AF333795

CHLORINE

Its Manufacture, Properties And Uses

EDITOR-IN-CHIEF

J. S. SCONCE

Technical Assistant to the Vice President
Research and Development
Hooker Chemical Corporation

ROBERT E. KRIEGER PUBLISHING COMPANY
MALABAR, FLORIDA

ORIGINAL EDITION 1962

REPRINT 1972, 1982

Printed and Published by
ROBERT E. KRIEGER PUBLISHING CO., INC.
KRIEGER DRIVE, MALABAR, FLORIDA 32950

Reprinted by arrangement with
VAN NOSTRAND REINHOLD COMPANY

Library of Congress Catalog Card Number: 62-20781
SBN 0-88275-075-5

Printed in the United States of America

CONTRIBUTORS

L. R. BELOHLAV
Purdue University

F. M. BERKEY
Monsanto Chemical Company

RALPH C. DOWNING
E. I. du Pont de Nemours & Co., Inc.

M. F. FOGLER
Allied Chemical Corporation

SHERWOOD N. FOX
Scientific Design Company, Inc.

P. E. HOCH
Hooker Chemical Corporation

ANITA S. KASTENS
Union Carbide Chemicals Company

MORTON S. KIRCHER
Hooker Chemical Corporation

RALPH LANDAU
Scientific Design Company, Inc.

EDMUND J. LAUBUSCH
The Chlorine Institute, Inc.

LORRAINE B. LOWRY
Stauffer Chemical Company

R. B. MACMULLIN
R. B. MacMullin Associates

E. T. MCBEE
Purdue University

CHARLES L. MANTELL
Newark College of Engineering

A. P. MUREN
Pittsburgh Plate Glass Company

H. D. PARTRIDGE
Hooker Chemical Corporation

L. W. PIESTER
Pittsburgh Plate Glass Company

B. H. PILORZ
Shell Chemical Company

DAVID J. PORTER
Diamond Alkali Company

WALTER H. PRAHL
Hooker Chemical Company

ALEXANDER REDNISS
Technical Enterprises, Incorporated

DAVID S. ROSENBERG
Hooker Chemical Corporation

E. D. SERDYNSKY
Dow Chemical Company

W. H. SHELTMIRE
Olin Mathieson Chemical Corporation

C. B. SHEPHERD
E. I. du Pont de Nemours & Company

PETER W. SHERWOOD
Ethyl Corporation

A. E. SKRZEC
Stauffer Chemical Company

D. L. TAYLOR
Hooker Chemical Corporation

EDWARD D. WEIL
Hooker Chemical Corporation

ALEX H. WIDIGER
The Dow Company

WILLIAM H. WILLIAMS
The Dow Chemical Company

GENERAL INTRODUCTION

American Chemical Society's Series of Chemical Monographs

By arrangement with the Interallied Conference of Pure and Applied Chemistry, which met in London and Brussels in July, 1919, the American Chemical Society was to undertake the production and publication of Scientific and Technologic Monographs on chemical subjects. At the same time it was agreed that the National Research Council, in cooperation with the American Chemical Society and the American Physical Society, should undertake the production and publication of Critical Tables of Chemical and Physical Constants. The American Chemical Society and the National Research Council mutually agreed to care for these two fields of chemical progress. The American Chemical Society named as Trustees, to make the necessary arrangements of the publication of the Monographs, Charles L. Parsons, secretary of the Society, Washington, D. C.; the late John E. Teeple, then treasurer of the Society, New York; and the late Professor Gellert Alleman of Swarthmore College. The Trustees arranged for the publication of the ACS Series of (a) Scientific and (b) Technological Monographs by the Chemical Catalog Company, Inc. (Reinhold Publishing Corporation, successor) of New York.

The Council of the American Chemical Society, acting through its Committee on National Policy, appointed editors (the present list of whom appears at the close of this sketch) to select authors of competent authority in their respective fields and to consider critically the manuscripts submitted.

The first Monograph of the Series appeared in 1921. After twenty-three years of experience certain modifications of general policy were indicated. In the beginning there still remained from the preceding five decades a distinct though arbitrary differentiation between so-called "pure science" publications and technologic or applied science literature. By 1944 this differentiation was fast becoming nebulous. Research in private enterprise had grown apace and not a little of it was pursued on the frontiers of knowledge. Furthermore, most workers in the sciences were coming to see the artificiality of the separation. The methods of both groups of workers are the same. They employ the same instrumentalities, and frankly recognize that their objectives are common, namely, the search for new knowledge for the service of man. The officers of the Society therefore combined the two editorial Boards in a single Board of twelve representative members.

Also in the beginning of the Series, it seemed expedient to construe rather broadly the definition of a Monograph. Needs of workers had to be recognized. Consequently among the first hundred Monographs appeared works in the form of treatises covering in some instances rather broad areas. Because such necessary works do not now want for publishers, it is considered advisable to hew more strictly to the line of the Monograph character, which means more complete and critical treatment of relatively restricted areas, and, where a broader field needs coverage, to subdivide it into logical subareas. The prodigious expansion of new knowledge makes such a change desirable.

These Monographs are intended to serve two principal purposes: first, to make available to chemists a thorough treatment of a selected area in form usable by persons working in more or less unrelated fields to the end that they may correlate their own work with a larger area of physical science discipline; second, to stimulate further research in the specific field treated. To implement this purpose the authors of Monographs are expected to give extended references to the literature. Where the literature is of such volume that a complete bibliography is impracticable, the authors are expected to append a list of references critically selected on the basis of their relative importance and significance.

AMERICAN CHEMICAL SOCIETY

BOARD OF EDITORS

PREFACE

In recent years a number of recommendations were made to the ACS Monograph Board concerning the need for a treatise on the chemistry and technology of chlorine, devoting main attention to the American industry.

After discussions with members of The Chlorine Institute and a number of others in this industry, it became evident that a text of this magnitude could not be covered accurately and currently by any one individual, and that the logical approach would be chapter contributions by experienced scientists currently at the peak of knowledge in their phase of the chlorine industry. Again with the help of The Chlorine Institute, authoritive chapter authors were obtained covering the properties, manufacture and major uses of chlorine. One notable exception is the manufacture of chlorine by electrolysis of molten salts. It was believed that this subject had been so fully and recently covered in a recent ACS Monograph, "Sodium—Its Manufacture, Properties and Uses," by Marshall Sittig that repetition in this text would be unnecessary.

The chapters were prepared to cover past history, chemical technology, commercial processes and end product uses. Particularly in the sections covering chemical technology and commercial processes the authors have had to work carefully between the zone of published information and confidential company technology. This was pointed out in our initial contact letters and it is a fine compliment to the willingness to publish of our American industry that acceptances were almost unanimous and that the chapters are so informative.

There are a few cases where certain chlorinated products have been covered in two different chapters, but these present different approaches to the same end product and are thus worthy of inclusion.

It is hoped that this will be a useful text and reference for those interested in the chlorine industry. I realize that in some cases the reader may be at variance with the technology or conclusions of the chapter's author. I am certain that they will welcome correspondence concerning these differences and may be able to further clarify and resolve these differences.

It is impossible to list all the names of those who have helped in one way or another in the development of this monograph, but all of them should have a warm feeling of accomplishment in being a part of our growing American technology.

Niagara Falls, N.Y.
August 1962   J. S. SCONCE

CONTENTS

1. DISCOVERY AND EARLY WORK

L. R. BELOHLAV and E. T. MCBEE
Purdue University

The history of elemental chlorine* is relatively short, embracing less than
two hundred years; yet the story of chlorine compounds reaches far back
into prehistoric times. To find the proper perspective for evaluating the
role of chlorine in modern society, it is important to examine briefly the
past of the basic raw material for chlorine, salt, or more specifically,
sodium chloride.

Long before any other mineral resources of the earth had been utilized
for human endeavors, salt was a determining factor in the organization of
primitive society. Thousands of years ago, Chinese, Egyptian and Hindu
scholars tried to establish the time and place of the discovery of salt, but
because it had been in use long before man learned to report events, their
efforts were unsuccessful. It is reasonable to assume that the first organized
search for and exploitation of natural salt resources coincides with the first
tentative formation of agricultural societies, presumably by Mesolithic
people before 6000 B.C.

Since all body fluids contain chloride (0.25 percent in human blood, 0.1
to 0.2 percent as hydrochloric acid in the gastric juices), the maintenance
of the proper chloride level is essential for life. As long as the human
race was restricted to a nomadic existence where man lived primarily on
milk or flesh which contain adequate amounts of sodium chloride, the addition of salt to the diet was not necessary. On the other hand, a cereal or
vegetable diet calls for a supplement of salt. Thus, the habitual use of salt,
the most common chlorine compound, became intimately related to the advance from nomadic to agricultural life. An adequate supply of salt was the
prerequisite for the transformation from the demanding life of the hunter
and gatherer to the more contemplative existence of the farmer. Readily
accessible sources of salt were an essential factor in setting the stage for the
"creative leisure" of man which was responsible for the rise of civilization
in historical times. It is undoubtedly no coincidence that the cradle of
many civilizations is found in coastal regions where recovery of sea salt

*Pronounced chlō · rine.

could be achieved conveniently. Thus, it is surprising that, in the face of the important role common salt came to assume in early society, "scientific investigation" of this commodity by our standards was taken up so late.

The first written records of rock salt are ascribed to Herodot (fifth century B.C.) who mentioned the rock salt deposits in Lybia. Since that area was generally known as *Ammonia*, the early scholars named the mineral *sal ammoniacum*. Because this name was later reserved for what we now refer to as ammonium chloride, much confusion as to the interpretation of early texts resulted. However, reference to the observed decrepitation of rock salt on heating and dissolving, as well as to its use for human consumption, by Pliny, Apicius, Arrian, Synesius, and Aetius settle the meaning of *sal ammoniaeum* in the early writings beyond all doubt.*

The purification of rock salt was first successfully attempted by Geber (eighth century). The processes of dissolution, filtration, and evaporation of the filtered liquid were apparently well known to him.

Rock salt already had been put to technical use by the Romans, especially, in one of the most important technological problems of the times —the separation of silver from gold. Pliny[1] (77 A.D.) described their manner of purifying gold by heating it with salt (iron or copper sulfate) and *schistos* (clay). This mixture will give off fumes of hydrogen chloride; however, the actual mechanism of this purification and the nature of the effluvia was not recognized or even further investigated at that time.

The Arab alchemists undoubtedly knew also the hydrochloric acid in its admixture with nitric acid, the *aqua regia*.

The preparation of dilute aqueous hydrochloric acid by the distillation of common salt with vitriol† was first mentioned by Basilius Valentinus.[2] Libavius[3] gives several prescriptions and includes alum as a possible ingredient of the starting mixture. Although these are the first records of the preparation of "hydrochloric acid," it is more than likely that it was made repeatedly by earlier investigators. The distillation of mixtures of various salts and earth was one of the "unit operations" of early alchemists and the necessary ingredients were readily accessible.

The powerful solvent action of this *spiritus salis* (occasionally referred to as *aqua caustica*) was well recognized by both Basilius Valentinus and Libavius. However, the manufacture apparently was attempted only rarely

---

*Ammonium chloride was most likely unknown before the Arab chemist Geber. For several hundred years the term *sal ammoniacum* was used rather indiscriminately for salt in general, and rock salt as well as ammonium chloride proper. The term *sal ammoniacum* (or sal armoniacum) was not reserved exclusively for ammonium chloride before the thirteenth century.

†"Vitriol" was the common name for all water soluble sulfates of bivalent heavy metals, in particular iron, copper, manganese, cobalt, and zinc.

and J. L. Glauber[4] describes it as the most expensive and most difficultly prepared acid of his time. Glauber improved on the method of preparing the acid from common salt by using oil of vitriol or alum, and he obtained a particularly strong acid by heating a mixture of hydrated zinc chloride with sand. This fuming acid was called *spiritus salis Glauberianus* until the end of the eighteenth century.

Glauber noticed that the acid could dissolve all metals except gold and silver; Boerhave knew about the insolubility of lead in hydrochloric acid. The formation of a precipitate on combining solutions of silver nitrate with hydrochloric acid was pointed out by Boyle. As a point of interest, it might be mentioned that with the overemphasis not infrequently accorded to new inventions, hydrochloric acid was actually recommended as a substitute for vinegar and lemon juice, for salting roasted meat, for preserving fruit, and as a basic ingredient for a soft drink prepared from the acid with honey and sugar.

With reference to its production from sea salt, the term "Meersalzsäure" was coined from the German translation for *spiritus salis* or *acidum salis* (Salzgeist, Salzäure). This term was translated into the French *acide marin* and adopted by the English as "marin acid." In 1787 A. L. Lavoisier suggested the name *acid muriatique* (from the Latin *muria*, brine); hence the English "muriatic acid."

Glauber emphasized the necessity for the presence of water in his preparation of hydrochloric acid. Boyle observed the evolution of the gas upon contacting alkali salts of this aqueous acid with strong sulfuric acid and Hales[5] ("Vegetable Staticks," 1727) noticed the formation of a readily water soluble gas on heating ammonium chloride with oil of vitriol. It was left to Priestley,[6] however, to collect hydrogen chloride gas over mercury and to describe its properties; he called it marine acid air.

There can be little doubt that the corrosive, suffocating greenish-yellow fumes of chlorine actually have been observed ever since the thirteenth century when the knowledge of *aqua regia* became widespread among alchemists. Van Helmont[7] noted that a *flatus incoercible* resulted whenever *sal armoniacus* (ammonium chloride) and *aqua chrysolca* (nitric acid) were heated together, although each of the components could be distilled separately without suffering decomposition. This *flatus incoercible* certainly contained chlorine. It is C. W. Scheele's merit, however, to have started a closer examination of this gas. He described it in much detail in his treatise on pyrolusite.[8] On digesting pyrolusite (manganese dioxide) with hydrochloric acid, he noticed an odor reminiscent of *aqua regia*. In investigating the source of this odor, he collected the evolved gas and tested it with several agents. He noticed the greenish yellow color of chlorine and discovered its bleaching effect on vegetable matter, the reaction with mercuric sulfide to give mercuric chloride, its suffocating effect upon insects and its attack upon even as noble a metal as gold. C. W. Scheele considered this gas to be muriatic acid freed of its combustible property (*phlogiston*). Accordingly, in the language of his time, he called it dephlogisticated marine acid air. B. Pelletier (1785) and W. J. G. Karsten (1786) observed the formation of yellow crystals on cooling moist chlorine. Until 1810, these crystals were considered to represent solid chlorine. H. Davy showed experimentally that these crystals are not formed by pure, dry chlorine and demonstrated the presence of water in chlorine hydrate. Faraday determined the quantitative composition in 1823; following a suggestion by Davy he examined the thermal decomposition in sealed vessels and thus obtained liquid chlorine.

The first observation that certain compounds can ignite spontaneously in an atmosphere of chlorine in the absence of air was published anonymously in 1786 in Crell's "Beiträgen zu den chemischen Annalen." The incandescence displayed by several metals and metal sulfides in contact with chlorine was discovered by Westrumb in 1789.

The bleaching power of chlorine discovered by Scheele was put to technical use by Berthollet (1785). At that time, large areas were needed for solar bleaching of textiles (*grassing or crofting*) and the possibility of achieving the same effect in a restricted space was received with enthusiasm. However, elemental chlorine proved totally unsuitable for this purpose. The halogen caused intense discomfort to the workers, corroded all metal parts and tendered the treated fabrics. Chlorine water was then tried, with little gain in operating advantages. Later, in 1789, in a large scale experiment at Javelles, chlorine absorbed in a potash solution (whence the name *Eau de Javelles* for potassium hypochlorite solution) was used. Through Watt this new technique was publicized in England and first used on an industrial scale by McPherson in Glasgow. Soon the potash solution was replaced by milk of lime. In 1798 in Glasgow, C. Tennant, was granted a patent for this new bleaching solution. Linen printers and cotton manufacturers were quick to seize the advantages of the new process. Jean-Antoine-Claude Chaptal (1756 to 1832) introduced the new bleaching process into the paper making industry.

A discussion of the gradual acknowledgment of chlorine as an element must begin with the first theoretical concept of hydrochloric acid. Becher advanced the view that the characteristic property of this acid—to "volatilize metals"—could be explained only by the presence of the hypothetical element "mercurial earth." Stahl, the propagator of the phlogiston theory disagreed and proposed the theory that all acids contained "primitive acid." He considered hydrochloric acid as a vitriolic acid (sulfuric acid in modern terminology) which became disguised or changed in its properties by certain admixtures; he did not define the proposed ingredients more specifically. Nevertheless, he averred his capability to convert nitric, sulfuric, and hydrochloric acid into each other. Surprisingly, this claim was even supported by other investigators (e.g. Potts, 1739). Demachy refuted this preposterous claim, and until 1774, the year of Scheele's famous investigation on manganese (pyrolusite), hydrochloric acid was considered a unique non-cleavable acid.

The discovery of chlorine required a restatement of the nature of hydrochloric acid and the theory of its relation to chlorine. Scheele realized that pyrolusite, in order to form salt, had to undergo a transformation similar to that of the metal calces (metal oxides) when converted to the state of a *regulus*. This process was called "phlogistication;" when, for example, manganese dioxide was heated with sulfuric acid, *phlogiston* was supposedly introduced into the oxide from the heat (according to Scheele, consisting of phlogiston and oxygen) and the oxygen subsequently set free. In the action of hydrochloric acid on pyrolusite, the same mechanism was assumed; the phlogiston taken up by the oxide could originate only from the acid, which, after losing this moiety appeared as chlorine. Scheele considered this experiment unambiguous proof of the presence of phlogiston in *marine acid* and accordingly termed chlorine *dephlogisticated marine acid.*

Shortly after, Lavoisier promulgated his views that all acids contain oxygen. For hydrochloric acid he could not provide direct experimental evidence but by analogy he postulated the acid as a compound of a hypothetical *radical muriatique* or *base muriatique* and oxygen. This view was generally adopted within the framework of the new theory. Berthollet redefined the relation between hydrochloric acid and chlorine (1785 to 1786). The relinquishment of oxygen by manganese dioxide on solution in acid was experimentally proven; in addition, Berthollet discovered the formation of oxygen and the decomposition of chlorine water on exposure to light as well as the simultaneous formation of hydrochloric acid. The conclusion was drawn that solar irradiation decomposes chlorine into its elements, oxygen and hydrochloric acid. The apparent acidity of chlorine was explained by the rather weak affinity of these parts. Chlorine was renamed *acide muriatique oxigéné*, abbreviated to *oxymuriatic acid* in English (Kirwan).

Around 1800 the most divergent views on the actual nature of the muriatic radical were aired. At that time, hydrogen and oxygen (Girdanner, 1795; Pacchiani, 1805; Curaudan, 1798), zinc (Armet, 1795) and sulfur (Lambe, Woodhouse 1797) were considered as the elements, and none of these rash theories attracted too much attention. General agreement was reached only on the assumption that chlorine contains more oxygen than hydrochloric acid.

More competent attempts to gain further insight into the nature of the muriatic radical were undertaken by W. Henry in 1800. Gaseous hydrogen chloride was subjected to the action of electric sparks. Substantial attack of the mercury, used to retain the gas, was noted and ascribed to the oxygen content of the hydrogen chloride without a conclusive result being reached. Henry's result seemed, in the light of his interpretation, to indicate that even carefully dried hydrogen chloride still retained a residual amount of chemically bound water. A series of new experiments was initiated by Gay-Lussac and Thenard (around 1809) to settle the question of the water content of hydrogen chloride. On reaction of hydrogen chloride with lead oxide they obtained water. The amount evolved was found to produce as much metal oxide as was necessary to neutralize the applied acid. The evolution of acid in the reaction between hydrogen chloride and metal was ascribed to a decomposition of the water contained by the metal. Finally, the explosive recombination of hydrogen and chlorine without simultaneous formation of water under solar irradiation was discovered. The adopted interpretation of these phenomena was most precisely defined by Berzelius as late as 1819.[9] The hypothetical element *muriaticum* forms, by combination with oxygen (one atom of *muriaticum* plus two atoms of oxygen), the hypothetical anhydrous muriatic acid; gaseous hydrogen chloride was considered a compound of equal number of molecules of hypothetical anhydrous muriatic acid and water and was named *murias hydricus,* an analogue to *nitras hydricus,* (hydrated nitric acid) and *sulphas hydricus,* (hydrated sulfuric acid); chlorine, named *superoxydum muriatosum,* was presumed to represent a combination of one atom of *muriaticum* with three atoms of oxygen. Accordingly, anhydrous chloride could be formed by reaction of chlorine on metals, whereupon the former disintegrated into anhydrous muriatic acid and oxygen, which oxidized the metal; anhydrous muriatic acid and metal oxide then formed the anhydrous metal chloride. As an alternative, metal chlorides could also be formed by the reaction between hydrated muriatic acid and metal oxide, whereupon the water content of the former was separated from it.

Emphatic attempts to prove directly the presence of oxygen in chlorine ensued. However, even passing the gas over red-hot coal, one of the most efficient reducing agents of the time, was of no avail. Only indirectly, by forming chlorides (which were believed beyond all doubt to contain oxygen), could the presence of oxygen be ascertained. It is interesting to note that Gay-Lussac and Thenard already had mentioned an alternative possibility of interpreting the observed phenomena by assigning to chlorine the status of an element; however, it was argued, the hypothesis of an oxidized muriatic radical offered a preferable explanation, retaining the appealing simplicity of Lavoisier's theorem of the oxygen content of all acids as well as the generally accepted idea that all salts are combinations of oxides with acid. Indeed, it can be justly stated that only the emphasis on maintaining a unified salt theory retarded the realization of chlorine as an element.

In 1808, H. Davy presented to the Royal Society his first investigations on (gaseous) hydrogen chloride. On reacting potassium with the gas he had obtained hydrogen; he also decided at first, concurring with his French colleagues, on the hydrate character of the gas. However, repeated negative attempts to verify experimentally the presence of oxygen prompted him eventually to suggest for chlorine an elemental character similar to that of oxygen. Actually, Davy referred to his postulate as Scheele's hypothesis, and indeed the new view agreed superficially with Scheele's when "phlogiston" is simply translated as "hydrogen."*

In 1810, Davy justified and defined his ideas in greater detail. He pointed out that the older theory demanded the needless postulation of many hypothetical compounds while the new view succinctly expressed experimentally established facts. The expected criticism of the French school was met beforehand by assigning to chlorine the ability to form salts with metals directly, without the intermediary action of oxygen or oxygen compounds. This, at the time, unique property of a chemical element led Schweigger in 1811 to propose for chlorine the name halogen—a name derived from the Greek *hals* (salt) and Greek *gennao* (to produce) and later adopted for the whole group of halogen elements. Davy himself suggested the name chlorine or chloric gas (from Greek *chloros,* "yellowish green") as a replacement for the old term, "oxymuriatic acid;" Gay-Lussac in 1813 introduced the word *chlore* into the French literature, which was modified to *chlor* in German.

Davy's ideas were not accepted without challenge. Several criticisms were based on experiments performed with inadequate precautions (e.g., Murray claimed the formation of water in the combination of dry ammonia with dry halogen chloride). More serious and undoubtedly carrying more weight were the objections of the "elder statesman" Berzelius. He admitted the plausibility of Davy's arguments if his experimental materials were considered exclusively. However, he believed that he demonstrated one fallacy in Davy's view by pointing out that his formulation required basically different constituents of otherwise quite similar compounds (chloride as opposed to sulfate and nitrate), while the presence of oxygen could be demonstrated readily. Even in his defense of an erroneous concept, his attitude deserves respect to this day, as he stated: "I demand relentlessly of each chemical theorem that it correspond to the rest of chemical theory and can be incorporated into it; in the opposite case, I have to discard it

---

*Actually, Scheele did not consider hydrogen and phlogiston identical, although he believed substantial amounts of phlogiston to be present in hydrogen.

TABLE 1-1. THE EARLY HISTORY OF CHLORINE

6000 B.C.	Establishment of agricultural societies.
3000 B.C.	Archaeological evidence for the use of rock salt.
400 B.C.	Written records on salt.
100 A.D.	"Industrial" use of salt in the purification of noble metals.
800	Purification of rock salt.
1300	Presumably first observations of chlorine gas.
1600	Preparation of dilute hydrochloric acid.
1650	Preparation of concentrated hydrochloric acid.
1772	Preparation of gaseous hydrogen chloride.
1773	Use of hydrogen chloride fumes as a disinfectant.
1774	Preparation of chlorine gas. Definition of chlorine as dephlogisticated muriatic acid.
1785	Observation of chlorine hydrate—use of chlorine as a bleaching agent—definition of chlorine as oxymuriatic acid.
1798	First patent on chlorine bleaching solutions.
1800	Decomposition of hydrogen chloride by electric sparks.
1801	Development of a chlorine generator for sanitation purposes.
1809	Consideration of element character for chlorine.
1810	Definition of chlorine as a chemical element—hydrate character ascertained for chlorine hydrate.
1813	Coinage of the term "chlorine".
1815	General acceptance of element character for chlorine.
1819	Hydrogen chloride defined as a hydrate of the hypothetical muriatic acid.
1823	Quantitative determination of the composition of chlorine hydrate—installation of disinfecting equipment using chlorine in hospitals.
1826	Chlorine water used in obstetric wards for the prevention of puerperal fever.
1831	Fumigation with chlorine during the great European cholera epidemic.

unless its irrefutable evidence necessitates a revolution in the theory incompatible with it." It was Berzelius' contention that the adoption of the "chloristic theory" would necessarily bring about a complete revision of the experimentally well founded theories of the composition of other simple salts. It was not until the representatives of the chloristic school had made it clear that exceptional status was claimed only for chlorine, until iodine had been discovered and its properties described by the new theory, and, finally, until the cyanides were analyzed and shown to contain no oxygen (Gay-Lussac 1815) that chlorine was unanimously granted the rank of a chemical element.

References

1. Plinius, "Naturalis Historia," **33**, 25.
2. Valentinus, Basilius, "Triumphwagen des Antimonii," Leipzig, 1604.

3. Libavius, A., "Alchymia," Francoforti, 1595.
4. Glauber, J. L., "Furnis Novis Philosophicis," 1648.
5. Saint Hales, "Vegetable Staticks," 1727.
6. Priestley, J., "Observations on Different Kinds of Air," Phil. Trans., 1772.
7. Van Helmont, J. B., "De Flatibus in Ortus medicinae," 1668.
8. Scheele, C. W., "Brunsten, eller Magnesia och dess Egenskapen, Kong," Stockholm, Vetenskaps Academiens Handlingen XXXV, 1774.
9. Berzelius, J. J., "Schrift über chemische Proportionen," 1819.

General References

General Sources on the history of chlorine:
1. Knopp, H., "Geschichte der Chemie," Vol. 3, Braunschweig, 1845. Scholarly discussion of the early history with numerous quotations from difficultly accessible texts.
2. "The Early History of Chlorine," Alembic Club Reprint No. 13, Edinburgh, 1905. Contains translated excerpts from papers by C. W. Scheele, C. L. Berthollet, Guyton de Morveau, J. L. Gay-Lussac and L. J. Thenard.
3. "The Elementary Nature of Chlorine," Alembic Club Reprint No. 9, Edinburgh, 1929. Containing excerpts from papers by H. Davy.
4. "The History of Salt," Morton Salt Company, 1959.
5. Schroeter, J., *CIBA Review 12,* No. 139, p. 2.
6. Kaufmann, D. W., "Sodium Chloride," ACS Monograph Series, p. 1. New York, Reinhold Publishing Corp., 1960. A history of sodium chloride.

2. PRODUCTION AND USE PATTERNS

DONALD L. TAYLOR
Hooker Chemical Corporation

Chlorine production in the United States had its origin before the turn of the century. In 1892, The Electro-Chemical Company began operating a small installation at Rumford Falls, Maine, utilizing a bell-jar type electrolytic cell. This plant was moved to Berlin, New Hampshire, in 1898 and is presently operated by the Brown Paper Company. In 1895, the S. D. Warren Company began chlorine production at Cumberland, Maine. The Mathieson Alkali Company, now Olin Mathieson Chemical Corporation, produced chlorine at Saltville, Virginia, for a few years, beginning in 1895. The Dow Chemical Company began commercial production of chlorine in 1896 at Midland, Michigan, to exploit the brines of that area. Mathieson built the first plant in Niagara Falls, New York, and started production in 1897. Other companies followed shortly after the turn of the century, namely, the Roberts Chemical Company in 1901, and the Development and Funding Company in 1905, both of which are now a part of Hooker Chemical Corporation. Other pioneers in the manufacture and sale of chlorine were the Pennsylvania Salt Manufacturing Company in 1903 (now Pennsalt Chemical Corporation) and the Warner Klipstein Company in 1915 (now part of the Food Machinery & Chemical Corporation). In the interim, a number of pulp manufacturers began captive production of chlorine, including the New York and Pennsylvania Company and West Virginia Pulp & Paper Company.

The major use for chlorine at the start of the century was in the manufacture of bleaching powder by passing chlorine over beds of hydrated lime, and the product was harvested manually. Chlorine was first liquefied by the Niagara Alkali Company (now merged into Hooker) at Niagara Falls in 1909. This was one of the turning points in the growth of the chlorine industry and led to many new uses. By 1910 there were eleven operating plants in the United States with a total installed capacity of about 200 tons of chlorine per day. A typhoid epidemic in Niagara Falls in 1912 is credited with sparking an early expansion of the industry in which chlorine began to be widely used for water purification.

Industrial development between 1910 and 1920 felt the effect of World War I. During this period, the United States made its first great strides in the chemical industry by becoming independent of the European chemical industry. Many new uses for chlorine were developed during this period. By 1920, the installed chlorine capacity in the United States had increased to 600 tons per day.

The period from 1920 to 1940 was marked by the boom in the 1920's and the depression of the 1930's. However, this period saw many new industries developing and an increase in expenditures by industry for research. Many of the larger chemical demands for chlorine, such as ethylene glycol, chlorinated solvents, vinyl chloride, etc., were developed during this period. By 1940, U. S. installed chlorine capacity had reached nearly 2,000 tons per day.

During World War II, civilian use of chlorine-derived chemicals was restricted, while new military uses were developed. The war period catalyzed many new uses for chlorine and the trend in civilian demand returned following the war. By the end of 1950 chlorine capacity had reached approximately 6,000 tons per day, and in 1960 it was estimated at more than 14,000 tons per day. If the trend in demand during the next ten years increases at the expected annual rate of 8 percent, a capacity of 30,000 tons per day will be required by 1970.

PRODUCTION AND CONSUMPTION

Statistics on the production and consumption of chlorine during the early part of this century are rather meager for this reason: that data cannot be considered accurate, as exemplified by the inconsistency between capacity and production for the period from 1900 to 1925.

The U. S. Bureau of Census began reporting chlorine production at five-year intervals from 1899 to 1919; in 1921, it started reporting every two years. Accurate reporting of data was not realized until the passage of the Fifteenth Census Act, approved June 18, 1929. According to Section 17, "Director of the Census is authorized and directed to collect and publish, for every second year after 1927, statistics of manufacturing industries." The Chlorine Institute, a trade organization of chlorine producers, was formed in 1924 and has played an important part in collecting statistics within the industry.

The following table contains a review of statistics available on chlorine production, production for sale, exports and imports at five-year intervals ending with 1955 and then each year through 1960. The total U. S. chlo-

TABLE 2-1. UNITED STATES CHLORINE STATISTICS

| | | —Tons per Year— | | | |
Year	Total Capacity*	Total Production	Production For Sale	Exports	Imports
1900	11,000	1,400	N A	N A	N A
1905	30,000	4,500	N A	N A	N A
1910	69,000	11,000	N A	N A	N A
1915	150,000	30,000	N A	N A	N A
1920	220,000	61,000	N A	N A	N A
1925	255,000	100,200	52,500	N A	N A
1930	380,000	205,000	N A	N A	N A
1935	445,000	330,000	207,380	5,803	N A
1940	690,000	605,000	N A	5,005	N A
1945	1,464,000	1,192,100	N A	4,170	6,827
1950	2,150,000	2,084,160	1,073,300	N A	N A
1955	3,970,000	3,421,100	1,867,000	30,000	N A
1956	4,030,000	3,797,700	2,039,000	19,000	N A
1957	4,180,000	3,946,670	1,986,000	39,350	10,030
1958	–	–	1,781,060	25,360	14,210
1959	–	4,347,120	2,025,900	29,360	16,750
1960	–	4,636,940	2,216,560	27,350	26,710
1970	12,000,000	10,000,000			

*Estimated chlorine capacities as of the end of year.

rine capacity has been estimated for those years. Also estimated are capacities and production for 1970.

The Gompertz curve (Figure 2-1) shows the trend in chlorine production compared with actual production for the period from 1919 to 1958. These calculations were made with the aid of an electronic computer, which indicates that if the past trend continues, the chlorine industry will double in size between 1960 and 1970.

It is of interest to estimate the apparent annual consumption of chlorine as pounds per person in the United States over a 70 year period:

1900	.04	1940	9
1910	.24	1950	28
1920	1.2	1960	52
1930	3.3	1970	98

END USE PATTERNS

During the span of the chlorine industry in the United States, the largest single use has been in the manufacture of pulp and paper. In the early period of the industry, the second largest use was in sanitation for sewage treatment and purification of water. Although the growth in demand for sanitation purposes has continued, it has dropped from approximately 15 percent of the total domestic demand in 1925 to the present 3 percent.

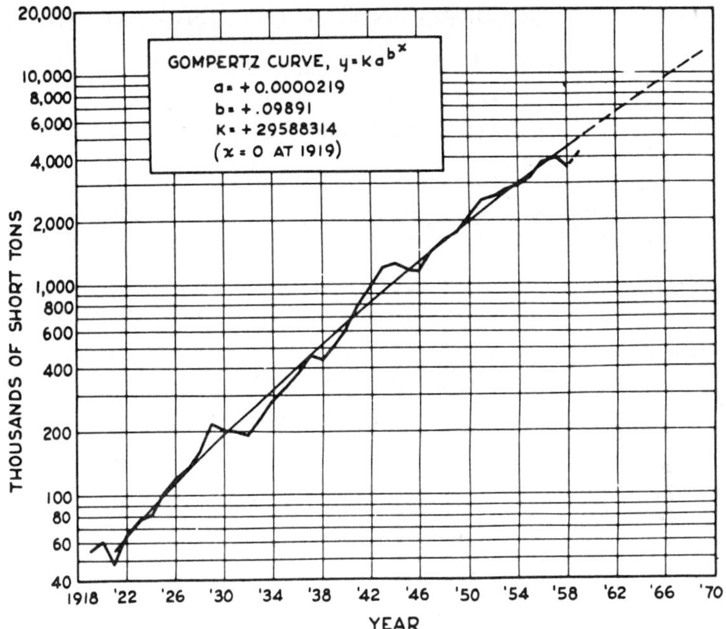

Figure 2-1.   U. S. chlorine production.

The use of chlorine in the manufacture of chemicals had its first impetus during World War I, but by 1925 this use accounted for only 17 percent of the U. S. demand. By 1930, however, 30 percent of the chlorine was being used for the manufacture of chemicals and since that time the chemical industry has been its own best customer. By 1940, 60 percent of the chlorine was being used by chemical manufacturers; 75 percent was being used by 1950 and in recent years, the demand has increased to 80 percent.

The following table shows the growth in demand for chlorine by various industries between 1925 and 1960.

Table 2-3 has been prepared to show the historical growth in demand for chlorine in chemical manufacture insofar as this is possible, using the information that is available. The demand for chlorine in tons per day is given for twenty major chemicals. These estimated demands are listed by five year periods from 1925 through 1960. The reasons for the variations above or below the trend line are due to economic situations experienced during the years covered in the Gompertz curve.

It will be noted that the largest single use for chlorine since 1950 has been for the manufacture of ethylene oxide and glycol. Some of the products are made by more than one process and may be made without the use of chlorine.

TABLE 2-2. U.S. CHLORINE END USE PATTERN—BY INDUSTRY

Estimated Percentage of Total Demand

	1925	1930	1935	1940	1945	1950	1955	1960
Pulp and paper	53	47	33	22	16	14	15	16
Chlorinated solvents	–	–	–	27	24	20	14	14
Automotive fluids	–	–	–	6	9	11	11	11
Plastics and resins	–	–	–	5	8	10	11	11
Insecticides and herbicides	–	–	–	1	6	10	8	6
Refrigerants	–	–	–	12	11	10	5	5
Other chemical uses	–	–	–	9	12	14	29	33
Chemicals–total	17	30	48	60	70	75	78	80
Sanitation	15	10	9	6	5	3	3	3
Textiles	10	8	6	5	4	3	–	–
All other uses	5	5	4	7	6	5	4	1

Estimated Tons per Day

	1925	1930	1935	1940	1945	1950	1955	1960
Pulp and paper	159	268	305	370	530	810	1340	1920
Chemicals	50	172	440	1010	2280	4350	6950	9600
Sanitation	45	57	83	100	165	175	265	360
Textiles	33	45	55	85	133	175	–	–
All other	14	28	37	115	200	290	360	120
Totals	301	570	920	1680	3308	5800	8915	12,000

Although uses for chemical products made from chlorine are discussed in other chapters of this book, it is of interest here to list the major end uses of the twenty products listed above:

(1) Benzene hexachloride: insecticides
(2) Carbon tetrachloride: manufacture of fluorocarbons for refrigerants and propellants
(3) Chloral: insecticides manufacture
(4) Chloroform: manufacture of fluorocarbons and drugs
(5) Chloro paraffins: lubricant additives and paints
(6) Dichlorobenzene: insecticide and organic solvent
(7) Ethyl chloride: tetraethyllead antiknock compounds
(8) Ethylene dichloride: tetraethyl lead antiknock compounds and vinyl chloride
(9) Ethylene oxide and glycol: antifreeze fluids and synthetic fibers
(10) Glycerine, synthetic: resins, tobacco manufacture and explosives
(11) Hydrochloric acid, synthetic: vinyl chloride and acid uses
(12) Methyl chloride: silicones manufacture; solvent and catalyst carrier
(13) Methylene chloride: paint remover, propellants and general solvent

(14) Monochloroacetic acid: herbicides, detergents, pharmaceuticals and permanent wave preparations
(15) Monochlorobenzene: phenol, aniline and DDT insecticide
(16) Perchloroethylene: dry cleaning
(17) Propylene glycol: polyester resins, cellophane
(18) Titanium: metal for special alloys
(19) Trichloroethylene: metal degreasing
(20) Vinyl chloride: plastic and resin products

Chlorine probably has more diverse uses than any other known chemical. It has hundreds of uses involved in the manufacture of products for food, clothing, shelter, health, transportation, communications, recreation and defense. One of the most unusual chlorinated products, from the standpoint of end uses, is monochloroacetic acid. It is used to make carboxymethyl cellulose which in turn is used as a detergent builder and as a filler in ice cream. Another use of this product is in the manufacture of thioglycollic acid for use in home permanents.

TABLE 2-3. CHLORINE DEMAND FOR MAJOR CHEMICAL PRODUCTS*

	1925	1930	1935	1940	1945	1950	1955	1960
Benzene hexachloride	–	–	–	–	–	89	86	30
Carbon tetrachloride	35	75	120	220	420	410	620	850
Chloral	–	–	–	–	50	125	194	300
Chloroform	3	6	5	8	20	40	98	190
Chloro paraffins	–	–	–	–	69	33	45	50
Dichlorobenzene	3	9	15	40	70	120	150	200
Ethyl chloride	–	–	–	–	110	170	260	300
Ethylene dichloride	–	–	–	–	60	100	170	400
Ethylene oxide and glycol	–	47	178	294	390	968	1450	2040
Hydrochloric acid-synthetic	–	–	25	55	165	310	440	520
Glycerine-synthetic	–	–	–	–	–	113	237	300
Methyl chloride	–	–	5	7	66	60	80	175
Methylene chloride	–	–	–	–	25	90	177	310
Monochloroacetic acid	–	–	–	–	–	30	55	55
Monochlorobenzene	8	10	40	95	230	350	390	570
Perchloroethylene	–	–	2	30	150	240	420	490
Propylene glycol	–	–	–	–	18	140	126	270
Titanium	–	–	–	–	–	–	63	80
Trichloroethylene	–	–	20	90	310	380	530	650
Vinyl chloride	–	–	–	2	105	285	510	610
Totals	49	147	410	841	2258	4050	6101	8390

*Estimated in Tons per Day

PRODUCERS AND LOCATION

The following map indicates the location of sixty-four chlorine plants in the United States. They are owned by thirty-five companies and are centered in the following major areas:

	Number of Plants	Estimated Capacity Tons/Day	Per Cent of U.S.
Texas	7	3940	26.5
Louisiana	6	1800	12.1
Michigan	4	1680	11.3
West Virginia	3	1490	10.0
New York	7	1150	7.7
Alabama	4	780	5.2
Ohio	3	720	4.8
Washington	3	630	4.2
	37	12,190	81.8
Total in U.S. (end of 1960)	64	14,900	100.0

Most plants are located where either cheap power or cheap salt is available. A combination of both factors, however, creates a favorable economic situation for supplying chlorine demands within a reasonable shipping distance.

It is estimated that between 70 and 75 percent of chlorine production is captive, the remainder being available for merchant sales.

CONCLUSIONS

To summarize, it may be stated that the chlorine industry will continue to grow at a healthy rate for many years to come; from all indications, it will double in size within the next ten years. In retrospect, it is evident that chlorine has played an important part in our industrial development. Although it is impossible to predict future developments, it is certain that chlorine will continue to contribute to the welfare of mankind.

Like all industries, the chlorine industry also has its problems. The ability to realize a fair return on invested capital has always been of major concern. Since chlorine is always made with a co-product, it is necessary to have a closely balanced demand for both products. In many chlorination processes, by-product hydrochloric acid is made and its sale is quite often a problem. Changing market demands from both a product and geographic standpoint has led to the construction of new plants over the years. Since chlorine is sold on an f.o.b. freight equalized basis, it is important that a chlorine plant supply an area as restricted in size as possible in order to minimize freight costs. With the expansion of the chlorine industry and the inflation in construction costs, it has been necessary to build larger plants.

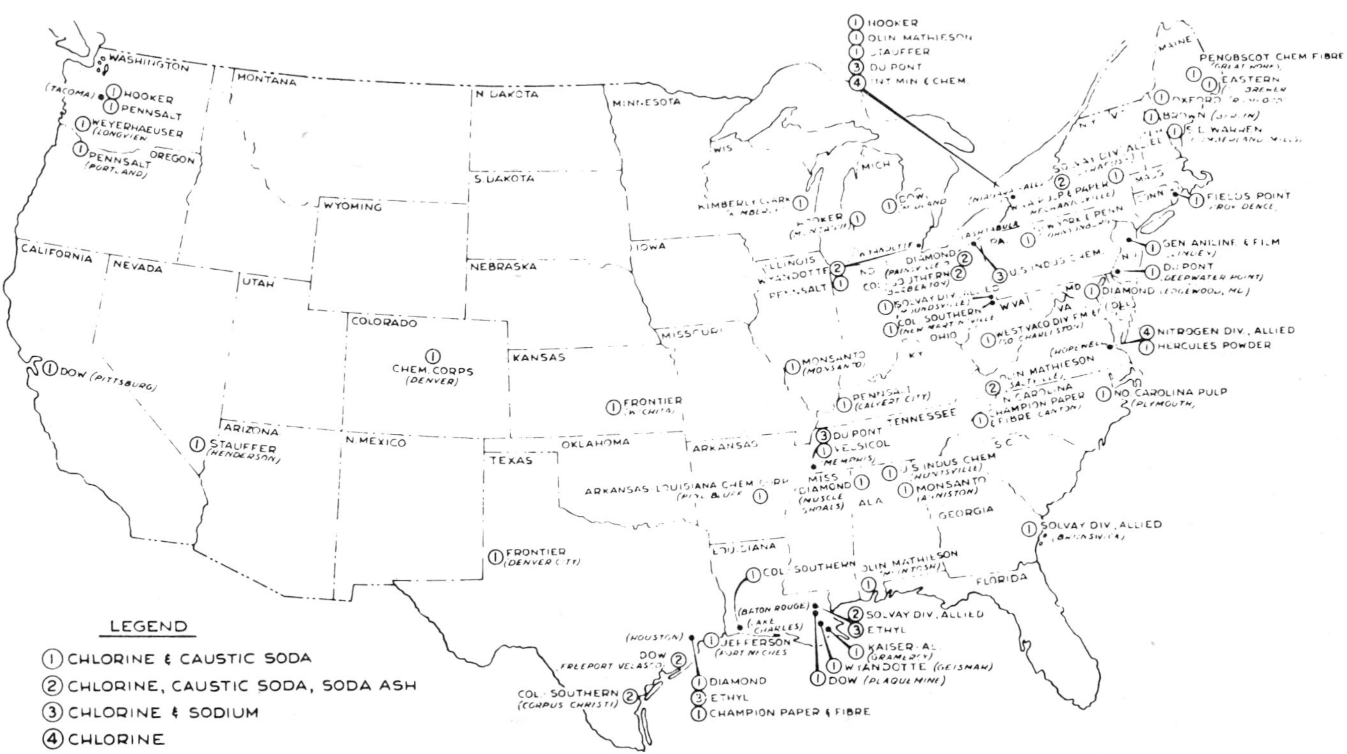

Figure 2-2. Chlorine producers in the United States (end of 1960).

TABLE 2-4. U.S. CHLORINE PRODUCERS

Producer	Location	Year Built
Allied Chemical & Dye Corp.	Baton Rouge, La.	1937
(Solvay Division)	Hopewell Va.*	1936
	Moundsville, W. Va.	1953
	Syracuse, N. Y.	1927
	Brunswick, Ga.	1957
Arkansas Louisiana Chemical Corp.	Pine Bluff, Ark.	1943
Chemical Corps, U.S. Army	Denver, Colo.	1943
Columbia-Southern Chemical Corp.	Barberton, Ohio	1936
	Corpus Christi, Tex.	1938
	Lake Charles, La.	1947
	Natrium, W. Va.	1943
Diamond Alkali Co.	Edgewood, Md.	1942
	Muscle Shoals, Ala.	1952
	Pasadena, Tex.	1938
	Painesville, Ohio	1928
Dow Chemical Co.	Midland, Mich.	1897
	Freeport, Tex.	1940
	Velasco, Tex.	1940
	Pittsburg, Calif.	1917
	Plaquemine, La.	1958
E.I. du Pont de Nemours & Co., Inc.	Deepwater Point, N. J.	1928
	Niagara Falls, N. Y.†	1898
	Memphis, Tenn.†	1958
Ethyl Corporation	Baton Rouge, La.†	1938
	Houston Tex.†	1952
Food Machinery & Chemical Corp.	South Charleston, W. Va.	1915
Fields Point Manufacturing Co.	Providence, R. I.	1921
Frontier Chemical Co.	Denver City, Tex.	1947
	Wichita, Kan.	1952
General Aniline & Film Corp.	Linden, N. J.	1956
Hercules Powder Co. Inc.	Hopewell, Va.	1939
Hooker Chemical Corp.	Montague, Mich.	1954
	Niagara Falls, N. Y.	1898
	Tacoma, Wash.	1929
International Minerals & Chemical Corp.	Niagara Falls, N. Y.	1916
Jefferson Chemical Co., Inc.	Port Neches, Tex.	1959
Kaiser Aluminum & Chemical Corp.	Gramercy, La.	1958
Monsanto Chemical Co.	East St. Louis, Ill.	1922
	Anniston, Ala.	1952
Olin Mathieson Chemical Corp.	MacIntosh, Ala.	1952
	Niagara Falls, N. Y.	1897
	Saltville, Va.	1951
Pennsalt Chemicals Corp.	Calvert City, Ky.	1953
	Portland, Ore.	1947
	Tacoma, Wash.	1929
	Wyandotte, Mich.	1898

TABLE 2-4 (*continued*)

Producer	Location	Year Built
Stauffer Chemical Co.	Henderson, Nev.	1942
	Niagara Falls, N.Y.	1898
U.S. Industrial Chemicals Co.	Ashtabula, Ohio†	1949
	Huntsville, Ala.	1943
Velsicol Chemical Corp.	Memphis, Tenn.	1943
Wyandotte Chemicals Corp.	Wyandotte, Mich.	1938
	Geismar, La.	1959

Pulp and Paper Mills

Producer	Location	Year Built
Brown Co.	Berlin, N. H.	1898
Champion Paper & Fibre Co.	Canton, N. C.	1916
	Pasadena, Tex.	1936
Eastern Corp.	South Brewer, Me.	1916
Kimberly-Clark Corp.	Kimberly, Wis.	1936
North Carolina Pulp Co.	Plymouth, N. C.	1952
New York & Pennsylvania Co.	Johnsonburg, Pa.	1905
Oxford Paper Co.	Rumford, Me.	1916
Penobscot Chemical Fibre Co.	Great Works, Me.	1916
S. D. Warren Co.	Cumberland, Me.	1916
West Virginia Pulp & Paper Co.	Mechanicville, N. Y.	1908
Weyerhauser Timber Co.	Longview, Wash.	1957

*Nitrosyl Chloride or salt process.
†Downs cells—metallic sodium

It is now felt that the smallest plant should have a capacity of 100 tons of chlorine per day to prove economical. Even that size may offer only a marginal profit and be justified on expected expansion. The cost of constructing new chlorine plants depends on the auxiliary facilities that have to be supplied along with the "battery limits" plant. A good average cost used by the industry for plant investment cost is $100,000 per ton day of chlorine. Such an investment is sizable, especially when considering the time factor involved in the construction, start-up problems and developing capacity demands.

Large sums have been spent by the industry in developing new uses for chlorine and will continue to be spent for some years to come. The industry is faced with the everlasting problem of obsolescence of processes and products. Typical examples are the new processes developed for the manufacture of ethylene oxide and glycerine without chlorine. If the automobile combustion engine is ever replaced by a turbine type engine, it would have a tremendous effect in reducing the demand for chlorine used in the manufacture of antiknock fluids, antifreeze compounds and existing lubricant additives.

In looking to the future, we may expect the demand for chlorine to continue growing. The abundance of salt in various parts of the country assures the industry of a cheap source of the major raw material. At this time, it appears that chlorine will continue to be one of the cheapest basic chemicals for use as a building block in the chemical industry. Based on present consumption, the chemical producer receives an average of $1.70 from each person in the United States for the chlorine consumed in the manufacture of the tremendous number of products used by the consumer in his daily life. The consumer has gained these benefits through the industry's continual advances in technology.*

General References

1. Arthur D. Little, Inc., *Service to Management* (issued quarterly).
2. *Chem. & Eng. News,* p. 38 (December 14, 1959).
3. *Chemical Week,* p. 116 (March 7, 1959).
4. The Chlorine Institute, Inc., "Chloralkali Producers in North America," (issued annually).
5. Foster D. Snell, Inc., *Chem. Market Abstracts* (issued monthly).
6. Gordon, J., *Chem. Week* p. 29 (April 4, 1953).
7. Hampel, C. A., and Ehlers, N. J., *J. Electrochem. Soc.,* 906 (October 1959).
8. Hampel, C. A., and Ehlers, N. J., 115th Meeting of the Electrochemical Society (May 4, 1959).
9. Hampel, C. A., and Ehlers, N. J., 117th Meeting of the Electrochemical Society (May 2, 1960).
10. Lamie, R. K., *Chem. Ind.,* 578 (April, 1948).
11. Lisman, C. E., Speech to the Chemical Market Research Association (February 25, 1954).
12. MacMullin, R. B., and Cole, J. C., *J. Electrochem. Soc.,* p. 550 (September 1958).
13. Murray, R. L., Ind. Eng. Chem., **41,** 2155 (October 1949).
14. National Production Authority, Alkali-Chlorine Conference, (March 15, 1951).
15. *Oil, Paint and Drug Rep.,* "Chemical Targets For The Sixties," p. 34 (December 28, 1959).
16. Sheets, T., Jr., *Chem. Eng. Progress,* 482 (October 1957).
17. Sommers, H. A., *Chem. Eng. Progress,* 508 (October 1957).
18. Stanford Research Institute, "Chemical Economics Handbook" (issued periodically).
19. Taylor, D. L., Speech to the Chemical Market Research Association (May 23, 1956).
20. Taylor, D. L., Speech to The Chlorine Institute (January 20, 1960).
21. U. S. Department of Commerce, "Facts For Industry" (issued annually).
22. U. S. Tariff Commission, "Synthetic Organic Chemicals, U. S. Production and Sales" (issued annually).

*The author expresses his sincere thanks for the assistance given to him in preparing data for this chapter by his associates, A. D. Kischitz and J. L. Olmstead.

3. PHYSICAL AND CHEMICAL PROPERTIES OF CHLORINE

Edmund J. Laubusch

The Chlorine Institute, Inc.

Chlorine is an element that is found in nature in the combined state only, chiefly with sodium as common *salt* (NaCl), *carnallite* ($KMgCl_3 \cdot 6H_2O$), and *sylvite* (KCl).[11,18] A member of the halogen family at standard conditions, it is a greenish-yellow gas which can be readily compressed into a clear, amber-colored liquid; it solidifies at about $-150°$ F under one atmosphere, and it crystallizes with water as chlorine hydrate, commonly $Cl_2 \cdot 8H_2O$, below $49.3°$ F.

Chlorine in commerce is a compressed, liquefied gas under pressure, about $1\frac{1}{2}$ times as heavy as water, packaged in steel containers (see Chapter 4). Liquid chlorine vaporizes under normal atmospheric temperature and pressure; the gas, about $2\frac{1}{2}$ times as heavy as air, has a characteristic penetrating and irritating odor. The physiological effects of chlorine are discussed in Chapter 4.

Properties herein noted are generally limited to those most frequently referred to, and are compiled from a number of sources believed to be representative and reliable. Data on magnetic, electrical and spectrographic properties, and on thermodynamic and chemical properties not herein included. have been summarized by Lange,[15] Mellor,[18] Gmelin,[8] and others.

ATOMIC AND MOLECULAR PROPERTIES

The atomic number of chlorine (number of excess positive charges on the atomic nucleus) is 17. Atomic chlorine, Cl, has an atomic weight of 35.457.[84] Molecular chlorine, Cl_2, has a molecular weight of 70.914.

Two isotopes of chlorine occur naturally, Cl^{35} and Cl^{37}, and at least five other isotopes have been artificially produced (see Table 3-1). Ordinary atomic chlorine consists of a mixture of about 75.4 percent Cl^{35} and 24.6 percent Cl^{37}.

Chlorine usually forms univalent compounds, but it also can combine with a valence of 3,4,5 or 7. In sodium chloride (NaCl) chlorine has a single negative valence; in sodium hypochlorite (NaClO) it is positive and univalent; in sodium chlorite ($NaClO_2$) it is trivalent; in sodium chlorate ($NaClO_3$) it is pentavalent; and, in sodium perchlorate ($NaClO_4$) it is heptavalent.

TABLE 3-1. ISOTOPES OF CHLORINE[11]

Isotope	Atomic Mass	Half-Life	Mode of Decay Radiation—Energy (mev)		
Cl^{33}	32.9860	2.8 sec	β^+ 4.1		
Cl^{34}	33.9826	33.0 min	β^+ 1.3	2.6	4.5
			γ 0.15	2.13	3.30
Cl^{35}	34.97867				
Cl^{36}	35.9784	4×10^5 yr	β^- 0.716		
			β^+		
			γ (?)		
Cl^{37}	36.97750				
Cl^{38}	37.981	38.5 min	β^- 1.11	2.77	4.81
			γ 1.6	2.15	
Cl^{39}		55.5 min	β^- 2.5		

β^- = beta particle
β^+ = positron
γ = gamma ray
mev = Million electron volts

PHYSICAL AND THERMODYNAMIC PROPERTIES

Boiling Point

The *boiling point* (or *liquefying point*) is that temperature at which liquid chlorine vaporizes under one atmosphere pressure. Variations in the value for chlorine are noted in the literature (see Table 3-2), but the value determined by Giauque and Powell[7] of $-29.29°F$ ($-34.05°C$) is popularly quoted in the industry.

Compressibility Coefficient

The compressibility coefficient of liquid chlorine (see Table 3-2), greater than for any other liquid element, represents the per cent decrease in volume corresponding to a unit increase in pressure, when the liquid is held at constant temperature.

Critical Properties

The critical data for chlorine reported by various scientists differ significantly (see Table 3-3).

TABLE 3-2. PHYSICAL AND THERMODYNAMIC PROPERTIES OF CHLORINE

Property	Value	Observer or Source
Boiling (liquefying) point	$-29.31°$ F $(-34.06°$C)	Rossini[27]
	$-29.29°$ F $(-34.05°$C)	Giauque and Powell[7]
	$-30.3°$ F $(-34.6°$ C)	Pellaton[22]
Compressibility coefficient—liquid at $20°$C $(68°$F)		Richards and Stull[25]
At 0.0 to 98.7 atm	0.0118 per cent/atm	
98.7 to 197.4 atm	0.0110 per cent/atm	
197.4 to 296.0 atm	0.0102 per cent/atm	
Critical density	35.77 lb/ft^3 (573 g/l)	Pellaton[22]
Critical pressure	1118.36 psia (76.1 atm)	Pellaton[22]
Critical temperature	$291.2°$ F $(144°$C)	Pellaton[22]
Critical volume	0.02796 ft^3/lb (0.001745 l/g)	Pellaton[22]
Density—Dry gas*	0.2003 lb/ft^3 (3.209 g/l)	Kapoor and Martin, computed[12]
	0.2006 lb/ft^3 (3.214 g/l)	Inter. Critical Tables[21]; Gmelin[8]
—Saturated gas at $0°$C†	0.7537 lb/ft^3 (12.07 g/l)	Kapoor and Martin, computed[12]
	0.799 lb/ft^3 (12.8 g/l)	Inter. Critical Tables[21]
—Liquid at $0°$C $(32°$F)†	91.67 lb/ft^3 (1468.4 g/l)	Kapoor and Martin, computed[12]
	91.64 lb/ft^3	Inter. Critical Tables[21]
—Liquid at $15.6°$C $(60°$F)‡	88.79 lb/ft^3 (11.87 lb/gal)	Kapoor and Martin, computed[12]
		Inter. Critical Tables[21]
Enthalpy—Dry gas*	232.74 Btu/lb	Kapoor and Martin[12]
—Saturated gas†	231.44 Btu/lb	Kapoor and Martin[12]
—Liquid†	116.39 Btu/lb	Kapoor and Martin[12]
Entropy—Dry gas*	0.73985 Btu/lb/$°$R	Kapoor and Martin[12]
—Saturated gas†	0.70222 Btu/lb/$°$R	Kapoor and Martin[12]
—Liquid†	0.46823 Btu/lb/$°$R	Kapoor and Martin[12]
Heat of fusion, solid chlorine (at melting point)	41.2 Btu/lb (22.9 g cal/g)	Inter. Critical Tables[21]
	38.86 Btu/lb (21.59 g cal/g; 1531 ± 1 cal/mol)	Giauque and Powell[8]

(*continued*)

TABLE 3-2 *(continued)*

Property	Value	Observer or Source
Heat of vaporization (at boiling point)	123.67 Btu/lb (68.7 g cal/g)	Kapoor and Martin[12]
	123.8 Btu/lb (68.8 g cal/g; 4878 ± 4 cal/mol)	Giauque and Powell[7]
	121.3 Btu/lb (67.4 g cal/g)	Inter. Critical Tables[21]
	123.55 Btu/lb	Ziegler[33]
Liquid-gas volume relationship	Wt of 1 vol liquid = wt of 457.6 vol gas at standard conditions*	Kapoor and Martin, computed[12]
Melting point (at 1 atm)	$-149.80°$ F $(-101.00°$ C$)$	Rossini[27]
	$-149.76°$ F $(-100.98°$ C$)$	Giauque & Powell[7]
	$-150.9°$ F $(-101.6°$ C$)$	Inter. Critical Tables[21]
Refractive index at 14 °C	1.367	Inter. Critical Tables[21]
Solubility (in water)	(See Fig. 3-5)	Winkler[31]
Specific gravity—Dry gas*	2.482	
—Saturated gas†	9.337	
—Liquid†	$1.468 \dfrac{0°}{4°}$ C	
Specific heat—liquid	0.236 Btu/lb/°F (0.236 g cal/g/°C) for liquid at equilibrium between 0°F and 100°F	Kapoor and Martin, computed[12]
		Kapoor and Martin, computed[12]
	0.233 between 0°F and 100°F	Ziegler[33]
	0.226 Btu/lb/°F (0.226 g cal/g/°C) between 32°F and 75°F (0°C to 24°C)	Inter. Critical Tables[21]
Specific heat—gas At constant pressure, C_p	0.115 Btu/lb/°F at 15 °C	Lange[16]
	0.115 Btu/lb at 15 psia between 50°F and 100°F	Kapoor and Martin, computed[12]
	0.122 Btu/lb at 100 psia between 100°F and 150°F	Kapoor and Martin, computed[12]
	0.116 Btu/lb at 14.7 psia between 50°F and 150°F	Ziegler, computed[33]

Property	Value	Reference
	0.128 Btu/lb at 100 psia between 50°F and 150°F	Ziegler, computed[33]
At constant volume, C_v	0.0849 Btu/lb/°F at 15°C	Lange[16]
	0.0848 Btu/lb at 15 psia between 50°F and 100°F	Kapoor and Martin, computed[12]
	0.0856 Btu/lb/°F between 50°F and 150°F	Nernst[83]
Specific heat ratio, C_p/C_v	1.355	Lange[16]
	1.329 ± .001	Partington[2]
Specific volume—Dry gas*	4.992 ft^3/lb	Kapoor and Martin[12]
—Saturated gas†	1.327 ft^3/lb	Kapoor and Martin[12]
—Liquid†	0.01091 ft^3/lb	Kapoor and Martin[12]
Surface tension	(see Table 3-6)	
Thermal conductivity—Gas		
At 0°C (32°F)	0.0042 Btu/hr/ft^2/°F/ft	Eucken and Hoffman[6]
At 55.5°C (132°F)	0.0049 Btu/hr/ft^2/°F/ft	
At 30°C (86°F)—Liquid	0.108 Btu/hr/ft^2/°F/ft	Smith, computed[28]
Vapor pressure at 0°C (32°F)	53.1155 psia (3.617 atm) (See Fig. 3-6)	Kapoor and Martin, computed[12]
Viscosity—Gas, at 20°C (68°F)	0.0000094 lb/sec/ft (0.000140 poises) (See Table 3-7 and Fig. 3-7)	Inter. Critical Tables[21]
—Liquid, at 20°C (68°F)	0.00023 lb/sec/ft (0.0035 poises)	Inter. Critical Tables[21]
Volume-temperature relationship	(See Fig. 3-2)	Kapoor and Martin, computed[12]

* At standard conditions: 32°F (0°C) and 14.696 psia (1 atm)
† Pressure of saturated gas and liquid at 32°F (0°C) is 53.155 psia (3.617 atm)
‡ Pressure of liquid at 60°F (15.6°C) is 85.61 psia

TABLE 3-3. CRITICAL DATA OF CHLORINE[33]

Critical Temperature (°C)	Critical Pressure (Atm.)	Critical Density (g/l)	Observer (Year)
148	–	–	Ladenburg[17] (1878)
141	83.9	–	Dewar[6] (1884)
146	93.5	–	Knietsch[14] (1890)
143.3	–	–	Estreicher[5] (1913)
144	76.1	573	Pellaton[22] (1915)

Pellaton's values,[22] included in Table 3-2, are generally considered most probable. The *critical density* is the mass of a unit volume of chlorine at the critical pressure and temperature. The *critical pressure* is the vapor pressure of liquid chlorine at the critical temperature. The *critical temperature* is the temperature above which chlorine exists only as a gas despite the pressure. The *critical volume* is the volume of a unit mass of chlorine at the critical pressure and temperature; it is the reciprocal of the critical density. Pellaton's values for critical properties have been used by Kapoor and Martin[12] in calculating the thermodynamic behavior of chlorine.

Density

Density represents the mass of a unit volume at specified temperature and pressure conditions. Values listed in Table 3-2, computed from Kapoor and Martin data[12] and the "International Critical Tables,"[21] differ to some extent.

The densities of gaseous and liquid chlorine, respectively, are graphically shown in Figures 3-1 and 3-3. Density decreases rapidly with increasing temperature, and the corresponding expansion of liquid chlorine in a shipping container (see Figure 3-2) is an important factor to be considered in filling chlorine containers and in handling liquid chlorine pipelines.

Enthalpy and Entropy

Enthalpy, H, and *entropy, S,* values for saturated and dry chlorine at certain conditions are given in Tables 3-4 and 3-5. Data are those computed by Kapoor and Martin[12] using a datum plane at absolute zero (At 0° F, H=O, S=O). (The properties of the saturated liquid at the boiling point, -29.29° F, referred to absolute zero, were based on the work of Giauque and Powell[7]). In the original work, thermodynamic properties of saturated liquid and vapor are given between −130°F and 291.2°F; properties of superheated gas are given at various temperatures between 0 psi and 1400 psi.

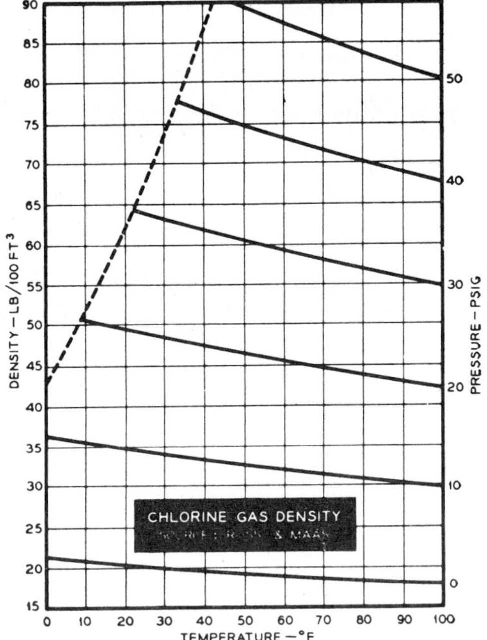

Figure 3-1

Heats of Fusion and Vaporization

The heat required to change one mass of solid chlorine to liquid (heat of fusion) or to change one mass of liquid to vapor (heat of vaporization) without change of temperature, according to various observers, is given in Table 3-2.

Liquid-Gas Volume Relationship

Liquid chlorine vaporizes rapidly at standard atmospheric conditions. One volume of liquid vaporizes into about 457.6 volumes of gas, or one pound of liquid forms about 5 cu. ft of gas.

Melting Point

The *melting point* (or *freezing point*) (see Table 3-2) is the temperature at which liquid chlorine solidifies under one atmosphere pressure.

Solubility in Water

Chlorine is only slightly soluble in water (see Figure 3-5), reaching its maximum solubility of approximately 1 percent at 49.3° F. Below 49.3° F,

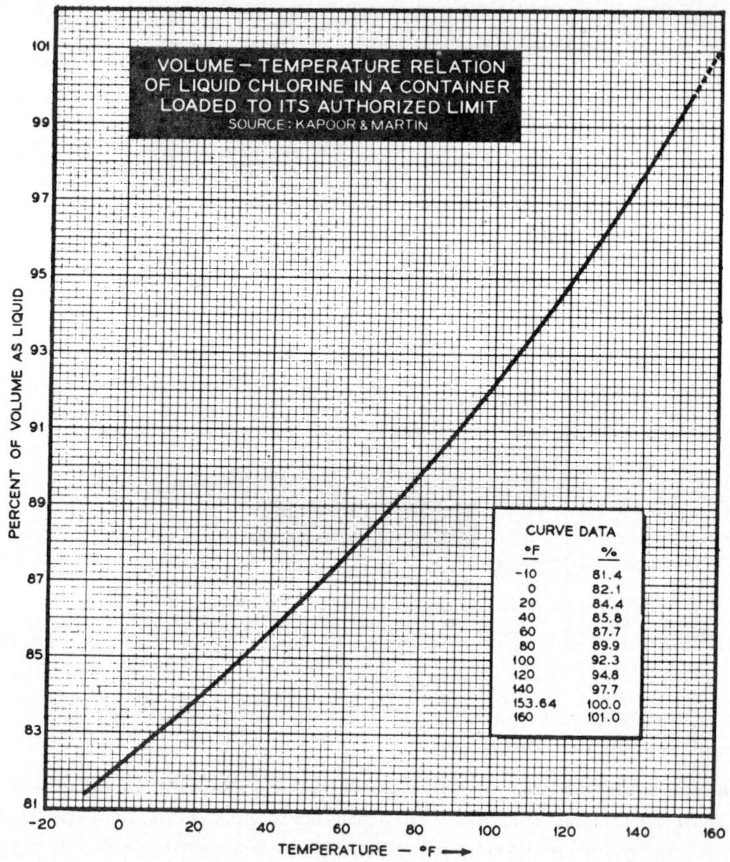

Figure 3-2

chlorine forms crystalline hydrates with water (frequently $Cl_2 \cdot 8H_2O$), referred to as "chlorine ice."   Above 49.3° F its solubility decreases with
temperature increases up to the boiling point of water at which point it is
completely insoluble.   (With organic solvents, solubility is frequently
greater, especially at low temperatures).

Specific Gravity

Dry chlorine gas at standard conditions of one atmosphere pressure and
at 32° F is about 2½ times heavier than air at the same conditions. The
values in Table 3-2 are based on an air density of 1.2929 g/l at standard

conditions. Liquid chlorine at 32° F and 3.671 atm. is about 1½ times heavier than water at its maximum density.

Specific Heat

The *specific heat* (see Table 3-2) is the amount of heat required to raise the temperature of chlorine by one thermal degree. Specific heat at

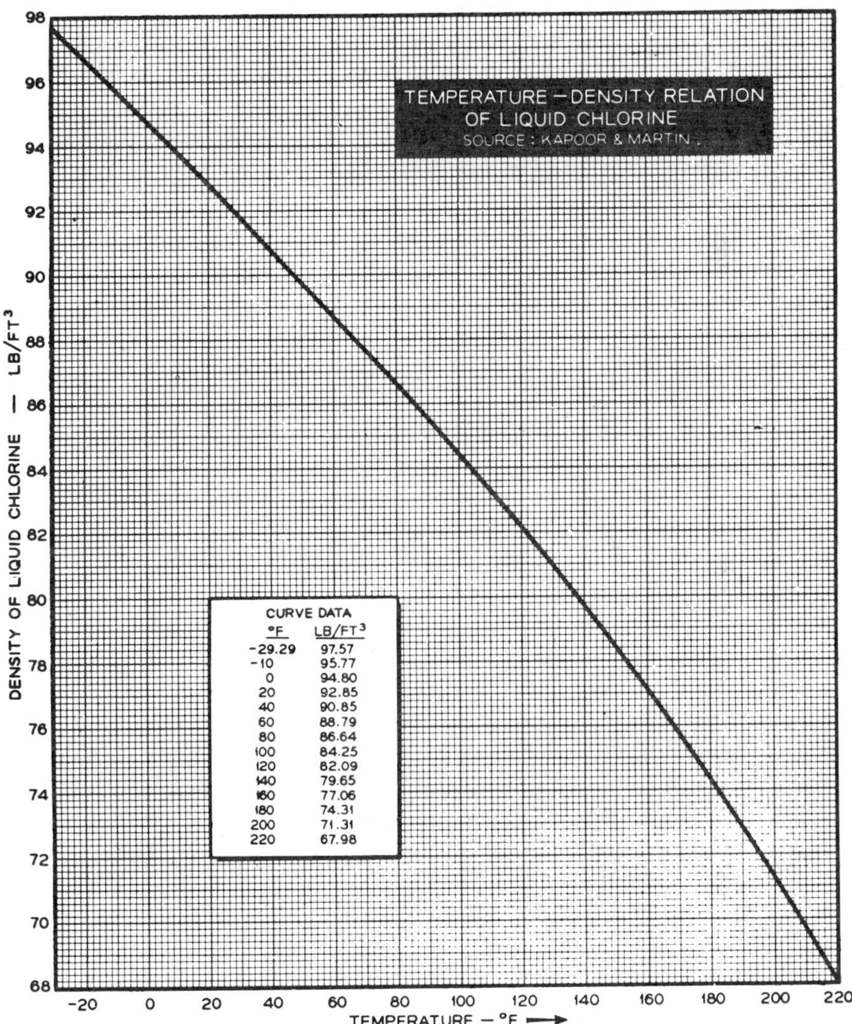

Figure 3-3

TABLE 3-4. THERMODYNAMIC PROPERTIES OF SATURATED CHLORINE [59]

Temp.	Pressure		Volume, ft^3/lb.		Enthalpy (H), Btu/lb			Entropy (S), Btu/lb/$^\circ$F			Temp.
$^\circ$F	Atmos.	psi	Liquid	Gas	Liquid	Vapori-zation	Gas	Liquid	Vapori-zation	Gas	$^\circ$F
-120	0.054607	0.80251	0.0094728	63.931	80.891	135.73	216.62	0.38200	0.39957	0.78156	-120
-100	.12006	1.7644	.0096278	30.739	85.697	133.00	218.70	.39574	.36976	.76551	-100
- 80	.23992	3.5258	.0097911	16.194	90.421	130.34	220.76	.40851	.34329	.75180	- 80
- 60	.44279	6.5073	.0099637	9.1996	95.087	127.72	222.81	.42048	.31955	.74002	- 60
- 40	.76444	11.234	.010146	5.5643	99.720	125.09	224.81	.43177	.29805	.72982	- 40
- 29.29	1.0000	14.696	.010249	4.3458	102.19	123.67	225.86	.43757	.28733	.72490	- 29.29
- 20	1.2473	18.330	.010341	3.5462	104.34	122.42	226.76	.44249	.27842	.72091	- 20
0	1.9396	28.504	.010548	2.3614	108.96	119.67	228.63	.45272	.26034	.71305	0
20	2.8946	42.539	.010769	1.6314	113.59	116.82	230.42	.46253	.24354	.70607	20
40	4.1696	61.276	.011007	1.1625	118.26	113.84	232.09	.47197	.22781	.69979	40
60	5.8251	85.606	.011263	0.85030	122.96	110.68	233.64	.48110	.21297	.69407	60
80	7.9246	116.46	.011542	.63565	127.71	107.32	235.03	.48995	.19885	.68880	80
100	10.534	154.80	.011846	.48389	132.54	103.70	236.24	.49858	.18529	.68387	100
120	13.721	201.65	.012181	.37387	137.46	99.783	237.24	.50704	.17213	.67917	120
140	17.558	258.03	.012555	.29229	142.51	95.483	237.99	.51539	.15922	.67461	140
160	22.119	325.05	.012976	.23053	147.74	90.710	238.45	.52370	.14638	.67008	160
180	27.481	403.87	.013458	.18285	153.21	85.329	238.54	.53208	.13339	.66547	180
200	33.729	495.67	.014024	.14533	159.01	79.151	238.16	.54065	.11998	.66064	200
220	40.952	601.83	.014710	.11519	165.29	71.882	237.17	.54960	.10576	.65536	220
240	49.244	723.69	.015588	.090340	172.26	63.022	235.28	.55921	.090072	.64929	240
260	58.711	862.81	.016827	.069012	180.38	51.529	231.91	.57007	.071598	.64167	260
280	69.464	1020.8	.019103	.048647	191.04	33.987	225.03	.58395	.045948	.62990	280
291.21	76.100	1118.37	.027960	.027960	207.77	00.000	207.77	.60582	.00000	.60582	291.21

Datum: H = 0; S = 0 for solid Cl_2 @ -459.69°F (0°R)

TABLE 3-5.   THERMODYNAMIC PROPERTIES OF SUPERHEATED CHLORINE [59]

Pressure, psia		0	60	120	180	240	300	360	420	480	540	Pressure, psia	
					TEMPERATURE, °F								
10	V	6.8782	7.7991	8.7171	9.6330	10.548	11.461	12.373	13.285	14.197	15.107	V	10
	H	229.30	236.07	242.94	249.92	256.99	264.13	271.33	278.58	285.85	293.13	H	
	S	.74325	.75708	.76961	.78107	.79163	.80142	.81054	.81907	.82706	.83459	S	
20	V	3.3997	3.8669	4.3312	4.7933	5.2539	5.7134	6.1720	6.6299	7.0873	7.5443	V	20
	H	228.94	235.76	242.68	249.69	256.79	263.95	271.17	278.43	285.71	293.01	H	
	S	.72338	.73732	.74992	.76143	.77203	.78185	.79100	.79954	.80755	.81508	S	
40	V		1.9001	2.1377	2.3731	2.6069	2.8395	3.0712	3.3022	3.5327	3.7627	V	40
	H		235.14	242.15	249.23	256.38	263.59	270.84	278.13	285.44	292.76	H	
	S		.71718	.72994	.74156	.75225	.76213	.77132	.77990	.78794	.79549	S	
60	V		1.2438	1.4062	1.5662	1.7245	1.8815	2.0376	2.1930	2.3478	2.5022	V	60
	H		234.50	241.60	248.75	255.96	263.22	270.51	277.83	285.17	292.52	H	
	S		.70506	.71800	.72974	.74051	.75046	.75970	.76832	.77639	.78396	S	
80	V		.91500	1.0401	1.1625	1.2831	1.4024	1.5207	1.6383	1.7553	1.8719	V	80
	H		233.83	241.04	248.27	255.54	262.84	270.18	277.53	284.90	292.27	H	
	S		.69620	.70933	.72120	.73206	.74208	.75137	.76003	.76813	.77573	S	
100	V			.82012	.92012	1.0182	1.1149	1.2106	1.3055	1.3998	1.4937	V	100
	H			240.46	247.78	255.11	262.47	269.84	277.23	284.63	292.02	H	
	S			.70245	.71446	.72542	.73550	.74485	.75355	.76168	.76931	S	
150	V			.52589	.59641	.66464	.73133	.79692	.86168	.92583	.98948	V	150
	H			238.94	246.50	254.01	261.50	268.99	276.47	283.94	291.40	H	
	S			.68940	.70181	.71303	.72331	.73279	.74160	.74982	.75751	S	
200	V			.37753	.43391	.48751	.53935	.58997	.63971	.68878	.73734	V	200
	H			237.30	245.16	252.87	260.52	268.12	275.70	283.25	290.77	H	
	S			.67946	.69237	.70390	.71438	.72401	.73293	.74124	.74900	S	
250	V				.33580	.38091	.42398	.46570	.50647	.54653	.58605	V	250
	H				243.72	251.68	259.50	267.23	274.91	282.55	290.14	H	
	S				.68461	.69650	.70722	.71702	.72607	.73446	.74230	S	
300	V				.26978	.30954	.34691	.38276	.41759	.45167	.48518	V	300
	H				242.19	250.44	258.45	266.32	274.11	281.84	289.50	H	
	S				.67786	.69019	.70118	.71116	.72033	.72882	.73673	S	
400	V				.18534	.21950	.25015	.27888	.30639	.33304	.35907	V	400
	H				238.69	247.75	256.24	264.44	272.48	280.39	288.21	H	
	S				.66591	.67946	.69111	.70150	.71096	.71967	.72773	S	
600	V					.12611	.15202	.17435	.19487	.21427	.23292	V	600
	H					241.13	251.26	260.38	269.03	277.39	285.56	H	
	S					.66116	.67506	.68662	.69681	.70601	.71444	S	
800	V						.10087	.12128	.13877	.15476	.16982	V	800
	H						245.10	255.80	265.30	274.24	282.82	H	
	S						.66091	.67447	.68567	.69550	.70436	S	
1,000	V						.066779	.088605	.10482	.11895	.13195	V	1,000
	H						236.41	250.49	261.26	270.92	279.99	H	
	S						.64543	.66331	.67600	.68663	.69599	S	
1,200	V						.028549	.065890	.081940	.095029	.10672	V	1,200
	H						209.81	244.10	256.83	267.42	277.08	H	
	S						.60796	.65205	.66707	.67872	.68868	S	
1,400	V							.048700	.065424	.077922	.088725	V	1,400
	H							236.05	251.96	263.75	274.08	H	
	S							.63967	.65845	.67142	.68208	S	

Datum: H = 0; S = 0 for solid Cl_2 @ -459.69°F (0°R)

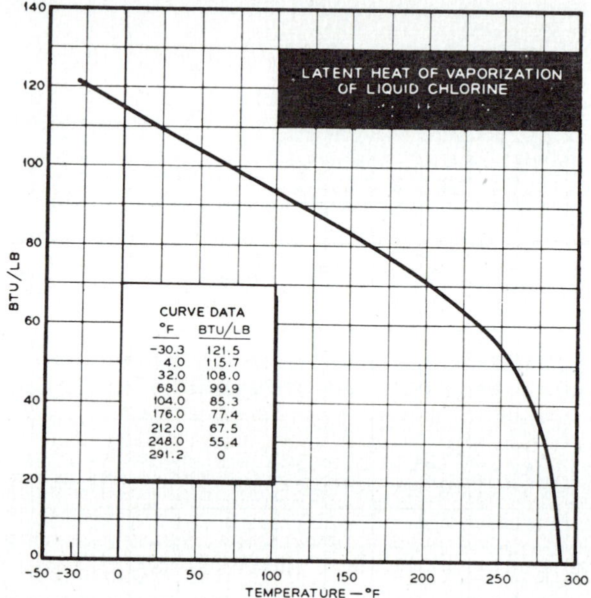

Figure 3-4

constant volume was obtained from Kapoor and Martin data (see Table 3-5).

Specific Volume

The *specific volume* (see Table 3-2) is the volume of a unit mass of chlorine at specified conditions of temperature and pressure.

Surface Tension (see Table 3-6).

Thermal Conductivity (see Table 3-2).

Vapor Pressure

The *vapor pressure* is the pressure of chlorine gas above liquid chlorine when they are in equilibrium. The relation between vapor pressure and temperature of chlorine is shown in Figure 3-6. Kapoor and Martin's data are based on the work of Giauque and Powell[7] between the melting and boiling points, and of Pellaton[22] between 32° F and the critical temperature.

Viscosity

Viscosity is a measure of internal molecular friction when a substance is in motion. The viscosity of chlorine gas is approximately the same as that

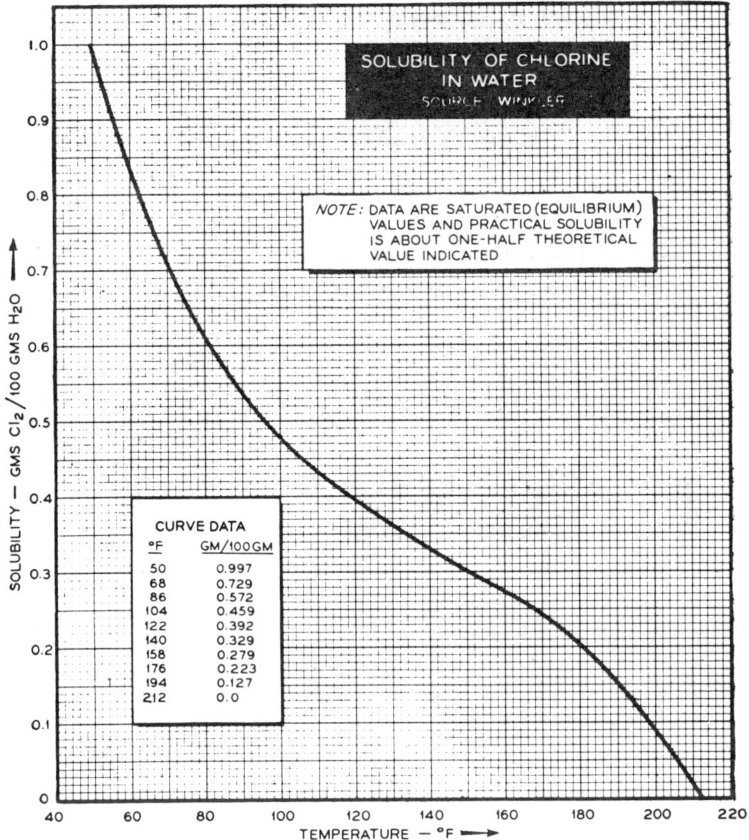

Figure 3-5

of saturated steam between 1 and 10 atm.; that of liquid chlorine is about one-third that of water at the same temperature. Hence, its friction loss in pipelines is definitely lower than that of water. (See Figure 3-7 and Table 3-7).

Volume and Temperature Relationship

The volume of liquid chlorine increases rapidly as its temperature increases. The volume-temperature relationship of liquid chlorine in a ship-

TABLE 3-6. SURFACE TENSION OF CHLORINE

Temperature	$°C$	-49.5	-28.7	0	10	20	28	50	100
	$°F$	-57.1	-19.9	32	50.0	68.0	82.4	122.0	212
Dynes/cm[19]		—	—	21.70	20.02	18.34	—	13.30	4.90
Dynes/cm[82]		29.28	25.23	21.90	—	—	16.99	13.39	—

ping container loaded to its authorized limit is shown in Figure 3-2. Chlorine confined in pipelines or containers that is subjected to a rising temperature increases in volume (and pressure).

BASIC CHEMICAL PROPERTIES

Chlorine is neither flammable nor combustible but it is a highly reactive material. It reacts under specific conditions with most elements, often with extreme rapidity. It reacts with many inorganic and organic compounds, in some cases with explosive violence and frequently with the evolution of heat.

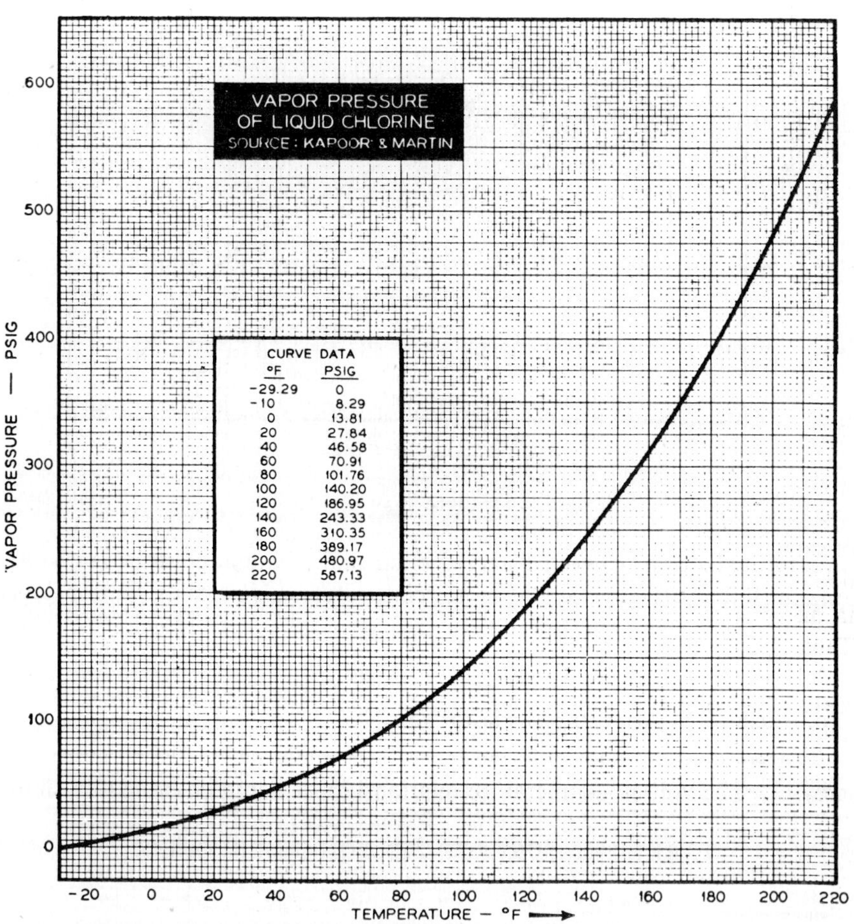

Figure 3-6

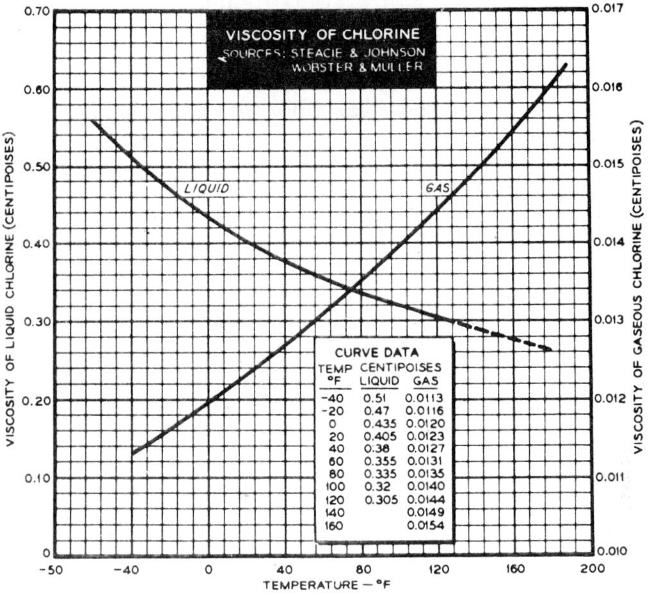

Figure 3-7

Reactions with Water

When chlorine is dissolved in water at temperatures between 49.3° F and 212° F, it reacts to form hypochlorous acid (HOCl) and hydrochloric acid (HCl) as shown (depending on the pH, some free chlorine may exist):

$$Cl_2 + H_2O \rightarrow HOCl + HCl \qquad (3\text{-}1)^{38,61}$$

$$HOCl \rightarrow H^+ + OCl^-$$

This reaction is the basis for the use of chlorine in municipal and swimming pool water treatment, and in the bleaching of paper pulps, textiles and other products.

At temperatures below 49.3° F, chlorine hydrate crystals ($Cl_2 \cdot 8H_2O$)

TABLE 3-7. APPROXIMATE VISCOSITY OF CHLORINE

Temper-ature		-53.1	-45	-35.3	0	20	50	100	150	200	250
	°C	-53.1	-45	-35.3	0	20	50	100	150	200	250
	°F	-63.5	-49	-31.6	32	68	122	212	302	392	482
Gas[30], cps.		—	—	—	—	.0133	.0147	.0168	.0188	.0209	.0228
Liquid[29], cps.		.569	.530	.494	.385	.345	.300	.249	—	—	—

are formed. This also can occur when liquid chlorine comes into contact with water.

Water is decomposed by chlorine at the boiling temperature or under the influence of light, and then decomposition is very slow. The reaction is as follows:

$$2\,Cl_2 + 2\,H_2O \rightarrow 4\,HCl + O_2 \qquad\qquad (3\text{-}2)^{[45,49]}$$

Chlorine reacts with water in the presence of charcoal at temperatures ranging from 0° C to 130° C as is shown:

$$2\,Cl_2 + 2\,H_2O + C \rightarrow 4\,HCl + CO_2 \qquad\qquad (3\text{-}3)^{[52]}$$

The formation of HOCl and HCl acids (Equation 3-1) explains why *moist* chlorine has a corrosive effect on metals which ordinarily are unaffected by gas or liquid chlorine.

Reaction with Gases

Ammonia. Chlorine reacts with ammonia (although it does not react directly with nitrogen) as follows:

$$12\,NH_3 + 6\,Cl_2 \rightarrow 9\,NH_4Cl + NCl_3 + N_2 \qquad\qquad (3\text{-}4)^{[44]}$$

In addition it may react in aqueous ammonia *solution* to form various chloramines or mixtures of chloramines, depending on conditions. Reactions are

$$NH_3 + Cl_2 \rightarrow NH_2Cl + HCl \qquad \text{(Monochloramine)} \qquad (3\text{-}5)^{[44]}$$

$$NH_3 + 2\,Cl_2 \rightarrow NHCl_2 + 2\,HCl \qquad \text{(Dichloramine)} \qquad (3\text{-}6)^{[44]}$$

$$NH_3 + 3\,Cl_2 \rightarrow NCl_3 + 3\,HCl \qquad \text{(Trichloramine or} \qquad (3\text{-}7)^{[44]}$$
$$\text{Nitrogen trichloride}$$

$$NH_3 + XCl_2 \rightarrow N_2 + N_2O + HCl \qquad\qquad (3\text{-}8)^{[44]}$$

Hydrogen. Chlorine reacts spontaneously with hydrogen, slowly in the dark but explosively in sunlight or at high temperatures, to form hydrogen chloride. The rate of reaction in an equivalent mixture of the two gases reportedly is increased markedly by the presence of oxygen or nitrogen chloride. Umland[78] reports that the lower explosive limit of hydrogen-chlorine mixtures varies from 3.1 percent to 8.1 percent, depending on pressure and other variables. The reaction is

$$H_2 + Cl_2 \rightarrow 2\,HCl \qquad\qquad (3\text{-}9)^{[44]}$$

Oxygen. Chlorine does not react directly with oxygen but several oxides of chlorine have been prepared by indirect means. These include chlorine monoxide (Cl_2O), chlorine dioxide (ClO_2), chlorine trioxide (Cl_2O_3), chlorine hexoxide (Cl_2O_6) and chlorine heptoxide (Cl_2O_7). For example, dry gaseous chlorine passed into ozonized oxygen at about 15°C yields chlorine hexoxide, i.e.,

$$Cl_2 + 2O_3 \rightarrow Cl_2O_6 \qquad (3\text{-}10)[76]$$

The photosynthesis of chlorine dioxide proceeds in the presence of light through such various reactions as

$$Cl_2 + 2O_2 \rightarrow 2ClO_2 \qquad (3\text{-}11)[40]$$

$$Cl + O_2 \rightarrow ClO_2 \qquad (3\text{-}12)[40]$$

$$Cl + 2O \rightarrow ClO_2 \qquad (3\text{-}13)[40]$$

The silent electrical discharge on a mixture of chlorine and oxygen, with a trace of nitrogen, yielding $N_2Cl_2O_{13}$, has been claimed.[58,68]

Hydrogen Sulfide. Chlorine reacts with hydrogen sulfide, and chlor-oxidation processes for odor control are frequently employed. Reactions are

$$H_2S + 4Cl_2 + 4H_2O \rightarrow H_2SO_4 + 8HCl \qquad (3\text{-}14)[2]$$

$$H_2S + Cl_2 \rightarrow 2HCl + S \qquad (3\text{-}15)[2]$$

Equation 3-15 is an example of the affinity of chlorine for hydrogen and the removal of hydrogen from some of its compounds.

Sulfur Dioxide. Dechlorination with sulfur dioxide is based on the reaction

$$SO_2 + Cl_2 + 2H_2O \rightarrow H_2SO_4 + 2HCl \qquad (3\text{-}16)[2]$$

Reactions with Metals and Other Elements (see Chapter 4).

Under certain conditions, chlorine reacts with nearly all metals to form metal chlorides which are usually quite soluble in water (cuprous, silver, plumbous, mercurous, platinous and thallous chlorides are only slightly soluble[68]), e.g.,

$$2Ag + Cl_2 \rightarrow 2AgCl \qquad (3\text{-}17)[60]$$

$$2Sb + 3Cl_2 \rightarrow 2SbCl3 \qquad (3\text{-}18)[60]$$

Temperatures encountered in normal uses of chlorine do not exceed 230° F. Below this temperature, copper, iron, lead, nickel, platinum, silver, steel and tantalum are chemically resistant to dry chlorine, gas or liquid. Certain copper and ferrous alloys, including "Hastelloy C," "Monel," and types 304 and 316 stainless steel also are resistant. The reaction rate of chlorine with most metals increases with extreme rapidity above certain temperatures, beyond which construction materials must be selected with great care; these temperatures are considerably lower if the metal or alloy is in finely-divided, powdered, sponge, or wire form.

Dry chlorine reacts with aluminum, arsenic, gold, mercury, selenium, tellurium, tin and titanium, e.g.,

$$Sn + 2\,Cl_2 \rightarrow SnCl_4 + 127{,}250\,cal \qquad (3\text{-}19)^{67}$$

At certain temperatures, potassium and sodium burn in chlorine gas. Carbon steel ignites at 483° F. Antimony, arsenic, bismuth, boron, copper, iron, phosphorus, and certain of their alloys, in powdered, sponge, or wire in finely-divided form, ignite spontaneously in chlorine.

Moist chlorine, primarily because of the hydrochloric and hypochlorous acids formed through hydrolysis, is very corrosive to all common metals. Gold, platinum, silver and titanium are resistant. At temperatures below 300° F, tantalum is totally inert to wet (and dry) chlorine. Moist (and dry) chlorine reacts readily with mercury.

Chlorine oxidizes iron in solution (e.g., in potable water treatment) as follows:

$$Cl_2 + Fe \rightarrow FeCl_2 \qquad (3\text{-}20)^2$$

$$Fe - 2\,e \rightarrow Fe^{++}$$

$$Cl_2 + 2\,e \rightarrow 2\,Cl^-$$

Dechlorination with activated carbon is based on the reaction

$$C + 2\,Cl_2 + 2\,H_2O \rightarrow 4\,HCl + CO_2 \qquad (3\text{-}21)^{52}$$

Treatment of phosphorous-containing waste waters is based on the reactions

$$2\,P + 3\,Cl_2 \rightarrow 2\,PCl_3 \qquad (3\text{-}22)^2$$

$$PCl_3 + 3\,H_2O \rightarrow H_3PO_3 + 3\,HCl \qquad (3\text{-}23)^2$$

Chlorine displaces bromine and iodine from their corresponding halide salts, and these reactions (see Table 3-8) are utilized in the commercial preparation of elementary bromine and iodine. Chlorine gas, led into iodine until it liquefies, forms monochloro-iodine

$$Cl_2 + I_2 \rightarrow 2\,CII \qquad (3\text{-}24)[51]$$

Chlorine passed over iodine until a red-yellow solid is formed yields iodine trichloride

$$3\,Cl_2 + I_2 \rightarrow 2\,ICl_3 \qquad (3\text{-}25)[51]$$

Reactions with bromine and fluorine also have been claimed, e.g.,

$$Cl_2 + F_2 + H_2O \rightarrow 2\,HF + 2\,HOCl \qquad (3\text{-}26)[49,68]$$

Reactions with Inorganics

Chlorine reacts with a great many inorganic compounds. The products of some of these reactions have considerable commercial significance. The preparation of soda and lime bleaches through the reaction of alkalies and alkaline earth metal hydroxides are prime examples:

$$2\,Ca(OH)_2 + 2\,Cl_2 + 2\,H_2O \rightarrow Ca(OCl)_2 \cdot 4\,H_2O + CaCl_2 \qquad (3\text{-}27)[35]$$

$$2\,CaO + 2\,Cl_2 + 4\,H_2O \rightarrow Ca(OCl)_2 \cdot 4\,H_2O + CaCl_2 \qquad (3\text{-}28)[51,64]$$

$$2\,NaOH + Cl_2 \rightarrow NaOCl + NaCl + H_2O \qquad (3\text{-}29)[35,68]$$

These hypochlorites are powerful oxidizers and have extensive commercial application.

Reactions with a variety of other inorganic compounds are summarized in Table 3-8.

Reactions with Organics

There are thousands of chlorine-containing organic compounds of commercial importance that are made by chlorinating hydrocarbons. Most involve addition or substitution reactions, and some involve decomposition of the hydrocarbon molecule (see Equation 3-35). In most instances the reaction is very rapid and gives off heat. It is in the organic and petrochemicals field that chlorine is most widely used as a chemical intermediate.

TABLE 3-8. SOME REACTIONS* OF CHLORINE WITH INORGANIC COMPOUNDS[60] (See Also Ref. 68)

Material	Reaction	Reference
Ammonium chloride	$3\ Cl_2 + NH_4Cl \longrightarrow NCl_3 + 4HCl$	44
Ammonium chloride	$3\ HOCl + NH_4Cl \longrightarrow NCl_3 + 3\ H_2O + HCl$	55
Calcium borate	$3(Ca_2B_4O_7 \cdot 3H_2O) + 6\ Cl_2 + 9\ H_2O \longrightarrow 12\ H_3BO_3 + 5\ CaCl_2 + Ca(ClO_3)_2$	69
Calcium carbonate	$CaCO_3 + 2\ Cl_2 \longrightarrow Cl_2O + CaCl_2 + CO_2$	53
Calcium carbonate	$14\ Cl_2 + 10\ CaCO_3 + 4\ H_2O \longrightarrow 8\ HClO + 9\ CaCl_2 + Ca(ClO_3)_2 + 10\ CO_2$	72
Calcium carbonate	$2\ Cl_2 + H_2O + CaCO_3 \longrightarrow CaCl_2 + 2\ HClO + CO_2$	72
Carbon monoxide	$CO + Cl_2 \rightleftharpoons COCl_2$	43
Hydrocyanic acid	$Cl_2 + 3\ HCN + CHCl_3 \longrightarrow C_3N_3Cl_3 + (CH_3Cl + HCl)$	57
Hydrogen peroxide	$H_2O_2 + Cl_2 \longrightarrow 2\ HOCl$	48
	$H_2O_2 + Cl_2 \longrightarrow 2\ HCl + O_2$	79
Iron bicarbonate	$2\ Fe(HCO_3)_2 + Cl_2 + Ca(HCO_3)_2 \longrightarrow 2\ Fe(OH)_3 + CaCl_2 + 6\ CO_2$	
Iron sulfate	$6\ FeSO_4 \cdot 7\ H_2O + 3\ Cl_2 \longrightarrow 2\ Fe_2(SO_4)_3 + 2\ FeCl_3 + 4\ H_2O$	
Manganous sulfate	$MnSO_4 + Cl_2 + 4\ NaOH \longrightarrow MnO_2 + 2\ NaCl + Na_2SO_4 + 2\ H_2O$	
Mercuric oxide	$HgO + 2\ Cl_2 \longrightarrow HgCl_2 + Cl_2O$	39
	$2\ HgO + 2\ Cl_2 \longrightarrow HgO \cdot HgCl_2 + Cl_2O$	36,50,73
	$2\ HgO + 2\ Cl_2 + H_2O \longrightarrow Hg_2OCl_2 + 2\ HClO$	37,41
	$2\ Cl_2 + 2\ HgO + O_2 \longrightarrow 2\ Cl_2O + 2\ HgO$	77
Nitric oxide	$2\ NO + Cl_2 \longrightarrow 2\ NOCl$	46
Potassium arsenite	$K_3AsO_3 + H_2O + Cl_2 \longrightarrow KH_2AsO_4 + 2\ KCl$	68
Potassium chlorite	$KClO + 2\ Cl_2 + 2\ H_2O \longrightarrow KClO_3 + 4\ HCl$	71
Potassium cyanide	$2\ Cl + KCN \longrightarrow CNCl + KCl$	63
Potassium hydroxide	$6\ KOH + 3\ Cl_2 \longrightarrow 5\ KCl + KClO_3 + 3\ H_2O$	34
	$6\ KOH + 3\ Cl_2 \longrightarrow 3\ KCl + 3\ KOCl + 3\ H_2O$	34
Potassium iodide	$Cl_2 + 2\ KI \longrightarrow 2\ KCl + I_2$	65
Silicon hydride (Silane)	$SiH_4 + 4\ Cl_2 \longrightarrow SiCl_4 + 4\ HCl$	65
Silver chlorate	$2\ AgClO_3 + Cl_2 \longrightarrow 2\ AgCl + 2\ ClO_2 + O_2$	62
Silver chloride	$Ag_2O + 2\ Cl_2 + H_2O \longrightarrow 2\ AgCl + 2\ HClO$	68,75
Silver fluoride	$8\ AgF + 4\ Cl_2 + 4\ H_2O \longrightarrow 6\ AgCl + 2\ AgClO + 8\ HF + O_2$	54
	$4\ AgF + 2\ Cl_2 + Pt \longrightarrow 4\ AgCl + PtF_4$	54

Silver nitrate	$Cl_2 + 2\ AgNO_3 \longrightarrow 2\ AgCl + 2\ NO_2 + O_2$	70
Silver carbonate	$Ag_2CO_3 + 2\ Cl_2 + H_2O \longrightarrow 2\ AgCl + CO_2 + 2\ HClO$	74,75
	$6\ Cl_2 + 3\ Ag_2O + 3\ H_2O \longrightarrow 6\ AgCl + 6\ HClO$	
	$6\ HClO + 3\ Ag_2O \longrightarrow 3\ H_2O + 6\ AgClO$	
	$6AgClO \longrightarrow 4AgCl + 2AgClO_3$	
Sodium bisulfite	$NaHSO_3 + Cl_2 + H_2O \longrightarrow NaHSO_4 + 2\ HCl$	
Sodium chlorite	$2\ NaClO_2 + Cl_2 \longrightarrow 2\ NaCl + 2\ ClO$	80
Sodium hypochlorite	$Cl_2 + NaOCl + H_2O \longrightarrow 2\ HOCl + 2\ ClO_2 + NaCl$	47
Sodium iodide	$Cl_2 + 2\ NaI \longrightarrow 2\ NaCl + I_2$	45
Sodium metabisulfite	$Na_2S_2O_5 + 2\ Cl_2 + 3\ H_2O \longrightarrow 2\ NaHSO_4 + 4\ HCl$	2
Sodium silicate	$Na_2O \cdot SiO_2 + Cl_2 \longrightarrow NaOCl + NaCl + SiO_2$	2
Sodium cyanide	$2\ Cl_2 + 4\ NaOH + 2\ NaCN \longrightarrow 2\ NaCNO + 4\ NaCl + 2\ H_2O$	42
	$3Cl_2 + 6\ NaOH + 2\ NaCNO \longrightarrow 2\ NaHCO_3 + N_2 + 6\ NaCl + 2\ H_2O$	
	$5Cl_2 + 10\ NaOH + 2\ NaCN \longrightarrow 2\ NaHCO_3 + 10\ NaCl + N_2 + 4\ H_2O$	
Sodium sulfite	$Na_2SO_3 + Cl_2 + H_2O \longrightarrow Na_2SO_4 + 2\ HCl$	
Sodium thiosulfate	$2\ Na_2S_2O_3 + Cl_2 \longrightarrow Na_2S_4O_6 + 2\ NaCl$	81
	$Na_2S_2O_3 + 4\ Cl_2 + 5\ H_2O \longrightarrow H_2SO_4 + Na_2SO_4 + 8\ HCl$	68
Tin Chloride	$SnCl_2 + Cl_2 \longrightarrow SnCl_4$	68

*Some reactions are not spontaneous but occur only under very specific circumstances.

Chlorine Addition. Chlorine adds directly at points of unsaturation to both sides of a double bond, e.g.,

$$+ 3\,Cl2 \rightarrow$$

(*Benzene, aromatic*)   (*Benzene hexachloride, saturated*)

$$C_6H_6 + 3\,Cl_2 \rightarrow C_6H_6Cl_6 \qquad (3\text{-}30)^{59}$$

Chlorine Substitution. Under controlled conditions chlorine reacts with hydrocarbons to form chlorinated hydrocarbons and hydrogen chloride, e.g.,

$$+ Cl_2 \rightarrow \qquad\qquad + HCl$$

(*Benzene*)   (*Monochlorobenzene*)

$$C_6H_6 + Cl_2 \rightarrow C_6H_5Cl + HCl \qquad (3\text{-}31)^2$$

Similarly, the reactions with methane and its chlorine derivatives are

$$CH_4 + Cl_2 \rightarrow CH_3Cl + HCl \qquad (3\text{-}32)$$

$$CH_3Cl + Cl_2 \rightarrow CH_2Cl_2 + HCl \qquad (3\text{-}33)$$

$$CH_2Cl_2 + Cl_2 \rightarrow CHCl_3 + HCl \qquad (3\text{-}34)$$

$$CHCl_3 + Cl_2 \rightarrow CCl_4 + HCl \qquad (3\text{-}35)$$

Hydrocarbon Decomposition. When turpentine, chiefly *a*-pinene, catches fire in chlorine, the reaction products are carbon and hydrochloric acid. The reaction is as shown:

$$C_{10}H_{16} + 8\,Cl_2 \rightarrow 10\,C + 16\,HCl \qquad (3\text{-}36)$$

References

Physical and Thermodynamic Properties

1. Brauns, D. H., "Empirical Relation between the Atomic Dimensions and the Melting and Sublimation Points of the Noble Gases, Halogens, and Elements of the Sulfur Group," Natl. Bur. Standards Res. paper, RP915, Washington, D. C. (1936).

2. Chlorine Producers. Miscellaneous Chlorine Safety Bulletins and Pamphlets.

3. Dewer, J., *Phil. Mag.* **5** (18), 210 (1884); (Ziegler, Ref. 33).

4. Eastman, E. D., "Specific Heats of Gases at High Temperatures," *U. S. Bur. Mines Tech. Paper No. 445,* Washington, D. C. (1929).

5. Estreicher, T., and Schnerr, A., *Krakauer Acad.* (Cracow), 345 (1910); (Ziegler, Ref. 33).

6. Eucken and Hoffman., *Z. Physik. Chem.*, **B5,** 422 (1929); (Ziegler, Ref. 33).

7. Giauque, W. F., and Powell, T. M., "Chlorine: The Heat Capacity, Vapor Pressure, Heats of Fusion and Vaporization and Entropy," *J. Am. Chem. Soc.,* **61,** 1970 (1939).

8. Gmelin. "Gmelins Handbuch der anorganischen Chemie," 8. Auflage. System-Nummer 6: Chlor. Verlag Chemie G.m.b.H., Berlin (1927).

9. Hulme, R. E., "Thermodynamic Properties of Superheated Chlorine," *Chem. Eng.,* **56** (11); (1949).

10. Hulme, R. E., and Tillman, A. B., "Thermodynamic Properties of Chlorine," *Chem. Eng.,* **56** (1); (1949)

11. Hodgman, Charles D., *et al.,* "Handbook of Chemistry and Physics," Ed. 38, Cleveland, Ohio, Chemical Rubber Publishing Co., (1956).

12. Kapoor, Rajendra, and Martin, Joseph J., "Thermodynamic Properties of Chlorine," *Eng. Res. Inst.,* Univ. of Michigan (1957).

13. Kiess, C. C., "A New Description and Analysis of the Arc Spectrum of Chlorine," Natl. Bur. Standards Res. paper RP570, Washington, D. C. (1933).

14. Knietsch, R., "On the Properties of Liquid Chlorine," *Ann. Chem., Justus Liebigs,* **259,** 100 (1890).

15. Ladenburg, R., *Ber. deut. chem. Ges.,* **11,** 818 (1870); (Ziegler, Ref. 33).

16. Lange, A., "Ueber einige Eigenschaften des verflussigten Chlor," *Z. angew. Chem.,* **13,** 683 (1900).

17. Lange, N. A., "Handbook of Chemistry," Sandusky, Ohio, Handbook Publishers, Inc. (1956).

18. Mellor, J. W., "A Comprehensive Treatise on Inorganic and Theoretical Chemistry," Vol. 2, New York, Longmans, Green and Co. (1927).

19. Marchand, J., *Chim. Phys.,* **11,** 575 (1913); (Mellor, Ref. 18).

20. Morris, J. Carrell., "The Mechanism of the Hydrolysis of Chlorine," *J. Am. Chem. Soc.,* **68,** 1692 (1946).

21. National Research Council. "International Critical Tables," New York, McGraw-Hill Book Co. 1926, 1928, 1929, 1930.

22. Pellaton, M., "Constantes Physiques du Chlor," *J. Chim. Phys.,* **13,** 426 (1915).

23. Pickering, S. F., "A Review of the Literature Relating to the Critical Constants of Various Gases," Natl. Bur. Standards, Sci. Papers, No. 541, Washington, D. C. (1926).

24. Pickering, S. F., "Relations between the Temperatures, Pressures and Densities of Gases," *Natl. Bur. Standards Circ.,* No. 279, Washington, D. C. (1926).

25. Richards and Stull. *J. Am. Chem. Soc.,* **26,** 408 (1904).

26. Ross, A. S., and Maass, O., "The Density of Gaseous Chlorine," *Can. J. Research,* **18, B** 55–65, (1940).

27. Rossini, F. D., *et al.,* "Selected Values of Chemical Thermodynamic Properties," *Natl. Bur. Standards Circ.,* No. 500, Washington, D. C. (1952).

28. Smith., *Ind. Eng. Chem.*, **22**, 1246 (1930).
29. Steacie and Johnson, "The Viscosities of the Liquid Halogens," *J. Am. Chem. Soc.*, **47**, 756 (1925); (Mellor, Ref. 18).
30. Trautz and Ruf., *Ann. phys.*, **20**, 127 (1934); (Ziegler, Ref. 33).
31. Winkler, L., "Math. es Termezettudomany, Ertesito," **25** (1907).
32. Wobster and Muller, *Kolloid-Beihefte*, **52**, 162 (1941).
33. Ziegler, L., "Thermische Eigenshaften von Chlor.," *Chem. Ing. Tech.*, **22**, 229 (1950).

Chemical Properties

34. Angel, Gosta., *Chem. & Met. Eng.*, **33**, 460 (1926); (Ref. 60).
35. Avery, J. M., and Evans, R. F., *Chem. & Met. Eng.*, **52**, 94 (1945), (Ref. 60).
36. Balard, A. J., *Taylor's Scientific Memoirs*, **1**, 269 (1837); (Ref. 18).
37. Balard, A. J., *Ann. Chim. Phys.*, **57**, 225 (1834); (Ref. 18).
38. Bergstrom, F. W., *J. Am. Chem. Soc.*, **48**, 2318 (1926); (Ref. 60).
39. Billeter, O., *Helvetica chimica Acta*, **1**, 486 (1918); (Ref. 60).
40. Bowen., *J. Chem. Soc.* (London) **127**, 342 and 510 (1925); (Ref. 60).
41. Carius., *Ann. Chem., Justus Liebig*, **126**, 196 (1863); (Ref. 60).
42. Chamberlin, N. S., and Snyder, H. B., "Treatment of Cyanide and Chromium Wastes," *Proc. Conf. Ind. Health*, Houston, Texas (1952).
43. Chapman, D. L., and Gee, F. E., *J. Chem. Soc.* (London) **99**, 1726 (1911); (Ref. 60).
44. Chapman, D. L., and Vodder, L., *J. Chem. Soc.* (London), **95**, 140 (1909); (Ref. 60).
45. Clark and Iseley., *J. Ind. Eng. Chem.*, **12**, 11-17 (1920); (Ref. 60).
46. Coates, J. E., and Finning, A., *J. Chem. Soc.* (London) **105**, 2444 (1914); (Ref. 60).
47. De Mallman, M. (Ref. 18).
48. Fairley, T., *B. A. Repl.*, **57** (1874); (Ref. 18).
49. von Falckenstein, V., *Z. Elektrochem.*, **12**, 763 (1906); (Ref. 60).
50. Garzarolli, K., and Schacherl, G., *Ann. Chem., Justus Liebig*, **230**, 776 (1885); (Ref. 60).
51. GersH, R., *Ber. deut. chem. Ges.*, **6**, 770 and 1321 (1873); (Ref. 60).
52. Gibbs, H. D., *J. Ind. Eng. Chem.*, **12**, 539 (1920); (Ref. 60).
53. Gopuer., *Gazz. chim. ital.*, **20**, 113 (1919); (Ref. 60).
54. Gore, George., *Chem. News*, **23**, 13 (1871); (Ref. 60).
55. Hanson, W. H., *Cereal Chem.*, **9**, 360 (1932); (Ref. 60).
56. Hardie, D. W. F., "Electrolytic Manufacture of Chemicals from Salt," ed. 1, Oxford Univ. Press, 1959.
57. Hartley, W. N., Dobbie, J. J., and Landen, A., *J. Chem. Soc.* (London), **79**, 852 (1901); (Ref. 60).
58. Hautefeuille, P., and Chappius, J., *Compt. rend.*, **98**, 273 (1884); (Ref. 18).
59. Henninger, A., *Ber. deut. chem. Ges.*, **5**, 222 (1872); (Ref. 60).
60. Jacobson, C. A., "Encyclopedia of Chemical Reactions," Vol. 2, Reinhold Publishing Corp., New York, 1948.

61. Jakowkin, A. A., *Z. phys. Chem.*, **29**, 613 (1899); (Ref. 18).
62. King, F. E., and Partington, J. R., *J. Chem. Soc.* (London), **129**(1) 926 (1926); (Ref. 60).
63. Langolis, *Ann. Chem., Justus Liebig*, **1**, 383 (1861); (Ref. 60).
64. Lunge, G., and Landolt, L., *Chem. Ind.*, **8**, 337 (1885) and *J. Soc. Chem. Ind.*, **4**, 722 (1885); (Ref. 18).
65. Lunge, G., and Shappi, H., *Dingler's J.*, **237**, 63 (1880); (Ref. 60).
66. Lebeau, P., *Compt. rend.*, **143**, 425 (1906); (Ref. 18).
67. Mantell, C. L., *Chem. & Met. Eng.*, **33**, 477 (1926); (Ref. 60).
68. Mellor, J. W., *J. Chem. Soc.* (London), **79**, 225 (1901); (Ref. 60).
69. Moore, C. C., British Patent 20384 (1899); (Ref. 18).
70. Namias, R., *Gazz. chim. ital.*, **26**, 41 (1896); (Ref. 60).
71. Oettel, Felix., *Z. Elektrochem.*, **1**, 360 (1895); (Ref. 60).
72. Richardson, A., *J. Chem. Soc.* (London), **93**, 288 (1908); (Ref. 60).
73. Russell, E. J., *J. Chem. Soc.* (London), **77**, 367 (1900); (Ref. 60).
74. Stas, J. S., *Chem. News*, **1**, 173 (1867); (Ref. 60).
75. Stas, J. S., *Mem. Acad. Belgique*, **35**, 92 (1865); (Ref. 18).
76. Schumacher, H. J., and Steiger, G., *Z. anorg. Chem.*, **184**, 272 (1929); (Ref. 60).
77. Secay, C. H., and Cady, G. H., *J. Am. Chem. Soc.*, **62**, 1036 (1940); (Ref. 60).
78. Umland, A. W., *J. Electrochem. Soc.*, **101**, 626 (1954); (Ref. 60).
79. Weltzein, M. C., *Chem. News*, **14**, 1 (1866); (Ref. 60).
80. White, J. F., and Vincent, G. P., *Chem. & Met. Eng.*, **47**, 630 (1940); (Ref. 60).
81. Willson, Virgil A., *Ind. Eng. Chem.*, Anal. ed., **7**, 44 (1935); (Ref. 60).
82. Johnson and McIntosh. *J. Am. Chem. Soc.*, **31**, 1143 (1909).
83. Nernst., "Die theoretische und experimentellen Grundlagen des Neuen Warmersatzes," 2 Ayfl., Halle a.s.S. 60 (1924).
84. Wichers, E., *J. Am. Chem. Soc.*, 74 (1952).

The assistance of Robert L. Mitchell, Jr., Secretary-Treasurer of The Chlorine Institute, Inc., in the preparation of material included herein, is acknowledged with appreciation by the author.

4. SAFE HANDLING OF CHLORINE*

EDMUND J. LAUBUSCH

The Chlorine Institute, Inc.

Chlorine in commerce, classified as a nonflammable compressed gas, is a liquefied gas under pressure; the chlorine in containers has both a liquid and a gas phase. All containers used in the transportation of chlorine in North America, as well as all means of their transportation, are controlled by strict governmental or other regulations. Interstate and foreign shipments originating within the limits of the United States must comply with Interstate Commerce Commission (ICC) Regulations[9,10] regarding loading, handling and marking of containers; water shipments of chlorine also must comply with U. S. Coast Guard Regulations.[12,13] Shipments in Canada must comply with identical Regulations of the Board of Transport Commissioners for Canada (BTCC).[3] Similarly, all chlorine shipping containers used in interstate (and, in practice, intra-state) commerce must comply with ICC Specifications,† and shipping containers in Canada must comply with BTCC Specifications. The ICC and the BTCC use the services of the Bureau of Explosives of the Association of American Railroads[1,2] to carry out the requirements of the law.

Hazards associated with chlorine handling are attributable to its chemical reactivity, physical properties and toxicological character. Neither liquid nor gaseous chlorine is explosive or flammable, but both react chemically (often vigorously) with many inorganic and organic substances, usually

*Information herein is based on accident prevention experiences of various members of The Chlorine Institute, Inc.; it does not necessarily reflect the position of, or endorsement by the Institute. Not every acceptable safety procedure is included; abnormal or unusual circumstances may require modified or additional procedures. References to proprietary products are for illustrative purposes only and are not intended to be all-inclusive. Literature references are included to serve readers who require more detailed, original information about many aspects of the subject which, of necessity, are presented only in brief. Regulations cited are those that apply in the United States and Canada.

†Except tank barges, which must comply with U. S. Coast Guard Specifications.

with the evolution of heat; chlorine also supports combustion. *Dry* chlorine does not react with (corrode) many metals at ordinary temperatures, but it is very reactive (strongly corrosive) in the presence of moisture. As discussed in Chapter 3, chlorine reacts with water to form hypochlorous and hydrochloric acids—a condition which complicates fire fighting where escaping chlorine is involved.

The volume of liquid chlorine increases considerably with increasing temperature (see Figure 3-2). Suitable precautions must be taken to preclude a build-up of excessive pressure and possible hydrostatic rupture in containers, pipelines and other equipment containing liquid chlorine. One volume of liquid chlorine vaporizes into about 450 volumes of gas, a significant fact, and one that should be reckoned with in handling chlorine leaks. Moreover, chlorine is only slightly soluble in water, and escaped chlorine usually can not be absorbed satisfactorily by flushing or by water-absorption schemes.

Liquid chlorine is a skin irritant and can cause severe damage to body tissues. It vaporizes rapidly to gas at normal atmospheric pressure and temperature conditions. Chlorine gas, in low concentrations, irritates the mucous membranes, the respiratory system and the skin; in extreme cases, the difficulty of breathing may increase to the point where death can result from suffocation (see Table 4-8).

CHLORINE CONTAINERS

Chlorine is shipped in steel containers of various designs, all of which are pressure-tested periodically as required by Regulations and equipped with one or more safety devices (depending on the type of container) which will be described later. The maximum filling density of all shipping containers is 125 percent. (Filling density is defined by ICC as "... the percent ratio of the weight of gas in the tank to the weight of water that the tank will hold .... at 60°F.")

Cylinders and Ton Containers

Construction. Chlorine cylinders (complying with ICC Spec. 3A480, 3AA480, 25, 3, 3BN480 and B.E. 25) are of seamless construction and have a capacity of from 1 to 150 lb; those of 100 and 150 lb predominate. The only opening permitted in cylinders is the valve connection at the top (see Figure 4-1). Approximate dimensions and weights of common cylinders are shown in Table 4-1.

Most cylinders are equipped with a Chlorine Institute Standard Cylinder Valve. Two types are in current use (see Figure 4-2). The valve outlet threads on both types are special, straight threads (conforming to connec-

Figure 4-1. Chlorine cylinders: foot ring type (left) and bumped-bottom type (right).

tion 660 of American Standard B57.1).[8] Both have a fusible plug safety device, designed to melt between 158° F and 165° F (70° C and 73.9° C), located below the valve seat to permit venting of the cylinder in the event of excessive pressures within resulting from fire and other high-temperature, increasing pressure situations.

Ton containers (complying with Spec. ICC 106A500, 106A500X and B.E. 27) are of welded construction and have a loaded weight of as much as 3700 pounds (see Figure 4-3). Ton containers have two convex-inward heads and sides that are crimped inward to form chimes (which provide a substantial grip for lifting clamps). Most containers have eight openings (for fusible plugs and valves). Approximate dimensions and weights of chlorine ton containers are shown in Table 4-1.

Ton containers have two identical operating valves, located on one head, which are similar to cylinder operating valves except that they contain no fusible safety plug and usually have a larger internal passage (see Figure 4-4). Each valve is connected to an eduction pipe, one terminating in the gas phase and the other in the liquid phase, depending on the position of the container (see Figure 4-3).

Most ton containers have six ($\frac{3}{4}$ or 1-in.) fusible metal safety plugs, three in each head. These are designed to melt within the same temperature range as the safety plug in the cylinder valve, and have the same purpose.

The valves on cylinders and ton containers are protected by a hood which should always be kept in place except during evacuation of chlorine.

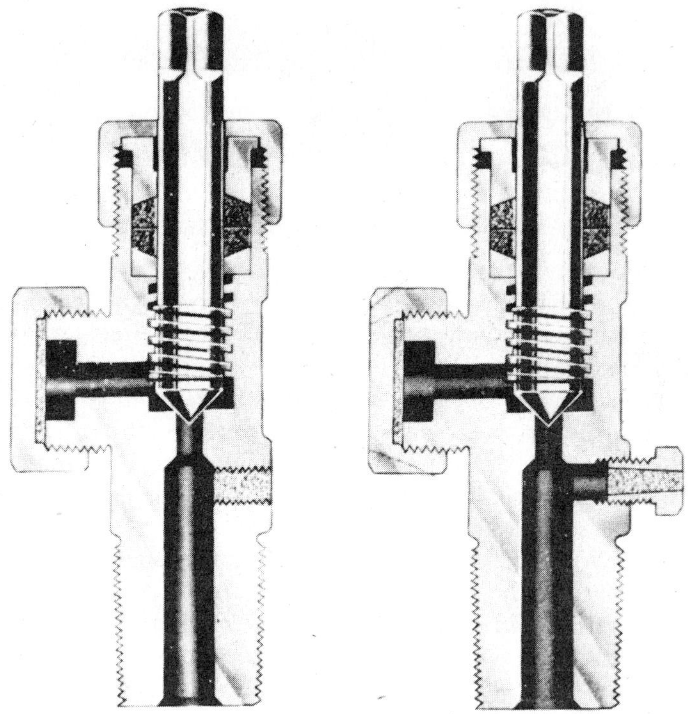

Figure 4-2. Chlorine Institute standard cylinder valves: poured type fusible plug (left) and screwed type fusible plug (right).

Shipping. Cylinders can be shipped by rail in carload or less than carload lots. Any number of cylinders can be shipped by truck in less than carload lots or in truckload quantities of various minimum weights.

Loaded ton containers may be shipped by rail only as part of a multi-unit (TMU) tank car. One or more ton containers may be transported by trucks or semi-trailers.

Moving. Loaded cylinders may be moved safely on a properly balanced hand truck having a clamp support at least two thirds of the way up the cylinder. If they are to be lifted and an elevator is not available, a crane or hoist having a special cradle or carrier should be used. Cylinders should not be lifted by means of the valve protection hood, nor should rope slings, chains or magnetic devices be used. Unloading docks preferably should be at truck or car-bed level.

Ton containers should be handled with a suitable lift clamp in combination with a hoist or crane having at least two tons capacity.

Cylinders and ton containers being trucked should be carefully chocked,

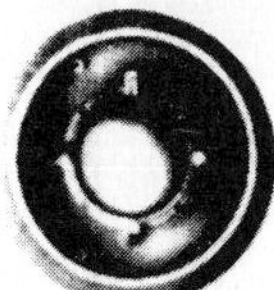 

Figure 4-3.   Chlorine ton container.

clamped, or otherwise suitably supported to prevent shifting and rolling. They should not be permitted to drop, and no object should be allowed to strike them with force.

Storage. Cylinders and ton containers, whether empty or full, should be stored in a dry, well-ventilated area, protected from external heat sources (such as steam pipes).   Fire-resistant storage areas are recommended and subsurface areas are to be avoided.   Full and empty units require spearate storage space.

Storage areas should be remote from elevators, gangways or ventilating systems because, in the event of a chlorine leak, dangerous concentrations of chlorine may spread rapidly.   Locations where heavy objects may strike or fall on containers should be avoided.   The storage area should be separate from that in which other compressed gas containers are stored, and should contain no turpentine, ether, anhydrous ammonia, finely divided metals, hydrocarbons or other flammable materials.

Cylinders should be stored in an upright position.   Ton containers should be stored on their sides, above the ground or floor, on steel or

TABLE 4-1.   DIMENSIONS AND WEIGHTS OF CYLINDERS AND TON CONTAINERS[7]

Capacity (lb)	Weight Class —	Empty Weight[a] (lb)	Outside Diameter (in.)	Over all Height[b] or Length (in.)
100	Heavy	80–115	8¼–8½	53–59
100	Light	63–79	8¼–8½	53–55
100	Heavy	95–105	10½–10¾	40–43
100	Light	63–76	10½–10¾	39½–43
105	Heavy	85	10¼–10½	41¼
105	Light	72–77	10¼–10½	40–41
105	Light	72–77	8¼–8½	57–58
150	Heavy	120–140	10½–10¾	53–56
150	Light	85–105	10¼–10¾	53–56
2000	—	1300–1650	30	79¾–82½

Notes: [a]Weight includes protection hood and valve(s).
   [b]Height to top of valve protection hood; height to center line of valve outlet is about 3½ in. less.

concrete supports; they generally should not be stacked or racked more than one high. Storage should be arranged to facilitate moving and frequent inspection, with minimum disturbance to other units.

Using. Cylinders and ton containers should be used in the order received. Damaged units should be reported promptly to the supplier; repairs or alterations should not be made by consumers. Connected units should be properly secured to prevent shifting.

Cylinders normally should be emptied in the gas phase, standing secured in an upright position. If it is necessary to empty them in the liquid phase, they should be partially inverted and clamped securely on a rack set at an angle of about 60 degrees to the horizontal. The dependable, continuous discharge rate of chlorine gas from a single 100 or 150-lb cylinder, without sweating under normal temperature (70°F) and air circulation conditions, is about 1¾ lb/hr against a 35 psi back pressure; for liquid, the dependable discharge rate under the same conditions is about 200 lb/hr. If sweating can be tolerated these rates can be doubled; for short periods they may be greatly exceeded.

Ton containers set in a horizontal position, with the valves in a vertical plane, deliver gas from the upper valve and liquid from the lower valve.

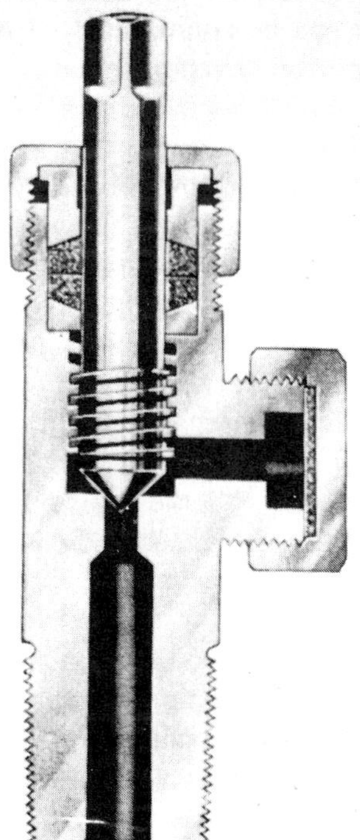

Figure 4-4. Chlorine Institute ton container valve.

When emptied in the liquid phase, a vaporizer (evaporator) is used. The dependable, continuous discharge rate of gas from a ton container, without sweating and under conditions above described for cylinders, is about 15 lb/hr; for liquid it is about 400 lb/hr. As for cylinders, these rates may be greatly exceeded under certain conditions.

The flow of chlorine gas from any chlorine container depends on the internal pressure which, in turn, depends on the temperature of the liquid chlorine. In order to withdraw gas, liquid must be vaporized. This tends to reduce its temperature and thereby its vapor pressure. At low discharge rates, when sufficient heat can be obtained from the surrounding air, the pressure in the container normally will remain constant, and uniform flow can be maintained. At high discharge rates, however, the temperature and pressure within the container will fall due to the cooling effect of vaporization in the container, and the rate of flow will gradually diminish. At excessive discharge rates, the liquid will be cooled sufficiently so that frost may form on the outside of the container. The insulating effect of the frost causes a further decrease in the rate of discharge. Discharge rates may be increased by forced circulation of room-temperature air around the container. Immersion of chlorine shipping containers in hot water, or the application of direct heat, should not be permitted.

If the gas discharge rate from a single cylinder or ton container will not meet demand requirements, two or more can be connected to a manifold and discharged simultaneously, or a vaporizer (evaporator) can be used. When discharging through a manifold, care must be taken that all containers are at the same temperature, particularly when connecting a new container to the manifold. If there is a difference in the temperature of the liquid chlorine, it will be transferred by distillation from the warm to the cool container, and the cooler container may become completely filled with liquid. Should this occur and the container valve then be closed, hydrostatic pressure may cause bursting. For this reason, extra precautions must be observed when closing valves of containers connected to a manifold. Connection of cylinders or ton containers discharging liquid chlorine to a manifold is not recommended.

A flexible connection between the container and the piping should be used; copper tubing, suitable for 500 psig service ($\frac{3}{8}$ in. OD $\times$ 0.035 in. wall), is recommended. A clamp and adapter connector (see Figure 4-5) is preferred; if a union connector is used, the threads on the connector must match the valve outlet thread. (Valve outlet threads are straight threads (1.030 in—14 NGO-RH-EXT), not standard taper pipe threads). A new gasket (see Table 4-6) should be used when making a connection.

Valves should be open one complete turn (counter clockwise) with a $\frac{3}{8}$-in. square box wrench not over 6 in. long. If the valve is difficult to

Figure 4-5. Yoke and adapter type connection.

open, the packing nut may be struck with the heel of the hand, no other implements ought to be used.

Container contents can be determined accurately only by use of suitable scales. The weight of the full container should be recorded and the empty weight determined by subtracting the specified weight of the contents. When the container is empty, the valves should be closed, lines should be disconnected, and the valve should be tested for leaks. If no leak is evident, the valve outlet cap and valve protection hood should be applied and the container removed. (If leaking occurs which cannot be stopped, the chlorine supplier should be notified promptly). The open end of the disconnected line also should be plugged immediately to prevent entry of moisture.

When chlorine is being absorbed in a liquid, proper precautions must be taken to prevent suck-back of the liquid into the container when it becomes empty (due to a partial vacuum created); a barometric leg and/or vacuum breaking device should be used.

Multi-Unit Tank Cars

A multi-unit tank car (TMU) consists of a specially constructed underframe on which 15 ton containers are mounted (see Figure 4-6). ICC and BTCC Regulations provide that TMU cars be consigned for delivery and unloading on a private track, except when a private track is not available within a reasonable distance. Containers may be removed on carrier tracks if special permission is obtained. Pertinent ICC or other regulations should be followed. Ton containers should be handled as outlined previously.

Dimensions and weights of TMU cars are shown in Table 4-2.

Figure 4-6.   Multi-unit chlorine tank car (TMU).

Single-Unit Tank Cars

Construction.   Single-unit tank cars (complying with Spec. ICC 105A300, 105A300W, 105A500, 105A500W, 105 or ARA V) in the United States and Canada are of 16-, 30-, or 55-ton capacity (see Figure 4-7). All except a few experimental cars have 4 in. corkboard insulation, protected by a steel jacket. At the center of the car, on top, there is a manway; this is the only opening permitted in the tank. Dimensions and weights of chlorine tank cars are shown in Table 4-2.

Five valves are mounted on the manway cover inside the housing (see Figure 4-8); four are angle valves, and the fifth (center) is a safety valve. Opposite each angle valve is an opening in the protective housing, protected by a cover, through which unloading connections are made.

Figure 4-7.   Single-unit chlorine tank car.

TABLE 4-2. DIMENSIONS AND WEIGHTS OF TANK CARS[7]

Car	Length Over Strikers[a]	Over-all Height[b]	Height to Valve Outlet[b]	Extreme Width[c]	Weight Empty (lb)	Weight Loaded (lb)
TMU	42'4"–47'0"	6'8"–7'6"	–	9'6"–10'1"	54,500–59,000[d]	84,500–89,000
16-ton	32'2"–33'3"	10'5"–12'0"	9'3¼"–10'0"	9'2"–9'6½"	42,000–51,000	74,000–83,000
30-ton	33'10"–35'11½"	12'4½"–13'7"	11'3"–11'9"	9'3"–9'10"	55,000–65,000	115,000–125,000
55-ton	38'9½"–43'0"	14'3"–14'10½"	12'6"–13'1½"	9'3"–10'7½"	76,000–94,000	187,000–204,000

[a]Add 2 ft. 6 in. for length over center line of coupler knuckles.
[b]Heights are for empty cars and are measured from top of rail; heights of loaded cars may be as much as 4 in. less.
[c]Width over grab irons.
[d]Weight for car with empty containers; underframe only weighs about 34,000 to 46,000 lb.
[e]Height to manway platform is 6 to 10 in. less than height to center line of valve.

Figure 4-8.   Valve arrangement and manway for single-unit tank car.

Most angle valves are of Chlorine Institute Standard Angle Valve design. Outlets, protected by pipe plugs, are 1-in. female, American Standard
taper pipe threads (see Figure 4-9). The two angle valves on the longitudinal center line of the car are for unloading liquid chlorine; the two on the
transverse center line terminate in the vapor space.

Under each liquid angle valve is an eduction pipe which is fastened to
the manway cover and extends to the bottom of the tank (see Figure 4-8).
At the top of each eduction pipe, immediately below the angle valve, is a
rising-ball, excess-flow valve (see Figure 4-10) and most of these are designed to close when the rate of flow of liquid chlorine exceeds about
7000 lb/hr. This safety device precludes the outward flow of liquid chlorine
if the angle valve is broken off or, under certain conditions, if the unloading line is severed.

The safety valve (see Figure 4-11) is designed to start discharging either at 225 psig (105A300 and 105A300W cars) or 375 psig (105A500 and 105A500W cars) pressure in the car.

Unloading. Unloading of single unit tank cars should be performed only by responsible, trained personnel who are thoroughly familiar with all pertinent details of ICC or other appropriate Regulations. All preparatory precautions regarding the location and positioning of the car[4] (on a private siding) should be observed before unloading is begun.

Unloading of single-unit tank cars should be done through a suitable metal connection (usually a prefabricated copper loop) which can accommodate the rise of the car as the springs depress or due to any other change in relative height of the car valve(s) and unloading line(s). If seamless, flexible metal ("Monel") hose is used, especial care should be taken to prevent entry of moisture and to ensure that the hose is in good condition so that no chlorine leaks occur. Unloading connections must be securely attached to unloading pipes before discharge valves are open. Cars should be attended at all times during unloading; if unloading is discontinued for any reason, all valves should be closed and unloading

Figure 4-9.   Chlorine Institute standard angle valve.

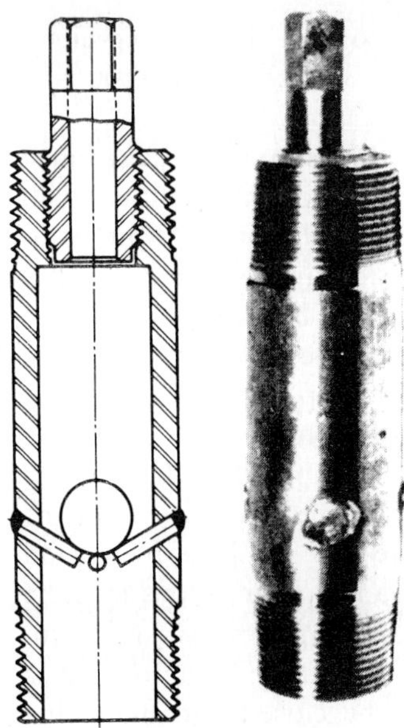

Figure 4-10. Excess flow-valve for single-unit tank cars.

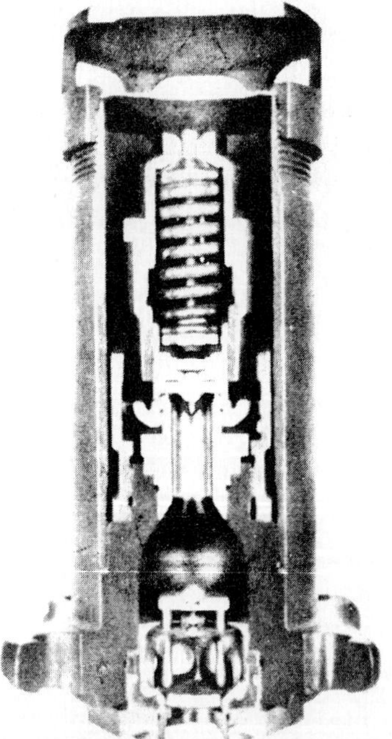

Figure 4-11. Chlorine Institute tank car safety valve.

connections should be carefully disconnected as later described, taking suitable precautions to keep moisture out of the chlorine system. When unloading is begun, chlorine pressure should be applied *cautiously* and the system should be tested for leaks. The liquid valve(s) should be opened *slowly* and *completely* (about two turns), otherwise the excess flow valve(s) will not function in the event that the unloading line(s) is severed.

Unlike cylinders and ton containers, single-unit tank cars are not designed to discharge chlorine gas. The liquid evaporation rate is limited by restricted heat transfer due to the car insulation. Processes requiring gaseous chlorine usually necessitate liquid withdrawal and subsequent evaporation. Liquid chlorine may be unloaded by its own vapor pressure. Decreased temperatures reduce the vapor pressure sometimes to a point where it seriously limits the unloading rate; in such cases, air padding on the car, supplied by the producer or consumer, may be necessary. When air padding is required, clean, dry, compressed air should be used, in accord with good practice.[4] Where air affects the process, an inert gas such as nitrogen can be used. Sometimes liquid chlorine discharge rate can be increased by keeping the car in a warm shed, but under no circumstances should heat ever be applied directly to a chlorine tank car.

The best indication that a tank car is empty is a rapid pressure drop. Scales are the preferred means of ascertaining the amount of chlorine remaining in a car during unloading operations, but some level gauge devices are reported satisfactory. After the angle valve has been closed, the chlorine unloading line should be emptied (before other valves are closed) to avoid trapping the liquid in the pipeline. Then, the valve outlet and the open end of the unloading line should be securely plugged. After testing the valves for leaks, the protective housing cover should be closed and the car should be returned promptly to the supplier as instructed. If it is necessary to return a partially-full car, the supplier should be consulted.

Tank Barges

Chlorine tank barges (see Figure 4-12) vary from 550 to 1100 tons capacity and comply with requirements of Class I arc-welded, unfired pressure vessels as defined by the U. S. Coast Guard Marine Engineering Regulations.[12] Tank barges are intended for unloading into chlorine storage containers. Valves and appurtenances are not uniform on all tank barges but, in general, safety and unloading procedures are similar to those applicable to single-unit tank cars. Unloading and handling of barges require individual study and evaluation.[5]

Tank Trucks

Chlorine tank trucks (complying with ICC Spec. MC 330) are used only infrequently in North America, although they are fairly popular in some

Figure 4-12.   Chlorine tank barge.

European countries. Like tank barges, they are designed for unloading into chlorine storage containers, rather than directly to a consuming process.

Storage Tanks

Because chlorine barges, unlike single-unit tank cars, are not trip-leased, storage facilities for chlorine must be provided. Bulk chlorine consumers remote from rail and water routes may receive chlorine shipments in tank trailers (although as previously noted, this means of shipment in the United States and Canada is practically nonexistant). In such cases, storage facilities usually are required.

For the average chlorine (tank car) consumer storage usually is not recommended for safety reasons including: (1) most consumers do not have constantly available personnel with the specialized training and experience required for operating and maintaining such installations; (2) in the event of a serious chlorine leak, most consumers do not have available means to dispose of the chlorine quickly; and (3) in the event of fire, storage tanks can not be moved from the fire area.

Where their use is clearly indicated, chlorine storage tanks should be located, designed, constructed, operated and maintained in accord with good industry practice (see References). Installations must be constantly under the supervision of trained, responsible personnel who thoroughly understand the properties of chlorine. Systematic inspection and maintenance of the entire facility is imperative.

General Safety Precautions

(1) Adhere to all ICC and other pertinent regulations or specifications.
(2) Permit only reliable, trained personnel to handle chlorine.

(3) Always keep valve protection hoods in place, except when containers are being emptied.

(4) Never tamper with fusible plugs or expose them to heat from any source.

(5) Never apply or expose cylinders or ton containers to heat for any purpose. Never place them in a bath of hot water or expose them to heat to increase gas discharge rates.

(6) Use only suitable and safe means to handle and move cylinders and ton containers.

(7) Store cylinders and ton containers away from flammable and other materials with which chlorine reacts, and in a clean, well-ventilated, fire-resistant area.

(8) Never mix chlorine and another gas in a shipping container.

(9) Never allow moisture or any liquid other than chlorine to enter a chlorine shipping container or a chlorine piping system. Suck-back of liquid from a chlorine-absorption system into a shipping container must be avoided.

(10) Never force connections that do not fit. Use only approved equipment to effect a valve opening or closure; never use a hammer for this purpose.

(11) Exercise especial care for manifolded systems. Do not manifold cylinders or ton containers being emptied in the liquid phase.

(12) Open operating valves on chlorine containers completely during unloading operations; never use them to control chlorine flow.

(13) Keep all equipment, tools, and implements used in connection with chlorine unloading and other handling operations free from oil, dirt and grit.

(14) Use only equipment suitable for chlorine service.

(15) Never ship a loaded container without the owner's consent. It is illegal to ship a leaking chlorine container or one that has been exposed to fire without permission of the AAR Bureau of Explosives.

(16) Report all difficult leaks and other irregularities to the chlorine supplier and be guided by his instructions. Inspect for them frequently and regularly.

ENGINEERING CONTROL OF HAZARDS

Building Design

General. Shipping containers and equipment containing chlorine preferably should be located indoors, in a suitable fire-resistant building. If a separate building is not provided, the chlorine containers and equipment should be located in an isolated room having floors and walls of fire-resistant construction. Standard fire walls may be needed to separate chlorine equipment from flammable materials. Subsurface locations should be avoided; if impossible, an adequate exhaust system should be provided.

There should be at least two means of exit from each separate room or building in which chlorine is stored, handled, or used. All exit doors should open out.

Careful consideration should be given to aids in handling containers to ensure their adequacy, e.g., height of ceilings for overhead hoists, or strength of floors for mechanical handling equipment.

Heating and Ventilating. If comfort heating is provided (or if the area is otherwise heated to increase chlorine discharge rates) care must be exercised to avoid overheating chlorine containers and equipment.

In some cases natural ventilation may be adequate; in others, ventilation by means of a suitable fan should be provided. A one to four minute rate of air change may be required in an emergency. Precautions must be taken to avoid discharging chlorine into areas where it can cause damage or injury.

If ducts are not necessary, a wall-type exhaust fan may be satisfactory if it can be located near the floor on an outside wall. Where ducts are required to bring air to the fan and carry it to a safe point of discharge outside the building, a pressure-type fan is indicated. Switches for all ventilating fans should be provided outside of chlorine rooms or buildings even when an inside switch is installed.

Chlorine gas is heavier than air and has a tendency to collect at floor level. The suction of ventilating fans should be located at or near floor level. Fresh air inlets should be located to provide cross ventilation and to prevent the development of a vacuum in the room. Multiple fresh air inlets and fan suctions may be necessary to exhaust air from some equipment areas.

Materials of Construction

At moderate temperatures, neither liquid nor gaseous chlorine attacks most engineering materials. At high operating temperatures corrosion may be a serious problem and only a few materials may be satisfactory. The presence of moisture, even in small amounts, creates severely corrosive conditions. Wet chlorine (not to be confused with liquid chlorine) is difficult to handle.

Before any material is chosen it should be evaluated under specific conditions of use, especially where operating temperatures and pressures exceed normal conditions or where other corrosive chemicals are present. Evaluations should include both physical (mechanical) and chemical (resistance) limitations. Some equipment that has been proved suitable for chlorine service is listed in Tables 4-3 to 4-6; space limitations preclude a complete listing.

Dry Chlorine, Gas and Liquid. Ordinary low carbon steel ($\frac{1}{2}$-in. minimum nominal pipe size) is the preferred material for handling dry chlorine up to about 350° F. Stainless steels are recommended for handling dry chlorine (types 304, 316, 317) at higher temperatures (up to about 600° F). Common steel construction materials suitable for dry chlorine at temperatures up to 300° F are listed in Table 4-3. Copper and copper alloy construction materials, which often give satisfactory service at temperatures up to about 400° F, are shown in Table 4-4. Other metals such as bronze, lead, nickel, nickel alloys, stainless steel and silver are suitable under some conditions, as are nonmetals such as ceramics, glass, rubber and certain plastics (see Table 4-5).

Wet Chlorine. Wet chlorine is very corrosive to all common construction materials. Gold, lead, nickel-molybdenum alloys, platinum, silver, tantalum and titanium are resistant. At low pressures, nonmetallic materials such as ceramics, glass, glass-lined steel, hard rubber and some plastics can be employed. For high pressures, common metals lined with resistant materials are suitable (see Table 4-5).

Piping Systems

Layout. Chlorine piping arrangements should be as simple as possible, having a minimum number of screwed or flanged joints (threaded joints may be used on pipe sizes through $1\frac{1}{2}$ in.) and a minimum number of loops and traps. Where it is necessary to bend pipe, it should be bent while hot to eliminate residual stresses.

Where possible, chlorine pipes should be located above ground so that leaks can be readily located and repaired, and inspection facilitated. They should be well supported (in a manner that provides for expansion and contraction), protected against temperature extremes, sloped to allow for drainage, and set at an elevation that leaves adequate clearance.

Shut-off valves at each end of a line, or valves at intermediate points in a long line, provide a means of isolating serious leaks. Liquid chlorine lines that are so isolated may require expansion chambers to avoid excessive hydrostatic pressures (liquid chlorine has a high coefficient of thermal expansion). Liquid chlorine should never be trapped between two shut-off valves unless the line is protected by a *suitable* expansion chamber.

Condensation or reliquefaction of gaseous chlorine may occur in pipelines that pass through areas where the temperature is below the temperature-pressure equilibrium indicated in the vapor pressure curve (see Figure

TABLE 4-3. STEEL CONSTRUCTION MATERIALS[6] SUITABLE FOR DRY CHLORINE AT PRESSURES UP TO 300 PSIG AND TEMPERATURES FROM −20°F to +300° F[a]

Equipment	Size and Description		Materials	ASTM Spec.
Pipe	½–1½ in. sch. 80 seamless	6 in. and up 2–4 in., schedule 80 seamless 6 in. and up, schedule 40 seamless	steel	A53 or A106, Grade A

Fittings	butt weld —sch. 80 seamless
	flanged —300# ASA, forgings
	screwed —2000# CWP, forgings
	socket weld—3000# CWP, forgings
		butt weld—seamless, thickness
		    to match pipe
		flanged —300# ASA, castings
		    or forgings
			steel	A234, Grade
				WPA or
				Grade WPB

Forged Flanges	rating—300# ASA type —screwed, socket weld or weld neck Sizes 1/2–1 in. only forged, 2 bolt-oval flange unions faced to manufacturer's standard rating—1500# CWP type —screwed	rating—300# ASA type —weld neck	steel	A106 or A181, Grade I or II

Valves	pattern—globe or angle
	rating —600# ASA
	design —O.S.&Y. forged body and
	    bonnet, bolted bonnet
	    and gland, renewable or
	    hard-faced seat
	end —screwed, socket weld or
	    flanged
		pattern—globe or angle
		rating —300# ASA
		design —O.S.&Y. forged or cast
		    body and bonnet,
		    bolted bonnet and
		    gland, renewable or
		    hard-faced seat
		ends —flanged
			steel body and
			  bonnet
			Monel or
			Hastelloy C
			  trim
			hard-facing
			  Colmonoy #5
				A106 or A181,
				Grade I or II

Valves for gas only	pattern—plug rating —300# ASA design—lubricated, non-lifting plug ends —screwed or flanged	pattern—plug rating —300# ASA design—lubricated, non-lifting plug ends —flanged	steel	A106 or A181, Grade I or II
Bolting	dimensions per ASA B18.2, threads per ASA B1.1 (thru 1 in.—UNC series; $1\frac{1}{8}$ in. up—8 N series) bolt studs—threaded full length bolts —heavy pattern nuts —semi-finished, heavy series		steel studs steel nuts	A193-B7 A354-BB A354-BC A194, Grade 1, 2, or 2H

[a] For temperatures below −20°F. See Ref. 6.

3-6). If necessary, this situation can be remedied by supplying properly controlled heat, by reducing the pressure, or by the use of noncombustible pipeline insulation.

Preparation for Use. Before any new piping installation is put into service it should be cleaned and dried thoroughly and checked for leaks.

Cutting oil, grease and other foreign material inside pipelines and fittings should be removed by methods such as flushing or pulling through each pipe length a cloth saturated with trichloroethylene or other suitable

TABLE 4-4. COPPER AND COPPER ALLOY CONSTRUCTION
MATERIALS[6] SUITABLE FOR DRY CHLORINE FOR
PRESSURES UP TO 300 PSIG AND
TEMPERATURES FROM
-20° F to $+300^\circ$ F

Equipment	Size	Description
Copper tube	3/16" to 3/4" OD	soft seamless copper tube 3/16" OD. Minimum wall 0.032" 1/4" to 1/2" OD. Minimum wall 0.035" 5/8" to 3/4" OD. Minimum wall 0.049"
	1/4" to 1½" (nominal)	copper water tube, type K, soft per ASTM-B88. (The OD of type K water tube is 1/8" larger than the nominal size.)
Fittings	3/16" to 3/4" OD	union type, with lead gaskets; fittings brazed to tubing. Three-piece flare type, Parker Triple-lok or equal. Straight bodies and nuts: brass barstock; shaped bodies: brass forgings; sleeves: copper silicon. Four-piece flareless type, Crawford Swagelock: brass bodies and nuts.
	1/4" to 1½" (nominal)	wrought copper solder joint fittings. (Joints should be made with a brazing alloy containing no tin.)

chlorinated solvent. Hydrocarbons or alcohol should not be used. New valves or other equipment received in an oily condition should be dismantled and cleaned before use.

Valves should be tested for seat tightness with 150 psi air before installation. For greatest safety, chlorine piping systems should be hydrostatically tested to 300 psig pressure before the system is dried.

Pipe lines must always be dried before use, e.g., by passing steam through the lines from the high end, allowing condensate and foreign matter to drain out. Steaming should be continued until the line is thoroughly heated, after which *dry* air should be blown through the lines until the

TABLE 4-5. GUIDE TO MATERIALS OF CONSTRUCTION
FOR CHLORINE SERVICE[11] TO APPROXIMATE
TEMPERATURE LIMIT INDICATED[a]

Material	Dry Chlorine	Wet Chlorine Gas	Uses
Carbon steel	300°F	unsatisfactory	pipe, fittings, valves, vessels and towers
Stainless steels (304,316,317)	600°F	unsatisfactory	pipe, fittings, valves, vessels and towers
High silicon irons:			
Durion	300°F	unsatisfactory	pipe, fittings, valves,
Durichlor	300°F	100°F (Durichlor with 3% Mo)	valve parts, pumps, pump parts
Tantalum	300°F	300°F	heat-transfer equipment, specialty parts, diaphragms, orifices
Titanium	unsatisfactory	200°F	pipe, fittings, valve parts, heat-transfer equipment
Nickel-Molybdenum alloys:			
Hastelloy C	1000°F	100°F	valves, valve parts, pumps, pump parts
Chlorimet 3	1000°F	200°F	valves, valve parts, pumps, pump parts
Nickel and alloys:			
Nickel	1000°F	unsatisfactory	pipe, fittings, valves, valve parts, vessels and towers, heat-transfer equipment
Monel	800°F	unsatisfactory	pipe, fittings, valves, valve parts, vessels and towers, heat-transfer equipment
Inconel	1000°F	unsatisfactory	vessels and towers
Lead	200°F	200°F (fair)	vessels and towers, packing gaskets, lining material
Copper and alloys: (copper, red brass, phosphor bronze)	400°F	unsatisfactory	pipe, fittings, valve parts, heat-transfer equipment
Silver	200°F	100°F	specialty parts
Platinum	600°F	200°F	diaphragms, specialty parts

(Continued)

TABLE 4-5 (*Continued*)

Material	Dry Chlorine	Wet Chlorine Gas	Uses
Glass	200°–300°F	200°F	pipe, fittings, valves, tower packing, heat-transfer equipment
Glass-lined steel	400°–600°F	200°F	pipe, fittings, valves, vessels and towers
Ceramics	200°F	200°F	pipe, fittings, valves, fume ducts, tower packing, vessels and towers
Hard rubber	Not usually recommended; unsuitable for liquid	150°F	lining material
Plastics:			
Fluorinated hydrocarbons	150°F	150°F	valve parts, packing, gaskets, lining material
Polyvinyl chloride	150°F (Not recommended for liquid)	50°F	pipe, fittings, fume ducts, lining material
Polyvinylidene chloride	150°F (Not recommended for liquid)	50°F	pipe, fittings, fume ducts, lining material
Impervious carbon	350°F	50°F	pipe, fittings, fume ducts, heat-transfer equipment

[a]Temperature limit represents safe value based on mechanical or corrosion resistance *or* value for which the material has been test evaluated.

dew point of the discharge air equals that of the entering air. After drying, the system should be filled with *dry* 150 psi air and tested for leaks by application of soapy water to the *outside* of joints. Small quantities of chlorine gas then should be introduced into the line, the test pressure built up with dry air, and the system tested for leaks.*

Containers, piping and equipment should be checked for leaks frequently, preferably at least daily.

Repairs. Repairs should not be attempted while chlorine equipment is in service. All chlorine piping and equipment that is to be repaired by welding should be thoroughly drained and/or purged of chlorine as the heat

---

*To find a leak, tie a cloth to the end of a stick; soak the cloth with strong (commercial 26° Bé) ammonia-water and hold it close to the suspected area. (Contact of ammonia-water with brass should be avoided). A white cloud of ammonium chloride will result if there is any chlorine leakage.

TABLE 4-6. EQUIPMENT SUITABLE FOR DRY CHLORINE AT PRESSURES UP TO 300 PSIG AND TEMPERATURES FROM −150°F. to +300°F.[5,7]

Equipment	Description	Material	Remarks
Fittings	see Table 4-4		
Flanges	see Table 4-4		
Fusible plugs	3/4 or 1 in.	fusible metal	plug designed to soften between 158°F. and 165°F.
Gaskets	flat gaskets 1/16 in. thick for pipe up to 1½ in.; 1/8 in. thick for pipe 2 in. and up	bonded asbestos fibre chemical lead, 2 to 4% antimony	per MIL-A-17472 for temperatures up to 150°F. maximum
Insulation (pipe)		foamglass or cork	foamglass preferred for warm pipelines
Pipe dope	for general use	linseed oil and white lead, linseed oil and graphite	applied to the male end only of threaded joints
	hard-setting	litharge and glycerin	for permanent joints
Pressure gages	diaphragm type, fluid-filled screwed or flanged	Monel, silver, tantalum Kel-F, or Kel-F coated nickel diaphragm; steel body	pressure range twice operating pressure; 250 psi for general purposes
Rupture discs	screwed, socket weld or flanged up to 1½ in.; flanged or butt weld for 2 in. and up	forged carbon steel flanges; silver, Kel-F coated nickel, tantalum, or impervious graphite discs	burst pressure not less than 40% above maximum working pressure; vacuum supports if necessary
Valve packing	ring or Chevron ring (only)	Teflon asbestos	— Garlock #7130 special or equal
Valves	see Table 4-4		
Welding rod	for gas welding for arc welding	ASTM-A-251 type GA60 ASTM A-233 type E-6010	

may cause piping to burn or even burst into flame. Purging may be done by flushing with water or steaming. Before returning equipment to service, it should be cleaned, dried and checked for leaks as previously described.

Equipment and tank cleaning or repairs should be performed only under the direction of thoroughly trained personnel who are fully familiar with the hazards involved and safeguards required.

General Safety Precautions

(1) Use equipment for chlorine service that has sufficient mechanical strength and is chemically resistant at the conditions of temperature and pressure for which it is employed. Special consideration is necessary in the case of wet chlorine gas; e.g., steel is unsuitable.

(2) Clean, dry and test all new and repaired equipment before placing it in service. Do not use hydrocarbons or alcohol for cleaning operations.

(3) Never attempt repairs of equipment or pipelines containing chlorine. Never attempt cutting or welding repairs of equipment or pipelines until all chlorine has been thoroughly purged therefrom.

(4) Immediately dry all pipelines or equipment into which water accidently has been introduced or which have been opened for repairs.

(5) Examine and test all chlorine containers, pipelines and other equipment for leaks frequently—preferably at least once a day.

(6) Never trap liquid chlorine in a pipeline unless it is protected adequately by a suitable expansion chamber.

HANDLING EMERGENCIES

Wherever chlorine is handled a potential risk is involved and a serious emergency might suddenly and unexpectedly occur. Emergency situations should be anticipated, plans established and persons trained to counteract them.

Preparation

Alkali Absorption. As a regular part of chlorine storage and use, provisions should be made for emergency disposal of chlorine from leaking cylinders or ton containers. Chlorine may be absorbed in solutions of caustic soda or soda ash, or in agitated hydrated lime slurries. Caustic soda is recommended as it absorbs chlorine most readily. The proportions of alkali and water recommened for this purpose are given in Table 4-7. A suitable tank to hold the solution should be provided in a convenient location. Chlorine should be passed into the solution through an iron pipe or rubber hose properly weighted to hold it under the surface; the container should not be immersed.

TABLE 4-7. RECOMMENDED ALKALINE SOLUTIONS FOR ABSORBING CHLORINE[7]

Container Capacity	Caustic Soda 100%	Water	Soda Ash	Water	Hydrated Lime[a]	Water
(lb-net)	(lb)	(gal)	(lb)	(gal)	(lb)	(gal)
100	125	40	300	100	125	125
150	188	60	450	150	188	188
2000	2500	800	6000	2000	2500	2500

[a]Hydrated lime solution must be continuously and vigorously agitated while chlorine is to be absorbed.

Emergency Kits. The majority of chlorine suppliers have emergency kits and skilled technicians to use them. These kits can be used to stop most leaks in a chlorine cylinder, ton container, tank car or barge tank, and can usually be delivered to consumer plants within a few hours in an emergency. Some consumers find it advisable to purchase kits and to train employees in their use; otherwise, some responsible person should be fully informed of the location and means of obtaining an appropriate kit and the assistance of a person knowledgeable of its use.

Employee Training. Employees should be instructed and periodically drilled on the locations, purpose, and use of emergency fire fighting equipment, fire alarms and emergency, crash shutdown equipment (such as valves and switches); personal protective equipment, including gas masks, showers, eye baths and similar water sources; and respiratory first aid and equipment.

Fire

In the event of fire, chlorine containers should be moved from the fire zone immediately; tank cars, barges, and trucks should be disconnected and pulled out of the danger area. If chlorine containers cannot be moved, water should be applied to cool them *providing no chlorine is escaping*. Only responsible, authorized persons should be permitted access to the affected area.

Handling Leaks

As soon as there is any evidence or indication of a chlorine leak immediate steps should be taken to correct the situation. Chlorine leaks never get better; they always get worse unless they are corrected promptly. If a leak cannot be handled promptly, assistance should be called for immediately; chlorine consumers should know where and how assistance can be obtained at any time.

General. Chlorine leaks should be investigated immediately by authorized, trained personnel equipped with *suitable* gas masks. All other persons should be kept away from the area until the cause has been isolated and the difficulty corrected. If the leak is extensive, an effort should be made to warn all persons in the path of the gas. Chlorine is heavier than air; therefore, persons should be instructed to keep above and upwind of the leak.

Use of Water. Water should never be used on a chlorine leak as it always makes the leak worse due to the corrosive effect. In addition, heat supplied by even the coldest water to a leaking container causes liquid chlorine to evaporate faster. A leaking chlorine container should not be immersed or thrown into a body of water as the leak will be aggravated due to the corrosive effect and the container may float when partially full, allowing gas evolution and dispersion at the surface.

Equipment and Piping Leaks. If a leak occurs in equipment in which chlorine is being used, the supply of chlorine should be shut off and the chlorine should be disposed of which is under pressure at the leak.

Leaks around valve stems usually can be stopped by tightening the packing nut or gland. If this does not stop the leak, the container valve should be closed, and the chlorine which is under pressure in the outlet piping should be disposed of. If a container valve does not shut off tight, the outlet cap or plug should be applied. In case of a valve leak on a ton container, the container should be rolled so the valves are in a vertical plane with the leaking valve on top; this is important.

If confronted with other container leaks, one or more of the following should be considered: (1) if the container is leaking chlorine, turn it so that gas instead of liquid escapes (the quantity of chlorine that escapes from a gas leak is about one-fifteenth the amount that escapes from a liquid leak through the same size hole); (2) apply appropriate emergency kit device, if available; (3) call the chlorine supplier for emergency assistance; (4) if practical, reduce pressure in the container by removing the chlorine as gas (not as liquid) to process or a disposal system. In some cases it may be desirable to move the container to an isolated spot where it will do the least harm.

Leaks in Transit. If a chlorine leak develops in transit in a populated area, it is generally advisable to keep the transporting vehicle moving until open country is reached in order to disperse the gas and minimize the hazards of its escape. Appropriate emergency measures should then be taken as quickly as possible.

If a motor vehicle is wrecked, leaking chlorine containers should be positioned, where possible, so that only gas escapes, and moved to a less hazardous area before attempting to stop the leaks. If a tank car is wrecked and chlorine is leaking, the danger area should be evacuated. Emergency clearing operations should not be started until safe working conditions are restored.

General Safety Precautions

(1) Know what to do in event of any emergency.
(2) Be thoroughly versed in the location, use and limitation of personal protective equipment, and in first aid for chlorine victims.
(3) Do not enter an area in which a chlorine leak has occurred unless adequately protected or until safety has been established.
(4) Only responsible, trained personnel equipped with a *suitable* respiratory protective device should investigate or attempt to repair any chlorine leak.
(5) Chlorine leaks never get better; they always get worse unless they are corrected promptly.
(6) If possible, position leaking chlorine containers so that only gas instead of liquid chlorine escapes.
(7) Keep upwind and above chlorine leaks.
(8) Never spray water or any other liquid on a *leaking* chlorine container for any purpose. Water may be used to cool non-leaking containers in a fire area if the containers cannot be moved.
(9) Never immerse leaking chlorine containers in alkali, water or other liquids.
(10) If a chlorine leak develops in transit, keep the vehicle moving.

HEALTH ASPECTS AND FIRST AID

Toxicity

Chlorine has a characteristic, sharply penetrating odor; concentrations above three to five parts per million (by volume) in air can be readily smelled by the normal person. At higher concentrations, the severely irritating effect of gas makes it unlikely that any person will remain in a chlorine-contaminated atmosphere unless he is unconscious or trapped. Liquid chlorine vaporizes to gas when exposed to normal atmospheric pressure and temperature.

Acute. As mentioned earlier, low concentrations of chlorine gas irritate the mucous membranes, the respiratory system and the skin. Large amounts cause irritation of eyes, coughing and labored breathing. If the duration of exposure or the concentration is excessive, general excitement of the person affected, accompanied by restlessness, throat irritation, sneezing and copious salivation, will result. The symptoms of exposure to high concentrations are retching and vomiting followed by difficult breathing; in extreme cases, the difficulty of breathing may increase to the point where death can occur from suffocation. The physiological effects of various concentrations of chlorine gas are shown in Table 4-8. Chlorine produces no known cumulative effects.

Liquid chlorine produces no known systemic effects, but when exposed to normal atmospheric pressure and temperature it vaporizes to gas which will produce the effects described above. Liquid chlorine in contact with the eyes, skin or clothing may cause serious burns.

TABLE 4-8. PHYSIOLOGICAL RESPONSE TO VARIOUS CONCENTRATIONS OF CHLORINE GAS[14]

Effect	Parts Chlorine Gas per Million Parts Air, by volume (ppm)
Least amount required to produce slight symptoms after several hours exposure	1
Least detectable odor	3.5
Maximum amount that can be inhaled for one hour without serious disturbances	4
Noxiousness, impossible to breathe several minutes	5
Least amount required to cause irritation of throat	15.1
Least amount required to cause coughing	30.2
Amount dangerous in thirty minutes to one hour	40 to 60
Kills most animals in very short time	1000

Chronic. A concentration of one part per million of chlorine gas may produce slight symptoms after several hours exposure, but careful examination of workers exposed daily to detectable concentrations reportedly has shown no chronic systemic effects. Local chronic effects due to chlorine have not been clinically demonstrated. Sensitization has not been a problem with chlorine.

Preventive Measures

Employee Selection. Chlorine is not a serious industrial hazard if workers are adequately instructed and supervised in proper methods of handling the chemical.

Asthma, bronchitis and other chronic lung conditions or irritations of the upper respiratory tract suggest that the person(s) in whom they are observed should not be employed where exposures to chlorine vapors might occur. Physical examinations, including a chest X-ray, are recommended for applicants and employees handling chlorine.

Personal Protective Equipment. Only respiratory protective equipment approved by the U. S. Bureau of Mines should be used.[15,16] It should be carefully maintained, inspected, and cleaned after each use and at regular intervals. Equipment used by more than one person should be sterilized after each use. No person should enter a chlorine contaminated area unless attended by an observer who can rescue him in the event of respirator failure or other emergencies.

The *industrial canister type mask,* with a full facepiece and a chlorine or all-purpose canister, is suitable for moderate concentrations of chlorine, provided sufficient oxygen is present. The mask should be used for a relatively short exposure period only. It may not be suitable for use in an emergency since, at that time, the actual chlorine concentration may exceed the safe 1 percent limit and the oxygen content may be less than 16 percent (by volume).

The wearer must leave the contaminated area immediately on detecting the odor of chlorine or on experiencing dizziness or difficulty in breathing; these are indications that the mask is not functioning properly, that the chlorine concentration is too high, or that sufficient oxygen is not available. Unless the presence of other gases requires the use of an all-purpose canister, the chlorine canister should be used.

Exceeding manufacturer's recommended limits on maximum non-use shelf life might be hazardous. Regular replacement of over-age canisters, even though unused, is recommended.

The *self-contained breathing apparatus,* with a full facepiece and a cylinder of air or oxygen carried on the body, or with a canister which produces oxygen chemically, is suitable for high concentrations of chlorine and is the preferred means of respiratory protection for the average chlorine consumer. It provides protection for a period which varies with the amount of air, oxygen, or oxygen-producing chemicals carried. Oxygen masks should not be used in a tank or other closely confined area where there may be danger of sparks or fire. In the case of oxygen-producing equipment entry into the affected area must be delayed a few minutes while the oxygen-generating reaction is starting.

The *positive pressure* (*blower*) *hose mask,* with a full facepiece and with air supplied through a hose from a remote blower, is suitable for high concentrations of chlorine provided conditions will permit safe escape if the air supply fails. The blower air supply must be free of air contaminants, and the intake preferably should be located at least six feet above the ground. The air intake for a gasoline-driven blower must not be near the engine exhaust.

The *compressed-air line mask,* with a full facepiece, a suitable reducing or demand type valve, an excess pressure relief valve and a filter, is suitable for high concentrations of chlorine provided conditions will permit safe escape if the air supply fails. Air is supplied through a hose from a source of compressed air. The compressed air supply must be free of air contaminants, particularly the harmful gases resulting from the decomposition of the compressor lubricating oil. The air intake for gasoline compressors must not be near the engine exhaust.

First Aid

Treatment. Prompt treatment of persons exposed to chlorine is of the utmost importance. Medical assistance should be obtained as soon as possible. Anyone overcome by or seriously exposed to chlorine gas should be moved at once to an uncontaminated area. *If breathing has not ceased,* the patient should be placed on his back, with head and back elevated, and kept warm, using blankets if necessary. Rest is essential. *If breathing apparently has ceased,* artificial respiration should be started immediately. The Nielson armlift-back pressure method is preferable. (If the Schaefer prone-pressure method is used, do not exceed 18 cycles per minute). A physician should be called immediately. If oxygen inhalation apparatus is available, oxygen should be administered by a person authorized for such duty by a physician. The instructions which come with the equipment must be followed carefully.

Stimulants rarely are necessary where adequate oxygenation is maintained, and any such drugs for shock treatment should be given only by a physician. Milk may be given in mild cases as a relief from throat irritation. Nothing should ever be given by mouth to an unconscious patient.

If liquid chlorine or chlorinated water has contaminated skin or clothing, the emergency shower should be used immediately. Contaminated clothing should be removed under the shower and the chlorine should be washed off with very large quantities of water. Skin areas should be washed with large quantities of soap and water. No attempt should be made to neutralize chlorine with chemicals. No salves or ointments should be applied for 24 hours.

If even minute quantities of liquid chlorine enter the eyes, or if the eyes have been exposed to strong concentrations of chlorine gas, they should be flushed immediately with copious quantities of running water for at least 15 minutes. No attempt should be made to neutralize with chemicals. The eyelids should be held apart during this period to insure contact of water with all accessible tissue of the eyes and lids. A physician, preferably an eye specialist, should be called at once. If a physician is not immediately available, the eye irrigations should be continued for a second period of 15 minutes. After the first period of irrigation is complete, it is permissible as a first aid measure to instil into the eyes two or three drops of 0.5 percent solution of pontocaine or other equally effective topical anesthetic. No oils or oily ointment should be instilled unless ordered by a physician.

The swallowing of liquid chlorine is extremely unlikely. However, if a person has swallowed chlorine *and is conscious*, he should immediately be made to drink copious amounts of lime water, milk of magnesia, or plain water if the others are not readily available; sodium bicarbonate should not be given. The victim may be expected to vomit spontaneously, but no attempt should be made to induce vomiting or to use a stomach tube. A physician should be called immediately.

Equipment. Properly designed emergency showers and eye baths should be provided in convenient locations.

Suitable equipment for the administration of oxygen and for automatic artificial respiration should be available. (Such equipment should be approved by the Council on Physical Medicine and Rehabilitation of the American Medical Association). If purchase of equipment is impractical, the location of the nearest equipment and experienced operator, as well as means of obtaining the same promptly if an emergency arises, should be determined. Such equipment is useless unless experienced operators are available whenever needed; special arrangements for night and week-end periods should be made.*

General Safety Precautions
(1) Make available a suitable gas mask to every person involved with chlorine handling.
(2) Use only respiratory protective equipment approved for chlorine service by the U. S. Bureau of Mines. Use only equipment suitable for the condition existing.
(3) Move all persons exposed to chlorine gas to an uncontaminated area. If breathing has ceased, start artificial respiration immediately and have someone call a physician at once.
(4) Never give anything by mouth to an unconscious person.
(5) Never attempt to neutralize chlorine with chemicals.
(6) Remove chlorine-contaminated clothing at once and wash affected skin areas with large quantities of soap and water. Never attempt to neutralize chlorine with chemicals.

*The assistance of Robert L. Mitchell, Jr., Secretary-Treasurer of The Chlorine Institute, Inc., and various representatives of U. S. chlorine producers in the preparation of material included herein, is acknowledged with appreciation by the author.

(7) Apply only such other first aid as may have been prescribed by a
physician pending his arrival.

References

1. Codes of Rules Governing the Condition of, and Repairs to Freight
and Passenger Cars for the Interchange of Traffic., Assoc. of Am. Railroads, Chicago 5, Ill.
2. Specifications for Tank Cars. Assoc. of Am. Railroads, Chicago 5, Ill.
3. Board of Transport Commissioners for Canada. Regulations for the Transportation of Explosives and other Dangerous Articles in Rail Freight and
Rail Express Service, including Specifications for Shipping Containers. The
Railway Assoc. of Canada, Montreal 25, Que. (Effective March 1, 1959).
4. "Consumer Air Padding of Chlorine Single Unit Tank Cars," The Chlorine
Institute, Inc., New York 17, N. Y.
5. "Handling of Chlorine Barges," The Chlorine Institute, Inc., New York
17, N. Y.
6. "Piping and Equipment for Use With Dry Chlorine," The Chlorine Institute,
Inc., New York 17, N. Y.
7. "Chlorine Manual," ed. 3, The Chlorine Institute, Inc., New York 17, N. Y.,
8. "American Standard" (B57.1), "Canadian Standard" (B96), Compressed Gas
Cylinder Valve Outlet and Inlet Connections. Compressed Gas Association,
Inc. (Sponsor), Pamphlet V-1, New York, N. Y.
9. Interstate Commerce Commission. ICC Regulations for Transportation of
Explosives and Other Dangerous Articles by Land and Water in Rail Freight
Service and by Motor Vehicle (Highway) and Water, including Specifications
for Shipping Containers. Issued as a tariff by T. C. George, Agent, Bureau of
Explosives, New York 7, N. Y.
10. Interstate Commerce Commission, ICC Regulations for Transportation of
Explosives and Other Dangerous Articles by Motor, Rail and Water, including
Specifications for Shipping Containers. Issued as a tariff by F. G. Freund,
Agent, American Trucking Associations, Inc., Washington 6, D. C.
11. "Chlorine," Olin Mathieson Chemical Corp., Baltimore 3, Md., 1959.
12. U. S. Coast Guard, "Marine Engineering Regulations and Material Specifica-
tions," U. S. Govt. Printing Office, Washington 25, D. C.
13. U. S. Coast Guard, Regulations Governing the Transportation or Storage of
Explosives or Other Dangerous Articles or Substances, and Combustible
Liquids on Board Vessels, Issued as a tariff by T. C. George, Agent, Bureau of
Explosives, New York 7, N. Y.
14. U. S. Depart. of the Interior, "Gas Masks for Gases Met in Fighting Fires,"
Bur. Mines Tech. Paper, No. 248 (1921); (Out of Print).
15. U. S. Depart. of the Interior, "Respiratory Protective Devices Approved
by the Bureau of Mines," *Bur. Mines Inform. Circ.,* No. 7885, Washington,
D. C. (1959).
16. U. S. Depart. of the Interior, "Bureau of Mines Approval System for
Respiratory Protective Devices," *Bur. Mines Inform. Circ.,* No. 7792 Washington, D. C. (1957).

Additional Selected References

1. Allied Chemical Corp., Solvay Process Divn., Solvay Emergency Devices for Stopping Chlorine Leaks, Allied Chem. Corp., New York 6, N. Y.

2. Association of American Railroads., *Circular 17 A*, Rev. 1, Assoc. of Am. Railroads, Chicago 5, Ill., (January 1956).

3. Chairs, Herbert, *et al.,* "Chlorine Accident in Brooklyn," *Occupational Med.,* **4,** 152 (1947).

4. Available Publications and Drawings, The Chlorine Institute, Inc., New York 17, N. Y.

5. Container Procedure at Chlorine Packaging Plants. The Chlorine Institute, Inc., New York 17, N. Y.

6. Excerpts of ICC Regulations and Specifications Applying to Chlorine Shipment in Cargo Tanks, The Chlorine Institute, Inc., New York 17, N. Y.

7. Literature on the Physiological Effects of Chlorine and Supplementary Bibliography, The Chlorine Institute, Inc., New York 17, N. Y.

8. Maintenance Instructions for Chlorine Institute Tank Car Safety Valves, The Chlorine Institute, Inc., New York 17, N. Y.

9. Maintenance Instructions for Chlorine Angle Valves, The Chlorine Institute, Inc., New York 17, N. Y.

10. Maintenance Instructions for Chlorine Institute 4 JQ Safety Valves, The Chlorine Institute, Inc., New York 17, N. Y.

11. Chlorine Institute, Inc. (The), "Recommended Specifications for Stationary Chlorine Storage Installations," *J. Am. Water Works Assoc.,* **46**(11) 1117 (November 1954).

12. Types and Locations of Chlorine Emergency Kits in North America, The Chlorine Institute, Inc., New York 17, N. Y.

13. Chlorine Producers, Miscellaneous Chlorine Safety Bulletins and Pamphlets.

14. Department of Scientific and Industrial Research, "Methods for the Detection of Toxic Gases in Industry-Chlorine," Leaflet No. 10, London, H. M. Stationery Office, (1955).

15. Hardy, G. C., and Barach, A. L., "Positive Pressure Respiration in the Treatment of Irritant Pulmonary Edema Due to Chlorine Gas Poisioning," *J. Am. Med. Assn.,* **128,** 359 (1945).

16. Heinemann, G., Garrison, F. G., and Haber, P. A., "Corrosion of Steel by Gaseous Chlorine: Effect of Time and Temperature," *Indus. Eng. Chem.,* **38,** 497 (1946).

17. Henderson, Y., and Haggard, H. W., "Noxious Gases and the Principles of Respiration Influencing Their Action," New York, Reinhold Publishing Corp., 1943.

18. Chemical Safety Data Sheet SD-80, Properties and Essential Information for Safe Handling and Use of Chlorine, Manufacturing Chemists' Association, Inc., Washington 9, D. C. (1960).

19. N.A.C.E. Committee Report, "A Bibliography of Corrosion by Chlorine," *Natl. Assoc. Corrosion Engineers,* Tech.; Unit Comm. T-5A on Corrosion in the Chemical Mfg. Industry (compiled by Task Group T-5A-4 on Chlorine), Pubn. 56-2. *Corrosion,* **12,** 59 (March, 1956).

20. "Respiratory Protective Equipment," Safe Practices Pamphlet No. 64. National Safety Council, Chicago 11, Ill.
21. Manual Covering ICC Regulations Applicable to Private Motor Truck Operations, Private Truck Council of America, Inc., Washington 5, D. C.
22. U. S. Dept. of Labor, Divn. of Labor Standards, "Chlorine, Controlling Chemical Hazards," Series No. 2, 1954.
23. The Code of Federal Regulations Title 46, Parts 98 and 146 to 149. U. S. Govt. Printing Office, Washington, 25, D. C.
24. Anonymous, "Chemical Resistance of Construction Materials and Nonmetals," *Chem. Eng.*, **53**, 120 (November, 1946).
25. Anonymous, "Corrosion by Chlorine and by Hydrogen Chloride at High Temperatures," *Ind. Eng. Chem.*, **39**, 839 (1947).
26. Anonymous, "Wet and Dry Chlorine vs. Materials of Chemical Plant Construction," *Chem. Eng.*, **54**, 211, 219, 213, 244, (1947).
27. Anonymous, Materials of Construction for Chlorine. *Chem. Eng.*, **57**, 115 (1950).

5. ELECTROLYSIS OF BRINES IN DIAPHRAGM CELLS

MORTON S. KIRCHER

Hooker Chemical Corporation

The diaphragm chlor-alkali cell is important as the principal twentieth century method of chlorine production. Approximately 75 percent of United States chlorine is produced in diaphragm cells as of 1961. Figure 5-2 shows growth in chlorine production by method of manufacture. The diaphragm cell is exceptionally simple in principal and in operation. Yet the processes involved in the diaphragm remain a challenge to the electrochemist.

The great growth in chlorine demand and the availability of new materials of construction have opened up opportunities for development so that during recent years more effort has been directed toward diaphragm cell development than at any previous time.

CHEMISTRY

In the operation of a typical diaphragm cell (see Figure 5-3) sodium or potassium chloride brine, nearly saturated and at a temperature approximately 60 to 70°C, is fed into the anolyte, which flows through the diaphragm into the catholyte where alkali is formed. Flow is continuous, with a differential head maintaining flow through the diaphragm. Chlorine gas is formed at the anode and hydrogen and alkali are formed at the cathode.

Modern cells operate with a cathode efficiency (hydrogen) very close to 100 percent and an anode efficiency (chlorine) of approximately 97 percent.

Anode Reactions

The principal anode reaction is

$$2\,Cl^- \rightarrow Cl_2 + 2\,e \tag{5-1}$$

This simple reaction represents approximately 97 percent of the anode discharge under good operating conditions; however, the following reactions representing minor losses are of particular significance since they

Figure 5-1.   An installation of modern diaphragm cells.

have a considerable effect on anode life, diaphragm life and purity of the products.

Chlorine formed at the anode saturates the anolyte and an equilibrium is established as follows:

$$Cl_2 + OH^- \rightarrow Cl^- + HOCl \tag{5-2}$$

$$HOCl \rightarrow H^+ + OCl^- \tag{5-3}$$

Reactions (5-2) and (5-3) indicate the importance of hydrogen ion concentration on the solubility of chlorine in the anolyte. Normally the pH of the anolyte is in the range of 3.0 to 4.0.

Hypochlorous acid and hypochlorite tend to react to produce chlorate according to the reaction used in electrolytic chlorate cells as shown:

$$2\,HOCl + OCl^- \rightarrow ClO_3^- + 2\,Cl^- + 2\,H^+ \tag{5-4}$$

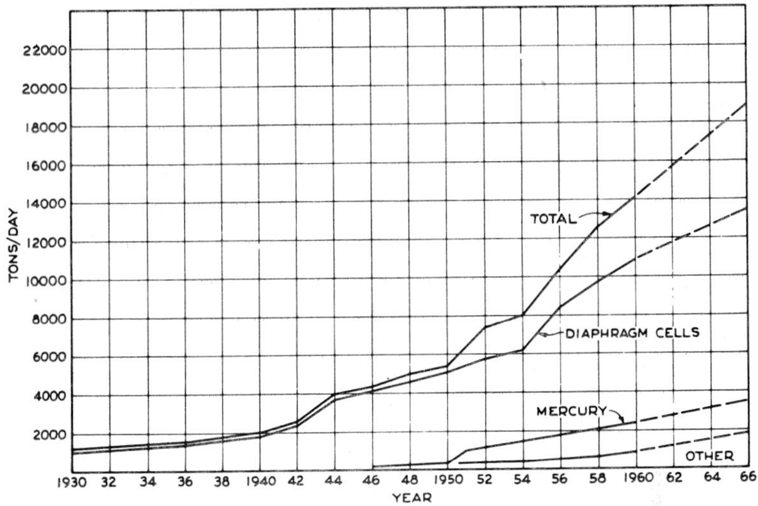

Figure 5-2.   Chlorine production in the United States by years and by
method of manufacture.

In addition to chloride ion, other anions may discharge at the anode.
Under usual operating conditions, OH^- discharge is the most important
reaction competitive with chloride discharge

$$4\,OH^- \rightarrow O_2 + 2\,H_2O + 4\,e \tag{5-5}$$

With graphite electrodes, which are the only type presently used in commercial production, Reaction (6) also occurs as shown:

$$4\,OH^- + C \rightarrow CO_2 + 2\,H_2O + 4\,e \tag{5-6}$$

The source of OH^- ions discharged at the anode is the cathode liquor,
from which the OH^- ions migrate under the influence of the electrical
potential across the diaphragm. Under steady state conditions the amount
of OH^- ion entering the anolyte must equal the amount used in other
reactions, e.g., chlorate formation plus the amount discharged at the anode.
Hence a certain pH is established in the anolyte which is sufficiently high to
cause the required rate of discharge at the anode.[23] The pH established
depends not only on the rate of migration of OH^- through the diaphragm
but also upon the characteristics of the anode material (the relative oxygen
and chlorine discharge potentials).

For a given rate of oxygen discharge, the anolyte pH with different anode
materials would tend to be in the following order:

Carbon anodes < untreated graphite < treated graphite < platinum.

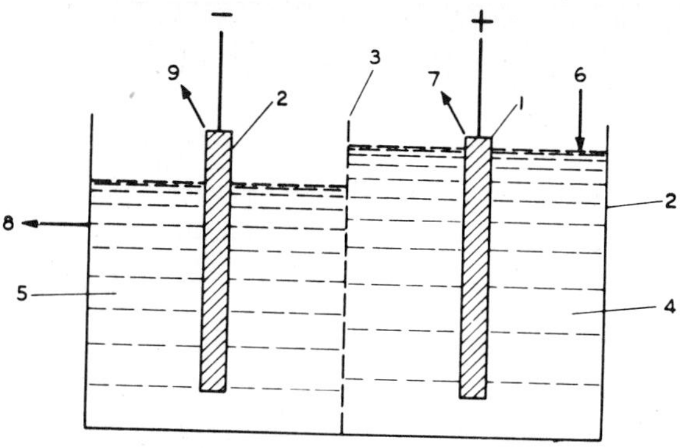

1. Anode
2. Cathode
3. Diaphragm
4. Anolyte-chlorine container for anolyte and chlorine
5. Catholyte hydrogen container for catholyte and hydrogen
6. Brine inlet
7. Chlorine outlet
8. Caustic outlet
9. Hydrogen outlet

Figure 5-3.   Diaphragm of percolating diaphragm chlor-alkali cell.

The reactions undoubtedly are much more complicated than represented. Peroxygen compounds, organic hydroxides, organic acids, and carbon monoxide probably are formed as intermediates. Traces of organic compounds resulting from anodic oxidation are always present to some extent in the caustic liquor from graphite anode cells; also small quantities of carbon monoxide, carbon tetrachloride, chloroform and hexachloroethane are always present in the chlorine. Vaaler[26] has studied the mechanism of oxygen and carbon dioxide discharge at chlorine cell anodes.

Presence of anions other than OH^-, e.g., HSO_4^-, ClO_3^-, OCl^-, are also factors in the anodic formation of oxygen and carbon dioxide. In small quantities such anions do not effect the total amount of oxygen formation because oxygen (or CO_2) discharge according to any mechanism also involves the formation of an equivalent quantity of H^+ ion. Hydrogen ion formed by such processes tends to neutralize an equivalent of hydroxyl ion and hence to reduce the extent of Reactions (5-5) and (5-6) correspondingly

Cathode Reactions

At the cathode, the primary reaction is discharge of hydrogen ion from the alkaline solution as shown:

$$2\,H^+ + 2\,OH^- \rightarrow H_2 + 2\,OH^- - 2\,e \qquad (5\text{-}7)$$

Discharge of hydrogen from the aqueous solution leaves an equivalent OH^- ion derived from water.

Reduction of HOCl to Cl^- also takes place at the cathode. However, for all practical purposes, the current efficiency of hydrogen discharge is 100 percent.

HISTORY

The first formation of chlorine by electrolysis is attributed to Cruickshank in 1800.[27] Commercial production of chlorine by oxidation of hydrogen chloride by manganese dioxide (Weldon process) and air (Deacon process) became important during the last half of the nineteenth century. Hydrogen chloride was derived from the Le Blanc Soda Ash process and the chlorine was converted to bleaching powder. By 1900, 150,000 tons per year of bleaching powder were being produced in England.[12]

British patents on chlorine-alkali cells to Cook in 1851, Watt in 1851 and Stanley in 1853 indicated an awareness of the possibilities of commercial electrochemical chlorine. The development of the dynamo about 1865 gave impetus to research on electrochemical processes which until this time had been quite academic. The observations of Watt indicate a remarkable insight into the problem which was not appreciated until nearly fifty years later. The theoretical work on electrolysis of brine by Fritz Foerster[8] and his students in Germany was outstanding and remains important today.

The first commercial production of chlorine was by the the Griesheim Company in Germany, in 1888. Actually the chlorine was a by-product of KOH manufacture. The first commercial production of chlroine in the United States and of caustic soda in the world was started by the Electrochemical Company of Rumford Falls, Maine, in 1892. Because there was no U. S. production of bleaching powder and cheap hydroelectric power was available, the infant U. S. electrochemical chlor-alkali industry was given a chance to survive competition with bleaching powder imported from England and Germany.[4,17,27]

The Griesheim cell,[7] illustrated in Figure 5-4, was based on the use of a porous cement diaphragm invented by Brauer (1886). The diaphragm was prepared by mixing portland cement with brine acidified with hydrochloric

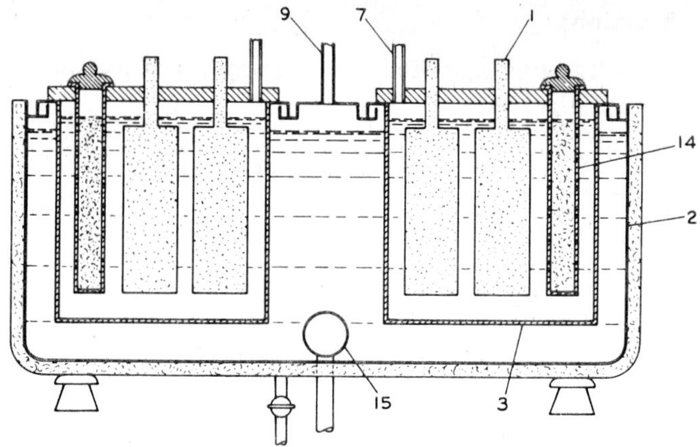

(Key to parts same as in Figure 5-3)

Special features:
   14. Porous container for salt (KCl)
   15. Heating tube

Figure 5-4.   Griesheim non-percolating diaphragm cell.

acid. After setting, it was soaked in water to remove soluble salts. A steel
tank served as cathode. Twelve anode compartments constructed of the
diaphragm material supported by steel angles were hung into the tank.
Carbon or magnetite was used for anodes. Porous pots were provided in
each anode compartment for addition of solid KCl. The cell was heated to
90° C by steam introduced in a central steel cylinder. The cell was operated
batchwise, KCl solution being introduced initially and electrolysis being
continued for about three days until a concentration of about 7 percent
KOH was obtained. The current efficiency was low (70 to 80 percent) but
the cell was simple, inexpensive and relatively large in capacity (2500
amperes).

The first cell employed at the Rumford Falls plant was not the cell later
known as the Le Sueur cell but a predecessor developed by Le Sueur and
first operated in 1890 on a pilot scale[4,17,27] (Figure 5-5). Le Sueur is credited
with making the first use of a percolating diaphragm which is the basis of all
diaphragm chlor-alkali cells in use today.   By permitting brine to flow into
the anolyte and through the diaphragm, continuous operation and much
higher current efficiency was obtained than with a non-percolating diaphragm such as in the Griesheim cell.

Le Sueur improved upon the structure of the cell, as shown in Figure 5-6,
about 1897. An installation of these cells with essentially the same design
is still in operation at the Brown Company in Berlin, New Hampshire.

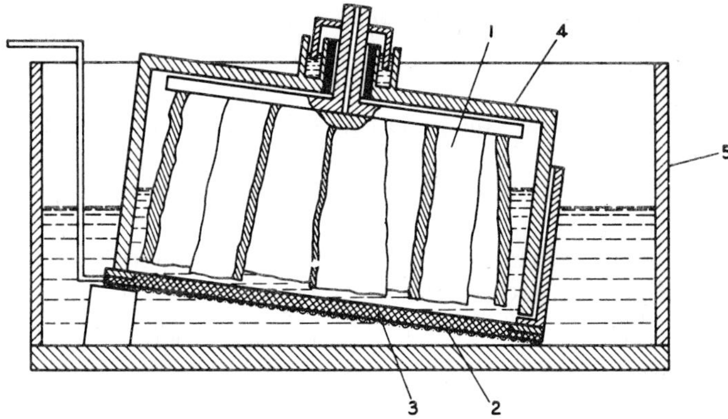

(Key to parts same as in Figure 5-3)

Figure 5-5.   Original Le Sueur cell.

The experience gained at the Electro-chemical Company plant influenced subsequent American cell design.   Curiously enough, however, all later American cells were vertical diaphragm cells.   Possibly the early American vertical cell designs were inspired by the Hargreaves-Bird cell[1] developed in England during the 1890's. This cell had a non-percolating diaphragm and reduced the back migration of hydroxyl by adding carbon dioxide and

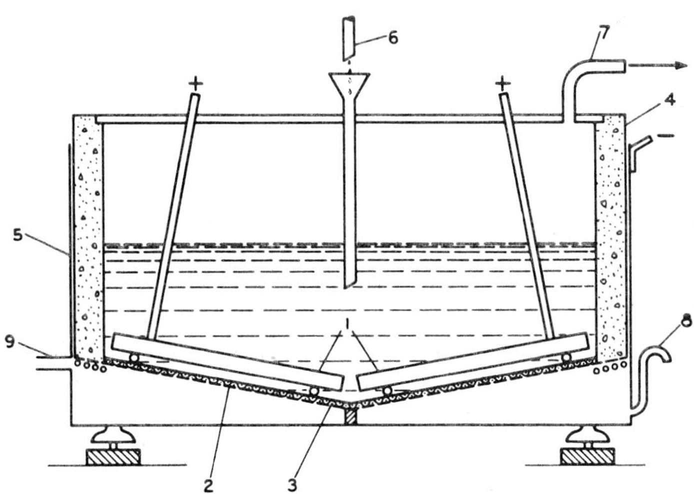

(Key to parts same as in Figure 5-3)

Figure 5-6.   Le Sueur cell.

steam to the cathode compartment to convert the caustic to sodium carbonate.

The H-B cell was the first commercial vertical diaphragm cell. Its construction, using a heavy concrete ring with anodes entering through the top and cathode plates bolted on each side, was typical of a whole class of commercial cells including Townsend, Giordani-Pomilio, Allen Moore, Buck-McRae, etc.

The Billiter cell developed in Germany is very similar to the Le Sueur cell although it was developed quite independently. During the 1920's the Billiter cell became the most widely used cell in the world. The simple flat cathode design is shown in Figure 5-7. Amperage capacities up to 12,000 amperes were achieved. By modifying the cathode to the corrugated shape shown in Figure 5-8, the I. G. Farbenindustrie increased cell capacity to 24,000 amperes.[13]

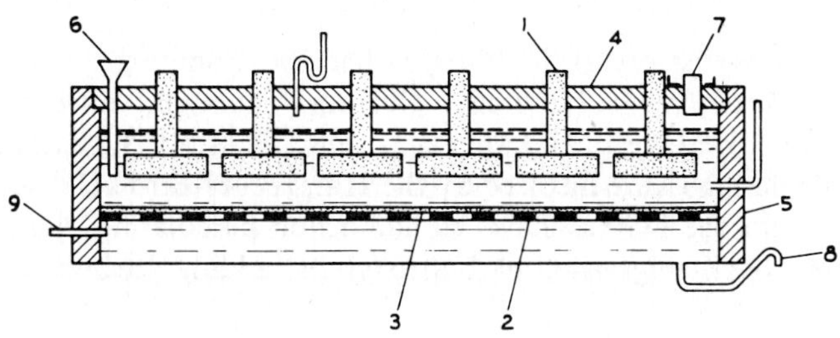

(Key to parts same as in Figure 5-3)

Figure 5-7.   Billiter cell, flat cathode.

One early cell should be mentioned as a matter of theoretical interest. This is the Aussig (Bohemia) Bell Jar cell which, although it had no diaphragm, operated in other respects as a diaphragm cell. As shown in Figure 5-9, the anodes were placed in an inverted vessel, a number of which were installed in a tank. Cathode rings surrounded the inverted bell jars. Brine was introduced into the anode compartment and flowed into the main tank. The difference in gravity of anode liquor from cathode liquor and the flow of liquor from anolyte to catholyte tended to prevent back migration of OH$^-$ ions. Allmand[1] gives extensive calculations on the flow rate required for a given current efficiency. The major drawbacks of this type cell are its very low capacity and the considerable distance between anode and cathode. As current density was increased evolution of gas bubbles tended to stir the respective compartments and cause mixing.

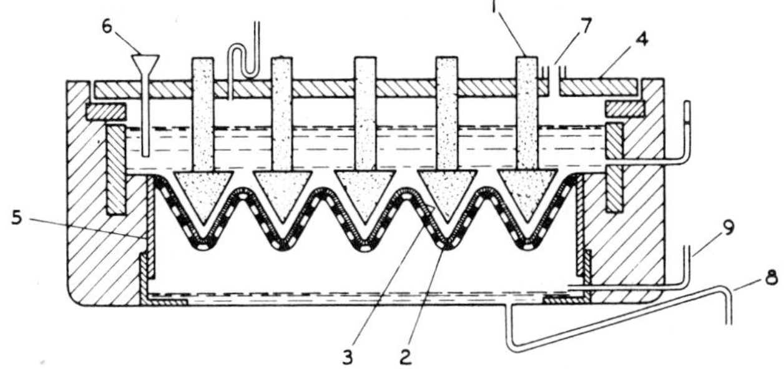

(Key to parts same as in Figure 5-3)

Figure 5-8.   Billiter cell, corrugated cathode.

A great deal of cell development was done in the United States in the years from 1890 to 1910. The following list of the number of U. S. patents issued by years is a measure of this activity.[22]

Period	No. of Patents
1883–1889	5
1890–1899	105
1900–1909	129
1910–1919	90
1920–1929	69
1930–1939	54
1940–1949	43

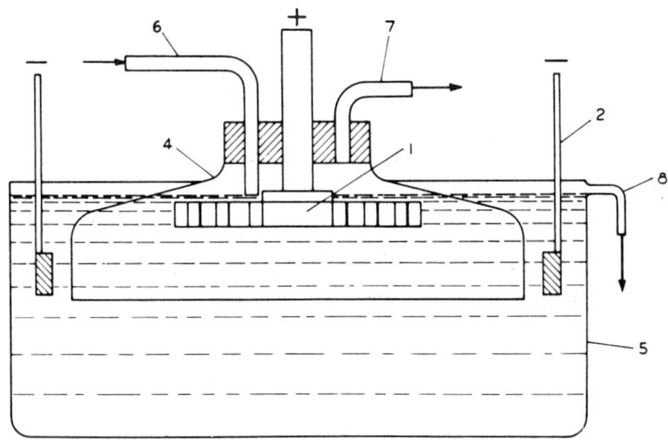

(Key to parts same as in Figure 5-3)

Figure 5-9.   Bell Jar cell.

Perhaps the most important discovery to the electrolytic chlorine-alkali industry was that of methods of producing synthetic graphite. H. Y. Castner and Chas. Vaughn found that baking carbon anodes in an electric furnace improved their performance while Acheson, working independently, discovered graphite surrounding silicon carbide in his electric furnace and then developed a method of producing it.

Following Le Sueur's pioneering work and the invention of synthetic graphite, scores of cells were invented. In the United States, three basic types were developed as follows:

(1) Rectangular Vertical Diaphragm cells including Allen-Moore, KML-Allen Moore, Townsend, Buck McRae and Nelson

(2) Cylindrical cells including Gibbs, Vorce and Wheeler

(3) Bipolar filter press cell, Dow

Claims were made for special features such as submerged diaphragms, unsubmerged diaphragms, special compounds to incorporate in the diaphragms, use of petroleum oil in the cathode compartment, etc. Although considered quite important at the time, in retrospect these differences are not of great significance.

In 1913 Marsh designed a cell with finger cathodes and side-entering anodes and cathodes. This increased the electrode area per unit of floor space. Asbestos paper was wrapped over this surface and sealed top and bottom with cement and putty. The putty joints provided a poor seal, hence the current efficiency of the Marsh cell never equalled that of cells with more simple construction. About 1928, in an attempt to overcome the disadvantage of the Marsh cell, Kenneth Stuart of Hooker Electrochemical Company developed a method of depositing asbestos fibre onto the cathode by immersing the cathode and applying a vacuum. The Marsh cell both before and after use of deposited diaphragms is shown in Figure 5-11.

Stuart made use of the flexibility of design permitted by the deposited asbestos diaphragm and designed the Hooker Type S cell, which has been further developed through various stages as the S–3, S–3A, S–3B and S–3C. These modifications have raised the amperage capacity from 5,000 to 30,000 amperes.

The importance of the Stuart invention may be gauged by the fact that more than 90 percent of the chlorine produced in diaphragm cells is produced in cells using deposited asbestos diaphragms.

Filter press cells which are used extensively for hydrogen-oxygen electrodyalsis and hydrogen-chlorine were designed for chlorine-alkali by several early workers including Kellner, Guthrie and Finaly. Although the filter press design is attractive from the standpoint of requiring a minimum of conductor material between cells, a minimum of floor space and low investment cost, the only commercial use of the filter press cell for chlorine-

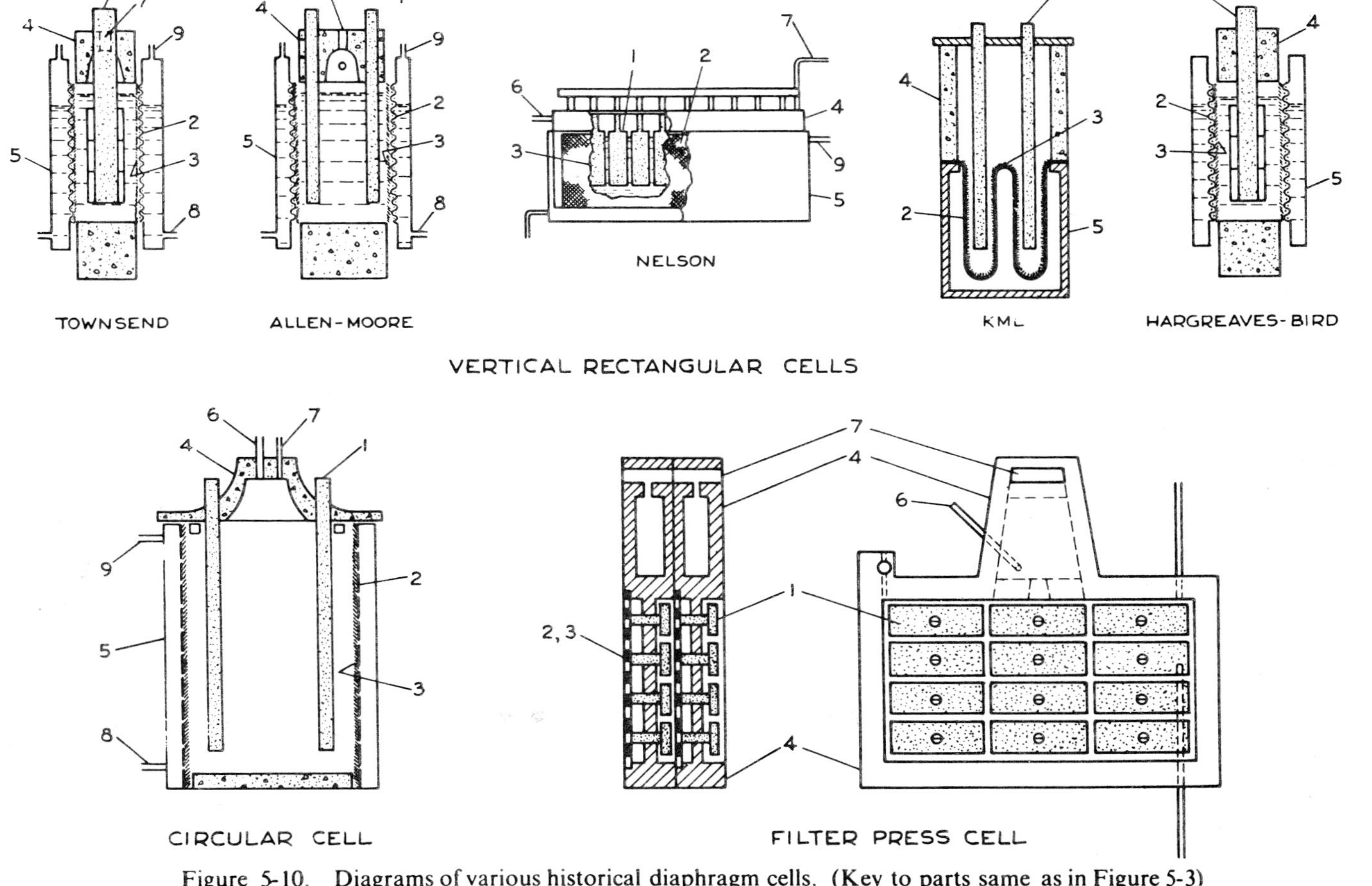

Figure 5-10. Diagrams of various historical diaphragm cells. (Key to parts same as in Figure 5-3)

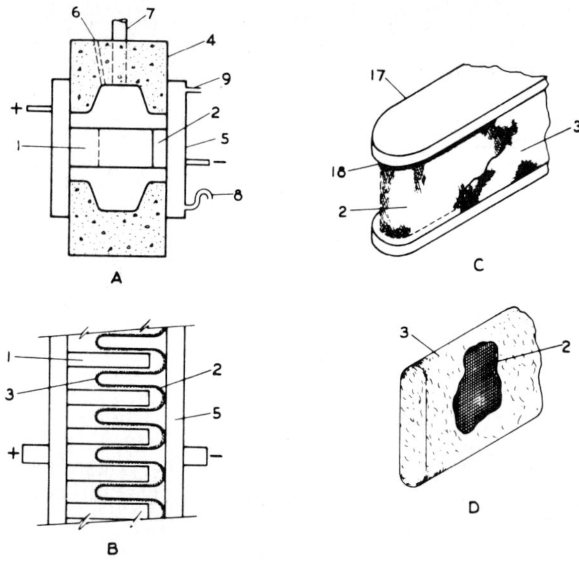

(Key to parts same as in Figure 5-3)

A. Vertical section
B. Plan, sectional diagram
C. Original cathode finger
   2. Cathode screen
   3. Asbestos paper diaphragm
   17. Concrete cap
   18. Putty seal
D. Deposited asbestos cathode finger
   2. cathode screen
   3. deposited asbestos diaphragm

Figure 5-11.   Marsh cell.

alkali has been by the Dow Chemical Company. The original patents were granted to Thomas Griswold, Jr. in 1911[10] and 1913.[11]

Over a period of fifty years, Dow has developed filter press cells through several stages characterized by simple, rugged, inexpensive construction.

The obvious disadvantages of the filter press cell are the large capacity of a single unit which must be removed for servicing, the multiplicity of joints which must be sealed, internal electrical leakage and necessity of shutting down a group of 50 cells upon failure of one cell.

The 1918 Ward design[28] of the Dow cell, shown in Figure 5-10, consisted of concrete filter press units with graphite rods extending through a concrete dividing wall, connecting cathode screen on the one side with graphite anode blades on the other.

The Hunter-Otis-Blue[15] design (1939) employed a graphite dividing wall cast into the concrete filter press frame with graphite anodes projecting vertically from the backing plate. Individual cathode fingers fabricated of steel wire screen were wrapped with asbestos paper and fastened to a screen backing plate which was connected by spring copper clips to the graphite backing plate of the adjacent cell unit. In 1947 cells of essentially the same design were converted to the use of the Stuart deposited asbestos diaphragm. Diagrams of this cell are shown in Figures 5-19 and 5-20.

A 1959 patent[18] describes an improved cell employing bipolar units which slip into grooves in a cell box. The cell box eliminates problems of sealing inherent in the filter press design. The cathode fingers of steel wire screen, and graphite anodes project vertically from a steel base plate forming an integral electrode unit. The extent to which this cell has replaced the older filter press design has not been published.

A modification of the flat finger structure used by Marsh and Stuart is the use of flattened tubes of steel screen open at both ends (rather than one), into the annular catholyte collecting chamber. This construction is somewhat cheaper than the construction used by Marsh and Stuart; however it has the disadvantage of not providing the same amount of circulation space for electrolyte provided in the Hooker Type S cell. The flattened tube design with both ends open is used in both Columbia-Hooker cells (Columbia-Southern Corporation) and the Diamond cell (Diamond Alkali Company). See Figure 5-12.

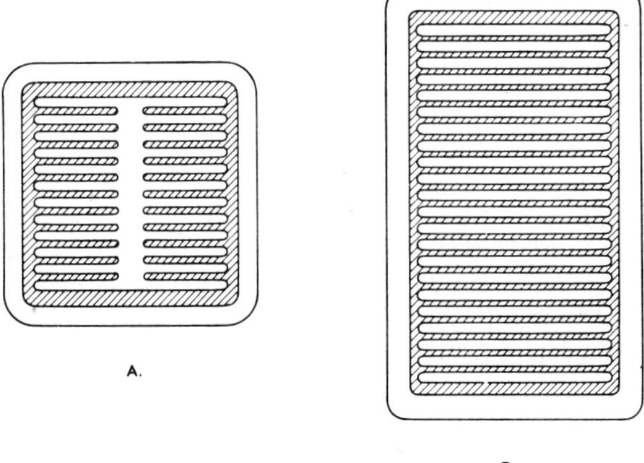

A. Finger type construction (Hooker type S cell)
B. Flattened tube-type construction (Columbia-Hooker
   Diamond, Hooker type T)

Figure 5-12.   Comparison of cathode construction.

PRESENT DIAPHRAGM CELLS

In the historical review it was noted that several hundred cell designs have been patented and that more than thirty types of diaphragm cells have been in use in the United States. During the past 10 years all new diaphragm cells have been of two basic types: The Stuart (Hooker) type and the Dow type. These types are similar, the only difference being that the Dow cell incorporates a multiplicity of unit cells into a filter press or box structure. Both types have (1) vertical graphite anodes, (2) steel screen cathodes and (3) deposited asbestos diaphragms. The Stuart type cell includes the Hooker Type S, S–3, S–3A, S–3B, S–3C and S–3M, Columbia-Hooker cells and Diamond cells. The Dow cell includes the Hunter-Otis-Blue cell design,[15] and the Lucas-Armstrong design.[18]

Hooker Type S Cell

The Hooker Type S cell has evolved in various models known as the S–1 (the original Type S cell) S–3, S–3A, S–3B, S–3M and S–3C cells. This description is based on the Type S–3C cell which is currently recommended[2] by Hooker for most installations. The cells are nearly cubical in shape and consist of three sections, bottom, middle and top, placed one upon the other. The anode assembly includes a concrete bottom in which is placed a casting of lead which contains the two flat copper anode connector bars and 128 vertically projecting graphite anode blades. The copper bars are 12 in. wide and $\frac{3}{4}$ in. thick at the ends and are tapered in the direction of diminishing current. The anode blades are $1\frac{1}{4} \times 6\frac{1}{4} \times$ 25 in. long. The lead casting is sealed into the concrete bottom by a mastic system of asphaltic material. It is a feature of the cell that the anode section which requires renewal least frequently is at the bottom and does not need to be disturbed when the diaphragm is renewed.

The cathode section consists of a steel frame with flanged top and bottom and with a copper conductor bar welded to the frame. A steel screen structure including 30 fingers is welded to the inside of the steel frame forming an integral cathode unit. The diaphragm is applied to this unit by immersing the cathode into a bath of asbestos slurried in cell liquor (the caustic liquor containing approximately 11.5 percent NaOH, 15 percent NaCl) and applying a vacuum according to a time schedule. The cathode with deposited diaphragm is placed over the graphite anodes, using a rubber cord gasket and putty for seal. Since the top of the cell is completely open, spacing between anode and cathode may be carefully adjusted. The concrete cell top is placed directly on the cathode with a seal of putty. The weight of the cell parts is sufficient to seal the cell so that no connectors are required to make the cell liquid-tight. Electrical connections from one cell to the next are made with L shaped copper connector bars slotted to provide for flexibility of assembly. Cells are removed for renewal from the circuit individually, using the portable jumper switch which operates on a mono-rail in back of the cells and which may service cells on either side of a double row. The jumper is applied without any interruptions of current to the circuit.

The cell is rated for operation at 30,000 amperes but may be operated within the range of 20,000 to 32,000 amperes. A sectional view of the cell is shown in Figure 5-13. Relationship of the cell parts may be seen more clearly by referring to Figures 5-14 and 5-15. Figure 5-14 shows the cell bottom with graphite anodes projecting upward from their anchorage in cast lead. The cathode section, with deposited asbestos fibre diaphragms, is placed on the bottom, fingers of the cathode alternating with the anode blades as shown in Figure 5-15.

Brine is introduced into the cell through a tantalum orifice placed in a glass tube set into a rubber stopper in the cell top. The head of brine on the orifices is controlled for the entire circuit and may be varied according to the current load and the strength of caustic which is desired.

This arrangement provides for a uniform feed of brine to each cell and a uniform rate throughout the life of the cell. The level of anolyte automatically adjusts itself in accordance with the porosity of the diaphragm. A cell with a new diaphragm will have a level approximately 5 in. above the top of the cathode. As the diaphragm becomes old and plugged with

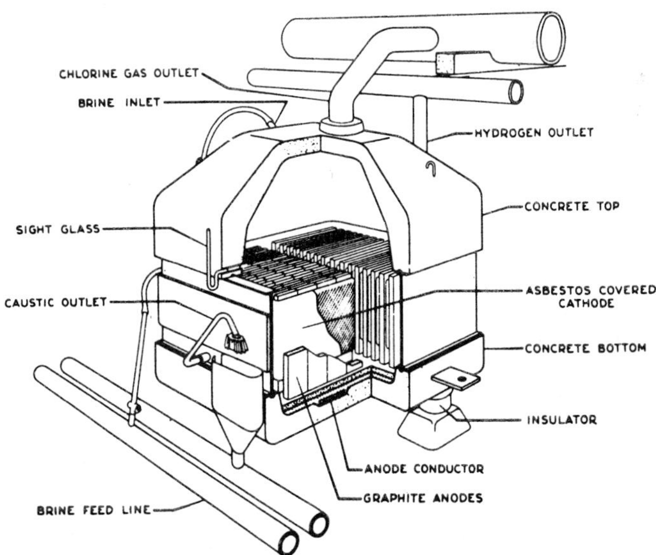

Figure 5-13. Sectional diagram of Hooker Type S-3C cell.

Figure 5-14.   Anode assembly in position (Hooker Type S-3A cell).

Figure 5-15.   Cathode with deposited diaphragm in position on anode (Hooker Type S-3A cell).

impurities, the level will rise to approximately 17 in. at which time the cell is shorted out for diaphragm replacement. The caustic liquor flows from the percolation pipe out of the cell whereas the hydrogen collected within the cathode is removed from the back of the cell through an insulator to the hydrogen header. Chlorine formed on the anodes bubbles up through the brine, accumulates at the top of the cell, and is withdrawn through a pipe to the chlorine header. The piping for brine, hydrogen and caustic liquor is generally constructed of steel whereas the chlorine line is usually constructed of ceramic, "Haveg" or "Hetron" polyester pipe.

An important feature of the Hooker cell design is that the cathode fingers do not extend completely across the cell but leave a central circulation space. During operation, the chlorine gas bubbles evolved from the anode set up a circulation causing the brine to rise in the space between the anode and diaphragm and to descend in the central aisle. It has been shown that up to 4 percent higher current efficiency may be obtained with this circulation when the anodes are new than when the circulation space is blocked off. A typical cell room of Type S cells is shown in Figure 5-16. Performance data are shown in Table 5-1.

The Columbia Hooker Cell

Columbia-Southern Chemical Corporation who are licensees of the Hooker type cell have designed a modification of the Hooker cell. This is

Figure 5-16.   Typical cell room, Hooker Type S-3A cells.

TABLE 5-1. TYPICAL OPERATING DATA HOOKER CELLS

	Type S-1		Type S-3B or S-3C			
Electrical Data						
Current, amperes	10,000	15,000	20,000	25,000	27,000	30,000
Current efficiency, percent	96.0	95.5	96.0	96.0	96.0	96.0
Voltage per cell, circuit average	3.75	3.25	3.50	3.75	3.85	3.95
Power, dc kwh/short ton Cl_2	2679	2335	2500	2680	2750	2820
Chlorine						
Production, lb/day/cell	672	1003	1344	1680	1814	2016
Caustic Soda						
NaOH Production, lb/day/cell	758	1131	1516	1895	2047	2274
Salt/caustic ratio, lb NaCl/lb NaOH	1.4	1.4	1.4	1.4	1.4	1.4
NaOH in cell liquor, percent	10.9	11.0	11.2	11.4	11.5	11.6
NaOH in cell liquor, gm/liter	135	136	140	143	144	146
Temperature of catholyte, °C	93	87	94	97	98	99.5
Evaporation req'd for 50% NaOH, lb H_2O/lb NaOH	5.7	6.0	5.7	5.3	5.0	4.8
Hydrogen						
H_2 Production, lb/day/cell	19.7	29.5	39.4	49.2	53.2	59.1
H_2, cu. ft/day/cell at 760 mm 0°C	3510	5265	7020	8775	9480	10,530
Graphite						
Pounds of graphite/short ton of Cl_2	6.7	7.0	6.3	5.3	5.3	5.3
Anode life/days	340	430	355	340	310	280

Diaphragm Life

2 or 3 diaphragms are used during the life of an anode

Typical Analyses

Chlorine: 97.5% Cl_2, 0.2 to 0.4% H_2, 1% CO_2, 0.5% O_2 by volume

Hydrogen: 99.8 to 99.9% H_2 by volume

Caustic: 0.1 part $NaClO_3$/100 parts NaOH

Brine feed: 26.6% NaCl, 322 gm/liter NaCl, 65-75°C

designed with fingers extending all the way across the cell as in the Diamond cell. Various size units have been designed up to 30,000 amperes in capacity. In other respects these cells operate essentially as the Hooker cell does.

Diamond Alkali D-3 Chlorine Cell

The Diamond Alkali D-3 chlorine cell is rated at 30,000 amperes and can be operated at 33,000 amperes for extended periods. A diagram of the cell is shown in Figure 5-17 while the cell room is shown in Figure 5-18. The cell is rectangular in shape and is 7 ft long, 3 ft 7 in. wide, and 4 ft 9 in. high. The base of the cell consists of a shallow, cast iron pan, housing flat copper grids. The copper grids and graphite anodes are imbedded in lead and the lead is sealed with protective coatings.

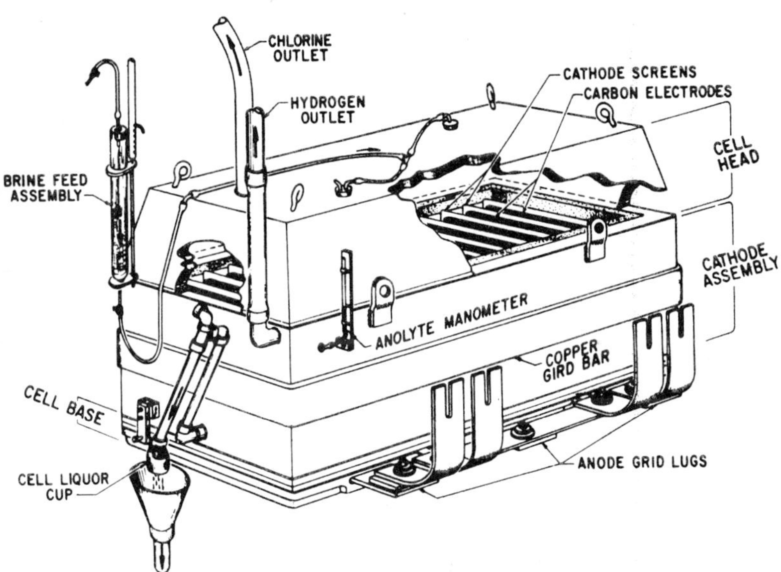

Figure 5-17.   Diagram of Diamond D-3 cell. *(Courtesy of Diamond Alkali Co.)*

The cathode assembly is a rectangular steel can. The inner section of this assembly consists of lateral rows of double metal screens, upon which a diaphragm of asbestos fiber has been deposited. These rows of screens fit between the anodes when the cathode is lowered into the base, thus forming the anolyte and catholyte sections of the cell. The cell is completed by placing a concrete head on top of the cathode assembly. This head contains

Figure 5-18.   A typical D-3 cell room. (This view shows a portion of the Diamond
   Alkali chlorine plant at Painesville, Ohio.) *(Courtesy of Diamond Alkali Co.)*

the chlorine outlet and the salt brine feed hoses. Electrical connections are
made to a copper bar around the outside of the cathode and to lugs which
extend from the anode grids.

In time, the anodes and diaphragm gradually wear out and must be
replaced. The D-3 cell renewal procedure involves removal of the cell from
the circuit and renewal in a separate area on a production line basis. When
a cell is removed, a rebuilt cell is substituted. Circuit operation during
removal and replacement is maintained by the use of jumpers which shunt
the current around the cell being replaced. This system results in a minimum loss of production and the convenience of mass production renewal
techniques carried out in a different area. Performance data are shown in
Table 5-2.

Dow Bipolar Cells

Dow had developed the bipolar filter-press type of cell through many
modifications. The filter-press cell provides a very compact unit with large

TABLE 5-2. PERFORMANCE DATA—DIAMOND ALKALI D-3 CELL

Cell amperes	27,000	30,000	33,000
Current efficiency	96.5	96.5	96.5
Current density, amp/in.^{2}	.751	.835	.920
Average cell voltage (inc. bus.)	3.68*	3.82*	3.94*
Power—KWHDC/ton Cl_2	2630	2720	2800
Graphite—lb/ton Cl_2	7.5	7.5	7.5
Average anode life, days	255	230	210
Average diaphragm life, days	115	115	110
Cell liquor temperature, °F	193	196	199
% NaOH in cell liquor	10.5*	10.5*	10.5*
$NaClO_3$ in cell liquor, gpl	0.07	0.07	0.07
Chlorine production, tons/day	0.91	1.01	1.11
NaOH production, tons/day	1.02	1.14	1.25
Anolyte temperature, °F	200	203	206

Weight of cell (empty) 5.6 tons
Weight of cell (full) 7.9 tons

*The D-3 Cell can be operated so as to produce cell liquor with a 11.2% NaOH content. This results in a cell voltage increase of .02 volt.

productive capacity for unit floor area. In addition, the investment cost is relatively low because the metal conductors between the anodes of one cell and cathode of the next cell are reduced to the minimum as is the container for electrolyte. The Hunter-Otis-Blue cell[15] consists of a series of abutting frames or unit cells each comprising one cathode and one anode, the anode of one unit cell being electrically connected within the cell frames to the cathode of the succeeding unit cell. As will be seen from Figure 5-19, the concrete frame (4) is cast around a graphite backing plate (1A); the graphite anodes (!B) are constructed integrally with the backing plate projecting at right angles to it. Spring copper contacts (15) are fastened to the reverse side of the backing plate make connections with similar spring contacts on the cathode of the adjacent cell. The cathode is a steel screen through which steel screen fingers are attached in such a way that they interleaf with the anode blades. The graphite backing plate of one cell and the cathode plate of the adjacent cell form the cathode compartment in which the caustic liquor and hydrogen are collected. The caustic liquor flows out from a hole in the bottom of each concrete filter press section and the hydrogen escapes from a similar hole at the top. As shown in Figure 5-20 the brine feed is made through a manifold which is an inverted concrete trough placed the full length of the filter press unit. Chlorine is collected from the individual unit cells in this trough. A similar trough collects the hydrogen. In one model, the cell operates at 10,000 amperes with 50 unit cells in a filter press unit. The individual concrete units are sealed with a ring of putty and are held together by tie rods extending the full length of the unit. Since a complete unit must be taken

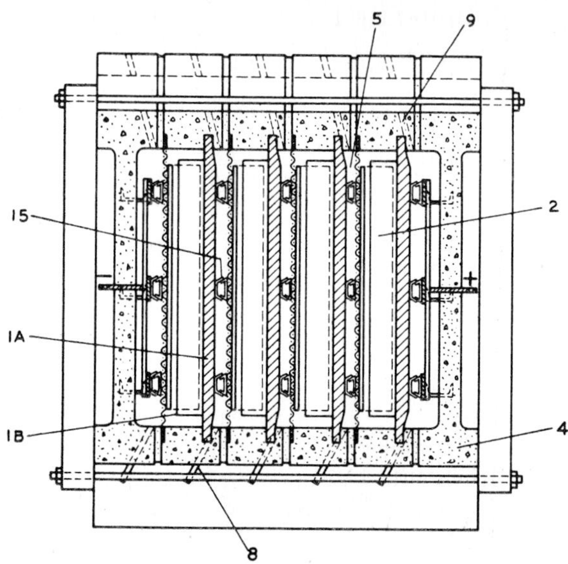

(Key to parts same as in Figure 5-3)

1A. Graphite anode backing plate
 B. Graphite anode
 2. Steel screen cathode finger
 3. Diaphragm (not shown)
 4. Concrete filter frame
 5. Cathode space for caustic liquor and hydrogen
 6. Brine inlet (not shown)
 7. Chlorine outlet (not shown)
 8. Caustic outlet
 9. Hydrogen outlet
 15. Spring clip connector between unit cells

Figure 5-19.   Dow filter press cell (Hunter-Otis-Blue).

apart for renewal of anode and diaphragms, the cells are arranged in the building so that up to 4 units may be operated in series making a total of up to 200 unit cells in a series.

One spare unit is provided for about 12 operating units and changes are made by moving flexible cables. The concrete unit cells are simply supported on treated wooden timbers. The caustic liquor is removed in individual streams from each unit cell. The rate of flow of the cell liquor may be controlled by the height of the outlet tube on each unit cell. The most serious disadvantages of this construction are that the complete unit of 50 cells must be dismantled in case any one of the individual units re-

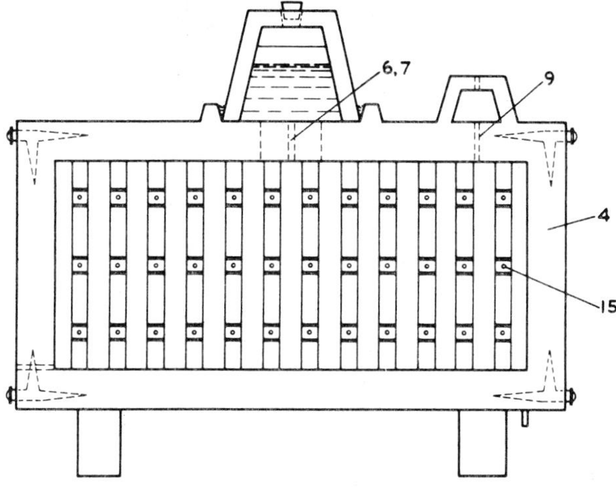

(Key to parts same as in Figure 5-3)

4. Concrete filter press frame
6,7. Brine inlet—chlorine outlet
9. Hydrogen outlet
15. Spring clip connector between unit cells

Figure 5-20.   Dow filter press cell (Hunter-Otis-Blue);
transverse vertical section.

quires repair, that a relative large unit must be shut down for renewal, and that leakage of current tends to occur within a given filter-press unit. Originally the cell was designed so that the cathode fingers could be wrapped with asbestos paper. At a later date the cell was adapted to use deposited asbestos diaphragms. An improved cell has been patented by J. L. Lucas *et al.*[18] using a solid box instead of filter-press frames. The anode of one cell and the cathode of the next are made into an integral unit which may be slipped into slots in the concrete box. To what extent this cell is used or how successful it has been has not been published. Figure 5-21 shows its construction. The concrete box (4) is shown with three unit cells. Presumably these cells may be increased to a number such as 20. The bipolar units are slipped into slots in the concrete with cathodes (2) inter-meshed with graphite anodes (1).

Figure 5-22 shows the detail of the construction of the bipolar unit. The main member of this structure is a steel plate (11) which is grooved to receive the anodes (1) and welded to the cathode structure which is fabricated of steel wire screen. The anode blades (1) are wedged into the grooves using the lead strips to make electrical contact and to prevent mechanical

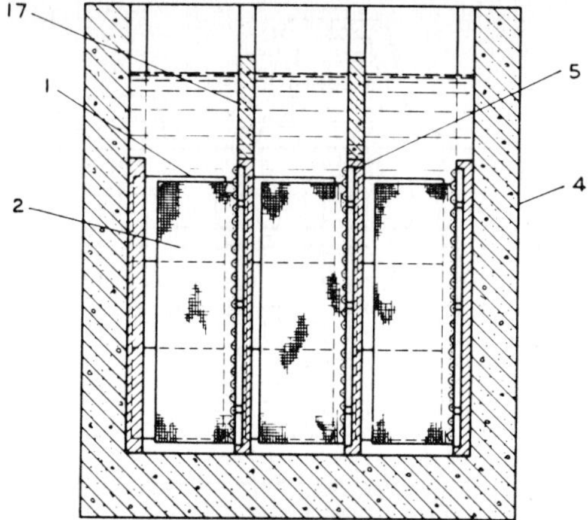

1. Anode
2. Cathode
4. Box container
5. Catholyte space
17. Intercell divider

Figure 5-21. Dow box cell (Lucas-Armstrong).

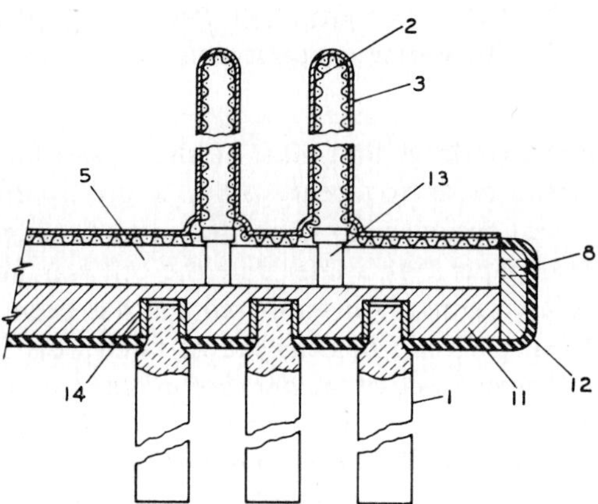

1. Graphite anode
2. Steel screen cathode
3. Deposited asbestos diaphragm
5. Container for catholyte and hydrogen
11. Steel base plate
12. Chlorine resistant, nonconductive coating
13. Integrally welded anode-cathode contact

Figure 5-22.   Dow cell: construction of bipolar
electrode, horizontal section (Lucas-Armstrong).

support. A coating of plastic material is applied over the surface of the steel plate.

Outlets for cell liquor and hydrogen are provided in the assembly. When the cell is assembled, the openings in the anode-cathode assembly line up with holes in the concrete container and a sealing material is used to completely seal the assembly in the concrete container. An advantage of this construction over that of the Hunter-Otis-Blue cell is that individual units may be replaced without dismantling a complete cell. Also the cell housing need not be moved when a cell is serviced. Further, when anodes and cathodes are installed, visual inspection of the parts can be made easily as they are interleaved.

DESIGN AND PERFORMANCE

Function of Diaphragm

The most primitive "diaphragm" cells were (1) the Bell-Jar type cell, in which the difference in density of anolyte and catholyte and the flow of anolyte were used to separate anolyte and catholyte and (2) non-percolating diaphragm cells, such as the Griesheim cell, in which the diaphragm was used simply to prevent mechanical mixing of anolyte and catholyte. Le Sueur is credited with combining the principles behind these cells to create a percolating diaphragm.

Maintenance of a flow of brine from the anolyte through the diaphragm into the catholyte to approximately equal the rate of OH^- ion migration from cathode to anode is the principle which has been applied in most diaphragm cells since 1892.

It has been assumed by various authors that the diaphragm acts as a filter bed and that the flow of brine and migration of ions can be calculated when the number of pores and average pore diameter is known. Observation of the diaphragms and of the effect of different variables on diaphragms indicate that this view is an over-simplification. There has been very little scientific study of the action of the chlorine- alkali cell asbestos diaphragm. It is thought that the effective diaphragm is a gel layer formed within the asbestos mat which is sensitive to pH and which tends to dissolve, precipitate and reform depending upon flow rate and salt content and pH of the flowing liquor. Preparation of a deposited diaphragm is shown in Figure 5-23.

It has been found that the pH of the anolyte flowing through the diaphragm has a major influence upon the rate of flow of anolyte at a given pressure. Normal anolyte pH during operation is approximately 3.0 to 4.0. In a cell without current, the flow of brine may be five times normal. If the pH is permitted to drop to 1.5 or below, the flow may decrease to one half of normal. Since the diaphragm separates a strongly alkaline solution from

Figure 5-23. Preparation of a deposited asbestos diaphragm. (Application of vacuum to the cathode according to a specified time-vacuum cycle forms a coating of asbestos on the cathode screen. The bath of asbestos slurry is prepared by mixing asbestos of certain graduated fiber length in caustic liquor from the cells).

a weakly acid solution, decrease of flow rate automatically allows the pH to rise, hence the diaphragms tend to be self-regulating.

An understanding of this effect has clarified several points which baffled early workers.* †

The relationship between flow rate of brine through the diaphragm and current efficiency was recognized early. Also the NaOH concentration, which depends upon flow rate and current density, was known to have a definite relationship to current efficiency. Sodium chloride concentrations of the brine feed, anolyte and catholyte were also known to have a definite effect on current efficiency.

*According to the concept that the diaphragm is a simple filter, an unsubmerged vertical diaphragm such as that used in the Vorce cell would tend to result in very high flow rates at the bottom of the cell and low flow rates at the top. Extremely poor performance would be expected, whereas such is not the case.

†In the early days when carbon anodes were used, either diaphragm life was very short or cell efficiency was very low. It is now known that carbon anodes tend to discharge OH^- ions from chloride solution much more readily than graphite, hence the anolyte pH tended to be low. Therefore the only practical method of operation was to permit a large amount of back migration of OH^- from the catholyte.

One of the most useful relationships in the study and control of diaphragm cell operation is the "percent decomposition" or salt-caustic ratio. This is the ratio of weight NaCl to weight NaOH in a given weight of caustic cell effluent. The percent of decomposition is directly calculated from the salt-caustic ratio as follows:

$$\frac{\text{Moles NaOH}}{\text{Moles NaOH + Moles NaCl}}$$

The salt-caustic ratio combines in one factor the effects of brine concentration, brine flow rate, current density and alkali concentration.

With a given cell type, a certain salt-caustic ratio tends to correspond to a certain current efficiency, independent of brine concentration, temperature, caustic concentration or current density. This empirical relationship greatly simplifies comparison of cell performance and has been surprisingly accurate in predicting performance.

Figure 5-24 shows the relationship between salt-caustic ratio as determined experimentally with a Hooker Type S cell operating at approximately 7700 amperes under a wide range of conditions. Curve *A* was derived from tests with NaCl brine, Curve *B* from tests with KCl brine. Curve *C* was proposed by A. H. Hooker[14] as the curve for a perfect diaphragm. (There does not appear to be sufficient theoretical justification for this assumption.) Curve *D* is from data by Moore and Curve *E* from data by Billiter.

A series of curves more-or-less parallel with Curves *A* and *B* would represent cells having performance characteristics better or worse than the particular Hooker Type S cells from which the data were taken.

Factors which are known to result in displacement from lines **A** and **B** are as follows:

(1) Diaphragm uniformity—greater uniformity: higher current efficiency (CE)

(2) Rate of internal circulation of anolyte within the cell—greater circulation: higher CE

(3) Current density—higher DC: slightly higher CE

At a salt-caustic ratio of 1.4, the following values of current efficiency are read or extrapolated from the graph for the various cells.

Townsend	96.6
Hooker Type S (mean)	96.5
Vorce	94
Allen-Moore	93.8
Marsh	92
Billiter	92

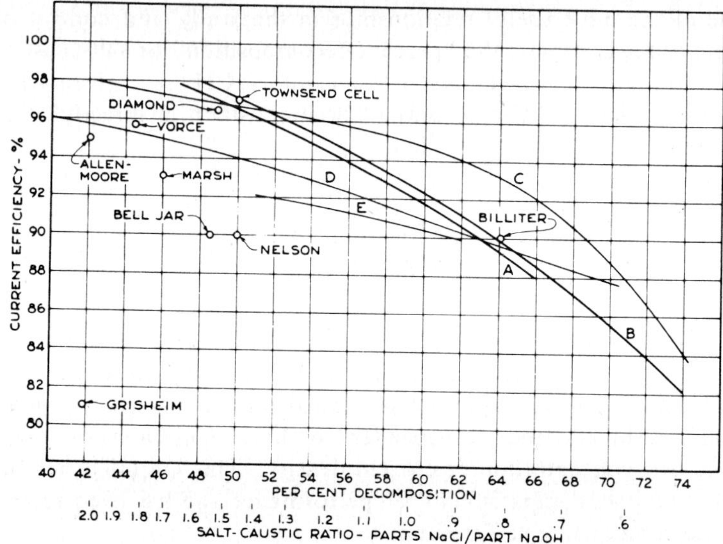

A. Experimentally determined curves, Hooker type S cell at 7700 amperes NaCl brine

B. Experimentally determined curve, Hooker type S cell at 7700 KCl brine

C. "Perfect diaphragm" curve proposed by A. H. Hooker

D. Moore data

E. Billiter cell data

(Other points as indicated in literature references)

Figure 5-24. Current efficiency as a function of per cent of decomposition.

Function of the Anode

It has been stated that the current efficiency of a diaphragm cell depends primarily upon the diaphragm and upon operating conditions which are summarized in the concept of "salt-caustic ratio" or percent decomposition. It has also been indicated that the structure, composition and performance of the diaphragm depend to a considerable extent upon operating conditions such as pH.

The anode discharges anions in accordance with their respective concentrations and in accordance with their respective discharge potentials at the anode surface. Oxygen and chlorine evolution are the principal anode reactions. Oxygen discharge is very complicated since several different processes may be taking place simultaneously.

The various anions which may participate in oxygen discharge are hydroxyl, chlorate, sulfate and hypochlorite. Any oxygen-producing discharge produces an equivalent of hydrogen ion. Hence in normal operation, it may be considered that all oxygen produced is derived from OH^- ion discharge. With graphite or carbon anodes, the OH^- ion discharge attacks the structure of the anode and tends to form organic oxygen compounds, carbon monoxide and carbon dioxide.

In a continuously operating cell, the rate of hydroxyl ion discharge must equal the rate of entry of hydroxyl ion into the anolyte, hence the ratio of oxygen to chlorine discharge is almost independent of the anode characteristics. Therefore when a certain rate of hydroxyl ion discharge is maintained, the anolyte pH tends to reach a certain level which depends upon the anode material, particular characteristics of anode material such as porosity, impregnation, etc. and concentration of other anions such as chloride, chlorate, sulfate. Empirically it has been shown by Murray and Kircher[23] that the oxygen current efficiency (including CO and CO_2) of the anode reaction is directly proportional to the reciprocal of the chloride concentration to the fourth power. The relationships of porous carbon, retort carbon, graphite, smooth platinum and impregnated graphite are shown in Figure 7 of Reference.[23] These were confirmed in recent work by L. E. Vaaler.[26] Figure 5-25 shows this relationship in the range of pH and NaCl concentrations normally found in chlorine-alkali cells. At an anolyte NaCl concentration of 270 gpl NaCl (4.6 molar) and an oxygen current efficiency of 4 percent (approximately 96 percent chlorine current efficiency) the anolyte pH's from the graph are listed in Table 5-3.

TABLE 5-3

	$\dfrac{1}{(H^+)(Cl^-)^4}$ at O CE = 4	$\dfrac{1}{(H^+)}$ at $(Cl^-) = 4.6$	$\log \dfrac{1}{(H^+)} = pH$ at O CE = 4 $(Cl^-) = 4.6$
X plain graphite	15	6740	3.83
Y impregnated graphite	20	9000	3.95
C impregnated graphite	28	12600	4.10
Z smooth platinum	32	14400	4.16

It will be noted that the pH at a given rate of oxygen discharge and given chloride concentration are in the order:

plain graphite < impregnated graphite < platinum

Carbon anodes tend to have pH values considerably lower than the range indicated and "pre-polarized" platinum electrodes considerably higher. Although the above relationships are typical, factors such as porosity of the

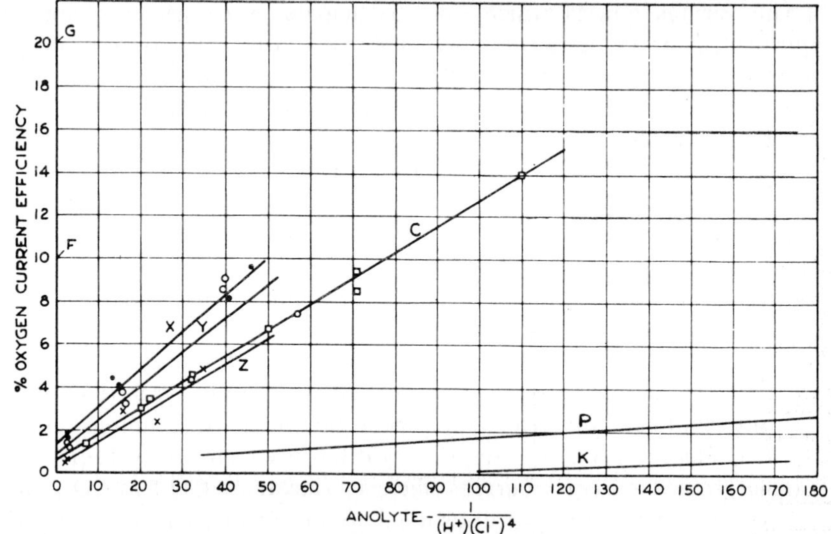

X. Plain graphite[26]
Y. Treated graphite[26]
Z. Platinum[26]
C. Treated graphite[23]
G. Porous carbon[8]
F. Retort carbon[8]
P and K. Platinum wire[8]

Figure 5-25.   Relationship between oxygen current efficiency and the hydrogen ion and chlorine ion concentration of the anolyte.

anode, length of operation of the anode, electrolyte velocity, etc. have an influence.

The relationship is useful since only one current efficiency determination is necessary to establish a curve for a given cell.   Having the curve established, the current efficiency may be estimated simply from a test of the anolyte for pH and chloride concentration.

Furthermore since the chloride ion concentration tends to remain fairly constant in good commercial practice, the anolyte pH and current efficiency may be correlated.

Figure 5-26 shows a relationship between anolyte pH and current efficiency.   A correlation has been found for Hooker cells according to anode and diaphragm ages.   All cells are classified arbitrarily in three groups as indicated in Figure 5-26.   In practice, if the anode age and diaphragm age are known and if the anolyte chloride concentration is approximately 270 gpl NaCl, the current efficiency may be estimated from the anolyte pH by reference to these three curves.

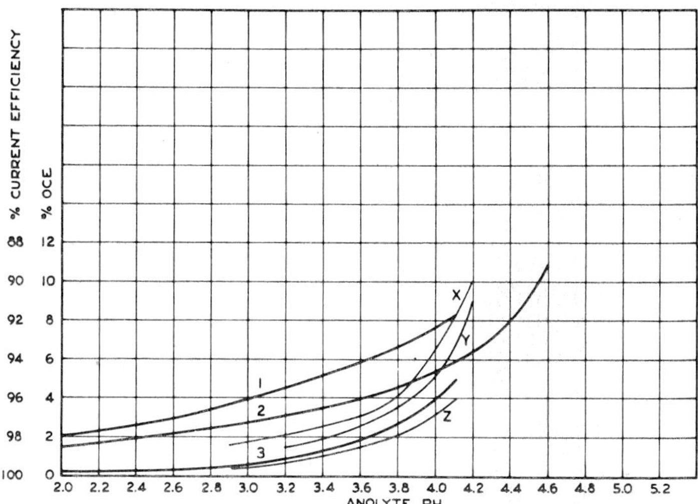

Hooker Type S cell—treated graphite, 270 gpl NaCl in anolyte.
1. New anode, entire life of first diaphragm
2. Old anode, first 30 days life of new diaphragm
3. Other cells

L. E. Vaaler data—270 gpl NaCl in anolyte[26]
X. Plain graphite
Y. Treated graphite
Z. Platinum

Figure 5-26.   Relationship between current efficiency and
anolyte pH.

As a matter of interest the three curves X, Y and Z from L. E. Vaaler's data (see Figure 5-25) are also plotted in Figure 5-26.

Figure 5-27 shows a typical breakdown of the losses of chlorine current efficiency expressed as a function of pH of anolyte.[23]

It will be noted that hydroxyl discharge (Reaction 5-5) is the principal loss under normal conditions.   Chlorate formation (Reaction 5-4) may become significant at high pH values.   This is the principal reaction in chlorate cells.   It is interesting to note that Reaction (5-4) in a chlorine cell at a pH of 4 takes less than one percent of the current efficiency; whereas in chlorate cell at a pH of 6.7 it takes up to 92 percent of the current.   Other than the difference in pH indicated, the only other difference affecting the reaction is time.   Chlorine cells are designed with a small anolyte volume whereas chlorate cells are designed with a large electrolyte volume to allow time for Reaction (5-4).

Since the major loss of current efficiency of the cell is "oxygen" discharge at the anode there is a definite relationship between current efficiency and

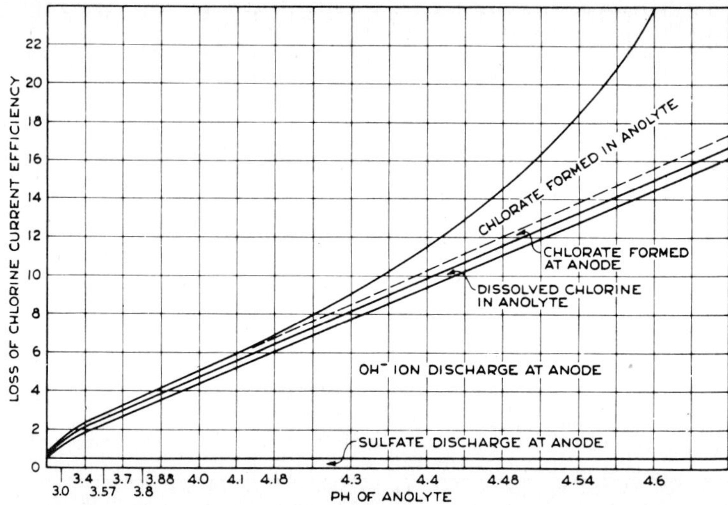

Figure 5-27.   Breakdown of losses of chlorine current efficiency as affected by anolyte pH.[23]

gas analysis. In fact, gas analysis provides the most practical method of determining current efficiency.[16]

In accordance with Reaction (1), formation of one gram mol of chlorine requires two faradays (faraday = 96,494 amperes seconds) whereas formation of a mole of oxygen according to Reaction (5-5) requires four faradays. Likewise formation of a mole of CO_2 according to Reaction (5-6) requires four faradays and formation of a gram mol of CO requires two faradays.

Assuming that the gases are "perfect" and that all of the gases produced from the cell are derived from anodic processes, the following relationship will hold:

$$\text{Chlorine current efficiency} = \frac{\text{faradays corresponding to chlorine}}{\text{Total faradays}}$$
$$\text{(CE)}$$

$$CE = \frac{\text{vol}\%\,Cl_2}{\text{vol}\%\,Cl_2 + 2(\text{vol}\%\,O_2 + \text{vol}\%\,CO_2)\,\text{vol}\%\,CO + \text{chlorine lost by}}$$
$$\text{solution in anolyte}$$

In applying this method for determination of current efficiency, correction must be made for CO_2 derived from Na_2CO_3 in the feed brine and for O_2 derived from air leakage (determined from the nitrogen content of the gas).

Function of Cathode

The cathode in modern cells is a steel wire mesh with wire approximately 1/16 in. diameter and 6 wires to the inch. The screen is flattened by hot rolling to provide a large surface and small holes. Older cells such as the Vorce and Nelson employed a perforated steel sheet. The asbestos paper or asbestos fibre is pressed tightly against, and is supported by the cathode. The pressure differential between anolyte and catholyte maintains the diaphragm, which has little mechanical strength *per se*, against the cathode support. With deposited asbestos diaphragms, the asbestos fibres tend to penetrate into the holes of the cathode structure and to lock the diaphragm to the screen.

Hydrogen bubbles form over the entire metal surface. Those formed on the face of the cathode toward the anode must penetrate minute spaces between the asbestos fibre and the metal surface to enter the cathode compartment. Some diffusion of hydrogen through the diaphragm into the anode compartment tends to occur and to result in an undesirable addition to the chlorine. Under good operating conditions hydrogen in the chlorine normally is less than 0.2 percent.

Contamination of the chlorine with hydrogen is objectionable principally because, after chlorine is removed from the cell gas by liquefaction, the residual gas may contain sufficient hydrogen to be in the explosive range.

The amount of hydrogen in the cell gas which is tolerable depends upon the particular chlorine handling system. Normally hydrogen up to 4 percent by volume is not likely to cause difficulty.

The causes of high hydrogen in the chlorine from diaphragm cells may be:
(1) Plugging of holes in cathode screen.
(2) Insufficient differential head between anolyte and catholyte.
(3) Rust spots on the diaphragm caused by allowing diaphragms to stand in a corrosive atmosphere too long before installation.
(4) Improper application of the diaphragm.

Effect of Variables on Cell Performance

The important independent variables concerned with cell performance are: current; surface areas of cathode, anode and diaphragm respectively; sodium chloride concentration of feed brine and rate of feed of brine. Less important independent variables are alkali content of feed brine, sulfate concentration of feed brine, physical characteristics of the anode, physical characteristics of the diaphragm and various items concerned with the geometry of the cell. As a matter of convenience, other variables which are dependent on the above are customarily used in correlating cell performance. These are current density, voltage, current efficiency, salt-caustic ratio of the caustic liquor and NaOH concentration in the caustic liquor.

Effect of Current Density on Voltage. Considering apparent areas only, the current density of anode, cathode and diaphragm are essentially equal on modern diaphragm cells. The voltage drop on a chlor-alkali cell is of considerable importance because the cost of power is one of the major items of operating costs. Cell design represents a compromise between low voltage-high investment and high voltage-low investment. The voltage drop may be broken down into the following components:

(1) The potential difference between the surface of the anode and the solution in contact with the anode (discharge potential) which causes chloride and hydroxyl ions to be oxidized.

(2) The potential difference between the solution in contact with the cathode and the cathode surface which causes hydrogen ions to be reduced.

(3) The potential difference between the diaphragm and the solution in contact with the anode required to cause current to flow through the anolyte.

(4) The potential drop across the diaphragm.

(5) The potential drop through the graphite anode.

(6) The potential drop through the metal conductors.

The conductor drop may be broken down further into many components including:

(1) Copper conductors between cells

(2) Contact resistance between joints

(3) Anode copper

(4) Contact between anode copper and lead

(5) Lead

(6) Contact resistance between lead and graphite

(7) Cathode copper

(8) Steel frame

(9) Wire screen and supporting steel

A classical study on voltage drop in a diaphragm chlor-alkali cell was made by Stender.[24] Articles on voltage balance in mercury cells have been made by Gardiner[9] and De Nora.[6] Figure 5-28 presents a voltage balance on a diaphragm cell of the Hooker type. The data presented are intended to show typical relationships rather than exact calculations. The total voltage is taken from a curve on Hooker S-3A cells showing relationship between voltage and current density at an average anode age. Curve *A* is the anode potential from Stender and *B* the cathode potential from Stender. Other curves are calculated for the condition in which the graphite

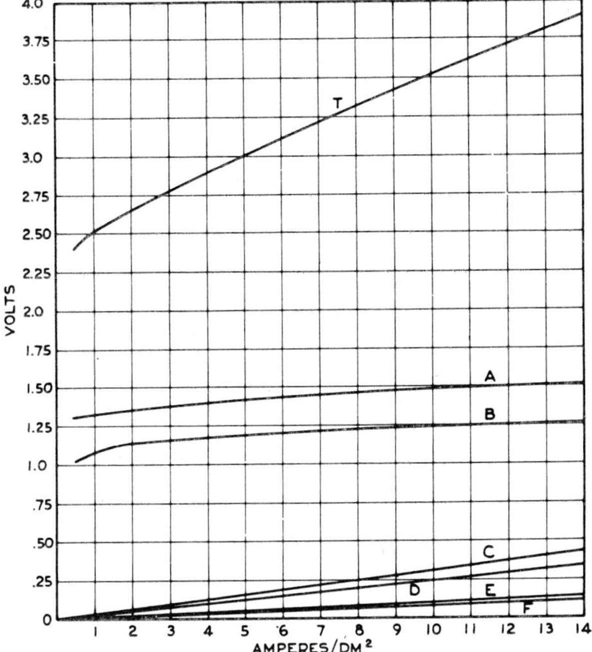

T . Total cell voltage (corresponds to average cell voltage
for Hooker Type S-3A cell)
A . Anode potential[24]
B . Cathode potential[24]
C . Anolyte (calculated for condition at half life of anode)
D . Diaphragm
E . Graphite anode (calculated for condition of half-life)
F . Metal conductors

Figure 5-28.   Voltage breakdown for typical
chlor-alkali cell.

blade is half consumed. It will be noted that voltage drop through the
brine, voltage drop through the diaphragm and voltage drop through the
graphite anode itself all have appreciable values. It will be noted further
that the total cell voltage increases at a higher rate than either anode or
cathode potential because of the additive effect of the various resistance
losses. In diaphragm cells both surfaces of the graphite anode are used and
are not adjustable. The distance between the anode and cathode increases
as it is worn. Hence the thickness and length of the anode blade and the
relative spacing between the anode and cathode are important factors in
the design of a cell. Figure 5-29 shows a family of curves representing the

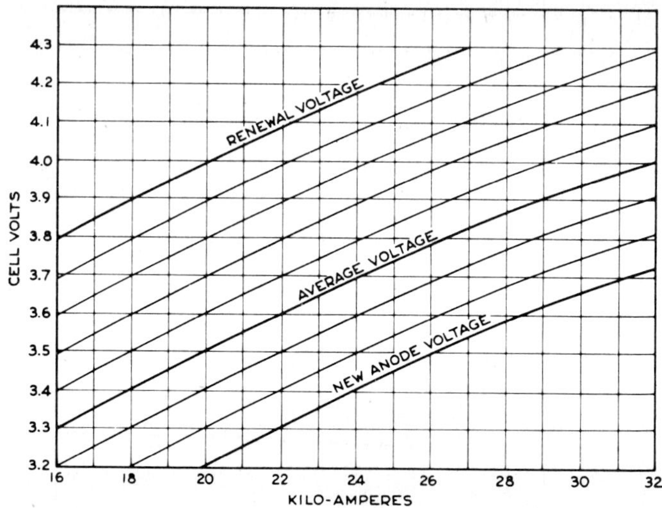

Figure 5-29. Voltage-current relationship (Hooker S-3 series cells).

voltage current relationship for the Hooker S-3 series of cells with the various lines representing different ages of anode.

Effect on Current Density on Current Efficiency. The relationship between current density and current efficiency is not as easy to determine as might be expected. Change in current density not only affects the rate at which products are formed at the electrodes but also affects the temperature of the cell and the characteristics of the diaphragm. Hence in comparing current efficiency at one current density versus the current efficiency at another density it is convenient to make the comparison on the basis of a given salt-caustic ratio in the cell liquor, recognizing that the diaphragm characteristics in one case are not exactly the same as those in the second case. Hence the curve in Figure 5-30 is not a relationship based on a given cell and a given diaphragm, rather it represents typical current efficiencies which may be obtained under typical operating conditions at various current densities at approximately the same salt-caustic ratio.

Sodium Chloride Concentration in the Brine. Sodium chloride concentration in the brine affects the sodium chloride concentration in the anolyte and also the salt-caustic ratio of the cell liquor produced. It has a significant effect upon graphite consumption, current efficiency and on the amount of water which must be evaporated to produce 50 percent caustic. The relationship between salt-caustic ratio and current efficiency has already been shown in Figure 5-24 and the effect of sodium chloride con-

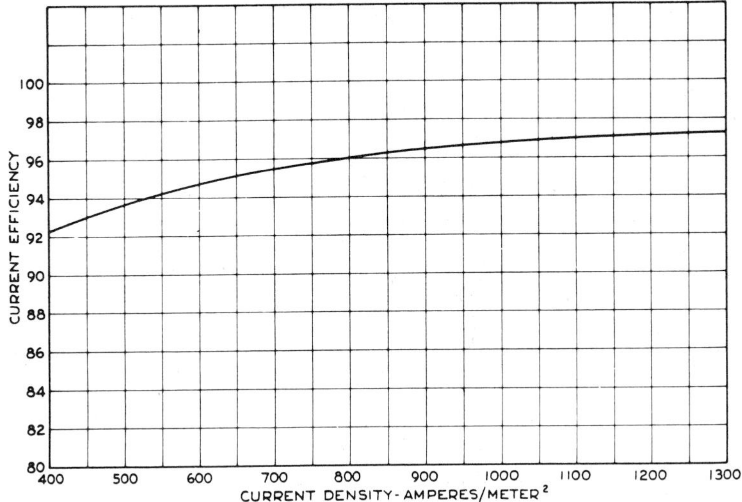

Figure 5-30. Current efficiency as a function of current density at a salt-caustic ratio of 1.4 in the caustic liquor (Hooker Type S cell).

centration in the anolyte on the pH of the anolyte and oxygen evolution at the anode have been shown in Figure 5-25. The relationship between sodium chloride concentration of the brine and the salt-caustic ratio and percent NaOH in the cell liquor is indicated in Figure 5-31. The temperature of the cell is an important variable in this relationship since the amount of water evaporated in the cell depends upon the cell temperature. Figure 5-32 which is derived from Figure 5-31 and Figure 5-24 indicates the effect of sodium chloride concentration of the brine on current efficiency at various caustic concentrations.

Sodium Hydroxide Concentration of the Cell Liquor. This depends, primarily, on the current on the cell and the rate of feed of brine and secondarily upon cell temperature. The concentration of NaOH obtained in diaphragm cells may range from approximately 8.5 to 13 percent. Modern cells tend to operate in the range of 11 to 12 percent. Whereas Figure 5-24 indicates the salt-caustic ratio which is required to obtain a certain current efficiency with a given cell, Figure 5-31 indicates the percentage of NaOH which would be obtained at such salt-caustic ratio with various concentrations of sodium chloride in the brine and with a certain indicated temperature.

Effect of Current Efficiency on Graphite Consumption. In the discussion on anodes it was indicated that the principal loss of current efficiency is the discharge of hydroxyl ions at the anode to form oxygen and carbon

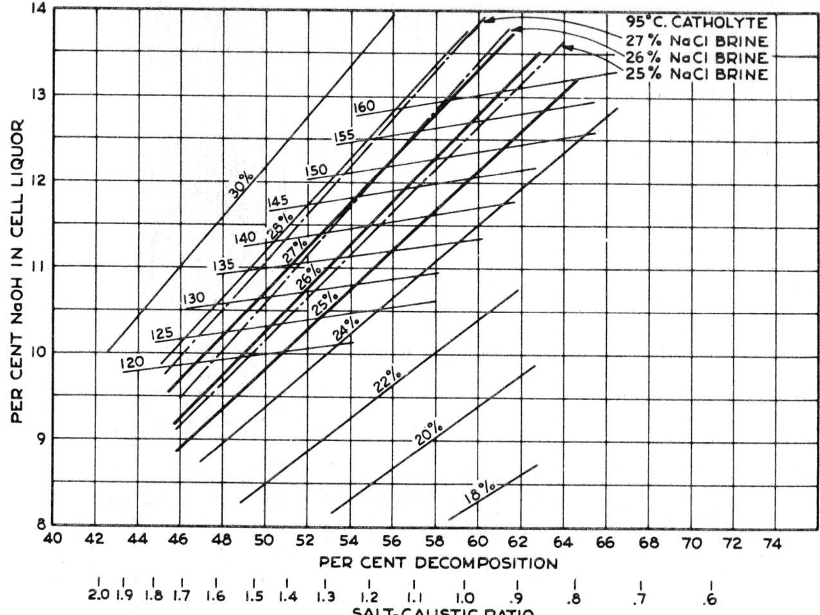

Figure 5-31. Effect of the NaCl concentration of brine on concentration of NaOH in catholyte and upon the salt-caustic ratio of catholyte. (Data correspond to operation at a temperature of 91° C. The effect of temperature is shown by three broken lines corresponding to operation at 95° C.)

Figure 5-32. Effect of NaCl concentration in the brine on current efficiency at various NaOH concentrations (derived from Figures 5-24 and 5-31 —based upon Hooker type S cells operating at 91° C.)

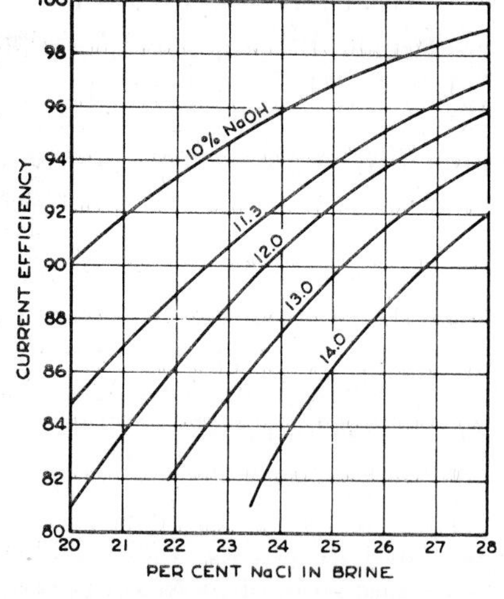

dioxide. Hence graphite consumption may be expected to be directly proportional to the loss of current efficiency. If the consumption of graphite is 6 lb/ton of chlorine at 96 percent current efficiency (4 percent loss of current efficiency), at 92 percent current efficiency (8 percent loss of current efficiency) the graphite consumption may be expected to be 12 lb/ton of chlorine.

Relationship between the Salt-Caustic Ratio and Water. The relationship between the salt-caustic ratio and the amount of water to be evaporated to produce caustic soda is an important factor since steam required for evaporation of the cell liquor to produce 50 percent caustic is one of the major items under the control of the cell operator. To obtain a high salt-caustic ratio, a high rate of brine feed, a high concentration of sodium chloride in the brine and a high cell temperature are favorable. From Figures 5-31 and 5-24 are calculated the pounds of water to be evaporated per ton of NaOH. The current efficiencies, salt-caustic ratio and graphite consumption are indicated. From these the balance which must be made to calculate an optimum salt-caustic ratio becomes apparent. The calculated values are indicated in Table 5-4.

TABLE 5-4

Salt-Caustic Ratio	NaOH Concentration* (%)	Current Efficiency†	Graphite lb/ton Chlorine	Pounds Water to Evaporate per Ton NaOH to Produce 50% NaOH
.80	15	89	16.5	7,720
.90	14.4	90.8	13.8	8,100
1.00	13.7	92.3	11.5	8,600
1.10	13.0	93.4	9.9	9,180
1.20	12.45	94.5	8.25	9,660
1.30	11.8	95.4	6.9	10,360
1.40	11.3	96.1	5.85	10,920
1.50	10.9	96.8	4.8	11,340
1.60	1 0.4	97.4	3.9	12,020

*From Figure 5-31 assuming 27.0% NaCl in brine feed and cell temperature of 95°C.
†From Figure 5-24 Curve *A*.

FLOW SHEET OF A TYPICAL CAUSTIC-CHLORINE PLANT USING DIAPHRAGM CELLS

A typical modern caustic-chlorine plant is shown in the schematic flow sheet of Figure 5-33 and also in photographs of various sections of the plant (Figures 5-34 to 5-39). Rock salt is dissolved in water (or brine is produced from underground salt deposits by water injection in a well) and

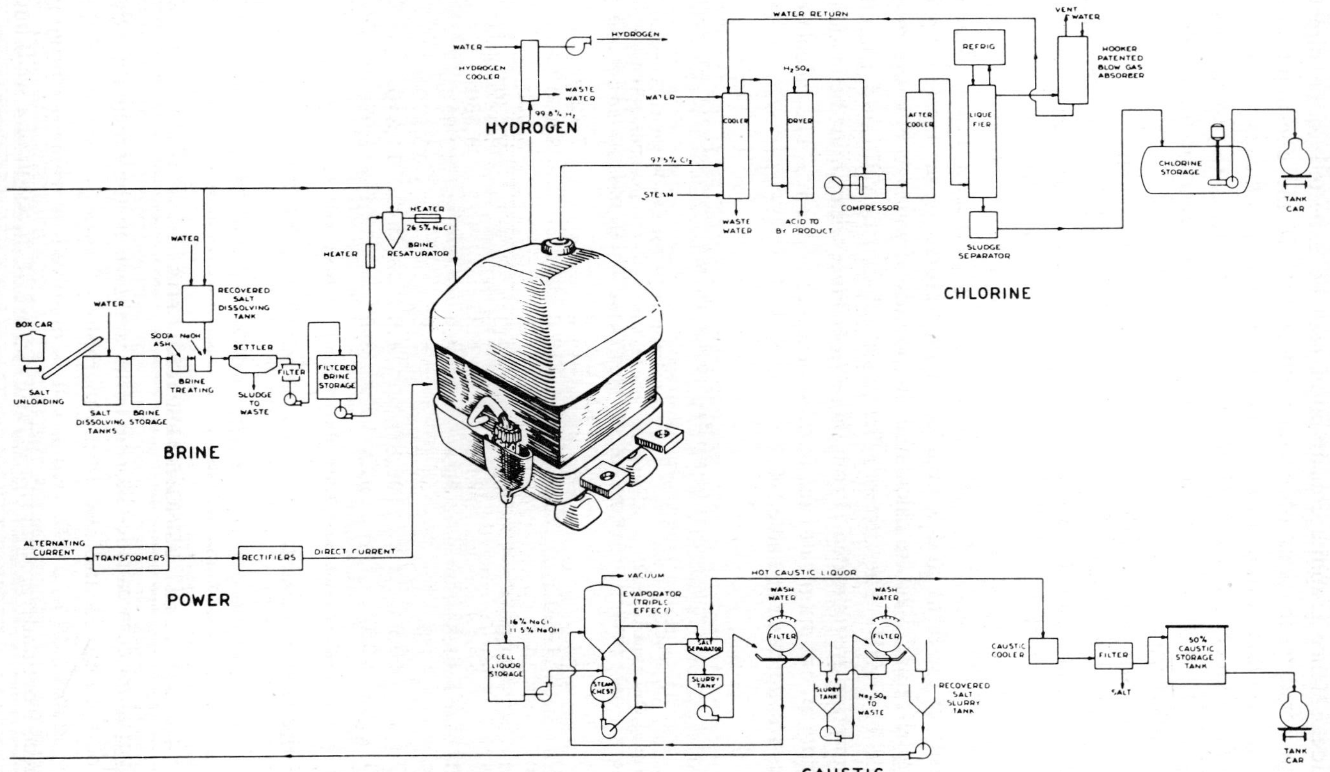

Figure 5-33.   Schematic flow sheet of typical caustic-chlorine plant using diaphragm cells.

Figure 5-34.　Unloading salt from ship to storage.

Figure 5-35.　Brine treating plant.

Figure 5-36. Silicon rectifiers.

Figure 5-37. Installation of diaphragm cells.

Figure 5-38.   Chlorine compressors.

the usual impurities, calcium and magnesium sulfates and chlorides are precipitated from the brine by addition of soda ash and caustic soda. Brine is clarified by settling and filtration. To increase the salt content to a practical maximum, brine is heated and saturated with purified salt, obtained from the evaporation of cell liquor, before going to the cells.

Direct current is supplied to the cells from an alternating current source by a mercury arc, silicon or other types of rectification equipment.

Electrolysis of the brine in the cell results in the formation of chlorine gas, hydrogen gas and caustic. The gases leave the cells saturated with water vapor. Caustic leaves the cell as cell liquor which contains about 11.5 percent NaOH by weight and about 16 percent NaCl by weight. The three products are withdrawn separately and processed in different departments.

Hydrogen gas is scrubbed with water sprays to cool and remove any traces of salt or caustic. It is then compressed for supplying various processes or for use as a fuel.

Chlorine cell gas is cooled by direct contact with water in a packed tower and then dried with sulfuric acid. The dried gas is compressed to 25 to 60 psi and is then liquified by refrigeration. The liquid chlorine is then transferred to tank cars, to process or to storage. Some processes can use the

Figure 5-39.　Caustic evaporation plant.

wet cell gases, other processes require dry cell gas and some need liquid chlorine. The "blow gas" from the chlorine liquifier contains non-condensable gases and some chlorine. The chlorine in the blow gas may be recovered by patented processes such as absorption in carbon tetrachloride (Diamond[3]); absorbtion in water (Hooker[3]) or may be absorbed in alkali for the production of hypochlorite or for disposal. In the Diamond process, chlorine is recovered from carbon tetrachloride solution by heating and stripping. In the Hooker process, chlorine is removed from the blow gas by absorption in water and recovered by using the chlorine-containing water for cooling the direct contact cell gas cooler. Before leaving the cooler, the water is acidified and heated with steam to strip it of chlorine.

Evaporation of the cell liquor until the NaOH content is 50 percent by weight is carried out in double or triple effect evaporators. Modern evaporators circulate the liquor rapidly in each effect through an external shell-and-tube heat exchanger. As the NaOH content increases, salt crystallizes and is separated from the caustic liquor by decantation and filtration. Salt is then washed free of caustic, dissolved and recycled to the brine system. Additional salt is removed from 50 percent caustic by cooling and settling. Standard 50 percent caustic containing only 1 percent salt is shipped as such in tank cars or may be evaporated in single effect evaporators to 73 percent caustic. Several processes can be used for further reducing the salt content of the standard liquor to convert it to rayon grade. The most widely used process for de-salting is the so called "DH process" of Columbia-Southern which employs liquid ammonia extraction of the impurities from the caustic.

Fused caustic, either solid or flake, is produced by evaporation and fusion of either 50 percent or 73 percent caustic. Fused caustic is usually shipped in metal drums.

FUTURE TRENDS

With the growth of the chlorine producing industry, diaphragm cells have increased in current capacity and in current density. It is to be expected that this trend will continue. The recent development and wide spread use of silicon rectifiers, which are economical in investment and efficient at considerably lower voltages than mercury arc rectifiers, will further increase the demand for high amperage units. There has been considerable activity on development of new types of anodes such as platinum plated titanium. In addition, there has been considerable research activity on development of synthetic ion exchange non-percolating diaphragms. It is to be expected that further work will be done in these directions. The great activity in the field of developing fuel cells may be of importance to the chlorine industry. For if the goal of commercial power in large quantities from fuel cells is attained, chlor-alkali cells would be an ideal load for such power because use of direct current generated at the site would eliminate transmission lines, transformers and rectifiers.

References

1. Allmand, A. J., and Ellingham, H. J. T., *"The Principles of Applied Electrochemistry,"* London, Arnold Co., 1931.
2. Anon., Hooker Chemical Corporation, Bulletin 20-A (1959).
3. Anon., *Chem. Eng.* **64**, (6), 154 (1957).
4. Barton, C. R., *Trans. Am. Inst. Chem. Engrs.,* **13**, 1 (1920).
5. Carrier, C. F., Jr., *Trans. Electrochem. Soc.,* **35**, 239 (1919).
6. De Nora, V., *Trans. Electrochem. Soc.,* **97**, 347 (1950).

7. Englehardt, V., *"Handbuch der Technischen Elektrochemie,"* Vol 2 (1), Leipsig Akademsche Verlagsgesellschaft, 1933.

8. Foerster, Fritz, Electrochemie, Wasseriger Losungen, 4th ed., Barth, 1923.

9. Gardiner, W. C., *Chem. Eng.,* **52**, (7), 110 (1945).

10. Griswold, T. Jr., U. S. Patent 987,717 (1911).

11. Griswold, T. Jr., U. S. Patent 1,070,454 (1913).

12. Hardie, D. W. F., *"Electrolytic Manufacture of Chemicals from Salt,"* Oxford University Press, 1959.

13. Hass, K., *"Ullmanns Encyklopadie der Technischen Chemie,"* Vol. 5, pp. 324-376 Munich, 1954.

14. Hooker, A. H., *Trans. Am. Inst. Chem. Engrs.,* **13**, 61, (1920).

15. Hunter, R. M., Otis, L. B., and Blue, R. D., U. S. Patent 2,282,058.

16. Kircher, M. S., Engle, H. R., Ritter, B. H., and Bartlett, *Trans. Electrochem. Soc.,* **100**, 448 (1953).

17. Le Sueur, E. A., *Trans. Electrochem. Soc.,* **63**, 187 (1933).

18. Lucas, J. L., and Armstrong, B. J., U. S. Patent 2,858,263.

19. Marsh, C. W., U. S. Patent 1,075,362 (1913).

20. Mantell, *"Industrial Electrochemistry,"* 3rd Ed., p. 421, New York, McGraw-Hill, 1950.

21. Moore, *Chem. Met. Eng.,* **23**, 1011, 1072, 1125 (1920).

22. Murray, R. L., *Ind. Eng. Chem.,* **41**, 2155 (1949).

23. Murray, R. L., and Kircher, M. S., *Trans. Electrochem. Soc.,* **86**, 83 (1944).

24. Stender, W. W., Zivotinsky, P. B., and Stroganoff, M. M., *Trans. Electrochem. Soc.,* **65**, 189 (1934).

25. Stuart, K. E., Lyster, T. L. B., and Murray, R. L., *Chem. Met. Eng.,* **45**, 354 (1938).

26. Vaaler, L. E., *Trans. Electrochem. Soc.,* **107**, 691 (1960).

27. Vorce, L. D., *Trans. Electrochem. Soc.,* **86**, 69, (1944).

28. Ward, L. E., U. S. Patent 1,365,875 (1921).

6. ELECTROLYSIS OF BRINES IN MERCURY CELLS

R. B. MacMullin

R. B. MacMullin Associates

HISTORY AND STATUS

The method of producing chlorine and caustic alkali in an electrolytic cell with a mercury cathode was simultaneously discovered by Hamilton Y. Castner, an American, and Karl Kellner, an Austrian, who, unaware of each other, applied independently for patents in 1892.* After an initial competitive fight, the two inventors united their efforts and sold their patent rights to Solvay *et Cie.* Industrial operations were entered upon under the name of Castner-Kellner.

The Castner rocking cell was first built at Oldbury, England. In 1894 the Mathieson Alkali Works, Inc. acquired the patent rights, and a demonstration plant was operated for a short time at Saltville, Virginia (1895 to 1897). These early cells, designed for 550 amperes, were first fitted with carbon anodes, which had a life that was short indeed. On a hunch that graphite might make a better anode, Castner invented a method of graphitizing his carbons.† This improvement and many others encouraged the owners to build a commercial plant at Niagara Falls in 1897. The success of this venture may be judged by the fact that the capacity of this plant was enlarged many times and, except for minor changes, the old Castner rocking cells were in continuous operation from 1897 to 1960. These were superseded in 1960 by the latest version of Olin Mathieson mercury cells having a rating of 100,000 amperes. Yet, the Castner rocking cell truly deserves a rewarding spot in the Valhalla of great American inventions.

In England, the Castner-Kellner process was established at Runcorn in 1897; in Belgium it was established at Jemeppe. Mercury cell development then took a different direction in Europe, beginning with the work of A. Brichaux and H. Wilsing of the Solvay Company who built the first "long"

---

*Actually a number of patents dealing with mercury cells issued prior to 1892, the first being that of Nolf, British Patent 4349 (1882).

†U. S. Patent 572,472.

cell at Osternienburg, Germany in 1898. The electrolyzer and the decomposer took the form of long parallel troughs, through which the mercury was circulated by lift pumps of various kinds. For many years platinum wire anodes were used and decomposer grids were of cast iron, although both were later replaced with graphite. In the meantime, there was a proliferation of diaphragm cell plants, and until the 1930's these dominated the chlor-alkali industry. With growing demands for rayon grade caustic, the mercury cell proponents seized the initiative.

In Germany, rapid progress was made in the I. G. Farbenindustrie plants during the 1930 to 1945 period. The various chlor-alkali plants exchanged information and shared the fruits of research and development. At the end of World War II, investigating teams of many countries had the unusual opportunity of studying the German plants, and it was generally conceded that Germany had reached a position of technical leadership in the mercury cell process for chlor-alkali.[36,58,59] This had a remarkably stimulating effect on other nations, and progress has been extremely rapid everywhere since 1945.

Prior to 1945 the usual size of a mercury cell was in the range of 12,000 to 14,000 amperes. During the war Germany was experimenting with larger cells rated at 28,000 amperes, and also with a rotating cathode cell rated at 40,000 amperes. Since 1945 cell sizes have been increased constantly, and as of 1960, cells rated at 100,000 to 200,000 amperes are on the market from a number of sources. That these larger cells are more economical than the small ones is quite clear. It is also evident that this trend is intimately related to the development of new types of rectifiers which are efficient and economical at the higher ampere loads and lower circuit voltages that go hand in hand with larger cells.

In the United States, the diaphragm cell process still dominates. In 1945, mercury cells only accounted for 4 percent of the U. S. production of chlorine; by 1959, this figure had been increased to 19 percent, and the trend is still up.[14] Elsewhere in the world the mercury cell is dominant. The contributions of Belgium, France, Germany, Great Britain, Italy, Japan and the United States to the new mercury cell technology are noteworthy.

General Description and Classification of Mercury Cells

The mercury cell to be discussed in this chapter is a device for the electrolytic decomposition of an alkali chloride salt, resulting in the production of chlorine, caustic alkali and hydrogen. It is comprised of two essential parts, an electrolyzer and a decomposer. In the electrolyzer an aqueous solution of the salt is electrolyzed, making use of an insoluble anode and a flowing mercury cathode. Chlorine gas is evolved at the anode, and alkali metal is deposited at the surface of the mercury cathode, in which it dissolves to form a liquid amalgam. In the decomposer, the amalgam is decomposed with water to form alkali hydroxide and hydrogen gas. A closed circulation of mercury through electrolyzer and decomposer is maintained by means of a pump or other device. Starting with sodium chloride which is the most typical example, the principal reactions are as follow:

In the electrolyzer:

At the anode,　　　　$Cl^- = \frac{1}{2} Cl_2 + e^-$

At the cathode,　　　$Na^+ + (Hg) + e^- = Na\,(Hg)$

Over-all,　　　　　$NaCl + (Hg) \xrightarrow{\text{1 Faraday}} Na\,(Hg) + \frac{1}{2} Cl_2$

In the decomposer:

At the anode,　　　　$Na(Hg) = Na^+ + (Hg) + e^-$

At the cathode,　　　$H_2O + e^- = OH^- + \frac{1}{2} H_2$

Over-all　　　　　$Na(Hg) + H_2O \xrightarrow{\text{1 Faraday}} NaOH + \frac{1}{2} H_2 + (Hg)$

Typically, direct current energy is applied to the electrolyzer at a voltage in the range of 4 to 4.5, with current efficiency in the range of 94 to 97 percent. Typically, the anode and cathode of the decomposer are short circuited, and no direct current energy is recovered.

It is seen that there is no direct connection between the brine solution in the electrolyzer and the caustic solution in the decomposer. For this reason, the caustic produced in the mercury cell is unusually pure. Moreover, the decomposer is usually operated to produce caustic liquor containing 50 percent NaOH and, under special conditions, 73 percent NaOH can be made. This is in sharp contrast to the diaphragm cell discussed in Chapter 5, which produces a caustic cell liquor containing 10 to 13 percent NaOH and an equivalent amount of undecomposed salt.

The principle of the mercury cell is illustrated in the simple sketch, Figure 6-1. Although all mercury type chlor-alkali cells operate on the same electrochemical principles, commercial cells differ greatly in design, capacity and results achieved. Mercury cells may be broadly classified as follows:

(1) Electrolyzers
    (a) Horizontal flowing mercury cathode;
       i Gravity flow
       ii Forced flow
    (b) Vertical falling film mercury cathode
    (c) Horizontal rotating mercury cathode

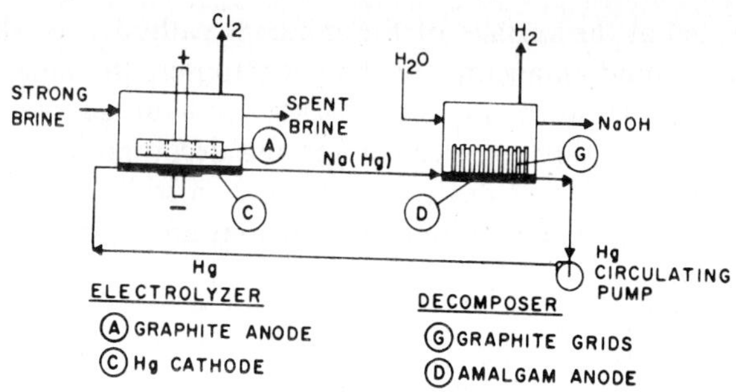

Figure 6-1.   The mercury cell principle.

    (d) Vertical rotating mercury cathode
    (e) Vertical jet stream mercury cathode
(2) Decomposers
    (a) Horizontal flowing amalgam anode, graphite grid cathode
    (b) Vertical towers, various types of graphite plate or packing serving as cathode
    (c) Rotating or vibrating cathodes

Further differentiation may be along the following lines:

Support for the mercury cathode
    (a) bare steel troughs, with varying degrees of rigidity
    (b) cement, stone, ceramic
    (c) various non-conducting coatings on (a) or (b), such as hard rubber, plastics
    (d) adjustability, various means for trough alignment
    (e) negative current connections, (various means for)

Anodes (orientation to match the cathode)
    (a) graphite, various styles of grooves and holes
    (b) metallic, such as Pt, Pt covered Ta or Ti
        i   smooth   *ii*   expanded

Anode supports and conductors
    (a) graphite, various treatments for
    (b) metallic, with various protective covers
    (c) adjustability, various seals for
    (d) single or dual support for anodes

Covers
    (a) rigid, various materials resistant to wet chlorine
    (b) flexible; rubber, plastics

Side walls
    (a) various designs
    (b) various protective coverings

Decomposers
    (a) horizontal, tubular or rectangular
    (b) horizontal, located alongside the electrolyzer
    (c) horizontal, located underneath the electrolyzer
    (d) staged, for concurrent or countercurrent operation
    (e) vertical, located at one end or the other of the electrolyzer
    (f) insulation and jacketing, various

Pumps
- (a) Archimedes spirals
- (b) bucket wheels
- (c) centrifugal, sump or tight suction
- (d) conical vortex
- (e) water lift

Cell arrangement
- (a) single deck, multi-deck, or stacked cells
- (b) horizontal U, vertical U
- (c) Aisles, or no aisles between cells

Diaphragm
- (a) vertical cells with or without diaphragm
- (b) cells where the diaphragm supports the mercury

Skimming of thick mercury
- (a) hand skimming
- (b) flushing devices
- (c) magnetic traps
- (d) continuous distillation of purge stream of mercury

Current breaks
- (a) for brine, various designs
- (b) for caustic, various designs

Cell switches
- (a) none at all
- (b) bars
- (c) gang type short circuiting switches for each cell
- (d) gang type short circuiting switches for a group of cells
- (e) mobile short circuiting switches
- (f) hand vs automatic switching
- (g) switches that isolate cell, and which may apply polarizing voltage.

Flow regulator
- (a) for brine
- (b) for decomposer feed water
- (c) for mercury

The above list is far from complete, yet it illustrates the fertility of invention in this field. A complete bibliography of the patent literature would be too voluminous to include in this book. However, some of the leading commercial cell types will be examined later in the chapter. For all of their differences in detail, the surprising thing is that they can all be recognized as mercury cells. Mercury cell design is still in a state of flux, and what the future may hold no one can safely predict.

GENERAL DESCRIPTION OF THE MERCURY CELL PROCESS

While the center of interest in any chlor-alkali plant is the cell house, and particularly the details of the cell itself, electrolysis is only one of many operations that are equally important. Any electrolytic process is concerned with the following irreducible functional parts:

Feed preparation　　　　　　　　Recovery of products
Electrolysis　　　　　　　　DC power supply

How these parts are related in a typical complete chlor-alkali plant is shown in Figure 6-2.

Just as there is almost infinite variety in mercury cells, so there is a great variety in the methods and equipment used in the other three departments. No two chlorine plants are ever entirely alike. In this flow sheet, therefore, functions are indicated and the major pieces of equipment are listed under each function.

Brine Dechlorination

In practically all mercury cell plants today, the depleted brine from the cells is dechlorinated, usually before resaturation. The reasons are:

(1) It is difficult to control iron removal in the presence of hypochlorite ion.

(2) If hypochlorous acid is not removed, the chlorate content of the brine will build up to the point where graphite consumption is adversely affected.

(3) Brine resaturation is less complicated and less nauseous if the brine contains no free chlorine.

The cell effluent is first acidified with HCl in sufficient amount to react with the HOCl present. Some chlorine gas is evolved at this point. Beyond this, practice varies according to whether there is a local outlet for sodium or calcium hypochlorite. For example, a captive chlorine plant serving a pulp mill bleachery must supply about one-half to two-thirds of the chlorine in the form of hypochlorite. In this case, the simplest method of brine dechlorination consists of blowing the brine with air in a packed tower. The chlorine in the tail gas is then scrubbed out with milk of lime or with dilute caustic soda.

Where the entire output of the chlorine plant is to be in the form of liquid, as in many merchant chlorine plants, it is customary to flash off the chlorine at reduced pressure. At $\frac{1}{2}$ atmosphere and a temperature in excess of 70° C, the available chlorine content is reduced to about 0.1 gpl. The apparatus consists of a packed section of rubber-lined tower, elevated high enough for the brine to discharge through a barometric leg. Vacuum may be produced by a water-sealed ceramic ring pump, in which case the seal water is indirectly cooled in order to remove the steam from the chlorine gas by condensation. The recovered chlorine, at atmospheric pressure, is vented to the cell-room chlorine manifold. Alternatively, the vacuum may be produced by pumping dechlorinated brine through a liquid type eductor.

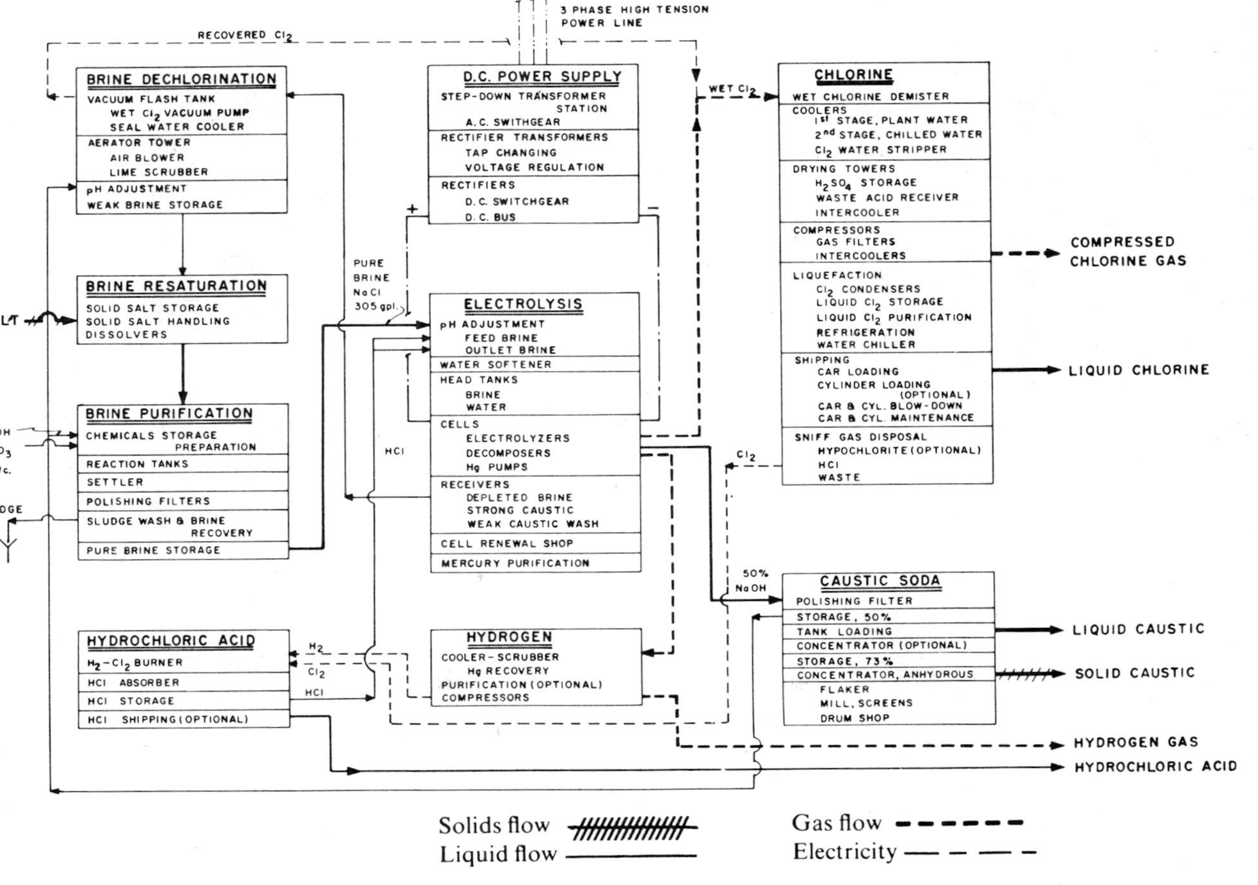

Figure 6-2. Functional parts of a typical complete mercury cell type chlor-alkali plant (major equipment for each part is listed).

The remaining available chlorine is removed by blowing compressed air through a column of the brine. The tail gas is either vented or killed in a milk of lime absorption tower. The dechlorinated brine contains less than 0.02 to 0.03 gpl available chlorine. Still another method of dechlorination is to pass the brine through a horizontal box fitted with porous diffuser plates, through which compressed air is blown.

It will be readily understood that either the aerator treatment or the vacuum treatment also evaporates water and hence cools the brine. This cooling may or may not be sufficient to balance out the heat developed by electrolysis. If not, additional heat exchangers or flash tanks are required.

The dechlorinated, but strongly acid brine is next neutralized or made slightly alkaline before it flows to the resaturator.

Salt

The process requires solid salt for resaturation of the partly depleted brine from electrolysis. If only brine is available, this must be evaporated first to produce solid salt. Typical sources of salt include:

Shaft mined rock salt

Vacuum pan salt

Solar pond salt

Salt recovered from the cell liquor evaporator in a diaphragm cell plant. To the above list we must add rock salt, for this is sometimes hydraulically mined with depleted brine from the cells, i.e., the brine resaturator is the salt cavity at the bottom of the brine well.[52]

Each salt listed has its own characteristics and must be handled accordingly. For example, coarse solar salt is non-caking and may be stored out-of-doors in a wet climate. Coarse rock salt cakes only on the surface of storage piles and remains generally free flowing. Fine rock salt and vacuum pan salt cake badly and must be stored dry, or under brine. In northern climates, all solid forms are stored indoors or under brine. The atmosphere around salt handling areas is very corrosive, and handling equipment must be of suitable material.

The chemical and screen analysis of the salt determines the kind of equipment required for resaturation and purification. Rock salt may be upgraded by using quick dissolvers, which leave considerable gypsum behind in the form of sand or sludge. On the other hand, suitably washed solar salt contains little gypsum and this practically all dissolves. The ideal raw material for the mercury cell process is, of course, highly purified recrystallized salt, which can be made with a purity of 99.99+ percent NaCl, either in vacuum pans or by the Richards process.[26,51]

Brine Resaturation

Depleted brine, dechlorinated and neutral to alkaline, containing 260 to 280 gpl NaCl, and at a temperature generally in the neighborhood of 50° to 60° C, is brought to saturation by forcing it upward through a bed of salt in the dissolving tanks. Conical bottom tanks are generally used, or if flat bottomed, the brine is distributed through horizontal spargers. Brick-lined, or concrete tanks are generally used. Alternatively, especially in Europe, the brine is made by hydraulically attacking the pile of salt in an open storage basin. Brine generally leaves the saturator hot and saturated and, to prevent crystallization downstream, it is generally diluted with a small bypass stream of weak brine.

Brine Purification

If the salt feed is of extremely high purity, the brine need only be filtered before returning it to the cells. Build-up of heavy metals can be prevented by purifying only a small purge stream or, if diaphragm process salt is used, the purge stream may be recycled to the brine treating system of the diaphragm cell process.

If rock or solar salt is used, the full stream of brine is treated for the removal of Mg, Fe and $CaSO_4$ equivalent to the pickup from the salt dissolved. Heavy metals, particularly V and Cr, must be controlled to close tolerances, as indicated later in this chapter.

Calcium sulfate may be knocked down with precipitated barium carbonate; Mg and Fe by caustic soda. If the SO_4 ion exceeds the Ca ion, calcium chloride may be added; if less, soda ash may be added. Heavy metals are usually sufficiently adsorbed by the precipitates formed: if not, the brine is deliberately treated with Ca and Fe to increase the volume of precipitates formed until control is obtained. The reactions are:

$$CaSO_4 + BaCO_3 \rightarrow CaCO_3 + BaSO_4$$

$$CaCl_2 + Na_2CO_3 \rightarrow CaCO_3 + 2\,NaCl$$

$$MgCl_2 + 2\,NaOH \rightarrow Mg(OH)_2 + 2\,NaCl$$

Treatment is carried out in a series of stirred tank reactors, the sequence of additions being $BaCO_3$, Na_2CO_3, NaOH. The brine is then clarified in a Dorr settler, and the overflow is made sparkling clear by filtering through sand or "anthrafilt." In many European plants, the brine from the reaction tanks is given a total filtration in a pressure leaf filter. Where salt is expensive, the sludges are sometimes washed to recover salt.

It is also possible to operate mercury cells with brine substantially saturated with $CaSO_4$ as described under Mathieson Cells, page 187. With this method, the alkaline brine will not dissolve any $CaSO_4$ in the saturator, hence no $BaCO_3$ is required to remove it. Since the solubility of $CaSO_4$ is greater in acid than in alkaline brine, by strongly acidifying the cell feed the tendency for gypsum to precipitate in the electrolyzer is offset. High calcium in the brine, however, results in some calcium metal in the amalgam, which may lower the capacity of the decomposer.

Still another method of calcium sulfate removal is practiced in some plants. A portion of the resaturated brine is treated with an excess of $CaCl_2$, to precipitate gypsum equivalent to the $CaSO_4$ picked up from the raw salt. This clarified stream then joins the main stream of brine, which is then treated with Na_2CO_3 and NaOH in the usual way to precipitate Ca and Mg. The brine so prepared will be low in Ca, but the Na_2SO_4 will average out at 7 to 10 gpl.

Electrolysis

Feed brine is acidified with HCl to a pH generally in the range 2.5 to 5, with a pH of 4 being common in many plants. The feed brine is distributed to the cells from a constant level head tank in or close to the cell house.

Cells are generally arranged in two banks, so that the electrical circuit is U-shaped. The two banks may be side by side in a one-layer cell house, or they may be above one another in a two-layer arrangement. The cell spacing, elevations and working aisle depend on the cell size and type, and will be discussed in more detail later.

The decomposers require a supply of soft water, but de-ionized or distilled water may be used if available. This is distributed to all the decomposers from a head tank. A flow divider is sometimes used, but more generally, each decomposer has its own manually adjusted flowmeter.

The following piping manifolds are commonly required:

Duty	Example of a Suitable Material
Brine feed	Rubber-lined steel
Depleted brine	Rubber-lined steel
Washout line	Rubber-lined steel
Plant water	Steel
Soft water	PVC
Chlorine gas	Rubber-lined steel, "Haveg"
Weak chlorine gas	Transite
Caustic soda	Nickel
Weak caustic wash	PVC
Hydrogen	Steel

Manifolds are electrically insulated from the cells by sections of nonconducting pipe. Current breaks are always provided for the strong caustic from the decomposers, and sometimes also in the brine connections.

Receivers below the cell deck are provided to collect depleted brine, caustic soda, etc. Acid is added to the anolyte receiver in sufficient amount to react with the HOCl—usually to a pH of 2.2 to 2.5. This causes evolution of some chlorine in the receiver, which is vented to the strong chlorine manifold.

The pressure in the electrolysis is maintained close to atmospheric, $\pm$ 15 mm. WG, by means of bypass suction control on the chlorine gas compressors. Hydrogen is always collected at a positive pressure.

For the currently popular horizontal mercury cells, which carry a total load of 40,000 to 180,000 amperes, the copper bus is carried in parallel lines from cell to cell, usually with the same spacing as the anodes. This saves copper. Where short circuiting switches are used, there is a switch on each bus, and the switches are ganged together for manual or motor-operated make or break.

Rectifier terminals are at one end of the cell room. At the opposite end, an empty bay usually is left for cell cover repair, refitting of new anodes, etc. Ventilation is very important in a mercury cell house, and 8 to 10 changes of air per hour are common. Regulations for mercury content of air in operating areas are very strict in most states. All floor drains and connections to sewers are trapped to catch accidental spillages of mercury. Cranes or monorail hoists are usually provided to lift and move covers, anodes, etc.

DC Power Supply

Direct generation of DC power is limited to those favored localities where cheap fuel or hydro power is immediately available. By and large, DC power for electrolytic plants is obtained by rectification of AC power. Rotary converters and motor-generator sets are now obsolete because of their cost and other technical reasons. Today, the choice is between the following rectifier types:

Mercury arc, pumped or pumpless
Mechanical
Semiconductor, germanium or silicon

The present trend is toward larger mercury cells with high current rating in order to minimize capital and operating costs. For a given plant capacity, the larger the cell rating, the fewer the cells and the lower the circuit voltage. For example:

	50	100	200
Cl_2, short tons per day	50	100	200
Cell rating, amperes	60,000	120,000	180,000
Number of cells	25–26	25–26	34
Circuit voltage at 4.3 volts/cell	112	112	146

The mercury arc rectifier operates efficiently only at voltages in excess of 500. Clearly, it is no longer considered favorable for plant capacities of less than about 600 tons per day of chlorine.

Fortunately, the mechanical and the semiconductor rectifiers operate at high efficiencies and down to 100 volts or even less. The development of semiconductor rectifiers for high kilowatt ratings is comparatively recent, but it is unmistakably clear that the semiconductors have numerous advantages over the mechanicals for service in mercury cell plants. For high ambient temperature conditions, silicon diodes appear to be preferable to the germanium diodes. It is important for mercury cell operators to remember, however, that semiconductor rectifiers cannot be safely overloaded, whereas mercury arc and mechanical rectifiers have frequently been pushed far beyond their design rating.

Hydrogen

The hydrogen gas collected from the decomposers is wet, contains some mercury vapor and also entrains caustic spray. Most of these are removed by cooling, either in a direct contact scrubber, or in a condenser. The mercury is removed. The gas then is 99.9 percent H_2, dry basis, and contains approximately 20 to 30 mg/m³ of Hg.

For some uses, the hydrogen may be further purified. The steps normally include catalytic removal of oxygen, drying, and removal of traces of mercury by deep cooling and filtering or by using activated adsorbents.

Caustic Soda

Most mercury cell operators prefer to make 50 percent NaOH in the decomposer. Some plants are limited to 40 percent; others are able to produce 60 percent. It is possible to produce 73 percent NaOH, but this requires ample decomposer dimensions and very careful attention to operating conditions. If 73 percent NaOH is desired, it is generally considered preferable to concentrate the 50 percent NaOH in a nickel evaporator. Solid forms of caustic soda are made by well known procedures, beyond the scope of this book.

It is customary to filter the caustic, if made in a tower decomposer, for the removal of graphite particles. These particles usually contain adsorbed mercury, and it is ordinarily worthwhile to recover the mercury from the sludge when a sufficient amount has accumulated.

Chlorine

The hydrogen content of the cell gas must be closely monitored. A cell to cell check is made if the cell room manifold is running abnormally high. This may indicate that certain cells need cleaning, or it may indicate improper operation in brine purification.

It has been found desirable to remove the mist of entrained brine from the wet cell gas, because such mist may carry right through the entire train of cooling and drying equipment to the compressors. If these are of the reciprocating type, dry salt fume gives trouble in the valves.

There is no essential difference in the method of drying and liquefaction of mercury cell chlorine and that of drying diaphragm cell chlorine. This method has been described in Chapter 5. There is one additional point, however: liquid chlorine made in a mercury cell plant generally contains less gum and organics than that made in a diaphragm cell plant, using oil treated graphite anodes. This difference may be a determining factor when liquid chlorine purification is being considered.

Hydrochloric Acid

A small amount of hydrochloric acid is required for the mercury cell process, as indicated on the flow sheet. If acid is not locally available, the required HCl is made from hydrogen and chlorine, as described in Chapter 26. Sniff gas from chlorine liquefaction is frequently used for this purpose.

PRINCIPLES OF OPERATION

The Mercury Cell Paradox

The principal over all reaction accomplished in the electrolyzer is:

$$NaCl \quad + \quad (Hg) \quad \xrightarrow{\text{1 Faraday}} \quad Na(Hg) \quad + \quad \tfrac{1}{2}Cl_2$$

Strong brine	*Mercury*	*Sodium amalgam* *cathode product*	*Chlorine gas* *anode product*

The free energy change for this reaction depends, of course, on the concentrations and hence the activities of the products and reactants, as well as on the temperature and pressure. For the usual range of conditions in the electrolyzer, $\Delta F = +70.4$ kg cal, corresponding to a reversible equilibrium potential of -3.05 volts; this means that electrical energy must be expended to bring about the reaction.

On the other hand, the possibility of decomposing water by a competing reaction is also present:

$$\tfrac{1}{2}\,H_2O(l) \xrightarrow{\ \text{1 Faraday}\ } \tfrac{1}{2}\,H_2\,(g) \quad + \quad \tfrac{1}{4}\,O_2\,(g)$$

Water in brine   *Cathode product*   *Anode product*

The free energy change for this reaction, under the same conditions (60° C) is $\Delta F = +28.35$ kg cal, corresponding to a reversible equilibrium potential of -1.23 volts.

Judging by the free energies only, the second reaction is overwhelmingly easier to accomplish than the first, i.e., from the thermodynamic standpoint, the amalgam-type chlorine cell is inoperable. This conclusion is absurd, since in practice the chlorine-producing reaction is favored over the hydrogen-producing reaction. To understand this paradox, we must examine the individual electrode reactions from the kinetic point of view. As a first step to understanding, it will be necessary to work out a material balance on the electrolyzer.

Cell Reactions, Current Efficiency and Recovery

There are many possible reactions that could take place in a mercury cell, as listed in Table 6-1. Some of these involve electron transfer at either anode or cathode, and some are purely chemical in nature. Not all the reactions are significant, and the problem is to select those reactions which best explain the observational facts listed in Table 6-2.

It must be borne in mind that conditions differ from plant to plant, from cell to cell, and from time to time. Hence no simple explanation will necessarily cover all conditions. We will, however, proceed to analyze a set of conditions which is typical of many plants, as shown in Table 6-3. Cell reactions will be written in over-all form, in order to more easily balance out the Faraday current between anode and cathode. The reactions selected will be the least in number, recognizing that these may summarize a set of intermediate mechanisms.

(1) For our typical plant, we assume a raw salt quality such that all brine is recycled through the purification step and that the treated brine contains an excess alkalinity of 0.12 gpl Na_2CO_3 and 0.04 gpl NaOH. The treated brine is then acidified to pH 4. This breaks up the carbonate and releases 0.05 gpl CO_2, which, however, remains dissolved in the brine because the solubility of CO_2 is of the order of 0.7 to 0.8 gpl. As the brine passes through the electrolyzer, the temperature rises, and the CO_2 is practically all purged out of the brine by the evolved Cl_2 gas. Thus, the incoming CO_2 appears in the cell gas, and only a part of the CO_2 in the cell gas results from the oxidation of graphite.

(2) The typical anolyte as it leaves the cell at 70° usually contains 0.5 to 0.6 gpl available Cl at a pH of about 3. To dechlorinate, HCl is added,

TABLE 6-1. COMPREHENSIVE LIST OF REACTIONS IN ELECTROLYZER OF A MERCURY CELL[42]

Anode Reactions:

$$2\ Cl^- \longrightarrow 2\ Cl^\circ + 2\ e^- \longrightarrow Cl_2$$

$$4\ OH^- \longrightarrow 4\ OH^\circ + 4\ e^- \longrightarrow 2\ O^\circ + 2\ H_2O \qquad \text{(alkaline)}$$

$$2\ H_2O \longrightarrow 4\ H^+ + 2\ O^\circ + 4\ e^- \qquad \text{(acid)}$$

$$2\ O^\circ \longrightarrow O_2$$

$$2\ O^\circ + C \longrightarrow CO_2$$

$$O^\circ + C \longrightarrow CO$$

$$4\ HSO_4^- \longrightarrow 4\ HSO_4^\circ + 4\ e^-$$

$$4\ HSO_4^\circ + 2\ H_2O \longrightarrow 4\ H^+ + 4\ HSO_4^- + 2\ O^\circ$$

$$6\ ClO^- + 3\ H_2O \longrightarrow 2\ ClO_3^- + 4\ Cl^- + 6\ H^+ + \tfrac{3}{2}\ O_2 + 6\ e^-$$

$$Cl^- + 3\ H_2O \longrightarrow ClO_3^- + 3\ H_2 + 6\ e^-$$

Cathode Reactions:

$$Na^+ + (Hg) + e^- \longrightarrow Na(Hg)$$

$$2\ H^+ + 2\ e^- \longrightarrow 2\ H^\circ \longrightarrow H_2 \qquad \text{(acid)}$$

$$H_2O + 2\ e^- \longrightarrow 2\ OH^- + H_2 \qquad \text{(alkaline)}$$

$$Cl_2 + 2\ e^- \longrightarrow 2\ Cl^-$$

$$HOCl + H^+ + 2\ e^- \longrightarrow H_2O + Cl^- \qquad \text{(acid)}$$

CHEMICAL REACTIONS (No electron transfer)

In the Bulk Electrolyte:

$$Cl_2 + H_2O \longrightarrow H^+ + Cl^- + HOCl$$

$$HOCl \longrightarrow H^+ + OCl^-$$

$$OCl^- + 2\ HOCl \longrightarrow ClO_3^- + 2\ Cl^- + 2\ H^+, \text{ or } 3\ HOCl \longrightarrow 2\ HCl + HClO_3$$

At the Anode:

$$HOCl + C \longrightarrow HCl + CO$$

$$2\ HOCl + C \longrightarrow 2\ HCl + CO_2$$

At the Cathode:

$$2\ Na(Hg) + 2\ HCl \longrightarrow 2\ NaCl + H_2 + (Hg)$$

$$2\ Na(Hg) + Cl_2 \longrightarrow 2\ NaCl + (Hg)$$

$$2\ H^\circ + ClO^- \longrightarrow H_2O + Cl^-$$

$$6\ H^\circ + ClO_3^- \longrightarrow 3\ H_2O + Cl^-$$

TABLE 6-2. ANALYTICAL DATA RELATED TO CURRENT EFFICIENCY

Cell Type	NaOH C.E. (%)	Cell Gas, Per Cent by Volume, Air Free, Dry		
		H_2	CO_2	O_2
De Nora, horizontal[57]	94–96	0.4–1.0	0.4–0.6	n.a.
vertical	95–97			
Solvay[6]	96+	0.3–0.9	n.a.	0.1
Krebs BASF[62]	94–95	0.2–0.4	0.6–0.7	n.a.
Uhde[18]	95–97	0.3	0.3	n.a.
Olin Mathieson[16,52]	95	0.23–0.9	0.3–0.6	n.a.

$NaClO_3$, Level ranges from 0.1–0.2 (Olin Mathieson);[16,52] 0.3 (Solvay);[6] others as high as 1 gpl. Negligible increase in $NaClO_3$ in one pass thru cell.

pH Brine feed 3–5, most plants, pH 4

  Anolyte, 2.5–3.5, most plants, about 3.0

Av. Cl. in anolyte:

  De Nora (Longview) 0.5–0.6 gpl; Solvay[6] 0.4 gpl

  I.G.F. (Germany)[56] 0.45–0.6 gpl; Uhde (Elworth)[34] 0.4 gpl

  Krebs (Shawinigan)[62] 0.3–0.45 gpl;

Av. Cl. in brine dechlorination:[35,58]

  After adding HCl, temperature approx. 70°C, 0.3 gpl

  After flashing to 1/2 atm., 0.1 gpl

  After aeration, 0.01–0.03 gpl

  pH rises 0.5–1.0 unit as Cl_2 is expelled.

usually to a pH of 2.5. Chlorine gas is released at once, leaving about 0.3 gpl available Cl dissolved in the brine. The only reaction which takes place instantaneously about 70° C is as follows:

$$HOCl + HCl \rightarrow Cl_2 + H_2O$$

Chlorate is not decomposed under these conditions to any appreciable degree.

  (3) Since the brine in the cell is acid, the principal source of HCl is from hydrolysis of chlorine,

$$Cl_2 + H_2O \rightarrow HOCl + HCl$$

However, the observed increase in HCl does not match the observed increase in HOCl. Hence, there must be a reaction which consumes HCl, thus permitting more HOCl to form according to the following equilibrium:[5,48]

$$K = \frac{(HOCl)(H^+)(Cl^-)}{(Cl_2)}$$

At 25° C, $K = 4.84 \times 10^{-4}$ and $\Delta H = +6.49$ kg cal. From the Van't Hoff equation, $d \log K = \dfrac{\Delta H}{4.5787} \, d\left(\dfrac{1}{T}\right)$, we calculate, at 70° C, $K = 2.03 \times 10^{-3}$. Since NaCl $= 270$ gpl $= 4.62$ N, and $\gamma = 0.83$, the activity $(Cl^-) = 3.83$. At pH 3, $H^+ = 0.001$.

Hence,

$$\frac{(HOCl)}{(Cl_2)} = \frac{2.03 \times 10^{-3}}{10^{-3} \times 3.83} = 0.53$$

But,

$$(HOCl) + (Cl_2) = \text{Total Av. Cl} = 0.00776 \, m = 0.550 \, gpl$$

Hence,

$$(HOCl) = 0.00776 \times 0.53/1.53 = 0.00269 \, m = 0.191 \, gpl$$

$$(Cl_2) \quad = 0.00776 - 0.00269 \quad = 0.00507 \, m = 0.359 \, gpl$$

The above calculated HOCl content of the anolyte closely approximates that observed in many Hg cell plants, agreeing roughly with the amount of Cl_2 liberated when the anolyte is acidified, as per paragraph 2 above.

(4) Since the current efficiency is only 95 percent, and current efficiency loss at the cathode is approximately equivalent to the H_2 in the Cl_2, this leaves about 4.5 percent loss due to chemical back-reactions at the cathode. Since brine containing free Cl_2 and HCl sweeps or diffuses

**TABLE 6-3. TYPICAL CONDITIONS ASSUMED
FOR MATERIAL BALANCE**

Treated brine alkalinity Na_2CO_3 0.12 gpl, NaOH 0.04 gpl

Cell brine

	Inlet	Outlet
Temperature	50°C	70°C
NaCl	305 gpl	270 gpl
Avail Cl_2	0.03 gpl	3.0 gpl
pH	4.0	3.0

Gas analysis (air free, dry)

H_2	0.5%
CO_2	0.5% or as calculated
O_2	–
CO	–
Cl_2	99.0%

Current efficiency, based on NaOH. 95.0%

past the cathode, it is logical to postulate back-reactions of these with the sodium in the amalgam and of the former much more than the latter because of its greater concentration. Back-reaction with HOCl is also possible, but need not be considered as an independent variable.

Taking the above reasoning into account, the postulated principal reactions are:

Electrolytic:

$$NaCl + (Hg) \xrightarrow{\text{1 Faraday}} \overset{\text{Cathode}}{Na(Hg)} + \overset{\text{Anode}}{\tfrac{1}{2} Cl_2} \tag{6-1}$$

$$\tfrac{1}{2} H_2O \xrightarrow{\text{1 Faraday}} \tfrac{1}{2} H_2 + \tfrac{1}{2} 0° \tag{6-2}$$

Chemical:

$$\tfrac{1}{4} C + \tfrac{1}{2} 0° \text{ (at the anode)} = \tfrac{1}{4} CO_2 \tag{6-3}$$

$$Cl_2 + H_2O \text{ (in the brine)} = HCl + HOCl \tag{6-4}$$

$$Na(Hg) + \tfrac{1}{2} Cl_2 \text{ (at the cathode)} = NaCl + (Hg) \tag{6-5}$$

$$Na(Hg) + HCl \text{ (at the cathode)} = NaCl + \tfrac{1}{2} H_2 + (Hg) \tag{6-6}$$

A material balance based on these six reactions now follows:

MATERIAL BALANCE

Basis: 1 liter of feed brine

NaCl decomposed,	0.60000 equiv.
Faradays at 95% CE* = 0.6000/.95 = 0.63157	
HOCl increase = .00282 − .00042 =	0.00240 equiv.
HCl formed via Reaction (6-4)	0.00240 equiv.
Initial HCl, pH 4	0.00010 equiv.
	0.00250 equiv.
Final HCl, pH 3	0.00100 equiv.
HCl consumed via Reaction (6-6)	0.00150 equiv.
Na in Na(Hg) = salt decomposed	0.60000 equiv.
Cl formed, all forms = decomposed	0.60000 equiv.
Brine out:	
Cl in Cl_2, 0.35/35.46 = 0.00988 equiv.	
Cl in HOCl, 0.20/70.92 = 0.00282 equiv.	
Cl in HCl 0.00100 equiv.	
0.01370 equiv.	

*current efficiency

Brine in:

Cl as HOCl, 0.030/70.9 = 0.00042 equiv.

Cl as HCl   0.00010 equiv.

                          0.00052 equiv.

Cl pick-up in brine   0.01318 equiv.   0.01318 equiv.

Cl as Cl_2 in cell gas   0.58682 equiv.

Cl_2 in cell gas   0.29341 moles

Cell gas at approximately 99% purity   0.29634 moles

H_2 in cell gas, 0.5% of 0.29634 =   0.00149 moles

H_2 from (6-6) = ½ of 0.00150   0.00075 moles

Leaving H_2 from Reaction (6-2)   0.00074 moles

0° from Reaction (6-2)   0.00074 moles

CO_2 eq. of O_2, Reaction (6-3)   0.00037 moles

CO_2 from brine = 0.050/44 =   0.00114 moles

Total estimated CO_2 in cell gas   0.00151 moles

Cell gas: Cl_2,   0.29341 moles 98.99%

            H_2,   0.00149 moles   0.50%

            CO_2,   0.00151 moles   0.51%

                   0.29641 moles (Air free, dry basis)

Total Faradays, 0.60000/.95	0.63157	F	100.000%
Reaction (6-2), 2 × 0.00074	0.00148	F	0.234%
Reaction (6-1), Total Na formed	0.63009	F	99.766%
Na in amalgam	0.60000	equiv.	95.000%
Na lost by Reactions (6-5) and (6-6)	0.03009	equiv.	4.766%
Na lost by Reaction (6-6)	0.00150	equiv.	0.238%
Na lost by Reaction (6-5)	0.02859	equiv.	4.528%

CHLORINE DISTRIBUTION

Cl_2 in cell gas   0.58682 equiv. 0.58682

Av. Cl in brine, 0.55/35.46   0.01549 equiv.

Total av. Cl   0.60231 equiv.

Cl_2 released on acidification, equiv. to

     OCl in brine   0.00564 equiv. 0.00564

Leaving in brine, 0.35 gpl av. Cl   0.00985

Released at 400 mm vac., 0.25 gpl av. Cl   0.00704 equiv. 0.00704

Leaving in brine, 0.1 gpl av. Cl   0.00281 equiv.

Gross Cl_2 recovered =   0.59950

Released in aerator, 0.07 gpl av. Cl.   0.00197

Leaving in brine,

     0.03 gpl av. Cl.   0.00084

     which recycles to cells.

HCl Requirements.

Brine Dechlorination:

HCl equiv. of HOCl =	0.00282 equiv.
Change pH from 3.0 to 2.5, .00316 − 00100 =	0.00216 equiv.
Sum	0.00498 equiv.

Acidification of brine feed:

Excess Na_2CO_3, 0.12 gpl/53 =	0.00226 equiv.
Excess NaOH, 0.04 gpl/40 =	0.00100 equiv.
Required to neutralize to pH 7	0.00326 equiv.
Change pH from 7 to 4,	0.00010 equiv.
Total HCl	0.00336 equiv.

Total HCl for Process: 0.00834 equiv.

		Anode *C.E.*
Total Faradays at 95% cathode C.E.	0.63157	
Cl_2 recovered from electrolysis	0.59950	95.00
Less Cl_2 to HCl burner	0.00834	1.32
Net Cl_2 recovered for use	0.59116	93.68

Cl_2 Yield on salt decomposed, 0.59116/0.60000 = 98.53%

NaOH Requirements.

Neutralize dechlorinated brine from pH 2.5 to pH 7   0.00316

Precipitation of Ca, Mg. This depends on the quality of the salt.

Assume 1.75 g raw salt per g Cl_2,

Net Cl_2 recovered = 0.59119 eq.	= 21.0 g
Raw salt = 1.75 × 21.0	= 36.8 g

Assuming a typical solar salt analysis,

Component	Wt. %	gm	Equiv. Wt.	Equiv.
$CaSO_4$	0.23	0.0845	68.00	0.00124
$MgSO_4$	0.12	0.0440	60.15	0.00073
$MgCl_2$	0.16	0.0598	47.60	0.00126

$$CaSO_4 + BaCO_3 = CaCO_3 + BaSO_4$$
$$MgSO_4 + BaCO_3 = MgCO_3 + BaSO_4$$
$$MgCO_3 + 2\,NaOH = Mg(OH)_2 + Na_2CO_3$$
$$MgCl_2 + 2\,NaOH = Mg(OH)_2 + 2\,NaCl$$

NaOH for pptn. Mg		0.00199 equiv.
NaOH excess, 0.04 gpl		0.00100 equiv.
Na_2CO_3 excess, .12 gpl	0.00226	
less Na_2CO_3 formed	0.00073	
Net Na_2CO_3, or NaOH equiv., about		0.00153 equiv.
Total NaOH for Ca and Mg removal		0.00452 equiv.
Total NaOH for process		0.00768 equiv.

		Cathode C.E.
Total Faradays at 95% cathode C.E.	0.63157	
NaOH recovered from decomposition	0.60000	95.00%
Handling and wash loss, 0.1%	−0.00060	− .10
NaOH to process	−0.00768	− 1.22
Net NaOH make	0.59172	93.68%
Yield on salt decomposed, approx.	98.62%	

The Cathode Reactions

Deposition of Na on a Mercury Cathode:

$$Na^+ + (Hg) + e^- = Na(Hg)$$

The directly measured emf, E_{Na}, of this half cell, referred to the hydrogen electrode, and for various concentrations of NaCl, Na in amalgam, and temperatures, as reported by Sugino and Aoki,[56] is given in Table 6-4.

The sodium overvoltage η_{Na}, at cathode current densities in the commercial range of 0.10 to 0.50 amps/cm², is reported to be quite small, i.e., about 0.01 to 0.02 volt.[16,53] Taking conditions typical of some commercial mercury cells, the total deposition potential V_{Na} is shown in Table 6-5. In horizontal mercury cells, the flow of brine and mercury is always co-current, for the reason that powdered graphite from the anodes is more easily floated down stream to an end box where such accumulations may be skimmed or flushed out of the electrolyzer.

Simultaneous Evolution of Hydrogen:

$$H^+ + e^- = \tfrac{1}{2}H_2, \quad \text{or}$$

$$H_3O^+ + e^- = \tfrac{1}{2}H_2 + H_2O$$

The standard emf E_H is by definition zero. However, the actual emf E_H depends upon the pH of the solution at the cathode interface:

$$E_H = -2.3026 \frac{RT}{F} \log \frac{(H_2)^{1/2}}{H^+} \qquad (6\text{-}7)$$

Since, in the main, the hydrogen is evolved at 1 atmosphere, and since pH $= -\log H^+$,

$$E_H = -1.9843 \times 10^{-4} \cdot T \cdot pH \qquad (6\text{-}8)$$

TABLE 6-4. EQUILIBRIUM POTENTIAL OF SODIUM AMALGAM IN CONTACT
WITH SODIUM CHLORIDE SOLUTION

(All Voltages are Negative, and Refer to Hydrogen Electrode)[21]

pH, 2.5-3.5 Temperature, °C.

Amalgam Wt. % Na	NaCl 300 gpl			NaCl 250 gpl			NaCl 200 gpl			NaCl 150 gpl		
	40°	60°	80°	40°	60°	80°	40°	60°	80°	40°	60°	80°
0.489	1.850	1.846	1.844	1.859	1.854	1.854	1.871	1.868	1.867	1.884	1.880	1.878
0.398	1.836	1.831	1.828	1.848	1.844	1.840	1.859	1.855	1.851	1.872	1.869	1.871
0.296	1.816	1.812	1.805	1.829	1.823	1.818	1.833	1.833	1.826	1.851	1.850	1.844
0.237	1.810	1.802	1.800	1.819	1.812	1.809	1.831	1.825	1.822	1.844	1.837	1.832
0.203	1.801	1.798	1.795	1.812	1.809	1.806	1.822	1.820	1.818	1.834	1.831	1.829
0.148	1.787	1.781	1.774	1.797	1.790	1.784	1.808	1.802	1.796	1.821	1.814	1.808
0.099	1.768	1.761	1.752	1.779	1.771	1.762	1.789	1.783	1.773	1.802	1.795	1.785
0.051	1.750	1.740	1.731	1.760	1.752	1.743	1.773	1.764	1.758	1.785	1.777	1.769
0.010	1.692	1.686	1.710	1.705	1.695	1.684	1.719	1.707	1.697	1.728	1.719	1.710

TABLE 6-5. SODIUM DISCHARGE POTENTIAL IN A TYPICAL CELL

	Inlet End	Outlet End
Temperature, °C	50	75
NaCl, gpl	305	270
Na in Hg, wt. %	0.01	0.15
Cathode current density, amps/cm^2	0.40	0.40
E_{Na}, volt	− 1.69	− 1.78
η_{Na}, "	− 0.01	− 0.02
V_{Na}, "	− 1.70	− 1.80

While the pH in the bulk of the brine will be about 3, due to the hydrolysis of chlorine with which the brine is saturated, the pH at the mercury interface is bound to be higher because of the depletion of H^+ ions by the very reactions under consideration.

The overvoltage of H_2 on a mercury cathode has been variously reported, and such figures must be used with discrimination. As a first approximation, the percent H_2 in the Cl_2 gas evolved in the electrolyzer is a measure of the hydrogen ion current density. Thus, if this gas contains 0.5 percent H_2, the hydrogen ion current density will be about 0.5 percent of the total current density, the balance being carried by the sodium ion.

The actual hydrogen overvoltage on mercury depends on many factors including H^+ current density, local pH, temperature, presence or absence of supporting electrolyte, and particularly on the degree of contamination of the mercury surface with heavy metal impurities. The available data are summarized in Table 6-6. Presumably, these results refer to the purest mercury obtainable. The coefficients refer to the Tafel equation[9] as follows:

$$\eta = \frac{2.3026\,RT}{\alpha F} \log \frac{i_0}{i} \qquad (6-9)$$

TABLE 6-6.   PARAMETERS FOR HYDROGEN EVOLUTION ON MERCURY CATHODE IN AQUEOUS SOLUTIONS,[9] AVERAGE VALUES

Solution	pH	$i_0(20°C)$	$\alpha(20°C)$	ΔH^*
1N HCl	0.1	2×10^{-12}	0.49	24.0
0.1N HCl	1.1	$5{-}9 \times 10^{-13}$	0.48–0.51	21.3
0.002N KOH	11.3	8×10^{-18}	0.59	
0.02N KOH	12.3	5×10^{-18}	0.64	16.0
0.1N KOH	12.9	4×10^{-16}	0.62	17.0
0.1N NaOH	12.9	3×10^{-15}	0.58	

and the activation energy, ΔH^*, is used to estimate the change in i_o with temperature according to the equation[8]

$$\log \frac{i_o, T_1}{i_o, T_2} = \frac{-\Delta H^*}{2.3026\ R} \left(\frac{1}{T_1} - \frac{1}{T_2} \right) \qquad (6\text{-}10)$$

It will be noted that the i_o's are markedly lower for alkaline conditions than for acid conditions. The actual pH at the interface *is* not an independent variable, but it is probably somewhere in the range of 4 to 7. For this range an i_o at 20° C of 10^{-17} is probable; $\alpha = 0.60$ and ΔH^* is taken as 19 kg cal. From these data the hydrogen overvoltage at 50° C and 70°C is derived as follows:

$$\text{At } 50°\,C,\ \eta_H = -1.684 - 0.108 \log i_H \qquad (6\text{-}11a)$$

$$\text{At } 70°\,C,\ \eta_H = -1.710 - 0.114 \log i_H \qquad (6\text{-}11b)$$

Since the total discharge potential for hydrogen V_H must equal that for sodium V_{Na} at the respective current densities of each, there is now a means of estimating the pH at the interface. In a properly working commercial cell, with reasonably pure brine and mercury, with graphite anodes low in heavy metals and with an acid feed brine in the range of pH 4 to 5.5, the average percent H_2 in the chlorine should be about 0.5 percent by volume. Since the conditions are more favorable for H_2 discharge at the outlet end (where impurities tend to collect), and less favorable at the inlet end, actual conditions may be approximated by assuming 0.75 percent H_2 in the gas evolved at the outlet end, and 0.25 percent H_2 in the gas evolved at the inlet end, with an average composition of 0.50 percent H_2 as observed. Calculations on the H_2 discharge potential on the cathode in a typical cell are summarized in Table 6-7.

The effect of contamination by heavy metals is of much concern to cell room operators since they lower the overvoltage of hydrogen and thus cause excessive hydrogen in the chlorine. One must, or should, operate outside the explosive limits, as shown in Figure 6-3. In practice, it is not usual to tolerate more than 1 percent H_2 in the Cl_2, because the chlorine gas is usually liquefied or absorbed in a chlorination reaction. In these operations, the percent H_2 in the remaining gas soon rises into the explosive range so that the lower the initial percent H_2, the greater the range of safe operation.

The effect of metallic impurities on mercury cell operation has been extensively studied by Angel *et al*.[2,3,4] Okada *et al*,[44,45] Czernotsky,[11] and others.[64] The results are summarized in Table 6-8 in the order of decreasing effect on hydrogen evolution. The effect of some acids and colloidal materials is given in Table 6-9. Combinations of two or more metals are likely

TABLE 6-7. HYDROGEN DISCHARGE POTENTIAL ON AMALGAM
CATHODE IN TYPICAL CELL

	Inlet End	Outlet End
No inpurities:		
Temperature, $^\circ$C	50	75
Local % H_2 in Cl_2	0.25	0.75
Total i_c, amp/cm^2	0.40	0.40
i_H, amp/cm^2	0.001	0.003
$V_H = V_{Na}$ (Table 6-5), volt	-1.70	-1.80
η_H, Equation (6-11), volt	-1.36	-1.42
E_H, by diff., volt	-0.34	-0.38
pH at interface, Equation (6-81	5.3	5.6
Impurities present:		
Local % H_2 in Cl_2	2.5	7.5
i_H, amp/cm^2	0.01	0.03
pH at interface, add 1	6.3	6.6
E_H, Equation (6-8), volt	-0.40	-0.45
η_H by diff., volt	-1.30	-1.35

to have worse effects than do the same metals taken separately, e.g., Mg
and Fe form a synergistic pair. The maximum allowable concentration of
metallics in the brine feed, for 0.3 percent H_2 in the Cl_2 gas made, operating
at 0.40 amps/cm² and 80°C, is[11] as follows:

Metal	mg/l
Ca	100.0*
Mg	1.0
Fe	0.3
Ti	0.1
Mo	0.001–0.010
Cr	0.001–0.010
V	0.001–0.010

*See Brine Purification for operation with brine saturated with $CaSO_4$.

While the exact mechanism of the catalytic effect of impurities on hydrogen evolution is not fully understood, it has been observed[64] that certain
impurities reduce the ability of the freshly deposited sodium to mix with the

TABLE 6-8. EFFECT OF METALLIC IMPURITIES ON HYDROGEN
EVOLUTION AT THE CATHODE OF A MERCURY CELL[2, 11, 44, 45]

Very harmful,	V, Mo, Cr, Ti, Ta, (Mg + Fe)
Moderately harmful,	Ni*, Co*, Fe*, W
Slightly harmful,	Ca, Ba, Cu*, Al*, Mg*, graphite
No effect,	Ag, Pb, Zn, Mn

(*Known to act as a Promoter)

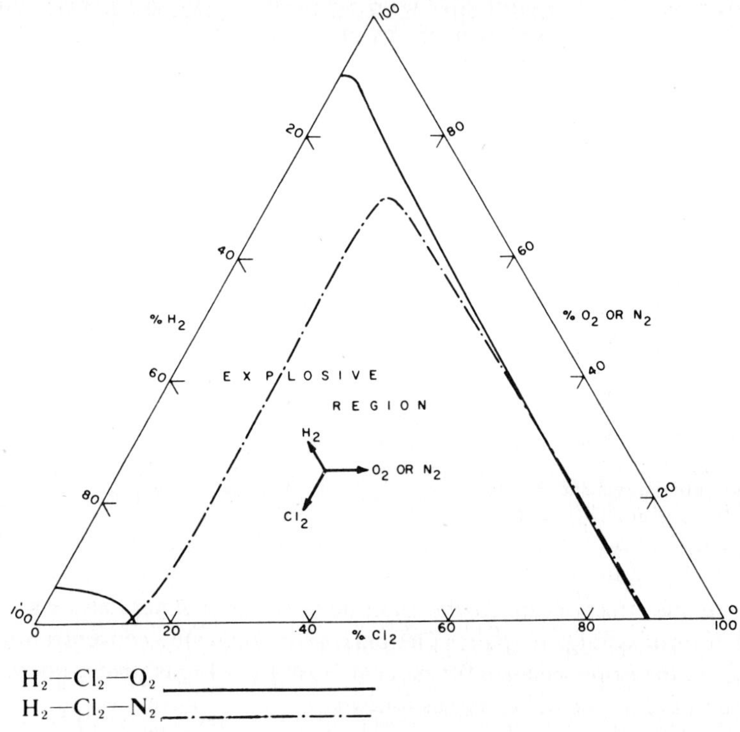

H_2—Cl_2—O_2 ——————————
H_2—Cl_2—N_2 —·——————·——·—

Figure 6-3.   Explosive limits for H_2—Cl_2—O_2 or H_2—Cl_2—N_2.

bulk of the mercury. This would cause an increase in sodium activity at the mercury-brine interface, raise the sodium deposition potential, and thus favor hydrogen evolution.

The sources and means of elimination of metallic impurities will be treated elsewhere. To show the effect of such impurities on the hydrogen evolution at the cathode as well as the effect on the pH at the cathode boundary, data of Table 6-7 have been plotted in Figure 6-4. Both normal

TABLE 6-9. EFFECT OF NON-METALLIC IMPURITIES ON
HYDROGEN EVOLUTION AT THE CATHODE
OF A MERCURY CELL[2]

No effect, SO_4, Br, I, ClO_3, F
Slightly beneficial, borate—tends to deactivate V weakly.
Beneficial,   silicate—forms colloidal silica, which deactivates
                              V, Mo.
                              pyrophosphate (deactivates V by forming complexes).
                              stannate—forms colloidal stannic acid which de-
                              activates V.

and abnormal conditions are shown. For the abnormal case it is assumed that there are sufficient impurities present to produce 5 percent H_2 in the chlorine, which is ten times that representing normal conditions. Since the H^+ ions are being used up ten times as fast at the metal boundary, the pH should be roughly one unit higher. Thus, impurities tend to produce a distinctly more alkaline condition at the mercury surface, in spite of the strong acidity in the bulk of the brine. This higher pH in turn raises the evolution potential, which to some degree counteracts the hydrogen current density and thus stabilizes the cathode reaction.

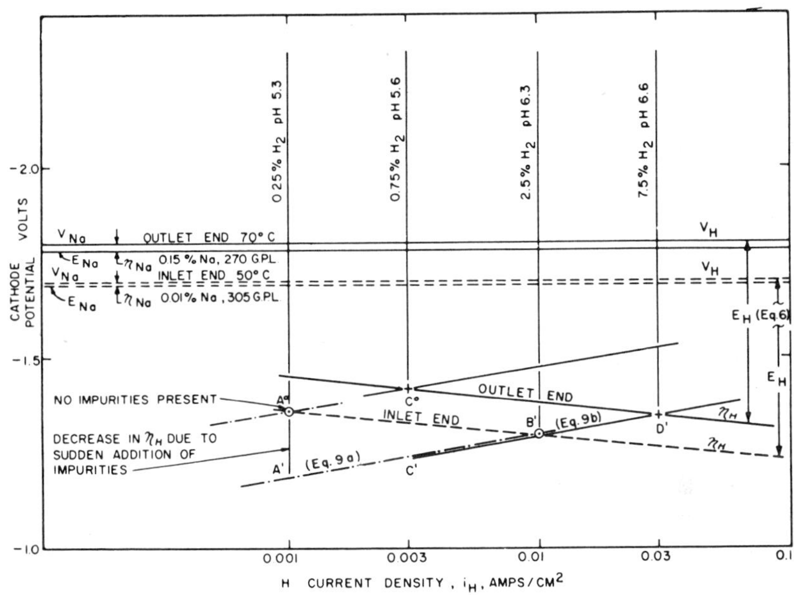

Figure 6-4. Typical mercury cell, cathode process; cathode current density = 0.40 amp/cm²; electrolyte pH = 2.0 to 3.0

The total potential for sodium discharge, $\quad V_{Na} = E_{Na} + \eta_{Na}$
The total potential for hydrogen discharge, $\quad V_H = E_H + \eta_H$
The total cathode current density, $\quad i = i_{Na} + i_H$

$V_{Na} = V_H$ at their corresponding current density.

At Inlet End, with minimum impurities in brine, Hg and graphite, the cell operates with overvoltage shown at point A°, 0.25 percent H_2, pH 5.3 at interface.

Assume impurities are suddenly added to the brine, lowering overvoltage to point A'. The increased hydrogen evolution raises the i_H along line A'B' according to Equation (6-9), at the same time raising the pH at the mercury surface. The reversible H potential E_H then increases, following the line A°-B', according to equation (6-8). These two lines cross at B', indicating a 10-fold increase in i_H and hence a 10-fold increase in percent of H_2 in the Cl_2, which would stabilize at 2.5 percent H_2. The corresponding pH is about one unit higher, stabilizing at pH 6.3.

Similarly, *at Outlet End*, if normal operations show 0.75 percent H_2, η_H is at point C°. Adding impurities drops η_H to C', and then η_H climbs along line B'D, while E_H follows along C°-D'. Stabilization is reached at the intersection D'.

It is fortunate indeed that the hydrogen overvoltage on mercury is as high as it is. The usual explanation for this[33] is that the reactions proceed stepwise, the first step being deposition of atomic hydrogen, which requires the additional free energy of dissociation of 48.57 kg cal.,

$$H^+ + e^- \rightarrow H°, \qquad E° = -2.10 \text{ volt} \tag{6-12}$$

The recombination reaction depends on a statistical probability, and apparently is catalyzed by the presence of metallic impurities on the mercury surface.

$$2 H° \rightarrow H_2, \quad \text{or} \tag{6-13a}$$

$$H° + H^+ + e^- \rightarrow H_2 \tag{6-13b}$$

The back-reactions taking place at the cathode, previously discussed, depend on the rate of diffusion and/or convection of dissolved chlorine and hydrochloric acid through the boundary layers to the cathode itself. The loss in current cathode efficiency due to these reactions is of the order of 3 to 5 percent. If a horizontal porous diaphragm be interposed between the mercury cathode and the anode above it, the loss due to back-reaction can be nearly eliminated.[49] An interesting laboratory prototype is described by Potter and Bisio.[50] However, such diaphragms are not practical and are never used on industrial horizontal type mercury cells.

The exact mechanism of the back-reactions may not be wholly as pictured by Reactions (6-5, 6-6). Other possible mechanisms are listed in Table 6-1.

The Anode Reactions

Evolution of Cl_2 Gas at Horizontal Graphite Anode

$$Cl^- \rightarrow \tfrac{1}{2} Cl_2 + e^- \tag{6-14}$$

The equilibrium emf, E_{Cl}, of this half cell, referred to the hydrogen electrode, and of various concentrations of NaCl and temperatures, as reported by Sugino and Aoki,[56] is given in Table 6-10 and 6-11.

TABLE 6-10. EQUILIBRIUM POTENTIAL OF GASEOUS CHLORINE
IN CONTACT WITH SODIUM CHLORIDE SOLUTION[56]

| | Temp (^{o}C) | | | |
NaCl Concentration (g/l)	10	40	60	80
119.2	1.370	(1.355)	(1.349)	(1.344)
230.3	1.354	(1.336)	(1.329)	(1.321)
150	–	1.350	1.343	1.338
200	–	1.341	1.335	1.327
250	–	1.333	1.326	1.317
300	–	1.326	1.318	1.309

[a] Potentials are volts, referred to hydrogen electrode.
[b] The parenthesized values were obtained by interpolation and extrapolation.

The chlorine overvoltage on a graphite electrode has been reported
variously, and published results are frequently discordant and must therefore be used with discrimination. There are two principal reasons for this:

(1) The variable nature of graphite itself.

(2) The manner in which it is used as an anode.

(1) Manufactured graphite anodes may differ as to source of the carbon,
the processing steps to form the baked carbon anode shape, the graphitiza-
tion and the machining. For use in mercury cells, active graphite anodes
are not oil treated, although the stems, or posts, are usually oil treated
graphite. As a result, one finds a range of properties as follows:

Density	1.55 to 1.75 gm/cm³
Porosity	30 to 20 percent
Permeability	indicative of pore structure
Hardness	indicative of degree of graphitization
Homogeneity	skin effects
Anisotropism	properties may vary in relation to axis of extrusion of green electrode
Impurities	may affect overvoltages

The above properties are important factors in anode life[29] and, to a lesser
extent, they also affect chlorine overvoltage and current efficiency.[28]

(2) Chlorine overvoltage is chiefly dependent on current density and temperature, as in most electrode reactions. The use of graphite as anode ma-

TABLE 6-11. VALUES OF ENTROPY CHANGE BETWEEN GASEOUS
CHLORINE AND CHLORIDE ION[56]

NaCl Concentration (gpl)	dE/dt, (V/^{o}C)	$\overline{\Delta S}$, (cal/mole, ^{o}C)
150	– 0.000325	– 15
200	– 0.000375	– 17
250	– 0.000425	– 20
300	– 0.000425	– 20

terial in a horizontal mercury cell involves a number of other factors which can have a profound effect on the overvoltage.

(a) *Age in Electrolytic Use.* In the first week or two, the pore structure of virgin graphite opens up until a steady state profile is obtained. The greatly increased surface lowers the true current density and hence the overvoltage, all other conditions remaining equal. Overvoltage on pre-electrolyzed graphite is of the order of 40 percent of that measured on virgin graphite.[63]

(b) *Orientation.* Overvoltage is always higher on a horizontal anode than on a vertical anode. The reasons are complex and are partly due to the greater shear velocity (due to the gas lift effect) on the vertical surface and partly due to the greater coverage of the horizontal surface with adsorbed chlorine.[20,46,47]

(c) *Grooves and Holes.* These are generally machined into the anode slab to provide channels or vents for the escape of chlorine gas that collects on the underside of the anode. Here again the effect is complex. Lower cell voltages result because of (1) lower percentage of dispersed gas bubbles in the brine between the electrodes, and (2) lower overvoltage on the surface of the anodes because of greater effective area and less blinding by adsorbed chlorine.[17,47]

(d) *Anode-cathode Gap.* The smaller the gap, the greater is the percentage of gas in the electrolyte, and also the greater is the overvoltage on the surface of the anode. The effect is not large when adequate vents are provided, as mentioned above.[20,46,47]

(e) *Forced Circulation.* Not a factor in present styles of commercial cells. However, prototypes indicate considerable reduction in overvoltage when brine is forced at high velocity between the electrodes.[15,19,43]

(f) *Surface-active Agents.* Numerous additives have been found that change the surface tension on the graphite in such a way that chlorine gas is released more easily. Cell voltages have been lowered as much as 0.2 volts with trace additions (0.02 percent) of a wetting agent in the brine.[25,61]

Numerical Values of the Chlorine Overvoltage on Graphite. The classical work of Stender *et al.* at the State Institute of Applied Electrochemistry, Leningrad, USSR,[55] did not cover the range of conditions found in modern mercury cells. More recently, comprehensive studies on mercury cell applications have been made at Kyoto University.[47] Pretreated and pre-electrolyzed graphite was used throughout, but the source of the original graphite and its physical properties were not stated. It is presumed that pre-electrolysis was done after machining the anodes. Some of the anodes tested are shown in Figure 6-5. The overvoltages for these are listed in Table 6-12 and are plotted in Figure 6-6.

Even these results must be used with caution. Figure 6-7 is taken from

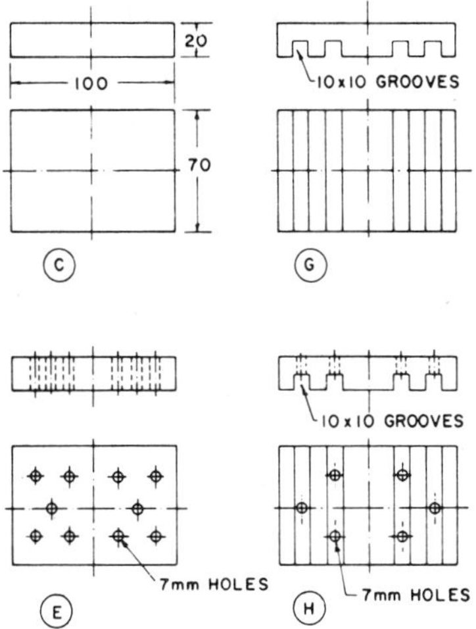

Figure 6-5. Various styles of
anodes tested.

TABLE 6-12. CHLORINE DISCHARGE POTENTIAL AND OVERVOLTAGE ON VARIOUS GRAPHITE ANODES[47]

C, E, G, H refer to horizontal anodes shown in Figure 6-5.

V = Total discharge potential, η = overvoltage, volts, Temp. 55°C

Current Density amp/cm^2	C Plain Surface V	η	E Holes, 7 mm dia. V	η	G Grooves, 10 × 10 mm V	η	H Holes and Grooves V	η
0.000*	1.283	–	1.283	–	1.313	–	1.313	–
0.005	1.300	0.017	1.300	0.017	1.320	0.007	1.320	0.007
0.010	1.330	0.047	1.320	0.037	1.330	0.017	1.325	0.012
0.020	1.365	0.082	1.350	0.067	1.350	0.037	1.340	0.027
0.050	1.430	0.147	1.380	0.097	1.410	0.097	1.385	0.072
0.100	1.510	0.227	1.450	0.167	1.470	0.157	1.440	0.127
0.200	1.750	0.467	1.600	0.317	1.580	0.267	1.510	0.197
0.500	2.810	1.527	2.140	0.857	1.830	0.517	1.680	0.367

*Reference emf at bottom center of anode.

an earlier paper by the same authors.[46] Comparison of the 55°C curve of Figure 6-7 with the *C* curve of Figure 6-6, reveals considerably higher overvoltages for the latter, possibly because the plain anode *C* was larger in area than the former plain anode.

Again, the grooves and/or holes, as machined in virgin anodes, do not long hold their shape and dimensions. As the anode wears back, the sharp rectangular grooves become rounded valleys. The holes become larger.

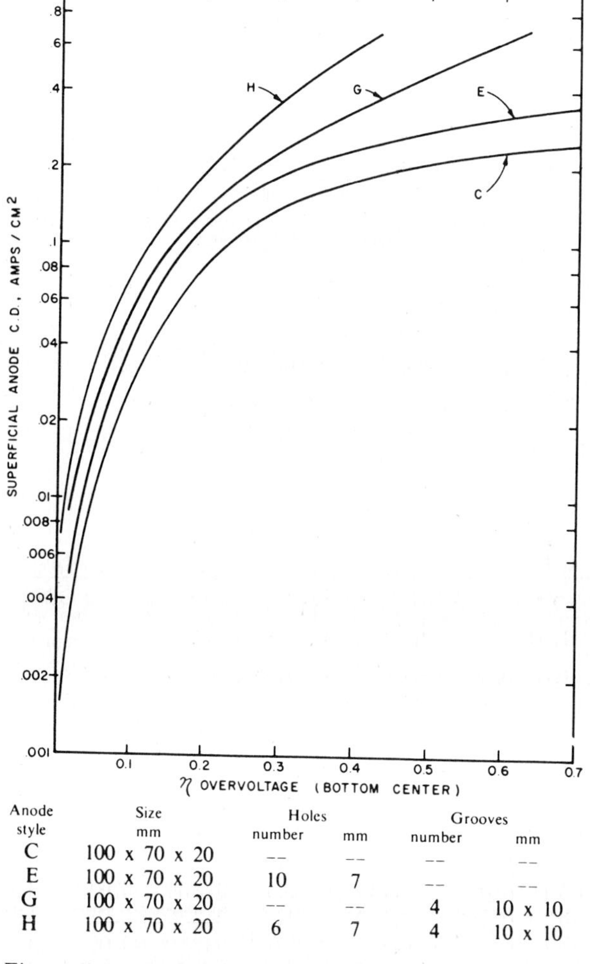

Anode style	Size mm	Holes number	mm	Grooves number	mm
C	100 x 70 x 20	--	--	--	--
E	100 x 70 x 20	10	7	--	--
G	100 x 70 x 20	--	--	4	10 x 10
H	100 x 70 x 20	6	7	4	10 x 10

Figure 6-6. Overvoltage of chlorine on various styles of graphite anodes; horizontal pre-electrolized graphite; 55°C.

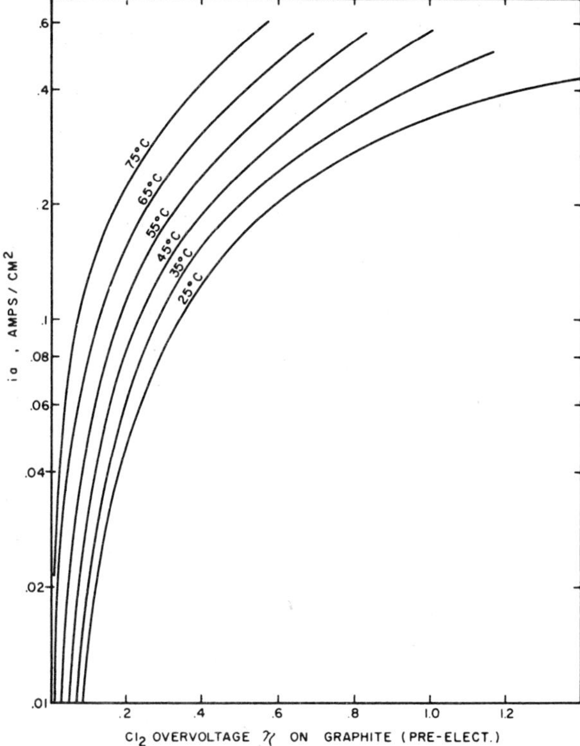

Figure 6-7.   Chlorine overvoltage on plain horizontal
anodes at various temperatures.

The bottom surface loses its flatness. The texture of the graphite becomes
very rough. There are no comparable data on chlorine overvoltage on serv-
iceable anodes of this kind.

The same Japanese authors[46] offer constants for the Tafel equation, as
shown in Table 6-13, which are valid at current densities up to 0.005 amps/
cm.² Making the assumptions that a layer of adsorbed chlorine blinds a
substantial percentage of the surface of the anode at higher current densities and that the Tafel equation can be applied to the remaining active
surface, the Japanese calculate the percentage of effective area as shown in
Figure 6-8. For this explanation to be quantitatively true, then, at apparent commercial current densities of 0.3 to 0.5 amps/cm², the true local
current density must be impossibly high to account for the observed overvoltages, (i.e., at 65°, i = 0.4 amps/cm², η = 0.52 volt, the local current
density i' is calculated to be 5×10^{12} amps/cm²). Nevertheless, it is quali-

TABLE 6-13.   TAFEL CONSTANTS, Cl_2 EVOLUTION
ON GRAPHITE[46]

$\eta = a + b \log i$

$\eta = Cl_2$ overvoltage, volts; i = current density, amps/cm^2

°C	a	b
25	0.179	0.058
35	0.158	0.053
45	0.142	0.051
55	0.095	0.043
65	0.072	0.036

tatively true that the adsorbed chlorine gas layer must have a profound effect on overvoltage.

A different approach was recently suggested by Ksenzhek and Stender.[32] Recognizing the porous nature of graphite and that specific surface is on the order of 0.8 to 1.5 m^2/g. for dense graphite; or 2.5 to 6 m^2/g. for a porous grade, they suggest that electrolysis takes place predominantly in the pores (near the active surface). With such an enormously augmented surface, the local current density is small enough for the Tafel equation to

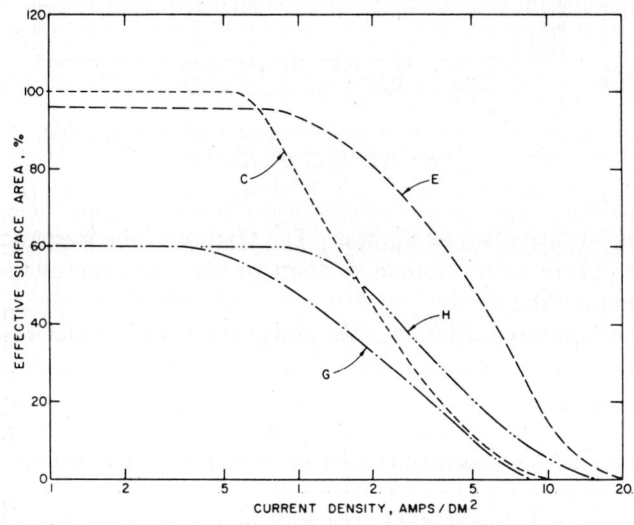

Conditions: C. Horizontal anode—plain
E. Holed anode (nonuniformly)
G. Grooved anode
H. Grooved and holed

Figure 6-8   Effective surface area of graphite anodes.

apply. They suggest an $i_o = 5 \times 10^{-6}$ amps/cm^2, and a constant, $b = 0.15$, for the Tafel Equation. This is equivalent to saying

$$\eta = +0.80 + 0.15 \log i' = a + b \log i' \qquad (6\text{-}15)$$

Since $\quad b = \dfrac{2.303\,RT}{\alpha\,F} = 0.15$

Then if $\quad T = 293°\,K, \alpha = 0.39$

Now $\quad i'$ is related to i by the equation

$$\frac{i}{i'} = d \cdot \rho \cdot s \qquad (6\text{-}16)$$

where $\quad d$ = effective depth for electrolysis, cm

$\qquad \rho$ = density of graphite in active layer, g/cm^3

$\qquad s$ = specific area, cm^2/g of graphite in this layer.

Thus, if $d = 0.1$ cm, $\rho = 1.65$ g/cm^3, $s = 10,000$ cm^2/g,

$i/i' = 1,650$. If $i = 0.40$ amps/cm^2, $i' = 2.4 \times 10^{-4}$ amps/cm.^{2}

According to Equation (6-15), at 20°C, $\eta = 0.26$ volts. The actual overvoltage is approximately 0.38 volts (small vertical electrode).

The energy of activation is given as approximately 8 kg cal from which we calculate the Tafel equation at 70° C to be

$$\eta = .760 + 0.175 \log i \qquad (6\text{-}17)$$

and at 70°, $\eta = 0.12$ volts. The actual overvoltage is approximately 0.12 volts (small vertical electrode.)

Thus it is seen that this hypothesis gives the right order of magnitude for the overvoltage on a vertical electrode. However, the overvoltages on a horizontal electrode, according to the Japanese, are much higher than on a vertical electrode.

This suggests that the true explanation of overvoltage probably should combine both the effects of augmented surface due to electrolysis in depth and reduction of surface by adsorbed chlorine.

Evolution of Oxygen at a Graphite Anode. The oxygen discharge at the anode may be written variously:

$$2 H_2O \rightarrow O_2 + 4 H^+ + 4 e^- \tag{6-18}$$

for which $\Delta F^\circ = 54.635 \text{ kcal}, E^\circ = -1.229 \text{ volt}$

$\Delta S = 15.57 \text{ cal}$

$E = -1.229 + 0.0592 \times \text{pH, at } 25^\circ C. \tag{6-19}$

$$\frac{dE}{dT} = \frac{\Delta S}{NF} = 0.000169 \text{ volts}/^\circ C \tag{6-20}$$

Temp. ($^\circ$C)	E_o	Coefficient of pH
25°	-1.229	0.0592
45°	-1.226	0.0632
55°	-1.224	0.0652
65°	-1.222	0.0672
75°	-1.220	0.0692

$$C + 2 H_2O \rightarrow CO_2 + 4 H^+ + 4 e^- \tag{6-21}$$

for which $\Delta F^\circ = 19.12 \text{ kcal}, E = -0.207 \text{ volt}$

$$H_2O \rightarrow O^\circ + 2 H^+ + 2 e^-; E^\circ = -2.42 \text{ volt} \tag{6-22}$$

followed by, $2 O^\circ \rightarrow O_2 \tag{6-23}$

or, $2 O^\circ + C \rightarrow CO_2 \tag{6-24}$

Whatever the mechanism of the reaction, oxygen is always found in the chlorine gas when Pt anodes are used and CO_2 when graphite anodes are used.[60] For cells fitted with graphite anodes, operating at a pH of about 2.5 to 3.0, the percent CO_2 in the chlorine is generally in the range of 0.3 to 0.6 percent, depending on brine strength, CO_2 in the feed, temperature and quality of the graphite. The corresponding current efficiency loss is 0.6 to 1.2 percent, uncorrected for CO_2 in the feed.

The overvoltage of oxygen on graphite and on Pt in acid solutions is not known with any certainty. The values usually quoted are for graphite only for 1 N KOH. For Pt, we take Hoar's parameters,[9]

$i_o = 9.10^{-12}, \alpha = 0.6, \text{temperature } 20^\circ C$

$\eta_o = 1.035 + 0.0938 \log i$

If $i_a = 0.40, CO_2 = 0.4\%, i_o = 0.0032 \text{ a/cm}^2$

$\eta_o = 0.80 \text{ v. at } 20^\circ C, 0.70 \text{ v. at } 50^\circ C, 0.65 \text{ v. at } 70^\circ C.$

At current densities in this range, η for graphite is about 0.10 to 0.12 volt less than for Pt. Hence for graphite,

η_0 = 0.69 v. at 20°, 0.60 v. at 50°, 0.53 v. at 70°

Now the discharge potential of chlorine must equal that of oxygen at the respective current densities of each. The anode voltages, as they appear in a mercury cell operating under the same typical conditions listed for the cathode reactions, are listed in Table 6-14.

TABLE 6-14. ANODE POTENTIALS IN TYPICAL MERCURY CELL

	Inlet end	Outlet end
NaCl in brine, gpl	305	270
pH	3.0	3.0
Temperature, °C	50	70
Total anode C.D., amps/cm^2 *	0.40	0.40
% CO_2 in anode gas	0.4	0.4
Oxygen C.D., amps/cm^2	0.0032	0.0032
Chlorine discharge:		
E_{Cl}	1.322	1.317
η_{Cl} (Model H, holes and grooves)	0.345	0.229
V_{Cl}	1.667	1.546
Oxygen discharge:		
E_0	1.032	1.026
η_0 (est'd)	0.600	0.530
V_0	1.632	1.556

*Note: Anode current density is usually 5 to 15% higher than cathode density when based on the superficial bottom area of the anodes. Taking into account the vertical surfaces of the anodes, which also are active and contribute to the conductance, the real anode current density is substantially the same as the cathode current density.

Miscellaneous Anode Reactions. The chlorine gas from graphite anodes in a mercury cell will contain traces of carbon monoxide. At the usual pH of 2 to 3, the oxygen is of the order of 0.03 percent, and the CO about 1/17th of the CO_2, or 0.02 percent.[60]

With platinum anodes, the oxygen is of the order of 0.15 percent or less, corresponding to a current efficiency loss of 0.3 percent or less. There will be no CO_2 except that entering as carbonate in the treated brine, which stays in solution on pre-acidification of the cell feed.

Voltage Drop in the Electrolyte

The resistivity of sodium chloride brine is shown in Table 6-15 and Figure 6-9.

The resistance of the electrolyte between the electrodes is increased by

TABLE 6-15. RESISTIVITY OF SODIUM CHLORIDE BRINE

(Estimated from ICT data by RBM)

ρ, ohm cm.; temperature $^\circ$C

NaCl	Temperature ($^\circ$C)			
(g/l)	25	50	75	100
200	4.83	3.07	2.23	1.67
225	4.57	3.00	2.10	1.57
250	4.37	2.82	2.00	1.47
275	4.29	2.74	1.93	1.42
300	4.11	2.67	1.88	1.35
320	4.07	2.64	1.83	1.32

the dispersed gas bubbles (mostly chlorine and water vapor). The following relation holds:[12]

$$\rho/\rho_0 = (1 - \epsilon)^{-1.5} \tag{6-25}$$

where ρ = resistivity with gas present

             ρ_0 = resistivity, no gas present

             ϵ = void fraction of dispersed gas.

The relation is shown in Figure 6-10.

It is generally difficult to independently estimate ϵ except in the simplest of cases. Instead, ρ/ρ_0 is measured and ϵ is calculated from it. Obviously, ϵ must depend on electrode orientation, being less for vertical than for horizontal anodes. The more vents and grooves provided, the less ϵ. E increases with current density, and decreases with electrode spacing.

The Japanese work[46] previously referred to in connection with chlorine overvoltage also includes a study of these variables. Their results are shown in Figure 6-11. In this plot, the letters refer to the anodes illustrated in Figure 6-5. The electrode spacing (nearest plane of anode to surface of mercury) was 10 mm, and the temperature was 55°C. At commercial densities in the range of 0.3 to 0.5 amp/cm^2, horizontal flat anodes give ϵ = 25 percent voids, ρ/ρ_0 = 1.5 or greater. With the most elaborate design of anode, with grooves and holes, we may take ϵ = 0.06 and ρ/ρ_0 = 1.10

The electrode spacing in a cell is continually increasing as the anode wears. Periodically, an adjustment is made to restore the desired spacing. Thus, the actual IR drop, and the average IR drop, is a matter of cell maintenance and local policy. The closest practical approach is about 4 mm. The wear amounts to about 0.14 mm per day at 0.4 amps/cm.^{2} Assuming an adjustment every month in modern cells, the average gap would then be 4 + 15 × .14 = 6 mm. Most older plants will average a little

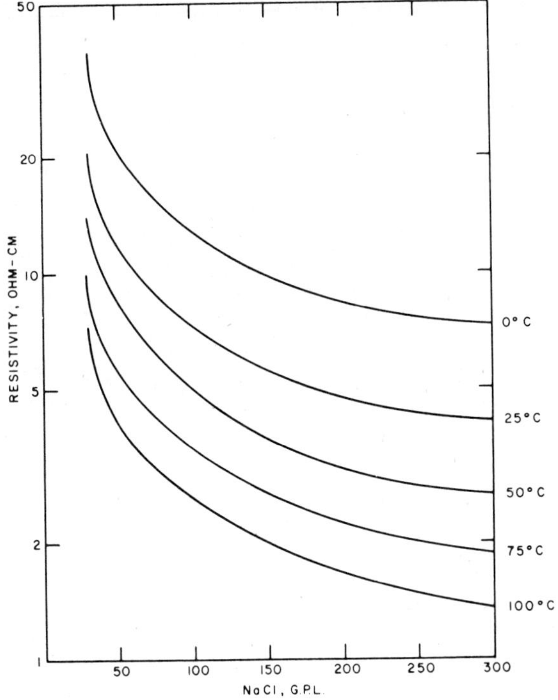

Figure 6-9.   Resistivity of sodium chloride brine.

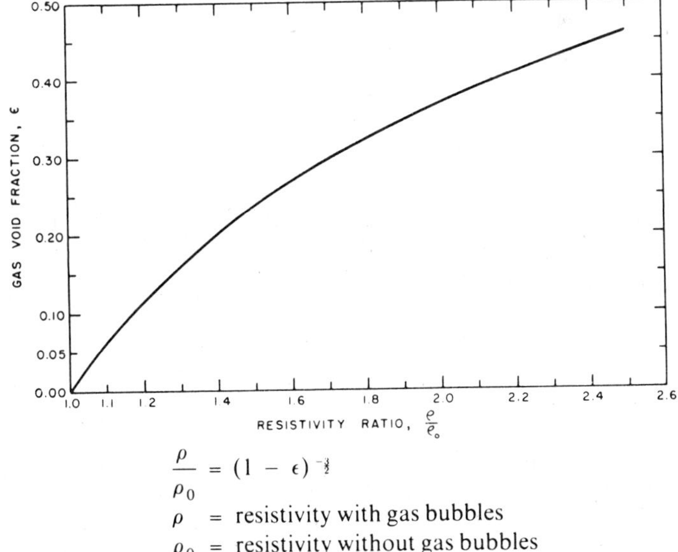

$$\frac{\rho}{\rho_0} = (1 - \epsilon)^{-\frac{3}{2}}$$

ρ = resistivity with gas bubbles

ρ_0 = resistivity without gas bubbles

Figure 6-10.   Effect of dispersed gas on resistivity of brine.

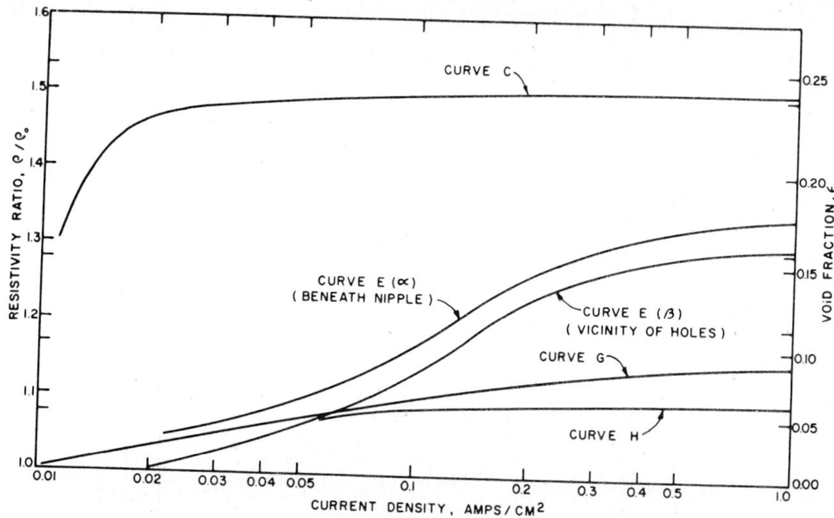

Figure 6-11. Effect of current density on gas dispersion in brine.

more than this—perhaps as much as 10 mm. The gap can be adjusted to compensate for any difference in cell voltages in order to maintain approximately constant current to each anode. The voltage drop in the anolyte of our typical cell is given in Table 6-16.

TABLE 6-16. VOLTAGE DROP IN ELECTROLYTE IN TYPICAL CELL

For "typical" cell operating at 0.4 amps/cm²

	Inlet	Outlet
NaCl, gpl	305	270
Temperature, °C	50	70
ρ_0, ohm cm	2.64	2.08
ε, (type H anode)	0.06	0.06
ρ/ρ_0	1.10	1.10
ρ, ohm cm	2.90	2.29
Cathode discharge potential (see Table 6-5), volt	1.70	1.80
Anode discharge potential (see Table 6-14), volt	1.67	1.55
	3.37	3.35
Gap, cm	0.50	0.66
R, ohms on 1 cm²	1.45	1.50
IR drop, volt	0.58	0.60
Total voltage, excluding conductors	3.95	3.95

Voltage Loss in the Electronic Conductors. The ohmic drop in the various conductors and contacts between conductors depends a great deal on the cell design. In the average typical cell, operating at 0.4 amps/cm², ohmic drop will be of the order of 0.25 to 0.30 volt; contact drop will be 0.10 to 0.15 volt; and total drop will be 0.35 to 0.45 volt.

Today cells are mounted side by side, as close as possible. Copper bus runs in parallel lines down the cell row, and vertical runs are minimized. The voltage drop in the cell to cell copper is included in the above figures.

Over-all Voltage Balance

The various voltages making up the total voltage for our typical cell are summarized in Table 6-17.

TABLE 6-17. OVER-ALL VOLTAGE BALANCE, TYPICAL CELL

Cathode C.D., 0.4 amps/cm²	Inlet	Average	Outlet
Reversible EMF, cathode	1.69	1.74	1.78
Reversible EMF, anode	1.32	1.32	1.32
Reversible EMF, total	3.01	3.06	3.10
Polarization, cathode	0.01	0.01	0.02
Polarization, anode	0.35	0.29	0.23
Polarization, total	0.36	0.30	0.25
Ohmic, electrolyte	0.58	0.59	0.60
Ohmic, conductors	0.28	0.28	0.28
Ohmic, contacts	0.12	0.12	0.12
Ohmic, total	0.98	0.99	1.00
Total cell/cell voltage ±.05	4.35	4.35	4.35

Voltage at Other Current Densities. The reversible emf remains constant. The overvoltage can be taken, over a moderate range, as the sum of a constant and variable portion. Within a range of ± 25 percent, the total cell voltage is approximately

$$V_{\mathrm{avg}} = 3.20 + 2.87\,i_{\mathrm{ac}}$$

where i_{ac} = means current density, amps/cm.²

i_{ac}, amps/cm²	0.20	0.30	0.40	0.50
V, volts	3.78	4.06	4.35	4.63

Energy Consumption

At 95 percent true current efficiency, and 98.5 percent recovery of cell product after internal uses for brine purification are deducted, the net current efficiency is 93.5 percent.

Theoretically, one short ton (2000 lb) requires 685,840 ampere hours of d-c current. At 93.5 percent recovery, it requires 735,000 ampere hours.

D. C. Energy per 2000 lb of net Cl_2 Recovered:

i_{ac},				
DC KWH/ton Cl_2,	0.20	0.30	0.40	0.50
	2774	2978	3192	3400

These results are compared with those claimed for various commercial cells in Table 6-18.

TABLE 6-18. ENERGY CONSUMPTION, VARIOUS MERCURY CELLS

Type	CD amps/cm^2	DC Kw-h per ton Cl_2
BASF—Krebs 80	0.3570	3,250
De Nora 18 SGL	0.3905	3,175
Mathieson E-8	0.4875	3,250
Solvay V-100	0.5440	3,130
Uhde 20	0.4062	3,120
Calculated example	0.40	3,192

Variations in Electrolysis

Horizontal mercury cells with graphite anodes have reached a high state of perfection, and commercial installations, as of 1960 are nearly all of this type. Much effort has also been devoted to the development of the vertical mercury cell, and also to the search for a better anode than graphite. We will now examine these variations from the electrochemical viewpoint.

Vertical Versus Horizontal Orientation. At high current densities, it is extremely difficult to get rid of the chlorine gas evolved on the under side of the horizontal anode. High overvoltage on the graphite and high ohmic voltage drop in the brine due to dispersed chlorine bubbles are the consequences. Anode current densities have reached 0.3 to 0.6 amps/cm², but it is unlikely that they will be pushed much further because of increased power consumption. This limitation could be lifted by forcing brine and mercury at high velocity through the anode-cathode gap, as proposed in some recent patents;[19,43] however, there may be serious practical limitations to designing such cells. As yet, no commercial cells have adopted this principle.

As far as the anode is concerned, a vertical orientation is much better. The overvoltage on graphite is much lower (see Figure 6-12). If 0.30 volt is

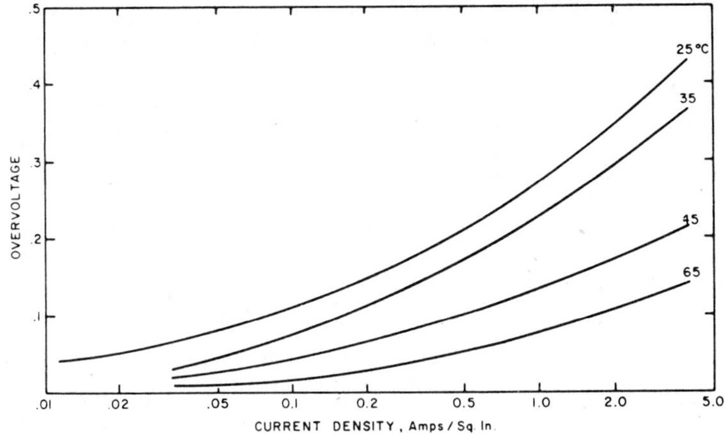

Saturated NaCl, 5% DEC., pH feed 4.0, pH anolyte 2.2–3.5
Estimated equilibrium potential *vs* N.H.E.

25° C	1.325
45° C	1.310
65° C	1.294

Figure 6-12. Overvoltage of Cl_2 on pre-electrolyzed graphite in vertical position.[46]

permissible in a horizontal anode at 0.40 amps/cm², then a vertical anode could be operated at 3.0 amps/cm² at this same overvoltage.

With a vertical arrangement, which includes a return path for circulation of brine within the cell as in Hooker or Diamond diaphragm cells, the gas lift effect will promote an enormous circulation of brine past the anode. The higher the current density, the greater is the circulation. This places a limitation on the percentage gas voids in the anolyte so that high ohmic voltage drop due to dispersed gas becomes a negligible factor, even at high current densities.

With these two obvious advantages the question might be asked: Why hasn't the vertical cell been adopted? This question can be answered with another: How does one maintain a mercury cathode in the vertical position? It can be done; the Honsberg rotating cathode does it.[59] The vertical mercury jet cathode cell does it.[40,41] The Hooker M cell does it.[30] The De Nora "Fluent" amalgam cell does it.[13] There are difficulties, and most of these have been mechanical ones.

In order for mercury to flow as a sheet smoothly down a fixed vertical surface, and not fly off, three basic conditions are necessary: (1) the steel cathode must be kept fully amalgamated everywhere, at all times, (2) some restraining frictional force must be provided to prevent the down-flowing amalgam from accelerating to the point where it flies off the cathode, (3) the cathode must be protected from the high velocity stream of rising anolyte, induced by the gas lift effect of the chlorine.

Respecting the first condition, the mercury feed to the top of the cathode must contain a certain amount of sodium in order to keep the cathode covered at the top. In other words, the mercury from the decomposition must be mixed with a bypass stream in controlled proportion. To overcome the tendancy to form hypochlorite, the cell should be fed with a more acid brine.

Respecting the second condition, the cathode is made of a wire screen, and the mercury is distributed to it as a very thin sheet through a very narrow slot, through which the screen is inserted. A stable fall velocity is achieved, and there is no limit to the height of the cathode on this score.

Respecting the third condition, a porous diaphragm or screen must be hung between cathode and anode. Perforated polyvinyl sheet appears to be suitable. The flow of the catholyte (actually this is the same as anolyte but free from chlorine bubbles) must be suitably controlled. The back reactions of chlorine and HCl with the sodium amalgam should be appreciably less with the diaphragm present.

It would appear that these three conditions have been satisfied in the De Nora "Fluent" amalgam cell. The chief drawback to such a cell is that anode adjustment is difficult, if not impossible. If graphite anodes are used, the gap continually increases, as it does in the Hooker or Diamond type diaphragm cell. So far, with graphite anodes, a reasonable average voltage can be achieved only at lower current densities (about 0.2 amps/cm^2).

"Permanent" Anodes. The term "permanent" is relative. Chlorine cells of all types have been built around the properties of graphite. The best graphite lasts 300 to 400 days in a mercury cell. As it wears, the gap increases and must be adjusted periodically to keep cell voltage from rising to prohibitive levels. It is obvious that an anode which held its dimensions throughout its life would be a boon to any cell, but particularly to the vertical mercury cell.

Platinum was used as an anode for many years in the horizontal mercury cells of Solvay *et Cie* in Belgium.[6] The corrosion rate is on the order of 0.3 to 0.6 gms. of Pt per ton of chlorine made.[7] However, the capital cost of Pt wire anode assemblies was very high—even though the Pt loss was low enough to be economical.

No other anode materials have yet been discovered that approach graphite and platinum in performance.

The idea of cladding a base metal with platinum, for use as an anode, is very old. The idea of cladding tantalum, or titanium, is relatively recent.[10,28] Ta and Ti appear to be completely inert to chlorinated brines and hypochlorites. These metals can not be used as anodes, however, because a nonconducting layer of oxide forms which will not pass anodic current. If, however, sufficient Pt is deposited on the surface, the Pt spots can act as anodes, while the exposed parts are inert. Pt can be electroplated on either Ta or Ti, and films ranging in thickness from 0.125 to 5.0 microns appear to be suitable for use on such a composite anode. Both continuous sheet and expanded metal sheet have been proposed.

We now turn to the electrochemical consequences of using, once again in history, a platinum anode in a mercury cell.[65] Reliable data on overvoltages on smooth Pt in brine, under cell conditions, are not available as yet. The older data of Knobel[31] are shown in Table 6-19.

TABLE 6-19. OVERVOLTAGES AT 25°, AS REPORTED IN I.C.T.[31]

Chlorine			Oxygen		
i amp/cm^2	Smooth Pt	Graphite	i amp/cm^2	Smooth Pt	Graphite
.05	0.047	0.188	.001	0.72	0.53
.10	0.054	0.251	.005	0.80	0.71
.20	0.087	0.298	.010	0.85	0.90
.50	0.161	0.417	.020	0.92	0.98
			.050	1.16	1.01
			.100	1.28	1.09

From the data of Table 6-19 it appears that the chlorine overvoltage on Pt is in an order of magnitude lower than that on graphite, while the overvoltage of oxygen is of the same order of magnitude for Pt and graphite. If these relations hold for conditions as they exist in the operating mercury cell, then the substitution of Pt for graphite—all other conditions remaining equal—should result in (1) lower cell voltages, and (2) lower current efficiency loss due to oxygen evolution. The second conclusion was substantiated recently by Vaaler.[60] In a diaphragm type test cell with brine 260 gpl, pH 3, and temperature 80° C, the current efficiency losses on Pt and plain graphite were 0.57 percent and 1.84 percent respectively

It was previously mentioned that Ta and Ti, if present in the brine, could have a moderately harmful effect on the hydrogen overvoltage. The use of these metals as component parts of anodes in a mercury cell raises the question of contamination, which may be slowly accumulative. Only long-term use will settle this point.

A possible difficulty with using platinized titanium anodes in a mercury cell is the damage to the anode that would result from accidental contact with mercury i.e., the danger that Pt would be transferred from anode to cathode, particularly during off-load periods.

Some commercial chlor-alkali cells of both diaphragm and mercury type have been fitted out with platinized titanium anodes for test. It will probably take several years before all factors are fully understood.

The Decomposer

Reactions and Kinetics. Caustic soda (or potash) is produced in the decomposer by reacting the sodium (or potassium) amalgam produced in the electrolyzer with water. The over-all reaction is

$$Na(Hg) + H_2O \rightarrow NaOH + \tfrac{1}{2}H_2 + (Hg) \tag{6-26}$$

This reaction is extremely slow if only the mercury surface is available for hydrogen evolution because of the high overvoltage. If now we make the amalgam anodic and provide a cathode of some convenient material which does not amalgamate with mercury then, on completing the electrical circuit, sodium will dissolve at the mercury surface to form sodium ion and hydrogen gas will be deposited at the cathode leaving hydroxyl ion. The electrode reactions are:

$$\text{Anode:} \quad Na(Hg) \rightarrow Na^+ + e^- \tag{6-27}$$

$$\text{Cathode:} \quad H_2O + e^- \rightarrow OH^- + \tfrac{1}{2}H_2 \tag{6-28}$$

The cathode is generally graphite which is in metallic contact with the amalgam, but otherwise immersed in the caustic solution. The decomposer acts like an internally shorted battery, and no attempt is made to recover energy from the battery.

The electromotive force corresponding to Reaction (6-26) may be expressed as follows:[24]

$$E = E_0 + \frac{RT}{F}\ln\frac{YW}{X} \tag{6-29}$$

where X is the activity of the NaOH at concentration x wt percent
        Y is the activity of the Na(Hg) at concentration y wt percent
        W is the activity of the water in the NaOH solution.
For the reaction

$$Na^\circ + H_2O \rightarrow NaOH + \tfrac{1}{2}H_2 \tag{6-30}$$

$$\Delta F + -43.49\,\text{Kcal.,} \quad E^\circ = +1.885\,\text{volt}$$

MacMullin[37] gives the potential of the half cell as follows:

$$Na°, Na(amalgam, activity\ Y) \tag{6-31}$$

$$E = E° + \frac{RT}{F}\ln Y, \text{ where } E° = -0.755 \text{ volt}$$

Adding the $E°$'s for Reactions (6-30) and (6-31), we obtain the following for Reaction (6-29):

$$E° = 1.885 - 0.755 = 1.130 \text{ volt}$$

The activity Y for sodium amalgam[37] is shown in Figure 6-13 while the activity X for NaOH solution[1] is shown in Figure 6-14. The activity of water in NaOH solution[9] is shown in Figure 6-15.

From the above data, the emf of Reaction (6-26) has been calculated as shown in Table 6-20.

Because of the emf, when the amalgam-caustic-graphite couple is shorted, current begins to flow. Soon after, the surfaces become polarized, and the current density settles down at some value such that the sum of the overvoltages plus IR drops balance the emf.

The kinetics of the decomposer reactions have been studied by Hine *et al* at Kyoto University and some of their data are reproduced here.[21,22,23,24]

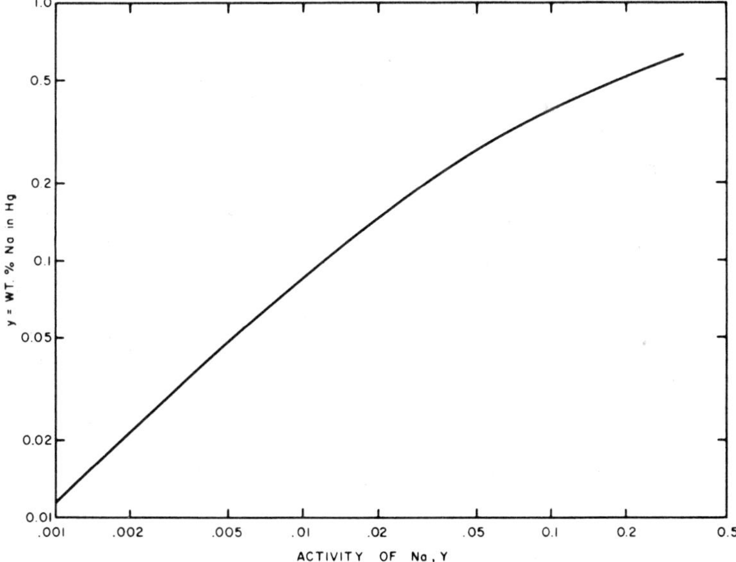

Figure 6-13. Activity of sodium in sodium amalgam at 25° C. (*Data: Richards & Conant Recalc. by C. Pruiss of R.B.M.A., 1959*)

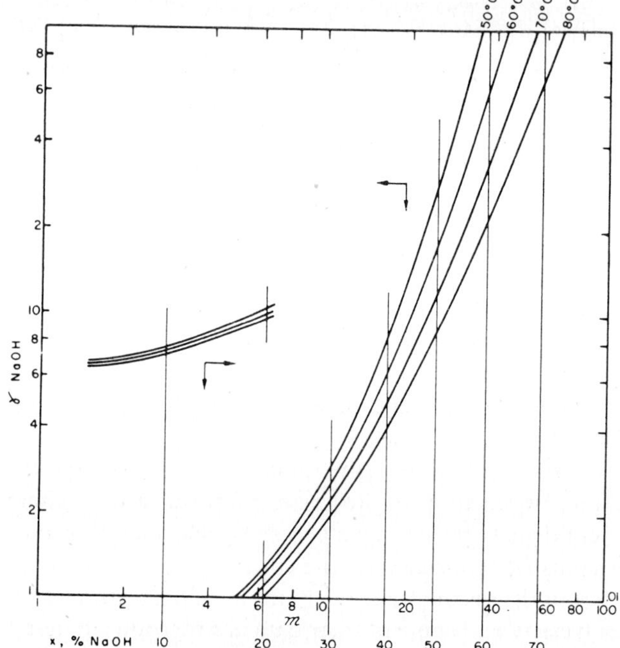

Figure 6-14.   Activity coefficient of NaOH in aqueous
solution.[1]

Figure 6-16 shows the anodic overvoltage curve of sodium amalgam at 80° C and at various concentrations of NaOH.   The overvoltage is quite small and has a minor influence on the kinetics of decomposition.

Figure 6-17 shows the cathodic hydrogen overvoltage curve for 40 per-

TABLE 6-20.   EMF OF THE DECOMPOSER REACTION

Temperature, 80° C

Weight % Na in Amalgam	Weight % NaOH						
	10	20	30	40	50	60	70
0.001	0.788	0.748	0.696	0.637	0.575	0.507	0.434
0.01	0.859	0.818	0.766	0.707	0.645	0.577	0.504
0.05	0.913	0.872	0.821	0.761	0.700	0.631	0.557
0.10	0.939	0.898	0.846	0.787	0.725	0.656	0.584
0.15	0.955	0.914	0.862	0.804	0.741	0.673	0.599
0.20	0.967	0.927	0.875	0.816	0.754	0.685	0.612
0.25	0.978	0.937	0.886	0.827	0.761	0.696	0.622
0.30	0.988	0.947	0.895	0.836	0.775	0.713	0.633

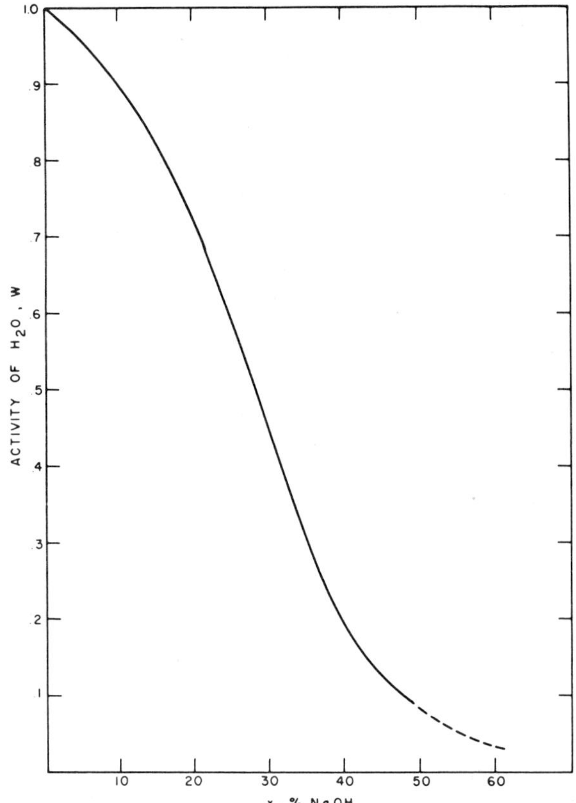

Figure 6-15.   Activity of H_2O in aqueous NaOH.[55a]

cent NaOH at various temperatures. The concentration for NaOH has only a slight effect on the overvoltage, being perhaps 10 percent higher than that for 10 percent NaOH.

The electrical conductivity of caustic soda is shown in Figure 6-18. The *IR* drop in the caustic adds to the overvoltage in limiting the short circuiting current, although the path of this current is indeterminate. Whatever the path, the higher the temperature, the lower is the resistance. Conductivity is at a maximum in the range of 20 to 25 percent NaOH.

The manner in which the local short circuiting current reaches a limiting value is shown in Figure 6-19 for a typical condition near the top of a countercurrent decomposer and for another near the bottom. It appears that the local current density is on the order of 0.25 amps/cm², at 80°C. To get at the total current flowing in any zone, one needs to know the effec-

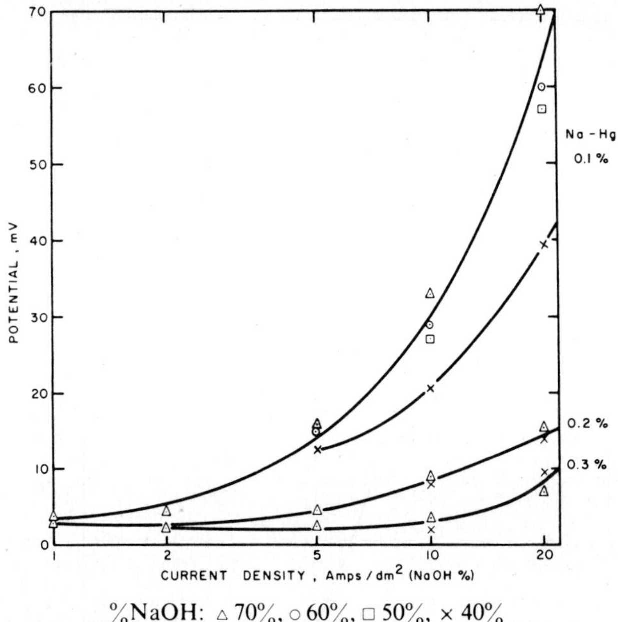

%NaOH: △ 70%, ○ 60%, □ 50%, × 40%

Figure 6-16.   Anodic polarization curve of amalgam
at 80° C.

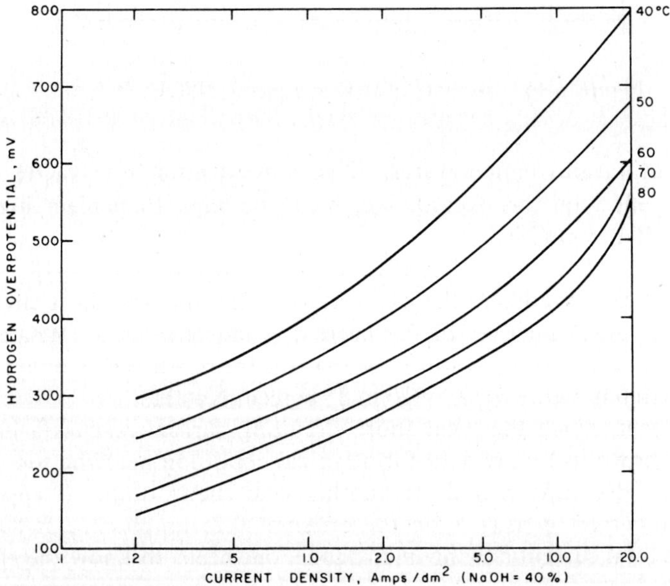

Figure 6-17.   Hydrogen overpotential on graphite cathode.

tive area of the graphite packing, which indeed is less than the total area. The integrated current, zone by zone, must equal the Faraday equivalent of the rate of production of caustic soda. For the detailed mathematical analysis of decomposer operation see the recent work by Hine, Yoshizawa and Okada.[24] A number of important conclusions can be drawn:

(1) For every type and size of packing there is an optimum mercury flow rate for maximum mass transfer. For 10 mm graphite spherules, this flow rate is about 400 kg. mols of mercury per hour and square meter. This

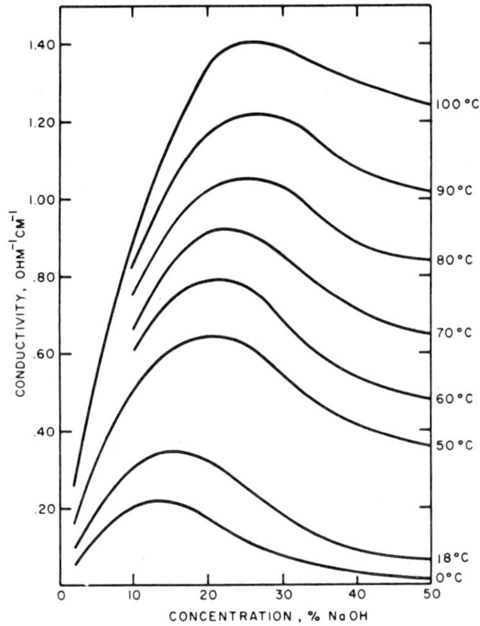

Figure 6-18. Electrical conductivity of caustic soda.

corresponds to an effective area for hydrogen deposition of 170 m^2/m^3, which is about 25 percent of the total area of packing.

(2) The higher the temperature, the smaller is the height of a transfer unit (HTU), and the fewer the number of transfer units (NTU) required.

(3) The higher the caustic strength made, the larger the decomposer volume that is required. 50 percent NaOH is usually made at 80° C; 73 percent is made at 110° C.

(4) Steel exhibits a lower overvoltage than graphite and thus permits higher driving force for decomposition. There is some tendency for steel to amalgamate, and this would slow down, or even stop, decomposition. Graphite loaded with 5 percent iron seems to be a good packing material.

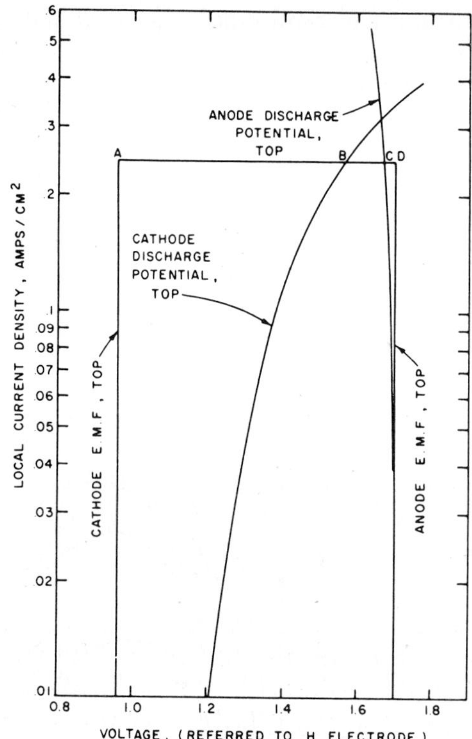

AB — Cathode overvoltage
BC — IR drops
CD — Anode overvoltage

Figure 6-19.   Polarization *vs* current density in the decomposer.

(5) Graphite packing sometimes becomes inactive. This is usually due to high calcium in the brine which carries over as calcium in the amalgam and precipitates as lime on the packing. Various activating treatments include washing, steaming, acid rinse and ferric chloride soaking.

(6) Horizontal decomposers (with graphite grids) are usually run countercurrent. A single stage vertical decomposer is not very effective as a countercurrent device. The mercury must flow down, but it is difficult to make the caustic flow up evenly because the heavier product caustic tends to gravitate downward and thus backmix with the feed water. The decomposer can be compartmentized, and the stages run countercurrent to the mercury.

Recovery of Power. Hamilton Castner, the inventor of the mercury cell, was the first to try this. He connected the positive and negative poles of his D.C. source to the anode of the electrolyzer and to the cathode of the decomposer respectively. This failed because the current efficiency of the electrolyzer was considerably less than that of the decomposer and soon led to the oxidation of the mercury.

There have been many schemes proposed and patented to overcome the above difficulty. None have met with favor because for two reasons:

(1) Separate circuits are required to keep electrolyzer and decomposer chemically in balance. This adds greatly to capital cost and operating labor.

(2) The driving force for decomposition is reduced by the emf recovered, and thus much larger decomposers are required as well as more mercury. The added costs do not justify the saving in power.

The use of sodium amalgam in fuel cells has been investigated by Yeager.[66] An oxygen cathode is used so that the reactions in the decomposer are as follows:

$$\text{Anode:} \quad Na(Hg) \rightarrow Na^+ + e^-$$

$$\text{Cathode:} \quad \tfrac{1}{4} O_2 + \tfrac{1}{2} H_2O + e^- \rightarrow OH^-$$

The cathode is a porous plate of metal or carbon which is activated to decompose hydroperoxide, the initial cathode product. In essence, the hydrogen, which is otherwise evolved in a short circuted amalgam decomposer, is burned nearly reversibly with oxygen to form water. The theoretical voltage available for recovering power is thereby increased by 1.23 volts.

It would appear entirely feasible to use an amalgam fuel cell in place of an ordinary decomposer and still make the usual 50 percent caustic soda— but with no hydrogen by-product. The net voltage of the electrolyzer–fuel cell combination could be in the range of 3.0 to 3.3 volts instead of the customary 4.2 to 4.5 volts. The balancing of current efficiencies in the two devices is mandatory, but presents no great difficulty.

Amalgam By-Products. In the amalgam process for the production of chlorine an aqueous solution of a chloride salt is electrolyzed. Chlorine gas is evolved at graphite anodes, and the metal of the salt in question is deposited on a flowing mercury cathode. The free metal dissolves in the mercury to form an amalgam, which is withdrawn from the electrolysis compartment. This amalgam can be utilized in a number of ways, which may be classified as follows:[37]

(1) Production of free metals, e.g., sodium, potassium, lithium, zinc, lead, etc.

    (a) Distillation of mercury from the amalgam, leaving free metal.
    (b) Electrodeposition.

(2) Production of caustic alkalies by decomposition with water.

(3) Production of metal alcoholates by decomposition of the amalgam with a suitable anhydrous alcohol.

(4) Production of various metallic salts by direct union of the metal dissolved in the amalgam with other elements and acidic compounds.

    (a) Sodium sulfite by reaction of sodium with sulfur.
    (b) Sodium hydrosulfite by reaction of sodium with sulfur dioxide.
    (c) Sodium chlorite by reaction of sodium with chlorine dioxide.
    (d) Sodium nitrite by reaction of sodium with nitrogen peroxide.

(5) For organic reductions:

    (a) Nitro to azo to hydrazo (aromatics)
    (b) Quinone to hydroquinone.
    (c) Oxalic acid to glyoxylic acid.
    (d) Ketones to pinacones.

(6) Amalgam metallurgy:

(a) Separation of various metals in their pure form from their mixed electrolytes, e.g., lead, copper, zinc, from the chlorides.

(b) Reduction of metallic chlorides to produce other metals, e.g., reduction of titanium and zirconium chlorides to produce titanium and zirconium.

The greatest use today of the amalgam type chlorine cell is for the production of caustic soda. Caustic potash can be made in the same equipment with no changes in design and with substantially the same results. Cell voltage on potash is generally 0.1 volt lower and current efficiency 1 or 2 percent lower. There are many mercury cell plants making caustic potash throughout the world.

For making other products, the amalgam reactor takes many forms, e.g., in making sodium sulfide, the reaction is

$$Na_2S_x + 2(x\text{-}1)\,Na(Hg) \rightarrow x\,Na_2S + (Hg) \qquad (6\text{-}32)$$

The decomposition of the amalgam with water is to be avoided. The best type of reactor for this purpose is a simple rubber-lined tower fitted with nonmetallic packing. It is important that the sodium not be wholly stripped from the mercury; otherwise HgS is formed. Hence, in actual installations, the polysulfide reactor is generally followed by a conventional decomposer which strips the amalgam of the remaining sodium before the mercury is returned to the electrolyzer. Installations of this type have operated at Marathon, Ontario, and at Tavazzano, Italy.

Sodium hydrosulfite is made by reduction of sodium sulfite solution with sodium amalgam at a controlled pH in the rage of 5 to 7. The principal reaction in the amalgam reactor is as follows:

$$4\,NaHSO_3 + 2\,Na(Hg) \rightarrow Na_2S_2O_4 + 2\,Na_2SO_3 + 2\,H_2O + (Hg) \qquad (6\text{-}33)$$

and in a second reactor in closed circuit with the above, the reaction

$$Na_2SO_3 + SO_2 + H_2O \rightarrow 2\,NaHSO_3 \tag{6-34}$$

A variety of reactor designs have been tried and there is a successful plant operating in the Netherlands.

For a more detailed and comprehensive discussion of amalgam by-products see MacMullin.[37]

MODERN MERCURY CELLS

There are a number of modern commercial types of mercury cell which are competitive not only with each other but also with other types of cell such as the diaphragm types described in Chapter 5. The word "modern," however, has meaning only in terms of the 1960 frame of reference. Magnificent installations of each type have been built or are currently being built, many of which have been described in the literature and to which reference will be made.

It is only natural that there should be differing opinions, even among experts, as to the relative merits of specific features of the various cells. The choice of cell depends on many factors, which have to be evaluated for each situation. Some of these factors are:
(1) The size of the operation
(2) The kind of salt available, and the delivered cost
(3) The kind of power service available, steady and off-peak, and its cost.
(4) The propinquity of skilled labor, mechanical, electrical and instrument services.
(5) The purpose of the chlor-alkali plant: merchant, captive or mixed.
(6) Limitations of space.
(7) Price and delivery of components and terms.
(8) Royalty charged and services rendered by licensor.
(9) Lastly, but importantly, evaluation of past performance of the various cell installations.

The article by H. A. Sommers[54] in which mercury cell types as of 1957 are critically compared is richly rewarding to read; see also R. B. Mac-Mullin, "Trends in Chlor-Alkali Technology" (1957).[38]

Brief and impartial descriptions of some of the leading commercial mercury cells now available (1960) follow. Structure and performance data on these are summarized in Table 6-21.

Mathieson Mercury Cells

The McIntosh, Alabama, chlor-alkali plant of Olin-Mathieson Chemical Corporation, employing the Mathieson Model E-8—a 30,000 ampere cell

TABLE 6-21. CHARACTERISTICS OF SELECTED SIZES OF VARIOUS MERCURY CELLS AVAILABLE IN 1960

		Mathieson		Solvay		Uhde	
		E-8	E-11	V-100	V-200	10 Sq. M.	20 Sq. M.
1	Rated, kiloamps	30	70	96	160	40	80
2	Maximum, kiloamps	34	100	110	190	50	100
3	Cathode c.d. (rated) amps/cm^2	0.48	0.47	0.53	0.53	0.40	0.40
4	Cathode c.d. (max.) amps/cm^2	0.63	0.67	0.61	0.60	0.50	0.50
5	Avg. cell voltage, rated	4.50	4.08	4.34	4.34	4.15	4.15
6	Avg. cell voltage, at max. amps	4.75	4.43	4.52	4.56	4.40	4.40
7	Current eff., (NaCl) %	95	95	96 +	96 +	95–97	95–97
8	Current eff., (KCl) %			93 +	93 +	92–94	92–94
9	*D. C. energy, rated, kwh/ton Cl_2	3250	2940	3130	3130	3000	3000
10	*D.C. energy, max. amps, kwh/ton Cl_2	3430	3200	3260	3280	3180	3180
11	Power to Hg pump, kw.	0.75	0.75	0.60	0.65	0.28	0.50
12	Cathode width, m	0.615	1.22	1.29	1.29	0.32	1.25
13	Cathode length, m	10.2	12.2	13.36	23.26	10.87	16.0
14	Cathode area, m^2	6.30	15.0	18.00	30.00	10.0	20.0
15	Slope, mm/m	3	6	6	6	4 to 5	4 to 5
16	No. anodes, width of cell	1	5	4	4	2	3
17	No. anodes, total	22	50	108	180	74	96
18	No. stems per anode	2	2	1	1	1	1
19	Anode dim., W × L × H, mm.	455 × 610 × 76	250 × 1220 × 102	315 × 509 × 84	315 × 509 × 120	450 × 290	400 × 500
20	Adjustable anodes	yes	yes	yes	yes	yes	yes
21	Anode seal	flex.	flex	graph/graph	graph/graph	packed	packed
22	Electrolyzer bottom	bare steel	bare steel	bare steel	bare steel	bare steel	bare steel
23	Electrolyzer walls	rub/steel	rub/steel	rub/steel	rub/steel	rub/steel	rub/steel
24	Electrolyzer covers	rub/steel	rub/steel	rub/steel	rub/steel	rub/steel	rub/steel
25	Decomposer type	vertical	vertical	horizon.	horizon.	horizon.	horizon.
26	Decomposer location	low end cell	low end cell	under cell	under cell	at side	horizon.
27	Decomposer lining	none	none	none	none	none	side or under
28	Hg pump, type	centrif.	centrif.	centrif.	centrif.	cone-cent.	none
29	Hg sump	yes	yes	yes	yes	none	centrif.

30 Short circuit switches	1 for 2 cells	1 per cell	none	none	1 per cell	1 per cell
31 No. of units	11	10	27	45	4	8
32 Overall width, m	0.98	1.35	1.59	1.59	1.55	2.06
33 Overall length, m	12.2	14.2	16.55	24.43	12.38	17.55
34 Cell spacing, ctrs., m	1.29	1.88	1.80–2.00	1.80–2.00		
35 Min. bldg. dimensions 1 bank cells, width, m	15.5	17.0	20	30		
36 Min. bldg. dimensions 1 layer cells, height, m	9	9	9	9		
37 Wt. anode graph., 1 cell, lb.	1400	5000	4560	8150	2500	5000
38 Wt. nipples graph., 1 cell, lb.	none	none	2250	3820	300	600
39 Wt. decomposer graph., 1 cell, lb.	300	800	750	1370	200	400
40 Anode life, (NaCl) months	7	10	17	24		
41 Anode life, (KCl) months						
42 Wt. mercury, 1 cell, lb.	3100	6100	4850	8150	2920	5840
43 *Hg loss, lbs/ton Cl_2	0.5	0.5	0.3	0.3	0.3–0.4	0.3–0.4
44 *Graph. cons., lbs/ton Cl_2 (NaCl)	5.3	4.8	3–4	3–4	4–5	4–5

*Per short ton (2000 lb)

TABLE 6-21. CHARACTERISTICS OF SELECTED SIZES OF VARIOUS MERCURY CELLS AVAILABLE IN 1960 (*Continued*)

		Krebs-BASF		De Nora Horizontal		De Nora
		P-50	P-100	14×3	18×6	FAC-24
1	Rated, kiloamps	40	80	60	150	60
2	Maximum, kiloamps	60	100	80	200	72
3	Cathode c.d. (rated) amps/cm^2	0.35	. 0.42	0.42	0.41	0.30
4	Cathode c.d. (max.) amps/cm^2	0.525	0.475	0.50	0.49	0.35
5	Avg. cell voltage, rated	4.15	4.20	4.20	4.20	4.30
6	Avg. cell voltage, at max. amps	4.40	4.35	4.50	4.55	4.60
7	Current eff., (NaCl) %	94–96	94–96	94–96	94–96	95–97
8	Current eff., (KCl) %	94–96	94–96	92–94	92–94	94–96
9	*D.C. energy, rated, kwh/ton Cl_2	2950	3000	3000	3000	3030
10	*D.C. energy, max. amps, kwh/ ton Cl_2	3225	3200	3250	3250	3250
11	Power to Hg pump, kw.	1.2	1.6	0.8	1.5	1.5
12	Cathode width, m	1.25	1.40	1.07	2.12	0.87
13	Cathode length, m	9.40	15.10	11.2	14.32	
14	Cathode area, m^2	11.4	21.0	11.95	30.40	20.22
15	Slope, mm/m	2 to 3	2 to 3	4	4	
16	No. anodes, width of cell	2	2	3	6	3
17	No. anodes, total	48 or 96	120	42	108	72
18	No. stems per anode	1	1	2	2	1
19	Anode dim., W × L × H, mm.	600 × 250 or 600 × 360	690 × 250	342 × 790 × 70	342 × 790 × 70	280 × 1000
20	Adjustable anodes	yes	yes	yes	yes	no
21	Anode seal	packed	packed	flex.	flex.	fixed
22	Electrolyzer bottom	rub/steel	rub/steel	plas/conc.	plas./conc.	
23	Electrolyzer walls	rub/steel	rub/steel	stone	stone	rub/steel
24	Electrolyzer covers	rub/steel	rub/steel	flex. rubber	flex. rubber	rub/steel
25	Decomposer type	horizon.	horizon.	vertical	vertical	vertical
26	Decomposer location	under cell	under cell	high end cell	high end cell	front cell
27	Decomposer lining	rubber	rubber	none	none	rubber
28	Hg pump, type	centrif.	centrif.	centrif.	centrif.	centrif.
29	Hg sump	yes	yes	yes	yes	yes
30	Short circuit switches	1 per cell	1 per cell	1 per cell	1 per cell	1 per cell
31	No. of units	4	4 × 2	14	18	
32	Overall width, m	1.60	1.80	1.40	2.55	1.8

33	Overall length, m	11.15	16.60	12.9	16.0	1.9
34	Cell spacing, ctrs., m	2.00	2.30	1.83	2.88	2.14
35	Min. bldg. dimensions 1 bank cells, width, m	13	21	16.5	22.0	6.1
36	Min. bldg. dimensions 1 layer cells, height, m	10	11	7.0	7.0	7.0
37	Wt. anode graph., 1 cell, lb.	3200	5200	3060	7884	none
38	Wt. nipples graph., 1 cell, lb.	incl. above	incl. above	320	1060	none
39	Wt. decomposer graph., 1 cell, lb.	375	600	800	2200	800
40	Anode life, (NaCl) months	12 to 14	12 to 18	10	10	
41	Anode life, (KCl) months	8 to 12	8 to 12	8	8	
42	Wt. mercury, 1 cell, lb.	3200	5200	3800	8580	2800
43	*Hg loss, lbs/ton Cl_2	0.1–0.2	0.1–0.2	0.2–0.4	0.2–0.4	0.3–0.5
44	*Graph. cons., lbs/ton Cl_2 (NaCl)	5–6	5–6	4–5	4–5	none

*Per short ton (2000 lb)

—has been adequately described.[52] The latest design, Model E-11, rated at 100,000 amperes maximum, is being installed at Niagara Falls, New York (1960). The main features of the E-11 are shown in Figure 6-20 and data on the E-8 and E-11 are summarized in Table 6-21.

The electrolyzer consists of a steel plate to which hard rubber-covered steel channels are bolted to form sides. The ends are closed by inlet and outlet end boxes which are cast iron, lined with hard rubber. The elec-

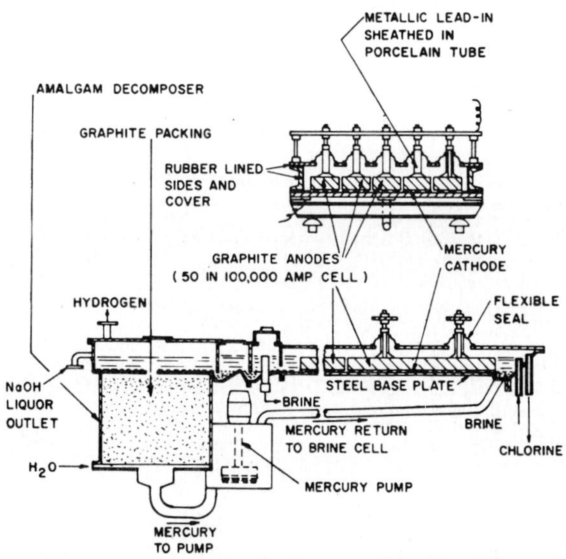

Figure 6-20.   Olin-Mathieson E-11 mercury cell.

trolyzer cover is a steel plate, covered with hard rubber on all surfaces exposed to chlorine. The cover rests on a soft rubber gasket positioned between it and the flanges of the side channels and end boxes; C-clamps are used to provide a gas-tight closure.

Fifty graphite anodes, each $4 \times 9 \times 48$ in., are supported from the distributing bus bars. Each anode has two metallic lead-in posts protected with porcelain sleeves. To promote the flow of chlorine away from the lower surface, slots are cut into the graphite and holes are drilled into each slot from above.

The decomposer is a vertical tower packed with lumps of broken graphite. The packing is held between screens compressed with jack screws. A distributor plate spreads the amalgam evenly over the top of the packing. The amalgam enters the top through a double seal to insure a minimum of chloride ion entering from the electrolyzer. Purified water enters the bottom of the decomposer below the packing and overflows as 50 percent sodium hydroxide above the packing. Sodium amalgam is decomposed in contact with the graphite packing and sodium hydroxide solution to form sodium hydroxide and hydrogen. Hydrogen is piped from a connection on the top of the decomposer. Mercury collects in a well in the bottom of the decomposer and flows to a sump-type centrifugal pump. This pump delivers mercury to the inlet end of the electrolyzer in a steel pipe at such a rate that 0.2 percent sodium is picked up by the mercury in passing through the electrolyzer.

Brine for the Mathieson process may contain calcium sulfate if the pH is adjusted to 2.5 by the addition of hydrochloric acid. The flow of concentrated brine to a cell is observed on a rotameter and is properly adjusted so that the brine leaving the cell contains 260 to 280 grams per liter of sodium chloride and the temperature does not exceed 85°C. Weak brine from the cells is dechlorinated, concentrated by contact with solid salt, treated with sodium hydroxide at pH 10, settled and filtered for return to the cells.

Mathieson mercury cell process U. S. patents include the following: 2,328,665; 2,334,354; 2,336,045; 2,423,351; 2,428,584; 2,511,466; 2,627,501; 2,732,284; 2,787,591; 2,837,408; 2,845,344; 2,872,393.

Solvay Mercury Cells

Solvay *et Cie*, Belgium, and Electrochemical Processes, Inc., USA, offer two models, the V-100 and the V-200, rated at 96,000 and 160,000 amperes respectively. These two models offer essentially the same features, the V-200 being nearly twice as long as the V-100. An excellent description of the V-200 was given in a paper by A. G. Basilevsky.[6] Solvay cells have been installed at three plants of Allied Chemical Corporation, the older S-60's at Syracuse, New York, and Moundsville, West Virginia and the V-100's at Brunswick, Georgia. The V-100's are operating at Rosignano, Italy (two levels); Linne-Herten, Netherlands (four levels); and Hallein, Austria (two levels); the V-200's are operating at Tavaux, France (four levels).

The main features of the V-200 are shown in Figure 6-21, as mounted in a multi-level cell house. Cells are mounted side by side with no aisle space between them. They are serviced from the end aisles, or from removable walkways positioned over the tops of the cells. The electrolyzer consists of a bare steel plate to which hard rubber covered members are attached to form side walls. There are 15 covers, of hard rubber-lined steel and each cover holds 12 anodes, 4 rows wide by 3 rows long. Each graphite anode —315 mm × 509 mm × variable thickness—is supported by one graphite

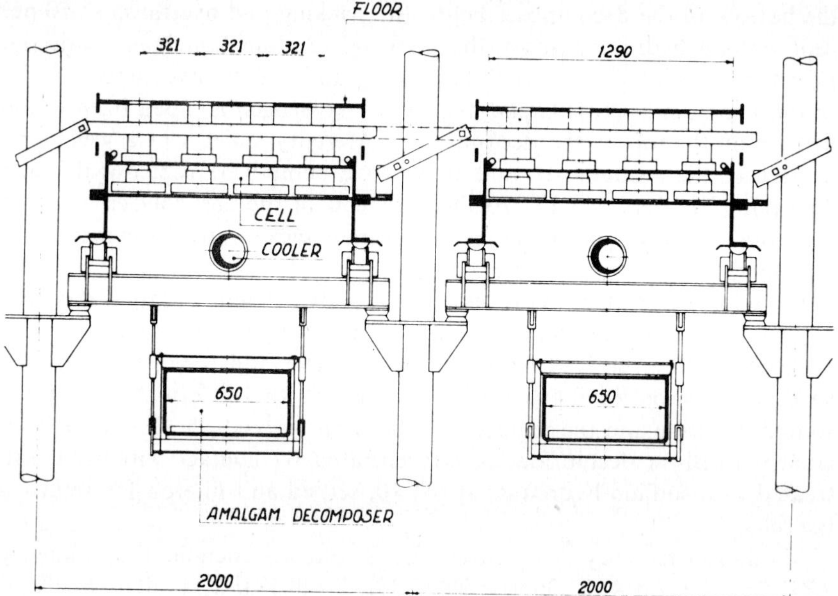

Figure 6-21.   Cross section of Solvay cell.

nipple 110 mm. in diameter, which fits tightly in a graphite ring bushing fastened to the cover; the bushing carrys the current to the positive bus. Anodes are adjusted by the operator, from the walkway above, by inserting a tool which engages the graphite nipple, and forces it down until the anode touches the mercury, after which the nipple is retracted a few millimeters (U. S. Patent 2,617,762).   A robot tool has been designed to assist the operator. Adjustments are usually made weekly, under full load, and require about 2 man hours per cell.

Covers are supported by and roll on tracks mounted on the level side walls of the electrolyzer.   Pneumatic seals are used to seal the covers to each other and to the cell walls.   At regular periods, usually of six to eight weeks, the cover at the mercury inlet end is lifted out, these anodes being thin. All covers are moved one space back, and another cover fitted with fresh anodes is dropped into place at the amalgam outlet end of the cell. At the same time, the cell bottom may be cleaned if necessary.   This operation, including making and breaking all seals and electrical connections, is said to take not more than two hours.   The operation is conducted while the cell is on full load, the chlorine being sucked from both ends into the weak gas headers.

No short circuiting switches are provided, although a cell may be safely short circuited in an emergency by repositioning the cross-over bus on the 15 inter-cell bus tabs. This takes two minutes at the most.

The decomposer is a rectangular steel duct mounted below the electrolyzer and protruding somewhat at each end. Graphite grids, which link to each other, can be fed in at the upper end and removed from the lower end. The grids can also be shaken in order to dislodge solid accumulations. Hydrogen passes out through a reflux condenser which returns condensed water and mercury to the decomposer. It is said that 73 percent NaOH can be made in this cell without application of additional heat. It is also claimed that mercury pumps can be changed in a few minutes under full load.

The cells operate at high current density (0.5 to 0.6 amps. per cm^2) and high temperature. Mercury is further economized by running a high sodium content in the amalgam (0.5 to 0.6 percent Na), which in turn requires a rather high gradient in the trough. These conditions necessitate exceptional brine purity and rigid specifications on graphite. The cell works best on continuous load with no power interruptions whatever. In spite of the high current density, the cell to cell voltage is not excessive because ohmic voltage drops in electrolyte, conductors and contacts have been minimized.

The Solvay mercury cells are covered by the following patents: U. S. Patents 2,550,231, 2,551,248, 2,617,762, 2,648,630, 2,704,743 and 2,757,076; Belgium Patent Application 437,658 and Belgium Patents 525,162, 525,535, 537,328 and 549,489.

Uhde Mercury Cells

Hoechst-Uhde Corporation offers four sizes of mercury cells rated according to cathode area at 5, 10, 20 and 30 square meters. Nominal cell loads are 25,000, 50,000, 100,000 and 150,000 amperes, respectively, but these cells will operate at substantially greater loads. Each size is available in two styles, i.e., with decomposer either alongside or underneath the electrolyzer. In the United States, Uhde 20 m^2 cells have been installed by Columbia-Southern Chemical Corporation at Natrium, West Virginia (1958) and by Hooker Chemical Corporation at Niagara Falls, New York (1961). The cell was developed in Germany, and various prototypes may be seen in the plants of Farbenfabriken Bayer A.G. and Farbwerke Hoechst A.G. Uhde 5 and 10 m^2 cells are in operation at Elworth,[34] Sandbach, England; Morón, Venezuela; and Suzan, Brazil. Uhde cells are also installed at Monthey, Switzerland; St. Gobain, France; and Oulu, Finland. The type of construction is shown in Figure 6-22.

The Uhde electrolyzer has a bare, accurately machined steel bottom of heavy, rigid welded construction, with rubber-covered detachable side walls and a rubber-lined cover. Each anode is supported by one paraffin treated graphite nipple that extends through an adjustable packed seal mounted on the cover. A device with a handwheel permits easy and accurate anode

Figure 6-22.   Cross section of Uhde cell.

adjustment under full load. In the Uhde 20 m² cell, there are 96 anodes, 3 rows wide by 32 rows long. The anodes are approximately 400 mm wide by 500 mm long by 80 to 100 mm thick, and they are slotted and drilled for chlorine relief. Bus bars are in parallel groups of 8 at the cell cross-overs and an 8-gang, spring loaded cut-out switch is provided for each cell. The switch can be manually operated, but automatically closes in case of interruption in mercury circulation.

The electrolyzer trough is provided with a two-compartment rubber-lined end box for washing of amalgam and manual skimming of graphite particles. A wash box is also provided at the mercury inlet end.

The decomposer is unlined and is provided with graphite grids. The mercury circulating pump is of the low-head, conical vortex type. Flow of water in the decomposer is normally counter-current, and pure caustic soda is produced in the range of 50 to 63 percent NaOH. Higher NaOH concentrations can be produced by using co-current water flow.

Uhde cells are customarily mounted with a working aisle between adjacent cells. Cross-over bus passes beneath such aisles.

Uhde cells attain minimum voltage drop through careful design and use of the anode adjusting device. Rugged construction and interchangeability of parts result in low maintenance cost.

Krebs—B.A.S.F. Cells

The Krebs *et Cie* engineering group (Paris, Zurich, and Brussels) have, since 1951, cooperated with Badische Anilin & Soda-Fabrik A.G. in the development and installation of mercury cells of various sizes and types. Since the B.A.S.F. originals have been described elsewhere, we take note of two new types, the Z-20 and Z-40, and the P-50, P-100 and P-125. The number following the letter is the nominal kiloampere rating. Various prototypes may be seen at the Ludwigshaven works of B.A.S.F., and the P-100's are being installed in a new 275 ton per day chlorine plant in

Russia. Krebs cells, B.A.S.F. cells, and various combinations are in operation in many countries, e.g., B.A.S.F. 40 kiloampere cells at Shawinigan Chemicals, Ltd., Canada; Krebs—B.A.S.F. cells at Montrose Mexicana S.A., Mexico and Kuhlmann S.A., France. The latest type of construction is shown in Figure 6-23.

The Krebs electrolyzer is a rubber lined steel trough formed of relatively light gauge steel plate. The trough, with its leveling screw, is mounted on a frame which is integral with the building structure. The bolted covers are rubber lined steel, the peripheral joint being a soft rubber gasket. Each graphite anode is supported by one graphite nipple, which passes thru a packed rubber lined seal in the cover. Each anode is provided with a regu-

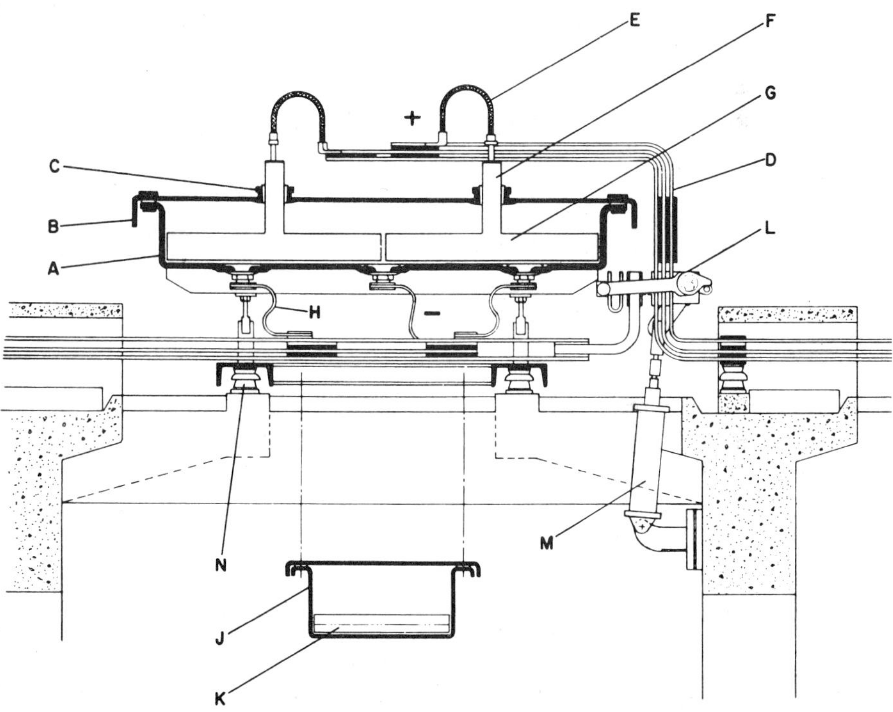

A. Rubber coated electrolytic compartment	G. Graphite anode block
B. Rubber coated cover	H. Cathode busbar connector
C. Anode support and gas seal	J. Horizontal trough type decomposer
D. Anode busbar	K. Graphite grids
E. Flexible anode cable	L. Cell shorting switch
F. Graphite anode post	M. Air operator for shorting switch
	N. Insulator

Figure 6-23.   Krebs B. A. S. F. cell (type P. 100-125).

lating device with high current transfer capacity. Anodes are arranged two abreast, and in the P-100, there are 120 anodes, 690 × 245 mm in all. The anodes are machined for chlorine relief, and the nominal current density is in the range of 0.4 to 0.5 amps/cm.²

Cathodic current is carried out through flat head steel contact plates inserted in the electrolyzer trough, the upper surface being slightly depressed below the surface of the rubber so that the steel always remains covered with mercury. Inter-cell copper bus is now carried across horizontally with the shortest possible runs. Each cell can be short circuited, and the P-100, for example, has four double gang switches which are air operated.

The Krebs decomposer is a steel trough, generally rubber-lined, mounted under the electrolyzer, and it is provided with slotted graphite grids. An iron-free caustic is thus made. The mercury circulating pump, of the centrifugal type, is mounted in a sump at the lower end of the decomposer. Mercury level and flow rate is automatically regulated.

The principal feature of the Krebs B.A.S.F. cell is the hard rubber lining of both electrolyzer and decomposer, which gives the maximum protection against iron pick-up, a potential source of trouble in mercury cell operation. The life of the rubber is said to be satisfactory and repairs are not difficult.

De Nora Horizontal Cells

Oronzio De Nora, Impianti Elettrochimici, of Milano, Italy, offer horizontal type mercury cells in five basic sizes, known as the 14 × 2, 14 × 3, 14 × 4, 14 × 5, and 18 × 6, rated at 40, 60, 80, 100, and 150 kiloamperes respectively. The present De Nora cell has been evolved through a series of changes since it was first introduced in America in 1950. Basically, however, the principle is unchanged—all De Nora cells have had a protected electrolyzer trough and a vertical decomposer. In the United States, De Nora cell installations are located at Muscle Shoals, Alabama; Plymouth, North Carolina; Calvert City, Kentucky; Longview, Washington and Deer Park, Texas. There are at least 60 De Nora plants located in 25 different countries, ranging in capacity from a few to several hundred tons of chlorine per day. The latest type of construction is shown in Figure 6-24.

The De Nora electrolyzer is a welded steel trough with a smooth concrete bottom and with granite stone side walls, also grouted in concrete. Cathode contacts are steel T bars welded to the steel trough and they run longitudinally down the cell. The tops of the T's are slightly depressed below the surface of the concrete. The concrete bottom is surfaced with a special plastic, which is exceptionally resistant to both chlorinated brine and sodium amalgam. The plastic adheres well to the concrete and, in case of damage, is easily patched.

Graphite anodes are suspended from a semirigid steel frame which is mounted above the cell on eight supporting steel studs. The cell code number refers to the arrangement of the anodes, e.g., the 18 x 6 means there are 6 anodes abreast and 18 rows, or 108 anodes total. These anodes are 31 x 13½ x 2¾ in. and each is supported by two 4-in. graphite nipples. The nipples in turn are supported by copper rods which are hung securely from the frame.

The cell cover is a flexible duplex sheet of rubber, fastened at the periphery to the cell flange and supported by the tops of the graphite anode nipples. The joint where the copper lead is screwed into the graphite nipple is closed by a washer. To adjust the anode-cathode gap, the supporting frame is warped down on two opposite studs at a time until the anodes touch the mercury; then it is backed off a few millimeters. The whole adjustment operation takes about 10 minutes per cell. The rubber cover is, of course, not permanent, but lasts as long as one set of graphite, after which it is discarded.

The De Nora electrolyzer is provided with a wash compartment at each end. At the amalgam end there is a quick-flushing device which is used to remove powdered graphite or thick mercury, which always accumulates behind the lower mercury seal in any mercury cell.

The De Nora decomposer is a steel tower in which baskets of hard graphite packing are fitted. Single stage decomposers are most common, but two stages are provided if caustic stronger than 60 percent is desired. The decomposer is steam jacketed, although steam is only used if the activity is not up to par. Mercury flows out of the decomposer into a sump tank which is large enough to hold all the mercury in the system should the pump be stopped. A centrifugal pump returns the mercury to the upper end of the electrolyzer. A device is provided to give signals in case of pump failure, or to close the short circuiting switch, if desired.

Cells are mounted side by side with no aisles between them. Operators can walk on the cell flanges or on a catwalk mounted on the flanges. Cell to cell copper is spaced the same as the anodes and takes the shortest possible route. The short circuiting switch on the 18 x 6 therefore has 18 switches ganged together, and is air operated.

The older De Nora cells, with stone bottoms and rigid covers, have in many cases been revamped to include the newer features just described. The new cells are particularly easy to operate and can be abused in situations where interruptions and power outages are forced on the chlorine plant operators by circumstances outside their control.

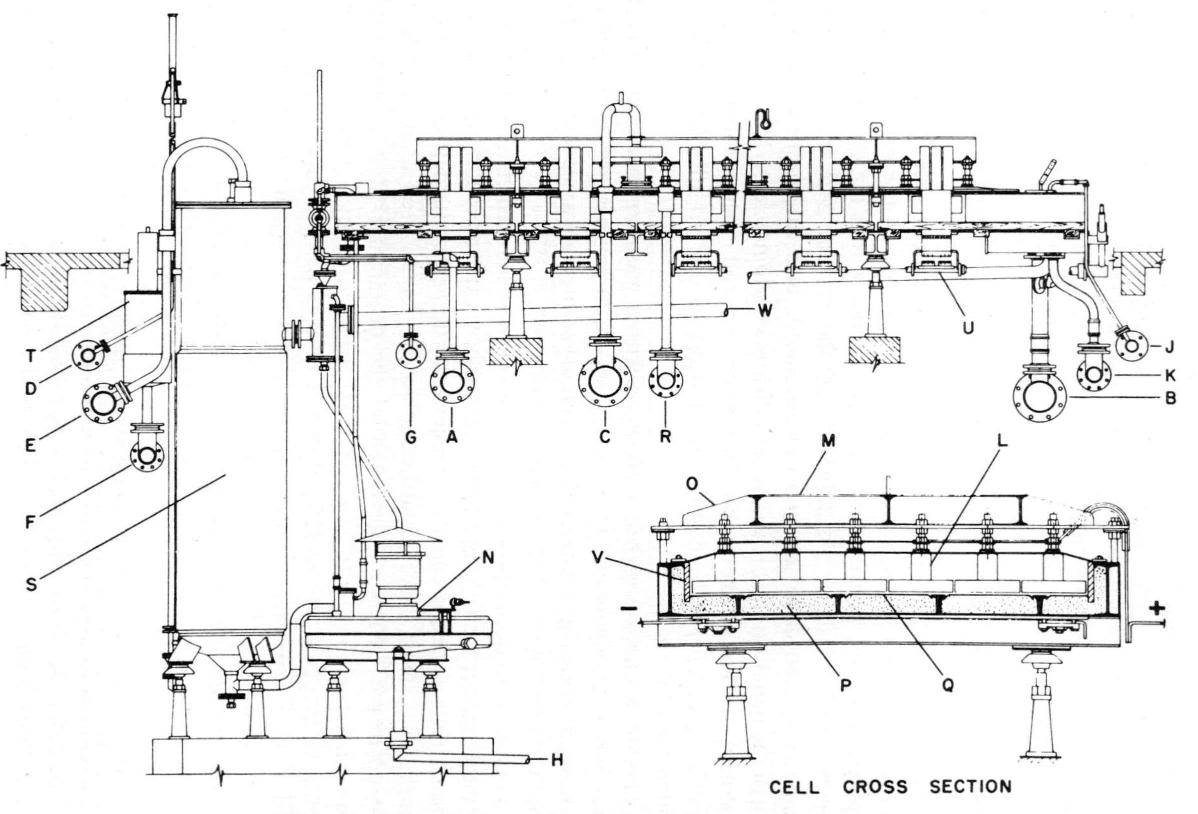

Figure 6-24. De Nora 18 × 6 amalgam type chlorine cell.

De Nora has designed and built the prototype of an amalgam cell, called the ODN, rated at 400 to 500 kiloamperes. For chlorine plants with a capacity in excess of 400 tons per day, this design offers promise of greater economy in capital and operating costs.

De Nora Vertical Cell

Oronzio De Nora, Impianti Elettrochimici, of Milano, Italy, now offers a commercial version of the long-sought vertical cathode mercury cell. It is known as the De Nora Fluent Amalgam cell, or FAC-24 and has a rated capacity of 60 kiloamperes. Although no commercial installations exist as of 1960, prototypes have been thoroughly tested.

The FAC consists of a plurality of alternating rectangular anode and cathode compartments, mounted vertically and bolted together into an assembly as shown in Figure 6-25. Each anode compartment is a rubber-lined steel frame, in which is mounted a "permanent" metallic anode and a plastic fabric diaphragm, or baffle. Positive current leads are through the top of this frame. Feed brine enters the frame at a lower corner and depleted brine and chlorine gas leave the frame at the opposite upper corner. These passages are connected to common corner ducts as in a filter press. A passage is also provided at the bottom of each frame, through which the amalgam flows to a collecting box and thence to the decomposer.

Each cathode compartment, on the other hand, is formed by a U-shaped spacer, provided with passages for the flow of fresh and spent brine and amalgam which match those of the anode compartments. Cathodes are amalgamated steel wire screens. Mercury is distributed uniformly along the top of these screens through slotted copper clad steel headers. The equipment is designed in such a manner as to keep an amalgam pool at the bottom of the cathode compartments in order to allow contact with the cathode screens and to prevent the escape of brine into the amalgam outlet.

A. Pure brine feed header	O. Rubber cover
B. Depleted brine collection header	P. Concrete grout.
C. Strong chlorine header	Q. Corrosion resistant membrane
D. Pure water feed header to decomposer	R. Diluted chlorine header
E. Hydrogen collection header	S. Decomposer
F. Caustic collection header	T. Current breaker
G. Pure water feed header to inlet end	U. Cell shorting switch
H. To mercury trap and sewer	V. Stone side facing
L. Anode post and plate assembly	W. Amalgam return
M. Anode support structure	Y. Quick flushing device
N. Mercury pump and sump	

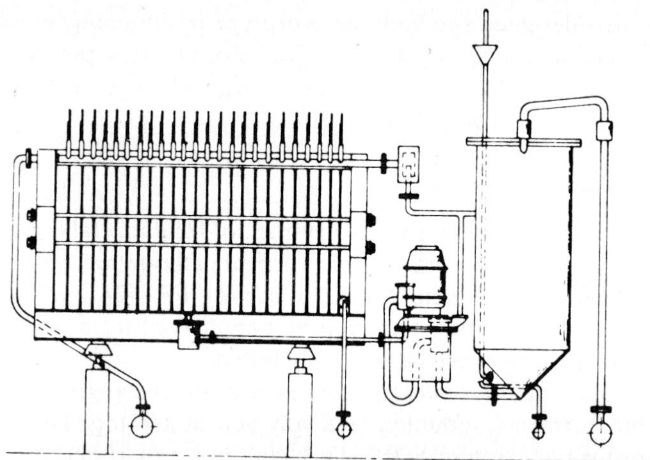

Figure 6-25.   De Nora fluent amalgam cell FAC-24.

Amalgam flows by gravity from the cells to the sump of the mercury pump where it is mixed with the sodium-free mercury coming from the decomposer. The mixture is then pumped partly to the cell and partly to the decomposer. The distribution of mercury is made in such a way that the mercury flowing back to the screen cathodes contains about 0.01 percent Na.

The "permanent" anodes used in the FAC cell are platinized titanium treated in a special way to prevent amalgamation of stray mercury with the platinum. Electrolyzer disassembly has been greatly simplified, and cell outage time is estimated at two days per year. The decomposer used is similar to that used on De Nora's horizontal cells. The principal advantage of the fluent amalgam cell is the economy of building space as compared to that of horizontal mercury cells.

Miscellaneous Types

There are many chlor-alkali plants scattered around the world that employ mercury cells of various types, designed, developed or improvised by local genius, and not widely known. Some radically different types of mercury cells have come to light, chiefly through patent literature, and it is quite possible that one or another of them may become a contender for commercial applications in the future. On the basis of intrinsic merit or novelty, the following are noteworthy:

Vertical Rotating Cathode Cell with Adjustable Gap. This cell, with an adjustable anode-cathode gap, is an attempt by Japanese electrochemists to overcome the shortcomings of the Honsberg cell.[60]   A 20,000 ampere prototype has been tested and is reported to be successful.   Adjustment can be made under full load.

Horizontal Forced-Flow Cathode Cell. This is also known as the "slot-type" mercury cell, in which the mercury and brine are forced through the restricted anode-cathode gap at high velocity.[19]   For an American version of the slot-type cell, it is stated that if the brine velocity corresponds to a Reynold's number in the range of 8,000 to 20,000, current density of the order of 1.26 amps/cm^2 is permissible without any sacrifice of current efficiency[43]—even when unpurified brine is used as the electrolyte.   A slot-type cell fitted with platinized titanium anodes is described.[28]

Horizontal Rotating Cathode Mercury Cell. A German arrangement of a multi-deck cell, combined with a centrally located decomposer, has as its chief advantage compactness and economy of floor space.[39]   Far more interesting is a current Japanese development which achieves forced flow by means of centrifugal force.[15]   The cell may be single or multi-deck and the decomposing chamber is located beneath each deck.   Using a current density of 1.0 amp/cm^2 and impure brine, current efficiencies in the range 96 to 98 percent were achieved.   A 25,000 ampere prototype with a single disc, 1.8 m diameter has been tested.   The mercury inventory was only 250 kg.

Vertical Jet Cathode Mercury Cell.   This is also a Japanese innovation.[40,41]   Cathode current densities of the order of 10.0 amps/cm^2 are possible and impure brine may be used.

References

1. Åkerlöf, G., Kegelis, G., *J. Am. Chem. Soc.,* **62,** 620 (1940).
2. Angel, G., and Lundén, T., *J. Electrochem. Soc.,* **99,** 435, 442 (1952); **100,** 39 (1953).
3. Angel, G., Lundén, T., Dahlerus, S., and Brännland, R., *J. Electrochem. Soc.,* **102,** 124 (1955).
4. Angel, G., Brännland, R., and Dahlerus, S., *J. Electrochem. Soc.* **102,** 246 (1955); **104,** 167 (1957).
5. Barr, Lars, *J. Electrochem. Soc.* **101,** 497 (1954).
6. Basilevsky, A. G., Paper presented at Electrochem. Soc. Meeting, May 1959.
7. Billiter, J., "Elektrolyse der Nichmetalle," Wien-Springer Verlag (1954).
8. Bockris, J. O'M., and Potter, E. C., *J. Electrochem. Soc.* **99,** 169 (1952).
9. Conway, B. F., "Electrochemical Data," Elsevier Publ. Co. (1952).
10. Cotton, J. B., *Platinum Metals Review,* No. 2, 45 (1958).
11. Czernotzky, A., *Acta Chim. Acad. Sci. Hung.,* **18,** 167 (1959).
12. De La Rue, R. M., and Tobias, C. W., *J. Electrochem. Soc.,* **106,** 827 (1959).
13. De Nora, V., Paper presented at Electrochem. Soc. Meeting, May, 1959.
14. Ehlers, N. J., and Hampel, C. A., Paper presented at Electrochem. Soc. Meeting, May 1960.

15 Fujioka, S., and Yoshida, S., U. S. Patent 2,916,425 (1959).

16. Gardiner, W. C., *Chem. Engr.*, (November 1947).

17. Gardiner, W. C., and Sakowski, W. J., Presented at Electrochem. Soc. Meeting, May 1959.

18. Grosselfinger, F. B., Private Communication (1959).

19. Heller, and Saunders: Canadian Patent 476,519.

20. Hine, F., Yoshizawa, S., and Okada, S., *J. Electrochem. Soc. Japan,* **24,** 370 (1956).

21. Hine, F., Okada, M., Yoshizawa, S., and Okada, S., *J. Electrochem. Soc. Japan,* **27,** 134 (1959).

22. Hine, F., and Yoshizawa, S., *J. Electrochem. Soc. Japan,* **27,** No. 7-9, E176 (1959).

23. Hine, F., Okada, M., and Yoshizawa, S., *J. Electrochem. Soc. Japan,* **27,** 419 (1959).

24. Hine, F., Yoshizawa, S., and Okada, S., *J. Electrochem. Soc.* preprint, May 1960 Meeting.

25. Hirsh, B. W., Canadian Patent 465,365 (1950).

26. Hopper, C. M., and Richards, R. B., U. S. Patent 2,876,182 (1959).

27. Imperial Chemical Industries, Ltd., Belgium Patent 573,978 (1958).

28. Inoue, T., and Sugino, K., *J. Electrochem. Soc. Japan,* **27,** 145 (1959).

29. Johnson, Neal J., *J. Electrochem. Soc.,* **86,** 127 (1944).

30. Kircher, M. S., U. S. Patent 2,762,765 (1956).

31. Knobel, Max, "Intern'l Critical Tables," Vol. 6, p. 339.

32. Ksenzhek, O. S., and Stender, V. V., *Trudy Chetvertovo Soveschaniya Elektrokim* (Moscow), 823 (1956).

33. Latimer, Wm., "Oxidation Potentials," 2nd Ed. Princeton, N. J., Prentice-Hall Inc. (1952).

34. Lewis, G., *Ind. Chemist,* December 1956, p. 535; and *Chem. Age,* December 1956, p. 359.

35. MacMullin, R. B., Fiat Final Report No. 732 (1946).

36. MacMullin, R. B., *Chem. Ind.,* July, 1947.

37. MacMullin, R. B., *Chem. Eng. Prog.,* **46,** 440 (1950).

38. MacMullin, R. B., "Trends in Chlor-Alkali Industry," Paper presented at AIChE Meeting, Baltimore, Md., 1957.

39. Messner, E., West German Patent 1,020,965 (1957).

40. Mizuno, S., Japan Patents 1315 (1956), 10,061 (1956).

41. Mizuno, S., Toshima, S., *J. Electrochem. Soc. Japan,* **27,** 391 (1959).

42. Murray, R. L., and Kircher, M. S., *J. Electrochem. Soc.,* **86,** 83 (1944).

43. Neipert, M. P., Blue, R. D., and Houser, H. E., U. S. Patent 2,898,284 (1959).

44. Okada, S., and Yoshizawa, S., *Memoirs Faculty Engr., Kyoto Univ.* XIV, No. 4, (October 1952).

45. Okada, S., and Yoshizawa, S.: Memoirs Faculty Engr., Kyoto Univ., 1954.

46. Okada, S., Yoshizawa, S., Hine, F., and Takehara, Z., *J. Electrochem. Soc. Japan,* **26,** 165 (1958).

47. Okada, S., Yoshizawa, S., Hine, F., and Takehara, Z., *J. Electrochem. Soc. Japan,* **26,** 211 (1958).

48. Opferman and Hockberger, "Technik u. Praxis der Papierfabrikation," Vol. 3, p. 63 (1935).

49. Philippov, T. S., and Drozin, N. N., *J. Appl. Chem. USSR,* **21,** 630 (1948).

50. Potter, Chas., and Bisio, G. L., *J. Electrochem. Soc.,* **101,** 158 (1954).

51. Richards, R. B., *Chem. Engr.,* (March 1952).

52. Sanders, H. J., Gardiner, W. C., and Wood, J. L., *Ind. Engr. Chem.,* **45,** 1824 (1953).

53. Sheludyakov, L. N., Saltovskaya, L. A., and Stender, V. V., *J. Appl. Chem. USSR,* **26,** 160 (1953).

54. Sommers, H. A., *Chem. Eng. Progr.,* **53,** 409, 596 (1957).

55. Stender, W. W., Stroganov, M. M., and Zhivotinsky, *Trans. Electrochem. Soc.,* **65,** 198 (1934); **68,** 521 (1935).

55a. Stokes and Robinson, *Ind. Eng. Chem.,* **41,** 2036 (1949).

56. Sugino, T., and Aoki, K., *J. Electrochem. Soc. Japan,* **27,** 26 (1959).

57. Thomas, L. R., *Ind. Chem.,* 203, (May 1954).

58. U. S. Chlorine Industry Team, FIAT Final Report No. 816, (1946), PB 33221.

59. U. S. Chlorine Industry Team, FIAT Final Report 824 (1946).

60. Vaaler, L. E., *J. Electrochem. Soc.,* **107,** 691 (1960).

61. Walde, H., West German Patent 1,022,194 (1958).

62. Zundel, M., and Hartmann, P., Paper Presented at Montreal (March 4, 1959).

63. Suzuki, O., Ikeda, A., and Abe, S., *J. Electrochem. Soc., Japan,* **27,** 121 (1959).

64. Yoshizawa, S., and Nishida, M., *J. Electrochem. Soc. Japan,* **52,** 184, 234 (1949).

65. Wranglen, Gosta, *Tek. Tid.,* (May 13, 1960).

66. Yeager, Ernest, Technical Report 14, Contract Nonr 2391(00), Project NR 359-277, Office of Naval Research, December 1960.

7. ELECTROLYSIS OF HYDROCHLORIC ACID SOLUTIONS

F. M. BERKEY

Monsanto Chemical Company

INTRODUCTION

Hydrogen chloride and hydrochloric acid are produced by one of four rather basic processes:

(1) Salt and sulfuric acid in Mannheim and other types of furnaces.

$$NaCl + H_2SO_4 \rightarrow HCl + NaHSO_4$$

$$NaCl + NaHSO_4 \rightarrow HCl + Na_2SO_4$$

(2) Burning chlorine in slight excess of hydrogen.

$$H_2 + Cl_2 \rightarrow 2\,HCl$$

(3) Hargreaves process.

$$4\,NaCl + 2SO_2 + O_2 + H_2O \rightarrow 2\,Na_2SO_4 + 4\,HCl$$

(4) By-product of organic syntheses.

$$C_6H_6 + Cl_2 \rightarrow C_6H_5Cl + HCl$$

$$C_6H_5Cl + Cl_2 \rightarrow C_6H_4Cl_2 + HCl$$

As practiced commercially, most of these processes produce hydrochloric acid solution. Demands for anhydrous HCl are usually supplied today by Process (2). Process (4) results in substantially dry HCl gas that is contaminated with organics; the common practice is to cool it to remove organics and then absorb it as a commercial grade of approximately 32 percent HCl.

The rapid growth of the chlorine industry has been due largely to demand for chlorine for organic chlorinations, with the resultant production of by-product HCl.

Table 7-1 shows a marked decrease in production from salt and sulfuric acid, a modest increase from burning hydrogen and chlorine and a very marked increase in by-product HCl. Figure 7-1 shows these same trends graphically and more impressively. From available information, one must conclude that this trend will continue or even be accelerated.

Processes that utilize HCl have not been developed fast enough to keep up with the supply. Shipments cannot be made economically over long distances because of freight charges on a material that is roughly two-thirds water. This situation makes it desirable to consider means of producing chlorine from HCl.

TABLE 7-1. PRODUCTION OF HYDROCHLORIC ACID
THOUSANDS OF TONS—100% HCl

Year	From salt and sulfuric acid	From chlorine and hydrogen	By-product and other	Total
1950	165.1	111.6	342.2	618.8
1951	174.99	123.0	397.5	695.6
1952	162.4	140.6	380.7	683.7
1953	180.0	152.9	440.5	773.5
1954	150.4	157.8	455.2	763.4
1955	161.1	160.0	517.1	838.2
1956	149.5	181.6	575.3	906.3
1957	138.9	182.4	626.3	947.7
1958	107.0	159.1	582.3	848.5
1959	100.0	184.3	751.6	1036.0

Source: Bureau of Census Industry Division

HISTORY OF HCl ELECTROLYSIS PROCESSES

Until recent years, the electrolysis of hydrochloric acid had not been studied very diligently. Only one type of cell has been developed that can be designated as a true HCl electrolysis unit. This cell is referred to as the Bitterfield type (a German development) or the De Nora type (an Italian development).

There are at least two other developments which produce chlorine from HCl indirectly by electrolysis of metallic chlorides produced by treating metals with HCl. In 1930 Westvaco suggested equipment for the electrolytic recovery of chlorine from hydrochloric acid;[3] in 1950 Roberts of Westvaco reported chlorine recovery from HCl by electrolysis of copper chloride.[4] In 1958 Schroeder of Seattle University proposed a process for electrolytic recovery of Chlorine from HCl by electrolysis of nickel chloride.[5] However neither has been used in a commercial installation. These processes may be identified as:

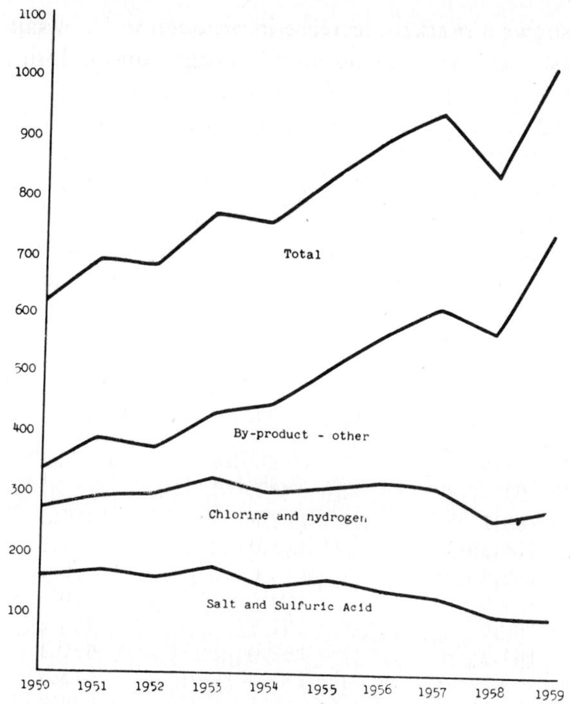

Source: Bureau of Census Industry Division

Figure 7-1. Hydrochloric acid production (thousands of short tons).

(1) Westvaco process for electrolysis of copper chloride.

(2) Schroeder process for electrolysis of nickel chloride.

Although chlorine recovery processes have not yet found wide commercial application, it is desirable to include development work in any discussion of HCl electrolysis, even though it has not reached the commercial stage. The Westvaco and Schroeder work fall in this area, being types of indirect electrolysis for the recovery of chlorine from HCl; both of these processes will be discussed later in the chapter.

Direct Electrolysis of HCl

Development of this process was conducted initially at Bitterfield in the I. G. Farben Industrie plant. This work was covered by German patent application J 7335, dated October 15, 1942, and made by Messner and Hoelmann. It was not published because of conditions following the last world war and it was only after the war that the development work by

G. Messner[6] and his collaborators became known and was taken up by De Nora in 1956,[7] resulting in the realization of a technically sound process and an electrolyzer of simple, rugged construction.

This first HCl electrolysis cell was constructed with vertical bipolar electrodes and with anodes of lump graphite. Extensive study was made of diaphragm materials and resulted in a diaphragm which gave relatively long life in resisting the attack of muriatic acid and free chlorine at temperatures of 70 to 80° C.

The details of manufacture of the diaphragm have never been published. I. G. Farben Industrie operated three technical-size HCl electrolyzers. The last and most improved electrolyzer contained thirty bipolar elements and was operated continuously for fifteen months. The HCl feed acid contained 2 to 8 percent H_2SO_4, and graphite consumption was approximately 0.8 kg/ton chlorine. This early work indicated some advantages from the electrolytic standpoint of improved conductivity when feed acid contained H_2SO_4.

After World War II a team of representatives of the U. S. Chemical Industry visited this installation at Wolfen, and Fiat Final Report, No. 832, dated June 19, 1946, covers their observations. This report indicated that much progress had been made in developing a practical unit. The essential problem was one of improving materials of construction.

Oronzio De Nora, Impianti Elettrochimici, of Milan, Italy, had successfully developed its design of water electrolyzers and the only basic problem which had to be solved in order to apply this know-how to HCl electrolyzers was, again, that of materials of construction. This study was started in 1950 and sufficient progress was made to justify a more active program which was initiated in 1956 under the direction of Dr. G. Messner, one of the original developers of the first units. An experimental cell was built in the De Nora plant in Milan to further the studies. The first commercial installation in the United States was made in 1958 for Monsanto Chemical Company at Anniston, Alabama.

How extensive a study of the process was made by others following publication of the Fiat Report, No. 832, is not easy to establish, but work was done by Dow Chemical Company, Midland, Michigan, and the result of this investigation is summarized in the following statement: "In 1950 The Dow Chemical Company built four experimental cells about 25 percent larger than those described in Fiat Report, No. 832, entitled 'Hydrochloric Acid Electrolysis at Wolfen.' These cells were operated for three months on an experimental basis without difficulty and substantiated the data presented in the Fiat Report. If such cells were needed, the Fiat Report would be a satisfactory source of estimate."

DE NORA PROCESS OF HYDROCHLORIC ACID ELECTROLYSIS

The detailed discussion of the De Nora Process in the following pages represents the development of the process and equipment as of 1960. The basic principle probably will not change but, as in the case of all new commercial applications of a process, there will surely be many improvements as new materials of construction become available and more operating experience is accumulated.

As mentioned, in a number of chemical processes, i.e., the chlorination of organic materials, hydrogen chloride is obtained as a by-product. However, possibilities of utilizing this by-product at the site are often quite limited so that there is a disposal problem due to the high cost of neutralization if stream pollution is to be avoided.

Therefore, a process that allows the chlorine value to be recovered from the by-product HCl is of special interest because with such a process any problem concerning neutralization or stream pollution disappears and the chlorine can be used again in the process by which hydrochloric acid is produced.

Actually, regeneration of chlorine from hydrogen chloride may be achieved by four methods:

(1) Oxidation of HCl by oxygen in the presence of a catalyst, e.g., the Deacon Process using a liquid catalyst, studied by I. G. Farben, Industrie Oppau.
(2) Oxidation of HCl by NO_2.
(3) Electrolysis of HCl with production of chlorine and hydrogen.
(4) Electrolysis of metallic chloride solutions.

However, of these four methods, the third is to be preferred, in general, because of the simplicity of equipment and operation as well as the relatively low installation cost. Moreover, this electrolysis process is characterized by great flexibility. Whenever necessary, the load on the electrolyzer can be varied over a wide range to meet the fluctuation in the quantity of HCl available.

De Nora Electrolyzers

Many laboratory and pilot plant tests have been carried out by De Nora to achieve a satisfactory unit design of commercial size.

The De Nora electrolysis unit (Figure 7-2), specially designed for hydrochloric acid, is of the filter-press type. It is composed of a plurality of single elements, i.e., the electrolytic cells proper.

All De Nora cells have vertical electrodes and vertical diaphragms. Both anodes and cathodes consist of graphite as shown in Figure 7-3 in vertical and horizontal cross sections. Each graphite plate acts as anode on one side and as cathode on the other side.

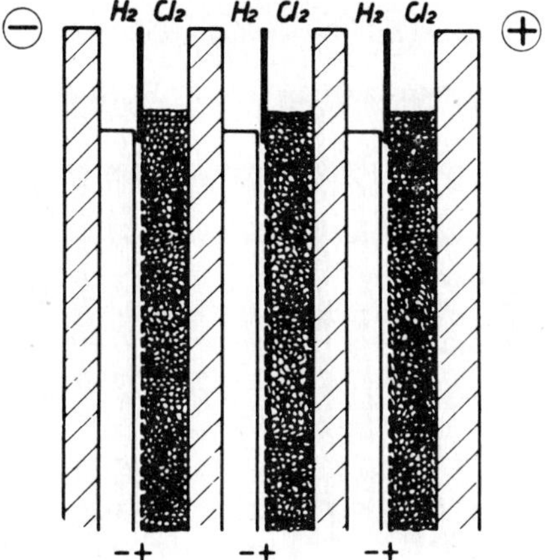

Figure 7-2.   HCl-cell schematic, vertical
section.

The single elements are assembled together (Figure 7-4) between two
steel end-plates[1] which are clamped against the series of cell frames[4] by
means of two capstan screws operated by the hand-wheels.[9] Each unit may
consist of 40 electrolytic cells, or more, according to the capacity required.

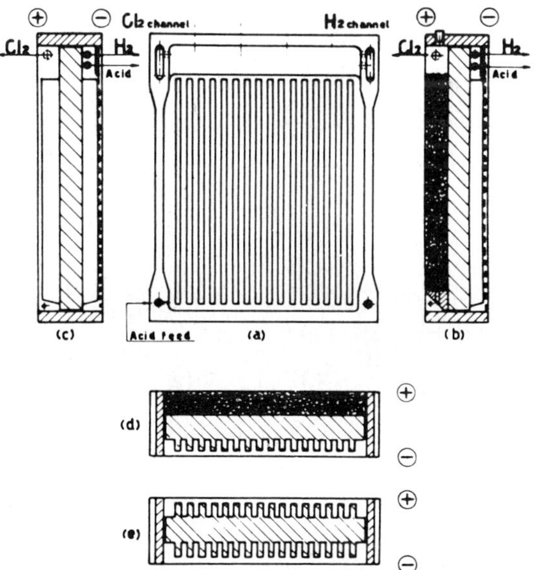

Figure 7-3.   Bipolar cell element from the cathode
side and vertical section.

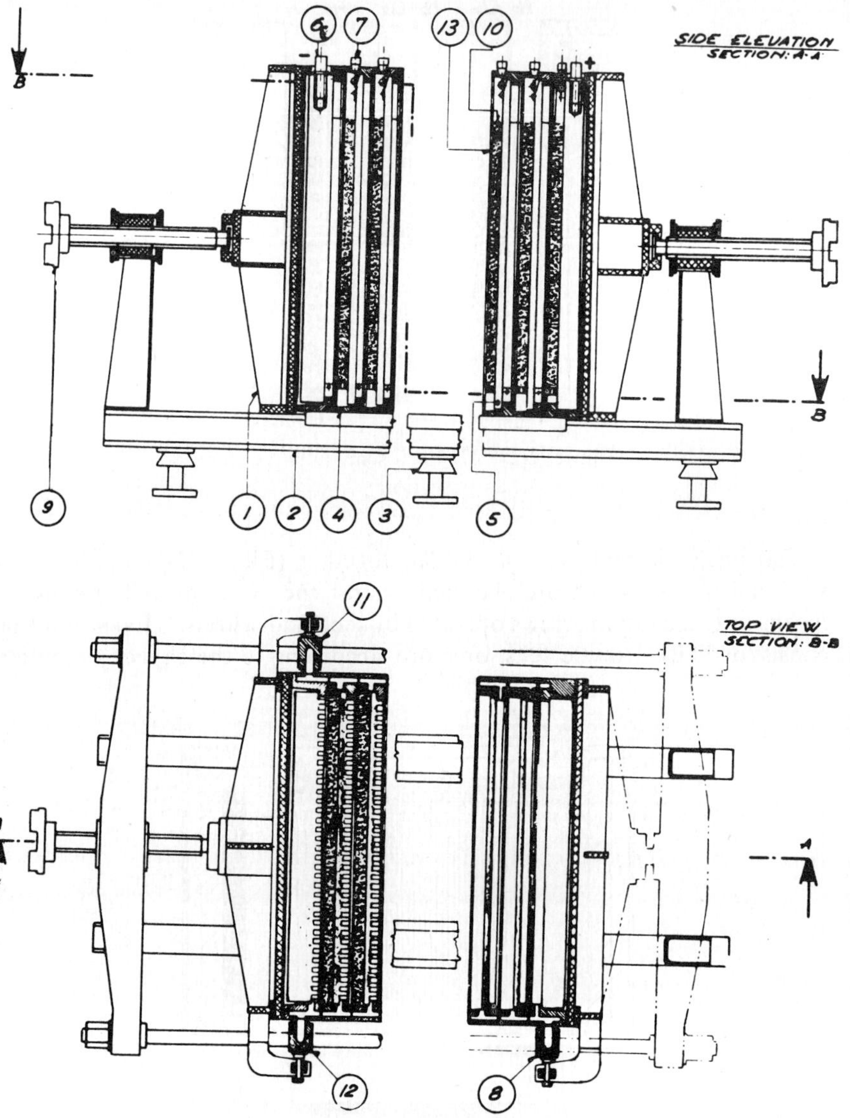

Figure 7-4.   Assembly of elements in electrolyzer.

The standard assembly is composed of 40 cells. The assembly rests on two I-beams,[2] which are covered with acid-resisting material, supported by porcelain insulators[3] and installed over a concrete tank with acid resistant lining. Design as well as construction are such that assembly and disassembly for maintenance work can be carried out quickly and simply.

Each cell in the assembly consists of a frame[4] of acid-proof synthetic material. Each frame holds a graphite plate[5] cemented to it. All of the graphite plates, with the exception of the two end ones, act as bipolar electrodes. This means that *each* plate performs as an anode on one side and as cathode on the other side; hence, external connections are required only at the two ends of the cell assembly. The D. C. voltage to be applied across the electrolyzer is equal to the voltage across each cell times the number of cells in the assembly. Assuming a nearly constant outlet acid containing 18 percent HCl, the cell voltage depends mainly upon the load. In the case of a load of 1500 amperes, the single cell voltage is 2.3 volts, corresponding to a DC power consumption of 1900 KWH/1000 kg, equivalent of 1750 KWH/2000 lb of chlorine. The end-plates are provided on top with a set of graphite stems,[6] which are screwed into the plates and connected to the electric leads (bus bars).

Each electrode plate[5] has on its cathodic side a plurality of parallel grooves from bottom to top, while on the anodic side the plate surface is smooth. An acid-resistant diaphragm[13] is cemented on the cathodic side of each element-frame, in such a way as to cover the cathodic face of the electrode plate. The free space left between the diaphragm and the opposite anodic face of the next plate is filled with graphite particles charged from the top through filling holes which are normally provided with stoppers.[7]

Anodes

The anodic side of the graphite plate does not function directly as anode, but rather the layer of graphite lumps in front of it do this.

Experience has shown that electrochemical attack on the bipolar graphite plate itself is negligible. Complete discharge of the Cl-ions takes place on the lump graphite, i.e., at the surface of the layer of graphite lumps closest to the cathode. This is because Cl-ion concentration in the electrolyte is very high, approximately double that in the brine.

Figure 7-5 shows on the left side pieces of graphite which have not yet been used in the HCl cell; these pieces have been cut from a graphite rod and the saw tooth marks are visible. At the right side there are graphite pieces which have been manufactured in the same manner but which have already functioned as anodic mass in a hydrochloric acid cell. It can be seen that the anodic attack has taken place almost entirely on the surfaces directly facing the cathode, even though the side surfaces are also in the electrolyte and their distance to the cathode is only a few millimeters more than that of the anodically attacked fron surfaces. This demonstrates that the layer of graphite lumps is sufficient to protect the bipolar graphite plate of the electrode against anodic attack.

Figure 7-5. Used and unused pieces of cut graphite.

In practical application cheaper scrap graphite pieces are used. The cut pieces illustrated were chosen and used merely to show the anodic attack more clearly.

This type of anode offers five advantages:

(1) Graphite lumps are the cheapest kind of electrode graphite.

(2) No time-consuming dismantling is necessary to renew this type of anodic mass as it is simply replaced through openings provided in the frames.

(3) As the graphite lumps slide down continuously because of their own weight, they automatically keep a continuous minimum distance between the electrodes and thereby a constant minimum cell voltage.

(4) The spaces between the loose graphite lumps serve as inlet passages for the feed hydrochloric acid. The acid is fed into each anodic space at the bottom of the layer of graphite lumps through holes in a grill and it flows up through the spaces, thus covering the complete anode surface uniformly.

(5) The spaces between the graphite lumps allow the chlorine bubbles formed at the anodic front to move upward into the chlorine collecting on top of the graphite lumps. The direction of movement of the chlorine bubbles and of the fresh acid is the same, i.e., from the bottom upwards so that the movement of gas and liquid do not conflict but rather assist one another.

The unhindered movement of chlorine gas and fresh acid which is promoted by this type of anode is very important for the operation of the cell because:

(a) On the one hand, any stoppage in the free flow of the chlorine gas would decrease the effective electrolysis-surface and would thereby increase the effective current density and cell voltage.

(b) On the other hand a blocked and thereby uneven supply of fresh acid to the anode surface would result in local depletion of HCl and *hence to an* increased graphite attack. In cases of very high local depletion, e.g., below 5 percent HCl in the electrolyte, even the bipolar graphite plates may be attacked.

Another type of graphite anode is available for the electrolysis of pure hydrochloric acid in which, besides the Cl ion, there are present practically no other ions such as SO_4 and PO_4; thus the anodic attack is kept very low. In this case cells can be supplied with bipolar elements that have ribs on both sides, as shown by the schematic drawings (c) and (e) of Figure 7-3. The movement of acid and chlorine gas is excellent, as it is in the case of graphite lumps.

In principle, anodes of platinized titanium could also be used in the electrolysis cell, but at present the price of this new type of electrode is still very high.

Cathodes

In Figure 7-2 there are "graphite ribs" on the cathodic sides of the bipolar graphite plates which protrude and on which the hydrogen is formed. The ribs are placed vertically in the electrolyte space and occupy about 50 percent of the cathode space. The spaces between the graphite ribs permit the hydrogen bubbles to rise and also let the electrolyte flow upwards to the overflow.

Figure 7-3 shows a view of the front of the cathode. At the upper end of the cathode the graphite ribs have been eliminated because at this point hydrogen gas and depleted acid are collected and have to move horizontally to the hydrogen outlet or to the depleted acid overflow.

Hydrogen is formed on the graphite under overvoltage conditions which depend not only on the temperature but also on the current density. It has been found that the overvoltage of hydrogen on graphite can be lowered by covering the cathode surface with metal, e.g., antimony, iron, nickel, copper, silver or platinum. However, because of the high operating temperature of 90 to 95° C this overvoltage is lowered so much that such means do not seem economically worthwhile.

Diaphragms

For separation of hydrogen and chlorine, a diaphragm is used which on one side rests directly on the ribs of the graphite cathode; on the other side are the anodic graphite lumps, (or between ribs in the case of rib anodes). Thus the distance of the electrodes is only as great as the width of the diaphragm, i.e., about 2 millimeters. The diaphragm consists of PVC cloth which has been specially treated in order to have unusually high resistance to hydrochloric acid and chlorine. Experience indicates that these diaphragms should have a life of at least three years. If the feed acid is free from solids which could clog the diaphragm pores, it can be expected that the hydrochloric acid cells will have to be opened only for renewal of the diaphragms. Care should be taken that no separate phase of chlorinated hydrocarbons is fed into the cell with the feed acid as a weakening of the PVC of the diaphragms might result. It has been observed, after working for several months with a hydrochloric acid which contained chlorinated hydrocarbons that the diaphragms have not been damaged; nevertheless, this is not a recommended operating practice.

Electrolyte

Normally the cell is fed with clear, filtered, concentrated hydrochloric acid. The concentration of this acid is not of great importance. Impurities in solution, e.g. up to several percent of $NaCl$, KCl, Na_2SO_4 have no damaging influence. H_2SO_4 may be present up to about 10 percent without effecting the HCl electrolysis; in fact, it improves the conductivity of the electrolyte but slightly increases the consumption of graphite lumps. If the acid to be electrolyzed contains more than traces of SO_4, anodes of graphite lumps are preferred.

Small amounts of chlorinated hydrocarbons, as long as they are in solution in the feed acid, have not deleterious effect on either the electrolytic process or the durability of the diaphragms. Even larger amounts of toluene, for example, or chloral have no deleterious effect. Figure 7-6 shows views of a hydrochloric acid cell in which the course of the electrolyte can be partly followed. The concentrated hydrochloric acid feed is measured by a rotameter so that the depleted acid leaves the cell with an HCl content of 18 to 19 percent. After the rotameter, the fresh acid is split between two branches of PVC pipe and enters the two ends of the acid

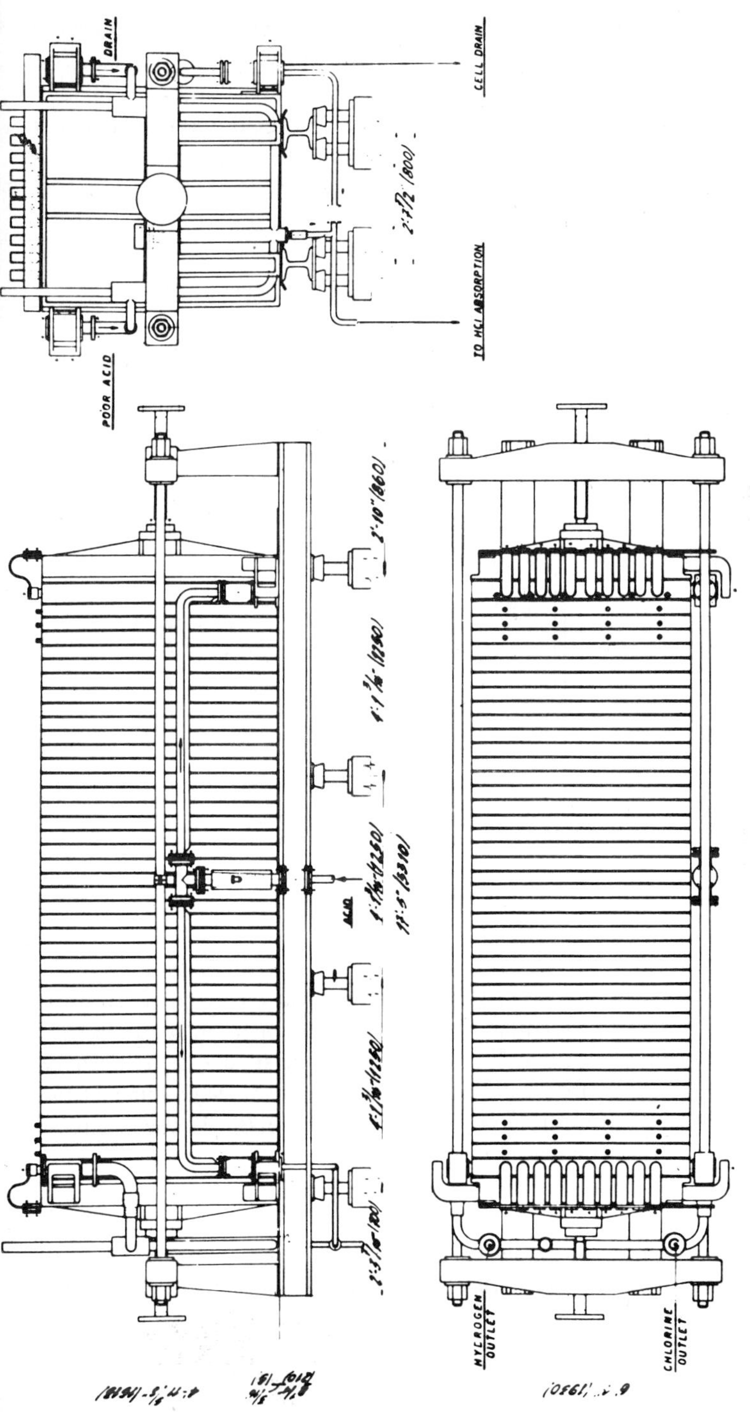

Figure 7-6.   The course of the electrolyte.

distributing channel. From this channel small openings lead into the acid distributors of the various anode spaces. The electrolyte enters the cathode spaces through the diaphragms and flows out of the cathode spaces through overflows. The main reason for feeding fresh acid into the anode spaces is to prevent a noticeable excess of HCl in the cathode spaces and a noticeable depletion of HCl in the anode spaces during electrolysis, due to the different migration velocities of H ions and Cl ions (see Figure 7-7).

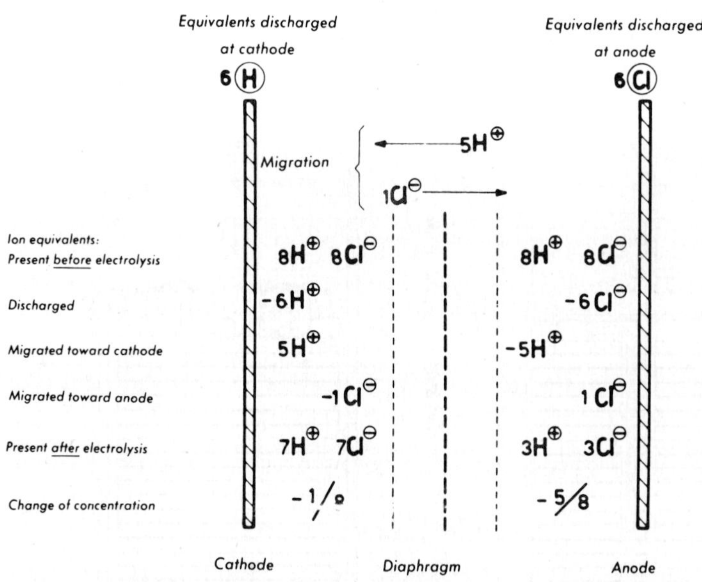

Figure 7-7. Alteration of HCl—concentration in the cell compartments (current: 6 equivalents).

Such a feeding prevents an increased attack on the graphite anodes which would have been due to the lower concentration of chlorine ions. An additional reason for feeding the concentrated acid into the anode spaces and taking the depleted acid out of the cathode spaces is the absence of active chlorine in the outflowing cathode acid. In this way some of the chlorine dissolved in the anolyte is carried into the cathode space and is there eliminated by cathodic reduction; however this amount is so small, because of the high operating temperature of the cell, that in spite of it a total current yield of 93 to 95 percent is obtained. Furthermore, it can be assumed that part of the cathodically reduced chlorine of the anolyte is saving a small amount of electric energy by corresponding depolarization of the cathode. The choice of the HCl concentration in the electrolyte is determined by:

(1) Concentration of chlorine ions—especially in the anode space; the higher the concentration of chlorine ions the lower is the discharge of OH ions and, therefore, the lower the consumption of the anode. For this reason it is advisable to keep the HCl concentration as high as possible.

(2) Electrical conductivity of the electrolyte; at a low HCl concentration, it increases at first together with the concentration; it reaches a maximum and then decreases rapidly with further increasing HCl concentration, as is shown in Figure 7-8 for various temperatures. The maximum of the conductivity declines at a content of about 18.5 percent HCl, at the operating temperature of the cell. For this reason

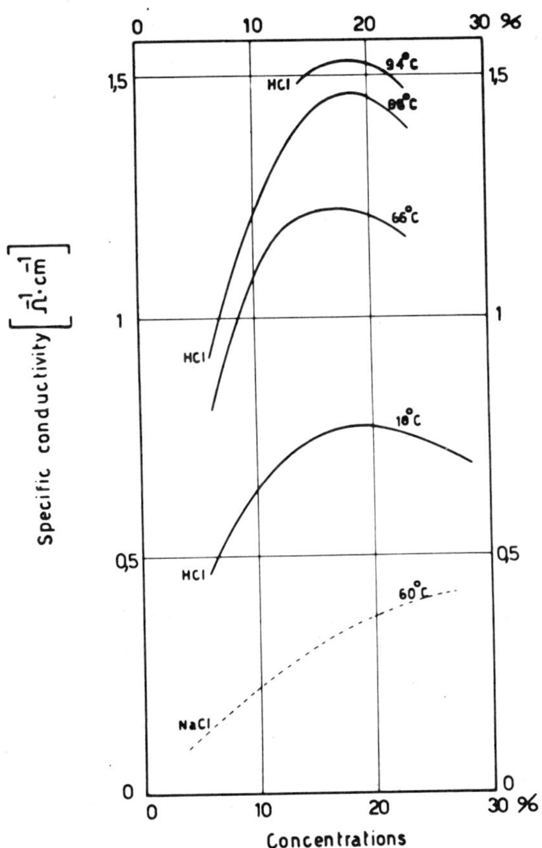

Figure 7-8. Conductivity of muriatic acid and NaCl brine at different concentrations and temperatures.

the depleted acid concentration should be kept between 18 and 19 percent HCl.

Effect of Impurities in the Electrolyte

Organic compounds in small amounts do not effect either the diaphragm or the electrolysis process itself, except methylene chloride which swells the fibers of the PVC cloth of the diaphragm.

The heavier organic compounds do not attack, i.e., even large amounts of chloral or toluene. As it is normally easy to separate the coherent phases of the organic compounds from the aqueous phase of the muriatic acid, there is practically no difficulty with by-product acid of any origin.

Inorganic compounds do not attack the diaphragm.

In the presence of HF there is, of course, no discharge of F ions.

In the presence of H_2SO_4, the electrolytic conductivity increases at small HCl concentrations and the consumption of anodic graphite also increases; however, with as much as 10 percent H_2SO_4 does not effect the diaphragm.

Construction of the Cell

The cell consists of a number of electrode plates assembled as is shown in Figure 7-2. Each electrode plate consists of graphite and is cemented into a rectangular frame of "Haveg" (see Figure 7-3). The "Haveg" frame has four openings. The two large upper openings are for collection and discharge of chlorine, hydrogen and depleted acid. The two smaller lower openings are for feeding the concentrated hydrochloric acid. If the feed acid is concentrated enough only one of the two feed channels is used.

All of these openings are equipped with adequate holes leading from the feed channel to the inside of the anodic electrode spaces. Hydrogen gas and depleted acid leave through two openings, one above the other, which lead into the same collecting channel. The position of the lower hole determines the height of the overflow, and thereby the minimum level of electrolyte in the electrode spaces. The level of electrolyte in the cell is of importance to the separation of Cl_2 and H_2. Above the level of the electrolyte Cl_2 and H_2 are collected and are kept separate by the graphite plate of the bipolar electrode on the one side, and by the Haveg separating wall, shown on the upper ends of (b) and (c) in Figure 7-3 on the other side.

Within certain limits the capacity of a hydrochloric acid cell can be determined by the number of bipolar elements which are assembled to form one block. At 1500 amperes, the capacity of a bipolar electrode of 1.5 sq. m active surface, corresponding to 16.1 sq ft, is 44 kg or 97 lb of chlorine per day. The standard De Nora cell of 40 elements produces, therefore $44 \times 40 = 1760$ kg, or about 3900 lb of Cl_2 daily. The cell, as a whole, is shown on Figure 7-6.

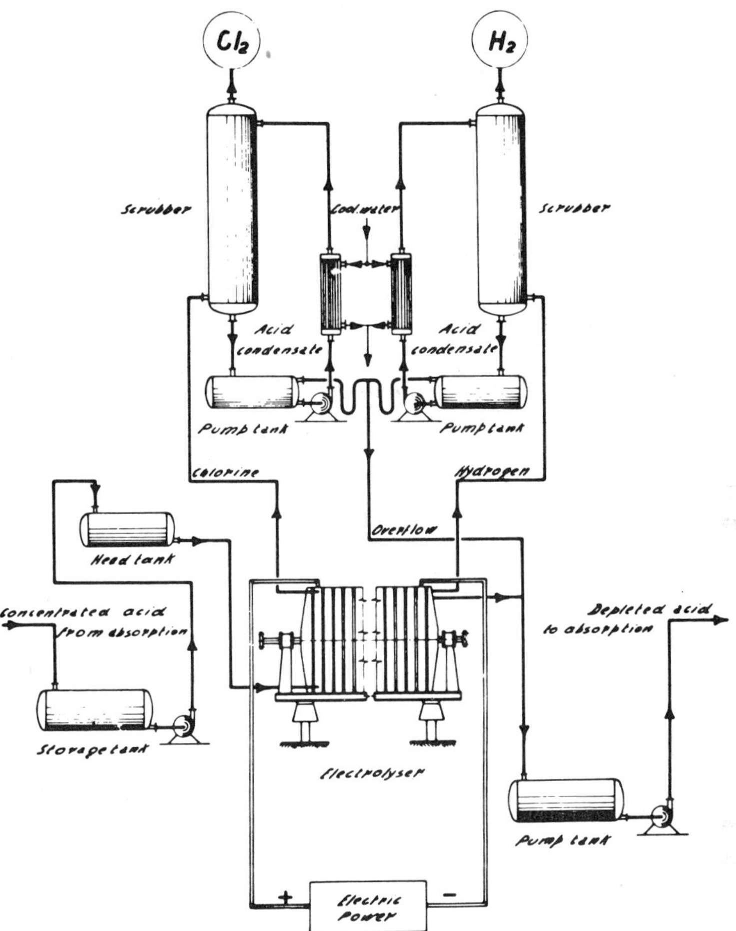

Figure 7-9.   Scheme for a De Nora plant.

The various electrodes with their "Haveg" frames are assembled as a filter press. As the "Haveg" frames are manufactured with completely uniform surfaces, only a low pressure is required to make the cell absolutely gas and liquid tight. The two electrodes at both ends of the cell function only on one side. The direct current is introduced into the two end electrodes through ten round graphite rods, each one by means of copper devices. The cell itself does not contain metal in any place.

This type of construction guarantees that no appreciable corrosion will occur in the cell. The iron compression device is almost wholly rubber covered so that acid vapors of outside origin cannot corrode it and substantially no protective painting is required. The unit may be built in the open but a roof is recommended for operating and maintenance during inclement weather.

Operation of the HCl Cell

When the current is switched on a cold electrolyzer the voltage is a few volts higher than when thermal equilibrium has been reached. As temperature rises the voltage of a single element decreases from about 2.6 volts to about 2.3 volts when the temperature reaches 90°C. When starting a cell with new diaphragms, the voltage of the elements is also 0.2 to 0.3 volt higher than normal for several days. When the "formation" (preparation of the pores) of the diaphragms is terminated, which takes in about one week, the voltage of the elements decreases to 2.3 volts and thereafter remains almost unchanged during the life of the diaphragms. It is therefore advisable when a cold electrolyzer with new diaphragms is started to operate at 70 to 80 percent of rated capacity for a few days.

Figure 7-10 shows a typical shutdown record of a 35-element electrolyzer at equilibrium temperature.

Figure 7-11 shows the volt-ampere diagram of a 35-element electrolyzer under equilibrium conditions.

Products

Chlorine and hydrogen of high purity are produced. After being passed through washing towers showered with cooled condensate produced from the gases themselves they contain only traces of HCl while water vapor content depends on the final temperature of the gases. Figure 7-9 is a schematic layout of a plant for electrolysis of hydrochloric acid. H_2 as well as Cl_2 pass through scrubbing towers. Condensate is collected from each tower and cooled in a graphite cooler and then fed again to the corresponding tower in which, by cooling of new gas, new condensate is formed. Surplus condensate overflows automatically and is combined with the depleted acid from the electrolysis cells. In this way, approximately the same volume of water is kept continuously in circulation.

Condensate from the chlorine and hydrogen is dilute hydrochloric acid, the HCl concentration of which depends on the average operating temperature of the cells. At an average cell temperature of 90°C (194°F) and an outlet concentration of the depleted acid of 18 percent HCl, the condensate from chlorine as well as from hydrogen would contain about 8 percent HCl.

In practice, however, it is much more diluted and contains about 3 percent HCl. Probably the reason for this is that a partial condensation takes place in the pipes between the electrolyzer and the scrubbing tower. It is advisable to install the chlorine and hydrogen gas outlet pipes with a

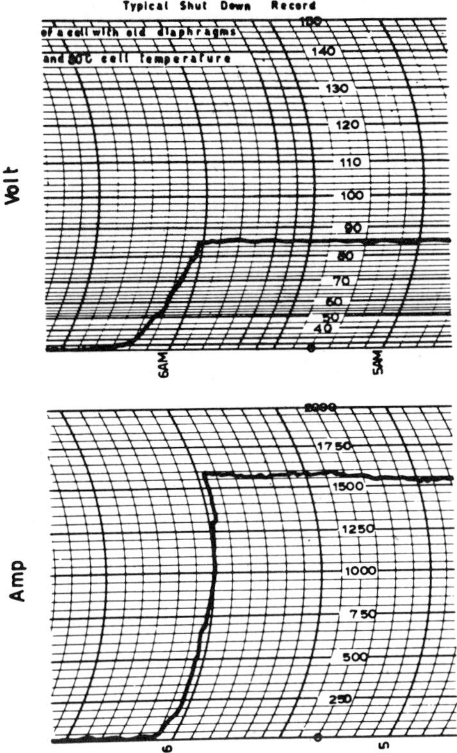

Figure 7-10. Hydrochloric acid cell (35 elements); typical shutdown record of a cell with old diaphragms and 80°C cell temperature.

slope toward the cells so that the pre-condensate richer in HCl is combined with the depleted acid of the cells and not with the main condensate in the cooling-washing towers. It is possible in this way to obtain a main condensate of very low HCl content and thereby also low in partial pressure of H_2O and HCl. Cooled Cl_2 and H_2 coming from the cooling-washing towers contain only traces of HCl and water vapor. Under normal circumstances these traces of HCl in the chlorine gas are of no consequence. In the case of hydrogen, however, even a trace of HCl can be of importance, e.g., if it is used for NH_3 synthesis, for hydrogenation of food products or for heating purposes. In these cases before drying, the hydrogen should be passed through a scrubbing tower using dilute NaOH or $Ca(OH)_2$. Its consumption is so low, that it practically can be disregarded from a cost standpoint. In this way hydrogen is obtained which is completely free

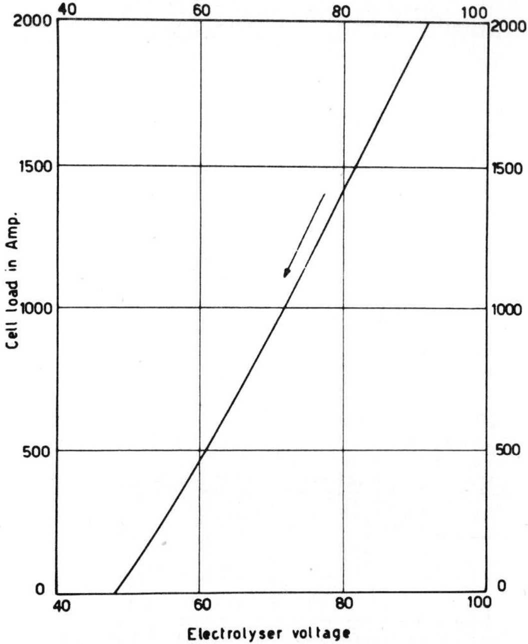

Figure 7-11.   Volt-ampere diagram of a HCl unit
with 35 elements.

of HCl and Cl_2. Gases from the electrolysis cells contain
  (1) Chlorine gas (dry basis) by volume: CO_2 0.05–0.4%

                  (dependent upon acid purity)

            H_2 Less than 0.1%

            O_2 Less than 0.01%

  (2) Hydrogen gas (dry basis) by volume: Cl_2 Less than 0.1%

If the cells and the connecting pipes are under a slight pressure, the air content of the chlorine and hydrogen is extremely low.

Current yield and consumption of energy are as follows:

          Current yield at 1000 amp: 92%

                  at 1500 amp: 94%

                  at 2000 amp: 95%

Consumption of energy of the 40-element cell unit:

$$\frac{94 \text{ volts (including copper bars)} \times 1500 \text{ amps} \times 24 \text{ hr}}{1760 \text{ kg } Cl_2 \text{ (per 24 hr)}}$$

$$= 1930 \text{ KWH per 1000 kg } Cl_2 \text{ or } 1750 \text{ KWH per short ton } Cl_2$$

An installation with 12 standard 40-cell electrolyzers for a rated capacity of 24 short tons of chlorine per day is shown in Figures 7-12, 7-13, and 7-14.

The 12 electrolyzers are divided into 2 electrical systems with 6 electrolyzers each. Each system has, at the normal load of 1600 amperes, a voltage of about 560 volts.

TABLE 7-2. STANDARD UNIT, TYPE AC—150D 40C TECHNICAL AND DIMENSIONAL CHARACTERISTICS*

Current rating	1600 amps
Voltage across unit	93 v.
Power consumption, DC, side	140 KW
Muriatic acid (33% HCl)	140 GPH
Number of elements	40
Nominal electrode size	2480 sq in.
Current density at rated capacity	0.0 amps/sq in.
Over-all dimensions:	
length (including capstan)	18 ft
width	6 ft
height	8 ft
Over-all floor space required, approximately 21 × 10 ft.	
Daily capacity	
Chlorine	4,000 lbs.
Hydrogen	20,000 cu ft.

*Each unit is installed over an acid-proof basin, approximately 8 in. high.

TABLE 7-3. UTILITY AND LABOR REQUIREMENTS

Based on the production of:	
Cl_2	2000 lb
H_2 (STP)	10,000 cu ft
Requirements	
HCl (as muriatic acid 100% basis)[a]	2060 lb
Electric energy	
DC, for electrolysis	1750 KWH (about 1900 KWH-AC)
AC, for auxiliary power	40 KWH
Broken graphite (for anode packing)	2.5 lb
PVC diaphragms (3 years life)	0.6 sq in.
Cooling water at 15°C (for Cl_2 and H_2)	10,000 gal
Labor[b]	1 man-hour

[a] On the assumption that HCl gas is supplied to the absorption and electrolysis process in the dry state. Otherwise, an additional amount is to be supplied in order to provide for the discarded HCl lost in the condensate and to prevent water accumulation in the cycle. (2 to 4% HCl is present in the condensate.)

[b] Based on a daily plant capacity of 24 short tons Cl_2 using one man per shift.

TABLE 7-4. MANUFACTURING COST (BG TYPE CELL)

(Based on a chlorine recovery of 24 short tons/day corresponding to 8,640 short tons/year)

	Dollars per Day	Dollars per ton Cl_2
Direct costs		
Muriatic acid	0	0
Electricity ($0.01/KWH)	456	19.00
Materials (graphite, PVC twill)	24	1.00
Cooling water ($0.01/1000 gal.)	2.40	0.10
Operating labor ($2.50/hour)	60	2.50
Maintenance (5% of investment)*	83	3.47
Indirect costs		
Supervision and overhead	40	1.66
Fixed costs		
Depreciation (10% of investment)*	166	6.90
Property taxes and insurance (1.25% of investment)*	21	0.87
Total manufacturing cost	778.15	35.50
Credit		
By-product H_2—240,000 cu ft/day		

*Based on $25,000 per daily ton Cl_2 capacity for electrolysis area.

WESTVACO PROCESS ELECTROLYSIS OF COPPER CHLORIDE

This process has never been applied commercially but development was carried through the pilot plant stage and represents very constructive work in the field of chlorine recovery from HCl by indirect electrolysis. The detail of this work is well presented in the September 1950, issue of "Chemical Engineering Progress" and only portions of this detailed report will be repeated here.

The process involves the electrolysis of a solution of a polyvalent metallic chloride to produce chlorine and a reduced metallic chloride. The reduced metallic chloride is oxidized with air and hydrogen chloride to regenerate the oxidized metallic chloride, which is recycled to the electrolytic step. The basic reactions involved in this two-step process are illustrated by the following equations:

$$2\,CuCl_2 \xrightarrow{\text{Electrolysis}} 2\,CuCl + Cl_2$$

$$2\,CuCl + 2\,HCl + \tfrac{1}{2}\,O_2 \rightarrow 2\,CuCl_2 + H_2O$$

Combining these two equations

$$2\,HCl + \tfrac{1}{2}\,O_2 \rightarrow Cl_2 + H_2O$$

This will be recognized as the familiar Deacon reaction.

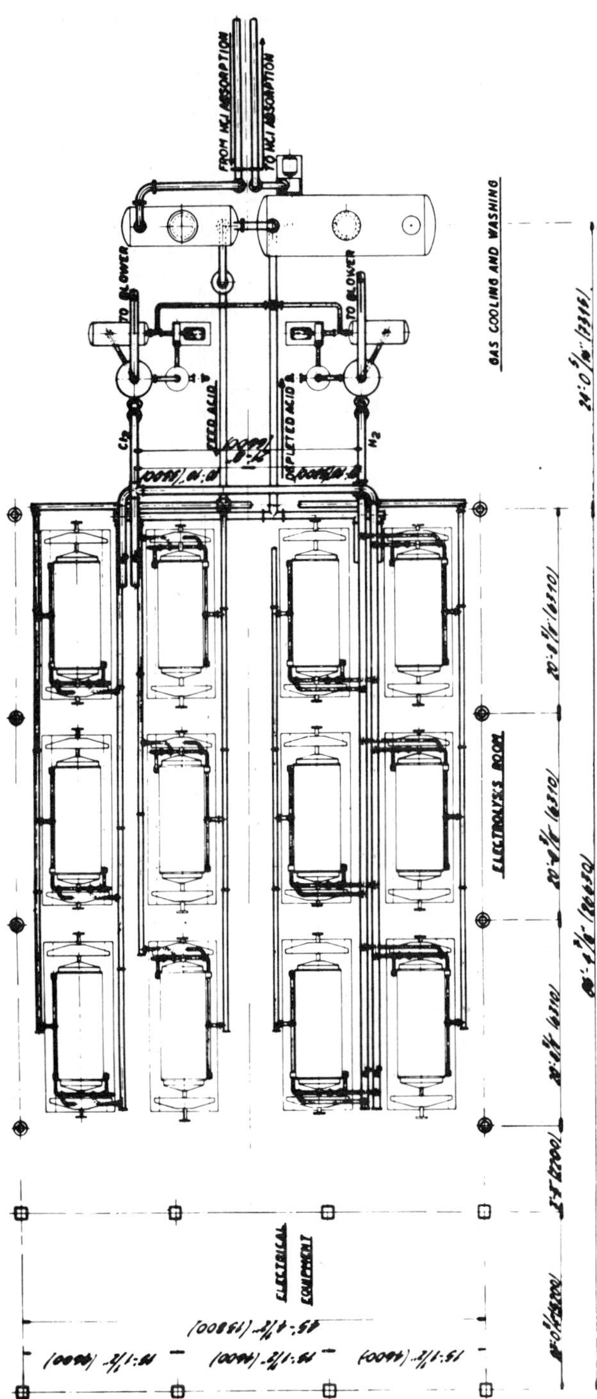

Figure 7-12. Typical arrangement of two circuits of electrolyzers.

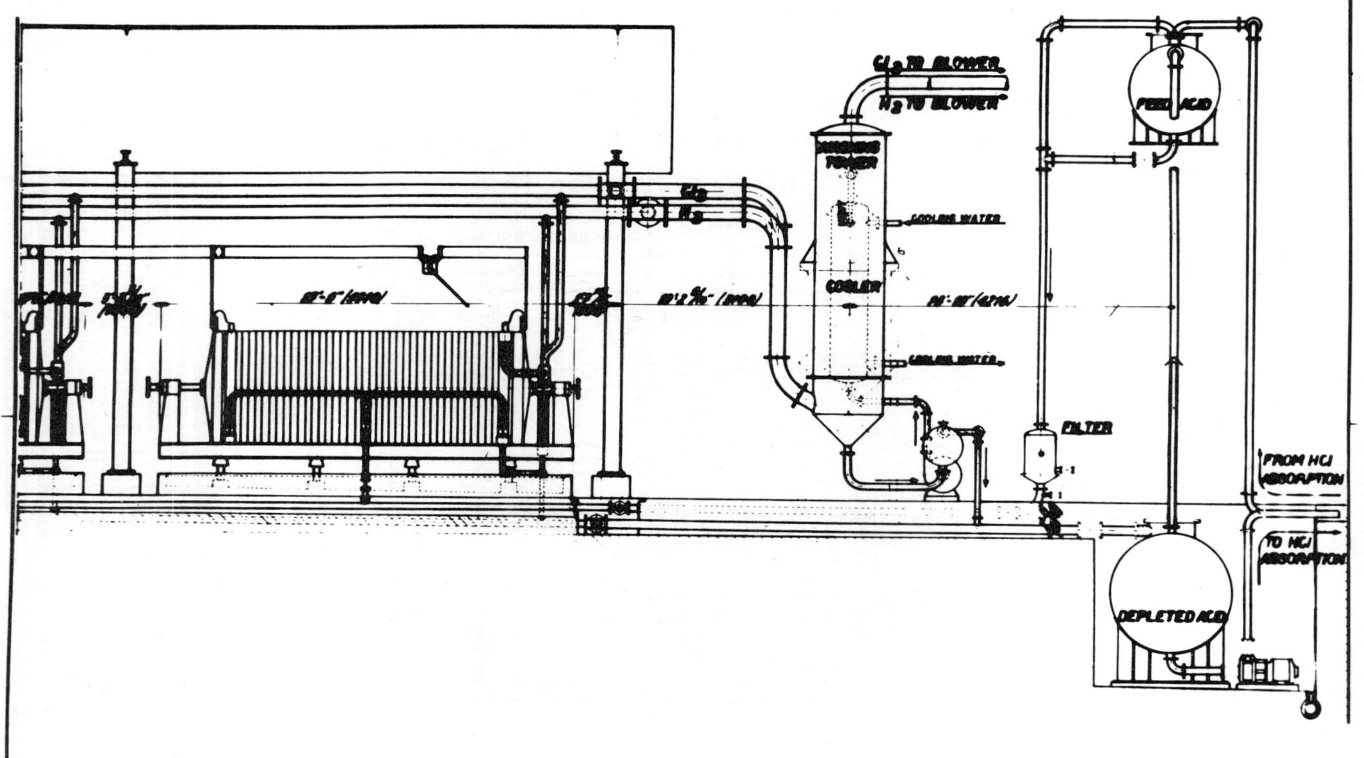

Figure 7-13. Ancillary equipment arrangement.

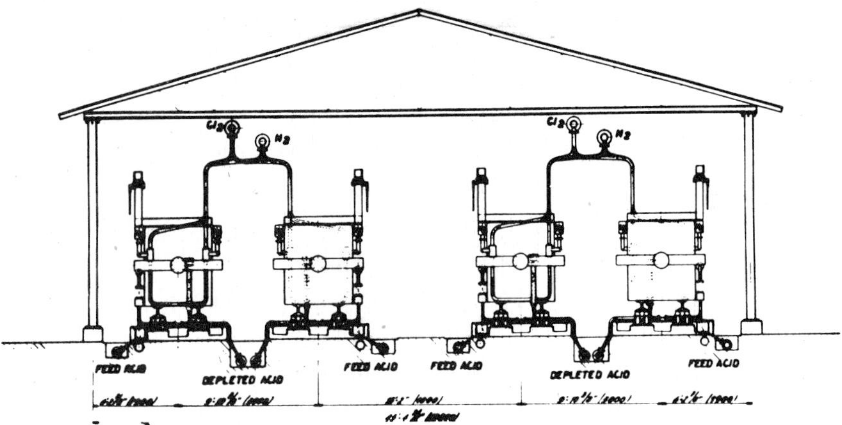

Figure 7-14.   Hydrochloric acid electrolysis (front elevation).

Although chemically equivalent to the Deacon reaction, the metallic chloride process has the advantage of producing chlorine in a pure form. The gas is not contaminated by the formation of an equilibrium mixture as in the Deacon reaction, and the nitrogen content of the oxidizing air readily is kept separate from the chlorine. The voltage and therefore the power requirements are less than for direct electrolysis. No hydrogen is generated to contaminate the chlorine or to create a potential explosion hazard. The absence of a second gas in the electrolytic step simplifies the cell and eases the materials of construction problem. Patents covering the process have been granted to F. S. Low[8] and other patent applications are pending.

As in the development of any process, much work was required to reach a suitable cell design. The results of all this work are shown in Figures 7-15 and 7-16 and explained in considerable detail.

Cell Design

The all-important practical details involved in designing a commercial cell, e.g., provisions for delivering the electrical load to the graphite electrodes, for feeding and distributing the electrolyte and removal of catholyte (cuprous chloride liquor), for escape of chlorine without back pressure, and for sealing of the various joints against penetration by the corrosive electrolyte, each presented problems which have been satisfactorily solved, with the possible exception of the last one. Numerious cell designs are, of course, possible. A compact and effective design is shown in Figures 7-15 and 7-16.

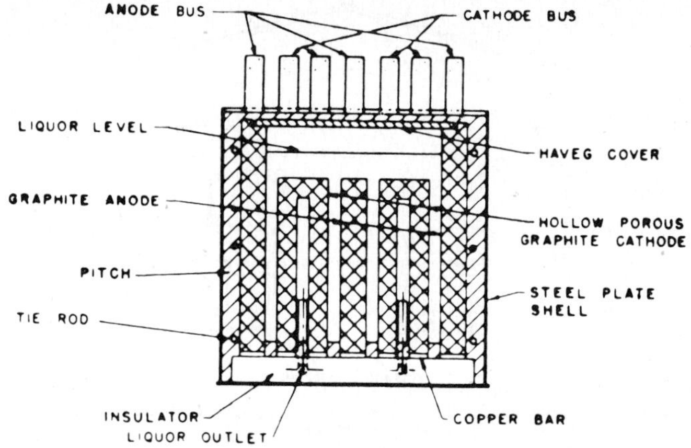

Figure 7-15. Copper chloride cell section through short dimension.

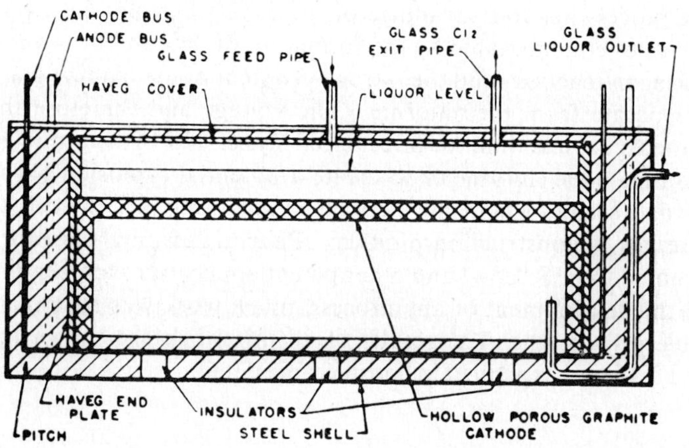

Figure 7-16. Copper chloride cell section through long dimension at center of hollow cathode.

An important feature of the cell is the hollowed-out monolithic cathode which is 12 in. high and 36 in. long. These cathodes are alternated with graphite anodes, the number employed depending upon the desired capacity of the cell. When using two cathodes and three anodes, as shown in the figures, 10 sq. ft of active cathode surface is available, giving the cell a capacity of 4000 amperes at 400 amperes per square foot. The electrodes are held in the proper relative positions by two grooved Haveg end-plates, one at each end of the electrode assembly, and these are held in place by tie rods running the length of the cell as shown. The bottom edges of the electrodes are lightly plated with copper, and multiple heavy flexible copper straps are soldered to the plated surface at intervals along the length of each electrode to distribute the electrical load. These flexible straps are connected to the solid copper bus work shown in the figures. This relatively elaborate distribution system is necessitated by the high current density at which the cell is operated.

The electrode assembly rests upon ceramic insulators in a sheet steel cell container. A molten sealing material such as asphaltic pitch is poured into the container in such a way as to rise around the bottom edges of the electrodes, thereby sealing the lower joints exposed to the electrolyte. This first portion of sealing material is allowed to solidify, after which additional sealing material is poured into the space between the electrode assembly and the cell container, filling this space and flowing over the "Haveg" cell cover which is supported by the outermost anodes. This seals the remainder of the joints in the cell. The electrolyte is fed to the cell through a glass pipe extending through the cover at the center. Electrolyte flows through the walls of the porous cathodes into the hollowed out portions, and from there through glass or Haveg pipe imbedded in the sealing material to a constant level over-flow device. Chlorine escapes from the cell through a second glass pipe mounted in the cell cover.

This completely sealed construction would not be feasible where periodic rebuilding or repair of the cells was required. However, the graphite electrodes and the "Haveg" end-plates give every indication of having an indefinite life, and with an adequate sealing material no periodic rebuilding should be necessary to realize a reasonable cell life. At present, no entirely satisfactory sealing material has been found, although a variety of asphaltic materials and related substances have been tested. After a few months of operation the sealing materials tested have failed due to attack by the electrolyte, permitting the electrolyte to reach the copper-graphite connections. The lower portions of the electrodes have been effectively impregnated with "Halowax" (a chlorinated naphthalene) to prevent penetration of the electrolyte through the pores of the graphite to the copper contacts. Failure occurs only as a result of penetration through the surrounding sealing material. It is anticipated that a satisfactory sealing material will eventually be found.

Figure 7-17 shows a flow sheet as developed for a plant using this process and Tables 7-5 and 7-6 give operating characteristics and process requirements estimated for such a plant.

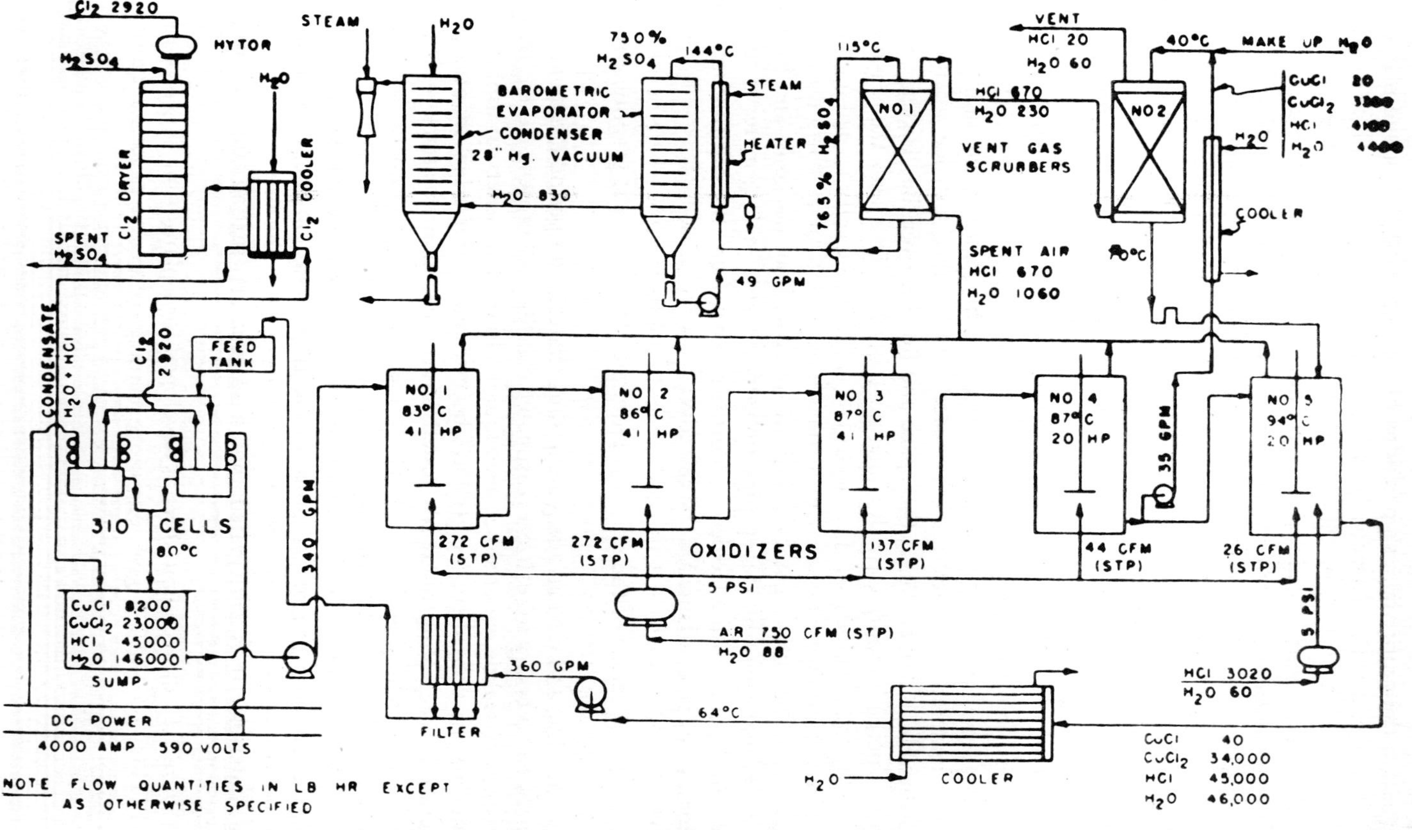

Figure 7-17.   Flow sheet for copper chloride process to produce 35 tons of chlorine per day.

TABLE 7-5. OPERATING CHARACTERISTICS OF 4000 AMPERE
COPPER CHLORIDE CELL

Feed rate, gal/hr	70
Feed density, gal/(Hr)(sq ft)	7.0
Feed composition: Cupric chloride, %	15
Hydrogen chloride, %	20
Feed temperature, °C	64
Cell temperature, °C	80
Current, amperes	4000
Cathode current density, amp/sq ft	400
Current efficiency, %	82
Voltage across cell	1.80
Power, KWH/lb Cl_2	0.80
Chlorine production, lb/hr	9.6
Effluent liquor composition: Cuprous chloride, %	3.7
Cupric chloride, %	10.2
Hydrogen chloride, %	20.2

TABLE 7-6. PROCESS REQUIREMENTS FOR COPPER
CHLORIDE PROCESS

Based on Chlorine Production of 35 Tons/Day

Direct supervision, foreman	1
Operating labor, men/shift	3
Maintenance labor, Avg. man-hours/day	40
Maintenance materials—Cost Approx. equal to maintenance labor	
Hydrogen chloride, ton/ton Cl_2	1.04
Cupric chloride, lb/ton Cl_2	3.6
Sulfuric acid, lb/ton Cl_2	62
Pure water, 1000 gal/ton Cl_2	0.3
Cooling water, 1000 gal/ton Cl_2	82
A.C. Power, KWH/ton Cl_2	155
D.C. Power, KWH/ton Cl_2	1600
Steam, lb/ton Cl_2	2500

SCHROEDER PROCESS ELECTROLYSIS OF NICKEL CHLORIDE

This process has not been developed beyond the laboratory stage but is of interest as it is another approach to the production of chlorine from HCl by indirect electrolysis of a metallic chloride.

As in the case of the other indirect electrolysis process, it is felt that a description of the salient points, while not as fully developed, should be presented and Mr. Schroeder may be contacted for details of his laboratory work.

The proposed process utilizes the reactions

$$NiCl_2 \xrightarrow{\text{Electrolysis}} Ni + Cl_2 \tag{7-1}$$

and

$$Ni + 2\,HCl \rightarrow NiCl_2 + H_2 \tag{7-2}$$

The net reaction is hence

$$2\,HCl \rightarrow H_2 + Cl_2$$

The process uses two cells (or sets of cells). In one cell nickel deposits on the cathode and chlorine is released at the anode. In the other cell, which does not require current, the nickel is dissolved by hydrochloric acid. The nickel chloride produced in the second cell is fed to the first cell. After a time there is a heavy deposit of nickel in the first cell while the second cell has been depleted of nickel. The roles of the cells are then reversed.

The electrolysis of nickel chloride solutions is best carried out in a cell with a carbon anode separated from a nickel or carbon cathode by about $\tfrac{1}{4}$ in. If a carbon cathode is used it becomes coated with nickel during the process. The temperature of the solution should be above 150° F and the pH of the solution between 1.3 and 3.5.

Table 7-7 shows current efficiency as a function of current density.

TABLE 7-7. CURRENT EFFICIENCY

Current Density (amps/sq. ft)	Cathode Efficiency (%)	Anode Efficiency (%)
25	95	93
50	97	95
200	99	97

The major loss of current efficiency is due to the reaction:

$$Cl_2 + 2\,e \rightarrow 2\,Cl \tag{7-3}$$

which occurs at the cathode but lowers both cathode and anode efficiency.

The chlorine produced contains about 1 percent oxygen plus carbon dioxide. The production of these gases accounts for the anode efficiency being about 2 percent below the cathode efficiency. The low oxygen production (and the consequent low anode consumption) is due to the low pH and the high chloride content of the electrolyte.

The Corrosion of Nickel by Hydrochloric Acid

In the light of previously published data[9] on the rate of corrosion of nickel by hydrochloric acid, Reaction (7-2) above does not seem a practical means of reforming nickel chloride from hydrochloric acid and deposited nickel. Nickel is built up at the rate of 0.125 in. per day if it is deposited by a current of 100 amperes per square foot. The maximum recorded rate of corrosion of nickel is 0.0157 in. per day, and this is in a solution much too acid to be fed to Reaction (7-1).

Reaction (7-2) became a practical part of the chlorine producing scheme when it was discovered that the addition of nickel chloride to a hydrochloric acid solution caused the rate of corrosion of nickel to increase by as much as 9 times. Table 7-8 shows some of the corrosion rates of nickel in hydrochloric acid.

TABLE 7-8. RATE OF CORROSION OF NICKEL IN HYDROCHLORIC ACID-NICKEL CHLORIDE SOLUTION

HCl (g)	H_2O (g)	$NiCl_2$ (g)	Temp. (°F)	Corrosion Rate (in./day)	(mdd)*	Source
0.5	99.5	0.0	Boiling	0.00083	1,875	Ref. 9
1.0	99.0	0.0	Boiling	0.00186	4,200	Ref. 9
1.0	99.0	83.0	238	0.0173	39,000	This work
5.0	95.0	0.0	Boiling	0.0157	35,400	Ref. 9
10.0	90.0	0.0	218	0.0163	36,700	This work
10.0	90.0	67.0	220	0.034	76,500	This work

*mdd—milligrams per square decimeter per day

Figure 7-18 is a flow diagram and material balance for a plant which will produce 1.7 tons of chlorine per day. It requires two cells of the type described below. A solution of nickel chloride is fed to the chlorine producing cell. Chlorine is released and nickel is deposited on the cathodes. The chlorine goes to a cooler and sulfuric acid drier (neither shown). It may then be compressed and liquified, if desired. The partially depleted nickel chloride solution is cooled and sent to the absorber where it is brought into contact with gaseous hydrogen chloride. The amount of hydrogen chloride required is 2 percent in excess of the theoretical for the chlorine produced. (In the event that aqueous hydrochloric acid is to be used the depleted nickel chloride stream will be sent to an evaporator where water equal to that in the acid feed will be removed. No absorber will be required.) The absorber is run hot enough to purge from its top any hydrocarbons which contaminate the hydrogen chloride. The solution from the bottom of the absorber goes to the hydrogen cell (which requires no current) where the hydrochloric acid dissolves the nickel which has

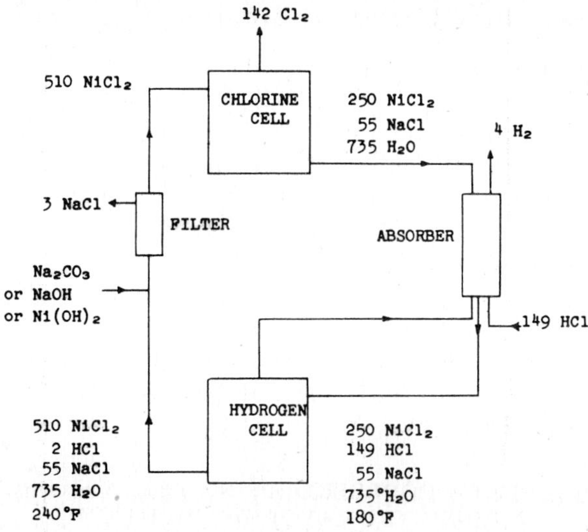

Figure 7-18. Flow diagrams and material balance. (Quantities shown in pounds per hour. Water leaves with hydrogen and chlorine. It may be made up at any convenient place, but this is not shown.)

previously been deposited. The gas from the hydrogen cell is routed to the absorber to recover the small amount of hydrogen chloride leaving the hydrogen cell cooled; it is then neutralized with caustic soda, soda ash or nickel hydroxide. The sodium chloride produced by neutralization builds up in the circulating solution till it reaches about the concentrations shown and thereafter precipitates and can be filtered out. The sodium chloride which remains in the circulating system does no harm. The salt filter will also remove, as ferric hydroxide, any iron in the system.

Operation continues in this manner until most of the nickel has been removed from the nickel dissolving (hydrogen) cell and a layer of nickel has been accumulated in the chlorine producing cell. Then the roles of the cells are reversed. A one to three day cycle is suggested.

Cell Design

The cell consists of a cubical Haveg lined tank measuring 5 feet on each side. At one end of the tank is a graphite anode, at the other end is a graphite cathode. Between the anode and cathode is an assembly of graphite plates measuring 5×3 ft $\times \frac{1}{4}$ in. thick, spaced $\frac{1}{4}$ in. apart. These plates act as bipolar electrodes. There are thus 120 cells in series in each large cell. There is a total of 1800 square feet of cathode area and

1800 square feet of anode area. No diaphragm is required. The current density proposed is 27 amperes per square foot since this current density will deposit nickel at the rate at which it can be dissolved. At this current density each cell will require 415 amperes at 205 volts. Each cell will produce 1.7 tons of chlorine per day while in chlorine service, or an average of 0.85 tons per day per cell since each cell is in chlorine service only half the time.

Alternate for Nickel in the Process

Cobalt or zinc can be substituted for nickel in the above process. Cobalt is in all technical respects as satisfactory (it actually corrodes about 25 percent faster than nickel) but is more expensive. Zinc can be used but a greater amount of electricity would be required. No other satisfactory metal was discovered.

Table 7-9 shows estimated utility and chemical consumption for the plant described. The figures given do not include the cost of compressing or liquefying the chlorine.

TABLE 7-9. UTILITY AND CHEMICAL CONSUMPTION
PER TON PER CHLORINE

Utilities

Electricity	1250 KWH
Steam	none
Fresh water	100 gal
Cooling water	3000 gal

Chemicals

Hydrochloric acid (100%)	2100 lb
Sulfuric acid (for drying)	25 lb
Soda ash (optional)	60 lb

Table 7-10 shows that there are now 88 producing points in 25 states and in several of these locations more than one process is used. The geographic distribution of these plants assures a supply of acid in all areas of the United States.

The steady increase in total tonnage without a corresponding development of new uses presents a serious problem of oversupply. Much study is being given to the development of new uses for hydrochloric acid either as gas or in solution to solve this imbalance between supply and demand. At the same time more and more interest is being shown in processes for recovering chlorine from HCl and it is predicted this will be the major means of solving the oversupply problem.

TABLE 7-10. PRODUCTION AND SALES OF HYDROCHLORIC ACID BY PROCESS

(Hydrochloric Acid Plants in United States)

	Salt-Acid	Hydrogen-Chlorine	By-Product	Total
Massachusetts	1	–	1	2
New Hampshire	–	1	–	1
New Jersey	3	–	7	10
New York	–	3	5	6 *
Pennsylvania	–	–	1	1
Alabama	–	–	2	2
Delaware	1	–	–	1
Georgia	–	1	2	3
Kentucky	1	3	3	5 *
Louisiana	2	2	3	5 *
Maryland	–	–	1	1
Tennessee	–	–	1	1
Texas	1	3	8	11 *
Virginia	1	1	–	2
West Virginia	–	–	6	6
Illinois	1	–	2	3
Indiana	1	–	2	3
Kansas	–	1	–	1
Michigan	–	2	6	7 *
Missouri	–	–	1	1
Ohio	2	1	4	6 *
California	–	2	3	5
Nevada	–	–	1	1
New Mexico	1	–	–	1
Washington	–	2	1	3
Total	15	22	60	88

*Some plants use more than one process. The total shown is the unduplicated count.

Source: Bureau of Census, Industry Division

Table 7-11 gives the estimated end uses at the present time. It must be realized, however, there are large quantities of acid now being wasted because these outlets are inadequate for the existing supply.

CONCLUSION

Which electrolysis process will prove the most attractive economically perhaps will not be established for some time. None of the processes are competitive with present operations of large chlorine producers. Considering developments of the various processes up to this time, the electrolysis of hydrochloric acid seems to be the most practical solution for the recovery of chlorine from hydrochloric acid which has no market and in addition requires an expenditure for its disposal. (The capital investment required for such an installation is approximately one-third of the investment required for a new chlorine plant of medium size.) Hydrochloric acid

TABLE 7-11. HYDROCHLORIC ACID END USE

	(%)
Chemical manufacture	23
Petroleum industry	30
Metals industry	23
Glucose & food products	12
Miscellaneous	12
	100

Source: Bureau of Census, Industry Division

electrolysis requires a rather low capital investment and has the additional advantage that it can be built in small units which do not require any higher investment per ton than large units. Operating labor costs will be higher for a small capacity plant unless the unit can be operated by present employees. Nevertheless it is a practical method of handling as little as a few tons per day of hydrochloric acid while a chlorine plant of this capacity would not be considered. An initial small capacity unit can be arranged and additions can be made at a minimum cost when required.

The practicability of this process has been proven by a commercial size unit in operation since September 1958, and no unusual difficulties have developed, as is often the case with new processes. Thus, any company with a surplus of hydrochloric acid that is either not saleable, or is saleable at a very low profit margin, or any company faced with a disposal problem or an additional cost of neutralization before disposal, cannot afford not to investigate this process; in most cases it will be found economical and practical, while operation of this plant is simple and does not require much labor or maintenance. The process is available in the United States by virtue of a working agreement between Oronzio de Nora, Milan, Italy, and Monsanto Chemical Company, Engineering Sales Department, St. Louis, Missouri.

Production of by-product hydrochloric acid is increasing every year. This growth will continue as the chlorine industry grows, and chlorine production increase is variously estimated at 6 to 10 percent per year. Along with the normal growth in the production of chlorine by the electrolysis of sodium or potassium chloride there is an equivalent production of quantities of sodium or potassium hydroxide. To date there usually exists a surplus of sodium hydroxide when normal chlorine demands are satisfied. Likewise, the normal chlorinations result in the surplus of hydrochloric acid for which there is no market. In many areas problems arising from disposal of waste acid are becoming more serious every year.

These combined problems of excess caustic soda and surplus hydrochloric acid can be solved by the electrolysis of hydrochloric acid, thereby providing chlorine without the surplus caustic as well as eliminating the disposal problem of acid.

References

(1) Fiat Report No. 833.

(2) H. I. Walter, U. S. Patent 2,878,105 (1959).

(3) Westvaco Chlorine Products, Inc., New York, U. S. Patent 1,746,542.

(4) *Chem. Eng. Progress,* p.p. 456-463, September 1950.

(5) David W. Schroeder, Seattle University, Seattle, Wash.

(6) Fiat Final Report No. 832.

(7) Oronzio de Nora, Impianti Elettrochimici, Milano, Italy.

(8) Low, F. S., U. S. Patent 2,468,766 (1949); U. S. Patent 2,470,073 (1949); British Patent 611,908 (1948); Canadian Patent 458,795 (1949): Canadian Patent 458,796 (1949).

(9) Friend, W. Z., and Knapp, B. B., *Trans. Am. Inst. Chem. Engrs.,* **39,** 731 (1943).

8. THE SALT PROCESS FOR CHLORINE MANUFACTURE

M. F. FOGLER

Allied Chemical Corporation

HISTORICAL BACKGROUND FOR SALT PROCESS

The commercial production of chlorine, in the form of bleaching powder, was first developed using the Deacon Process,[4] wherein chlorine was made by the oxidation of HCl with air, catalyzed either by MnO_2 or by Cu_2Cl_2. The over-all reaction for this process is represented by the following equation:

$$4\,HCl + O_2 \rightarrow 2\,Cl_2 + 2\,H_2O \tag{8-1}$$

This original commercial process was subsequently replaced by the electrolytic process involving the electrolysis of brine to give caustic soda and chlorine. While the electrolytic reaction was known in the eighteenth century, it was not until 1890 that caustic and chlorine were actually produced in this way for industrial consumption.[4] It is interesting to note that the original process was developed primarily for caustic manufacture and that chlorine was at that time only a by-product. At present, chlorine is the more valuable of the two products. The first American production of electrolytic chlorine was at Mumford Falls, Maine, in 1893.[4] The over-all reaction in the electrolytic process can be represented by the following equation:

$$2\,NaCl + 2\,H_2O \rightarrow Cl_2 + H_2 + 2\,NaOH \tag{8-2}$$

The electrolytic process is undoubtedly an ideal one when both products are in balanced demand and can be sold at fair prices. Caustic soda is also produced by causticizing soda ash, however, and the caustic market is not always able to absorb the entire output of both processes. In times when interest was devoted primarily to chlorine production, without the simultaneous production of caustic soda, other processes have been studied for possible replacement of the electrolytic process. Just such a condition prevailed at the time when the salt process was first introduced into industrial use in the United States.

In the 1920's, synthetic ammonia was made in industrial quantities for the first time by the fixation of atmospheric nitrogen. This major advancement in the chemical industry in America was coupled with ammonia oxidation processes to make nitric acid, the nitric acid being reacted with soda ash to form sodium nitrate. Sodium nitrate, at that time, was used extensively in the American fertilizer industry. Until the development of synthetic ammonia processes in the United States, we had been dependent on Chilean nitrate of soda.

While nitric acid was first manufactured for subsequent reaction with soda ash to produce sodium nitrate, its availability for reaction with salt, a cheaper source of sodium ion, to give chlorine and sodium nitrate hastened the development of the salt process in the United States. The early 1930's saw the first commercial installation of the salt process at Hopewell, Virginia, by the Solvay Process Company. The over-all reaction in the salt process is as follows:

$$3\,NaCl + 4\,HNO_3 \rightarrow NOCl + Cl_2 + 2\,H_2O + 3\,NaNO_3 \qquad (8\text{-}3)$$

Eventual changes in the fertilizer industry have led to the decline of fertilizer as a sodium nitrate outlet, while ammonium nitrate has largely taken its place. Ammonium nitrate is manufactured by the following reaction:

$$NH_3 + HNO_3 \rightarrow NH_4NO_3 \qquad (8\text{-}4)$$

The salt process still remains, however, as a method of producing sodium nitrate and chlorine, as well as nitrogen tetroxide (see Equation 8-8).

To complete the picture of the processes and products with which chlorine has become associated, it would be well to consider the history of soda ash manufacture in the United States.

The principal process for the manufacture of soda ash is the Solvay Process. While the Le Blanc process for soda ash manufacture was in universal use before the Solvay process, no Le Blanc plant was ever built in the United States.[4] The Solvay process can be represented by the over-all equation:

$$CaCO_3 + C + O_2 + 2\,NaCl \rightarrow Na_2CO_3 + CO_2 + CaCl_2 \qquad (8\text{-}5)$$

The soda ash thus produced can be upgraded to caustic soda by the lime-soda process represented by the equation:

$$Na_2CO_3 + Ca(OH)_2 \rightarrow 2\,NaOH + CaCO_3 \qquad (8\text{-}6)$$

The salt process is the final ingredient in the picture of the combined products and their related positions in the chlorine industry; this process is represented by Equation (8-3). The resultant gaseous products of chlorine and nitrosyl chloride can be recycled in either of two fashions. *First,* nitrosyl chloride can be reacted with soda ash to form salt, sodium nitrate, nitric oxide and carbon dioxide. The salt is recycled for further manufacture or reaction and the nitric oxide can be recycled for either nitric acid or sodium nitrate manufacture while the sodium nitrate is drawn off as a product and the carbon dioxide is wasted. This reaction is represented as follows:

$$3\,NOCl + 2\,Na_2CO_3 \rightarrow NaNO_3 + 3\,NaCl + 2\,CO_2 + 2\,NO \qquad (8\text{-}7)$$

The *second* process for handling or recycling nitrosyl chloride is represented as follows:

$$2\,NOCl + O_2 \rightarrow N_2O_4 + Cl_2 \qquad (8\text{-}8)$$

The additional chlorine can be drawn off as a product, the nitrogen tetroxide being taken either as product or recycled for acid for sodium nitrate manufacture.

To summarize the over-all picture as presented above, consider the attached Table 8-1. The table presents the entire picture of what the chemical process industry refers to as the soda ash—caustic soda—chlorine industry. In expanding this industrial complex to include the modern concepts, one must add sodium nitrate, ammonium nitrate, and nitrogen tetroxide. It will be noted that the basic raw material for all products is either salt or ammonia. (Historically salt has been obtained from natural sources in the United States and more recently, due to favorable economic conditions, some of it has been imported). Salt in turn can be upgraded to any one or more of the following products: sodium carbonate, sodium hydroxide, chlorine, and sodium nitrate. Ammonia can be upgraded to nitric acid and then to either ammonium nitrate, sodium nitrate and nitrogen tetroxide.

This general background of what has been termed the "chlorine" industry represents, in part, the complex vertical and horizontal integration that American chemical industry has undergone in arriving at diversification of products, raw materials, and intermediates.

PROCESS DEVELOPMENT

While commercial salt processes are limited to the use of sodium chloride, the salt process can be considered as the reaction of the chloride of an alkali metal with nitric acid to give chlorine and the nitrate. The fact that

TABLE 8-1. THE "CHLORINE" INDUSTRY—ITS MATERIALS, PRODUCTS AND PROCESSES.

Product	Process Name	Principal Raw Material	Over-all Reaction
Na_2CO_3	Solvay	NaCl	$CaCO_3 + C + O_2 + 2\ NaCl \longrightarrow Na_2CO_3 + CO_2 + CaCl_2$
NaOH & Cl_2	Lime-soda	Na_2CO_3	$Na_2CO_3 + Ca(OH)_2 \longrightarrow 2\ NaOH + Ca\ CO_3$
	Electrolytic	NaCl	$2\ NaCl + 2\ H_2O \longrightarrow 2\ NaOH + H_2 + Cl_2$
HNO_3	NH_3 oxidation	NH_3	$NH_3 + 2\ O_2 \longrightarrow H_2O + HNO_3$
NH_4NO_3	Neutralization	HNO_3, NH_3	$NH_3 + HNO_3 \longrightarrow NH_4NO_3$
N_2O_4, Cl_2 & $NaNO_3$	Salt process	NaCl, HNO_3	$3\ NaCl + 4\ HNO_3 \longrightarrow 3\ NaNO_3 + Cl_2 + NOCl + 2\ H_2O$
	a) W/NOCl oxidation	O_2	$2\ NOCl + O_2 \longrightarrow N_2O_4 + Cl_2$
	b) W/NOCl neutralization	Na_2CO_3	$3\ NOCl + 2\ Na_2CO_3 \longrightarrow 2\ NO + NaNO_3 + 3\ NaCl + 2\ CO_2$

chlorides are changed to nitrates by means of nitric acid has been known for a long time. Theoretically, one can trace its conception to the preparation of *aqua regia* in the eighth century, by the distillation of a mixture of niter, salt and sulfuric acid.[3]

In general, salt processes can be characterized by the following points:[2]

(1) On treating an alkali chloride with an excess of nitric acid, nitrates are obtained along with gaseous products such as chlorine, nitrosyl chloride and possibly other nitrogen oxychlorides.

(2) The mixture of gaseous products of the above reaction may be allowed to react with oxygen of the air to chlorine and nitric acid.

(3) The oxides of nitrogen can be handled in various ways. For example, as in an early process, they can be contacted with concentrated sulfuric acid wherein the nitrogen oxides are retained by the sulfuric acid and unabsorbed chlorine and HCl pass to a water wash with the HCl going to solution and the chlorine to product.

Historically, the chlorides studied for commercial development have been principally limited to the sodium chloride as given in Equation (8-3). This process has been known for a long time and was tried with various modifications on a commercial scale in England in the nineteenth century.[1] Many early investigators are credited with developments in this process. Among them are Watt and Tebutt, McDougal and Ramson, Goldschmidt, Lunge and Pelet, and, in 1847, Dunlop who patented a process for making chlorine from salt and nitric acid. In his patent, Dunlop proposed to treat the gases containing chlorine, nitrosyl chloride and nitric oxide with sulfuric acid to decompose the nitrosyl chloride into chlorine and nitric oxide. The nitric oxide would be absorbed by the sulfuric acid to form nitrosyl sulfuric acid. Chlorine leaving the absorber was washed with water to separate any HCl, and the chlorine was then sent to product. This process was operated for some time but was eventually abandoned due to the enormous quantities of sulfuric acid used.

After the abandonment of Dunlop's process, subsequent developments involved process studies using HCl as well as the alkali metals, but it was not until the twentieth century that a commercial process again came on the scene, involving sodium chloride and nitric acid. While several studies had been made for the production of a nitrate and chlorine from an alkali metal chloride, the process developed by Allied Chemical* is the only one, to the author's knowledge, that has been tested and proven commercially. The schematic flow diagram, as shown in Figure 8-1, represents the original process as installed commercially by Allied Chemical in 1936 at Hopewell, Virginia.

*See Patents cited under Technical Considerations.

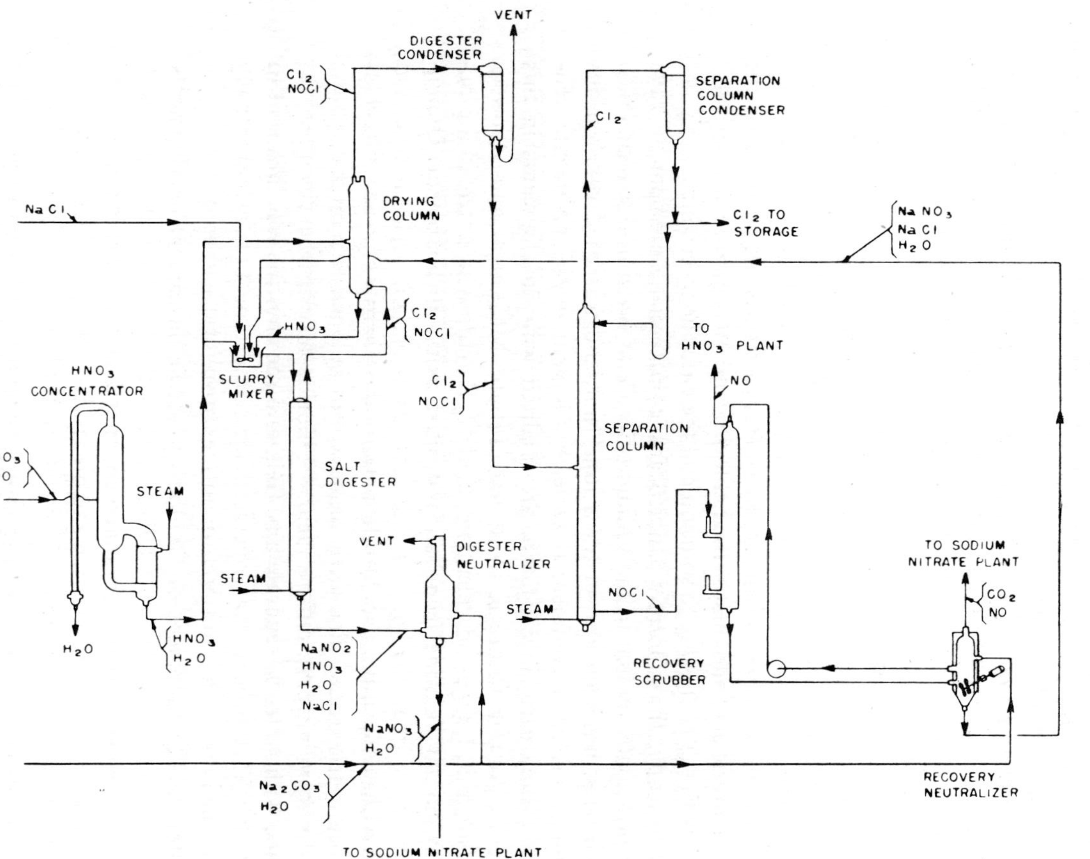

Figure 8-1. Chlorine by salt process with NOCl neutralization.

In this process, dilute nitric acid, which may be an aqueous acid containing 55 percent or less nitric acid, is first concentrated by evaporation of water. The resultant acid, (about 63 to 66 percent nitric acid) is then mixed with sodium chloride. From the mixer, the slurry passes into the top of the digester where it is heated and reaction takes place. The crystals gradually pass into solution and react to produce nitrosyl chloride and chlorine, in equal molar quantities, and sodium nitrate. The solution then passes down the column where nitrosyl chloride and chlorine are stripped to a point below which corrosion is negligible. The solution from the base of the digester enters the neutralizer where unconverted nitric acid is reacted with soda ash. The nitrosyl chloride-chlorine gaseous mixture, along with water vapor and some nitric acid, enters a drying column where the gas is scrubbed with refrigerated nitric acid. From the digestion system the gas passes to the separation system.

The nitrosyl chloride-chlorine gas from the salt digestion system is liquefied with refrigerated brine. The liquid nitrosyl chloride-chlorine mixture then runs by gravity into a separation column where it is separated, the nitrosyl chloride coming off the bottom as liquid and the chlorine off the top as gas. Gaseous chlorine is liquefied and sent to storage.

Liquid nitrosyl chloride from the separation column is sent to the recovery scrubber where it is contacted by an aqueous solution of soda ash, sodium nitrate and salt circulating from the recovery neutralizer. In this neutralizer system only a third of the nitrogen in the nitrosyl chloride is converted to sodium nitrate; the other two-thirds are converted to nitric oxide which is returned along with residual carbon dioxide to either acid or sodium nitrate manufacture. The stream from the recovery neutralizer, containing salt, sodium nitrate and water, is recycled back to the salt digestion system.

The over-all equation for the Salt process is as follows:

$$6\,NaCl + 12\,HNO_3 + 2\,Na_2CO_3 \longrightarrow$$
$$3\,Cl_2 + 10\,NaNO_3 + 2\,CO_2 + 2\,NO + 6\,H_2O \qquad (8\text{-}9)$$

This process, as originally installed, was used to convert salt, nitric acid and sodium carbonate entirely to chlorine and sodium nitrate. This process has come to be known as the salt process with nitrosyl chloride neutralization and, as shown in the flow sheet illustrated in Figure 8-1, is essentially the process employed by Allied Chemical until the 1950's.

The post-World War II period saw a trend toward higher analysis nitrogen fertilizers, resulting in a decline in demand for sodium nitrate as a fertilizer. The process with nitrosyl-chloride neutralization, as just described, gave a theoretical ratio of about four tons of sodium nitrate for every ton of chlorine, with approximately 40 percent of the sodium nitrate being derived from soda ash. Economics at that time indicated that the future of the salt process depended on the ability to make chlorine with by-product sodium nitrate being derived from sodium chloride alone. This meant that ways had to be found to convert the nitric oxide to nitric acid rather than soda ash-based sodium nitrate so that the nitric acid could be upgraded to ammonium nitrate, which was becoming an important fertilizer nitrogen outlet.

In effect, process improvements at this point revolved about the handling of the by-product nitrosyl chloride. Rather than neutralize this nitrosyl chloride to salt and sodium nitrate, development studies were undertaken to find a means of converting the nitrosyl chloride to chlorine and some nitrogen compound that could be recovered for further processing and use. Just such a process was developed and installed in 1953 at the Hopewell plant. This process is illustrated in Figure 8-2 and involves the oxidation of nitrosyl chloride with oxygen to form nitrogen tetroxide and chlorine as follows:

$$2\,NOCl + O_2 \rightarrow N_2O_4 + Cl_2 \tag{8-8}$$

In this process, nitrosyl chloride from the separation column, shown also in Figure 8-1, is vaporized and oxidized with oxygen, cooled, condensed and distilled; the nitrogen tetroxide is recovered as liquid product and the chlorine, with unreacted nitrosyl chloride, is sent back to the separation column. The nitrogen tetroxide can be stored and sold as such or recycled to produce nitric acid. In addition, a 50 percent increase in chlorine manufacture is effected by eliminating the recycling of salt.

The two processes, developed in Figures 8-1 and 8-2, represent the current status of the salt process for chlorine manufacture in the United States. At present, virtually all the sodium nitrate produced by this process is derived from the cheap sodium ion source—the chloride—while all unreacted nitric acid is recovered.

PROCESS DETAILS

When considering how this process is carried out on an industrial scale, it becomes obvious that corrosion is one of the major operating problems. Indeed, it is just such a reason that undoubtedly delayed the commercial realization of the salt process, the ability to handle *aqua regia*, or variations thereof, having been an industrial stumbling block for some time. But development of improved corrosion resistant materials from year to year and their application to appropriate steps in the process has led to considerable improvement. However, because corrosion is such an important problem,

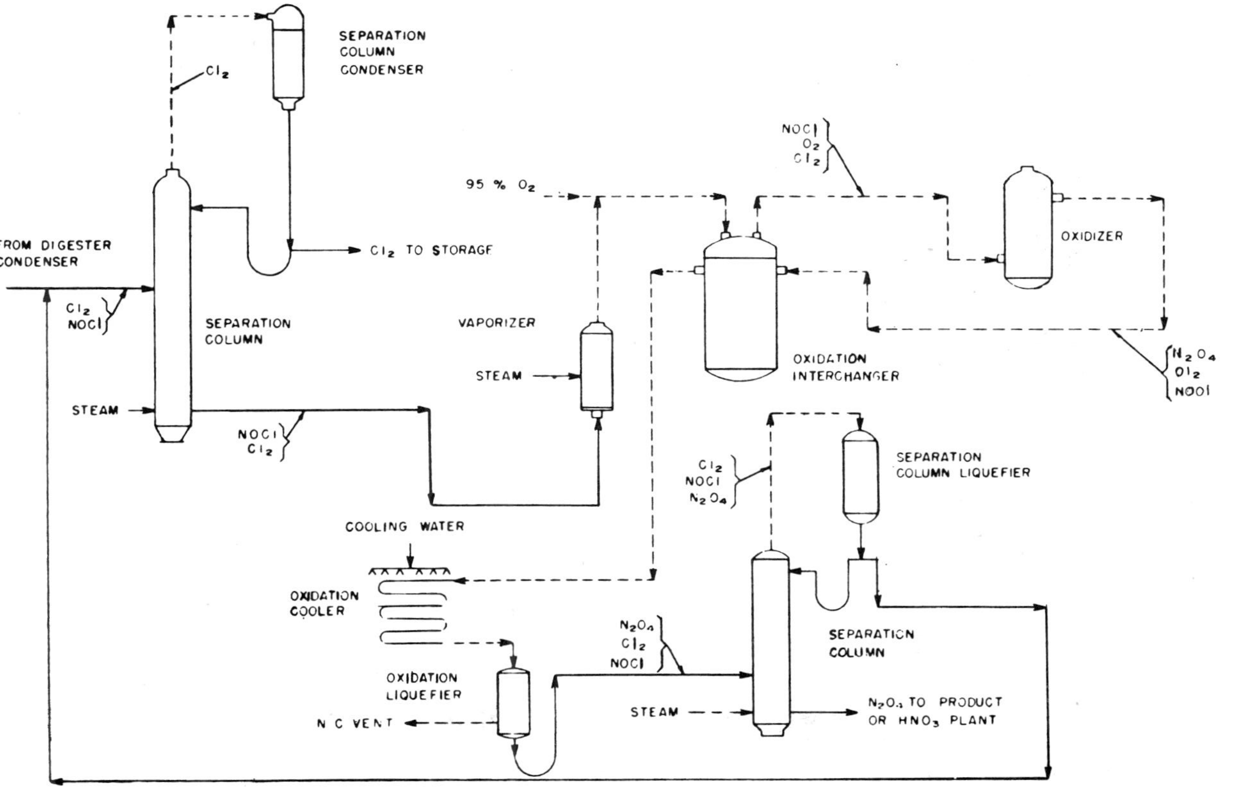

Figure 8-2.   NOCl oxidation.

in discussing the process details which follow, it has been deemed advisable to divide the process into the major units and then discuss operating details and corrosion problems together.

Salt Digestion

Quite naturally, a reaction mixture containing nitric acid and a chloride, and the gaseous products of reaction of the acid and the chloride are highly corrosive. It is of great importance, therefore, to provide an apparatus for the treatment of such a reaction mixture which will be fluid-tight and at the same time corrosion resistant. It is also of importance to provide an apparatus which is sufficiently rugged mechanically to withstand the stresses to which it is subjected in operation.

Many materials of construction, which are suitably resistant to corrosion, do not lend themselves to the construction of a single piece of apparatus of sufficient size for commercial operation; instead, it is necessary to build up the apparatus from sections or parts joined together. The joints between these sections, even though closed by a corrosion resistant cement, represent points at which a fluid might escape from the interior of the apparatus.

The salt digester used for the commercial process at Hopewell is comprised of a reaction vessel open at the bottom and provided with a closure for the top. The vessel and closure are constructed of chrome-iron alloys and acid-resistant brick. A shell, open at the bottom, surrounds the vessel, leaving a space between the inner wall of the shell and the outer wall of the vessel. A plate abuts both the bottom of the reaction vessel and the bottom of the shell. This plate is also constructed of a chrome-iron alloy.

In the space between the reaction vessel and the surrounding shell there is maintained a mantle of liquid which is noncorrosive towards the material of which the shell is constructed. The liquid mantle fills the space between the vessel and shell and is maintained under a pressure exceeding that within the vessel, so that any flow of fluid through the vessel is an inward one, rather than one permitting fluid to escape from the inside of the vessel. In such an arrangement the liquid forming the mantle is usually water and nitric acid or an aqueous nitrate solution.

However, it is not necessary that the mantling liquid fill the space between the reaction vessel and shell or that one of the above-mentioned liquids be used. One or both walls bounding the space between the vessel and shell may be washed with a liquid which will absorb and dilute fluids escaping into this space from the vessel. In such an arrangement the mantling liquid is under a pressure lower than the operating process pressure and is preferably water or dilute solution of an alkaline metal, e.g., sodium carbonate which will neutralize acidic fluids escaping from the inner vessel. The reaction vessel is a vertical tower containing an acid-resistant packing material, mounted on chrome-alloy metal plate.

For reaction in this vessel, salt suspended in 63 to 66 percent nitric acid, in a mole ratio of about 1 to 1.7–1.9, is first introduced into a slurry mixer which operates at about 40°C, a portion of the acid having previously passed through the drying column. The slurry is then fed to the salt digester. The slurry mixer is a chrome alloy tank.

After introduction of the salt-nitric acid mixture into the salt digester, reaction is carried on at about 75 to 125°C and at substantially atmospheric pressure. The nitrosyl chloride-chlorine gas mixture from the digester passes to the drying column, countercurrent to a stream of 63 to 66 percent nitric acid which has been previously cooled by refrigerated brine to about –10 to 0°C. The gas leaving the scrubber will then be dry enough to be handled in corrosion-resistant equipment.

Gas from the drying column passes into the digester condenser where it is liquefied at about atmospheric pressure by cooling with brine to about –15 to –20°C. The digester condenser is so elevated that the static head of the column of condensed nitrosyl chloride-chlorine mixture is used to force the liquid into the separation column. The digester condenser is also a chrome-alloy vessel.

Separation

Leaving the digester condenser, and under its own static head, the liquid nitrosyl chloride-chlorine mixture enters the separation column in which the chlorine is separated from the nitrosyl chloride by fractionation. The column operates at super-atmospheric pressure at a range of temperatures from 0 to 35°C. Reflux and chlorine product are condensed by refrigerated brine.

Liquid chlorine from the product condenser is then passed to storage, weighing and sales.

The original process, depicted in Figure 8-1, is no longer industrially important in the United States from the point of the separation column on; rather, the nitrosyl chloride is now oxidized, resulting in nitrogen tetroxide and chlorine as shown on Figure 8-2.

Nitrosyl Chloride Oxidation

In the nitrosyl chloride oxidation process, the nitrosyl chloride is drawn from the bottom of the separation column, vaporized and heated, along with an oxygen stream, in an oxidation interchanger to about 100 to

200°C. The oxygen and nitrosyl chloride mixture passes to an oxidizer wherein the nitrosyl chloride is oxidized to nitrogen tetroxide and chlorine. The resultant reaction-product mixture leaves the oxidizer at about 200 to 400°C, and it passes first to the oxidation interchanger for interchange with the feed stream, and then to an oxidation cooler where it is cooled to about 25 to 35°C. Next, the product mixture is passed to an oxidation liquefier where it is condensed with brine at a temperature of about –10 to –20°C, and then to a separation column, operating at about 15 to 60°C. There the nitrogen tetroxide and chlorine are separated, the nitrogen tetroxide passing as the bottom effluent and the chlorine, with traces of unreacted nitrosyl chloride, passing overhead. The entire nitrosyl chloride oxidation system operates at pressures up to 8 atmospheres.

As indicated previously the present salt process consists of salt digestion, chlorine recovery, and nitrosyl chloride oxidation to nitrogen tetroxide and additional chlorine. The over-all balance, expressed in terms of raw material usage and product distribution, is presented in Table 8-2.

TABLE 8-2. THEORETICAL RAW MATERIAL—PRODUCT
DISTRIBUTION SALT PROCESS

(Over-all Equation)

$$6\ NaCl + 8\ HNO_3 + O_2 = 6\ NaNO_3 + 4\ H_2O + 3\ Cl_2 + N_2O_4$$

Component	Tons of Component
Raw materials	
$NaCl$	1.65
HNO_3	2.37
O_2	0.15
	4.17
Products	
$NaNO_3$	2.40
Cl_2	1.00
N_2O_4	0.43
H_2O	0.34
	4.17

Technical Considerations

Most of the technical considerations embodied in the salt process for chlorine manufacture can be found in the patent literature issued through the United States patent office as well as through foreign patent offices. The subsequent material is primarily a listing of these patent references, which are arranged by classes in the following manner:

1) Patents devoted to the over-all salt process with specific references to sodium chloride as the alkali metal chloride being used as the raw material.
2) Patents involving the handling of nitrosyl chloride as produced in the salt process reaction.
3) Patents involving use of alkali metals other than sodium.
4) Patents designating materials of construction.
5) Miscellaneous.

Class 1. Among the most important patents issued in recent years on the salt process using sodium chloride, are those listed in the footnotes and arranged numerically.*

The main hindrance to the early development of the salt process lay in the fact that complete conversion of salt to elemental chlorine could not be accomplished. As described in United States Patent 2,181,559, this drawback was overcome by the then novel invention therein described. By controlling both the concentration of nitric acid and the salt-nitric acid feed ratio, virtually complete conversion of the sodium chloride to the elemental chlorine can be accomplished. By heating the proper reaction mixture with steam, complete evolution of gaseous nitrosyl chloride and chlorine can be accomplished. Maintenance of the reactor at the boiling point of the reaction mix provides for complete recovery in the gaseous state of all chlorine and nitrosyl chloride formed. These exit gases, containing some water and nitric acid, were then brought into contact with refrigerated nitric acid of the feed concentration for recovery of the vaporized water and nitric acid. Such a process dried the resultant chlorine and nitrosyl chloride mixture to the point that it could be handled in corrosion resistant materials.

Class 2. The nitrosyl chloride-chlorine separation involves no unusual processing techniques; ordinary distillation techniques are used to obtain the desired separation.

The degree of interest in means of disposing of the nitrosyl chloride is best illustrated by the large number of patents issued on the subject.†

*U. S. Patents: 1,036,611; 1,036,833; 1,965,400; 2,092,383; 2,138,016; 2,148,429; 2,148,793; 2,181,559; 2,212,835; 2,215,450; 2,225,685; 2,241,613; 2,269,000; 2,296,763; 2,535,989; 2,793,102.
British Patents: 310,230; 517,174.
German Patents: 630,652; 636,981.
†U. S. Patents: 2,004,663; 2,025,391; 2,038,083; 2,064,978; 2,130,519; 2,138,017; 2,159,528; 2,208,112; 2,210,439; 2,211,531; 2,215,451; 2,228,273; 2,240,668; 2,247,470; 2,258,771; 2,258,772; 2,261,329; 2,268,999; 2,296,328; 2,296,762; 2,297,281; 2,309,919; 2,320,257; 2,731,329.
German Patents: 526,476; 589,072.

While volumes have been written on the subject, the simple fact remains that the nitrosyl chloride oxidation process (described in U. S. Patent 2,130,519) is the only known commercial process for conversion of nitrosyl chloride to economically important products. It should be recalled, however, that low cost oxygen has been available only since World War II, and this factor undoubtedly delayed the commercialization of the nitrosyl chloride oxidation modification of the original salt process.

Class 3. While commercial salt processes have been limited to sodium chloride usage, considerable study has been made on other alkali and alkali earth metals, principally potassium chloride. Listed below are the chief patents issued on these studies.‡

Class 4. One patent** has been issued to designate materials of construction for the process.

Class 5. Several related patents are of interest for possible variations of the salt process.††

Market and Trends

In predicting the future of the salt process, the past history developed in the first section of this chapter must be recalled.

It must also be recalled that the original process became available just before sodium nitrate began its decline as a commercial fertilizer. The arrival of the nitrosyl chloride oxidation phase of the process has been too recent to permit re-evaluation of the present salt process in the light of existing markets. A very recent factor, lending impetus to the salt process, has been the selection of nitrogen tetroxide as the Titan II storable oxidizer.

The future role of nitrogen tetroxide in the missile program will have an important effect on the economics of the salt process.

Another variation of the salt process has recently received closer scrutiny. Such a variation embodies the use of potassium chloride rather than sodium chloride and results in the manufacture of potassium nitrate and chlorine. Operation is similar with one exception: minor amounts of soda ash now used must be replaced with potassium carbonate to prevent a product mixture of potassium nitrate with sodium nitrate as an impurity. Potassium carbonate is a relatively high cost raw material. The total tonnage requirement however is small. One American company has announced plans for construction of such a process for completion in early

---

‡U.S. Patents: 1,604,600; 1,899,123; 1,930,664; 2,057,957; 2,916,353.

**U.S. Patent: 2,201,423.

††U.S. Patents: 1,310,943; 1,658,519; 1,932,939; 2,072,947; 2,174,574; 2,287,555; 2,343,635; 2,347,073; 2,692,818; 2,761,761.

British Patents: 310,230.

1961. Several foreign countries are known to be actively studying the patent files for possible use of a potassium chloride-based salt process. Potassium nitrate provides two forms of fertilizer material—potassium and nitrogen; however it has the disadvantage of a very high K_2O to N ratio.

In summary, several factors have been presented in this text which will play an important role in the future of the salt process: health of the soda ash, caustic, the chlorine industry, demand for nitrogen tetroxide, adaptation of other alkali metal chloride such as potassium chloride, and the ever-increasing availability of newer corrosion-resistant materials of construction.

The process has been proven to be technically feasible and the future of its commercial use is dependent upon the factors herein listed.

References

1. DeJahn, F. W., *Chem. & Met. Eng.,* p. 537, October 1935.
2. Dominik, V., *Chem. & Ind.,* **18,** pp. 24–32, July 1927.
3. Mellor, J. M., "Modern Inorganic Chemistry," 8th Ed., 2nd impression, p. 623, 1934.
4. Shreve, "Chemical Process Industries," 1st Ed., p. 271 ff., 1945.

9. HCl OXIDATION PROCESSES

ALEXANDER REDNISS

Technical Enterprises, Incorporated

INTRODUCTION

Chlorine demands in the eighteen and nineteenth century were relatively small, and original supplies were produced by the oxidizing action of MnO_2 on HCl, without satisfactory manganese recovery as follows:

$$MnO_2 + 4\,HCl \rightarrow MnCl_2 + Cl_2 + H_2O$$

Weldon developed the first practical process in 1866 for decomposition of $MnCl_2$ and for producing MnO_2 for recycle, by the action of excess $Ca(OH)_2$ and air on $MnCl_2$. In both cases, however, the theoretical yield of chlorine is 50 percent. Weldon's modification of the MnO_2 oxidation enjoyed wide use until displaced by the Deacon process.[1]

The literature of the nineteenth century is replete with varied processes for producing chlorine by heating metallic chlorides in the presence of air e.g., $CaCl_2$, $MgCl_2$, $FeCl_3$, $NiCl_2$, Cr_2Cl_6, NH_4Cl and others. By-product HCl was available in large quantities from the LeBlanc soda ash process. In all of these processes air was the basic oxidizer, if only to recover the metallic oxide for recycle. The labor and heat costs of these processes were very high and led to their displacement by the Deacon Process, which enjoyed wide usage in Europe but was never commercially practiced in U.S.A.

However, the Deacon process was responsible for the world's chlorine for a good many years. And this process, i.e., direct catalytic oxidation of HCl by air, was the first large scale gas-phase catalytic process used by the chemical industry. Its problems were many, but the results were generally good. As long as the principal use of chlorine was for bleaching powder, the process had economic utility. However, liquid chlorine was very costly to produce because of dilution of the gas (about 10 percent) and lack of suitable solvent absorption processes.

The last European Deacon process plant closed around the time that mass markets for liquid chlorine developed in the 1920's thus succumbing to the superiority of the electrolysis of NaCl solutions.

As of 1960, the only chlorine produced industrially from HCl in the United States is manufactured by Allied Chemical Corporation at Hopewell, Virginia, where nitric acid is employed as the oxidant on both NaCl and HCl. This is covered in Chapter 8.

HCl CATALYTIC OXIDATION PROCESSES

As mentioned in Chapter 7, in recent years the needs for chlorine have increased at a faster rate than those for caustic soda. These needs, in turn, have created a real demand for chlorine production from by-product HCl and metal chlorides because electrolysis of salt produces more caustic soda than can be conveniently sold. This situation has been aided by the availability of tonage oxygen of about 95 percent purity at very low prices, resulting from efficiencies obtained in new processes of liquid air fractional distillation under high pressure. At the same time, new materials of construction have become available as follows: (1) plastics based on organo fluorine compounds; (2) polymers of furfural, ethylene, propylene and vinyl chloride; (3) newer HCl resistant metals such as tantalum, titanium, zirconium and others; (4) glass linings on steel of improved thermal shock resistance; (5) impervious graphite bonded by furan or phenolic resins; (6) improvements in glass and ceramics and pure silica for piping and structural materials. The list is very long indeed. These advances in technology and technical materials have stimulated re-evaluation of old processes. A great deal of research and development of HCl oxidation processes has been undertaken in all of the principal industrial countries. It has taken the following major directions:

(1) Improvements in the Deacon process using air or oxygen which are most promising indeed.[2,6,7,8,9,12,13]

(2) Chlorine liberation processes by action of oxygen on metallic chlorides especially ferric chloride.[1,6,10,11,12,13,18]

(3) Utilization of the Deacon process, with simultaneous chlorination of organics.[13,27,28,29]

(4) Oxidation of HCl to SO_3HCl, and catalytic decomposition to SO_2, chlorine and H_2SO_4.[3,19,20,21,22,23,24,25,26]

DEACON PROCESS AND MODIFICATIONS

The Deacon process, wherein air oxidizes HCl in vapor phase, over a hot copper containing catalyst, was patented in the U. S. A. in 1868.[13] Lunge[1] gives a very good and detailed description of the process as practiced about 1911. The principal difficulties of fugitive copper chlorides, low chlorine composition in exit gas, limited available materials of construction with high maintenance and poor catalyst activity, led in the 1920's to its displacement by electrolysis methods—which produce a practically pure chlorine at lower cost and at the same time produce caustic soda.

Today, however, due to glutted local markets, disposal of HCl is accompanied by serious problems and considerable expense in many areas of the world.

At the same time, the basic simplicity of the Deacon process, with its mildly exothermic reaction coupled with low electrical power and thermal needs, has attracted many chemists who have developed improvements, so that today it is again ready to take its place producing chlorine from by-product HCl from organic substitution chlorinations.

The reaction takes place in the vapor phase over a copper base catalyst as follows:

$$4\,HCl + O_2 \xrightarrow{\;(450-650^\circ\,C)\;} 2\,Cl_2 + 2\,H_2O$$

$$146 \qquad 32 \qquad\qquad 142 \qquad 36$$

$$\Delta F^\circ = -27{,}460 + (8.65\,T \log T) - 0.00229T^2 + 0.263 \times 10^{-6}\,T3 + 80T$$
(with 95% O_2)

There are no side reactions, or competing reactions; the problems are principally to develop operating conditions which best balance the increased rate of reaction at higher temperatures against higher yields at lower temperatures. This reaction is reversible and cannot proceed to completion from right to left but reaches an equilibrium which varies according to temperature and other conditions. Figure 9-1 gives the plot of equilibrium composition of gases, at both 450° C and 650° C for 95 percent oxygen.

Figure 9-2 gives a plot of the thermodynamic data with a plot of ΔF° and log Kp against temperature. Figure 9-3 gives adiabatic degrees of completion of oxidation as plotted against temperature or reaction, with various feed ratios and feed temperatures. A study of these curves indicates the favorable theoretical possibilities of this method of producing chlorine.

Airco Process

The Air Reduction Company has developed a number of improvements of the basic Deacon process in recent years using both air and oxygen.[5,14] The net result indicates that the process is practical today for by-product HCl. A product gas was produced at one ton per day level, at about 27

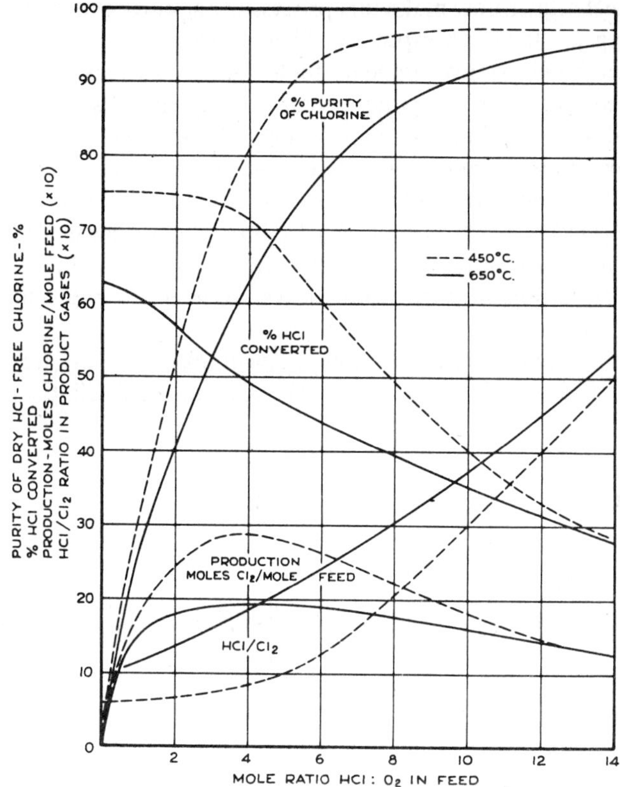

Figure 9-1. Equilibrium composition of gases in HCl oxidation process (purity of oxygen, 95%).

percent Cl_2 by volume with air, and about 90 percent Cl_2 with 95 percent oxygen. Conversions per pass with high $HCl:O_2$ ratios ran about 35 percent with air and they ran about 52 percent per pass for oxygen with low $HCl:O_2$ ratios. Over-all yields can be on the order of 90 to 95 percent of theory. Production costs using air under 1960 U.S.A. conditions are calculated to be between $15 and $45 per short ton of liquid chlorine, including all manufacturing and capital costs, depending primarily on the size of the plant and the availability of unsalable by-product HCl gas or solution. Economically, the minimum size unit is one producing about 15 tons per day. The problem of catalysis has been solved by the use of improved copper base catalyst bodies with at least one year of useful activity, utilizing rare earths as promotors and accelerators in a special reversing flow reactor which makes the reaction self-sustaining without the addition of external heat. New methods, materials of construction and engineering techniques now make this process economically practical with air or with oxygen—the latter especially when a large capacity oxygen plant is available on the site which can be useful for other purposes.

The advent of selective organic solvents for chlorine has further advanced the possibilities of commercial exploitation. Major American companies, now employ these methods on snift gas—using carbon tetrachloride as a solvent. Other and better solvents are also available, but data is not currently published.[14,15,16,17] A plot of solubility of chlorine in CCl_4 solution against temperatures given in Figure 9-4 shows the possibilities of employing CCl_4 for scrubbing either snift gases from electrolytic chlorine production or Deacon product gases with air—since both gases generally have about the same chlorine composition. Solvent losses of CCl_4 are claimed to be on the order of $3.00 per ton by the Diamond Alkali Company and

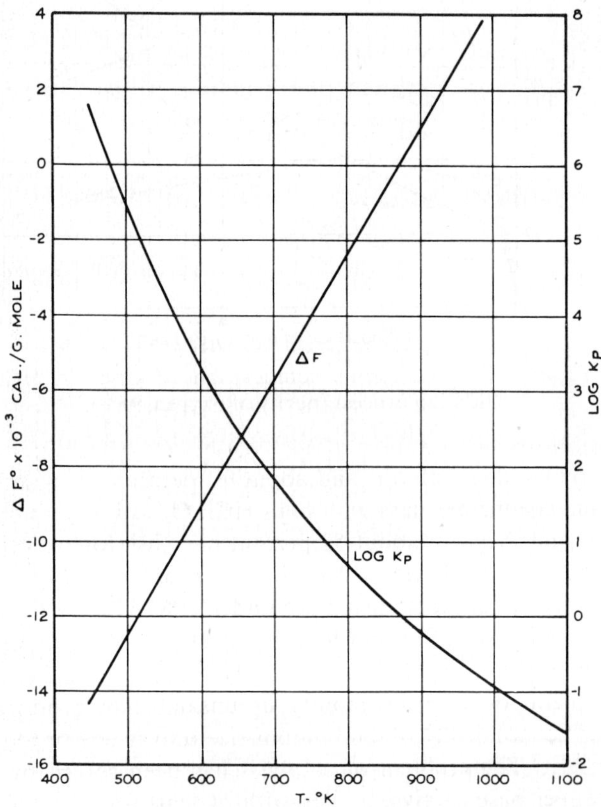

Figure 9-2. Thermodynamic data for reaction $4HCl + O_2 \rightleftharpoons 2H_2O(g) + 2Cl_2$.

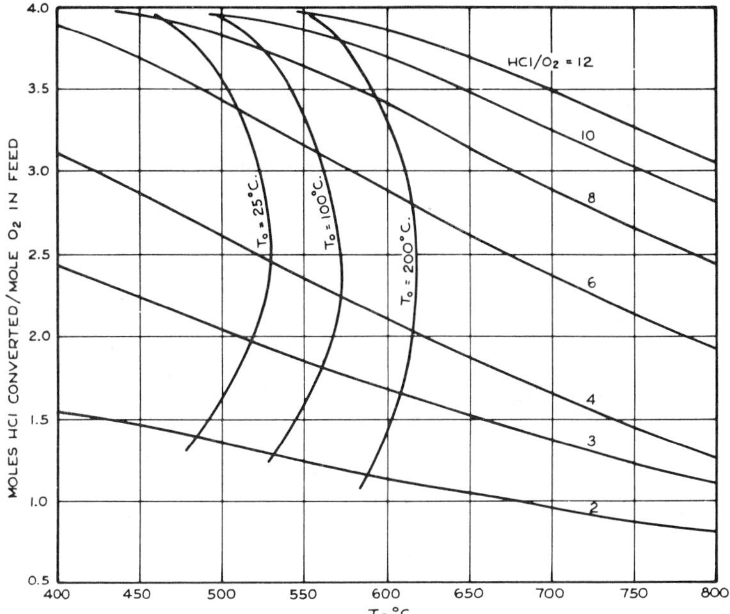

Figure 9-3.   Adiabatic equilibrium degrees of completion of oxidation of HCl for various feed ratios and temperatures (purity of oxygen, 95%).

much reduced by Hooker Chemical Corp. Process, using an additional high boiling solvent such as hexachlorobutadiene.

Sulfuric acid dehydration (and/or adsorption dehydration) of gas streams of chlorine-containing gas, allows the utilization of carbon steel equipment, with relatively small and predictable maintenance costs. Lowering capital costs in this manner makes the adsorption-dehydration processes of more economic industrial utility.

A flow sheet of the Airco process for oxygen is shown in Figure 9-5. The air system is similar to that of the Deacon process but equipment is considerably larger due to the larger gas flow. The functions of the equipment may be described as follows:

(1) The catalytic reactor consists of a U-shaped steel, lead and "foam-sil" lined catalyst container with both gaseous feed and discharge at the bottom. Periodic reversal of flow controls reaction temperature, preheats the feed and contains the copper, and the catalyst mass acts as an efficient heat regenerator. Gases leave the reactor slightly below 200° C.

(2) In the cooler, product gas is scrubbed by cold 33 to 36 percent

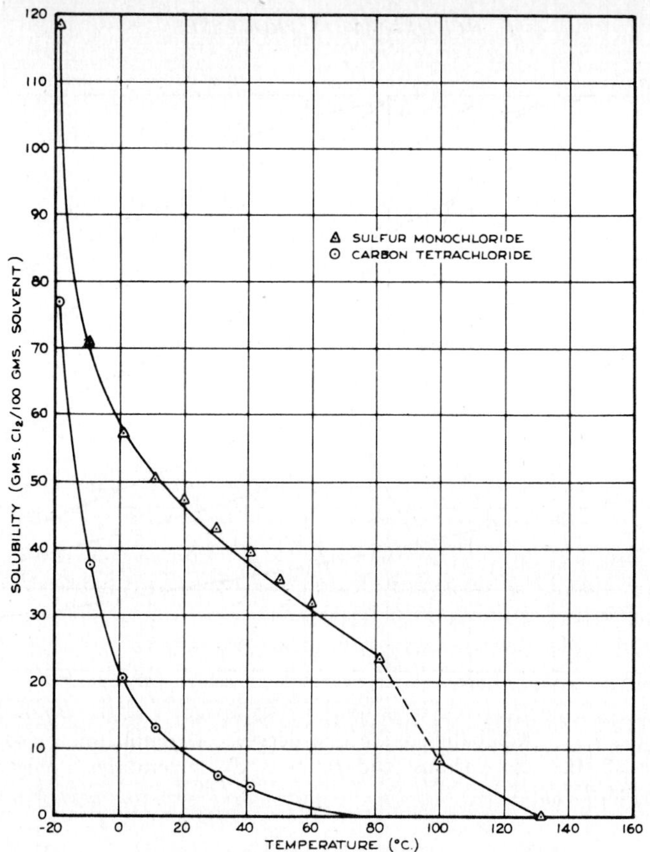

Figure 9-4.   Solubility of chlorine (atmospheric pressure).

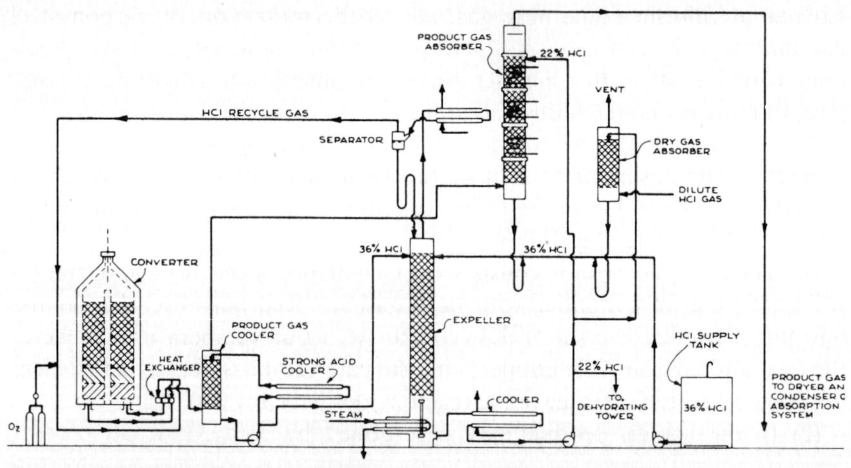

Figure 9-5.   Flow sheet—Air Reduction Co. process for oxidation of HCl with O₂

HCl solution in water.   Substantially all the water of reaction condenses to produce concentrated HCl solution, which is stripped of its HCl content above a H_2O-HCl azeotrope in the expeller described below.   This cooler is of chlorine and HCl resisting plastic construction.

(3) In the absorber, the cooled reactor gas stripped of most of its water, is absorbed in cold 20 to 22 percent HCl solution in a falling film unit of karbate and plastic construction.   It leaves as a 33 to 35 percent solution, which is then stripped in the expeller.   The absorbing liquid (20 percent HCl) is the cooled "foots" of the expeller.

(4) The expeller or stripper is a karbate packed-tower which is fed strong, cool HCl solution which is stripped to produce a "product" of 98 percent HCl gas and a foots as close as practical to 20 percent solution. The foots are cooled and recycled to the absorber.

The water build-up is discharged from the system by stripping HCl as gas, in a separate small system using H_2SO_4 or strong $CaCl_2$ solution.

(5) Dehydrating Tower.   In large plants, water is condensed first by chilling and secondly by sulfuric acid scrubbing, but in small plants the effluent gas from the absorber goes to a ceramic or polyester fiber-glass dehydration tower fed with 98 percent H_2SO_4.   The gas leaves the tower containing about 10 ppm of water, and the H_2SO_4 is left at about 70 percent concentration.   Corrosion at this water level is very small.

(6) The dry gas is then fed to a liquifaction system when oxygen is used and most of the snift gas recycled—or it is fed to an absorption-desorption system when air is used, although absorption-desorption may be used with oxygen also.   In either case, chlorine emerges as a pure liquid from the liquifaction system, under pressure and reduced temperature.

The Airco work has shown that the catalyst reactor successfully contains and controls the heat of reaction, that sufficient heat is generated to maintain continuous reactor operation, that yield with air and oxygen as well as chlorine composition of product gas is reasonably close to theory, and that temperatures and copper chloride volatility can be controlled within practical limits.   These and long catalyst life are important reasons for future utility of these modifications of the basic Deacon process.

Fluid-Bed Deacon Processes

The patent literature contains several fluid bed processes which should be promising because of ease of maintaining temperature and catalyst activity.   Copper volatility is reduced by partially condensing the water of reaction and feeding the solution back into the fluid bed.   This also helps remove part of the exothermic heat.   This method has a good many unsolved mechanical and corrosion problems, but portions of the catalyst can be removed for reactivation and more can be inserted without shutting down the operation.   The so-called hot spot, used to advantage in the Airco process, is minimized but the whole bed must be held at a higher average temperature.   More will be heard about this method in the future.[7,9,12,13]

Stripping HCl Solutions

Processes for stripping HCl solutions for recycle or feed fall generally into three principal systems as follows:

(1) *Use of Sulfuric Acid.* Contact in a tower between a 33 percent HCl solution and strong sulfuric acid causes a great local evolution of heat because of dilution of sulfuric acid. The solubility of HCl in sulfuric acid at elevated temperatures is practically zero, and thus dry 100 percent HCl leaves the top of the tower. The diluted sulfuric acid, leaves at the base of the tower at between 50 and 60 percent concentration as desired. This can be recycled after concentrating by evaporation or by SO_3 absorption from contact gas.

(2) *Use of Metallic Chlorides.* In this method, the 33 percent HCl solution is contacted in a tower by hot concentrated metal chloride solutions which take up its water by metal chloride dilution. Calcium chloride or Lithium chloride are preferred materials. The solubility of HCl in these hot solutions is very small, and HCl passes out the top as a gas—generally containing 1 to 2 percent water. The diluted hot metal salt is pumped to a submerged combustion unit to evaporate absorbed water, cooled a little, and recycled to the contact tower. When $CaCl_2$ is used, it absorbs at 55 percent and is evaporated at 50 percent by weight.

(3) *Use of Distillation.* In this process, 33 percent HCl is fed to a packed distillation tower equipped with a reboiler built of impervious graphite. The vapor pressure of HCl above its solutions increases sharply with increase in temperature and HCl gas passes out of the top of the tower with a water content in equilibrium with the temperature of the HCl feed solution; this will generally be between 1 and 5 percent water. From the bottom of the tower, the constant boiling solution (HCl-20 percent) leaves hot. It is cooled and is ready for refortification.

Any of the above systems may be used to obtain a gas feed for a Deacon or a Metal Salts process.

PRODUCTION OF CHLORINE FROM HCl BY USE OF METAL CHLORIDES

The utilization of metal chlorides decomposed by oxygen into chlorine and metallic oxides has been the subject of a great deal of work, from the mid-nineteenth century until today. Lunge[1] gives analyses and gives many references to early work. It is the subject of many patents.

The work of Grosvenor and Miller[10] is especially significant in this regard. This patent describes a fixed bed system in which the following reactions are alternately conducted.

$$Fe_2O_3 + 6\,HCl \xrightarrow{250-300°\,C} 2\,FeCl_3 + 3\,H_2O$$

$$2\,FeCl_3 + 1\tfrac{1}{2}\,O_2 \xrightarrow{475-500°\,C} Fe_2O_3 + 3\,Cl_2$$

In this process a catalytic mass is prepared by impregnating a porous carrier, such as calcined diatomite rock, with a mixture of metallic chlorides, the latter constituting between 25 percent and 50 percent of the final mass. The metallic chlorides are in the ratio of 30 parts iron, 5 parts cadmium, 26 parts potassium. This mass is dried and the chlorides converted to oxides by a stream of air or oxygen passing over the mass and held at 450° to 500° C until the chlorine evolution stops.

The ferric oxide is the working substance in the reaction with HCl to form $FeCl_3$. KCl is used as a vapor pressure depressant to reduce the amount of volatile $FeCl_3$ coming over in the chlorine stream. In this function it makes the process workable, as described below in the Oppau liquid molten catalyst process. The addition of cadmium as an accelerator to speed up the reaction is useful. This process has been the subject of intensive laboratory and pilot plant investigation.

A minimum of two fixed beds are utilized, one to conduct the chlorination of Fe_2O_3 with HCl at 250 to 300° C, and a second to conduct the oxidation at 450 to 500° C. They alternate positions in each cycle. HCl is fed into the reactor preheated to between 250 and 300° C, at which temperatures the rate of reaction is fast and complete. The HCl is passed in until it appears in the discharged effluent gas, containing water of reaction. Since this is only slightly exothermic, the feed must be preheated in order to maintain the temperatures needed and still avoid the problem of $FeCl_3$ volatility.

When the mass is converted to chlorides preheated nitrogen is passed in to sweep out HCl gas and, when finished, preheated oxygen is brought in to oxidize the ferric chloride to oxide and chlorine. Since this reaction, also, is only slightly exothermic, it is necessary to add heat and recover part of the exit heat. This leads to engineering complications in materials of construction, in addition to dilutions when cycles are shifted. The problem is partially overcome by the use of multiple beds, with an exit from the oxidizer preheating a chlorinator to reaction temperature of 500° C. This method eliminates some of the problems of the Deacon process, e.g., the separation of water of reaction and unreacted HCl from the product chlorine gas. It does, however, have more serious problems of materials of construction for heat transfer at relatively high temperatures with substantial quantities of heat to add or remove and $FeCl_3$ volatility. It produced a chlorine gas of 28 to 30 percent of composition with oxygen and about 12 to 15 percent (volume) with air. By the use of 3 to 5 reactors connected in series, the product gas can be brought to a maximum concentration of 65 to 70 percent thus consuming most of the excess oxygen. Concentrations above 70 percent have proved to be impractical perhaps due to reaching an equilibrium. Modern chlorine solvents should be used to separate pure chlorine, as liquid or gas, from its inert admixtures.

Grosvenor-Miller Process as a Continuous Deacon System

It is visualized that a 3 to 5 bed system can be used to produce chlorine up to 70 percent on a continuous system. A description of the three bed system will serve to illustrate the procedure. Reactor I is laden with chlorides at 250 to 300° C and is fed oxygen. In this bed the $FeCl_3$ is converted to chlorine and Fe_2O_3, and the gas leaves containing about 30 percent chlorine, with most of the balance unreacted oxygen. This effluent gas is passed into Reactor II which is maintained as a mixed bed of Fe_2O_3 and $FeCl_3$ at 500° C, to serve as both chlorinator and oxidizer. Here both HCl and O_2, with excess HCl, are passed in so that HCl reacts with Fe_2O_3. The chlorine does not react and with excess HCl passes through Reactor II. The effluent gas, containing chlorine, HCl, steam, and excess oxygen is passed into Reactor III (maintained at 250 to 400° C) which was previously Reactor I and thus oxide laden. Reactor III strips the HCl from the gas, chlorine does not react and passes through, carrying with it steam and excess oxygen, until it becomes saturated with chlorides. At this point the functions of Reactor III are transposed to those of Reactor I, and gas flow is III to II to I since I is now oxide laden.

A product gas composition approaching a maximum of 70 percent Cl_2 could be obtained in this way. This can easily produce liquid chlorine in an absorption-desorption system.

Dow Moving Bed Process

The Dow Chemical Company has done a great deal of development work on most phases of HCl oxidation to chlorine, as befits the major chlorine producer for captive use.

Two patents[6,12] have been issued for utilization of moving beds, one using a Deacon catalyst (copper chlorides) with partial water condensation to return vaporized copper to the fluid bed, and the other utilizing the Grosvenor-Miller principle, applied to a moving bed of catalyst pellets.

This technique of superimposing three connected moving, vertical beds so gravity would flow solid materials from one to the next below, and material from the lowest bed was removed from recycle by air is good, if a sticky catalyst can be avoided. The lowest bed is the oxidizing bed at 500°C where oxygen is introduced. The chlorine leaves the top of this lower bed, containing chlorine and excess oxygen. The patent claims a chlorine purity as high as 99.7 percent Cl_2 with that expected in commercial application of this process to be 90 percent. We doubt that 90 percent can be attained commercially because of a suspected equilibrium at about 70 to 80 percent. The circulation of catalyst from the bottom by a normal temperature air-lift cools the catalyst somewhat before dropping it to the top bed.

In the top circular bed, the returned catalyst can be cooled or heated, as desired, by passing air upwards through this bed. This is done to provide temperature control in the middle bed (350 to 400°C), which is the chlorination bed. Here HCl gas is fed to the bottom of the middle bed and passes upwards through downward moving catalyst, and the steam generated by the reaction is conducted out from the top of the bed.

The catalyst can be removed, heated or cooled as desired in this system. This provides a better chance to condense any vaporized $FeCl_3$ and return it to the bed. Catalyst can be removed continuously from the system, and replaced with active catalyst. This is an excellent approach to the recovery of chlorine from HCl, utilizing techniques commercially practiced in the oil industry. The process was developed in a 4 inch diameter reactor and then carried through a pilot plant stage at about one ton per day. The flow sheet in Figure 9-6 describes the process.

This process was the basis of a commercial plant built by Hercules Powder Company in Brunswick, Georgia, and designed for production of about 50 tons per day of chlorine from by-product HCl resulting from chlorinated camphenes. This daily capacity would normally be smaller than minimum normally economic in an oil refinery operation. Such a unit would probably operate at about 30 to 60 psig pressure most effectively. It utilized 95 percent oxygen to give maximum chlorine content which reached 80 percent but generally ran between 60 to 70 percent. The plant capacity, for a variety of reasons, was about 25 to 30 tons per day. Air utilization would have necessitated higher pressures and larger volume moving beds which were more costly than the oxygen plant. No recycle of oxygen was used. The flow sheet in Figure 9-6 gives an outline of this plant. A great deal of money and three years were spent to perfect this unit, but trace corrosion problems peculiar to Hercules feed gas were encountered, necessitating a feed gas purification installation to overcome very high maintenance costs in the remainder of the plant. Abrasion, sticking of catalyst and volatilization problems of $FeCl_3$ or $CuCl_2$ were also encountered. Finally the plant was closed and sold when pipeline chlorine became available there at a low price. It did prove that a moving

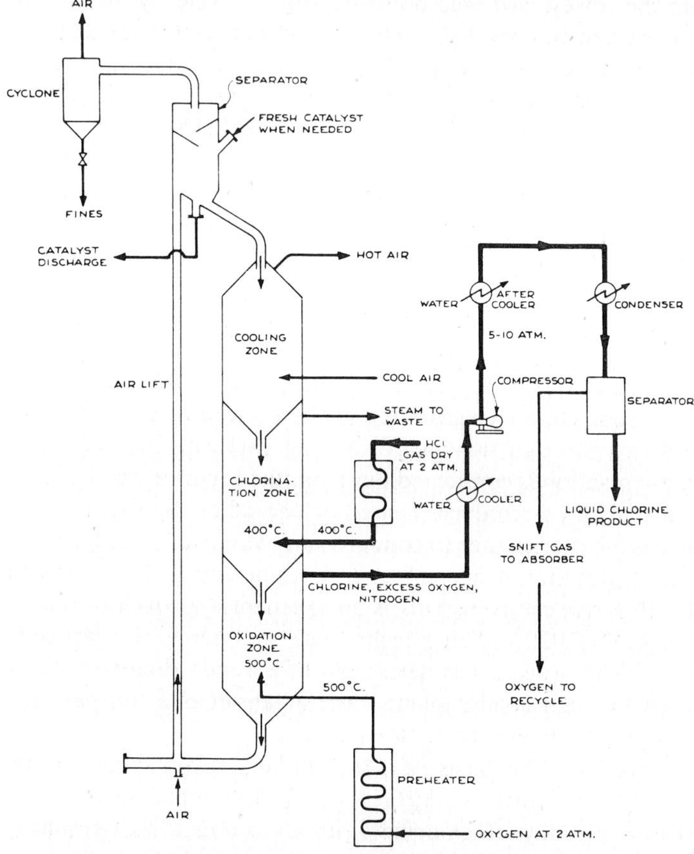

Figure 9-6. Flow sheet—Dow-Hercules system for chlorine production from HCl.

$$Fe_2O_3 + 6\,HCl \rightarrow 2\,FeCl_3 + 3\,H_2O$$

$$2\,FeCl_3 + 3/2\,O_2 \rightarrow Fe_2O_3 + 3\,Cl_2$$

bed process will work, and has the basic elements needed for a commercially successful plant. It is regrettable that this heroic effort was stopped, just when success was within reach.

Several other patents utilizing fluid bed are covered below.[7, 13]

Other Catalysts

The use of oxides of chromium and manganese as Deacon catalysts and as reactive oxides for the formation of metal chlorides with HCl with subsequent reaction with oxygen, or heat alone, to discharge chlorine.[11, 13]

Fixed beds have been discussed. Their activity is not as fast or complete as that of copper plus promotors, and generally temperatures are as high or higher.

III DEACON PROCESSES WITH MOLTEN METALLIC CHLORIDES

The process developed by I.G. Farbenindustrie at Oppau in a large pilot plant during the period 1939 to 1944 utilized a molten mixture of ferric chloride and potassium chloride.[18] In this process, the molten chlorides were the contact mass in a column about 4 meters high, with preheated feed, HCl and 98 percent oxygen; hot gas of about 28.3 percent HCl, 21.5 percent Cl_2, 29.4 percent O_2; $11.6 N_2$, was yielded, using an $HCl:O_2$ ratio 100 percent higher than the theoretical and with oxygen recycle.

The large oxygen excess drives the reaction to about 60 percent conversion of HCl per pass and the product gas after drying, containing 28 percent chlorine (volume), is absorbed in cold S_2Cl_2. The S_2Cl_2 liquid is pumped to the stripper where chlorine is released at a pressure of seven atmospheres, passing through a packed column in which it is washed by a reflux of liquid chlorine and freed of S_2Cl_2 or SCl_2 by condensation.

The process is generally described by the flow sheet in Figure 9-7 and by Rule.[18] Solubility of chlorine in S_2Cl_2 is given in Figure 9-4 in the section describing the Deacon process. Some details are as follows:

(1) *The reactor.* This is a vertical steel shell lined with a thick layer of acid and insulating brick, filled with a 4 meter height of molten salts. The

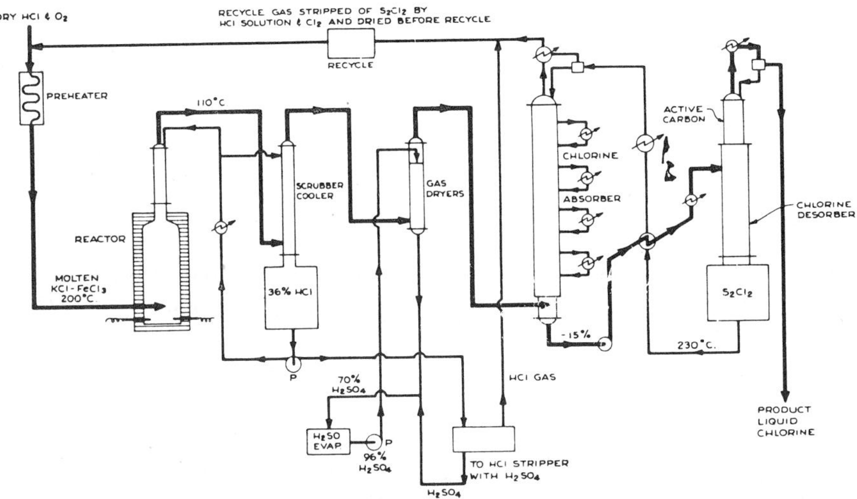

Figure 9-7. Flow sheet—I. G. Farbenindustrie process for oxidation of hydrogen chloride using molten $FeCl_3$.

HCl and O_2 feed, with the recycle stream, are preheated to 200° C and passed through this reactor. The gases are dried with H_2SO_4 to avoid excessive corrosion of the preheater and feeding system. Graphite electrodes extending into the molten salts are used whenever exothermic heat is not available to keep the bath molten. The reaction temperature is maintained at 450° C. The HCl : O_2 mole ratio was 2, and 98 percent oxygen feed was used.

The gases pass into a dephlegmator located on top of the reactor in order for vaporized $FeCl_3$ to reflow. It is visualized that at this point major difficulties will be encountered.

(2) *Cooler-Condenser-Dryer.* The gases from the dephlegmator are led to a packed tower fed with cold 36 percent HCl solution at 20° C in order to cool the gases and to condense most of the water of reaction as a strong HCl solution, which is stripped by sulfuric acid and recycled to reactor feed, or sold. The dilute (70 percent) sulfuric is reconcentrated to 96 percent for recycle.

This scrubber is a rubber lined, tile protected, packed tower. The cold HCl solution passes in at the top and leaves at the bottom. Part of the bottoms is cooled and recycled to the top of the tower and part sent to the stripper to recover their HCl content for recycle and also to discharge water of reaction from the system. The gases leaving the top of the scrubber have a composition of about 27.8 percent (volume) Cl_2, 1.2 percent H_2O, 32.2 percent O_2, 29.6 percent HCl and 9.2 percent N_2.

This gas is dried with 96 percent H_2SO_4 in two packed towers in series and leaves dry.

(3) *Absorption.* The dry gas is compressed by a Nash-type unit using S_2Cl_2 as a sealant. The gas enters at the bottom and rises. Recycled, stripped S_2Cl_2 at –20° C is fed to the top of the absorption tower, and then passes downward through several external recirculated cooler sections to remove the substantial heat of solution and reaction.

$$S_2Cl_2 + Cl_2 \rightarrow 2SCl_2$$

The temperature of the saturated liquid leaving the bottom of the tower is about –15° C, and solubility of chlorine is about mole for mole, but HCl remains insoluble and passes through the tower. This tower is built of carbon steel or low chromium steel.

The cold gases leaving the top of the tower are recycled after clean up to the reactor. They would have an approximate composition of 45.3 percent (volume) O_2, 41.7 percent HCl, 12.9 percent N_2, 0.1 percent S_2Cl_2.

Such gas contains enough S_2Cl_2 to form sulfates in the molten salts catalyst to raise its melting point to the point where it solidifies. Therefore it is stripped in a small packed tower sprayed with 50 percent H_2SO_4 plus a little chlorine and then it must be dried in another sulfuric acid drying tower. The gas composition from this tower has been analyzed as about the same as that entering, except the S_2Cl_2 has been replaced by about 0.4 percent Cl_2.

(4) *Desorption.* The saturated liquid from the bottom of the absorber, is heated to about 200° C and flashed into a packed tower operating at 7 atm gauge. This tower has four parts, i.e., a heavy steel still pot, a stainless steel bottom tower section, a steel middle section, and a steel top section filled with active carbon.

The function of the carbon is to absorb sulfur chlorides and a small reflux of liquid chlorine continuously washes the carbon free enough of sulfur so that the vapor chlorine emerges sulfur free, from the top of the desorber.

The hot liquid S_2Cl_2 from the bottom of the tower is continuously removed, cooled, refrigerated to –20° C and recycled to the absorber.

The pure chlorine gas leaving the top of the tower passes to a condenser. Since it is at 7 atm pressure, it can be liquified with cold water.

This process, was piloted in Germany. It has good potential to develop into a commercial process, but many small problems remain to be solved— especially those of materials of construction. The handling of S_2Cl_2 with traces of water is by no means easy and S_2Cl_2 traces in process streams must be avoided.

A later German Patent (857,796) in 1952 adds water to reduce the melting point of the molten catalyst mass which is useful for extended shutdowns.

It would be interesting to run this reaction under HCl rich conditions (large $HCl:O_2$ ratio) which may still produce a gas rich enough in chlorine to compete with the Deacon process, when organic adsorbants are used. This process is indeed worthy of further study.

The Shell Development Company has patented[9] the use of $PbCl_2$ as an activator and promotor in a liquid melt, similar to the Oppau process described above. They used mixtures such as 40 mol percent $CuCl_2$, 30 percent KCl, 10 percent NaCl and 20 percent $PbCl_2$, and held the fluid at 275° C to 375° C. This can be employed in a one- or two-step process to produce chlorine from HCl.

In the one-step process, operated similarly to the Oppau process, the HCl and O_2 are introduced together into the fluid melt. The HCl and O_2 react with lower valence chlorides to produce higher valence chlorides, and simultaneously the reduction of higher valence occurs to release chlorine and lower metal valence chlorides for reuse.

Although no real data is given in the patent, it seems promising as a potential method either to produce chlorine or to produce chlorine and utilize it *insitu* for oxychlorination of organic compounds.

This fluid melt has another attractive possibility—namely, use of a two stage process. In this case, HCl and air (or oxygen) are used to oxidize the metal chlorides in the melt from lower to higher valence in one reactor without substantial chlorine liberation. The H_2O of reaction, and inerts of air are discharged to waste. The fluid catalyst can then be pumped to a second reactor where heating of the catalyst mass evolves dry chlorine with a minimum of contamination if done continuously. Of course, this makes it necessary to use additional heat since the exothermic reaction takes place separately and loses some of the attractiveness of the true Deacon process. However, the advantages of lesser chlorine purification may overcome this disadvantage. The handling of the fluid melt also is advantageous, avoiding the complications of a fluid bed. The presence of new materials of construction should make this type of reaction more practical commercially than before.

Fluid melts have inherent advantages over solid contact or solid mass catalysts in that temporary mistakes do not poison the catalyst. Heat exchange is made easier and contact can be easily controlled. The problems of higher cost, greater difficulty of containment and control of chloride volatility should be solvable with sufficient effort.

USE OF THE DEACON PROCESS WITH A CHLORINE ACCEPTOR

In theory, all HCl oxidation processes form atomic chlorine at relatively high temperature where the rate of activity is very rapid. A most practical utilization of the chlorine recovered would be *insitu* chlorination of organic materials to take up chlorine as fast as it is formed and drive the reaction to completion.

Laboratory tests with Deacon catalysts indicate quantitative uptake of chlorine and practically zero HCl in the exit gas stream.

A number of patents have been issued utilizing methods of oxychlorination. Most of the major chemical companies interested in chlorinations have been and are working in this field. No statement, however, can be made as to current commercial utilization. Cass[27] was one of the early workers in this field. He covers the utilization of the Deacon reaction with ethylene as follows:

$$CH_2{=}CH_2 + HCl + \tfrac{1}{2}O_2 \rightarrow CHCl{=}CH_2 + H_2O$$

$$CH_2{=}CH_2 + 2\,HCl + O_2 \rightarrow CHCl{=}CHCl + 2\,H_2O$$

$$CH_2{=}CH_2 + 3\,HCl + 1\tfrac{1}{2}O_2 \rightarrow CHCl{=}CCl_2 + 4\,H_2O$$

$$CH_2{=}CH_2 + 4\,HCl + 2\,O_2 \rightarrow CCl_2{=}CCl_2 + 4\,H_2O$$

In these processes, HCl and ethylene are passed into the hot fixed-bed catalyst mass, at a lower temperature than that for chlorine recovery, simultaneously with O_2 or air. The above reactions take place rapidly because of the availability of reactive atomic chlorine proximate to the chlorine acceptor. These reactions can be run so that the yield of chlorinated hydrocarbons on HCl is practically quantitative. The hydrocarbons formed are a mixture of the various chemicals shown above. The ratio of one to the others is variable according to temperature, pressure, activity of the catalyst, ratios of feed, etc. The products of the reaction can be separated by distillation or other techniques. Potentially, this can be a most practical way to produce vinyl chloride, trichloroethylene or perchloroethylene, or other compounds with a double bond. Also these reactions apply to most other hydrocarbons, both aliphatic and aromatic.

Johnson and Chernaivsky[28] have patented this same process, using a fluid bed, continuously removing about 0.5 percent of the catalyst per hour with the effluent gases and quenching partially in cold HCl solution. This solution is used partly as feed to the fluid bed to control temperature and catalyst feedback and also partly as feedstock for a 55 percent $CaCl_2$ stripper operation to produce substantially pure HCl gas for recycle.

Oxychlorination processes should be applicable to production of a wide range of chlorinated hydrocarbons.

The commercially successful Raschig process for phenol produces chlorobenzene by passing HCl and air, with an excess of benzene, through the catalyst mass. The chlorobenzene is then hydrolyzed in vapor phase to phenol, stripping HCl for recycle.   The production of aromatic chlorine derivatives by oxychlorination should have a good future because of new materials of construction now available.[29]

Du Pont[30] is currently operating a pilot plant at Orange, Texas, using HCl and Oxygen in a fluidized bed reactor to produce 40 percent carbon tetrachloride, 40 percent chloroform, 15 percent methylene chloride and 5 percent methyl chloride.

CHLORINE FROM HCl USING SO_3 AS AN OXIDANT

The use of SO_3, readily available in sulfuric acid contact gas, is an interesting possibility for future recovery of chlorine from HCl *via* chlorosulfonic acid (HSO_3Cl). Enough has been done on this process to know that it works. The principal current obstacle is separation of SO_2 and Cl_2 in the final step. Johnstone[2] ably summarizes the state of the art and gives kinetics of the principal reactions involved.

$$HCl\,(g) + SO_3\,(g) \xrightarrow{\;70\text{-}100^\circ C\;} HSO_3\,Cl\,(l)\ \ \Delta H_{289} = 140.2\ \text{kg cal}$$

$$2\,HSO_3Cl \xrightarrow[\text{HgCl}_2]{160°\,C} SO_2Cl_2\,(g) + H_2SO_4\,(l)$$

$$\Delta H_{298} = 6190\,cal;\ \Delta F° = 6190 = 16.27\,T$$

$$SO_2Cl_2\,(l) \rightarrow SO_2 + Cl_2\,(g)\ \Delta H_{298} = 7760\,cal;\ \Delta F° = 7760\text{--}22.67\,T$$

$$SO_2Cl_2\,(g) \rightarrow SO_2 + Cl_2\ \Delta H_{298} = 11,100\,cal;\ \Delta F° = 10,500\text{--}461\,T$$

$$\log T - 16.08\,T$$

Chlorosulfonic acid is prepared in large scale commercial production from contact plant gas and dry HCl. There are many different systems employed, but in general both gases are passed to the base of a packed tower, with a slight excess of HCl. Flowing down the tower is cooled chlorosulfonic acid, which dissolves the SO_3 and exposes it to HCl for reaction. Enough circulation must be maintained to conduct away the substantial heat of reaction and sustain a temperature of about 100° C. The vent gasses are cooled and discharged. The product leaves with the cooling chlorosulfonic acid in circulation.

The catalytic decomposition of chlorosulfonic acid with mercuric chloride is conducted under pressure. The curves in Figure 9-8 show the decomposition of chlorosulfonic acid with 3.1 percent $HgCl_2$, and the effect

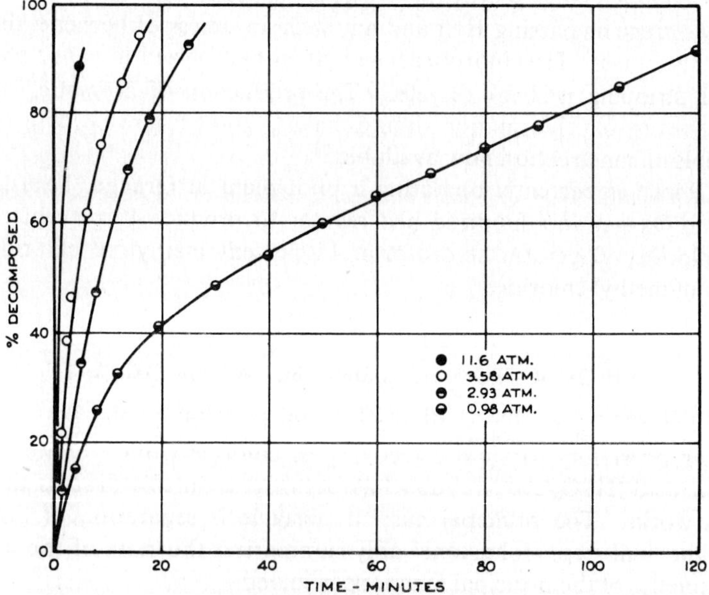

Figure 9-8.   Catalytic decomposition of chlorosulfonic acid in presence of mercuric chloride (3.1%).

of pressure on conversion rate. The rate of reaction between 3 and 11 atmospheres appears fast enough to be utilized in a continuous production of sulfuryl chloride in liquid phase. The difficulty seems to be the method of separation of catalyst from the batch. Table 9-1 is a tabulation of results obtained with several runs, and the calculation of the second-order rate

TABLE 9-1. SECOND-ORDER RATE CONSTANT FOR REACTION

$$2\ HSO_3Cl(1) \qquad H_2SO_4(1) + SO_2Cl_2(g)$$

Run	Temperature ($^\circ$C.)	Pressure of Boiling (atm)	Initial Conc. $HgCl_n$ (wt. %)	Range % Undecomposed	k
4	156	0.98	0.49	73.7-63.8	0.72
5	156	0.98	0.95	70.8-57.5	0.94
6	156	0.98	1.43	73.5-57.1	1.43
8	156	0.98	3.1	66.4-57.2	1.40
8	159	0.98	3.1	45.4-32.8	1.61
9	156	0.98	6.5	68.9-56.7	1.51
9	159	0.98	6.5	44.5-34.1	1.55
11	199	3.57	3.2	53.5-38.3	12.4
11	205	3.57	3.2	38.3-29.5	16.1
11	209	3.57	3.2	29.5-19.0	25.5
12	204	3.52	3.2	54.6-32.3	8.4
13	158	varying	3.1	39.3-21.3	1.12
15	145	varying	2.8	73.8-52.0	0.41

constant for the reaction. This was run by Kallal[19] in a continuous glass apparatus with indications that it could be made a continuous operation if operational problems were solved. Indications also are that steel equipment could be used. However, more bench scale work is required.

In the decomposition of sulfuryl chloride to SO_2 and Cl_2 activated charcoal is used as a catalyst. Figure 9-9 gives equilibrium decomposition for the reaction $SO_2Cl_2 \rightarrow SO_2 + Cl_2$. This reaction is visualized as one to be conducted in the vapor phase at elevated temperatures. The exit gas could be cooled sufficiently to remove most of the unreacted SO_2Cl_2 and the balance removed by scrubbing with cold sulfuric acid or in an activated carbon tower. The mixture of SO_2 and Cl_2 can then be passed through a solvent, such as CCl_4, in which SO_2 is barely soluble and Cl_2 is soluble. Several methods have been published for this separation, utilizing metal chlorides, tin, aluminum, and zirconium. The boiling points (SO_2 at -10° C and Cl_2 at -33.7° C) are far enough apart to permit separation in a short column, but formation of azeotropes prevents the use of this method.

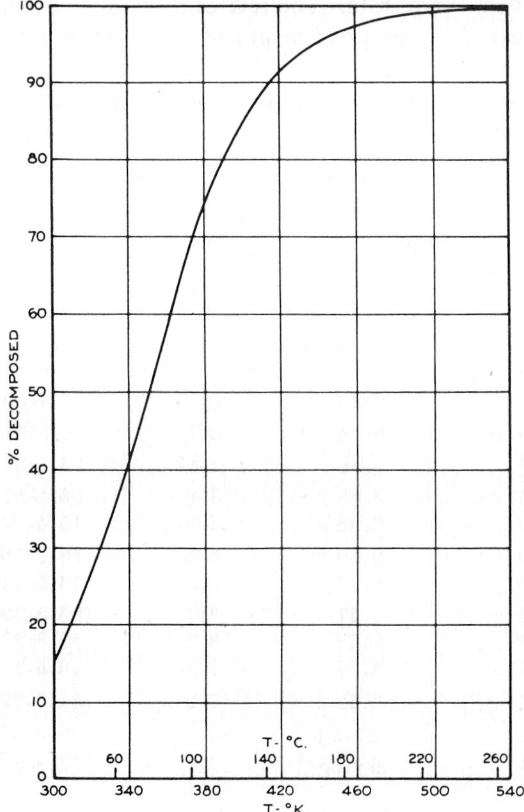

Figure 9-9. Equilibrium   decomposition   of
sulfuryl chloride.

Newer organic solvent systems for separation of the two gases should make
the process under discussion commercially interesting.

The over-all reaction of the three steps is as follows:

$$SO_3 + 2\,HCl \rightarrow H_2SO_4 + Cl_2 + SO_2$$

All three products are salable in relatively pure form. It is felt that this
process has great potential merit, but a good deal of bench and pilot plant
study will be required before commercial utilization is possible.[19, 20]

References

Sulfuric Acid and Alkali

1. Lunge, G., II; 263, III 525 D. Van Nostrand & Co. (1911).
2. Johnstone, H. F., *Chem. Eng. Progress,* **44,** 657 (1948).

3. Johnstone, H. F., "Production of Chlorine from Salt and Sulfur," Report to Office of Production Management (January 1942).

4. Hou, T. P., "Manufacture of Soda," ACS Monograph 65, 2nd ed., New York, Reinhold Publishing Corp., 1942.

5. Balcar, F. R. (to Air Reduction Co.), U. S. Patent 2,204,172 (1940); U. S. Patent 2,312,952 (1943); U. S. Patent 2,271,056 (1942); U. S. Patent 2,447,834 (1948).
Miller, H. S. (to Air Reduction Co.), U. S. Patent 2,204,733 (1940).

6. Davis, C. W., and Ehlers, F. A. (to Dow Chemical Co.), U. S. Patent 2,547,928 (1951).

7. Belchetz, A. (to M. W. Kellogg Co.), U. S. Patent 2,602,021 (1955).

8. de Jahn, F. W., U. S. Patent 2,330,114 (1943).

9. de Benedictes, A., and Luton, Jr., D. B., (to Shell Development Co.), U. S. Patent 2,448,255 (1948).
Johnson, A. J., and Cherniavsky, A. J. (to Shell Development Co.), U. S. Patent 2,746,844 (1956).

10. Grosvenor and Miller, (to Grosvenor Lab.), U. S. Patent 2,206,399 (1940).

11. Banner and Perrin, (to Diamond Alkali Co.), U. S. Patent 2,678,259 (1954).

12. Pye, D. J., and Joseph, W. J., (to Dow Chemical Co.), U. S. Patent 2,577,808 (1951).

13. Additional Patents References on Chlorine From HCl

Deacon, H.	U. S. Patent	85370	(1868)
Deacon, H.	U. S. Patent	141333	(1873)
Deacon, H.	U. S. Patent	165802	(1876)
Deacon, H.	British Patent	1927	(1876)
	U. S. Patent	*Number*	
Mond, L.	U. S. Patent	529130	(1896)
Goldschmidt	U. S. Patent	934400	(1909)
Kipper	U. S. Patent	1,355,020	(1918)
Kipper	U. S. Patent	1,512,225	(1924)
Cannon	U. S. Patent	1,355,105	(1920)
Muller	U. S. Patent	1,810,055	(1931)
Peir	U. S. Patent	1,845,058	(1932)
Mitchell, J. A.	U. S. Patent	1,979,280	(1934)
	U. S. Patent	2,051,962	(1936)
Gensch	U. S. Patent	1,984,369	(1934)
Odell	U. S. Patent	1,984,380	(1934)
Osborne, *et al*	U. S. Patent	2,020,431	(1935)
Calcott	U. S. Patent	2,034,896	(1936)
Drefus	U. S. Patent	2,075,889	(1937)
Barr	U. S. Patent	2,256,969	(1940)
Rudbach	U. S. Patent	2,270,903	(1942)
Morey	U. S. Patent	2,288,320	(1942)
Rosenstein	U. S. Patent	2,299,427	(1942)
Hemminger	U. S. Patent	2,303,047	(1942)
Thomas	U. S. Patent	2,304,128	(1942)
Halt	U. S. Patent	2,340,878	

Blaker	U. S. Patent	2,351,094	(1944)
Roetheli	U. S. Patent	2,376,190	(1945)
Walk	U. S. Patent	2,387,378	(1945)
Reeves	U. S. Patent	2,411,592	(1946)
Simpson	U. S. Patent	2,412,917	(1946)
Thompson	U. S. Patent	2,415,152	(1947)
Patterson	U. S. Patent	2,416,019	(1947)
Gorin	U. S. Patent	2,418,402	(1947)
Gorin	U. S. Patent	2,418,930	(1947)
Gorin	U. S. Patent	2,444,289	(1948)
Murphee	U. S. Patent	2,436,870	(1948)
Richardson	U. S. Patent	2,451,870	(1948)
Beck	U. S. Patent	2,464,480	(1949)
Sawyers, R. H.	U. S. Patent	2,642,339	(1955)
Reynolds	U. S. Patent	2,783,286	(1956)

14. "Chlorine Without Caustic, Economic Now?', *Chem. Week* (July 6, 1957).
15. U. S. Patents for Organic Absorption Processes

		Number	
Bouchard, F. J.	U. S. Patent	2,393,229	(1946)
Newbauer, F. A., *et al.*	U. S. Patent	2,540,905	(1951)
Hulme, R. E.	U. S. Patent	2,765,873	(1956)
Hooker, T., Gee-rong, E. J., and Maude, A. H.	U. S. Patent	2,841,243	(1958)

16. *Chem. Week,* p. 66 (September 17, 1958); *Chem. Processing,* p. 178 (September 1958).
17. *Ind. Eng. Chem.,* **49,** 87A (April 1957).
18. Rule, K. C., FIAT Final Report No. 833 (July 6, 1946).
19. Kallal, R. J., B.S. Thesis, U. of Illinois (1946).
20. Edwards, W. A., BIOS Final Report 243.
21. Jackson, K. E., *Chem. Rev.,* **25,** 81 (1939).
22. Johnson, R. S., M.S. Thesis, Univ. of Illinois (1946).
23. Kudryavtsev, S. A., *J. Applied Chem.* (USSR) **10,** 1260 (1940); *J. Phys. Chem.* (USSR) **13,** 1340 (1939).
24. Ruff, O., *Ber.,* **34,** 3509 (1901).
25. Sanger, C. R., and Riegel, E. R., Proc. *Am. Acad. Sci.,* **47,** 673 (1912).
26. McKee, R. H., and Saals, C. M., *Ind. Eng. Chem.,* **16,** 351 (1924).
27. Cass, O. W. (to E. I. du Pont de Nemours & Co.), U. S. Patents 2,308,489 and 2,327,174 (1943).
28. Johnson, A. J., and Chernaivsky, A. J. (to Shell Development Co.), U. S. Patent 2,542,961 (1951); U. S. Patent 2,644,846 (1953); U. S. Patent 2,746,844 (1956).
29. Prahl, Walter, U. S. Patent 1,963,761 (1934); U. S. Patent 2,035,917 (1936); U. S. Patent 2,156,402 (1939).
30. *Chem. Eng.,* **332,** 6 (1961).

10. PULP BLEACHING AND PURIFICATION

H. deV. Partridge

Hooker Chemical Corporation

THE HISTORY OF CHLORINE IN PULP BLEACHING

The first commercial use of chlorine for bleaching cellulose mentioned in the literature appears to have been that made in Russia about 1830, where it was used to bleach straw and esparto. The straw was disintegrated with caustic soda or caustic potash and then treated with chlorine. Both sodium hypochlorite and chlorine water are mentioned. By 1874 chloride of lime (probably hypochlorite) had been substituted for chlorine water. These steps were followed by pressure cooking in stationary cookers with caustic soda to produce defibration in preparation for bleaching.

Attempts to improve the physical properties of straw and esparto pulps led to defibering techniques involving a water soak or boil, instead of the alkaline stage, followed by exposure of the wet fibre to gaseous chlorine. The fibers were draped on racks in a chamber and exposed to chlorine gas. This treatment followed by a caustic extraction produced a defibered material of a light gray color easily bleached white with hypochlorite.

The first mention of the use of elemental chlorine in the United States was by Watt and Burgess who obtained U. S. Patent 11343 in 1854.

This combination of chlorination with chlorine gas or chlorine water, followed by alkaline extraction and hypochlorite bleach was well known by 1874. However, as so often happens, the procedure was ahead of the times and it was not until after World War I, with liquid chlorine available in tank cars, that the three-stage process began to be used for sulphite pulp, and with additional stages (up to six or seven) was adaptable to bleaching Kraft pulps to high whiteness.

Prior to 1918, the various cellulose pulps were bleached mainly in a single stage process using calcium hypochlorite which was prepared from bleaching powder purchased from a chemical supplier. Charles Tennant of Glasgow is credited with first producing bleaching powder in 1799 by absorbing chlorine gas in dry, hydrated lime. An excess of lime was necessary to give the powder stability. Bleach solution was made by dissolving this powder in water and allowing the excess solids to settle out. Between 1919 and 1930, with liquid chlorine available for bleach making in the mills, a rapid expansion in hypochlorite bleaching took place.

Batch chlorinators were also developed in which a weighed amount of liquid chlorine was introduced. This system eventually gave way to continuous chlorination; the chlorine liquid was vaporized in a suitable evaporator heated with hot water or steam and metered continuously as a gas into the stock ahead of a pump or mixer.

About 1930, Rue and Sconce[1] of Hooker Electrochemical Company introduced a new concept into the pulp bleaching field by the development of a Hooker two-phase chlorination system. In this process, excess chlorine was injected into the pulp with continuous vigorous agitation in specially designed in-line mixers. The stock then passed into a small tower with 3 to 5 minutes retention time, which is sufficient to react 80 to 90 percent of the chlorine added.

At this point, milk of lime was injected just ahead of another set of in-line mixers, and the pH was carried rapidly over to the alkaline side (pH 8 to 9). The bleaching was then continued in a second 1-hour retention tower to a low residual test.

This system is still in use in at least one very large Kraft mill in the United States and in many smaller mills, particularly for bleaching de-inked stock. This system does an excellent job for a small capital investment.

The successful use of continuous acid chlorination depended not only on an adequate supply of chlorine at a reasonable cost but also very heavily on the development of suitable materials of construction for the pipelines, mixers, towers, and washers.

The advent of rubber-lined pipe, pumps, towers, and rubber-covered washers made large-scale, continuous chlorination possible. Other resistant materials which have been developed and found suitable are Saran, PVC, glass-lined and fiberglass reinforced polyester resins such as Hooker's "Hetron" 72.

Chlorine Dioxide in Pulp Bleaching

As early as 1923 a German chemist, Erick Schmidt, who used a solution of chlorine dioxide to dissolve out the encrusting matter from wood, discovered that chlorine dioxide was an excellent delignifying and bleaching agent for cellulose. He found that it would attack the lignin without appreciably affecting the cellulose or hemicellulose fractions and obtained patents on bleaching pulp by this method. However, its hazardous nature and high cost in those early days delayed its development for nearly 25 years.

In the 1930's, Mathieson Chemical Corporation developed a process for manufacturing chlorine dioxide which was relatively cheap; they converted it into sodium chlorite and placed this chemical on the market around 1940. During the same period a great deal of research was done on bleaching with chlorites, which behave in a similar manner to ClO_2 when activated with chlorine, acid, or hypochlorite. Small quantities of specialty pulps were bleached with sodium chlorite but, despite reduction in price, sodium chlorite is still too costly for large-scale use in pulp bleaching.

In the search for a cheaper means of generating chlorine dioxide, attention was directed to the reduction of sodium chlorate with suitable reducing agents in a strongly acid medium; in 1946, three processes based on the reduction of sodium chlorate with sulphur dioxide in strong sulphuric acid were introduced into pulp mills in Canada and Sweden. These were the Rapson-Wayman process, installed at the Kipawa Mill of Canadian International Paper Company in June 1946; the Holst process, installed at the Husum Mill of Mo och Domsjo A/B in July 1946; and the Persson process installed at the Shutsgar Mill of Stora Kopparbergs A/B in Sweden during August 1946.

The next installations on this continent were made in 1950 at the Hawkesbury Mill of C.I.P. Company and the Natchez Mill of International Paper Company. In 1951, the Harmac Mill of MacMillan & Bloedel installed the only Holst plant on this continent, and in 1952, the Riegelwood Mill of Riegel Paper Company installed the first Solvay chlorine dioxide plant.

Since then, a very rapid increase in the number of plants, has taken place until nearly all bleached kraft mills and many bleached market sulphite pulp mills have installed chlorine dioxide plants. In all there are more than sixty plants on this continent, and more installations are being actively considered in many of the smaller mills.

Of the units in operation today, a large majority are Mathieson systems, with Solvay units in second place. There is one operating Day-Kesting unit, four Hooker R-2 plants, and a number of R-2 plants under construction.

PULP BLEACHING PROCEDURES AND STAGES

Pulp bleaching may be defined as the process of decolorizing and/or removing colored or color producing materials from cellulose fibres by the use of oxidizing or reducing substances, along with alkalies or acids. The color producing material may be degraded or altered so that it becomes soluble in the treating liquor and can be washed out, or it may be chemically altered so that it becomes more nearly colorless.

The choice of a bleaching agent depends on its ability to produce the desired improvement in brightness without undue degradation of the cellulose

 CHLORINE

and on its cost. Of the oxidizing agents available for pulp bleaching, elemental chlorine, hypochlorites and, in the last 10 years, chlorine dioxide have taken the major share of the market because of their effective purification and bleaching action combined with low cost.

The active agents in the chlorination of lignin are the molecules or ions present in a dilute aqueous solution of chlorine. These depend upon the pH of the solution and on the temperature of the reaction mixture, and may be represented by the following equation.

$$Cl_2 + H_2O \rightleftharpoons HCl + HOCl$$
$$H^+ + Cl^- \quad H^+ + OCl^-$$

At a pH of 2 the equilibrium is forced to the left, and the reaction is largely that of chlorination, with some oxidation due to the presence of small amounts of HOCl.

As the pH rises the equilibrium shifts producing more hypochlorous acid and a more oxidative reagent. As pH reaches 6 to 8 mixtures of hypochlorous acid and hypochlorite ion are largely present, and oxidative degradation of the cellulose is very severe. However, as pH rises above 9, hypochlorous acid is largely replaced with hypochlorite ion. The oxidation potential drops, and the damage to cellulose is minimized. Therefore, for maximum chlorination with minimum attack on cellulose, low pH and low temperature are required, and for maximum oxidation with minimum attack on cellulose, a pH of not less than 9 and preferably 10.5 is required. The pH range from 5 to 8.5 must be avoided in any bleaching operation with elemental chlorine.

The earliest bleaching operations with sulphite pulp were carried out batchwise using lime bleach without any particular attention being paid to the pH. A certain degree of delignification was obtained along with rather severe degradation of the cellulose. With the advent of liquid chlorine, the process developed into a two-stage operation consisting of acid chlorination in dilute solution followed by Bellmer bleaching at low consistency with alkaline hypochlorite.

As the need for chemical cellulose of high purity developed, a caustic extraction stage after the chlorination was introduced—strength of caustic being dictated by the *alpha* cellulose content required. Until very recently, sulphite paper and dissolving grade pulp were almost universally bleached with this combination.

As a result of pressure to improve the strength characteristics of sulphite pulps while maintaining brightness levels of 92 or 93 G.E., a chlorine dioxide stage has recently been added to the chlorination, caustic extraction and hypochlorite sequence.

The bleaching of sulphate pulps proved to be a more difficult task than that of sulphite pulp, and until the development of chlorine dioxide bleaching, it was next to impossible for a mill to deliver consistently 85 G.E. brightness, even with multiple hypochlorite stages. With a five stage system (chlorination, caustic extraction ClO_2, caustic extraction and ClO_2) 88 to 90 brightness can be obtained consistently, and 90 to 92 by careful control of the conditions.

A single-stage hypochlorite treatment on Kraft pulp can be expected to produce a pulp with 60 to 65 G.E. brightness and the three-stage (chlorination, extraction, hypochlorite) sequence gives 70 to 75 G.E. brightness. The latest development in Kraft pulp bleaching has been the elimination of the hypochlorite stage leaving only chlorination (C), caustic extraction (E), and chlorine dioxide (D) in a CEDED sequence, which gives 88 to 92 G.E. brightness. In the search for still higher brightness levels for Kraft pulps (91 to 92 +) other, more expensive bleaching agents than ClO_2, have been suggested as additional stages, e.g., peroxides and peracetic acid.

The above serves as a general introduction to the subject of pulp bleaching. A more detailed discussion of the various stages follows.

Pulp Chlorination

Acid chlorination is almost always the first step in bleaching wood pulp, whether for paper or for chemical cellulose. The treatment consists of injecting the required quantity of chlorine gas into the unbleached stock, either directly or through a water injector, mixing chlorine and stock intimately to assure uniform treatment of all fibres, and retaining the slurry until chlorine absorption is almost complete. The treatment is done at low consistency 3 to 4 percent, and usually at the prevailing water temperature. Chlorination is usually carried out in an upflow tower with a 60 to 90 minute retention time at the prevailing tonnage rates.

This tower has generally been sized to provide the proper time at the lowest seasonal water temperatures and it has been assumed that no appreciable degradation results from chlorination for these same times under summer conditions when temperatures may rise as high as 110° F. However, recent work by Rapson and Duncan[2] has shown that this assumption is incorrect. The effect of increasing temperature of chlorination on pulp viscosity for both sulphite and Kraft pulps is shown in Figure 10-1. The viscosities shown are those of pulps which have had a 1 percent caustic extraction treatment at 15 percent consistency for 2 hours at 50° C. The large drop in viscosity at the higher temperatures indicates that considerable degradation of the cellulose has taken place.

Figure 10-2 shows the rate of consumption of chlorine by Kraft pulp at several temperatures and under plant conditions (dotted line). Figure 10-3

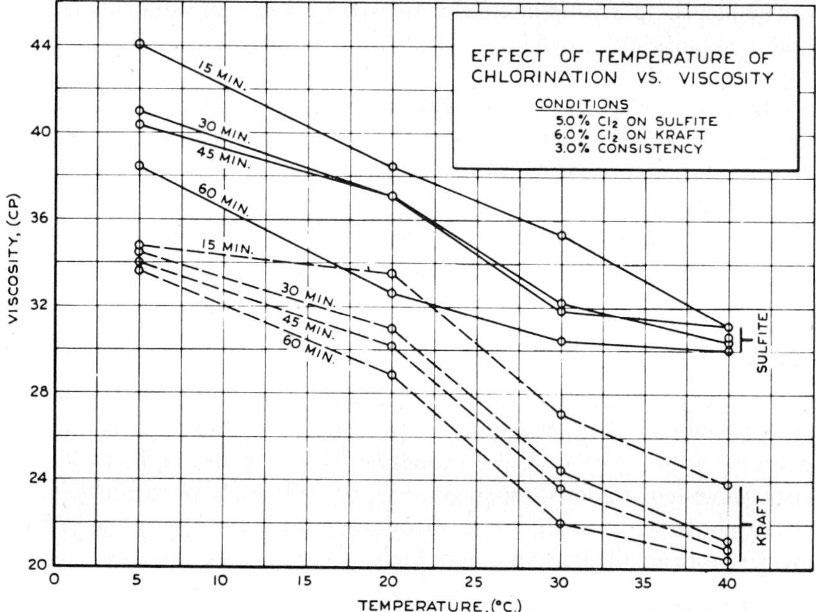

Figure 10-1

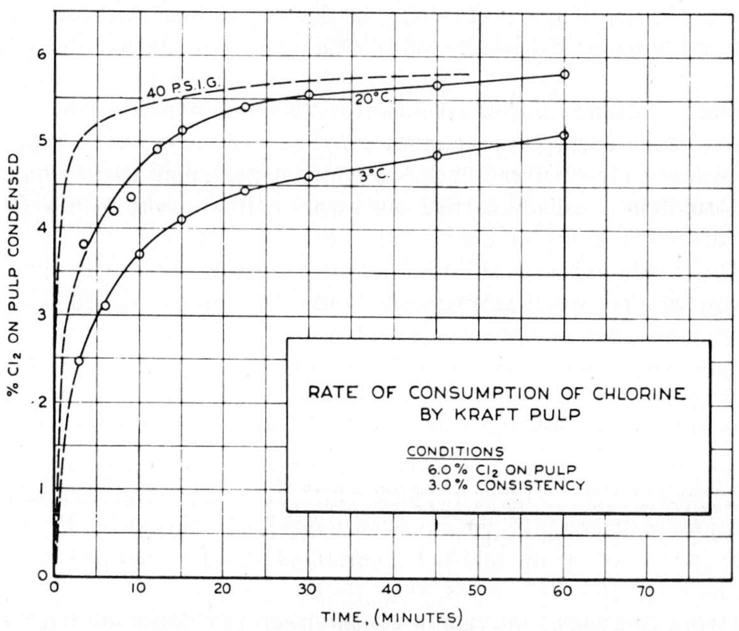

Figure 10-2

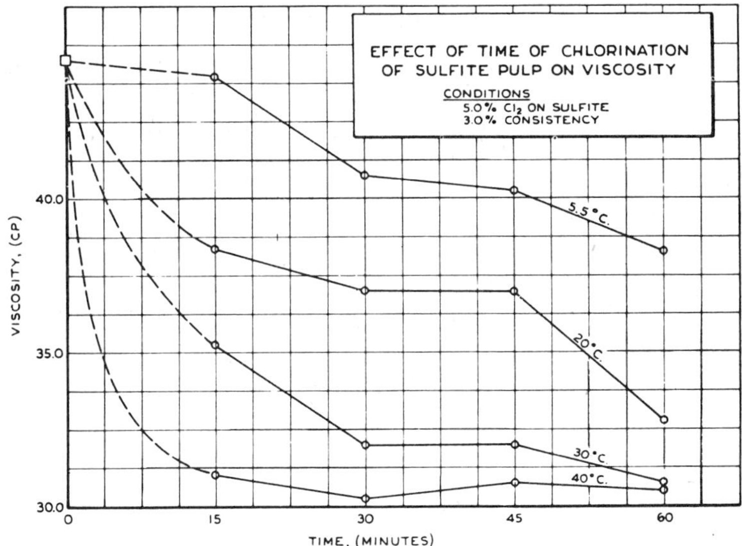

Figure 10-3

shows the effect of time of chlorination of sulphite pulp on viscosity and the rapid rate of degradation at the higher temperatures. Figures 10-4 and 10-5 show the effects of time and temperature on the Kappa Number of the pulp. Since the Kappa Number is proportional to lignin content, the curves show that for sulphite pulps the temperature of chlorination has little effect on the Kappa Number and that even at 5° C lignin removal is essentially complete after 15 minutes and further action is largely attack on the carbohydrate fractions.

With Kraft pulp, however, there is a greater dependence of Kappa Number on temperature and it appears that 20° C (68° F) is the highest temperature allowable for conserving viscosity and that the retention time need not be more than 20 to 30 minutes to affect lignin removal.

Figure 10-6 shows the effect of time and temperature on the hot caustic solubility of Kraft pulp. Solubility in 7.14 percent hot caustic soda solution is a measure of the carbonyl group content of pulp. The data again shows that high temperature increases hot alkali solubility and therefore presumably increases color reversion, yield loss, and chemical usage. By keeping the temperature down to 20° C, hot alkali solubility is kept at a minimum; if higher temperatures are to be used, the time should be shortened to minimize carbonyl group formation.

With the above considerations in mind it should be possible to arrange retention towers to provide variable chlorination times to meet prevailing

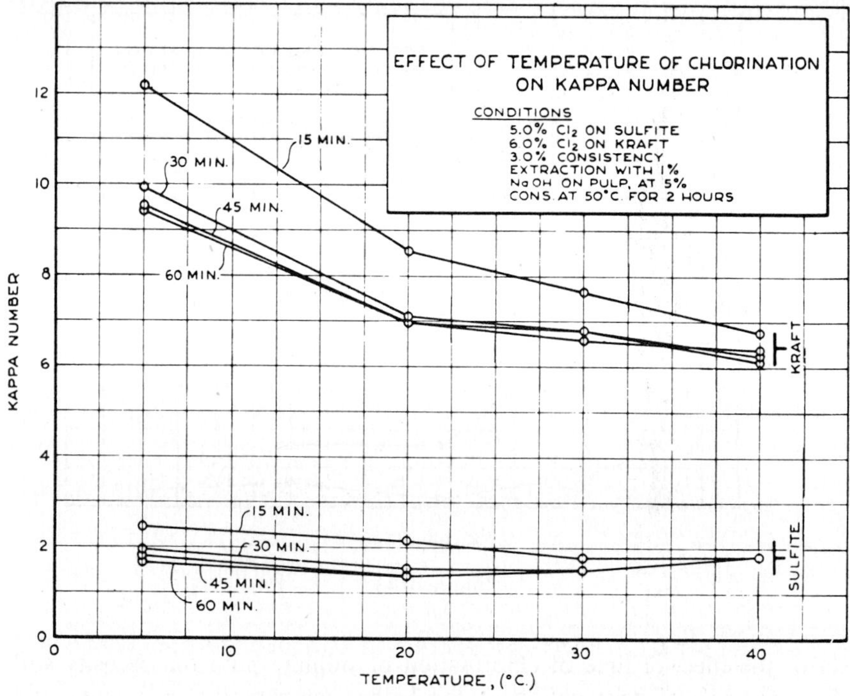

Figure 10-4

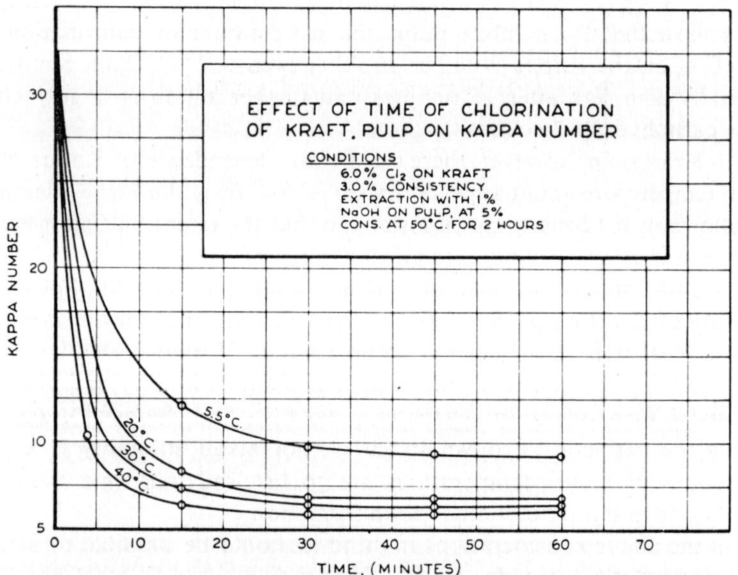

Figure 10-5

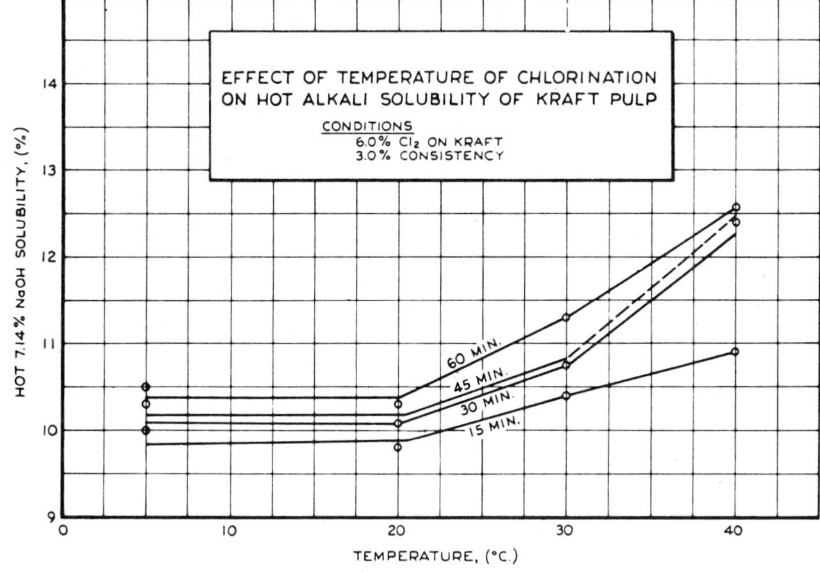

Figure 10-6

temperature conditions. Table 10-1 shows the optimum retention time for a given chlorination temperature.

Redox Control of Pulp Chlorination. The course of the reaction between pulp and chlorine in an aqueous medium can be followed by measuring the oxidation-reduction potential between chlorine and chloride and this potential can be used to control the addition of chlorine to the chlorination stage.

Oxidation potential can be calculated from the Nernst equation:

$$E = E_0 - \frac{0.059}{n} \log \frac{Ar}{Ap}$$

E = oxidation potential in volts

E_0 = oxidation potential at unit ratio of reactant to product

n = number of electrons exchanged

Ar = activity of reactant

Ap = activity of product

For pulp chlorination this equation can be written as:

$$E = E_0 - 0.059 \log \frac{\text{residual chlorine}}{\tfrac{1}{2}\,\text{chlorine consumed}}$$

TABLE 10-1. OPTIMUM RETENTION TIMES
PULP CHLORINATION

| Temperature | | Chlorination |
(°C)	(°F)	Time (min.)
5	41	120
10	50	80
20	68	40
30	86	20
40	104	10

In order to use oxidation-reduction potential to control the chlorine addition, a number of conditions must be met. These are well described in papers by Duncan[3] and Seymour.[4]

(1) Mixing of chlorine and pulp must be thorough and prolonged so that uniform reaction with fibres can be achieved and a representative sample of liquor obtained for the Redox potential measurement.

(2) Sufficient time (1 to 3 minutes) must be allowed to elapse so that 80 to 90 percent of the Cl_2 has reacted. At this point, the acidity due to the HCl formed by the reaction has lowered the pH sufficiently, to depress the hydrolysis of Cl_2 to HOCl and the rate of the various reactions has slowed. The Redox potential under these conditions becomes proportional to the chlorine-to-chloride ratio.

(3) It is recommended that for sulphite pulps, the ORP be measured on the 3 percent stock without dilution. Most Kraft bleachery installations, however, dilute to 1 percent to minimize plugging in the sample line before measuring the potential.

(4) If dilution to 1 percent is used at the measuring electrode, this dilution should be accurately controlled.

(5) Electrode design and dimensions are important and better results are obtained with larger area electrodes on 3 percent undiluted stock.

With due regard to the above principles, installations are now in use which operate smoothly, have shown substantial savings in chlorine and bleach liquor and have brought about substantial improvements in pulp uniformity and physical characteristics.

A successful installation for ORP controlled chlorination is shown in Figure 10-7. It consists of a chlorine-water injector. The control valve on the chlorine line is operated by the ORP controller which responds to the chlorine-chloride ratio at a point in the retention tower.

The chlorine-water injector of advanced design discharges the chlorine gas in the form of small bubbles into the stock stream. The mixture of stock, chlorine water, and chlorine gas then passes through a series of in-

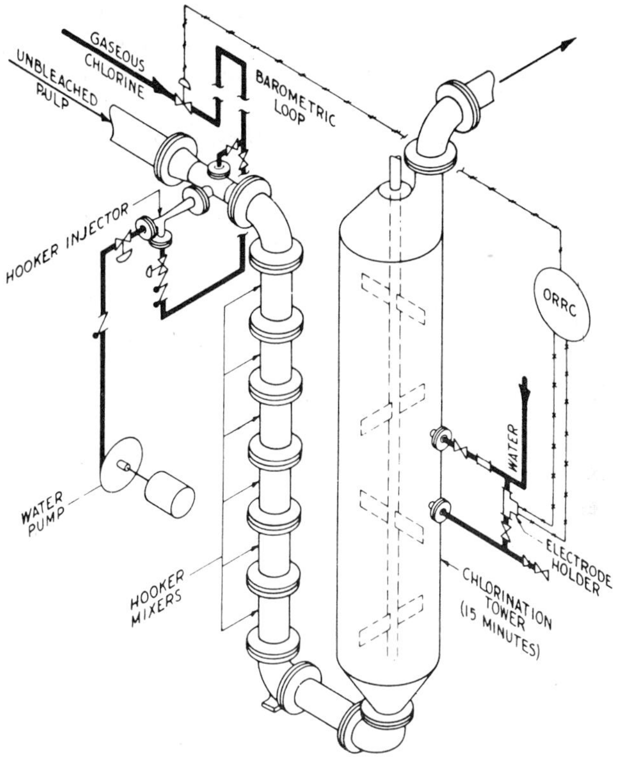

Figure 10-7.   Chlorine injector, mixer, and redox control.
(*Hooker Chemical Corp.*)

line mixers (Figure 10-8) which disperse the chlorine gas evenly throughout the pulp mass. The stock then passes to a 5-minute retention tower where mixing and reaction continue.

The sample for the ORP measurement is withdrawn from the tower, through a water aspirator. Its potential is measured at 3 percent stock density before dilution and after 3-minutes reaction time.

After the 5-minute pre-retention tower the stock can be pumped to a suitable retention tower. Retention towers should be constructed with takeoffs at different levels so that changes in chlorination time can be made to compensate for changes in chlorination temperatures.

The above detailed discussion of pulp chlorination and the equipment recommended raises the question of what is to be gained from better pulp chlorination. The following are the most easily discerned advantages.

Chemical Savings. A number of pulp mills have installed Hooker chlorine-water injectors and in-line mixers and have found savings in bleaching chemicals either in chlorine for chlorination or in hypochlorite or chlorine dioxide in later bleaching sequences. One mill[5] has reported a saving of \$0.50 to \$0.75 per ton of pulp. Others estimate the savings at 10 to 15 lb per ton of chlorine. Another west coast mill reported a slightly

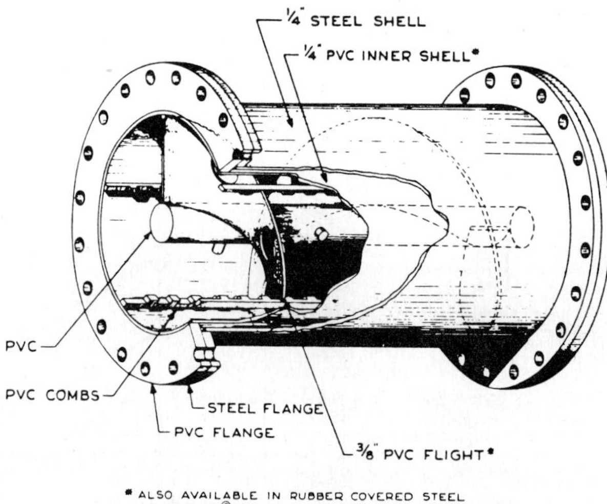

AVERAGE TONNAGE	MIXER DIAMETER	REMARKS
90	12	MINIMUM MIXERS REQUIRED KRAFT - 6 SULFITE - 3
120	14	
200	16	ΔP = APPROXIMATELY 2 LBS. TO 3 LBS. PER MIXER
300	18	O.A. LENGTH OF EACH MIXER = 3'-0"
400	20	MAXIMUM TEMPERATURE = 140°F.
500	22	STOCK CONSISTENCY = 3-3½%

Figure 10-8.   In-line mixer. (*Hooker Chemical Corp.*)

increased usage of chlorine but a substantial savings in ClO_2 in later bleaching stages. One large southern mill reports savings of about 10 pounds of chlorine per ton of pulp after installation of a specially designed external mixer with several minutes retention time and very vigorous agitation.

Improved Brightness with less Chemical Usage.   Under-chlorinated pulp fibres resulting from uneven chlorine distribution can not subsequently be brought to maximum brightness without severe treatment with attendant degradation. Improved chlorine mixing makes close ORP control possible with savings in chlorine usage, and also results in improved strength properties due to reduction in degradation during the chlorination stage, and reduced chemical requirements in later stages.

Caustic Extraction

The caustic extraction stage or stages in multistage bleaching are essential to obtaining the highest brightness levels with minimum strength loss and color reversion. For sulphite paper pulps, this stage consists of a very mild treatment with 0.5 to 1.0 percent NaOH based on pulp, at 80 to 100°F for perhaps 1 hour. The treatment is mild to conserve the hemicelluloses which contribute to paper strength. For kraft paper pulps the treatment must be more drastic in order to dissolve the products of the chlorination. The percent of NaOH will vary from 1.5 to 3.5 at about 140°F for about 2 hours. If a second caustic extraction stage is used between bleaching stages it will probably be a much milder treatment, e.g., 0.25 to 0.5 percent NaOH at 140°F for 1 hour. In any case, the pH of the caustic liquor at the end of the stage should not be less than 10.

Caustic extraction under these conditions breaks caustic sensitive linkages in the cellulose molecule and removes carbonyl groups which contribute heavily to color reversion.

Hypochlorite Bleaching

Hypochlorite bleaching with either sodium or calcium base liquor is now almost universally carried out at high density in continuous towers—either downflow or upflow. The reasons for this are the savings of steam and chemicals possible with high density and the capital savings realized by using smaller retention towers.

The problems connected with high density bleaching are largely those of providing sufficient mixing to ensure a uniform distribution of chemicals to every fiber in the pulp mass. The most effective means for doing this consists of diluting the incoming high density stock with the required solution and then thickening the stock again, leaving the proper amount of chemical intimately mixed with the stock at the desired treatment temperature. These installations, however, have been largely displaced for economic reasons by single or double shaft mixers which mix the dense stock with treating chemicals or steam and depend on diffusion to distribute the liquor evenly throughout the lumps of fibers.

pH in hypochlorite bleaching is extremely important. If the pH of the hypo stage is 6 to 8, the bleaching action is rapid and serious degradation occurs. This is manifested by rapid decrease in viscosity, increase in soda solubility, and copper number, indicating strength loss and later serious color reversion.

As the pH of the bleach liquor increases, the oxidation potential drops, degradation is lessened, and bleach consumption slowed down. Since the rate of consumption of bleach decreases with increasing pH of the bleach liquor, a compromise must be struck between pH and bleaching temperature to accomplish the required bleaching in the available reaction time.

The effects of pH on pulp properties of a sulphite and sulphate pulp bleached with chlorine are shown in Figures 10-9 and 10-10. The graphs

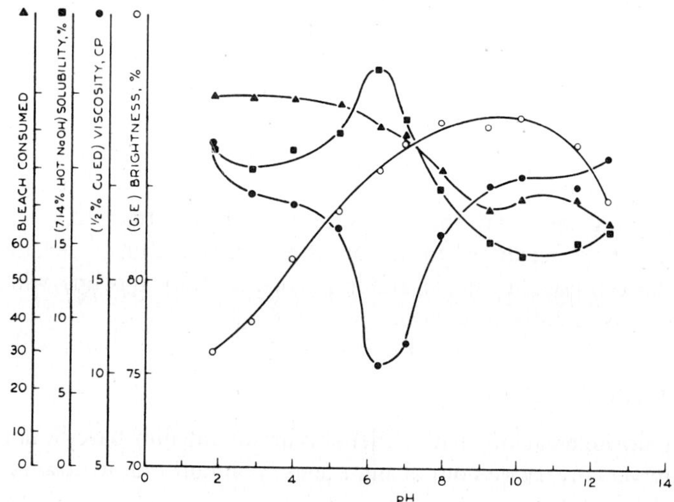

Figure 10-9. Effect of pH on bleaching chlorinated and extracted pulp with chlorine.

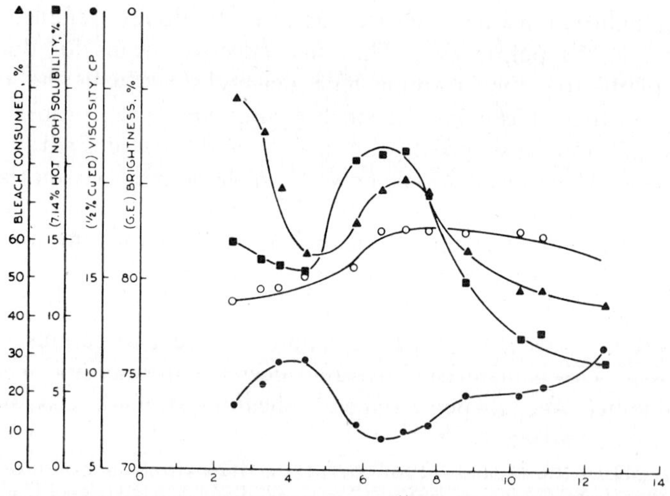

Figure 10-10. Effect of pH on bleaching partially purified sulfate pulp with chlorine.

show the effect of pH on a final hypochlorite stage and vividly illustrate the degrading effects on viscosity and the increase of hot soda solubility caused by bleaching in a pH range of 5 to 8 for both sulphite and Kraft pulps.

It is interesting to note that bleaching at pH > 10 decreases the hot caustic solubility below that of the original pulp sample, which suggests that at the high pH, hypochlorite bleaching should improve color stability. This is confirmed by mill experience.

Residual Bleach. Color reversion will take place in the bleaching tower in the absence of some residual bleach. It is not clear whether the bleach prevents the color formation or bleaches it out as fast as it develops. However, it is known that the formation of carbonyl and carboxyl groups is enhanced by the presence of water, and Rapson has shown that this reversion in aqueous solutions at various pH's actually takes place.[6]

It is therefore important to operate the hypochlorite bleach stage or stages to provide a definite residual chlorine content, and to maintain at least pH 9 at the end of the stage and preferably pH 10.

Chlorine Dioxide Bleaching

Chlorine dioxide bleaching is carried out at high density (11 to 13 percent) and high temperature (140 to 180° F) for 3 to 6 hours in one or two stages, depending upon the brightness required. The reaction vessel is generally either an upflow-downflow tower or an upflow tower. In either of these vessels the stock at about 12 percent consistency is pumped through a steam mixer which heats it, a chemical mixer where ClO_2 solution is added and dispersed, and into the bottom of the tower.

In an upflow tower the ClO_2 and stock pass into the bottom of the tower and move slowly to the top of the tower during a period of several hours. The ClO_2 is held in the stock under pressure of the stock in the tower until it has had time to react with the non-cellulosic constituents of the pulp.

In an upflow-downflow tower, the stock mixed with ClO_2 solution passes into the bottom of a pre-retention tube where the fastest part of the reaction takes place under the hydrostatic pressure prevailing in the tube. When the stock reaches the top of this tube it discharges into a larger downflow tower. There is some flashing off of unreacted ClO_2 at this point, and the ClO_2 released is generally picked up in water in a small packed tower and returned to ClO_2 storage. The stream of cold liquid should not be returned to the tower as it creates a cold zone in which bleaching practically ceases.

Chlorine Dioxide Towers. Of the two main types of ClO_2 towers available it is a matter of personal choice which type is used. Each one has certain features which are attractive.

In upflow towers the stock from the previous washer, at about 12 percent consistency is transferred to an external double-shaft mixer where steam to heat the stock to a tower temperature of 140 to 180° F is added. The stock then passes to the high consistency pump which moves it against the full pressure of the pulp in the retention tower.

The stock then passes through the radial flow mixer which is integral with the bottom of the tower or through an external mixer. The chlorine dioxide as an aqueous solution containing 6 to 10 gpl ClO_2 is introduced through two or more nozzles and is mixed with the stock by the rotating blades. The mixer in one case also acts as a diffuser to give uniform distribution of the stock in the tower and to prevent channeling.

The main difficulties with these towers is a lack of uniform mixing of ClO_2 with the stock and channeling. Both these problems are aggravated by higher consistencies, and it is often necessary to compromise to obtain good mixing. These mixers do not generally do a good mixing job at more than 12 percent consistency.

The Impco Upflow-Downflow tower is also very extensively used in ClO_2 bleaching stages. In this tower the stock from the previous washer, usually adjusted to the desired entering pH, is dropped into a double-shaft mixer where steam to heat the stock to tower temperature (140 to 180° F) is added. The stock then passes to the high consistency pump, which operates against the full pressure of stock in the pre-retention tube.

The high consistency pump moves the stock through a single-shaft mixer where the aqueous solution of ClO_2 is added and into the bottom of the pre-retention tube. This tube is of varying diameter depending on the rated capacity of the unit. The newest units are employing pre-retention tubes as large as 10 ft in diameter × 65 ft for a 600 ton per day unit. With this size tube the greater part of the reaction takes place before the stock enters the downflow tower, and flashing off of unreacted ClO_2 is minimized. Uniform distribution of the stock is obtained by building the pre-retention tube with a 60° cone bottom.

One of the advantages of the upflow-downflow tower is the variable retention time which can be achieved. However, the mixers used with this tower suffer from the same defects as the radial-flow mixers and the evidence suggests that for proper mixing of ClO_2 and stock, the consistency should be less than 12 percent. There is also considerable evidence that improved mixing can be accomplished with the Impco mixer by reducing the clearance between the blades and the casing. This is based on unpublished data from Dryden Pulp Company, the pioneer in this development.

Materials of Construction for Chlorine Dioxide Stages. Chlorine dioxide solutions at high temperatures are extremely corrosive and much work has gone into the development of suitable materials of construction.

The towers themselves are tile-lined and pointed with special cements which stand up satisfactorily.

From the point of injection of the ClO_2 solution to the tile-lining of the tower, Titanium is rapidly becoming the first choice. "Hastelloy C" mixers have shown variable life, and it is now customary to line the mixer with thin titanium sheet and cover the mixer blades with a layer of titanium. Recently a solid titanium mixer was installed which was the largest single titanium casting ever made. The pipe from mixer to tower may be titanium lined or, as in a recent installation, solid, welded titanium pipe. In other instances, fiberglass reinforced polyester pipe is being used successfully.

The ClO_2 washers are generally constructed of Type 316 or 317 stainless steel and are protected from corrosion by destroying the ClO_2 with SO_2 or by making the stock alkaline with caustic soda before it reaches the washer vat.

Sampling Points. For proper control of the ClO_2 stage sampling points are essential. In the upflow tower it is simple to obtain samples of the stock leaving the tower for residual ClO_2 and pH measurements and difficult to obtain uniform samples of the stock immediately after mixing of the ClO_2 and stock. It is even more difficult to obtain representative samples of stock from the upflow-downflow tower, and most of these towers are operated without adequate testing.

Conditions for Chlorine Dioxide Bleaching. Chlorine dioxide is most effective as a bleaching agent for cellulose at pH 6.0; for most effective bleaching it should be kept between 5.0 and 6.5 from the beginning to the end of the bleach. As the pH drops below 5, the brightness decreases for a given amount of ClO_2 at a given consistency, temperature, and retention time. As the pH increases over 6.5, under the same conditions, the brightness drops off rapidly. Chlorine dioxide bleaching has very little effect on viscosity below pH 7.0. Above pH 7, the viscosity drops off rapidly which indicates attack on the cellulose. Hot caustic solubility remains low up to pH 6.5, but increases rapidly as pH increases. This indicates an increase in aldehyde and/or ketone groups with attendant increased color reversion. Figure 10-11 shows the effects of pH on the bleaching of Kraft pulp with chlorine dioxide. On the basis of laboratory findings, the ideal conditions for the chlorine dioxide stage are to mix pulp, ClO_2 solution, and a buffer, in such a way that they enter the reaction tower at pH 6.0 and leave after the reaction at the same pH. However, the use of buffers is too costly, and pH must be adjusted with caustic soda or acid to bracket the pH 6.0 range.

The best compromise would be to adjust the pH at the point of entry to the tower so that the residual at the exit from the tower will be all chlorine dioxide with no chlorite present, as determined by a neutral and acid

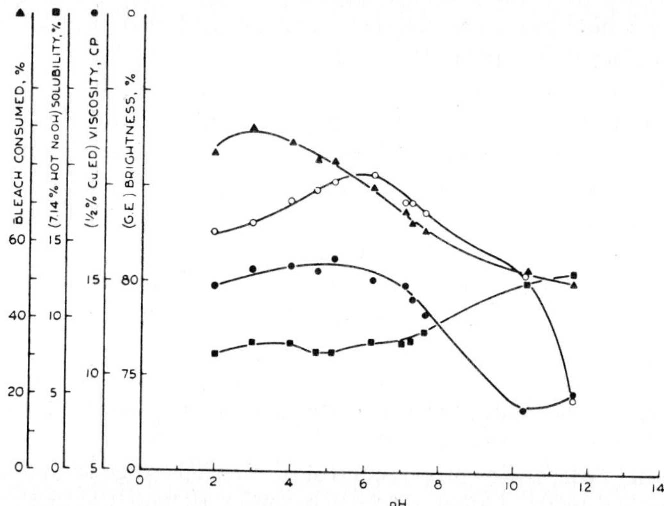

Figure 10-11.   Effect of pH on bleaching partially purified sulfate pulp with chlorine dioxide.

titration of the residual. It is not possible to specify the conditions as each situation differs in Kappa Number, amount of ClO_2 used, acidity of the ClO_2 solution, and retention time and temperature. But a starting pH of 7 with a gradual drop at the end of the stage to 3.5 is a reasonable compromise. In order to accomplish this, ClO_2 bleaching is generally done in two stages if more than 0.5 percent ClO_2 is to be applied. By introducing two-stage ClO_2 bleaching, 1 to 3 points higher brightness can be obtained than by a single stage. If a caustic extraction stage is added between the stages another point or two of brightness with improved brightness stability can be obtained.

The usual amount of chlorine dioxide applied to a softwood kraft in a sequence is 0.9 to 1.2 percent for 89 to 90 G.E. brightness. If the hypochlorite stage is eliminated, the ClO_2 may increase to 1.6 to 1.8 percent.

Residual ClO_2. A residual of ClO_2 at end of bleaching stage is absolutely essential to prevent color reversion in the tower. It has been shown[6] that color reversion takes place in the tower itself in the absence of a bleaching agent. The extent of the reversion depends upon the extent of carbonyl group formation produced by previous bleaching sequences. It is most severe following hypochlorite bleaching at pH 7.

Temperature of Treatment. Temperatures used in these stages in industry vary from 140 to 180° F, with 160° F being the preferred temperature.

Bleaching Sequences

The sequence and number of stages employed in pulp bleaching depends upon the type of stock being bleached and the degree of whiteness required. The simplest bleaching sequences consist of hypochlorite only (H) or chlorination followed by hypochlorite (C-H) and are now used mainly for bleaching de-inked ledger stock and de-inked stock containing some ground wood. The Hooker two-phase chlorination system (C-H) is also valuable for these applications.

In the Hooker system, the stock is first treated with chlorine for a short period (about 5 minutes) in a pre-retention tower and then made alkaline by injection of caustic or milk of lime and pumped to a retention tower to exhaust the alkaline bleach. In this system 70 to 80 percent of the chlorine is used up at a low pH and the remainder as hypochlorite bleach. This system uses more alkali for neutralization of HCl but is low in capital investment.

For sulphite pulps the sequence chlorination, caustic extraction, and hypochlorite bleach is almost universally used up to 86 G.E. brightness. In the last two years, the demand for higher brightness combined with equal or better strength has caused some mills to introduce a ClO_2 stage into this sequence either after the hypochlorite stage (CEHD) or before it (CEDH). These sequences make possible 92 G.E. brightness with excellent strength characteristics. The sequence CEDH can be accomplished without washing between the last two stages and is the subject of Canadian and U. S. Patents.[7]

For Hardwood Kraft pulps, the three-stage CEH sequence, is used for brightnesses up to 80 to 82. For Softwood Kraft pulps, it becomes imperative to use a minimum of 4 stages for 80 to 83 brightness and 5 (CEDED) or 6 (CEHDED) stages to obtain 88 to 90 brightness with maximum strength retention. The newest bleacheries for paper pulps on this continent are being built for the 5 stage CEDED sequence with no hypochlorite stage. For dissolving pulps, a hypochlorite stage is maintained to provide viscosity control. This will be discussed under bleaching of dissolving pulps.

PURIFICATION OF DISSOLVING PULPS

The purpose of the reagents used in purifying dissolving pulps is somewhat different than that of reagents used in bleaching paper pulp in that the stages are designed to remove hemicelluloses and non-cellulosic materials as completely as possible. The chlorination is essentially the same as for a paper pulp and serves the same purpose, that of removing as much

lignin as possible. With some hardwood pulps with high resin contents
the amount of chlorine is limited to keep the formation of insoluble
chlorinated resins to a minimum. The caustic extraction is much more
drastic than for paper pulps and treatment may consist of from 6 to 15 percent NaOH based on dry pulp at 200 to 235° F for 1 to 4 hours.
The amount of caustic and the temperature of the extraction depends upon
the *alpha* cellulose level desired in the finished pulp. The upper limit with
hot caustic extraction is about 96 percent. Surfactants are generally added
to this stage to remove as much of the solvent extractable fraction as possible. The most effective are the non-ionics of the nonyl phenol-polyoxy-
ethylene type. They are added in amounts from 0.05 to 0.5 percent, based
on O.D. pulp.

Hypochlorite Bleach

The hypochlorite stage is operated for both bleaching and viscosity
control and sometimes precedes the caustic extraction stage so that the
short-chain length fractions formed by the bleaching can be dissolved out,
leaving only the higher DP fractions of the cellulose in a narrower DP
distribution. The stage operates at high pH (10 to 11) and at temperatures
up to 160° F. The rate of change of viscosity is followed by rapid CED
or cuprammonium viscosity tests so that the bleaching can be controlled
within narrow viscosity limits. For viscose pulps, the rate of alkaline
depolymerization of the alkali cellulose is strongly dependent on the pH
at which the pulp was bleached with hypochlorite. Alkali cellulose from
pulp bleached at a pH of 6.5 to 8.0 ages much faster than alkali cellulose
from pulp bleached at a pH of 10+. This again appears to be connected
with the presence of carbonyl groups.

Chlorine dioxide bleaching of dissolving pulps is used where maximum
bleaching is required with minimum viscosity reduction. The special pulps
for cellulose acetate and for tire yarn manufacture fall in this category.
There is no essential difference in the operation of this stage for paper or
dissolving pulps. Chlorine dioxide treatment usually consists of two stages
separated by a mild caustic extraction stage.

Cold Caustic Extraction Stage

This special stage is used to prepare the highest grades of dissolving pulp
for both tire cord and acetylation. *Alpha* cellulose levels of 99 percent are
obtainable at much higher yields than with hot caustic treatments. Treatment consists of extracting the cellulose, usually after the first hot-caustic
stage, with 6 to 8 percent caustic soda solution or white liquor at 20 to
40° C for up to 1 hour and removing the caustic liquor by filtration and

displacement washing. The strong caustic liquor is recycled and subsequently recovered in the cooking liquor recovery system. The strong swelling produced increases the sensitivity of the pulp to degradation by hypochlorite and makes chlorine dioxide bleaching essential.

MULTISTAGE BLEACHING OF KRAFT PULP

Table 10-2 shows the laboratory results obtained on a northern softwood kraft pulp in the laboratory with various combinations of chlorination, caustic extraction, hypochlorite and chlorine dioxide treatments.

TABLE 10-2. LABORATORY BLEACHING OF MILL KRAFT PULP WITH ClO_2

Cl_2 (%)	NaOH (%)	ClO_2 (%)	Hypo (%)	NaOH (%)	ClO_2 (%)	G.E. Brightness Before Aging	After Aging	P.C. No	CED Viscosity, (cps)
4.5									
4.5	2.0					35.0	34.5	–	17.6
4.5	2.0	1.0				74.8	69.8	2.29	17.4
4.5	2.0	0.5	0.2			81.4	75.6	1.81	16.0
4.5	2.0	0.5	0.5			82.6	75.6	2.11	14.8
4.5	2.0	0.5	1.0			86.7	79.6	1.59	12.4
4.5	2.0	0.5	–	–	0.5	85.6	80.3	1.21	16.6
4.5	2.0	0.5	–	2.0	0.5	86.9	81.6	1.09	17.2
4.5	2.0	1.0	–	2.0	0.5	91.0	87.2	0.49	16.0
4.5	2.0	1.0	1.0	–	0.5	90.3	84.2	0.96	14.7

The treatments applied were as follows:

Chlorination:	4.5 percent Cl_2 on pulp, 25° C, 3 percent consistency, 1 hour
Caustic extraction:	2 percent NaOH on pulp 60° C, 12 percent consistency, 2 hours
Hypochlorite:	Percent NaOCl as shown, 40° C, 6 percent consistency, 3 hours
Chlorine dioxide:	Percent ClO_2 as shown, 70° C, 6 percent consistency, 3 hours

The data indicates that it is possible to reach 70 to 75 G.E. with a three-stage CED combination without viscosity loss with a 5 point reversion* and a high PC number.

A G.E. range of 81 to 86 is possible with a four-stage CEDH sequence using 0.2, 0.5, and 1.0 percent hypochlorite in the last stage. Reversion is again 5 to 6 points and viscosity drop is considerable with increasing hy-

*The reversion test used in this work was 18 hours in an oven at 105° C which is more severe than the usual 1 hour at 105° C.

pochlorite. The same brightness range 85 to 86 is obtainable with two stages of ClO_2 with 0.5 percent in the first stage and 0.5 percent in the second stage with a reversion of about 5 points but with considerably less viscosity drop for 86 point brightness. 90 G.E. brightness is obtainable with a hypochlorite stage between the two ClO_2 stages, but reversion is again over 6 points and viscosity drops considerably. 90 to 91 G.E. brightness is obtainable with a CEDED sequence, with the lowest reversion 4 points and minimum viscosity drop. This is the preferred sequence for brightness, strength characteristics, color stability, and cost. It is interesting to note that the caustic extraction between the chlorine dioxide stages raised the G.E. brightness level from 85.6 to 86.9 or a total of 1.3 points for the same 1 percent ClO_2 usage. This is a very inexpensive way to gain brightness and stability. As has been mentioned earlier in this chapter, many of the latest bleachery installations have been made with this sequence.

CHEMICAL USAGES IN MILL BLEACHERIES

At the 1961 summer meeting of the Canadian Pulp and Paper Association, a symposium on the chemical cost of bleaching was held at which chemical usage data was presented by a number of Canadian Mills. The data was based on yearly inventory figures calculated on oven-dry bleached pulp. A few of these figures for mills using different woods and bleaching sequences are given in Table 10-3.

The other striking fact about these data is that the mills trying to produce high brightness with four-stage bleacheries are using as much chemical for 85 to 87 brightness as 5 and 6 stage bleacheries are for 90 G.E. brightness.

Comparing mills B and G, Mill B with CEDED is producing 87.5 G.E. brightness from 22 K number pulp for less than Mill G which is producing 87 G.E. using CEHD sequence and 20 K number.

These comparisons are difficult to make accurately because of the variations in material costs in different locations and do not take into account relative costs of steam, labor, maintenance, overhead, and such things as wood yield and pulp physical properties. They do show, however, the importance of careful economic evaluation of the bleaching process and the continuing evolution of bleaching processes resulting from sound economic and technical studies.

PREPARATION OF CHEMICAL BLEACHING AGENTS

Chlorine Handling

The subject of safe handling of chlorine and the technique and equipment for vaporizing chlorine are thoroughly discussed in an earlier chapter in

TABLE 10-3. BLEACHING CHEMICALS USED IN VARIOUS MILLS

Mill	Woods Used	Average K number	Bleaching Chemicals Used, lb/O.D. Ton Bleached						Average G.E. Brightness
			C	E	H	D	E	D	
A	50% lodge pole pine 50% spruce	22	170 (25.4)	105 (24)	23 28[a] (.97)[b]	13.5 (1.03)	10 –	7.0 (1.6)	90.1
B	67% spruce 33% Jack pine	22	125 (11.5)	67 (pH 10.5)	– –	23 (3.6)	11	6 (2.9)	87.5
C	Spruce-fir	18	99 (6.4)	88.9 58 (.35)[c]	14[d] (.78)	24[d] (1.68)	–	–	85.3
D	Jack pine & white spruce	25.5	199	126	–	19.5	32	19.5	89.6
E	65% Western Hemlock 15% Cedar, larch spruce balsam	18.5	150	100	–	24	10	8	90–92
F	Hemlock, spruce, pine	22.0	138	67	–	26 4.4[e]	17.7	17 6.7[e]	89.5
G	Northern softwoods	20.0	81	86	78 (85)[a]	19.8	–	–	87
F	50% Maple: 40% birch +	13.5	78	24	–	20 4.4[e]	18	13.1 0.7(SO$_2$)	89.5
C	Mixed eastern hardwoods	13.0 18[f]	52	56 34[c]	58 6	21			86.5

[a]NaOH with hypochlorite for pH control
[b]Figures in brackets are residual(s) in lb/ton
[c]Pounds Cl$_2$ as hypochlorite added to caustic extraction
[d]ClO$_2$ stage precedes hypochlorite stage (CEDH)
[e]Pounds/ton NaOH for pH control
[f]Pre-bleach with hypochlorite before chlorination for pitch control.

this monograph and will not be repeated here. It is assumed in the discussions on bleach making and chlorination which follow, that an adequate supply of dry chlorine gas at a uniform temperature and pressure is available for use.

Calcium Hypochlorite System

Calcium hypochlorite bleach liquor of the desired composition is made by reacting a clarified milk of lime slurry and evaporated gaseous chlorine in a specially designed reactor (Figure 10-12). In recent years there has been a rapid change-over from batch to continuous systems. Hooker has

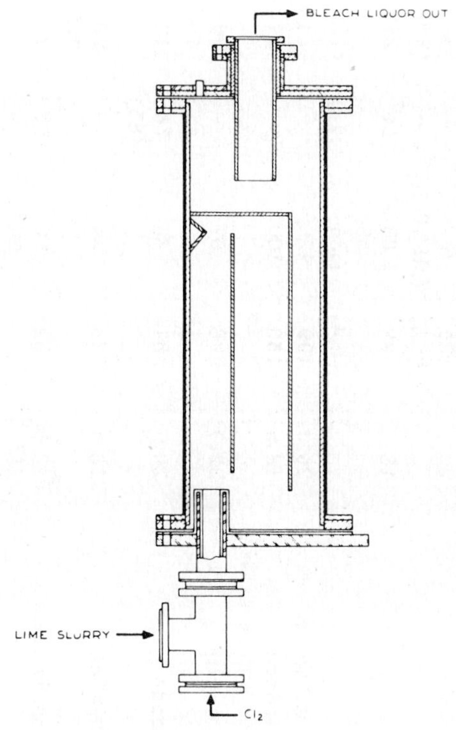

Figure 10-12.   Continuous calcium hypochlorite reactor (*Hooker Chemical Corp.*).

pioneered in this field and offers fully automatic, continuous systems, with oxidation-reduction potential (ORP) controlled, for both sodium and calcium hypochlorite manufacture.[8]

These systems have been engineered to provide the maker of bleaching solutions with the following major advantages:

(1) Lower cost installations and reduced space requirements, through the elimination of large settling and storage tanks.

(2) Limited operating attention required.

(3) Lower chemical costs because of complete utilization of chlorine added.

(4) Improved stability and uniformity of bleach liquor.

Previous research has established the suitability of oxidation-reduction potential (ORP) for the control of hypochlorite production.[9] Laboratory, pilot plant and actual full-scale, continuous operation over a long period have confirmed the suitability of this method of control for industrial applications. The continuous methods described herein incorporate the features and improvements resulting from several years of continuous operation in a number of pulp mills.[10,11]

The Hooker Calcium Hypochlorite Bleach-liquor system, which is automatic, consists of efficient continuous equipment for preparing a milk of lime slurry of definite controlled composition. Lime solids and dirt are removed from the milk of lime slurry continuously and returned to the clarifier. This pre-classification is carried out in a centrifugal cleaner of the Dorrclone type which removes some of the load from a similar unit that classifies the bleach liquor after chlorination. Classification with subsequent recycle of the unreacted lime permits maximum utilization of the lime and chlorine going into the system. The flow diagram in Figure 10-13 indicates a general layout of equipment. The description is divided into three parts: preparation of lime slurry, bleach preparation and control, and bleach liquor clarification.

Lime Slurry Preparation and Control. Lime slurry is obtained from an existing slaker which is not shown in Figure 10-13. Because of the heat of formation of calcium hypochlorite, the lime slurry must be below 80° F before entering the chlorinator.

Reaction Control. As shown in Figure 10-13 bleach liquor is metered from storage to the bleach plant through a flow recorder controller. As demand varies, the level controller on the hypochlorite storage tank adjusts the flow of dilute lime to the reactor. The Redox controller quickly compensates for the change and alters the flow of chlorine gas to maintain the selected potential of the finished bleach liquor. Selection of a Redox potential set point establishes the ratio of excess dissolved lime to available chlorine in the finished liquor. Available chlorine content will then vary only with the concentration of lime slurry entering the reactor. With close control of lime dilution, a uniform bleach liquor product is assured. Typical properties of chemically pure calcium hypochlorite bleach-liquor solutions for various oxidation reduction potentials are shown in Figure 10-14. Properties of commercial bleach liquors will vary according to the analysis of the lime and will have to be determined in each case.

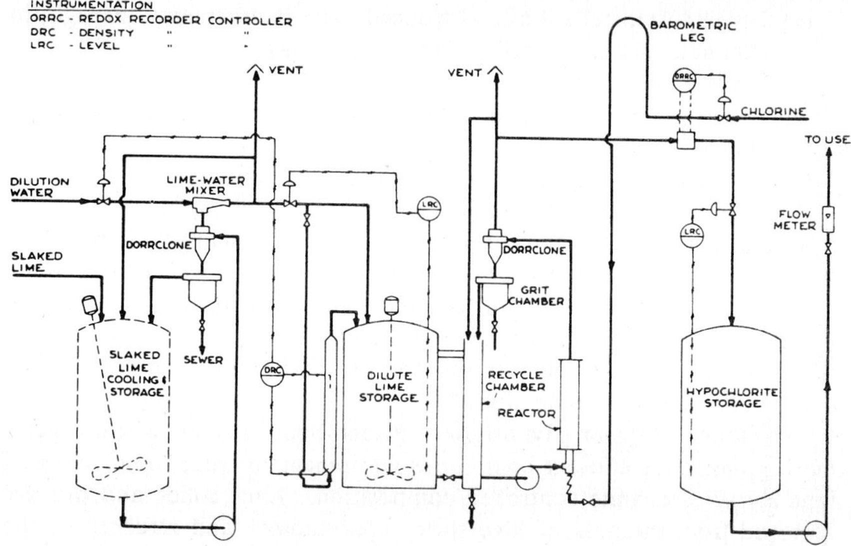

Figure 10-13.   Automatic bleach liquor production, calcium hypochlorite (*Hooker Chemical Corp.*).

Safety Interlocks.   Proper design of any continuous system using a hazardous chemical such as chlorine must include automatic shut-down provisions for safety reasons. Low pressure switches should be installed on the water, chlorine and caustic supply lines. These switches should be connected in series so that de-energizing any one of the switches shuts off all the feeds.

Bleach Liquor Clarification.   After chlorination the bleach liquor containing suspended solids is passed through a 6 in. Dorrclone which removes particles larger than 50 microns. This semi-clarified liquor contains about 0.2 percent of suspended solids less than 50 micron particles size. The liquor has a translucent appearance and may be used without further clarification. No decrease in pulp brightness or increase in ash content of pulp bleached with this liquor has been found.

Sodium Hypochlorite System

The same reactor may be used for the sodium hypochlorite as for the calcium hypochlorite system. The auxiliary equipment for preparing dilute caustic is much simpler than that required for preparing lime slurry, and no centrifugal clarification of the liquor is required.

Figure 10-15 is the flow diagram for continuous sodium hypochlorite production with Redox control of chlorine flow. Selection of a Redox po-

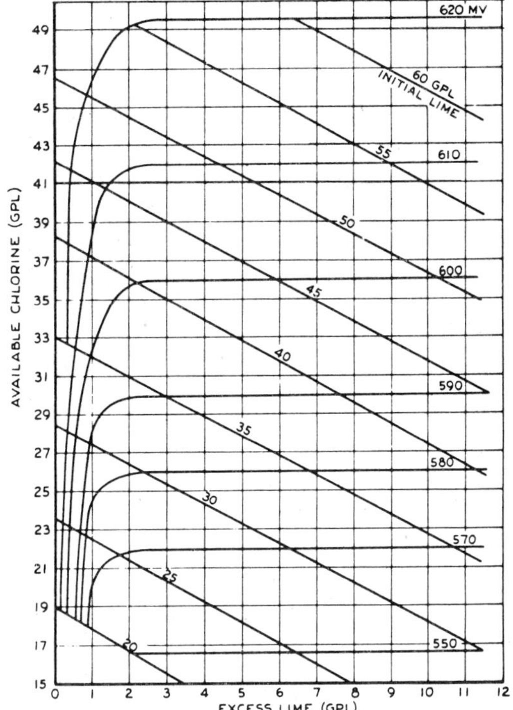

Figure 10-14. Oxidation potentials for various calcium hypochlorite solutions (based on solutions made from reagent grade lime).

tential set point establishes the ratio of excess caustic soda to available chlorine in the finished bleach liquor and is based upon original caustic strength. The Redox controller maintains the potential of finished bleach liquor passing through the electrode cup by adjusting the flow of chlorine to the injector. Available chlorine content will then vary only if the concentration of the dilute caustic entering the reactor varies. Thus, with close control of caustic dilution, a uniform bleach-liquor product, is assured.

Dilute caustic soda of constant composition is prepared from liquid 50 percent caustic, which is pumped along with dilution water into a mixing tee and reaction coil. The water flow is regulated by a level recorder-controller in the dilute caustic storage tank and the 50 percent caustic addition is regulated with a conductivity recorder controller in the line after the mixer reactor. The flow of dilute caustic soda to the chlorination reactor is controlled by a level recorder-controller in the bleach-liquor storage

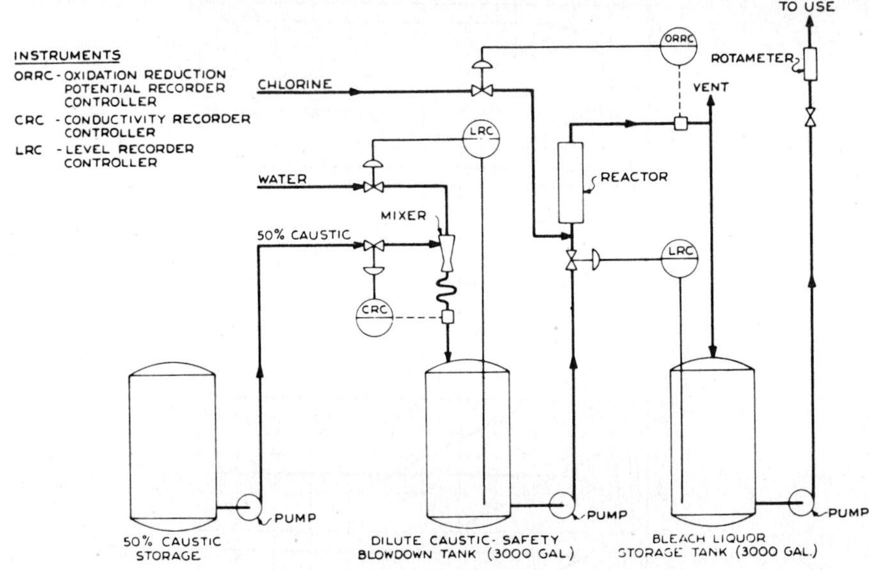

Figure 10-15. Sodium bleach liquor manufacture—redox control of chlorine. (*Hooker Chemical Corp.*).

tank, which senses any change of bleach liquor usage. The chlorine flow is quickly adjusted to any change in dilute caustic flow by the oxidation-reduction recorder controller.

Figure 10-16 shows the properties of sodium hypochlorite bleach solutions versus Redox potential. The same safety interlocks are required with the sodium as with the calcium system.

Chlorine Dioxide Solution Preparation

Chlorine dioxide is a gas at ordinary temperatures. It has an intense greenish-yellow color, a density of approximately 2.4 and a pungent, irritating odor. It is more irritating and toxic than chlorine.

The boiling point of liquid ClO_2 is 11°C and its melting point –59°C. Pure gaseous chlorine dioxide decomposes at a measurable rate at temperatures above 30°C, and from 50 to 60°C it decomposes explosively. Chlorine dioxide is usually handled in dilute mixtures with air in the range of 8 to 12 percent by volume. Spontaneous decomposition can and does occur at this dilution in the presence of iron rust, grease, and many organic particles. Direct sunlight causes decomposition. At this dilution the explosions are relatively mild and are referred to in the industry as "puffs." To take care of "puffs," generators, storage tanks, and bleaching towers are equipped with explosion lids.

Chlorine dioxide solutions for use in pulp bleaching are prepared by passing ClO_2 gas, diluted with air, through an absorption tower. The ClO_2 is absorbed in chilled water at a strength of 6 to 10 gpl. Chlorine dioxide solutions are very corrosive and must be stored in suitably lined

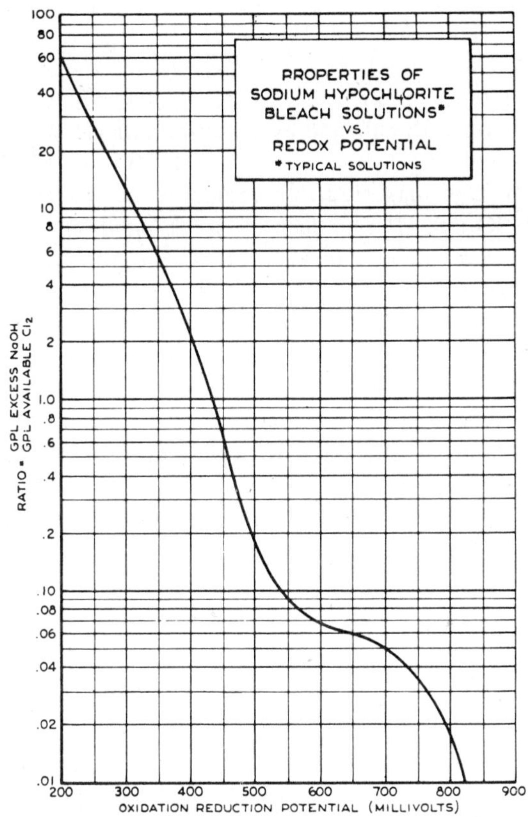

Figure 10-16

equipment. Suitable materials for storage vessels are glass-lined steel, fiberglass polyester, and tile-lined steel. Piping pumps valves and lines may be made of glass, glass-lined, titanium, PVC, Saran, "Kel F," "Teflon," and "Karbate." Because of its unstable nature, ClO_2 is always generated at the point of use.

Generation of Chlorine Dioxide. The evidence is strong that the mechanism of the formation of chlorine dioxide from sodium chlorate is the same for all processes. Rapson[12] suggested, "that the fundamental reactions producing chlorine dioxide from chlorates are:

$$HClO_3 + HCl \rightarrow HClO_2 + HClO \qquad (10\text{-}1)$$

$$HClO_3 + HClO_2 \rightarrow 2ClO_2 + H_2O \qquad (10\text{-}2)$$

and that HClO may react with HCl to produce chlorine

$$HClO + HCl \rightarrow Cl_2 + H_2O \qquad (10\text{-}3)$$

or with another reducing agent to produce HCl, which is then available for further reaction with chlorate by Reaction (10-1). Although it is possible that other reducing agents may reduce $HClO_3$ to $HClO_2$ to enter into Reaction (10-2), it is suggested that reduction of the $HClO_2$ so formed, stepwise, to HClO and HCl by such reducing agents is rapid, and that the reducing agent serves mainly to keep HClO reduced to HCl and thereby minimize the formation of chlorine. Thus, a delicate balance among the rates of these various reactions determines the rate of production of chlorine dioxide and the ratio of chlorine to chlorine dioxide produced."

In studying these reactions, Rapson found that chloride would reduce chlorate to chlorine dioxide and chlorine in a yield of 99 percent under the following conditions: (a) the generator solution is ION H_2SO_4, (b) the chlorate is low 0.1 M, and (c) the chloride is also low .02 to .04 M. This process is called the R-2 process and is covered by a U. S. and Canadian Patents.[13]

$$NaClO_3 + NaCl + H_2SO_4 \rightarrow ClO_2 + \tfrac{1}{2} Cl_2 + Na_2SO_4 + H_2O \qquad (10\text{-}4)$$

$$NaClO_3 + 5NaCl + 3H_2SO_4 \rightarrow 3Cl + 3Na_2SO_4 + 3H_2O \qquad (10\text{-}5)$$

98 to 99 percent of the sodium chlorate goes by reaction (10-4) and only 1 to 2 percent by Reaction (10-5). Excess chloride favors Reaction (10-5), but since it takes 5 moles of NaCl to 1 mole of $NaClO_3$ to cause this reaction, an excess of salt will only displace the equilibrium slightly. In practice, 5 mole percent excess chloride is provided to make the reaction go more nearly to completion.

In processes using other reducing agents, the basic reactions are believed to be the same, i.e., the reducing agent reduces hypochlorite or chlorine to chloride to react with the chlorate to form more chlorine dioxide, as in the following equations:

$$HClO + H_2SO_3 \rightarrow HCl + H_2SO_4 \qquad (10\text{-}6)$$

$$HClO + CH_3OH \rightarrow HCl + HCOH + H_2O \qquad (10\text{-}7)$$

$$HClO + HCOH \rightarrow HCOOH + HCl \qquad (10\text{-}8)$$

There are five principle methods of generating chlorine dioxide which are now in use throughout the world: (1) The Mathieson process, (2) The Solvay process, (3) The Day-Kesting process, (4) The Hooker R-2 process, (ER-2 Process in Canada), and (5) The Persson process. Several other processes are still in limited use and are mentioned here primarily because of their historical importance in the development of ClO_2 generation; they are the Rapson-Wayman process,[14] which was the first continuous process to be used commercially in North America, and the Holst process, used in Sweden.

All these processes depend on the reduction of sodium chlorate with a reducing agent in strong acid medium. All except the Holst process are continuous.

The Mathieson Process. (Figure 10-17). In this process separate streams of 46 percent sodium chlorate solution and 66° Bé sulfuric acid are con-

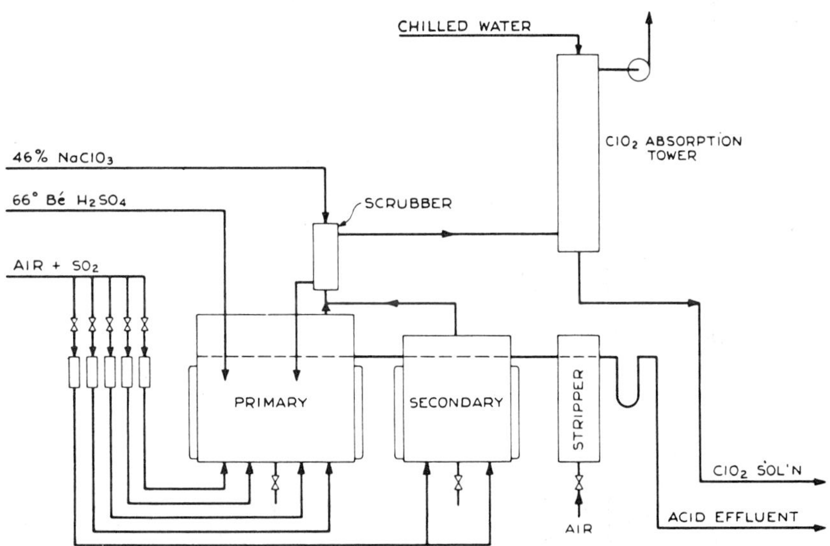

Figure 10-17. Mathieson ClO_2 generator.

tinuously fed into the bottom of a jacketed reaction vessel. Air for dilution mixed with sulfur dioxide is blown into the bottom of the vessel through sparger plates. The ClO_2 gas generated passes from the top of the vessel into the bottom of a packed absorption tower down which chilled water flows. The aqueous ClO_2 solution passes to a storage tank for subsequent use. The effluent from the generator contains 9 to 10 N sulfuric acid, sodium sulfate, and small amounts of chlorate, chloride, ClO_2 and Cl_2.

The Mathieson generator is usually equipped with a smaller secondary generator, a stripper for removing the last of the dissolved ClO_2 and a scrubber which is mounted in the gas line from the generator. The incoming chlorate feed trickles or is recirculated, in recent units, over the packing, scrubbing HCl and H_2SO_4 from the ClO_2 gas stream and returning them to the generator. With these latest units, yields of 91 percent are obtainable.

The Solvay Process. (Figure 10-18). This process is similar to the Mathieson process in that sodium chlorate solution and 60° Bé sulfuric acid are fed into the bottom of the first of two jacketed reaction vessels in

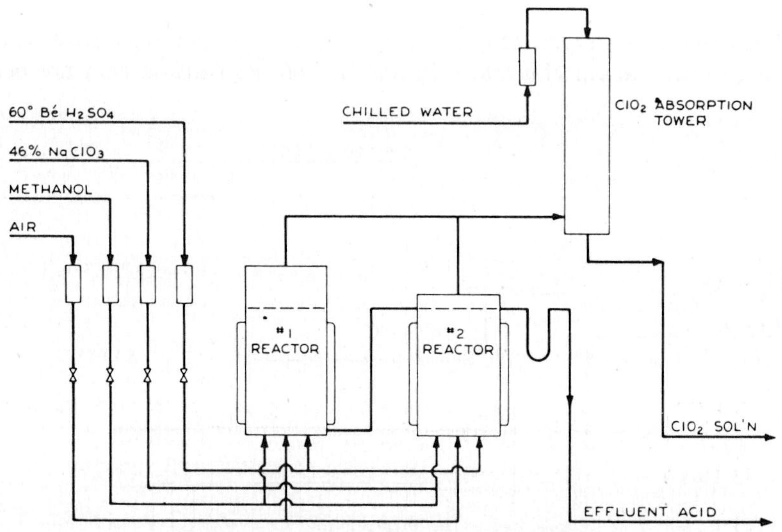

Figure 10-18. Solvay ClO_2 generator.

series. A feed of methanol acts as the reducing agent in this case and air for dilution of the ClO_2 and agitation of the liquid is blown into the bottoms of the two vessels. The vessels have about one third of the volume of a Mathieson generator of the same rated ClO_2 capacity. The ClO_2 is absorbed in a packed tower which is similar to that used for the Mathieson system.

The Day-Kesting Process. (Figure 10-19). This process was patented initially in Germany and is in commercial use in Germany, Sweden, and England. The same process was independently developed and patented in the United States and assigned to Brown Company, who constructed a plant to use the process and began operation in 1957.

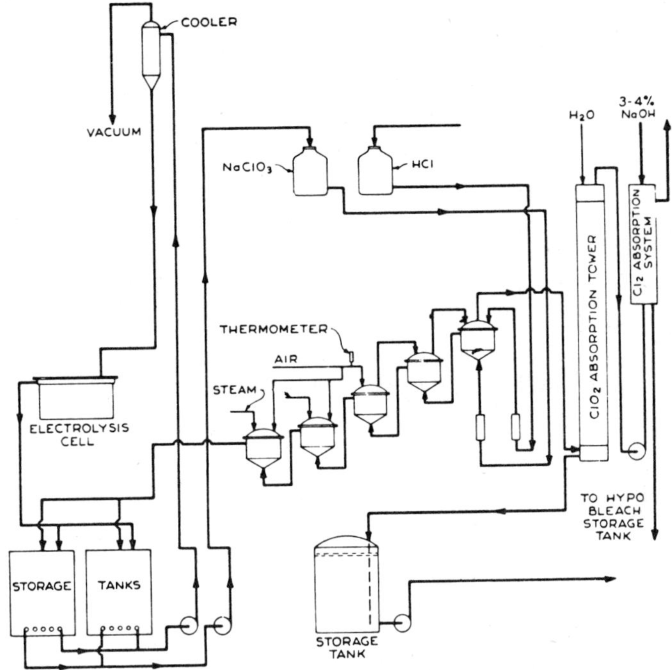

Figure 10-19.   Day-Kesting ClO₂ generator.

In this process the sodium chlorate is reduced with hydrochloric acid. In order that the reaction may proceed according to Equation (10-9) and that the undesired reaction (10-10) may be kept at a minimum, only a portion of the chlorate is reduced.

$$NaClO_3 + 2HCl \rightarrow ClO_2 + \tfrac{1}{2}Cl_2 + NaCl + H_2O \qquad (10\text{-}9)$$

$$NaClO_3 + 6HCl \rightarrow 3Cl_2 + NaCl + 3H_2O \qquad (10\text{-}10)$$

The concentration of the chlorate is kept high at all times and that of the hydrochloric acid is kept low. This is made practical by integrating the chlorine dioxide producing plant with a plant for the electrolytic production of sodium chlorate to which the chlorate solution which has been partially reduced to chloride can be sent for reoxidation to chlorate. The mixture of chlorine and chlorine dioxide produced in the process is separated into a chlorine dioxide-rich fraction and a chlorine-rich fraction by countercurrent contact with a stream of water which preferentially dissolves the chlorine dioxide. The solubility of chlorine dioxide in water is shown graphically in Figure 10-20.

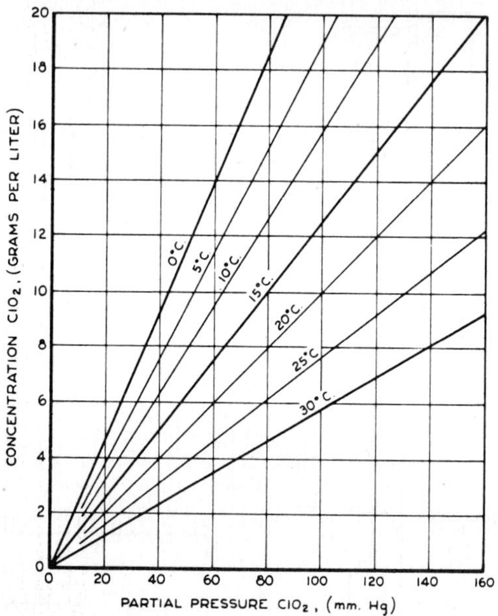

Figure 10-20.   Solubility of ClO_2 in water.[18]

The required hydrochloric acid may be purchased, or it may be made by burning chlorine, separated from the chlorine dioxide, plus purchased chlorine with hydrogen from the chlorate electrolytic cells.   Whether hydrochloric acid is purchased or made will depend on the relative cost, locally, of chlorine and hydrochloric acid.

The R-2 Process.[15]   (Figure 10-21).   This process, covered by U. S. Patent 2,863,722 issued to W. Howard Rapson, is assigned to Hooker Chemical Corporation in the U. S. A., and the equivalent Canadian Patent 543,589 is assigned to Electric Reduction Company in Canada.   The process consists of feeding a solution containing both sodium chlorate and sodium chloride in essentially equimolar ratio, and concentrated 66° Bé sulfuric acid into a reaction vessel.   Air is blown into the bottom of the vessel to provide rapid agitation, stripping, and dilution of the ClO_2 produced.   In this process ClO_2 and Cl_2 are evolved in what is nearly a 2 to 1 ratio.   The ClO_2 is absorbed in a conventional tower with chilled water.   About 25 percent of the chlorine evolved (12 to 15 percent based on ClO_2) dissolves in the ClO_2 solution.   The remainder passes through the tower and is picked up as either calcium or sodium hypochlorite in the second, smaller tower or suitable absorption system.   The yield of ClO_2 by the R-2 process is 95 to 96 percent of the theoretical under mill conditions.[16]   This process is cur-

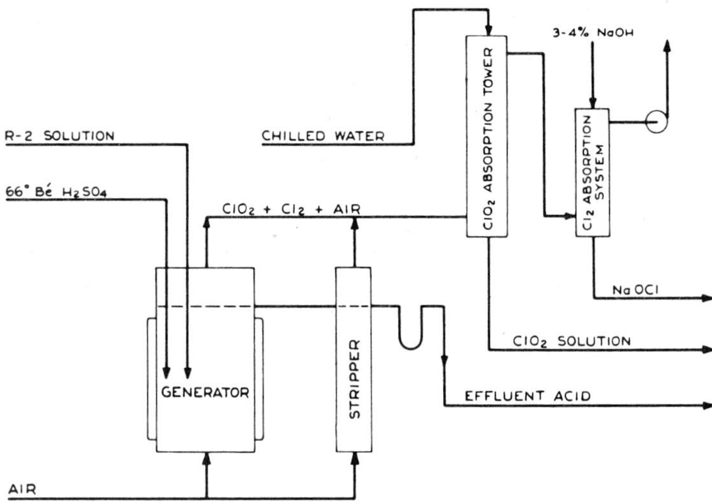

Figure 10-21.   Hooker R-2 ClO₂ generator.

rently operating in four mills in the U. S. A. and is being installed in a number of other mills. It produces ClO_2 at the lowest cost of any process in mills where full use can be made of the chlorine, salt cake, and effluent acid. Advantages of the R-2 process are as follows:

(1) The process gives the highest yield of ClO_2 of any commercial process.
(2) The process is extremely simple to operate.
(3) The process responds immediately to a change in production rate and can be started up from full shutdown to full steady production in less than 20 minutes.
(4) The ClO_2 production is strictly proportional to chlorate fed at all production rates.
(5) The process can be run in any existing ClO_2 generator and will produce at up to 10 times the rated capacity of the existing generator.
(6) The process can be fully automated to run off a float valve in the ClO_2 storage tank.

Chemical Cost of ClO_2—Various Processes

A comparison of the Mathieson, Solvay, and R-2 processes is given in Table 10-4. In this comparison, a yield of 95 percent is taken for the R-2 process and 86 percent for the other two processes. These yields are based on average operating data from many mills.

The gross chemical costs are 21.4, 20.2 and 20.7 cents/lb of chlorine dioxide for the Mathieson, Solvay, and R-2 Processes respectively.

In a Kraft mill, cooking pine wood and converting the tall oil soap to crude oil and having a hypochlorite stage in the bleachery, the salt cake, chlorine and sulphuric acid may all be usable at full value. If all the effluent acid is usable, the total credits for the R-2 system are 8.8 cents/lb ClO_2 and for the Mathieson and Solvay processes they are 3.9 cents/lb respectively. These calculations show that ClO_2 can be produced by the R-2 process for 11.9 cents/lb, a savings of from 4.4–5.6 cents/lb of ClO_2.

If all the effluent acid cannot be used for tall oil soap splitting, the residual acid can be neutralized with soda ash and added to the kraft recovery system as salt cake makeup. Salt cake made this way often is slightly cheaper than the purchased material so that the sulphuric acid has a significant value—perhaps 0.5 cents/lb. To the chemical costs of ClO_2 must be added the capital, operating, and maintenance costs which are generally about 2.5 cents/lb of ClO_2.

TABLE 10-4. CHEMICAL COST COMPARISONS OF VARIOUS PROCESSES PER LB OF ClO_2

Reactant	R-2 Process lb cents/lb cents	Mathieson Process lb cents/lb cents	Solvay Process lb cents/lb cents
$NaClO_3$	$1.66 \times 9.0 = 14.9$	$1.85 \times 9.0 = 16.7$	$1.85 \times 9.0 = 16.7$
H_2SO_4	$4.75 \times 1.1 = 5.2$	$1.3 \times 1.1 = 1.4$	$2.5 \times 1.1 = 2.8$
$NaCl$	$0.9 \times 0.7 = .6$	–	–
SO_2	–	$0.85 \times 4.5 = 3.8$	–
Methanol	–	–	$0.2 \times 3.5 = 0.7$
Gross chemical cost	20.7	21.4	20.2
Credits			
Na_2SO_4	$2.3 \times 1.7 = 3.9$	$1.2 \times 1.7 = 2.0$	$1.2 \times 1.7 = 2.0$
Cl_2	$0.4 \times 3.5 = 1.4$	–	–
H_2SO_4	$3.2 \times 1.1 = 3.5$	$1.7 \times 1.1 = 1.9$	$1.7 \times 1.1 = 1.9$
Total credits	8.8	3.9	3.9
Net chemical cost	11.9	17.5	16.3

Addition of chloride to improve yield of ClO_2[17] in a number of SO_2 type units, particularly those being forced beyond their rated capacity, the yield of chlorine dioxide from chlorate has been increased from 80 to 83 percent to 93 to 94 percent of the theoretical by the addition of 10 mole percent of sodium chloride. This chloride may be dissolved in the chlorate solution, or it may be added to the generator as a separate feed of saturated brine. Since the latter method introduces extra water into the system, extra sulphuric acid will have to be used to maintain proper acidity in the generator.

In the second situation where more ClO_2 is required than the generator is rated for, additional chlorine dioxide can be generated by adding equimolar quantities of sodium chlorate and sodium chloride. The extra ClO_2 is then produced by the R-2 reactions. Additional sulphuric acid must be added to take care of the extra sodium added with the sodium chloride and sodium chlorate. A number of SO_2 type generators are operating in the United States and Canada with the addition of 10 to 15 mole percent sodium chloride both to improve the yield and to produce extra ClO_2.

References

1. Rue, John, D., and Sconce, J. S., Hooker Bulletin 200, Page 21.
2. Duncan, E. P. and Rapson, W. H., *Pulp & Paper*, (October-November 1960).
3. Duncan, E. P., *Canadian Pulp & Paper Ind.*, (January-February 1960).
4. Seymour, George W., *TAPPI*, **40** (June 1957).
5. Schlumberger, Allan A., "Pulp Chlorination Improvements at St. Helens," *TAPPI*, **44** (April 1961).
6. Rapson, W. H., Paper at National TAPPI February 1961.
7. Rapson, W. H., (to Canadian International Paper Company), U. S. Patent 2,587,064; Canadian Patent 470,478.
8. Hooker Chemical Corporation, Bulletin 251, U. S. Patent 2,889,199.
9. Pye, D. J., *J. Electrochem. Soc.*, **97** (August 1950).
10. "Continuous Calcium Hypochlorite at Crofton," *TAPPI*, **43** (October 1960).
11. "Continuous Calcium Hypochlorite at St. Helens," *Paper Trade Journal* (May 2, 1960).
12. Rapson, W. H., *TAPPI*, **39** (8) 554 (August 1956).
13. Rapson, W. H. (to Hooker Chemical Corporation), U. S. Patent 2,863,722; Rapson, W. H., (to Electric Reduction Company of Canada) Canadian Patent 543,589.
14. Rapson, W. H., and Wayman, M., U. S. Patent 2,481,241 (1949); Canadian Patent 466,816 (1950).
15. Rapson, W. H., *TAPPI*, **41**, 181 (1958); U. S. Patent 2,863,722.
16. Partridge, H. D., and Rapson, W. H., "Mill Trials of R-2 Process," Hooker Bulletin 262; *TAPPI*, **44** No. 10 (1961).
17. Rapson, W. H., (to Hooker Chemical Corporation) U. S. Patent 2,936,219 (May 1960); Rapson, W. H. (to Electric Reduction Company of Canada) Canadian Patent.
18. Haller, J. F., and Northgraves, W.W., *TAPPI*, **38**, 199 (April 1955).

11. ETHYLENE AND PROPYLENE OXIDES AND GLYCOLS

Anita S. Kastens

Consultant, Union Carbide Chemicals Company

Chlorine was essential to the establishment of the petrochemicals industry in the 1920's. The derivative ethylene chlorohydrin made feasible commercial production of the industry's first building block, ethylene oxide, and the industry's first major product, ethylene glycol. Later propylene chlorohydrin made possible the equivalent three-carbon products—propylene oxide and propylene glycol. Since then, ethylene oxide and ethylene glycol, at least, have grown to heavy chemical size, with combined 1960 production almost 2.8 billion pounds. The industry derived from this beginning accounts for approximately 30 percent or 56 billion pounds of total U. S. chemical manufacture (some 185 billion pounds) and more than 61 percent of the estimated 11.2 billion dollar chemical value of products.

Ethylene oxide is the first member of the class of *alpha* oxides known as olefin oxides, in which the first two carbon atoms on a straight chain are linked by an oxygen atom (CH_2—CH—). Ethylene glycol is the simplest

$$CH_2-CH- \atop \diagdown O \diagup$$

member of the glycol or diol family. It alone is more important than all the other glycols combined. Propylene glycol ranks second. Strictly speaking, a glycol is a compound having two hydroxyl groups attached to separate carbon atoms in an aliphatic carbon chain, $C_nH_{2n}(OH)_2$. Common usage has extended the glycol nomenclature, however, to compounds having two hydroxyl groups attached to carbon chains interrupted by oxygen (e.g., diethylene glycol, dipropylene glycol, polyethylene glycols) or sulfur (thiodiglycol). Like ethylene glycol and propylene glycol, "glycols" of this sort are derivatives of ethylene oxide and propylene oxide and as such will be discussed here under the headings "Derivatives of Ethylene Oxide" and "Derivatives of Propylene Oxide."

Ethylene oxide is useful primarily as a chemical intermediate. Polyethylene glycols, acrylonitrile, ethanolamines, glycol ethers, surface-active agents, in addition to ethylene glycol, are its major derivatives. Propylene oxide is the starting material for comparable propylene-derived compounds. The major outlet for ethylene glycol by far is as permanent-type antifreeze, although it is also utilized as an intermediate and as a humectant, as a component of hydraulic fluids, as engine coolant for airplanes, and so on. Propylene glycol has applications similar to those of ethylene glycol, but because of its low order of toxicity, it is also used in drugs, cosmetics, and foods.

The source of much of the information in this chapter is "Glycols," by Curme and Johnston.[12] For references and further information, the reader is referred to this A.C.S. Monograph which comprehensively covers the subjects discussed here. Polymers from ethylene oxide are thoroughly reviewed in another monograph, "Polyethers," by Gaylord.[15] Further information on ethylene oxide-based surfactants is contained in "Surface Active Agents" by Schwartz, Perry, and Berch.[20]

HISTORY

Both ethylene oxide and ethylene glycol were first prepared by Wurtz in 1859. He obtained ethylene glycol by saponifying ethylene glycol diacetate with potassium hydroxide. Then he treated the reaction product of ethylene glycol and hydrogen chloride (ethylene chlorohydrin) with potassium hydroxide to get ethylene oxide. Later he made the full circle by hydrating ethylene oxide to ethylene glycol. Carius was the first, however, to prepare ethylene chlorohydrin by reacting ethylene with aqueous hypochlorous acid (1863). The original work of these two men is the basis for the glycol production techniques in use today.

Ethylene oxide and ethylene glycol remained buried in the literature until the early 20th Century. The first indication of any commercial outlet came in 1904 when the glycol derivative ethylene glycol dinitrate was patented as a freezing point depressant in nitroglycerin dynamite. It was not until World War I, however, that the shortage of glycerol and the hazards of using nitroglycerin dynamite led to the manufacture of ethylene glycol for conversion into its dinitrate. In Germany Th. Goldschmidt A.-G. utilized alcohol as a source of ethylene for chlorination to ethylene dichloride, which was hydrolyzed to the glycol. In this country The Commercial Research Company produced some ethylene glycol by converting ethylene obtained by the Pintsch gas process to ethylene chlorohydrin and then glycol. Ethylene chlorohydrin was considered, however, more as a starting material for the production of mustard gas and was employed for this purpose both in the United States and in Germany. At the end of the War, interest in ethylene glycol still centered around ethylene glycol dinitrate, and a premature stab at marketing ethylene derivatives ended in 1920 when

Willard Dow failed to uncover peacetime markets for ethylene chlorohydrin, ethylene oxide, and ethylene glycol.

In the meantime, the first really adequate source of raw material ethylene was being developed at the Mellon Institute in Pittsburgh. Looking for a source of acetylene cheaper than calcium carbide, Curme discovered that gas oil could be cracked to acetylene, as well as to ethylene, propylene, and butylene. Abandoning acetylene, Curme and his co-workers determined optimum conditions for cracking gas oil to ethylene and developed commercial methods for converting it to ethylene oxide, ethylene glycol, ethylene dichloride, and other derivatives. In 1922 Carbide and Carbon Chemicals Corporation (now Union Carbide Chemicals Company) started small-scale production of these materials at Clendenin, West Virginia, and in 1925 large-scale manufacture at South Charleston, West Virginia. At the Charleston installation, chlorine was pumped from the adjacent Westvaco plant to supply the raw material for the essential intermediate ethylene chlorohydrin. Explosives and antifreeze were the sole outlets for the early glycol production.

In 1937 Carbide put into operation the first ethylene oxide units utilizing techniques for the direct catalytic oxidation of ethylene. Carbide's dependence on chlorine as a raw material was thereby eliminated. Dow, however, started commercial production in the late 1930's based on chlorine, caustic soda, and ethylene. Postwar producers (Jefferson, Wyandotte, and Mathieson) likewise utilized the chlorohydrin techniques. Still more recent entries into the oxide-glycol business, however (Allied, General Aniline), utilize direct oxidation methods, which basically are more economically sound.[16] Today, approximately 40 percent of the ethylene oxide produced in the United States is made via ethylene chlorohydrin, the balance directly from ethylene.[6] A unique Du Pont process, in operation since 1940, bypasses ethylene oxide and yields only ethylene glycol (Tables 11-1 and 11-2). Current oxide-glycol capacity is about 1.7 billion pounds.[6]

Worldwide demand for ethylene oxide and ethylene glycol has resulted in the construction of production facilities, too numerous to mention, all over the world, from Europe to Japan. In recent years many U. S. producers have been active in joint efforts with local companies in foreign operations. American chemical construction firms, such as Scientific Design, have also been successful at selling oxide-glycol package plants to foreign companies.

The 1961 price of ethylene oxide stands at $15\frac{1}{2}$ cents per pound in tank car quantities, of ethylene glycol at $13\frac{1}{2}$ cents. Ethylene oxide has dropped from a 19 cent high in 1952 to its current level after going through a low of $13\frac{1}{2}$ cents in 1954. The price of ethylene glycol has ranged from 60 cents a pound in 1925, through a low of $9\frac{1}{2}$ cents after World War II, up to 17 cents in 1951, and back to its current level in 1957. Increasing competition with expanded production is likely to maintain prices at current levels.[6]

Propylene Oxide-Propylene Glycol

Wurtz also prepared propylene glycol in 1859, by hydrolyzing propylene glycol diacetate. It was first produced commercially by Carbide in 1931, at South Charleston, West Virginia. The process was similar to that for the production of ethylene glycol through the chlorohydrin. Current propylene oxide-propylene glycol capacity is estimated at 373 million pounds (Table 11-3).[7]

COMMERCIAL PRODUCTION

Ethylene oxide is produced commercially in two ways—by the dehydrochlorination of ethylene chlorohydrin and by the direct oxidation of ethylene. Almost all ethylene glycol is produced by the hydration of ethylene

TABLE 11-1. U.S. ETHYLENE OXIDE CAPACITY,[a] 1960[6]

Company	Location	Direct oxidation	Chlorohydrin
		(million pounds)	
Union Carbide	Institute, W. Va.	150	–
	Seadrift, Tex.	210	–
	S. Charleston, W. Va.	30	90
	Texas City, Tex.	150	–
	Torrance, Cal.	50	–
	Whiting, Ind.	100	–
Dow Chemical	Freeport, Tex.	–	220
	Midland, Mich.	–	20
	Plaquemine, La.	–	60
Allied Chemical	Orange, Tex.	35	–
Calcasieu Chemical	Lake Charles, La.	60	–
General Aniline & Film	Linden, N. J.	60	–
Jefferson Chemical	Port Neches, Tex.	60	110
Olin-Mathieson	Brandenburg, Ky.	60	100
Wyandotte Chemicals	Geismar, La.	60	–
	Wyandotte, Mich.	–	30[b]
		1025	630

Total: 1,655 million pounds

[a]Does not include Carbide's Ponce, Puerto Rico plant (estimated capacity 100 million pounds per year) or Houston Chemical's Beaumont, Texas plant (capacity, 80 million pounds per year), under construction.
[b]Converted to propylene oxide.

TABLE 11-2. U.S. ETHYLENE GLYCOL CAPACITY,[a] 1960[6]

Producer	Capacity (million pounds, est.)
Based on ethylene oxide	
Union Carbide	755[b]
Dow Chemical	350
Jefferson Chemical	180
Olin Mathieson	100
Wyandotte Chemicals	90
Calcasieu Chemical	75
General Aniline & Film	35
Allied Chemical	35
	1620
Based on formaldehyde	
Du Pont	150
Grand total	1770

[a] Includes capacity for approximately 150 million pounds per year of propylene glycol and about 200 million pounds per year of higher glycols.
[b] Plus 100–120 million pounds per year at Ponce, Puerto Rico.

TABLE 11-3. U. S. PROPYLENE OXIDE-GLYCOL CAPACITY,[a] 1960[7]

Company	Location	Capacity, millions of pounds per Year
Celanese	Bishop, Tex.	8
Dow Chemical	Freeport, Tex.	100
	Midland, Mich.	15
	Plaquemine, La.	30
Jefferson Chemical	Port Neches, Tex.	40
Olin Mathieson	Brandenburg, Ky.	40
Wyandotte Chemicals	Wyandotte, Mich.	25
Union Carbide Chemicals	S. Charleston, W. Va.	115
Total		373

[a] Source: C & EN estimates

oxide. Only ethylene glycol manufactured by Du Pont's high-pressure, high-temperature reaction of formaldehyde, carbon monoxide, and hydrogen is the exception. This method accounts for less than 10 percent of total glycol production.

Chlorohydrin Method

An aqueous solution of chlorine (hypochlorous acid) is reacted with ethylene to form ethylene chlorohydrin. This in turn is hydrolyzed with caustic soda or milk of lime to ethylene oxide. Hydration to the glycol takes place in the presence of strong acid catalysts. These steps are indicated in the following equations:

$$CH_2CH_2 + HOCl \rightarrow CH_2ClCH_2OH$$

$$2\,CH_2ClCH_2OH + Ca(OH)_2 \rightarrow 2\,\underset{\diagdown O \diagup}{CH_2\!-\!CH_2} + CaCl_2 + 2H_2O$$

$$\underset{\diagdown O \diagup}{CH_2\!-\!CH_2} + H_2O \rightarrow HOCH_2CH_2OH$$

In practice, ethylene, chlorine, and water are introduced into a reaction tower in a continuous process. A concentration of about 5 percent ethylene chlorohydrin is maintained in the reaction liquid. The crude chlorohydrin solution is passed to a hydrolyzer containing an excess of 10 percent milk of lime where it is heated with live steam at 96 to 102°C at pressures between 60 and 80 mm Hg. Exit gases at 80°C contain about 70 percent water, 26 percent ethylene oxide, 3 percent ethylene dichloride, and 1 percent miscellaneous. The vapors are cooled, and the ethylene oxide recovered by fractionation. Maximum yield is 95 percent based on ethylene chlorohydrin, 80 percent based on ethylene. By-product ethylene dichloride is formed to the extent of 0.1 to 0.15 pound per pound of ethylene oxide, and β,β'-dichloroethyl ether to the extent of 0.07 to 0.09 pound per pound of ethylene oxide. Raw materials required to produce one ton of ethylene oxide are: ethylene, 1600 pounds; chlorine, 4000 pounds; lime (100 percent CaO), 3200 pounds; electricity, 180 kw-hr; steam, 24,000 pounds; and water, 60,000 gallons.[13]

Ethylene oxide is converted to ethylene glycol by reaction with an excess of water at elevated temperatures and pressures in the presence of strong acid catalysts. The crude solution is concentrated in evaporators and purified by distillation. Diethylene glycol and triethylene glycol are separated as by-products, the amounts formed being dependent on the ratio of water to oxide. Maximum yields of 92 to 95 percent ethylene glycol are feasible. Raw materials required to produce one ton of ethylene glycol via chlorohydrin are: ethylene, one ton; chlorine, 2.35 tons; milk of lime, 2.2 tons; NaOH, 80 pounds; steam, 24,000 pounds; electricity, 160 kw-hr; and water, 80,000 gallons.[13]

Direct Oxidation of Ethylene

Although Wurtz stated in 1859 that ethylene could not be combined directly with oxygen to form ethylene oxide, Lefort discovered a procedure in 1931 whereby oxygen could be added directly to the ethylene bond. Carbide developed his techniques into commercial production methods utilizing air as a source of oxygen (1937). More recently (1958) Shell Development-designed plants have employed purified oxygen. The relative merits of air versus oxygen have been discussed.[9]

No matter whether the raw material is air or oxygen, the basic reactions are the same:

$$CH_2CH_2 + \tfrac{1}{2}O_2 \rightarrow \underset{O}{CH_2{-}CH_2}$$

$$CH_2CH_2 + 3O_2 \rightarrow 2H_2O + 2CO_2$$

Because the reaction products are formed with the liberation of much heat, the reaction equipment must be designed for excellent temperature control.

In general, nonexplosive mixtures of ethylene, air (oxygen), and diluent (ethylene dichloride) are passed over a silver catalyst. Ethylene oxide is removed from the effluent gas containing the oxide, unreacted ethylene, carbon dioxide, and water by adsorption on activated carbon or absorption in water, methanol, or other solvent. Refined ethylene oxide is obtained by distillation.

Direct oxidation plants vary in construction detail and operating conditions.[10] Temperatures range from 150 to 500° C, with 220 to 280 apparently optimum. Pressures are in the range of 225 to 250 psi. Catalysts are generally silver coated on inert carriers prepared in various ways, i.e., by thermal treatment of silver salts of carboxylic acids. The catalysts may be fixed or fluid bed. Optimum ratio of oxygen to ethylene is between 1:1 and 2:1.

Yields for the direct oxidation with air run 55 to 65 percent. The raw materials required to produce one ton of ethylene oxide are: ethylene, 2500 pounds; air, 26,500 pounds; ethylene dichloride, 40 pounds; silver (catalyst replacement), 1.4 pounds; electricity, 1700 kw-hr; steam, 200 pounds; water, 50,000 gallons.[13] Yields of ethylene glycol produced through direct oxidation techniques run 50 percent. Raw materials for one ton of ethylene glycol are 1800 pounds of ethylene, 19,000 pounds of air, 36 pounds of ethylene dichloride, and one pound of silver.[13]

Details of the Jefferson ethylene oxide plant developed by Scientific Design Company, Inc. are shown in Figure 11-1. The charge is high purity ethylene and air. The methods are described as follows:[2]

Compressed air, ethylene, and recycle gas are mixed and fed to a multi-tubular, catalytic reactor. The temperature of oxidation is controlled by an organic cooling medium which is omitted on the process flow diagram. From the reactor the effluent gases, which contain

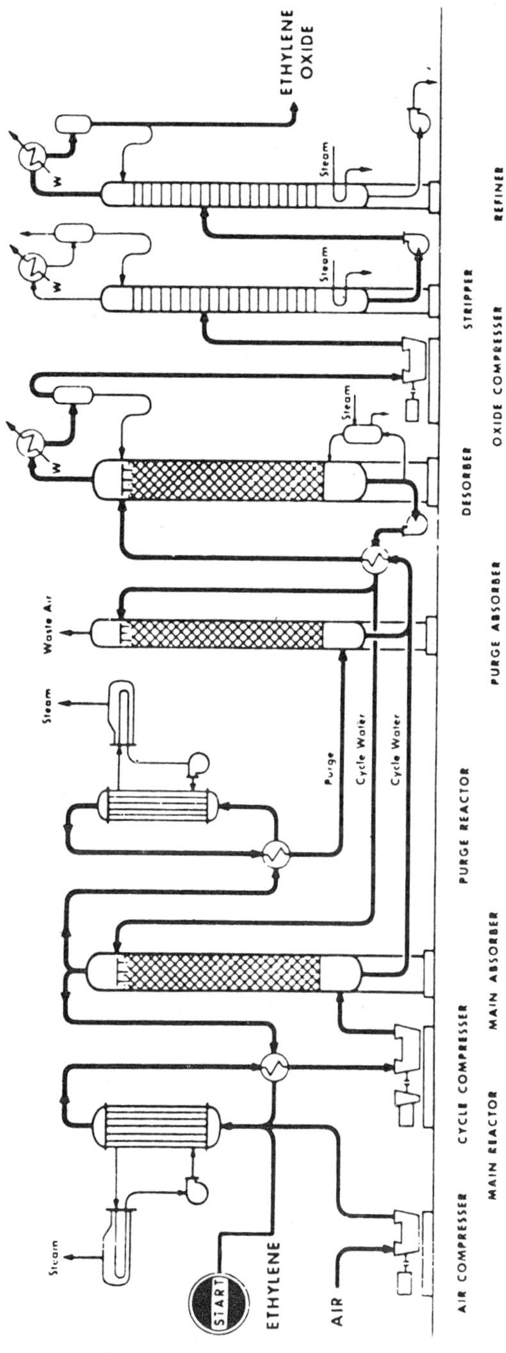

Figure 11-1. Jefferson Chemicals' ethylene oxide plant developed by Scientific Design Company, Inc.[2] *(Courtesy of Hydrocarbon Processing & Petroleum Refiner," copyrighted by Gulf Publishing Company, Houston, Texas).*

ethylene oxide, are cooled and further compressed. The cooling is accomplished first by recuperative exchange with the recycle gases, and finally by a water-cooled exchanger. The gases then pass to a scrubber where ethylene oxide is absorbed in a dilute aqueous solution. Most of the unabsorbed gases are returned to the reactor via the previously mentioned recuperative exchanger, thus completing a closed circuit. A portion is diverted to the secondary reactor, to purge accumulated inert gases. The secondary reactor operates to scavenge the remaining ethylene. From the secondary reactor the effluent gases are cooled as before. The ethylene oxide is absorbed in a scrubber, and the residual gases are discharged from the system.

The dilute solution of ethylene oxide is stripped by heat. A feed-bottoms exchanger is usually employed to save steam. The overhead vapor, which contains a substantial amount of water vapor, may be passed through a partial condenser for enrichment. This partial condenser is omitted on the flow diagram because it is not essential and because of space limitations. Oxide vapor is compressed and refined in a distillation train. There are several alternate arrangements for the distillation. As shown here, the carbon dioxide and other light ends are first stripped out of the oxide, and then pure oxide is distilled away from water and other heavier liquids. The pure ethylene oxide product is pumped as a liquid to storage tanks which are padded with inert gas.

The catalyst contains silver. Its long life and durability make catalyst costs a relatively unimportant item in the ethylene oxide manufacturing budget. The reactions involved are simple:

$$(1) \quad C_2H_4 + \tfrac{1}{2} O_2 \rightarrow C_2H_4O$$

$$(2) \quad C_2H_4 + 3 O_2 \rightarrow 2 CO_2 + 2 H_2O$$

Operating conditions: oxygen and ethylene concentrations in the reaction gases are quite critical. In order to avoid explosion hazards, they must be maintained at a low level. Reaction temperatures range between 450 and 600° F. The reaction pressure may vary between 120 and 300 psig. Distillation conditions are usually determined by the available cooling water temperature.

Yields: the weight yield, expressed as pounds of ethylene oxide produced per pound of ethylene consumed approaches one hundred percent, in plants of the proper design.

A schematic diagram of the Shell process for ethylene oxide-ethylene glycol is shown in Figure 11-2. Oxygen, rather than air, is a basic raw material. The method is described as follows:

The process produces ethylene oxide by direct oxidation of ethylene with 95 percent oxygen. Product distribution between ethylene oxide and glycol can be varied over a wide range.

Charge:

Ethylene		Oxygen	
Ethylene	98% min.	Oxygen	90–95%
Methane	1% max.	Nitrogen	Balance
Ethane	1% max.	Argon	
Acetylene	10 ppm max.		

Products: Ethylene oxide (99.7 percent min.) and ethylene glycol (99.6 percent min.) Description: Ethylene and oxygen together with recycle gas are charged to a tubular isothermal reactor. The reaction takes place in the presence of a silver catalyst and proceeds according to the equation:

$$C_2H_4 + 1/2 \ O_2 = C_2H_4)$$

The heat of reaction is removed by a coolant, and the hot coolant is used to generate steam.

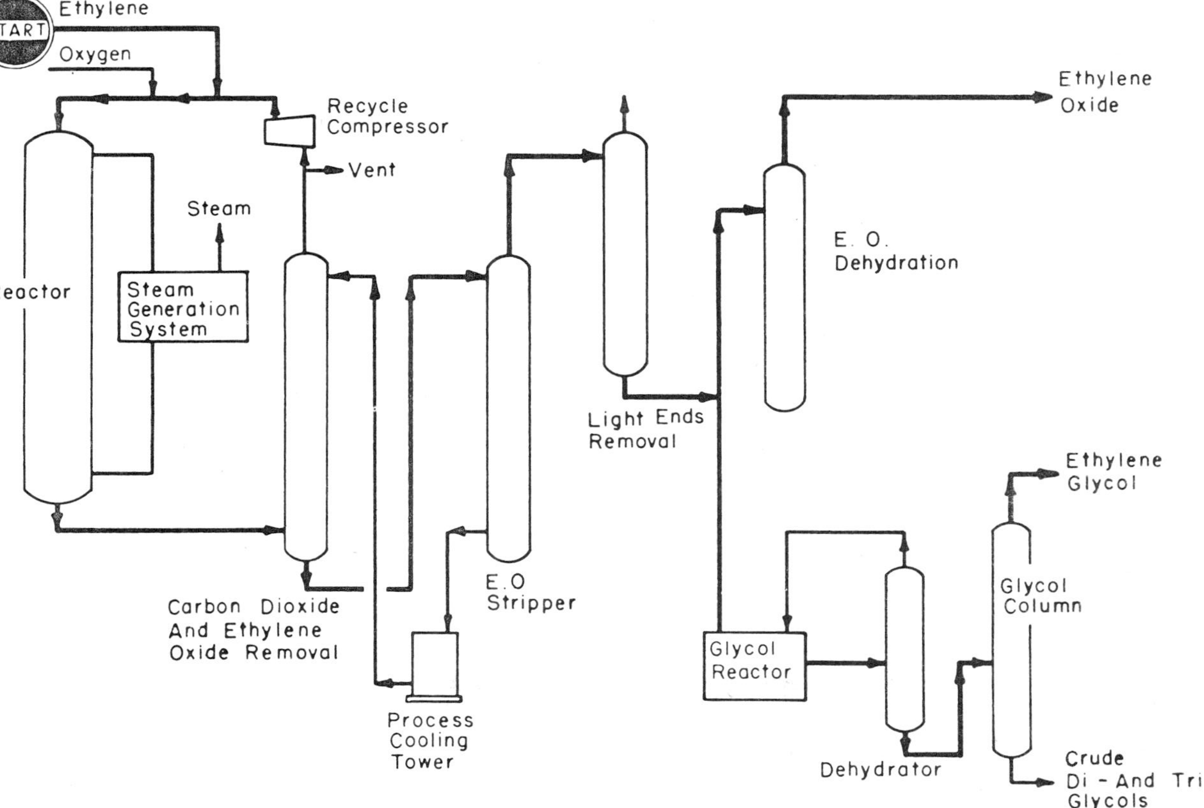

Figure 11-2.   Shell Development's process for making ethylene oxide-ethylene glycol.[3]   (*Courtesy of "Hydrocarbon Processing & Petroleum Refiner," copyrighted by Gulf Publishing Company, Houston, Texas*).

Ethylene oxide is recovered from the reactor effluent by absorption in water. Except for a small vent, the residual gases are compressed and recycled. The ethylene oxide is stripped from the "fat" water and distilled to remove light ends. The "lean" water is cooled in a process cooling tower before being returned to the absorber. Part of the raw ethylene oxide is dehydrated to produce finished ethylene oxide; the balance is reacted with excess water in a special reactor according to the equation:

$$C_2H_4O + H_2O = HOC_2H_4OH$$

The resulting aqueous solution of ethylene glycol is dehydrated to produce antifreeze grade ethylene glycol; industrial grade ethylene glycol and diethylene glycol also may be produced.

Economics

Direct oxidation is definitely the preferred modern technique for producing ethylene oxide. In recent years only Dow has added chlorohydrin capacity. Other chlorohydrin producers are even converting ethylene chlorohydrin units to propylene chlorohydrin units.[6] The advantages of the direct oxidation process are two: the cost of the raw materials is relatively small, and maintenance costs are much lower.

Plants for the production of ethylene oxide and ethylene glycol range in size from 20 to 200 million pounds annual capacity. Minimum economic size for a new plant is in the 30 million pound class. Oxidation plants in the 30 to 40 million pound range cost $100,000 to $150,000 per million pound annual capacity, depending on the size and design of the particular plant. In general, chlorohydrin plants cost about 30 percent less.[13]

Propylene Oxide—Propylene Glycol

The production of propylene chlorohydrin, propylene oxide, and propylene glycol is completely analogous to that of the ethylene equivalents. The reactions are carried out under nearly the same conditions; yields and efficiencies are also comparable. This is the reason ethylene chlorohydrin units can be converted to production of propylene chlorohydrin. A description of Jefferson Chemical's propylene oxide plant, converted from an ethylene oxide unit, is as follows:[4]

Chlorine and 90 percent propylene are reacted in dilute aqueous solution to yield propylene chlorohydrin, propylene dichloride, and other by-products. The reaction mixture is then contacted with lime slurry to produce a crude propylene oxide stream and a stream of calcium chloride brine, which is sewered. The crude propylene oxide stream is then successively fractionated to remove propylene dichloride, water and propionaldehyde.

Crude propylene dichloride is purified by passage in block operation through the expanded existing ethylene dichloride purification facilities.

Propylene oxide is also produced by the vapor-phase partial oxidation of aliphatic hydrocarbons such as propane and butane (Celanese). It is not separated or sold as such.

ETHYLENE OXIDE

Ethylene oxide (epoxyethane, oxirane, dimethylene oxide) is a colorless, essentially odorless gas above its boiling point of 10.7°C at atmospheric pressure. It is utilized primarily as a chemical intermediate, but also finds use as a fumigant, fungicide, and sterilizing agent. Because its vapors in all concentrations above 3 percent are flammable and explosive and because it is a highly toxic material in both gaseous and liquid states, ethylene oxide must be used with extreme caution. Its physical properties are listed in Table 11-4, end-uses in Table 11-5.

Ethylene oxide is a highly reactive material and reacts exothermally with all compounds having a labile hydrogen atom. This reaction permits the introduction of the hydroxyethyl group ($-CH_2CH_2OH$) into a variety of compounds. The general reaction may be written:

$$
\begin{array}{ccc}
CH_2 & & CH_2A \\
\diagdown & & | \\
\quad \diagdown O + HA \rightarrow & & | \\
\diagup & & | \\
CH_2 & & CH_2OH
\end{array}
$$

Reaction products include the following: with water—ethylene glycol, diethylene glycol, and triethylene glycol; with ethylene glycol—polyethylene glycols; with alcohols and phenols—ethylene glycol monoethers; with acids—ethylene glycol monoesters; with amines—ethanolamines; with hydrogen cyanide—ethylene cyanohydrin; and with cellulose—hydroxyethyl cellulose.

When ethylene oxide is used to excess in the reactions with long-chain alcohols, phenols, acids, and other reactants mentioned above, polymeric polyoxyethylene derivatives are obtained.[18] In this manner ethylene oxide reacts with hydrophobic substances containing active hydrogen atoms to make nonionic surface active agents. Since the water solubility and surface activity of the products are determined by the number of moles of ethylene oxide reacted with the hydrophobic structure, practically any hydrophobe-hydrophile balance can be obtained by varying the ratio of the reactants. They can be combined to obtain optimum detergency, wetting, lime soap dispersion, and emulsifying properties as desired. The most important nonionics contain 6 to 15 oxyethylene units, although there may be as many as one hundred.

TABLE 11-4. PHYSICAL PROPERTIES OF ETHYLENE OXIDE AND PROPYLENE OXIDE*

Physical Property	Ethylene Oxide	Propylene Oxide
Boiling point @ 760 mm	10.7°C	33.9°C
@ 50 mm	−44°C	−24°C
@ 10 mm	−66°C	−49°C
Δb.p./Δb.p., 740 to 760 mm	0.036°C/mm	0.036°C/mm
Coefficient of expansion @ 20°C	0.00161	0.00151/°C
Critical temperature	195.8°C	215.3°C
Explosive limits in air		
@ 760 mm (upper)	100% by vol.	21.5% by vol.
(lower)	3% by vol.	2.1% by vol.
Flash point		
Tag glass open cup	<0°F	−
Cleveland open cup	−	−35°F
Freezing point	−111.3°C	−104.4°C
Heat of combustion	312.55 ± 0.20 kcal/g mol	7796 cal/g
Heat of vaporization @ 760 mm	245 Btu/lb.	160 Btu/lb.
Molecular weight	44.06	58.08
Refractive index,		
n_D at 7°C	1.3597	−
n_D at 20°C	−	1.3657
Solubility in water @ 20°C	Complete	40.5% by wt.
Specific gravity (apparent) @ 20/20°C	0.8711	0.8304
ΔSp. gr./Δt	0.00140	0.00125/°C
Specific heat	0.268 cal/g/°C	0.51 cal/g/°C
Vapor pressure @ 20°C	21.2 psia (1095 mm)	445 mm
Viscosity (absolute)	0.32 cps (at 0°C)	0.38 cps (at 20°C)

*Tables 11-4, 11-7 and 11-8 from "Glycols," G. O. Curme, Jr. and F. Johnston, Reinhold Publishing Corp., New York, 1952.

DERIVATIVES OF ETHYLENE OXIDE

Approximately 65 percent of the ethylene oxide produced in recent years was converted to ethylene glycol. The rest was used for the production of nonionic surfactants, polyglycols, ethanolamines, acrylonitrile, glycol ethers, and miscellaneous items. Output of ethylene oxide derivatives projected to 1965 is shown in Table 11-6.[6]

Ethylene Glycol

Ethylene glycol (ethanediol-1,2) is a colorless, practically odorless liquid with a sweetish taste. It is relatively viscous and nonvolatile. It is com-

TABLE 11-5. CONSUMPTION OF ETHYLENE OXIDE[6]

(million pounds)

End Use	1950	1955	1960	1965
Ethylene glycol	315	600	900	1,120
Nonionic surfactants	25	90	125	180
Diethylene glycol	18	55	94	115
Triethylene glycol	7	15	24	30
Ethanolamines	26	65	105	145
Acrylonitrile	25	55	65	75
Glycol ethers	25	50	60	70
Hydraulic fluids, oxide polymers, dyestuffs, morpholine, others	12	30	67	115
Total:	453	960	1,440	1,850

pletely miscible with water and the lower aliphatic alcohols, aldehydes, and ketones, but is practically insoluble in hydrocarbons and similar compounds. Toxicologically, ethylene glycol is not hazardous in its conventional industrial applications. LD_{50} values for guinea pigs, rats, and mice have been reported at 7.35, 5.50, and 13.1 ml./kg., respectively. Physical properties of ethylene glycol are listed in Table 11-7, end-uses in Table 11-8.

Chemically, ethylene glycol can be esterified, etherified, dehydrogenated, oxidized, and dehydrated. Its individual hydroxyl groups react as the hydroxyl groups in monohydric alcohols, except when both hydroxyl groups are involved in one reaction; e.g., oxidation of ethylene glycol to glyoxal. Most ethylene glycol derivatives are prepared commercially through ethylene oxide, however, and will be discussed here as derivatives of the oxide.

While many new markets for ethylene glycol have developed since 1925,

TABLE 11-6. U. S. OUTPUT OF PRINCIPAL ETHYLENE OXIDE[6] DERIVATIVES

(million pounds)

Product	1954	1955	1956	1957	1958	Estimated 1960	1965
Ethylene glycol	638	888	1,027	1,199	1,116	1,225	1,520
Nonionic surfactants	155	180	240	285	297	350	490
Acrylonitrile	63	118	141	174	180	245	300
Ethanolamines	63	78	93	94	97	126	175
Diethylene glycol	36	75	92	113	110	130	160
Triethylene glycol	17	19	20	25	25	30	40
Polyethylene glycols	na	na	25	25	25	28	35

na: not available

TABLE 11-7. PHYSICAL PROPERTIES OF ETHYLENE GLYCOL, DIETHYLENE GLYCOL, TRIETHYLENE GLYCOL, AND PROPYLENE GLYCOL

Physical Property	Ethylene Glycol	Diethylene Glycol	Triethylene Glycol	Propylene Glycol
Boiling point				
@ 760 mm	197.6°C	245.0°C	287.4°C	187.4°C
@ 50 mm	123°C	164°C	198°C	116°C
@ 10 mm	89°C	128°C	162°C	85°C
Δb.p./Δp, 740 to 760 mm	0.043°C/mm	0.049°C/mm	0.054°C/mm	0.042°C/mm
Coefficient of expansion @ 20°C	0.00062/°C	0.00064/°C	0.00069/°C	0.00073/°C
Density (true) at 20°C	1.11336 g/ml	1.1161 g/ml	–	1.0363 g/ml
Flash point (open cup)	240°F	290°F	330°F	225°F
Freezing point	–13°C	–8°C	–7.2°C	–
Heat of combustion	283.3 kcal/mol	567 kcal/mol	–	435.9 kcal/mol
Heat of vaporization				
@ 760 mm	191 cal/g	150 Btu/lb	179 Btu/lb	170 cal/g
@ 10 mm	–	279 Btu/lb	–	–
Initial decomposition temperature	–	164°C	206°C	–
Molecular weight	62.07	106.12	150.17	76.09
Refractive index,				
n_D @ 20°C	1.4316	1.4472	1.4559	1.4326
$\Delta n_D/\Delta t$	0.00026/°C	0.00028/°C	0.00031/°C	0.00030/°C
Specific gravity (apparent)				
at 20/20°C	1.1155	1.1184	1.1254	1.0381
ΔSp. gr./Δt	0.00070/°C (0 to 40°C)	0.00071/°C (5 to 40°C)	0.00078/°C (5 to 40°C)	0.00073/°C (0 to 40°C)
Specific heat at 20°C	0.561 cal/g/°C	0.500 Btu/lb/°F	0.525 Btu/lb/°F	0.593 cal/g/°C
Surface tension	48.4 dynes/cm (@ 20°C)	48.5 dynes/cm (@ 25°C)	45.2 dynes/cm (@ 20°C)	72.0 dynes/cm (@ 25°C)
Vapor pressure at 20°C	0.06 mm	<0.01 mm	<0.01 mm	–
Verdet constant at 25°C				
and 5461 Å	0.01456 min./ gauss cm	0.01490 min./ gauss cm	–	0.01540 min./ gauss cm
and 5893 Å	0.01232 min./ gauss cm	0.01264 min./ gauss cm	–	0.01291 min./ gauss cm
Viscosity (absolute) at 20°C	20.93 cps	35.7 cps	47.8 cps	56.0 cps

TABLE 11-8. ETHYLENE GLYCOL END-USE PATTERN

End-Use	1950		1960*	
	million pounds	% of total	million pounds	% of total
Antifreeze	380	76.6	900	73.4
Cellophane**	20	4.0	47	3.8
Explosives	30	6.0	25	2.0
Synthetic fibers (chiefly polyesters)	0.5	0.1	40	3.3
Polyglycols, glycol ethers	15	3.0	45	3.7
Polyester, alkyd, ester gum resins	5.5	1.1	23	1.9
Exports	20	4.0	55	4.6
Hydraulic fluids, industrial coolants, inventories, misc.	26	5.2	90	7.3
Totals	497	100.0	1,225	100.0

*Estimated.
**The market for ethylene glycol in cellophane had essentially been entirely transferred to propylene glycol by 1961.

its undisputed claim to fame is as permanent type, nonvolatile antifreeze. Antifreeze sales in 1960 amounted to 120.9 million gallons. Of this total, ethylene glycol types accounted for 110.5 million gallons and methanol types for the remainder.[1] Roughly 5 million gallons were the "extended life" antifreezes. Suitably inhibited ethylene glycol protects automobile cooling systems by lowering the freezing point of water to any degree required by local climate. It has the edge over methanol in that it has a higher boiling point, is odorless, and does not affect automotive finishes.

The second largest outlet for ethylene glycol is in the production of polyester fibers, films, and to a small extent plasticizers. The ethylene glycol is reacted with dibasic acids. Such reaction products as "Dacron" polyester fiber and "Mylar" film are super polymers prepared by condensing terephthalic acid with ethylene glycol.

The third largest outlet for ethylene glycol is as an intermediate for ethylene glycol dinitrate, which is used as a freezing inhibitor in nitroglycerin explosives and dynamites. In practice, mixtures of glycerol and ethylene glycol are nitrated with sulfuric and nitric acids to form solutions of nitroglycerin in ethylene glycol dinitrate. Permissible explosives incorporate 10 percent by weight of these solutions, while gelatin dynamites incorporate as much as 36 percent. These products are used almost exclusively in mining, road building, and construction work.

Ethylene glycol finds a host of other important, but relatively small, applications. It is a coolant for piston-type airplane engines, as well as for machine guns and army tanks. Solutions of boric acid or its salts in ethylene glycol form the electrolyte in electrolytic condensers for use in refrigerators, television sets, radar equipment, and so on. Ethylene glycol is incorporated into formulations for hydraulic fluids, and acts as a softening agent for fibers, paper, and leather. Inhibited ethylene glycol is used for snow removal and as deicing fluid. Ethylene glycol is also the starting material for glyoxal, prepared by vapor phase oxidation of the glycol over a supported copper catalyst. Glyoxal is primarily useful for stabilizing rayon against shrinkage.

Diethylene Glycol, Triethylene Glycol, Polyethylene Glycols, "Polyox" Resins

Ethylene oxide reacts with water to form ethylene glycol, diethylene glycol, triethylene glycol, and the lower polyethylene glycols:

$$n\mathrm{C_2H_4O} + \mathrm{H_2O} \rightarrow \mathrm{HO(C_2H_4O)}_n\mathrm{H}$$

where n is 1, 2, 3, 4, 5, ... The amount formed depends on the ratio of water to ethylene oxide.

Commercially, diethylene glycol and triethylene glycol are produced in continuous operations as co-products in the manufacture of ethylene glycol. The polyethylene glycols, however, are more uniformly produced by batch techniques utilizing a caustic material as catalyst.[18] Ethylene oxide is added to the reactor until the molecular weight has gone to the desired level. For polyethylene glycols with molecular weights up to 1000, ethylene oxide with ethylene glycol or diethylene glycol are ordinarily the starting materials. For higher molecular-weight polyglycols, the charge is ethylene oxide and a polyethylene glycol such as polyethylene glycol 1000. Polyethylene glycols with molecular weights up to 10,000 are practical by these methods. However, high polymers of ethylene oxide with molecular weights above 100,000 ("Polyox" resins) may be produced using special catalytic systems such as zinc alkyls,[11] aluminum alkyls,[19] and aluminum alkoxides.[19]

Diethylene glycol, $\mathrm{HO(C_2H_4O)_2H}$, is a hygroscopic, practically odorless and colorless liquid. It resembles ethylene glycol in most ways, but has added solvent power due to the extra ether group in the molecule. Its physical properties are listed in Table 11-7.

Diethylene glycol finds widespread application as a solvent, hygroscopic agent, plasticizer, lubricant and conditioning agent. It is utilized as a softener for fibers and solvent for dyes. It is a coupling agent for soluble oils used for textile solutions, metal cutting, agricultural sprays, and polishes. It is effective for softening and moistening tobacco, cellophane, composition cork, glues, casein, paper, and synthetic sponges. It is used to absorb moisture from natural gas, and in foundry sand mixes it keeps the molten metal from reacting with either the moisture in the sand or the sand itself. It acts as an extractant in "Udex" units.

Diethylene glycol derivatives are usually prepared as secondary derivatives in the reaction of ethylene oxide with active hydrogen compounds. The alkyl ethers of diethylene glycol, marketed by Union Carbide Chemicals as "Carbitol" solvents, are excellent solvents for cellulose esters, dyestuffs, and woodstains. They are useful in the formulation of lacquer thinners, varnishes, enamels, printing inks as well as hydraulic fluids and soluble oil preparations. Mono- and diesters of diethylene glycol are useful as plasticizers and emulsifying agents. Polyester resins derived from diethylene glycol, maleic anhydride, and styrene are important in the field of reinforced plastics. A particularly good solvent for fats, waxes, and greases is "Chlorex" solvent or bis(2-chloroethyl) ether, in which the hydroxyl groups of diethylene glycol are replaced by chlorine. It is used to improve viscosity characteristics of lubricating oils, as an assistant in textile operations such as scouring, as a solvent in the purification of butadiene, and as a soil fumigant and chemical intermediate.

Triethylene glycol, $HO(C_2H_4O)_3H$, is a colorless liquid with a slight sweet odor. It closely resembles diethylene glycol in properties and uses. Its physical properties are listed in Table 11-7. Triglycol is incorporated into air conditioning systems designed for dehumidifying air and is useful as an air disinfectant in hospital wards, theaters, offices, and homes. It is a liquid desiccant for removing water from natural gas, and an extractant in "Udex" units. Various triethylene glycol esters are useful as plasticizers for compounding such materials as rubber and vinyl chloride resins. The di-2-ethylbutyric acid ester of triethylene glycol (sold as "Flexol" plasticizer 3GH), for example, is the plasticizer for the polyvinyl butyral resins used in the manufacture of safety glass.

Polyethylene glycols range from water-white liquids to wax-like solids. They are stable, nonvolatile, odorless materials. As the molecular weight increases, the freezing or melting range, specific gravity, flash point, and viscosity increase; water solubility, vapor pressure, hygroscopicity, and solubility in organic compounds decrease. Polyethylene glycols are completely water soluble and are miscible with many waxes, gums, oils, starches, and organic solvents.

Polyethylene glycols, alone or in blends, are used as lubricants, vehicles, solvents, binders, and intermediates in the rubber, food, pharmaceutical, cosmetic, agricultural, textile, paper, petroleum, and many other industries. Chemically, polyethylene glycols are typical glycols in their reactions. Like ethylene glycol, however, most of their derivatives are prepared commercially through ethylene oxide, except the esters. Polyglycols are reacted with fatty acids to obtain nonionic surface-active agents useful as emulsifying agents, detergents, and dispersants in the agricultural and textile industries.

"Polyox" resins are ethylene oxide-derived polymers with molecular weights ranging from 100,000 to 10,000,000. These polyethers are homologs of the lower molecular-weight polyethylene glycols, but differ in that they show the characteristics of true high polymers, i.e., the ability to be cold drawn.

"Polyox" resins are dry white powders that are completely water-soluble at room temperatures. Their solutions are viscous even at low concentrations, and those with higher molecular weight show unusual thickening effects. The resins are tough, thermoplastic, and highly crystalline materials that can be calendered, molded, extruded, cast, and heat-sealed. The dry resins, films, or coatings show low atmospheric pick-up except at very high humidities and are resistant to most greases and oils. They resist biological attack and show no biological oxygen demand. Their unusual combination of properties make them useful in textile warp sizes, paper coatings, detergents, aerosol hair sprays, toothpastes, water-soluble packaging films, adhesives, and in coagulating operations.

Glycol Ethers

Monoalkyl ethers of ethylene glycol are commercially produced by reacting ethylene oxide with alcohols, for example, ethanol:

$$CH_2-CH_2 + C_2H_5OH \rightarrow C_2H_5OCH_2CH_2OH$$
$$\diagdown O \diagup$$

High ratios of alcohol to ethylene oxide favor the formation of ethers of ethylene glycol, while lower ratios produce monoethers of the polyglycols. Ethylene glycol monoethers can also be prepared from the chlorohydrin or glycol with sodium hydroxide and a dialkyl sulfate.

The monoethers of ethylene glycol are colorless, mobile, almost odorless liquids, sold under the Union Carbide Chemical trade name "Cellosolve." Their boiling points and refractive indexes increase with molecular weight; their specific gravities decrease. They are miscible with most organic solvents and up to the butyl ether are completely miscible with water. They are important solvents, coupling agents, and chemical intermediates in such industries as protective coatings, textiles, metals, leather, and paper.

Dialkyl ethers of ethylene glycol are preferably prepared by reacting the sodium alcoholate of an ethylene glycol monoether with an alkyl halide or by reacting a dialkyl sulfate with a monoether. Dialkyl ethers are useful as inert reaction media and as mutual solvents.

Glycol Esters

Ethylene glycol mono- and diesters are prepared by reacting the glycol with either organic or inorganic acids, although ethylene oxide may be used as the starting material. Only ethylene glycol dinitrate (see page 325) and some fatty acid esters are industrially important. The fatty acid derivatives are useful as emulsifying, stabilizing, dispersing, wetting, foaming, and suspending agents.

Glycol Ether-Esters

Glycol ether-esters are usually prepared by the esterification of ethylene glycol monoethers with acids, acid anhydrides, or acyl halides. Ether-esters of acetic acid are the major commercial materials. Lower molecular-weight ether-esters are used as solvents. Higher molecular-weight products are employed as plasticizers, softening agents, and resin solvents.

Ethanolamines

The reaction of ethylene oxide with ammonia yields mixtures of ethanolamines—the proportion of mono-, di-, and triethanolamines formed depending on the ratio of ethylene oxide to ammonia. Excess oxide favors the formation of triethanolamine.

The chief outlet for the ethanolamines is in detergent manufacture. They are reacted with fatty acids for the production of anionic surfactants. The ethanolamines are also used in the scrubbing of carbon dioxide and hydrogen sulfide from refinery gases. They find other markets in cosmetics, textile processing, agricultural sprays, and emulsion cleaners.

Acrylonitrile

Ethylene oxide reacts with hydrogen cyanide to produce ethylene cyanohydrin for conversion into acrylonitrile. Synthetic fibers such as "Dynel" and "Orlon" are major outlets for acrylonitrile. Nitrile rubbers and plastics consume smaller amounts.

Dioxane

Dioxane is a powerful industrial solvent produced by the dimerization of ethylene oxide or by the dehydration of ethylene glycol. It has excellent solvent properties for cellulose nitrate, cellulose acetate, cellulose ethers, natural resins, vegetable and mineral oils, and oil-soluble dyes. A high order of toxicity has severely limited its commercial utility, however.

Hydroxyethylcellulose

Hydroxyethyl ethers of cellulose are formed by the action of ethylene oxide on cellulose. The number of ethylene oxide moles per glucose unit of cellulose can be varied by the amount of ethylene oxide used. These ethers are utilized as thickeners for resin emulsions and textile pastes, textile sizing agents, coatings for paper, and suspending agents for inert solids.

Thiodiethylene Glycol

Ethylene oxide reacts with hydrogen sulfide to form 2-mercaptoethanol and thiodiethylene glycol ($HOCH_2CH_2S\ CH_2CH_2OH$). Thiodiglycol, sold as "Kromfax" solvent, is primarily useful in preparing printing pastes of vat and basic dyestuffs, and as a solvent for inks used in printing.

Nonionic Surface-Active Agents

Ethylene oxide is condensed with a variety of high-carbon-content materials for the manufacture of nonionic surface-active agents.[14,15] Alkyl phenols, tall oil, and fatty or long-chain synthetic alcohols are the major co-reactants. Partially esterified polyhydric alcohols (such as sorbitol), C_{10}—C_{18} branched chain alkyl mercaptans, and polypropylene glycols (see page 332) are less prominent starting materials, as are higher alkyl amines and alkyl amides. Amine and amide condensation products, however, tend to be cationic. In general, as the length of the polyoxyethylene chains increases, water solubility of the product increases and oil solubility decreases.

Alkyl phenol-ethylene oxide polyethers are the largest group of nonionics. They usually contain 4 to 30 moles of ethylene oxide. Types containing up to 15 moles are liquids; above 15 moles they are waxes. The low molar ratio products act as water-in-oil emulsifiers and as detergents in nonaqueous media such as dry cleaning solvents. They are also sulfated for the production of high-foaming anionic surfactants. Higher mole-ratio products are widely used in low foaming household and industrial detergents. Industrial applications include their use as petroleum demulsifiers, insecticide and latex paint emulsifiers, and textile surfactants.

Tall oil-ethylene oxide adducts are polyoxyethylene esters of carboxylic acids. Tall oil is the source of fatty acids, rosin acids, and naphthenic acids. The low mole-ratio adducts are emulsifiers and textile lubricants. The 12 to 16 mole ethylene oxide adducts are household detergents, primarily for automatic washer products.

Alcohol-ethylene oxide condensation products are based on fatty and rosin alcohols, such as lauryl, oleyl, tallow, and cetyl alcohols, or synthetic alcohols, such as trimethylnonanol. Such adducts are used in raw wool scouring, dye leveling, kier boiling, liquid dishwasher detergents, and emulsifying operations.

Other ethylene oxide-based surfactants have miscellaneous specialty applications. For example, the partially esterified polyhydric alcohols are emulsifiers in cosmetic and pharmaceutical preparations. Fatty amine adducts are used as textile lubricants and softeners. The mercaptan condensates are detergent sanitizers and dairy cleaners. They also are employed for reducing air pollution.

DERIVATIVES OF PROPYLENE OXIDE

Propylene oxide is a colorless liquid at room temperature. Its physical properties are listed with those of ethylene oxide in Table 11-4. Like the lower oxide, propylene oxide is primarily a chemical intermediate. As shown in Table 11-9, propylene glycol, polypropylene glycols, and mixed ethylene oxide-propylene oxide derivatives are its major end products. They are produced by methods comparable to those for the ethylene equivalents.

Propylene Glycol

Propylene glycol has the same basic characteristics as ethylene glycol but is slightly more volatile and about three times as viscous at room temperature. Physical properties of the two may be compared in Table 11-7.

The applications for propylene glycol are similar to those for ethylene glycol. Propylene glycol is used in preference to ethylene glycol in food, drug, and cosmetic uses because it is nontoxic. It is incorporated in the

TABLE 11-9. CONSUMPTION OF PROPYLENE OXIDE[7]

(million pounds)

	Year	
End Use	1959	1960
Propylene glycol	130	155
Polypropylene glycols and oxide adducts for urethane foams	45	70
Polypropylene glycols for brake fluids	20	21
Surface active agents for detergents, etc.	12	13
Miscellaneous uses, including isopropanol amines and dipropylene glycol	25	28
Total:	232	287

Source: C & EN estimates

New and Non-Official Remedies for pharmaceutical products, and the Food and Drug Administration lists it as one of the few "generally recognized as safe" items for use in food products. LD_{50} values for small animals are close to 20 ml/kg. End uses for propylene glycol are shown in Table 11-10.

TABLE 11-10. CONSUMPTION OF PROPYLENE GLYCOL[7]

(million pounds)

End Use	Year	
	1959	1960
Polyesters	43	54
Plasticizers	6	6
Brake fluids	10	10
Cellophane	18	20
Export	15	15
U.S.P. grade	33	35
Miscellaneous	11	12
Total:	136	152

Source: C & EN estimates

Accordingly, propylene glycol is the coolant in refrigerating systems where toxicity may be a factor: in dairies, breweries, and food packaging plants and on ships, trains, and planes. It acts as a hygroscopic agent and preservative for tobacco, as a softening agent for cellophane, and as a solvent for inks used in printing food wrappers of paper and cloth. In foods, propylene glycol is a solvent for flavoring materials such as vanilla, preservative against mold growth in syrups, and humectant for such items as coconut. It is incorporated into many formulations for both pharmaceuticals and cosmetics. Like ethylene glycol, propylene glycol is a component of hydraulic fluids and an intermediate for resins and plasticizers of the polyester type. The resins are primarily used in reinforced plastics, the plasticizers in high-temperature applications.[5]

Polypropylene Glycols

Polypropylene glycols are more oil soluble and less hygroscopic than the comparable polyethylene glycols. They are colorless to light yellow, nonvolatile viscous liquids. The lower molecular-weight compounds are completely soluble in water. They are more compatible with vegetable oils and natural waxes and resins than are the polyethylene glycols. Chief outlet for polypropylene glycols are as urethane intermediates. They are also useful in hydraulic fluids and automobile radiator compounds. They are mold lubricants, anti-dusting agents, and antifoaming agents. Polypropylene glycol esters are used as petroleum emulsion breakers and as emulsifiers in insecticidal preparations. Ethers are employed as lubricants and hydraulic fluids.

Mixed Alkylene Oxide Derivatives

Ethylene oxide condenses with polypropylene glycols to form nonionic surfactants with the general formula

$$HO(C_2H_4O)_a(C_3H_6O)_b(C_2H_4O)_cH$$

They are useful as general purpose emulsifiers and detergents. Ethylene oxide condenses also with propylene oxide to mixed polyalkylene glycols. These polymeric products, sold under the trade name "Ucon," are lubricants in metal forming operations and for such items as industrial machinery and electric motors. They are widely used in formulations for automotive and industrial hydraulic fluids, as textile and rubber lubricants, and as oily bases for hair dressings.

References

1. Anon., *Chem. Eng. News,* 130 (September 4, 1961).
2. Anon., *Petroleum Refiner,* **38,** 247 (November 1959); *Petrochemical Industry,* 18 (September 1959).
3. Anon., *Petroleum Refiner,* **38,** 248 (November 1959); *Petrochemical Industry,* 11 (January 1959).
4. Anon., *Petrochemical Industry,* 22 (September 1959).
5. Anon., *Chem. Week,* 85 (April 29, 1961).
6. Anon., *Chem. Week,* **84,** 37 (April 25, 1959).
7. Anon., *Chem. Eng. News,* 20 (October 10, 1960).
8. Anon., *Chem. Week,* **82,** 74 (June 28, 1958).
9. Anon., *Petroleum Refiner,* **38** (11), 246 (1959).
10. Aries, R. S., and Schneider, H., "Encyclopedia of Chemical Technology," Vol. 5, p. 912, New York, The Interscience Encyclopedia, Inc., (1950.)
11. Bailey, F. E., Jr., Powell, G. M., and Smith, K. L., *Ind. Eng. Chem.,* **50,** 6 (1958).
12. Cume, G. O., Jr., and Johnston, F., "Glycols," New York, Reinhold Publishing Corp., 1952.
13. Faith, W. L., Keyes, D. B., and Clark, R. L., "Industrial Chemicals," New York, John Wiley & Sons, Inc., (1957).
14. Fine, R. D., *J. Am. Oil Chem. Soc.,* **35,** 542 (1958).
15. Gaylord, N., "Polyethers," Interscience, 1962.
16. Katzen, R., *Petroleum Refiner,* **39,** 167 (1960).
17. Landau, R., *Petroleum Refiner,* **32** (9), 146 (1953).
18. Malkemus, J. D., *J. Am. Oil Chem. Soc.,* **33,** 571 (1956).
19. Price, C. C. *et al., J. Polymer Sci.* (1958).
20. Schwartz, A. M., Perry, J. W., and Berch, J., "Surface-Active Agents," Vol. II, New York, Interscience Publishers, Inc., (1958).

12. CHLORINATED METHANES

RALPH LANDAU AND SHERWOOD N. FOX
Scientific Design Company, Inc.

INTRODUCTION

Chlorinated methanes are hydrocarbons based on methane (CH_4) in which one or more hydrogen atoms have been replaced by chlorine. The resulting four compounds are methyl chloride (chloromethane, CH_3Cl), methylene chloride (dichloromethane, CH_2Cl_2), chloroform (trichloromethane, $CHCl_3$), and carbon tetrachloride (CCl_4). Molecular weight, specific gravity and boiling point increase with a corresponding increase in the number of chlorine atoms (Table 12-1). With the exception of methyl chloride, which is a gas easily liquefied under pressure, the chlorinated methanes are liquid at ordinary temperatures and pressures.

Reference books on organic chemistry generally contain little on chlorinated methanes and other chlorinated hydrocarbons because historically their chief service to industry has rested on their specific physical properties rather than on their chemical properties. In recent years, however, their value as chemical intermediates has taken a sharp upturn in stride with increased industrial emphasis on materials such as silicones, chlorofluorocarbons and tetramethyllead. In fact, a typical characteristic of the use pattern of the chlorinated methanes with subsequent rises in production over the years, has been their ability to recover from obsolescence in certain major areas by finding new and more promising outlets. A striking example is the way carbon tetrachloride has lost ground as a dry cleaning solvent, once its chief application, only to gain new prominence as an intermediate in chlorofluorocarbon production.

With the exception of carbon tetrachloride, the price structure for commercial grades and quantities of the chloromethanes has remained relatively constant. Over the past decade, the list price of methyl chloride has risen from 11.5 cents to 12.75 cents per pound, freight equalized, in 1961. During the same period, methylene chloride prices have fluctuated somewhat, but the 1961 list price (12.25 cents per pound, delivered) is essentially the same as it was in 1952. The delivered list price for chloroform has remained con-

TABLE 12-1. PHYSICAL PROPERTIES OF THE CHLORINATED METHANES

	Boiling point (°F)	Freezing point (°F)	Specific gravity 68°/39°F	Pounds/gal 68°F	Molecular Wt.
Methyl chloride	− 10.7	− 143.7	0.920	7.68	50.5
Methylene chloride	104.2	− 142.1	1.326	11.07	84.9
Chloroform	142.3	− 82.3	1.489	12.43	119.4
Carbon tetrachloride	168.8	− 9.13	1.594	13.30	153.8

stant at 17 cents a pound. On the other hand, the list price of carbon tetrachloride has increased steadily from less than 8 cents in 1952 to 10.75 cents a pound, freight allowed, in 1961.

All of the chlorinated methanes are stable enough to be of commercial value in many applications. They do, however, decompose through oxidation, hydrolysis and pyrolysis. High temperatures, light and impurities will accelerate oxidation. Elevated temperatures and certain metals will step up rates of hydrolysis. All of the chlorinated methanes will break down in direct flame or on hot surfaces. Thermal decomposition, a strong factor in the decline of carbon tetrachloride as a fire extinguishing agent, can produce hydrochloric acid, phosgene and chlorine.

The chlorinated methanes are toxic and, as a general rule, humans should not be exposed to their fumes, even when the odor of these compounds is barely noticeable. The literature is filled with accounts of their toxicity and the anesthetic effects, in some cases, of their fumes, causing a delay of the symptoms associated with poisoning. For these reasons, adequate ventilation should be supplied to those exposed to the fumes. Contact with the skin should also be avoided since these compounds are excellent solvents and may dissolve the fats of the skin.

In recent years increasing demand for the chlorinated methanes as chemical intermediates has led to expanded production (Table 12-2). There are currently 14 prime producers at 19 locations (Table 12-3). Present process technology, involving six basic processes, is described in a later section.

METHYL CHLORIDE

Dumas and Peligot discovered methyl chloride in 1835, but it was not until 1874 that the Germans began making it in commercial quantities for use as a methylating agent in manufacturing dyes and organic derivatives.[4] Eventually, the thermodynamic and physical properties of the chemical drew attention to it as a refrigerant. This application dates back to 1884, when Crespin and Marteau built the first successful machines in Paris. By

TABLE 12-2. U.S. PRODUCTION OF THE CHLORINATED METHANES

Methyl Chloride Production, U. S.

Year	Number of Manufacturers	Thousands of Pounds
1954	5	33,147
1955	6	36,318
1956	6	41,564
1957	7	46,756
1958	7	43,532
1959	8	67,067

Methylene Chloride Production, U. S.

Year	Number of Manufacturers	Thousands of Pounds
1954	4	69,833
1955	5	73,963
1956	6	95,391
1957	7	94,175
1958	7	88,968
1959	7	112,740

Chloroform Production, U. S.

Year	Number of Manufacturers	Thousands of Pounds
1954	6	32,087
1955	6	40,396
1956	6	46,308
1957	8	57,403
1958	8	47,137
1959	9	70,717

Carbon Tetrachloride Production, U. S.

Year	Number of Manufacturers	Thousands of Pounds
1954	6	234,895
1955	5	287,371
1956	7	302,480
1957	8	317,839
1958	6	312,875
1959	7	367,847

SOURCE: U.S. Tariff Commission

1900 its use as a refrigerant was well established in Europe, especially in France and Germany. In the United States, however, little use was found for methyl chloride outside the laboratory until the advent of the household and small commercial refrigerator. U. S. commercial production, therefore, did not begin until 1920 despite early widespread use in Europe. The first American producer was the Roessler & Hasslacher Chemical Company. Following the acceptance of methyl chloride by a handful of refrigerating equipment manufacturers, American engineers became more familiar with it, and the variety of its applications steadily increased.

The raw material used initially for producing methyl chloride in Europe was the waste residue of beet sugar manufacture. Trimethylamine was dry distilled from the betaine present in these residues, the amine yielding methyl chloride and ammonium chloride when treated with hydrogen chloride. With the introduction of synthetic methanol and better methods of chlorinating methane, new and less expensive ways to produce methyl chloride emerged.

TABLE 12-3. PRINCIPAL U.S. PRODUCERS OF
CHLORINATED METHANES

Company	Location	Chlorides Produced			
		CH_3Cl	CH_2Cl_2	$CHCl_3$	CCl_4
Allied Chemical Corp. (Solvay Process Div.)	Moundsville, W. Va.	X	X	X	X
Ancon Chemical Co.	Lake Charles, La.	X	–	–	–
Ansul Chemical Co.	Marinette, Wis.	X	–	–	–
Brown Company	Berlin, N. H.	–	–	X	–
Diamond Alkali Co.	Belle, W. Va.	X	X	X	–
	Painesville, Ohio	–	–	–	X
Dow Chemical Co.	Freeport, Tex.	X	X	X	X
	Pittsburg, Calif.	X	X	X	X
	Plaquemine, La.	–	–	–	X
Dow Corning Corp.	Midland, Mich.	X	–	–	–
E. I. du Pont de Nemours & Co., Inc. (Electrochemical Dept.)	Niagara Falls, N. Y.	X	X	X	–
Ethyl Corp.	Baton Rouge, La.	X	–	–	–
FMC Corporation (Westvaco Chlor-Alkali Division)	S. Charleston, W. Va.	–	–	–	X
General Electric Co. (Silicone Prod. Dept.)	Waterford, N. Y.	X	–	–	–
Pittsburgh Plate Glass Co. (Chemical Division)	Barberton, Ohio	–	–	–	X
Stauffer Chemical Co.	Louisville, Ky.	–	X	X	X
	Niagara Falls, N. Y.	–	–	–	X
Vulcan Materials Co. (Frontier Chem. Co. Division)	Wichita, Kan.	–	X	X	X
(Kolker Chem. Div.)	Newark, N. J.	X	X	X	–

Physical and Chemical Properties

A gas under normal conditions, methyl chloride is heavier than air, colorless, and has a somewhat ether-like odor. It is stored and shipped under pressure as a water-white liquid. Chemically, it is more reactive than ethyl chloride but less reactive than methyl bromide and methyl iodide. Methyl chloride readily furnishes methyl groups, and methylations may be carried out through the Friedel-Crafts reaction, the Wurtz synthesis, the Grignard reaction, and by direct reaction with metals such as silicon. The last two, used in preparing organosilicon compounds as intermediates in the production of silicones, serve as important outlets for most of the methyl chloride produced today.

Methyl chloride has the highest thermal stability of the chlorinated methanes. In the presence of water, it hydrolyzes slowly to form hydrochloric acid. In contact with air at ordinary temperatures it is stable, but at higher temperatures it decomposes in direct contact with flame or very hot surfaces. Classed as moderately flammable, it is explosive in concentrations of about 8 to 18 percent in air. The products of decomposition are hydrogen chloride and phosgene, which though toxic are very irritating and thus afford adequate warning of their presence in harmful amounts. Methyl chloride fumes, on the other hand, are both toxic and anesthetic and do not give adequate warning to humans. A 2 percent concentration after two hours may be fatal.[10] The symptoms of methyl chloride poisoning by inhalation are similar to those of alcohol intoxication: giddiness, nausea and vomiting, and drowsiness passing to anesthesia. Symptoms may not develop for several hours after exposure, becoming progressively worse for several days until improvement begins or death occurs.

Beyond the normal precautions for moderately flammable, moderately toxic liquefied gases, no special or unique techniques need be employed in handling and storing methyl chloride. Most construction metals except aluminum may be used for plant equipment and containers.

Uses

At present, although methyl chloride is still used to a small extent as a refrigerant, its chief application is as a methylating agent, primarily in preparing intermediates in the synthesis of silicones. Fully three quarters of the methyl chloride consumed in 1960 went into silicones production. One reaction unites methyl chloride directly with silicon at about 360° C to form dimethyldichlorosilane, which is then hydrolyzed and condensed with other alkylchlorosilanes to form the resin. In another method, methyl chloride and magnesium are first combined as a Grignard reagent and then pumped into an ether solution of silicon tetrachloride to obtain dimethyldichlorosilane. In Friedel-Crafts reactions, methyl chloride is used to a lesser extent by the chemical industry to methylate benzene, toluene, xylene and other aromatics. Another minor outlet is in the manufacture of methyl cellulose.

The second major use of methyl chloride is in butyl rubber manufacture; here it is used as catalyst carrier, diluent and solvent. Less than ten years ago, almost half of the methyl chloride consumed went into this application; however, automobile inner tube production, the chief outlet for butyl rubber, has steadily declined since the introduction of the tubeless tire so that now only about 10 percent of the methyl chloride consumed is attributed to this use.

In addition to its limited use as a refrigerant, methyl chloride has less important applications as a solvent and propellant for certain high pressure aerosols used in greenhouses, as a low temperature extractant in the drug industry, and as a solvent for impurities in metal annealing salt baths.

A new and potentially huge application of methyl chloride is in the manufacture of tetramethyllead (TML). At least one major oil company has begun using this compound as a gasoline antiknock agent along with the more familiar tetraethyllead (TEL). Whether the other oil companies follow suit remains to be seen, but there is little doubt of the favorable market possibilities that such a development would bring to methyl chloride producers. TML is made by the Kraus-Callis reaction, the same process used to make TEL. Methyl chloride (ethyl chloride in the case of TEL) reacts with lead-sodium alloy to give sodium chloride and TML. Unreacted lead is recycled.

METHYLENE CHLORIDE

The history of methylene chloride is obscure, probably because its remarkable solvent power has only been recognized within the past 20 or 25 years. The earliest record of its preparation was in 1840 by Regnault, discoverer of carbon tetrachloride, who chlorinated methyl chloride in synthesizing the new compound.

Physical and Chemical Properties

The least toxic of the chlorinated methanes, methylene chloride is a clear, colorless, volatile liquid with a not unpleasant characteristic odor. It is miscible with alcohol, ether, and most organic solvents and is practically insoluble in water. It is relatively stable even in the presence of water, has no flash point, and does not support combustion. Combined with its excellent solvent activity, these properties make it especially useful in cleaning operations such as vapor degreasing, hot stripping and cold cleaning.

Chemically, methylene chloride behaves much like methyl chloride, though it reacts more slowly and less smoothly, and gives more complex product mixtures. As a result, its chief applications lie in those areas which make use of its physical rather than its chemical properties. Since methylene chloride will gradually hydrolyze to form hydrochloric acid, it should be kept dry to minimize losses and prevent corrosion. It can be used in storage systems made of most common construction materials, the one exception being aluminum.[5] No special handling or storing equipment or methods are required.

Methylene chloride will not form combustible mixtures under ordinary conditions, but vapors coming in contact with open flame or hot metal surfaces can form phosgene and hydrogen chloride. The maximum allowable concentration for methylene chloride fumes has been set at 500 ppm, the highest of any of the chlorohydrocarbons. Besides being toxic, the chemical is narcotic, and inhaling its vapors may produce nausea, dizziness, and headaches.[3] Prolonged exposure to high concentrations may result in unconsciousness or even death. Symptoms of poisoning disappear quickly in fresh air, however, with no permanent effects on the vital organs.

Uses

Paint strippers based on methylene chloride are claimed to remove more types of paint finishes from a greater variety of metals than any other stripping agent—a fact which accounts for the position of this compound as the nation's top industrial stripper in dollar sales.[13] Despite the disadvantages of high cost and high rate of evaporation, about one third of the methylene chloride consumed goes into paint and varnish strippers, both industrial and retail. Effective even at room temperature, methylene chloride strippers wrinkle rather than dissolve paint finishes. Because they don't react with aluminum under these conditions, their biggest outlet is in the aircraft field.

Two quite recent uses of methylene chloride, i.e., as an aerosol propellant and as a "safety solvent," account for another 30 percent of its total consumption. Aerosol producers use it in combination with certain propellants as both a solvent and a diluent; for example, it will dissolve 80 percent of its own weight of DDT while at the same time reducing the possibility of clogged nozzles. Added to petroleum solvents, methylene chloride effectively raises the flash point of the material, thereby increasing the safety factor of operations using these flammable solvents.

Methylene chloride's ability to dissolve cellulose acetate accounts for more than 15 percent of its total consumption, mostly in the manufacture of triacetate safety film used by the motion picture and television industries. On a smaller scale, its low boiling point, high solvency and nonflammability also make it useful to industry as an extractant for fine chemicals, alkaloids, antibiotics and fats and oils; and as a solvent for certain resins, many lacquers, ethyl cellulose, nitrocellulose, waxes and both synthetic and natural rubber. By adding an evaporation retardant such as a water seal, it can be used in degreasing and cleaning operations.

CHLOROFORM

Credit for discovering chloroform is shared equally by Liebig of Germany, Soubeiran of France and Guthrie of the U. S., all of whom, working independently of one another, simultaneously discovered this compound in 1831. Liebig prepared it from acetone and bleaching powder, Soubeiran and Guthrie from alcohol and bleaching powder.[111] Its first use was as a medicine in treating asthma. Then, in 1847, the Englishman Sir James Y. Simpson discovered the anesthetic property of this chemical that was to play such a critical role during the American Civil War and, years later, almost entirely supplant ether as the world's general anesthetic before fading into obscurity in this application.

Simpson described the anesthetic properties of chloroform in his famous pamphlet "New Anesthetic Agent as a Substitute for Sulphuric Ether in Surgery and Midwifery."[197] Chloroform took the medical world by storm, partly because of Simpson's writing efforts and partly because of certain advantages it held over ether. It was easier to manage, acted faster, and was more pleasant to take than ether. By 1870, it had become so popular in the operating room that Simpson was able to write to an American colleague, "Chloroform is the greatest triumph of all. For it has, if not entirely, yet nearly entirely, superseded the use of 'sulphuric ether.' "

But in 1890, after 20 years filled with records of accidents from the use of chloroform, surgeons all over the world began to discard it for ether. Dr. Edward R. Squibb, who was largely responsible for introducing anesthetics in American medicine, was of the opinion that the deaths attributed to chloroform were due to overdosage rather than to something intrinsic in the chemical. Even after he developed the first effective mask to administer ether safely, he did not give up his belief that chloroform was safe and efficient for short periods, and he continued to use chloroform anesthesia on his own family when he thought it appropriate.[36] With the appearance of more modern anesthetics, the market for chloroform eventually dwindled to almost nothing. However, its value as a solvent and chemical reactant more than made up for its abandonment as an anesthetic, so that today it is still an important chemical.

Physical and Chemical Properties

Chloroform is a clear, colorless liquid, volatile but not flammable, with a characteristic sweet odor. When exposed to air and light it decomposes slowly to form phosgene, hydrochloric acid, and chlorine. Its excellent solvent action is one of the properties which have made it useful to industry. Besides behaving as a typical alkyl chloride, it reacts with potassium dichromate and sulfuric acid to produce phosgene and chlorine, and with nitric acid to give small yields of chloropicrin. Chlorinated, it yields carbon tetrachloride; reduced with zinc and hydrochloric acid, it forms methylene chloride. Important commercially is its reaction with metal fluorides such as antimony trifluoride in producing chlorofluorocarbons.

Being reasonably toxic, chloroform's maximum allowable concentration for eight hours without serious effects has been set at 100 ppm. In the laboratory, the greatest hazard arises from the presence of phosgene, formed when chloroform has not been properly stored or has been stored too long. When stabilized (usually with ethyl alcohol or amylene), chloroform may be used with the common construction metals up to 248° F in the presence or absence of light and water.[8] Chloroform's high boiling point and low freezing point make unnecessary any need for special temperature requirements when storing. Pure chloroform does not act on most metals, so it requires no special handling or storing techniques. Stocks should be checked regularly for corrosion by hydrochloric acid which may have formed from contact with air, light, or moisture.

Uses

A decade or so ago, considerable amounts of chloroform were being used by the drug makers to extract and purify antibiotics. However, since 1951, when 38 percent of the chloroform consumed went into penicillin production alone, the use pattern has been altered considerably. Pharmaceutical uses have declined to less than 5 percent of total chloroform consumption, most of which goes into cough medicines, liniments and vitamin manufacture. Chloroform has even fewer applications as a freezing point depressant in carbon tetrachloride fire extinguishers and as a solvent in dye and perfume manufacture.

The main factor in the growth of chloroform use in the past few years has been the soaring demand for it as an intermediate in the production of chlorofluorocarbon refrigerants and fluorocarbon resins (mostly "Teflon"). In 1959, 90 percent of total consumption was attributed to these two applications. The refrigerant that contributes most heavily to chloroform's consumption is type "22" chlorofluorocarbon, monochlorodifluoromethane, in which two atoms of chlorine have been replaced by two atoms of fluorine through reaction with anhydrous hydrogen fluoride over an antimony halide catalyst. When type "22" chlorofluorocarbon is pyrolyzed under pressure, it polymerizes to the inert plastic tetrafluoroethylene, "Teflon."

CARBON TETRACHLORIDE

Carbon tetrachloride was first noted by Regnault about 1840 as the result of the reaction in sunlight between chlorine and chloroform. Later, Dumas produced it by chlorinating the methane in marsh gas. In 1843, Kolbe reported its preparation in a red hot tube by the action of chlorine on carbon disulfide. Hoffman, in 1860, also used carbon disulfide, but he chlorinated it with antimony pentachloride to form carbon tetrachloride and sulfur. Commercial production started near the turn of the century, but exact dates are not available. As early as 1900, Dow Chemical Company was selling carbon tetrachloride to the Laskin Soap Company as a nonflammable cleaning fluid.[111] But first full commercial production probably was by the Warner Chemical Company at Carteret, New Jersey, which made about a ton of the solvent in 1902.

A few years after Dow started a commercial operation at Midland, Michigan, in 1908, the industrial use of carbon tetrachloride as a solvent steadily increased. And when the dry cleaning industry began using it, production boomed. By 1935, about 65 percent of the total U. S. production was going into this outlet.[11] Only 15 years later, however, perchloroethylene had gained the attention of the dry cleaning industry, and by 1956 only 7 percent of the carbon tetrachloride used went to this industry. The advantages of perchloroethylene in dry cleaning lie in its lower toxicity, its lower corrosive action on common metals, and its lower volatility which permits a greater recovery in the small reclaiming units now so common in the shop of the independent dry cleaner. During the same period that carbon tetrachloride was losing ground as a dry cleaning solvent, demand for it was increasing because of totally new applications. In 1955, for example, about 57 percent of the 260 million pounds produced was used as an intermediate in the manufacture of industrial propellants and refrigerants.

Physical and Chemical Properties

The heaviest of the chlorinated methanes, carbon tetrachloride is a clear, colorless liquid with an ether-like odor similar to that of chloroform. Like chloroform, it has no flash point; unlike chloroform, it does not decompose in light. Its use as a fire extinguishing agent rests on its nonflammability and its failure to support combustion of carbonaceous materials. It is not suitable as an extinguisher for burning metals, however, since it yields most of its chlorine to the metal and violently accelerates the combustion. In contact with water, carbon tetrachloride slowly hydrolyzes to form hydrochloric acid and other products. Metals accelerate the rate of hydrolysis and are in turn attacked by the acid formed. In contact with open flame, carbon tetrachloride will decompose forming hydrogen chloride and, under certain conditions, phosgene. This does not materially affect its suitability as a fire extinguishing agent, however, since during most fires other gases such as carbon monoxide are present which lack at least the saving grace of being as easily detectable as hydrogen chloride and phosgene.

Carbon tetrachloride is toxic when inhaled or swallowed. Symptoms are headache, mental confusion, depression, fatigue, loss of appetite, nausea, vomiting, loss of coordination and sense of balance, and visual disturbances.[2] Poisoning may be insidious, with symptoms delayed for as long as eight days. Maximum allowable concentration for an eight-hour working exposure is 100 ppm. Carbon tetrachloride is much more hazardous to the drinking individual because the presence of even small amounts of alcohol in the blood synergistically intensifies the toxic action.[182] After prolonged contact with the skin, this solvent can cause poisoning with all the symptoms that accompany its inhalation and ingestion, and besides may inflame the skin. Other than providing adequate ventilation in work areas, particularly at floor level, no special handling or storing techniques are required for carbon tetrachloride. Pure and dry, it is not corrosive to most metals.

The demand for carbon tetrachloride is increasing. The biggest single outlet at present is as an intermediate in chlorofluorocarbon manufacture, a market that now takes at least 65 percent of all carbon tetrachloride consumed. Chlorofluorocarbons based on this compound can be made by the reaction of anhydrous hydrogen fluoride on carbon tetrachloride in the presence of a catalyst and under controlled conditions. The most important of these are type "12" (dichlorodifluoromethane) and type "11" (trichloro-monofluoromethane). Today, type "12" by itself accounts for more than half of the chlorofluorocarbon production in the United States;[17] it not only serves as an important aerosol propellant but also is in wide use as a refrigerant in all types of air conditioning and refrigerating equipment, both large and small.[21] Type "11" became popular when the low pressure aerosol container was introduced and is also one of the most widely used refrigerants in air conditioning and refrigerating equipment of the industrial type.

Thermoelectric cooling devices, presently limited in size because of low cooling capacity, seem to pose no immediate threat to the position of the halocarbons in the refrigerant market. And any immediate losses to thermoelectrics by these compounds will be more than compensated by the growing trend toward their use as blowing agents for urethane foam refrigerator insulation.

The role of carbon tetrachloride as a grain fumigant runs a weak second to its primary use, taking about 15 percent of total consumption. That this application is becoming increasingly large, however, is indicated by its position ten years ago when it accounted for only slightly more than 2 percent of the then lower total consumption of carbon tetrachloride. The value of the chemical as a fumigant hinges on its toxicity to a wide variety of pests and its nonflammability and nonexplosiveness. Often used in combination with carbon disulfide or ethylene dichloride, this fumigant is readily volatile, clean, and leaves no odor.

Carbon tetrachloride is still in fairly wide use as a fire extinguishing fluid, though in increasingly stiff competition with the "dry" types such as carbon dioxide. Both function in much the same manner, i.e., by excluding oxygen from burning surfaces, and both are nonconductors of electricity. Carbon dioxide, however, has no adverse effects on humans and animals when exposure is reasonable, and it has the added advantage of exerting a cooling effect as it is released from high pressure in the extinguisher. Probably no more than 6 percent of total carbon tetrachloride consumption now results from this application. Other miscellaneous uses are dependent primarily on the solvent power of the compound; e.g., cleaning operations, degreasing, extracting and spotting.

PROCESS TECHNOLOGY

In recent years, increases in consumption of the chloromethanes as chemical intermediates relative to the older solvent uses has led to expanded activity and a changing process pattern. As noted earlier, there are currently 14 prime producers at 19 locations. Shifts in demand toward the lower molecular-weight chlorides have led to increasing emphasis on processes which can economically produce a spectrum of chlorides.

There are at present six chloromethanes processes in use:

(1) Direct thermal chlorination of methane
(2) Intensive thermal chlorination of light
    hydrocarbons (carbon tetrachloride only)
(3) Photochemical chlorination
(4) Chlorination of carbon disulfide (carbon
    tetrachloride only)
(5) Methanol-HCl reaction (methyl chloride
    only)
(6) Haloform synthesis (chloroform only)

Four of these are directed primarily toward a single chloromethane; however, methyl chloride from the methanol process can be further chlorinated to produce higher chlorides. Chloroform can also be produced from carbon tetrachloride under strong reducing (e.g., electrochemical) conditions, but these reactions are only of historical interest.[30,42,52,53,67,100,177,188]

Estimated capacity for each process is presented in Table 12-4. Direct chlorination of methane, it can be seen, is now the dominant process. The advantages of this method include the ability to produce a distribution of

TABLE 12-4. ESTIMATED U.S. CAPACITY FOR CHLORINATED
METHANES BY PROCESS

Estimated Capacity (Million annual pounds)

Chloride Process	CH_3Cl	CH_2Cl_2	$CHCl_3$	CCl_4	Total
Direct (thermal and photochemical) chlorination of methane	50	100	70	90	310
Intensive thermal chlorination of light hydrocarbons				100	100
Carbon disulfide chlorination				210	210
Methanol-HCl process	165	35	10		210
Haloform synthesis			5		5
TOTAL	215	135	85	400	835

chlorides over wide ranges (which can be varied with market conditions) and the inherent raw materials economy, provided that provisions can be made for the disposal of by-product HCl. It may be anticipated that direct chlorination of methane will account for the major portion of future expansion.

While some chloromethanes facilities are located close to supplies of hydrocarbon gases, the majority are currently in cheap power areas where chlorine is readily available. Disposal of by-product HCl (which cannot be shipped economically over any great distance) probably will become one of the major factors in the future location of plant sites.

Chemical characteristics and technological details for each of the six processes are presented below. The first three of these (thermal chlorination of methane, intensive chlorination of light hydrocarbons, and photochemical chlorination) share similar bodies of basic information as well as a similar HCl by-product situation. Hence, some of their reference material overlaps, making it preferable to consider these three jointly.

Direct Thermal Chlorination of Methane

This route involves the direct reaction of methane and chlorine to produce all four chlorinated methanes in variable proportions. One mole of HCl is produced as the by-product for each mole of chlorine consumed. The direct chlorination process accounts for nearly 40 percent of all the chlorinated methanes produced. Excluding carbon tetrachloride (for which specific processes are available), this proportion exceeds 50 percent.

Advantages. One of the important advantages of this process is the ability to produce simultaneously all four chloromethanes in a wide range of distributions governed by variations in process conditions. The process also is characterized by relatively low raw materials costs. Chlorine (as purchased at list prices) accounts for well over half of the manufacturing costs. Since 50 percent of the chlorine feed to the process emerges as by-product HCl, disposal of this by-product at an attractive price is obviously of great importance to the over-all economics of chloromethanes manufacture. Most producers of chloromethanes are able to find uses for the by-product HCl of such a nature that the equivalent chlorine value of the HCl is essentially recovered. Major HCl uses include the combination with ethylene to form ethyl chloride and with acetylene to form vinyl chloride. Other uses for which HCl may have a high value include oil well activation, metal pickling, and the manufacture of chloroprene and chlorinated polyether monomers. Because of the difficulties in transporting HCl, it has in the past been necessary to locate the chloromethanes plant close to HCl consumers. The marketability of HCl, however, may be appreciably enhanced by new shipping techniques such as those involving the use of tank trucks for anhydrous liquefied HCl.[15] The importance of marketing HCl is underscored by the fact that by-product HCl accounts for two thirds of total HCl consumption.

The Chloromethanes Process. The formation of chloromethanes by direct chlorination proceeds by a free radical chain mechanism. Activation (or dissociation of chlorine) is initiated thermally by heating the reactants to between 275 and 325° C. Once begun, the reaction is generally self sustained by virtue of the high heat evolved, and it proceeds homogeneously in the absence of catalysts.

Upon activation, the system may be represented as taking place by consecutive bimolecular substitution reactions:

$$CH_4 + Cl_2 \rightarrow CH_3Cl + HCl \tag{12-1}$$

$$CH_3Cl + Cl_2 \rightarrow CH_2Cl_2 + HCl \tag{12-2}$$

$$CH_2Cl_2 + Cl_2 \rightarrow CHCl_3 + HCl \tag{12-3}$$

$$CHCl_3 + Cl_2 \rightarrow CCl_4 + HCl \tag{12-4}$$

The changes in free energy of these reactions are all negative and have absolute values in excess of 20 Kcal; thus the equilibrium, if any, of each of these reactions is displaced far to the right. Most writers class the reactions as irreversible. Hence, product distribution is controlled by reaction rates and the feed methane-to-chlorine ratio.[82,85,105,106,107,109,140,149,163,164,168,169] All four rate constants are required for complete design of the system, including reactor size. However, it may be shown that material balances may be completely specified from any three independent rate constant ratios, greatly simplifying design procedures. Rate constant ratios are not only relatively simple to determine experimentally, but, because the activation energies of all four reactions are nearly equal, the ratios are relatively insensitive to temperature. The reactor effluent is virtually chlorine-free and contains a substantial amount of methane, which must be recycled. Methane-to-chlorine ratios are generally far in excess of unity and increase in the direction of lighter distributions; higher chlorine concentration in the feed leads to higher yields of carbon tetrachloride. Production of methyl chloride only is kinetically impossible. In moving toward the lower chlorides, capital costs increase rapidly both because of higher quantities of recirculating methane and greater difficulties in separating the product from the reactor effluent. Carbon tetrachloride can be produced as the only chlorinated product, if desired. However, special techniques are required in moving toward the higher chlorides; e.g., higher temperatures and re-circulation of inert gas through the reactor. Process design details change considerably with distribution; the amount of recycle methane and the nature of the chlorides separation steps are particularly sensitive.

The chief side reactions include the formation of C_2 chlorides from C_2 impurities in the feed. These C_2 chlorides represent chlorine yield losses and purification problems. Table 12-5 points up the difficulties in separating C_2 chlorides by distillation; many of these have boiling points in the

TABLE 12-5. BOILING POINTS OF LIGHT HYDROCARBON CHLORIDES

	Boiling point (°C)
Methyl Chloride, CH_3Cl	− 23.7
Vinyl chloride	− 13.9
Ethyl chloride	+ 12.2
1,1-Dichloroethylene (vinylidene chloride)	31.7
Isopropyl chloride	35.4
Methylene Chloride, CH_2Cl_2	40.1
Allyl chloride	44.6
n-Propyl chloride	47.2
1,2-Dichloroethylene, *cis*	48.4
1,1-Dichloroethane (ethylidene chloride)	57.3
1,2-Dichloroethylene, *trans*	60.3
Chloroform, $CHCl_3$	61.3
1,1,1-Trichloroethane	74.1
Carbon Tetrachloride, CCl_4	76.0
1,2-Dichloroethane (ethylene chloride)	83.6
Trichloroethylene	87.0
Tetrachloroethylene	121.2

same range as the chloromethanes. Another side reaction, resulting from local overheating, is carbonization.

The chlorination reactions are highly exothermic, liberating over 40,000 Btu per pound mole of chlorine reacted. Hence, heat removal can be a problem. For light distributions, continuous preheat of the feed is required, while for moderate distributions the reaction is essentially self sustaining. In moving toward the heavy chlorides, heat evolution is so rapid that special methods of control are required; e.g., circulation of inert gas such as nitrogen or carbon tetrachloride.

An important feature of the chloromethanes process is separation of, first, HCl and, second, chloromethanes from reactor effluent prior to recycle. The first requirement may be met by efficient aqueous absorption. Recovery of organic chlorides from the HCl-free gas involves compression, cooling or refrigeration, and in extreme cases, absorption, the specific economic combination depending on desired product distribution. The difficulty of this recovery increases rapidly with increase in lighter chlorides in the product distribution.

Dry cell gas is technically suitable as chlorine feed, but liquefied chlorine is usually economically preferred. Freedom from oxygen and water is important also.

Natural gas stripped of C_2 and higher hydrocarbons is suitable for methane feed. It is also desirable to minimize the inerts content in order to reduce purge rates. The current tendency is to supply 99 percent-plus methane prepared by low-temperature fractionation.[16,19]

A number of review articles[9,70,112,137,207] relating to methane chlorination have appeared. The great bulk of the literature is understandably devoted to reactor design and operation. Reactor geometry and provision of heat transfer surface have received attention.[31,32,192] Considerable work has been devoted to mixing within the reactor and reactor hydrodynamics.[23,77,108,146,149,206,210] In recent years, the use of the fluidized bed as a reaction moderator has received attention.[123,124,155] Many other means of carrying out the reaction have been proposed.[6,7,122,135,161,184]

Recycle of reaction products for control purposes has been described in the old as well as recent literature.[47,66,121] Separation of products from reactor effluent is of considerable economic importance. The older literature proposes separate absorption of both organic chlorides and HCl[190] or total anhydrous liquefaction of products.[60] More recent detailed attention has been devoted to the technology of HCl recovery by absorption.[1,24,54]

While modern technology has demonstrated the economic superiority of thermal activation, the older literature makes reference to catalytic methods,[37,87,125,127,129,143,167] usually in efforts to increase the yield of one particular chloromethane over others.

Figure 12-1 is a flow diagram for a typical modern chloromethanes

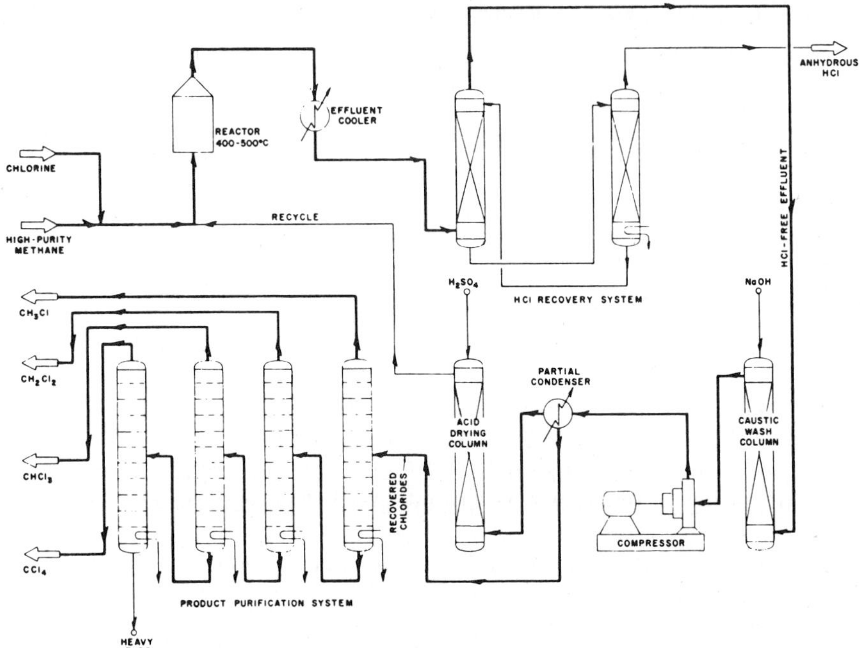

Figure 12-1. Chloromethanes by direct thermal chlorination of methane for typical intermediate product distribution.

process for an intermediate product distribution. The process sequence requires four main steps: (1) reaction, (2) HCl recovery, (3) chlorides recovery, and (4) chlorides refining. High-purity methane, chlorine, and recycle methane (containing most of the required chlorides recycle) are premixed and fed to the reactor. The reactor effluent, containing organic chlorides, HCl, excess methane, and only traces of chlorine, is cooled and fed to the HCl recovery system. The first column in this system is an absorber designed for efficient HCl removal. The bulk of the absorbing liquor is HCl azeotrope (about 20 percent by weight HCl). The rich acid is thus above the azeotrope and allows stripping of anhydrous HCl. The second column distills off anhydrous HCl and produces the required azeotrope in the bottoms.

The HCl-free gases from the absorber are washed with caustic soda to remove final traces of HCl and are then ready for chlorides recovery. For the intermediate product distribution under discussion, compression, cooling, and drying with sulfuric acid are an economical combination; for other distributions, different combinations of unit operations may be required.

The refining system may be composed of four distillation columns in conventional sequence.

In-process materials are highly corrosive. Hot, wet HCl is a particularly troublesome material, though moist organic chlorides can be equally difficult. Carbon steel and stainless steel are generally unsatisfactory for methane chlorination. Instead, complex schedules of alloy metals, ceramics, and plastics are required to give reasonably low maintenance costs.

Intensive Thermal Chlorination of Light Hydrocarbons

The intensive thermal chlorination process usually involves the reaction of chlorine with C_1 to C_3 hydrocarbons at conditions generally more severe than those necessary for directly chlorinating methane. The reaction products are carbon tetrachloride, CCl_4, and perchloroethylene, C_2Cl_4, the proportions of the two being controlled by recycle and certain process variables. HCl is a by-product. About 25 percent of carbon tetrachloride is currently produced by this process. The principal difference between this method and thermal chlorination of methane to produce mixed chlorocarbons is the higher temperature and the fact that excess chlorine is fed to the reactor.

Advantages. LPG (liquefied petroleum gas) is a suitable feedstock for the intensive chlorination process. The lower cost of this material relative to high-purity methane, coupled with lower chlorine consumption (less HCl is produced), makes raw materials costs highly favorable, provided by-product HCl can be disposed of effectively. A main characteristic of direct chlorination of methane is the ability to produce four products, while intensive chlorination produces only two; this may be an advantage for either, depending on the circumstances. Use of this process will probably continue to expand in cases where carbon tetrachloride is the desired chlorinated methane. HCl considerations relative to the intensive chlorination process are similar to those for the methane chlorination process, although it will be seen that the allowable propane feedstock leads to lower HCl production.

Reactions. Under the stringent conditions of intensive chlorination, most aliphatic hydrocarbons react with chlorine. The higher molecular weight materials (C_4 and heavier) tend to form chlorinated substitution or addition compounds. The lower hydrocarbons, however, react to produce primarily perchloroethylene and carbon tetrachloride. These lower hydrocarbons include both saturated and unsaturated compounds, although acetylene is not a satisfactory feed material. Many chlorinated hydrocarbons (e.g., methyl chloride, methylene chloride, ethyl chloride, ethylene dichloride, allyl chloride, and isopropyl chloride) will also react to form perchloroethylene and carbon tetrachloride; hence, intensive chlorination may serve as a means for disposing of by-product chlorocarbons from other sources. Using propane for illustration, these prime reactions may be written:

$$2\,C_3H_8 + 14\,Cl_2 \rightarrow 3\,C_2Cl_4 + 16\,HCl \qquad (12\text{-}5)$$

$$C_3H_8 + 10\,Cl_2 \rightarrow 3\,CCl_4 + 8\,HCl \qquad (12\text{-}6)$$

Under the usual conditions, both of these reactions are displaced far to the right thermodynamically, and their procedure is governed by rate processes. However, the reaction products are in equilibrium:

$$C_2Cl_4 + 2\,Cl_2 \rightleftharpoons 2\,CCl_4 \qquad (12\text{-}7)$$

the reaction being forced to the left by high temperatures.

The chief side reactions result in formation of hexachloroethane and hexachlorobenzene:

$$2\,C_3H_8 + 17Cl_2 \rightarrow 3\,C_2Cl_6 + 16\,HCl \qquad (12\text{-}8)$$

$$2\,C_3H_8 + 11\,Cl_2 \rightarrow C_6Cl_6 + 16\,HCl \qquad (12\text{-}9)$$

From an equilibrium viewpoint, these reactions are displaced far to the right and their course is controlled by reaction rates. Hexachloroethane is in equilibrium with the principal products:

$$C_2Cl_6 \rightleftharpoons C_2Cl_4 + Cl_2 \qquad (12\text{-}10)$$

$$C_2Cl_6 + Cl_2 \rightleftharpoons 2\,CCl_4 \qquad (12\text{-}11)$$

Hexachlorobenzene does not take part in any equilibrium transition; its formation is thus irreversible and controlled solely by the rate of reaction.[9]

Another undesirable side reaction involves carbon formation:

$$C_3H_8 + 4\,Cl_2 \rightarrow 3\,C + 8\,HCl \qquad (12\text{-}12)$$

This reaction is favored by high temperature and may take place explosively.

Hydrocarbon chlorination liberates over 40,000 Btu per mole of chlorine reacted. Reactions (12-5), (12-6), (12-8), (12-9), and (12-12) are thus highly exothermic. Temperature control and absorption of heats of reaction are usually effected mainly by diluting reaction gases, since surface cooling of the hot, corrosive reaction gases is difficult. Such diluents as nitrogen and HCl have been suggested, but recycle of the chlorocarbon products of reaction (perchloroethylene and carbon tetrachloride) is highly advantageous in that the materials are readily available in process and no foreign substances are introduced. Furthermore, the recycle may be used to influence the equilibrium (12-7) and consequently the product split.

Temperature Factors. Operating temperature is extremely important from the standpoint of product distribution, as reflected both in proportion of desired products and in by-product yield. High temperature, of course, favors the rates of all reactions. While this effect is beneficial in reactions (12-5) and (12-6), it is detrimental in (12-8), (12-9) and (12-12); this is particularly serious in the last two reactions, where the products are irreversibly formed. Temperature also has a major effect on equilibrium constant. Thus, the reactor temperature can control product split through equation (12-7), though this effect can also be controlled by product recycle. Similarly, the equilibria in (12-10) and (12-11), and consequently the hexachloroethane by-product yield, are influenced by temperature.

With all the pertinent factors considered, suitable reactor temperatures are generally in the range of 550 to 700°C and thus are considerably higher than those required for thermal chlorination of methane to mixed chloromethanes. However, reactor temperature is not the only important temperature factor. Because of the fundamental equilibrium nature of the processes, rapid quench of reactor effluent is required. This is in marked contrast to the chloromethanes process which is rate controlled. If the quench is not efficient, yields of perchloroethylene may be lower than desired as the equilibrium of reaction (12-7) shifts on slow cooling. More important, reactions (12-10) and (12-11) would result in high yields of hexachloroethane. The importance of quench cannot be overemphasized.

Amount of Chlorine Feed. The chlorine feed is supplied in large excess because of the exhaustive nature of the chlorination required. Deficiency in chlorine also leads to high yields of hexachlorobenzene. This, again, is in contrast to the chloromethanes process where the feed is characterized by a deficiency of chlorine. Ordinary cell gas (containing hydrogen) is technically suitable, but liquefied, dried chlorine is economically preferred. The chlorine should, however, be oxygen free; while phosgene will not form at reaction temperature, it can appear in the quenched effluent.

Feedstocks. Hydrocarbon gases in the C_1 to C_3 range are suitable feedstocks for the intensive chlorination process. Higher hydrocarbons are not suitable. C_3 feedstocks are preferable to the lower hydrocarbons in that they require less chlorine and produce less by-product HCl. This may be seen by inspecting the following representative reactions for carbon tetrachloride:

		Cl_2/CCl_4	HCl/CCl_4	
Methane	$CH_4 + 4\,Cl_2 \rightarrow CCl_4 + 4\,HCl$	4	4	(12-13)
Ethylene	$C_2H_4 + 6\,Cl_2 \rightarrow 2\,CCl_4 + 4\,HCl$	3	2	(12-14)
Ethane	$C_2H_6 + 7\,Cl_2 \rightarrow 2\,CCl_4 + 6\,HCl$	3.5	3	(12-15)
Propadiene	$C_3H_4 + 8\,Cl_2 \rightarrow 3\,CCl_4 + 4\,HCl$	2.7	1.3	(12-16)
Propylene	$C_3H_6 + 9\,Cl_2 \rightarrow 3\,CCl_4 + 6\,HCl$	3	2	(12-17)
Propane	$C_3H_8 + 10\,Cl_2 \rightarrow 3\,CCl_4 + 8\,HCl$	3.3	2.7	(12-18)

Furthermore, C_3's are readily available in the form of LPG at attractive prices.

In view of the diversity of variables, it is not surprising that a large number of component and process designs have appeared in the literature. Two recent articles of general interest in aliphatics chlorination are those of Nekrasova and Shuikin[158] and of Longiave.[137] Fink and Bonilla[78] discuss the CCl_4-perchloroethylene equilibrium and include valuable basic data.

The use of chlorinated hydrocarbons (which may be available as by-products of other chlorination processes) as feedstocks has been suggested. [49,147,148,150,151,173,204] Catalytic, as opposed to thermal, reaction has been considered. Hennig[74,75,114,117] proposes the use of activated carbon, while Grebe et al.[97] prefer molten metal chlorides; Fuller's earth has also been suggested,[65]

Reactor design has received substantial attention. Several two-stage reactors have been proposed.[20,22,64,159] Brown[38,48] is concerned with the mixing problem. Bijarvat et al.[35] propose a fluidized-catalyst reactor. A porous-plate reactor has also been suggested.[61]

It was pointed out above that product recycle is useful in controlling product split, and a number of patents have been concerned with this point.[40,59,76,113,157,204,205,208] The use of recycle for purposes of temperature control has also been considered.[63,160,175,194] Brown et al.[39] propose the use of aqueous HCl as a quench medium, a procedure that has considerable economic merit, while Heitz and Brown[113] prefer a perchloroethylene quench, partly because of its ability to remove hexachlorobenzene. Colton[55] discusses recovery of chlorine from reactor effluent.

Figure 12-2 is a schematic flow diagram for a typical carbon tetrachloride-perchloroethylene plant, following the patent of Heitz and Brown.[113] Any proportion of carbon tetrachloride and perchloroethylene may be produced.

Fresh chlorine feed together with recycle chlorine and hydrocarbon gas feed are introduced into a vaporizer where they are mixed with recycle chlorides in the vapor space. The chlorine is in 10 to 25 percent excess.

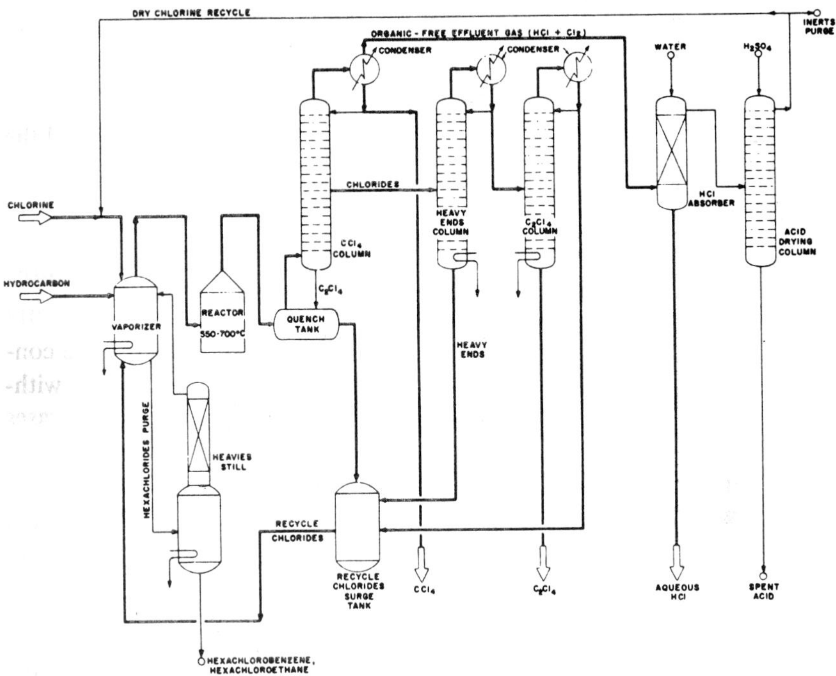

Figure 12-2. Intensive thermal chlorination of light hydrocarbons to CCl_4 and C_2Cl_4 with C_2Cl_4 quench.

The recycle chloride diluent rate (controlled by heat input to the vaporizer) is designed to control adiabatic reactor temperature at 550 to 700°C and is usually about 75 mol percent of the total stream; its composition also serves to control CCl_4-C_2Cl_4 ratio. The mixed gases, at about 1 atm. pressure, are fed to a refractory lined reactor where they are rapidly mixed with the contents, thereby heated to ignition, and reacted adiabatically. The reaction is self sustaining and supported by the considerable heat of reaction (after start-up), but is tempered and controlled by the diluent action of the recycle chlorides.

The reactor effluent is essentially free of unreacted hydrocarbon and consists mainly of carbon tetrachloride, perchloroethylene, HCl and excess chlorine. It is rapidly quenched by intimate contact with a liquid which is largely perchloroethylene. The rapid quench serves to preserve the equilibrium ratio attained in the reactor and prevents formation of undesirable by-products. It also serves to dissolve any quantities of hexachlorobenzene which may be present (these are small if reaction temperature is maintained below 650°C) and which might introduce difficulties due to deposition of solids on heat transfer surfaces and in vessels. Some heat economy is obtained from the hot reactor effluent, which serves to boil the contents of the quench tank and thus provide boil-up for the CCl_4 column. Hexachlorobenzene is purged from the quench tank by allowing liquid to overflow to the recycle surge tank.

The CCl_4 column, operating on boil-up from the quench tank, returns quench liquid rich in perchloroethylene as a bottoms stream. Fractionation results in an overhead which is largely free of perchloroethylene. The condenser yields carbon tetrachloride, the desired quantity of which is withdrawn as product, the remainder being used as reflux. Chlorides-free gases (HCl and chlorine) are also disengaged at the overhead. These gases are scrubbed with water to remove HCl, the effluent liquor forming the aqueous HCl by-product. They are subsequently dried with concentrated sulfuric acid and purged to maintain the balance of inert gases; the remaining chlorine (with some uncondensed chlorides) is recycled to feed.

A perchloroethylene-rich stream is removed as a side stream from the CCl_4 column and processed to produce perchloroethylene product. Heavy ends are first removed by distillation and returned to the recycle surge tank. The overhead from the heavy ends column is fractionated in the C_2Cl_4 column where the desired quantity of perchloroethylene product is removed as the bottoms and the overhead (largely carbon tetrachloride) is sent to recycle.

The recycle surge tank serves as a reservoir for three recycle streams: (1) the quench tank overflow—largely perchloroethylene with some hexachlorobenzene, (2) heavy ends from the heavy ends column, and (3) the C_2Cl_4 column overhead (largely carbon tetrachloride). Proper control results in a recycle mixture of the desired composition, and this is fed to the vaporizer where the vapors are mixed with the feed materials.

To prevent buildup of hexachlorobenzene in the system, liquid is withdrawn from the vaporizer and sent to the heavies still. Here the more volatile components are stripped off and returned to the vaporizer, while heavy liquid consisting of hexachlorobenzene and some hexachloroethane is purged from the still pot.

Another process (Figure 12-3) has been recently described,[12,14] which differs from the preceding chiefly by virtue of aqueous HCl quench, sequence of product separation, and chloride recycle procedure. By-product HCl may be recovered as the anhydrous gas. Chlorine, hydrocarbon, and several recycle streams are mixed and fed to the chlorination furnace. The chlorination process, sustained by the heat of reaction, proceeds readily at 500 to 650°C near atmospheric pressure without the aid of catalyst. From the furnace, the reaction products go to a quench vessel where they are suddenly cooled to 140°C.

The liquid quench mixture, consisting of carbon tetrachloride, perchloroethylene and 21 to 36 percent aqueous HCl is decanted to separate

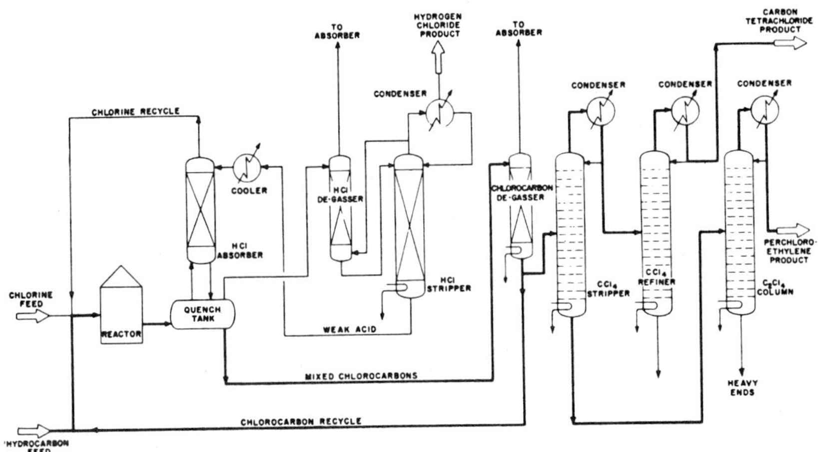

Figure 12-3.   Intensive thermal chlorination of light hydrocarbons to CCl₄ and C₂Cl₄ with HCl quench.

acid from the chlorocarbons. Part of the acid is cooled and returned to the HCl absorber. The chlorocarbons are stripped in a degassing column.

The mixed chlorocarbons from the stripping column are first fractionated in a carbon tetrachloride stripper. The crude carbon tetrachloride overhead is distilled and withdrawn as product. The remaining crude perchloroethylene is refined to yield high purity product, and the bottoms from the C_2Cl_4 column are recycled to the furnace.

Unquenched gases from the quench vessel pass to an HCl absorber where hydrogen chloride is absorbed in weak acid. The remaining gases from this absorption column are recycled.

The strong hydrochloric acid from the HCl absorber is also partially degassed to remove traces of dissolved chlorine and other contaminants, which are returned to the HCl absorber. The purified hydrochloric acid is then sent to a conventional HCl stripping system which yields anhydrous hydrogen chloride gas as the by-product of the process. Weak acid, close to the azeotropic composition, is recycled as scrubber liquor.

All equipment is of corrosion resistant alloy or nonmetallic materials of construction. In particular, equipment coming in contact with hydrogen chloride must be ceramic lined or made of impervious graphite or other suitably corrosion-resistant material.

Photochemical Chlorinations

In the reaction of methane or other light hydrocarbons with chlorine, prior activation is required to produce free radicals of monatomic chlorine. Catalytic methods were investigated some years ago but have long been abandoned as either unworkable or uneconomical. Thermal activation is presently most widely favored. However, photochemical activation was well regarded for a number of years until it was established that both high investment and high operating costs as well as low efficiencies are necessarily associated with the source of radiation. Nevertheless, the process is still used and is certainly of interest from a historical point of view.

Radiation of wave length below 5000 Å is required for initiation. The mercury arc lamp, which produces in the 3000 to 5000 A range, is the common source. The quantum yield from this source is generally quite low.

One of the attractive facets of the photochemical process is the generally mild reaction conditions allowable, particularly regarding temperature. Even liquid-phase chlorinations may be carried out, and the higher chloromethanes form more easily than in the thermal process. For this reason, photochemical reactors are sometimes used to finish off effluent from a thermal reactor, especially where large quantities of carbon tetrachloride are desired.[26]

The theory of photochemical chlorination to chloromethanes has been covered by Ritchie and Winning[178] and by Pritchard *et al.*[170] Hirschkind[118] outlines the technology of the process. Design of photochemical reactors has received major attention.[72,136,152,186,187,188,189,209] A comparison of vapor-phase and liquid-phase processes is available.[33,34] Several proposals have been made for processes giving high selectivity to carbon tetrachloride.[162,166,181] Brown and Davis[40] describe a photochemical-catalytic process operating on inexpensive propane. As in the case of thermal chlorination, recycle for reaction control has been proposed.[185,191]

Carbon Disulfide Chlorination

This process involves the reaction of carbon disulfide, CS_2—produced by thermal reaction of coke with sulfur and, more recently, by catalytic reaction of methane with sulfur—with chlorine. Sulfur chlorides always appear as intermediates and may, in some variations, actually serve as indirect chlorinating agents. Sulfur is a by-product and may be recycled to the carbon disulfide plant, although substantial purification is usually required. Carbon tetrachloride is the sole product; it may be reduced to chloroform, but this is not current industrial practice.[30,42,52,53,67,100,177,187] About 50 percent of current carbon tetrachloride production stems from the disulfide process.

If carbon disulfide production by the methane route is considered integrally with chlorination, the process, in effect, is a chlorination of methane with complete utilization of chlorine and no HCl by-product. Raw materials costs are therefore potentially low. However, sulfur conservation and the relatively extensive processing are serious drawbacks. Economic operation probably is contingent on integration with large-scale carbon disulfide facilities. Decreasing markets for rayon (one of the large outlets for carbon disulfide) may adversely affect this route to carbon tetrachloride in the United States.

The desired reaction for the carbon disulfide process is:

$$CS_2 + 2\,Cl_2 \rightarrow CCl_4 + 2\,S \tag{12-19}$$

The sulfur may be returned to the carbon disulfide plant where it can be reacted with carbon or methane:

$$C + 2\,S \rightarrow CS_2 \tag{12-20a}$$

$$CH_4 + 4\,S \rightarrow CS_2 + 2\,H_2S \tag{12-20b}$$

Actually, the sulfur may require considerable treatment before re-use; one treating method includes neutralization with soda ash, followed by filtration. The over-all reactions may be written:

$$C + 2\,Cl_2 \rightarrow CCl_4 \tag{12-21a}$$

$$CH_4 + 2\,Cl_2 + 2\,S \rightarrow CCl_4 + 2\,H_2S \tag{12-21b}$$

Both of these are highly attractive from a raw materials standpoint in that chlorine is completely utilized and there is no HCl by-product; for the methane process, hydrogen sulfide also can be converted readily back to sulfur (e.g., the Claus process, which proceeds *via* partial oxidation to SO_2, followed by reaction with H_2S). Unfortunately, however, the formation of sulfur chlorides complicates the procedure. Actually, reaction (12-19) yields sulfur monochloride and proceeds largely as follows:

$$CS_2 + 3\,Cl_2 \rightarrow CCl_4 + S_2Cl_2 \tag{12-22}$$

This reaction goes readily to completion in the presence of an iron catalyst, but the value in the sulfur monochloride, S_2Cl_2, must be recovered. This is made possible by using the sulfur monochloride to chlorinate additional carbon disulfide at higher temperatures:

$$CS_2 + 2\,S_2Cl_2 \rightleftharpoons CCl_4 + 6\,S \tag{12-23}$$

Reaction (12-23) is an equilibrium, and special techniques are required to force the reaction to the right. Furthermore, separation of residual carbon disulfide (for recycle) from carbon tetrachloride by distillation is reportedly difficult.

Sulfur dichloride, SCl_2, may also exist under process conditions, and higher chlorides have been reported. The interrelationship among these compounds may be represented by the S-Cl equilibria:

$$2\,S + Cl_2 \rightleftharpoons S_2Cl_2 \tag{12-24}$$

$$S_2Cl_2 + Cl_2 \rightleftharpoons 2\,SCl_2 \tag{12-25}$$

From (12-23) and (12-25), reversible chlorination of carbon disulfide by sulfur dichloride may be written:

$$CS_2 + 2\,SCl_2 \rightarrow CCl_4 + 4\,S \tag{12-26}$$

Similarly, (12-26) may be represented as producing various combinations of sulfur with sulfur dichloride, as:

$$CS_2 + 4\,SCl_2 \rightarrow CCl_4 + 2\,S_2Cl_2 + 2\,S \tag{12-27}$$

It is clear that many convenient representations of these reactions may be written, but all are derived from combinations of reaction (12-22) with the sulfur-chlorine equilibria, (12-24) and (12-25). Actual product distribution may be predicted from the equations of equilibrium.

The complexity and reversibility of the system can lead to a wide variety of process schemes. For example, Reilly [68,174] has suggested the use of excess sulfur chlorides and pressures of 1.5 to 10 atmospheres to force reactions (12-23) and (12-26) to completion at 50 to 70°C. Beanblossom[27] describes a single fractionating column with carbon disulfide and chlorine feed and carbon tetrachloride and sulfur products. Beanblossom and Scott[28,29] use sulfur chlorides to chlorinate carbon disulfide, remove sulfur equivalent to the net feed carbon disulfide by crystallization, directly chlorinate to complete reaction of carbon disulfide and remove residual sulfur, distill off the carbon tetrachloride product and, finally, recycle sulfur chlorides to the initial stage.

Industrially, both direct and indirect chlorination of carbon disulfide are employed. In a direct chlorination process (Figure 12-4), a mixture of carbon tetrachloride, carbon disulfide and sulfur chlorides, recycled from a subsequent process step, is directly contacted with excess chlorine at about 30°C over a divided iron catalyst, converting over 99 percent of the carbon disulfide to carbon tetrachloride. Some fresh carbon disulfide feed may also be diverted to the primary chlorinator for purposes of controlling downstream loads. It is essential that the reaction be complete but that over-chlorination of sulfur to sulfur monochloride be avoided to prevent appearance of carbon disulfide and sulfur monochloride in the distilled crude product. The required stringent control makes batch reactors desirable. Distillation of the reactor effluent yields a bottoms product (which is largely sulfur chlorides from which value is subsequently recovered) and an overhead product of relatively pure carbon tetrachloride. This material may be treated with a base to destroy sulfur chlorides (which form sulfites and chlorides) and dried. In the process shown, neutralization is carried

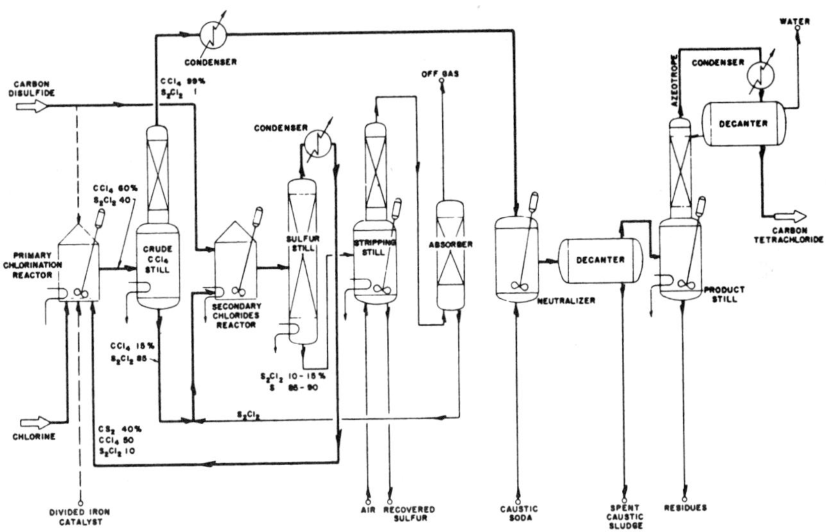

Figure 12-4. Carbon tetrachloride via carbon disulfide chlorination (direct chlorination process).

out with caustic soda and by subsequent drying by azeotropic distillation. In this latter operation, the CCl_4-H_2O azeotrope is taken overhead; condensation results in phase splitting, the water phase being rejected and the carbon tetrachloride phase serving to provide both the final product and reflux to the column. An alternate purification involves treatment with alcoholic caustic (which provides simultaneous neutralization and drying), followed by distillation. A second alternative is solid-liquid contact with alkaline dehydrating agents, e.g., caustic soda, silica gel or calcium chloride. The sulfur chlorides from the crude still bottoms are sent to a secondary reactor where fresh carbon disulfide feed is partially converted to carbon tetrachloride at about 50° C in accordance with reaction (12-23). Distillation yields an overhead (containing carbon tetrachloride and carbon disulfide) which serves as feed to the primary reactor. The bottoms product consists largely of molten sulfur with 10 to 15 percent sulfur chloride. The sulfur may be precipitated and possibly returned to the CS_2 plant, but, as shown here, air stripping may also be used to recover sulfur monochloride, which is then recycled to the secondary reactors. Yields of carbon tetrachloride are about 90 percent on carbon disulfide and 80 percent on chlorine. This process is currently in use in the United States, and a similar process has also been used by I. G. Farben in Germany.[138]

An indirect chlorination process is similar to the Beanblossom and Scott patents[28,29] and is shown in Figure 12-5. Fresh carbon disulfide feed reacts with sulfur monochloride in accordance with equation (12-26) under conditions such that the carbon disulfide is all but completely reacted. However, the reactor product contains a fraction of a percent of carbon disulfide as well as carbon tetrachloride and sulfur. A direct-chlorination polishing reactor is used to convert this remaining carbon disulfide and to facilitate the subsequent distillation, where crude carbon tetrachloride is removed overhead and molten sulfur containing some sulfur monochloride is the bottoms product. The crude carbon tetrachloride may be purified as in the direct chlorination process, and alcoholic caustic treatment with redistillation is shown on the flow sheet. Sulfur equivalent to the fresh carbon disulfide feed is separated from the bottoms product and may be returned to the CS_2 plant for processing. The residual sulfur stream is directly chlorinated to provide the sulfur chlorides chlorinating agent for the first reaction stage.

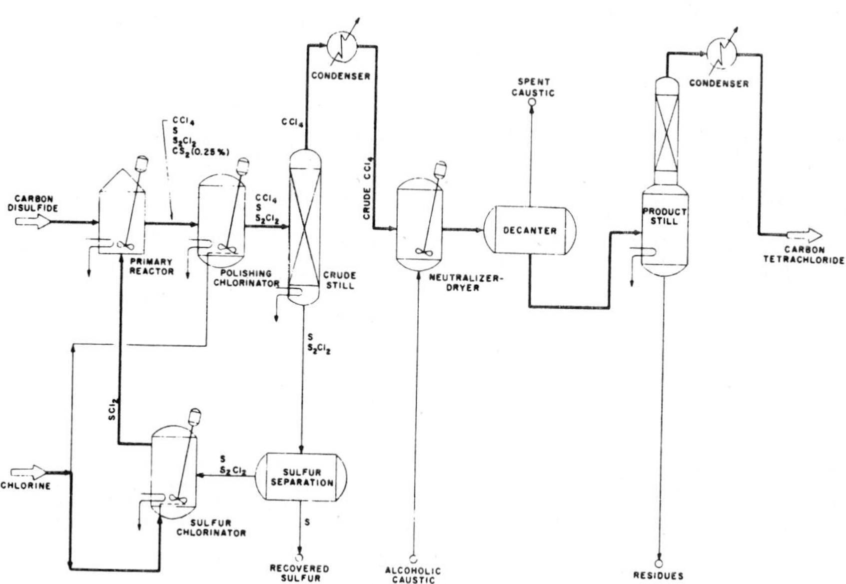

Figure 12-5. Carbon tetrachloride via carbon disulfide chlorination (indirect chlorination process).

While steel equipment is satisfactory at some points in these processes, materials more usually required include lead, nickel, and 18-8 stainless steel.

Methanol–HCl Process

Methanol, CH_3OH, reacts with HCl under mild dehydrating conditions to yield methyl chloride. There are no theoretical by-products. The methyl chloride product may be further chlorinated to produce the higher chloromethanes. The methanol route accounts for only 20 to 25 percent of the chloromethanes, but nearly 80 percent of all methyl chloride.

Though there are no by-products and HCl provides a cheap source of chlorine, the methanol requirement leads to high raw materials costs. The methanol route has, however, found favorable acceptance where methyl chloride is the only desired product. Typical is the captive production of methyl chloride by silicone and tetramethyllead producers. Not only does cost of the ultimate product tend to overshadow methanol charges, but plants are also of sufficiently small size to make raw materials charges less important relative to capital and labor-related costs.

The reaction between methanol and HCl is a simple substitution:

$$CH_3OH + HCl \rightarrow CH_3Cl + H_2O \qquad (12\text{-}28)$$

Dehydrating conditions are obviously beneficial, and, because of this, two molecules of methanol take part in a side reaction to produce dimethyl ether:

$$2\,CH_3OH \rightarrow CH_3OCH_3 + H_2O \qquad (12\text{-}29)$$

In the liquid phase, reaction (12-28) may be carried out catalytically at 1 atmosphere and 100 to 150° C. In aqueous solution, suggested catalysts include zinc chloride, ferric chloride, bismuth chloride and other metal chlorides,[58,62,83,84,119,120,130] with various stoichiometric ratios and reaction conditions. A fused zinc chloride system has also been suggested.[25] Harding[104] calls for hydrated metal chloride to minimize ether formation. Scott[183] uses amine hydrochloride catalysts. The liquid-phase reaction may also be carried out without a catalyst under pressure or by refluxing.[41,91] Holt and Daudt[120] directly couple an HCl generator (from salt and sulfuric acid) to their reactor.

The reaction also proceeds catalytically in the gas phase, but temperatures of 300 to 450° C are required.[43,153] Suggested catalysts include alumina gel, zinc chloride on pumice, cuprous chloride, activated carbon, pyridine hydrochloride and picolines. In one suggested procedure, an externally heated fixed bed reactor is operated at 350°C and 1 atmosphere with space velocities referred to standard conditions near 300 hr.$^{-1}$

Purification of reactor effluent is relatively simple, consisting of sucessive removal of excess HCl, residual methanol, and light and heavy ends. HCl may be removed by aqueous absorption. Recovery of the excess HCl by stripping may, however, introduce difficulties since HCl-watei forms an azeotrope at 20 percent by weight HCl. Residual methanol is readily removed by cold water.[200] Green and Heminger[98] propose to separate, purify, and dehydrate the product by distillation with sulfuric acid or alkyl sulfates.

An alternative preparation of methyl chloride from methanol with ammonium chloride instead of HCl has been proposed.[50,51,193] Reaction temperatures are much higher, and the advantages of this procedure are not immediately evident.

The methyl chloride product may be further chlorinated to yield higher chlorides, and this is currently being carried out industrially. The procedure does not differ materially from direct chlorination of methane, and the kinetic principles described above may be applied directly. The HCl produced in the direct chlorination may, of course, be used for the primary reaction with methanol. Several specific procedures for thermal chlorination of methyl chloride have been described.[46,101,128,199] Guinot[101] chlorinates over an activated carbon catalyst. Thomas and Hindley[199] use sulfur monochloride as the chlorinating agent.

Most of the current industrial processes are carried out continuously in the liquid phase using a zinc chloride catalyst. A flow sheet for this process, based on the above references, appears in Figure 12-6. Anhydrous HCl gas feed is sparged into the agitated reactor, maintained at 100 to 150°C under atmospheric pressure, and the volatile reaction products are continuously removed overhead. The methanol is all but completely reacted, and conversion of HCl is high. The gaseous reaction products (including some residual methanol, excess HCl, methyl chloride, and water) pass to a separator, where any entrained liquid is removed and returned to the reactor and then to the scrubbing system.

The product gases from the absorber are next subjected to cold water wash to remove residual HCl and methanol. Caustic scrubbing serves to remove the last traces of HCl, and a concentrated sulfuric acid wash serves to dry the methyl chloride product. Finally, the methyl chloride is compressed to 120 psia, its storage pressure, and is condensed.

Haloform Synthesis

The haloform synthesis as currently practiced involves the action of hypochlorite on acetone to yield chloroform. This is the oldest chloro-

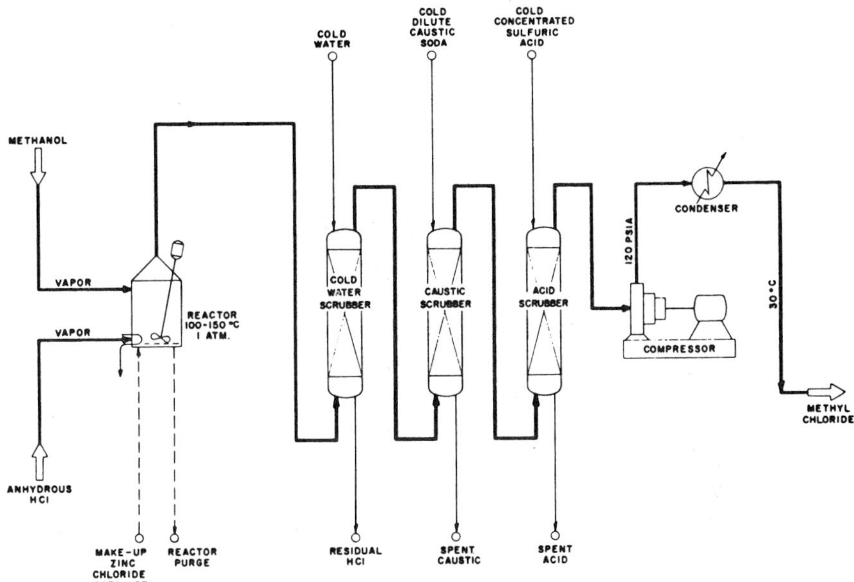

Figure 12-6.   Methyl chloride via methanol (liquid phase catalytic process).

methane process now in use, but it accounts for only a fraction of total production. Since each mole of acetone yields only one mole of chloroform, it is clear that raw materials charges are extreme. However, the process is specific for chloroform and may be convenient where small quantities are required for specialized use. There is little possibility that the haloform synthesis will receive large scale attention in the future, particularly in the U. S.

The haloform reaction involves the action of hypochlorite ion, OCl^-, on any of a group of oxygenated aliphatics, e.g., ethanol, acetone and acetaldehyde.[86] Hypochlorite can play both oxidative and chlorinative roles. Using acetone, the reaction may be written as follows:

$$CH_3COCH_3 + 3\,OCl^- \rightarrow CHCl_3 + CH_3COO^- + 2\,OH^- \qquad (12\text{-}30)$$

Considering the source of hypochlorite (solution of chlorine in base to form OCl^- and Cl^-, theoretical chlorine utilization is the same as for direct chlorination, but the Cl^- (as a salt) has no significant value. For ethanol, OCl^- is also consumed for oxidation:

$$CH_3CH_2OH + 4\,OCl^- \rightarrow CHCl_3 + CHOO^- + 2\,OH^- + Cl^- + H_2O \qquad (12\text{-}31)$$

and chlorine requirements are increased. The haloform reaction may also be carried out electrolytically on acetone-NaCl solutions, where the required OCl⁻ is effectively produced *in situ*.[196]

Industrially, acetone is contacted with a slurry of bleaching powder, $CaOCl_2$, at temperatures below about 65°C. Because the reaction is rapid, turbulent conditions are maintained for temperature control. Crude chloroform is distilled from the reaction mixture and is purified by agitation with about 50 percent of its weight of concentrated sulfuric acid. Neutralization with lime and redistillation yields the final product. The acetone required may be industrial material, or wood spirits may be used. Yields are 85 to 90 percent based on acetone and about 60 percent based on $CaOCl_2$. Figure 12-7 is a schematic diagram for this process.

Process variations suggested include the use of acetaldehyde[45] or secondary alcohols[165] and continuous flow reaction.[44]

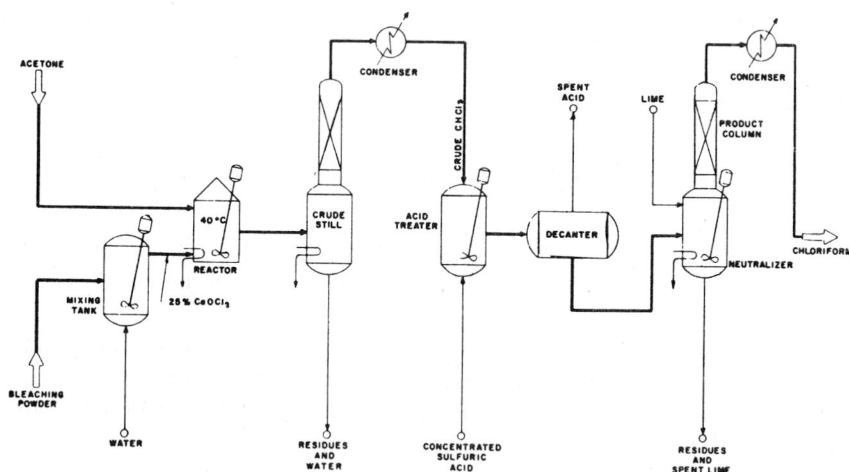

Figure 12-7.   Chloroform *via* haloform synthesis.

Other Process Activity

In addition to the six commercial chloromethanes routes described above, considerable effort has been expended in developing other processes. These are discussed below. It should be noted that none of these has achieved commercial success.

Much work has been done on processes which use copper chlorides as intermediates in hydrocarbon chlorination.[80,81,92,94,95,172,201]

$$CH_4 + 2\,CuCl_2 \rightarrow CH_3Cl + Cu_2Cl_2 + HCl \qquad (12\text{-}32)$$

The cupric chloride may be regenerated from the effluent by-products, frequently in molten form, by adding the net HCl requirement and reacting with air:

$$Cu_2Cl_2 + 2\,HCl + \tfrac{1}{2}\,O_2 \rightarrow 2\,CuCl_2 + H_2O \qquad (12\text{-}33)$$

The over-all reaction is:

$$CH_4 + HCl + \tfrac{1}{2}\,O_2 \rightarrow CH_3Cl + H_2O \qquad (12\text{-}34)$$

Riblett[176] has devised a scheme which couples chlorine recovery from HCl by-product (by combustion) with a direct chlorination. In more refined procedures, air for HCl combustion is fed directly to the reactor with chlorine:[134,142,202]

$$2\,CH_4 + Cl_2 + \tfrac{1}{2}\,O_2 \rightarrow 2\,CH_3Cl + H_2O \qquad (12\text{-}35)$$

Potassium, zinc or cupric chloride catalysts are used and yields of methyl chloride are high.

Some of the older literature[14,99,133,179,180] considers the combination of HCl and air as a chlorinating agent, thus carrying out reaction (12-34) *in situ* in a single step. Thode and Meissner[198] have revived interest in this procedure. Gorin[93] substitutes cupric oxychloride for air, and regenerates the salt externally. Recent activity[126,171] has led to high efficiency of HCl utilization.

Chlorinating agents other than HCl have also received attention. Moyer,[156] Fogler,[79] and Grebe *et al.*[96] use nitrosyl chloride, NOCl. Phosgene, $COCl_2$, is employed by Machalske[139] and by Haszeldine and Iserson;[110] a general discussion of phosgene processes is presented by Fink and Bonilla.[78] Sulfur compounds have also been considered. McKee and Sells[154] claim that the use of sulfuryl chloride, SO_2Cl_2, reduces over-all thermal effect. More recently, Verti and Salvi[203] investigated a sulfuryl chloride process. Other investigators have worked with sulfur trioxide-metal chloride systems,[57,88,89,90] with alcohol-sulfur chloride systems,[69] and with the action of chlorine on dimethyl sulfide.[102,103]

The direct reaction of carbon to form carbon tetrachloride has been widely reported. Fink and Bonilla[78] discuss the basic technology. In an early process, Matthews[144] reacts carbon and chlorine in the electric furnace in the presence of calcium chloride. Hennig[73,115,116] and Stock *et al.*[195] directly halogenate activated carbon. Machalske[139] reacts carbon with phosgene. Combes[56] and Maiofis *et al.*[141] use sulfuryl chloride, while Korshak *et al.*[132] use metal chlorides as chlorine carriers.

Various catalytic processes for halogenating hydrocarbons, as alternates to thermal or photochemical methods, were considered above.[37,65,74,75,87,97,114,117,125,127,129,143,167]

An interesting recent development is the synthesis of carbon dichloride, CCl_2, which shows promise as an intermediate.[18]

References

1. Allen, G. L., Jr., Heitz, R., and Henderson, G., U. S. Patent 2,402,978, to Dow Chemical Co. (1946).
2. Manufacturing Chemists' Association, Chemical Safety Data Sheet SD-3.
3. E. I. du Pont de Nemours & Co., Inc., Chlorine Products Bulletin S24-659.
4. E. I. du Pont de Nemours & Co., Inc., Technical Booklet, "Methyl Chloride."
5. Allied Chemical Corp., Solvay Process Division, Technical Pamphlet, "Methylene Chloride."
6. C. I. O. S. Report No. 26-11, "I. G. Farbenindustrie A. G., Hoechst-am-Main: Miscellaneous Chemicals," June 6, 1945.
7. Fiat Final Report No. 1299, "Manufacture of Methane Chlorination Products in I. G. Hoechst Plant," March 3, 1948.
8. U. S. Department of Commerce, Chemical and Drugs Industry Report, March 1949.
9. *Petroleum Refiner,* **33,** (12) 136 (December 1954).
10. *J. Royal Naval Medical Service,* **41,** 90 (1955).
11. *Ind. Eng. Chem.,* **48,** (11) 17A (November 1956).
12. *Chem. Eng.,* **65,** (9) 116 (May 5, 1958).
13. *Chem. Week,* **84,** (4) 49 (January 24, 1959).
14. *Petroleum Refiner,* **39,** (11) 283 (November 1959).
15. *Chem. Week,* **87,** (5) 65 (July 30, 1960).
16. *Chem. Eng.,* **67,** (12) 71 (June 13, 1960).
17. *Chem. & Eng. News,* **38,** (29) 92 (July 18, 1960).
18. *Chem. Week.,* **87,** 63 (August 13, 1960).
19. *Chem. Eng.,* **67,** (18) 98 (September 5, 1960).
20. *Chem. Wk.,* **87,** (11) 100 (September 10, 1960).
21. *Ind. Eng. Chem.,* **52,** (10) 41A (October, 1960).
22. *Chem. Eng.,* **68,** (1) 52 (January 9, 1961).
23. Ayres, E. E., Jr., U. S. Patent 1,717,136 to B. A. S. Co. (1929).
24. Ayres, E. E., Jr., U. S. Patent 1,831,474 to B. A. S. Co. (1931).
25. Backhaus, A., U. S. Patent 1,509,463, to U. S. Industrial Alcohol Co. (1924).
26. Badische Anilin- & Sodafabrik A. G., British Patent 688,004 (1953).
27. Beanblossom, J. E., U. S. Patent 2,287,225 to Hooker Electrochemical Co. (1942).
28. Beanblossom, J. E., and Scott, W. B., U. S. Patent 2,316,736 to Hooker Electrochemical Co. (1943).
29. Beanblossom, J. E., and Scott, W. B., U. S. Patent 2,316,737 to Hooker Electrochemical Co. (1943).

30. Bellone, A. F. S., U. S. Patent 1,627,881 to Societe Chimiques des Usines du Rhone (1927).

31. Bender, H., U. S. Patent 2,089,937, to Great Western Electro-Chemical Co. (1937).

32. Bender, H., U. S. Patent 2,170,801, to Dow Chemical Co. (1940).

33. Bender, H., U. S. Patent 2,200,254, to Dow Chemical Co. (1940).

34. Bender, H., U. S. Patent 2,200,255, to Dow Chemical Co. (1940).

35. Bijarvat, H. C., Patel, N. B., and Potnis, G. V., *J. Appl. Chem.* (London), **6,** 375 (1956).

36. Blochman, L. G., "Doctor Squibb," New York, Simon and Schuster, Inc., 1958.

37. Boswell, M. C., and McLaughlin, R. R., Canadian Patent 301,542 (1930).

38. Brown, D., U. S. Patent 2,839,589, to Chempatents, Inc. (Scientific Design Co., Inc.), (1958).

39. Brown, D., Colton, J. W., and Landau, R., U. S. Patent 2,746,998 to Chempatents, Inc. (Scientific Design Co., Inc.), (1956).

40. Brown, T. E., and Davis, C. W., U. S. Patent 2,377,669, to Dow Chemical Co. (1945).

41. Buc, H. E., and Gleason, A. H., U. S. Patent 2,153,170, to Standard Oil Development Co. (1939).

42. Byers, H. G., and van Arsdel, W. B., U. S. Patent 1,534,027, to Brown Co. (1925).

43. Carlisle, P. J., U. S. Patent 1,834,089, to Roessler & Hasslacher Chemical Co. (1931).

44. Carlisle, P. J., U. S. Patent 1,915,354, to E. I. du Pont de Nemours & Co., Inc. (1933).

45. Carlisle, P. J., Canadian Patent 349,740, to Canadian Industries, Ltd. (1935).

46. Cass, O. W., U. S. Patent 2,406,195, to E. I. du Pont de Nemours & Co., Inc. (1946).

47. Cevidalli, G. B., Pirani, R., and Zanetti, F., Italian Patent 469,574, to Montecatini Societa Generale per L'Industria Mineraria e Chemica (1952).

48. Chempatents, Inc. (Scientific Design Co., Inc.), British Patent 772,126 (1957).

49. Churchill, J. W., U. S. Patent 2,440,768, to Mathieson Alkali Works, Inc. (1948).

50. Churchill, J. W., and Thomas, R. M., U. S. Patent 2,755,310, to Olin Mathieson Chemical Corp. (1956).

51. Churchill, J. W., and Thomas, R. M., U. S. Patent 2,755,316, to Olin Mathieson Chemical Corp. (1956).

52. Coleman, G. H., and Hadler, B. C., U. S. Patent 2,095,240, to Dow Chemical Co. (1937).

53. Coleman, G. H., Hadler, B. C., and Zuckermandel, E. C., U. S. Patent 2,104,703, to Dow Chemical Co. (1938).

54. Colton, J. W., U. S. Patent 2,901,407, to Scientific Design Co., Inc. (1959).

55. Colton, J. W., U. S. Patent 2,909,240, to Scientific Design Co., Inc. (1959).

56. Combes, C., British Patent 25,688 (1902).

57. Cunningham, J. L., U. S. Patent 2,820,068, to Horizons, Inc. (1958).

58. Cone, L. H., and Daudt, H. W., U. S. Patent 1,983,542, to E. I. du Pont de Nemours & Co., Inc. (1934).

59. Consortium Fuer Electrochemische Industrie G.m.b.H., French Patent 844,300 (1939).

60. Curme, G. O., Jr., U. S. Patent 1,422,838, to Carbide & Carbon Chemicals Corp. (1922).

61. Dachlauer, K., and Schnitzler, E., U. S. Patent 2,156,039, to I. G. Farbenindustrie A. G. (1939).

62. Daudt, H. W., U. S. Patent 2,016,075, to E. I. du Pont de Nemours & Co., Inc. (1935).

63. Davis, C. W., Dirstine, P. H., and Brown, W. E., U. S. Patent 2,442,323, to Dow Chemical Co. (1948).

64. Degeorges, M. E., and Thizy, A., U. S. Patent 2,957,033, to Progil S. A. (1960).

65. Diamond Alkali Co., British Patent 701,244 (1953).

66. Diamond Alkali Co., Belgian Patent 537,777 (1955).

67. Dow, H. H., and Quoyle, W. O., U. S. Patent 1,311,329 (1919).

68. Dow Chemical Co., British Patent 484,888 (1938).

69. Drefus, H., British Patent 341,878 (1929).

70. Egloff, G., Schaad, R. E., and Lowry, C. D., Jr., *Chem. Revs.,* **8,** 1 (1931).

71. Ernst, O., and Wahl, H., German Patent 486,952, to I. G. Farbenindustrie A. G. (1922).

72. Farbenindustrie, I. G., A. G., British Patent 489,554 (1938).

73. Farbenindustrie, I. G., A. G., French Patent 826,875 (1938).

74. Farbenindustrie, I. G., A. G., British Patent 513,235 (1939).

75. Farbenindustrie, I. G., A. G., French Patent 836,979 (1939).

76. Farbenindustrie, I. G., A. G., Belgian Patent 449,037 (1943).

77. Farbwerke Hoechst A. G., British Patent 789,314 (1958).

78. Fink, C. G., and Bonilla, C. F., *J. Phys. Chem.,* **37,** 1135 (1933).

79. Fogler, M. F., U. S. Patent 2,174,574, to Solvay Process Div., Allied Chemical & Dye Corp. (1939).

80. Fontana, C. M., and Gorin, E., U. S. Patent 2,575,167, to Ethyl Corp. (1951).

81. Fontana, C. M., Gorin, E., Kidder, G. A., and Meredith, C. S., *Ind. Eng. Chem.,* **44,** 363 (1952).

82. Francis, A. W., and Reid, E. E., *Ind. Eng. Chem.,* **38,** 1194 (1946).

83. Frei, J., U. S. Patent 1,784,423, to E. I. du Pont de Nemours & Co., Inc. (1930).

84. Frei, J., U. S. Patent 1,824,951, to E. I. du Pont de Nemours & Co., Inc. (1931).

85. Fuoss, R. M., *J. Am. Chem. Soc.,* **65,** 2406 (1943).

86. Fuson, R. C., and Bull, B. A., *Chem. Revs.,* **15,** 275 (1934).

87. Garner, J. B., and Clayton, H. D., U. S. Patent 1,262,769, to Metals Research Co. (1918).

88. Giraitis, A. P., U. S. Patent 2,698,347, to Ethyl Corp. (1954).

89. Giraitis, A. P., U. S. Patent 2,698,348, to Ethyl Corp. (1954).

90. Giraitis, A. P., U. S. Patent 2,821,560, to Ethyl Corp. (1958).

91. Gleason, G. W., British Patent 486,453, to I. G. Farbenindustrie A. G. (1937).
92. Gorin, E., U. S. Patent 2,407,828, to Socony-Vacuum Oil Co. (1943).
93. Gorin, E., U. S. Patent 2,498,546, to Socony-Vacuum Oil Co. (1950).
94. Gorin, E., Fontana, C. M., and Kidder, G. A., *Ind. Eng. Chem.,* **40,** 2128 (1948).
95. Gorin, E., Fontana, C. M., and Kidder, G. A., *Ind. Eng. Chem.,* **40,** 2135 (1948).
96. Grebe, J. J., Bauman, W. C., and Robinson, H. A., U. S. Patent 2,366,518, to Dow Chemical Co. (1945).
97. Grebe, J. J., Reilly, J. H., and Wiley, R. M., U. S. Patent 2,034,292, to Dow Chemical Co. (1936).
98. Green, A. D., and Heminger, C. E., U. S. Patent 2,308,170, to Standard Oil Development Co. (1943).
99. Gremli, E., Austrian Patent 108,424 (1927).
100. Griswold, T., Jr., and Strosacker, C. J., U. S. Patent 1,101,025, to Midland Chemical Corp. (1914).
101. Guinot, H. M., British Patent 708,194 (1954).
102. Hallstein, A., German Patent 416,603, to Chemische Fabrik auf Aktien vorm. (1924).
103. Hallstein, A., German Patent 417,970, to Chemische Fabrik auf Aktien vorm. (1924).
104. Harding, E. A., U. S. Patent 1,816,845, to Roessler & Hasslacher Chemical Co. (1924).
105. Hass, H. B., and Marshall, J. R., *Ind. Eng. Chem.,* **23,** 352 (1931).
106. Hass, H. B., McBee, E. T., and Weber, P., *Ind. Eng. Chem.,* **28,** 333 (1936).
107. Hass, H. B., McBee, E. T., and Hatch, L. F., *Ind. Eng. Chem.,* **29,** 1335 (1937).
108. Hass, H. B., and McBee, E. T., Canadian Patent 374,241, to Purdue Research Foundation (1938).
109. Hass, H. B., and Weber, P., *Ind. Eng. Chem.,* Anal. Ed., **7,** 231 (1935).
110. Haszeldine, R. A., and Iserson, H., *Nature,* **179,** 1361 (1957).
111. Haynes, W., "American Chemical Industry," Vol. 1, New York, D. Van-Nostrand Co., Inc., 1954.
112. Hebberling, H., *Seifen-Oele-Fette-Wachse,* **76,** 211 (1950).
113. Heitz, R. G., and Brown, W. E., U. S. Patent 2,442,324, to Dow Chemical Co. (1948).
114. Hennig, B., U. S. Patent 2,160,574, to I. G. Farbenindustrie A. G. (1939).
115. Hennig, B., U. S. Patent 2,223,448, to General Aniline and Film Corp. (1940).
116. Hennig, B., German Patent 693,414, to I. G. Farbenindustrie A. G. (1940).
117. Hennig, B., German Patent 712,492, to I. G. Farbenindustrie A. G. (1941).
118. Hirschkind, W., *Ind. Eng. Chem.,* **41,** 2749 (1949).
119. Holt, L. C., and Daudt, H. W., U. S. Patent 1,983,542, to E. I. du Pont de Nemours & Co., Inc. (1934).
120. Holt, L. C., and Daudt, H. W., U. S. Patent 2,091,986, to E. I. du Pont de Nemours & Co., Inc. (1937).
121. Holzverkohlungs-Industrie A. G. in Konstanz, i.B., German Patent 477,494 (1929).

122. Ionescu, V., and Anghelescu, N., *Revista de Chimie* (Bucharest), **8,** 745 (1957).

123. Johnson, P. R., U. S. Patent 2,585,469, to Atomic Energy Commission (1952).

124. Johnson, P. R., Parsons, J. L., and Roberts, J. B., *Ind. Eng. Chem.,* **51,** 499 (1959).

125. Jones, G. W., Allison, V. C., and Meighan, M. H., *U. S. Bur. Mines Tech. Paper 255,* 1921.

126. Joseph, W. J., U. S. Patent 2,752,401, to Dow Chemical Co. (1956).

127. Kazuba, Z., *Przeglad Chem.* (Poland), **6,** 123 (1948).

128. Ketslakh, M. M., Rudkovskii, D. M., and Suknevich, I. F., *Zhur. Priklad. Khim. (U.S.S.R.),* **23,** 221 (1950).

129. Kiprianov, A. I., and Kusner, T. S., *Zhur. Priklad. Khim., (U.S.S.R.),* **8,** 673 (1935).

130. Klein, H., and Pfaundler, C., U. S. Patent 2,026,131, to I. G. Farbenindustrie A. G. (1935).

131. Kolker, L. A., U. S. Patent 2,847,484 (1957).

132. Korshak, K. V., Strepikheev, Y. A., and Velatova, L. F., *Zhur. Obschbi Khim., (U.S.S.R.),* **17,** 1626 (1947).

133. Krause, E., and Roka, K., U. S. Patent 1,654,821, to Holzverkohlungs-Industrie A. G., in Konstanz, i.B. (1928).

134. Krentsel, B. A., Topchiev, A. V., and Andreev, L. N., *Doklady Akad. Nauk. S.S.S.R.,* **112,** 73 (1957).

135. Lacy, B. S., French Patent 480,064 (1916).

136. Levine, A. A., U. S. Patent 1,975,727, to E. I. du Pont de Nemours & Co., Inc. (1934).

137. Longiave, C., *Chimica e industria (Milan),* **36,** 693 (1954).

138. Loveless, A. H., P. B. Report No. A-63613, U. S. Department of Commerce.

139. Machalske, F. J., U. S. Patent 808,100, to F. Darlington (1905).

140. MacMullin, R. B., *Chem. Eng. Prog.,* **44,** 183 (1948).

141. Maiofis, L. S., Schusterovich, G. M., and Bibikov, N. N., *Trans. State Inst. Appl. Chem., (U.S.S.R.),* **24,** 119 (1935).

142. Mamaliev, Y. G., Dementyeva, V. V., Kuliev, A. M., and Bakshiev, A. A., *Zhur. Priklad. Khim., (U.S.S.R.),* **12,** 1826 (1939).

143. Martin, L. F., and Lux, A. R., U. S. Patent 1,801,873, to Dow Chemical Co. (1931).

144. Matthews, J. M., U. S. Patent 835,307, to F. Darlington (1906).

145. Mattson, G. W., U. S. Patent 2,622,107, to Ethyl Corp. (1952).

146. McBee, E. T., and Hass, H. B., U. S. Patent 2,004,072, to Purdue Research Foundation (1935).

147. McBee, E. T., Hass, H. B., Chad, T. H., Welch, Z. D., and Thomas, L. E., *Ind. Eng. Chem.,* **33,** 176 (1941).

148. McBee, E. T., Hass, H. B., and Pierson, E., *Ind. Eng. Chem.,* **33,** 181 (1941).

149. McBee, E. T., Hass, H. B., Neher, C. M., and Strickland, H., *Ind. Eng. Chem.,* **34,** 296 (1942).

150. McBee, E. T., Hass, H. B., and Bordenca, C., *Ind. Eng. Chem.,* **35,** 317 (1943).

151. McBee, E. T., and Devaney, L. W., *Ind. Eng. Chem.,* **41,** 803 (1949).

152. McBee, E. T., and Devaney, L. W., U. S. Patent 2,473,162, to Purdue Research Foundation (1949).

153. McKee, R. H., and Burke, S. P., U. S. Patent 1,738,193 (1929).

154. McKee, R. H., and Sells, C. M., U. S. Patent 1,765,601 (1930).

155. Mndzhoyan, A. L., and Aroyan, A. A., *Izvest. Akad. Nauk Armyan. S.S.S.R., Khim. Nauki*, **11**, 45 (1958).

156. Moyer, W. W., U. S. Patent 2,152,375, to Solvay Process Div., Allied Chemical & Dye Corp. (1939).

157. Mugdan, M., and Wimmer, J., German Patent 680,659, to Consortium Fuer Electrochemische Industrie G.m.b.H. (1939).

158. Nekrasova, V. A., and Shuikin, N. I., *Uspekhi Khim.*, **22**, 179 (1953).

159. Neubauer, J. A., FIAT Final Report No. 1154, "High Temperature Chlorination of Methane at Huels," June 5, 1947.

160. Obrecht, R. P., and Bender, H., U. S. Patent 2,857,438, to Stauffer Chemical Co. (1958).

161. Paoloni, C., and Eusepi, A., Italian Patent 513,373, to Rumianca Societa per Azioni (1955).

162. Passler, F., U. S. Patent, 2,906,681, to Farbwerke Hoechst A. G. (1959).

163. Pease, R. N., *J. Am. Chem. Soc.*, **56**, 2388 (1934).

164. Pease, R. N., and Walz, G. F., *J. Am. Chem. Soc.*, **53**, 3728 (1931).

165. Phillips, M., U. S. Patent 1,359,099 (1920).

166. Pianfetti, J. A., and Timmerman, R. W., U. S. Patent 2,606,867, to Food Machinery & Chemical Corp. (1952).

167. Pie, P. F., Jr., U. S. Patent 2,280,928, to Darco Corp. (1942).

168. Potter, C., and McDonald, W. C., *Can. J. Res.*, **25**, 415 (1947).

169. Pritchard, H. O., Pyke, J. B., and Trotman-Dickenson, A. F., *J. Am. Chem. Soc.*, **76**, 1201 (1954).

170. Pritchard, H. O., Pyke, J. B., and Trotman-Dickenson, A. F., *J. Am. Chem. Soc.*, **77**, 2629 (1955).

171. Pye, D. J., U. S. Patent 2,752,402, to Dow Chemical Co. (1956).

172. Randall, M., U. S. Patent 2,547,139 (1951).

173. Reilly, J. H., U. S. Patent 1,947,491, to Dow Chemical Co. (1934).

174. Reilly, J. H., U. S. Patent 2,110,174, to Dow Chemical Co. (1934).

175. Reitlinger, O., U. S. Patent 2,429,963 (1947).

176. Riblett, E. W., U. S. Patent 2,334,033, to Process Management Co. (1944).

177. Richter, G. A., and van Arsdel, W. B., U. S. Patent 1,535,378, to Brown Co. (1925).

178. Ritchie, M., and Winning, W. I. H., *J. Chem. Soc.*, (London), **1950**, 3583.

179. Roka, K., British Patent 186,270, to Holzverkohlungs-Industrie A. G. in Konstanz, i.B. (1921).

180. Roka, K., German Patent 478,083, to Holzverkohlungs-Industrie in Konstanz, i.B. (1923).

181. Rommel, O., German Patent 857,955, to Badische Anilin-& Sodafabrik A. G. (1952).

182. Roueche, B., "The Incurable Wound," p. 68, Boston, Little, Brown & Co., 1957.
183. Scott, N. D., U. S. Patent 2,570,495, to E. I. du Pont de Nemours & Co., Inc. (1951).
184. Shaw, H., and Whitston, O., B.I.O.S. Final Report No. 851, "I. G. Hoechst Chlorinated Methanes Derivatives," 1949.
185. Skeeters, M. J., and Cooper, R. S., U. S. Patent 2,688,592, to Diamond Alkali Co. (1954).
186. Snelling, W. O., U. S. Patent 1,271,790 (1914).
187. Snelling, W. O., U. S. Patent 1,285,823 (1918).
188. Snelling, W. O., U. S. Patent 1,325,214 (1919).
189. Snelling, W. O., U. S. Patent 1,339,675 (1920).
190. Snelling, W. O., U. S. Patent 1,421,733 (1922).
191. Soell, J., and Runkel, C., German Patent 491,316, to I. G. Farbenindustrie A. G. (1931).
192. Solvay *et Cie,* British Patent 772,910 (1957).
193. Spencer, H., U. S. Patent 2,755,311, to Olin Mathieson Chemical Corp. (1956).
194. Stauffer Chemical Co., British Patent 820,850 (1959).
195. Stock, A., Lux, H., and Wustrow, W., *Z. anorg. allgem. Chem.,* **195,** 149 (1931).
196. Swann, S., Jr., *Trans. Electrochem. Soc.,* **69,** 296 (1936).
197. Swathmey, J. T., "Anesthesia," New York, D. Appleton Co., 1914.
198. Thode, E. F., and Meissner, H. P., *Ind. Eng. Chem.,* **43,** 129 (1951).
199. Thomas, E. B., and Hindley, F., U. S. Patent 2,715,146, to British Celanese, Ltd. (1955).
200. Thronson, E. A., and Mendolia, A. A., U. S. Patent 2,421,441, to E. I. du Pont de Nemours & Co., Inc. (1947).
201. Tizard, H. T., Chapman, D. L., and Taylor, R., British Patent 214,293 (1924).
202. Vdovichenko, V. T., Galenko, N. P., and Sarishvili, I. G., *Ukrain. Khim. Zhur.,* **23,** 110 (1957).
203. Verti, V., and Salvi, G., *Riv. Combustibili (Milan),* **1,** 64 (1947).
204. Warren, G. W., U. S. Patent 2,577,388, to Dow Chemical Co. (1951).
205. Warren, G. W., U. S. Patent 2,727,076, to Dow Chemical Co. (1955).
206. Wheeler, T. S., and Mason, J., British Patent 342,329, to Imperial Chemical Industries, Ltd. (1931).
207. Wilson, M. J. G., and Howland, A. H., *Fuel,* **28,** 127 (1949).
208. Wimmer, J., U. S. Patent 2,305,821 (1942).
209. Wintersberger, K., and Rommel, O., U. S. Patent 2,706,709, to Badische Anilin-& Sodafabrik A. G. (1955).
210. Zanetti, F., and Cevidalli, G., "A Type of Furnace for Chlorination of Methane," VII National Convention of Natural Gas and Petroleum, Taormina, Italy, April 1952.

13. TRICHLOROETHYLENE AND PERCHLOROETHYLENE

C. B. Shepherd

E. I. du Pont de Nemours & Company

HISTORY AND DEVELOPMENT

Michael Faraday discovered perchloroethylene* in 1820 when he de-chlorinated hexachloroethane by pyrolysis.[111] Forty-four years later, Fischer reacted hexachloroethane with zinc and dilute sulfuric acid and re-portedly formed trichloroethylene.[120] Fischer is usually credited with the discovery of trichloroethylene,† but in 1836 Laurent passed chlorine over ethylene dichloride for two days with gentle heating and obtained a product which was probably trichloroethylene, although he ascribed to it the formula $C_8H_2Cl_6$.[187]

The first commercial manufacture of trichloroethylene appears to have taken place in Germany in the early 1900's. Manufacture in Austria and England followed soon after. Requirements of the United States were imported from these countries until World War I conditions made a domestic supply imperative.[296] U. S. production, however, was small and intermittent until 1925 when a plant supplying 2.5 tons per day was started in Niagara Falls, New York, by the Roessler & Hasslacher Chemical Company (acquired by the Du Pont Company in 1930). Following the introduction of the vapor-degreasing process for cleaning metals with trichloroethylene in the early 1930's, Du Pont gradually increased production capacity and, by 1945, was producing trichloroethylene at the rate of 115,000 tons/year. Other major producers of trichloroethylene, in the order of their entrance into the market on a significant volume basis, are listed below:

*The correct name, approved by the International Union of Chemistry and by the American Chemical Society, is "tetrachloroethylene." Chemical industry prefers the name "perchloroethylene" (often written perchlorethylene) to avoid any possibility of its confusion with tetrachloroethane, an unstable, highly toxic solvent.

†Trichloroethylene is usually written as "trichlorethylene" in commercial usage.

	Year
Westvaco Chlorine Products Co.	1933, (discontinued in 1951)
Hooker-Detrex Corp.	1947
The Dow Chemical Co.	1948
Niagara Alkali Co.	1949
Columbia-Southern Chemicals Div.	
Pittsburgh Plate Glass Co.	1956

U. S. capacity in 1960 was an estimated 250,000 tons annually[5] with a total domestic production of about 176,000 annual tons. Imports of trichloroethylene have been a steadily increasing factor in the U. S. market since 1953 and in 1960 represented about 16 percent of U. S. consumption.

Because perchloroethylene is a co-product of so many chlorohydrocarbon syntheses, the history of its early manufacture is difficult to trace. The U. S. Tariff Commission's annual reports on "Synthetic Organic Chemicals, United States Production and Sales" mention it for the years 1923, 1924, and 1930, but the Du Pont Company apparently was the first continuous volume producer.[259] Its Niagara Falls, New York, plant began manufacturing perchloroethylene in 1932. Although perchloroethylene was first promoted for dry cleaning in 1933, its use in this field accelerated most rapidly after 1945, and dry cleaning now represents the chief outlet. Below are listed other major perchloroethylene manufacturers, in the order of their entry into the market.

	Year
The Dow Chemical Co.	1935
Columbia-Southern Chemicals Div.	
Pittsburgh Plate Glass Co.	1949
Diamond Alkali Co.	1950
Stauffer Chemical Co.	1954
Frontier Chemical Co. Div.	
Vulcan Materials Co.	1958

U. S. 1960 capacity has been estimated at 137,500 tons annually,[259] but is now believed to be considerably more than this. Domestic production of perchloroethylene in 1960 was about 105,000 tons. Imports of perchloroethylene have been increasing since 1958 and in 1960 accounted for approximately 11 percent of U. S. consumption.

In a span of thirty years, trichloroethylene and perchloroethylene have become commercial solvents whose combined production exceeds 280,000 tons and accounts for some 10 percent of U. S. chlorine consumption.[193] This growth stems from two major outlets: the vapor degreasing of metals with trichloroethylene and dry cleaning of fabrics with perchloroethylene.

Inasmuch as metal cleaning by vapor degreasing accounts for over 90 percent of the trichloroethylene market[5] this market follows closely metal-fabricating activity. Figure 13-1 compares the FRB index (1947–9 = 100) on U. S. metal fabricating, reconstructed for the years prior to 1947, with U. S. consumption of trichloroethylene. A 1953 estimate credited vapor degreasing with 15 to 20 percent of all metal cleaning.[262] As Figure 13-1 suggests, this share is increasing. Novel techniques for phosphatizing and painting metals from a trichloroethylene base, introduced late in 1959 (see under "Uses—Metal Finishing"), make possible completely anhydrous cleaning and finishing treatments for metals. The consumption trend of trichloroethylene is expected to continue to climb not only because of these new uses but also because of increased applications of vapor degreasing.

Trade estimates of U. S. consumption of perchloroethylene for dry cleaning place it at 75 to 76 percent of total consumption.[11,259] Figure

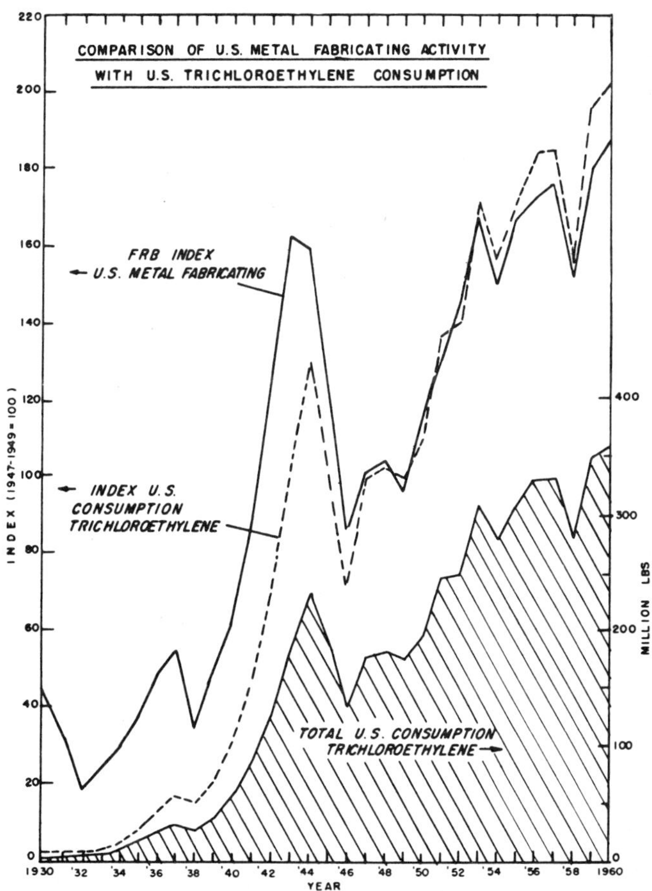

Figure 13-1

13-2 shows yearly dollar purchases of dry cleaning services and illustrates the increasing penetration of this market by the perchloroethylene process. Government allocations during World War II and its aftermath retarded normal growth of perchloroethylene dry cleaning for 6 to 7 years. However, after removal of supply uncertainties, dry cleaning with perchloro-

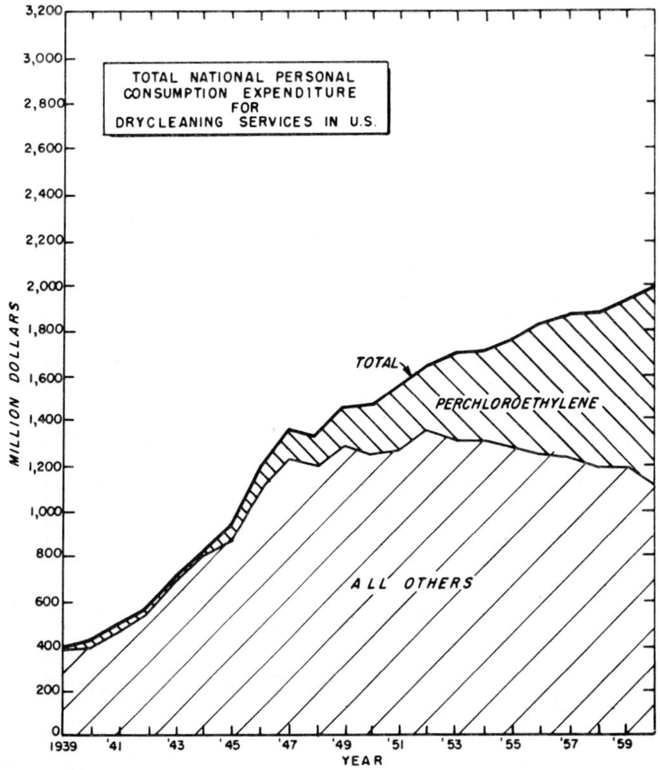

Figure 13-2. (Breakdown based on solvent process used).

ethylene advanced rapidly. Whereas the 1954 Census showed 33 percent of all dry cleaning plants to be using perchloroethylene,[11] spot surveys in heavily populated areas indicated the ratio probably exceeded 50 percent in 1959. Two factors should accelerate perchloroethylene sales for dry cleaning: (1) personal expenditures for dry cleaning are increasing at twice the rate of population growth,[193] and (2) the 1960 appearance of coin-operated perchloroethylene dry cleaning machines in self-service laundry centers is creating additional dry cleaning volume by attracting "marginal" articles that would not be sent to a professional cleaner.[7]

The following sections describe the manufacture and stabilization of trichloroethylene and perchloroethylene, their useful properties and the applications in which these serve, and the safety considerations and handling procedures attending their use.

MANUFACTURE

Although petroleum gases are now finding increasing use as raw material for the manufacture of trichloroethylene and especially perchloroethylene, these solvents continue to be classed as acetylene derivatives in recognition of the historical and dominant role of acetylene chlorination in their commercial production. The principal reactions are as follows:

$$CH{\equiv}CH + 2\,Cl_2 \rightarrow CHCl_2CHCl_2 \qquad (13\text{-}1)$$

Acetylene   *sym-Tetrachloroethane*

$$CHCl_2CHCl_2 \rightarrow CHCl{=}CCl_2 + HCl \qquad (13\text{-}2)$$

Trichloroethylene

$$CHCl{=}CCl_2 + Cl_2 \rightarrow CHCl_2CCl_3 \qquad (13\text{-}3)$$

Pentachloroethane

$$CHCl_2CCl_3 \rightarrow CCl_2{=}CCl_2 + HCl \qquad (13\text{-}4)$$

Perchloroethylene

sym-Tetrachloroethane is the prime intermediate. Patent literature describing means of controlling Reaction (13-1) dates back to 1903.[71] Important factors are feed pressures, reactor temperature, and catalyst. Relatively high-purity chlorine is used to minimize by-product formation and to avoid introduction of air in concentrations sufficient to form explosive compositions with the acetylene. Because vapor phase reactions form unwanted by-products, the chlorine and acetylene are usually reacted in tetrachloroethane solution at moderate temperatures. The two gases may feed into an agitated bath of tetrachloroethane in a steel or clad-steel reactor, or they may rise in a packed steel tower filled with the product.[50,203,306] Higher temperatures and excess acetylene favor the formation of dichloroethylene.[73]

Trichloroethylene is obtained from tetrachloroethane by treatment with inorganic HCl acceptors, notably lime, or by pyrolytic loss of HCl.

The chlorination of trichloroethylene to pentachloroethane can take place in the equipment used for Reaction (13-1) and under similar conditions. High-purity chlorine, moderate reaction temperatures, and a slight excess (to 5 percent) of trichloroethylene minimize chlorine losses.[50]

Removal of HCl from pentachloroethane by cracking or liming yields perchloroethylene.

Trichloroethylene

Of the 176,000 tons of trichloroethylene produced in the United States in 1960 approximately 90 percent was manufactured from acetylene via the cracking or liming of tetrachloroethane. The following discussion highlights tetrachloroethane liming and cracking operations and describes briefly other processes of lesser commercial importance.

Tetrachloroethane Liming. Until the late 1940's, nearly all the trichloroethylene produced was obtained by dehydrochlorinating tetrachloroethane with a slurry of lime, using methods based on a 1905 German patent.[72]

$$2\,CHCl_2CHCl_2 + Ca(OH)_2 \rightarrow 2\,CHCl{=}CCl_2 + CaCl_2 + 2\,H_2O \quad (13\text{-}5)$$

The lime slurry, which may contain from 5 to 25 percent calcium hydroxide, is normally in excess. The slurry is stirred and the reaction takes place in a vessel at 70° to 80° C, maintained by the heat of reaction. Live steam or cold water is added, as required, to control temperature. The reaction mixture passes to a stripper where the organic products are stripped from the aqueous phase by steam. The organics pass to a rectifier which removes high boilers, and then to a still which separates the trichloroethylene from low boilers to give a product which is over 99.9 percent pure. The residues from the stripping, rectifying, and distilling steps are redistilled for recoverable values as shown in Figure 13-3.

Tetrachloroethane Cracking. In the liming process, the HCl is lost in the form of dilute calcium chloride solution. The tetrachloroethane cracking process offers an economic advantage wherever chlorine values are recoverable and usable as HCl gas.[3]

Tetrachloroethane decomposes in the temperature range 263 to 282° C, by a homogeneous first-order reaction which forms trichloroethylene and hydrogen chloride.[24] However, commercial cracking operations usually take place at temperatures up to 600° C.[104,106] A variety of surface active materials with and without added catalytic agents have been employed to facilitate the pyrolysis. These include refractory materials[293] or pumice[70] at 400 to 500° C, activated carbon at 260° C[38] and 500° C,[70] fluidized sand,[123,162] thorium oxide below 390° C,[56] organic nitrogen bases at 150 to 200° C,[303] bone char at 300 to 310° C[305] and barium chloride at 300° C.[56,86,107,231] Conversions per pass of 90 percent and over are possible.[13,107] The addition of oxygen or air to high-purity tetrachloroethane vapors reportedly promotes dehydrohalogenation to trichloroethylene (93 percent at 400° C).[226]

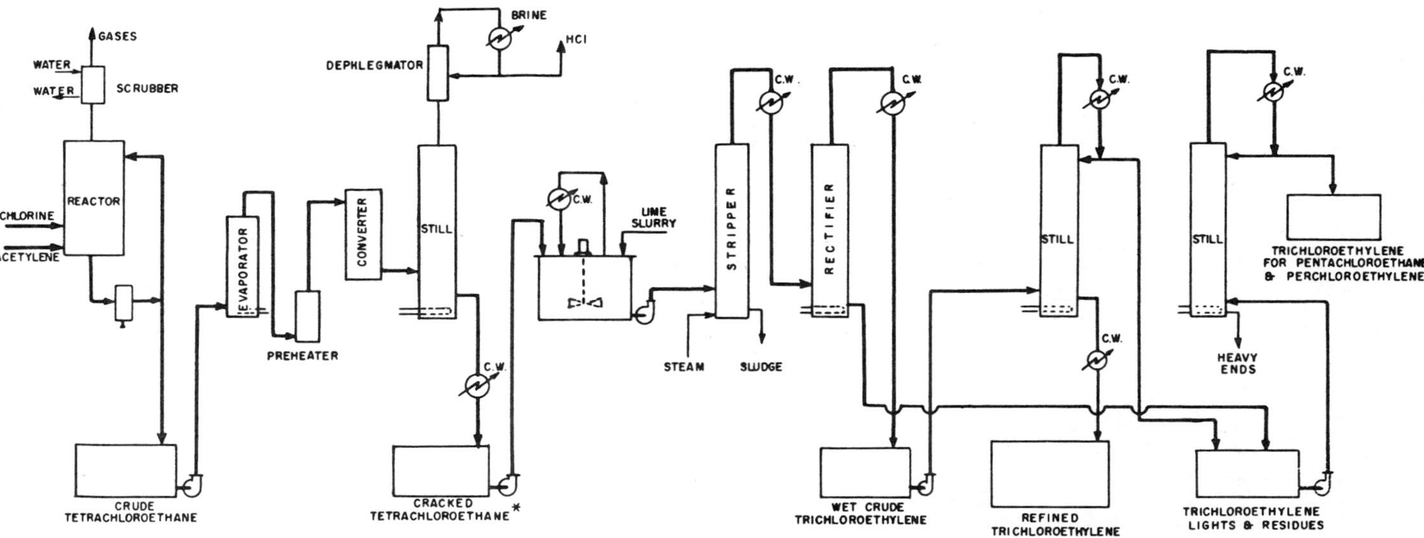

Figure 13-3. Trichloroethylene from tetrachloroethane.

Figure 13-4. Columns such as these at Du Pont's Niagara Falls, New York, plant refine trichloroethylene and perchloroethylene with a 2 to 3°F boiling range.

Because the instability of tetrachloroethane makes separation of pure trichloroethylene difficult, the product mix from the cracking operation normally goes through a liming step to convert the remaining tetrachloroethane to trichloroethylene (see Figure 13-3).[50]

Many variations of the pyrolytic process are cited in the journal and patent literature. Simple immersion of a heating element at 800 to 1100° C in tetrachloroethane causes trichloroethylene to flash distill in 74 percent conversion.[128] Tetrachloroethane may be cracked in the presence of an organic HCl acceptor. For example, reaction of tetrachloroethane with acetylene at 250° C over activated charcoal impregnated with metal halides produces trichloroethylene together with vinyl chloride.[307] In other processes the chlorination of acetylene and pyrolysis of the intermediates are carried out simultaneously or in stages in one operation. Both trichloroethylene and perchloroethylene are formed by reaction of acetylene and chlorine mixed with recycled chlorohydrocarbons at 350 to 450° C over fluidized charcoal impregnated with cupric chloride and barium chloride.[292] Alternatively, the exothermic chlorination reaction and the endothermic dehydrochlorination step are effected in one pass through two separate catalytic zones.[68]

Non-acetylenic Routes. The patent art is replete with processes for trichloroethylene based on hydrocarbon raw materials other than acetylene. Chlorination of ethylene dichloride in a bath of molten KCl and $AlCl_3$ at 400 to 480° C forms trichloroethylene along with dichloroethylene.[251] Similarly, ethylene dichloride vapor or ethylene gas, passed with chlorine at 300 to 500° C through a fluidized bed of fuller's earth, charcoal, or graphite, forms trichloroethylene as the principal product.[69,94]

Ethane and chlorine, passing with or without an inert gas through turbulent sand, yield a complex mixture containing trichloroethylene, perchloroethylene, ethylene, dichloroethylene, trichloroethane, and some high boilers.[124]

Oxidative chlorination of olefins or chlorolefins favors chlorine-substitution over chlorine-addition reactions. Thus ethylene reacts with HCl and oxygen at 375 to 490° C over an oxidation catalyst, such as copper oxide on pumice or Fe_2O_3 on fuller's earth,[252] to form vinyl chloride, dichloroethylene, trichloroethylene, and perchloroethylene. Under similar conditions, ethylene and a mixture of 1 mol Cl_2 with 0.5 mol O_2 react to form dichloroethylene and trichloroethylene which, on recycling, give a 75 to 80 percent yield of perchloroethylene.[52] Vinyl chloride undergoes a similar reaction.[54,55]

Direct oxidation of 1,1,2-trichloroethane at 400° C over copper oxide on fire brick forms trichloroethylene, but the simultaneous cracking of 1,1,2-trichloroethane gives considerable dichloroethylene as by-product.[53]

Perchloroethylene

Numerous routes are available for the manufacture of perchloroethylene. These may be classified into three broad categories: (1) the dehydrochlorination of pentachloroethane derived from acetylene, (2) direct processes based on acetylene or its chlorination products, and (3) the cracking of other chlorohydrocarbons. Some of the processes typical of these categories are described below.

Dehydrochlorination of Pentachloroethane. Houser and Bernstein studied the kinetics of pentachloroethane decomposition to perchloroethylene at 407 to 438° C in flow systems using helium or nitrogen as the carrier gas and providing contact times ranging from 6 to 35 seconds. They found the reaction to be autocatalytic and probably of a radical-chain nature.[156] Barton, who studied the reaction rate at 450° C in clean-walled glass tubes, found that it was enhanced by the presence of small quantities of either oxygen or chlorine.[23]

One of the earliest patents covering the preparation of perchloroethylene describes the dehydrochlorination of pentachloroethane at 280° C over bone black.[305] Other catalysts cited in the patent literature include aluminum chloride,[323] and barium chloride,[230,231] or barium chloride with cupric chloride[86] on activated carbon. Immersion of a heating element at 800 to 1100° C in pentachloroethane reportedly produces 97 percent perchloroethylene.[128]

In the manufacture of perchloroethylene by cracking pentachloroethane, the so-called "cracked penta" usually undergoes a liming treatment to complete conversion of the charge.

$$2CHCl_2CCl_3 + Ca(OH)_2 \rightarrow 2CCl_2{=}CCl_2 + CaCl_2 + 2H_2O \qquad (13\text{-}6)$$

Another method of dehydrochlorinating pentachloroethane is to react it with acetylene at 260° C in the presence of activated charcoal impregnated with metal chlorides.[26]

Direct Processes Based on Acetylene and Its Chlorination Products. The patent art cites a number of methods for the direct chlorination of acetylene to perchloroethylene, but none of these are known to be used commercially in the United States. Acetylene and chlorine can burn to form perchloroethylene along with other chlorohydrocarbons.[126] When intimately mixed in proper proportions, acetylene and chlorine react at 450 to 500° C without burning to give up to 95 percent yield of perchloroethylene.[60] Means of controlling the explosion hazard of this reaction involve the use of diluents such as inert gases or recycled product and by-products. The reaction may be conducted over activated carbon and other catalysts at 250 to 400° C in the presence of nitrogen or HCl[29,149,164] and in molten salt baths at 190 to 210° C using a chlorohydrocarbon diluent.[250] Feeding a mixture of acetylene, chlorine, and trichloroethylene to a reactor filled with tetra- or pentachloroethane results in a mixture of chloroethanes. The chloroethanes are limed to form trichloroethylene and perchloroethylene, and the trichloroethylene is recycled to the chlorination step.[299] Chlorination of acetylene can also be effected by chlorocarbons which release free chlorine under pyrolytic conditions. For example, perchloroethylene is formed by the reaction of acetylene with hexachloroethane at 200 to 400° C over activated carbon.[27,28,274]

Perchloroethylene may also be produced by direct reactions on intermediate chlorohydrocarbons derived from acetylene. Chlorination of trichloroethylene at 150 to 350° C over activated carbon and other catalysts gives perchloroethylene in over 95 percent yield.[85,122] In the presence of chlorination-cracking catalysts, chlorine reacts with tetrachloroethane vapors at 150 to 450° C to form perchloroethylene as the principal product.[25] Catalytic oxidation of tetrachloroethane with air or oxygen at 300 to 600° C produces perchloroethylene and water. Suitable catalysts include copper oxide[53] or metal halides on an inert carrier.[109,114,168,297]

Chlorination-Dehydrochlorination of Other Hydrocarbons and Their Chlorine Derivatives. Since the rise of the petrochemical industry, the direct manufacture of perchloroethylene by the chlorination-dehydrochlorination of hydrocarbon gases and their chlorine derivatives has assumed increasing importance. Processes based on hydrocarbons other than acetylene now account for approximately two thirds of the perchloroethylene produced in the United States. Propane,[12,42,43,51,58,59,102,146,194,204,233,291] propylene, ethane,[138,146,167,233] ethylene,[4,51,100,195,246,275,310] and methane,[84,146,153,163,183,233,255,324] or their mixtures as natural gas, are the hydrocarbons which today serve as the most common sources of perchloroethylene. The literature also describes processes based on ethylene dichloride,[95,249] and carbon tetrachloride,[61,101,153,181,290,311] and chloroform,[61,260] which are derived from these raw materials.

As a rule, the reactions take place in the vapor phase at temperatures ranging from approximately 200 to 800° C, depending upon the presence or absence of catalyst and/or diluent. Control of reaction conditions to avoid explosion hazards or undesirable by-products usually takes the form of variable volume relationships between the reaction gases; their passage over large-area contact surfaces such as fuller's earth, fire brick, silica gel, charcoal, and activated carbon; and/or the addition of diluents such as inert gases (nitrogen, HCl), or chlorohydrocarbon vapors—frequently in the form of recycled crude product or by-product. The choice of catalyst, if used on the contact surface, depends on the reaction intended. Most common are the metal chlorides, especially cupric chloride, for the addition of chlorine and/or the elimination of HCl.

The selection of raw materials (hydrocarbon or chlorohydrocarbon; chlorine or HCl and air) depends on their economic advantages at the plant site.† Figure 13-5 illustrates a typical system for the manufacture of perchloroethylene from gaseous aliphatic hydrocarbons. In the reactor, the heat of the chlorination reaction cracks the saturated chlorohydrocarbon

---

†See "Development of a Perchloroethylene Process" by J. J. Lukes (Diamond Alkali Co.) in *Chem. Eng. Prog.*, **54**, (3), 75 (1958), for a description of the complex factors entering such selection and the means used by Diamond to resolve them (and devise a process based on the chlorination of ethylene).

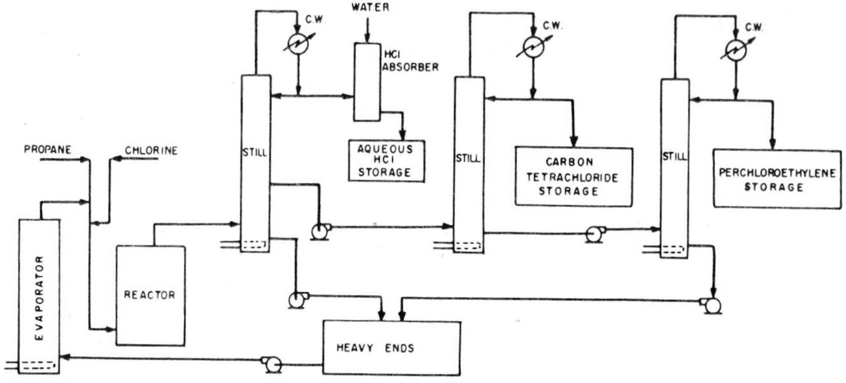

Figure 13-5. Perchloroethylene from propane.

formed, and the cracking gases pass to a quench column where HCl escapes, leaving a crude product mix which passes to the fractionating stills.[153]

Chlorination of ethylene dichloride at 300 to 500° C over coke or pumice gives perchloroethylene as the principal product.[249] Ethylene dichloride chlorinates and dehydrochlorinates to perchloroethylene when passed over a fluidized bed of sand, silicon carbide, clay, silica gel, or alumina gel, heated by burning natural gas in the air.[95]

Weiser and Wightman[311] and, more recently, Korshak and co-workers[181] have studied the thermal decomposition of carbon tetrachloride to perchloroethylene and chlorine. This is the basis of a process used to produce substantial quantities of perchloroethylene in this country during World War II,[153,290] but subsequently discontinued. Without catalysts and in the presence of a chlorine acceptor such as tetrachloroethane, acetylene or methane, the reaction takes place at 500 to 1000° C.[101]

An equimolar mixture of carbon tetrachloride and chloroform passed through a quartz reactor filled with inert material at 600 to 750° C, forms perchloroethylene.[61] Chloroform vapors decompose at 450 to 525° C to form perchloroethylene and HCl.[260]

Stabilization

Carlisle and Levine found that pure trichloroethylene does not hydrolyze readily and that it is stable towards light unless oxygen is also present.[49] Their unpublished work compares the action of light, moisture, metals, and air on trichloroethylene, perchloroethylene, *sym*-tetrachloroethane, and pentachloroethane (Table 13-1).

TABLE 13-1. DECOMPOSITION OF CHLOROHYDROCARBONS ON HEATING 24 HOURS AT THEIR RESPECTIVE BOILING POINTS

	Acidity, cc 0.01N HCl/25 ml solvent			
Test Conditions	C_2HCl_3	C_2Cl_4	$C_2H_2Cl_4$	C_2HCl_5
Dry solvent in nitrogen	10.2[a]	10.5[a]	248	16.8
H_2O-saturated solvent in nitrogen	7.7	12.5	66	12.3
Wet solvent in nitrogen	1.1	0	107	296
H_2O-saturated solvent, soft steel, N_2	11.0	11.6	11	–
Wet solvent, soft steel, nitrogen	5.6	–	–	–
Dry solvent in oxygen	820	782	1985	11050

[a]Acidity attributable to traces of air entrapped in the sealed tubes.

More recent work comparing the hydrolysis rates of chlorohydrocarbons at 25° C (Table 13-2) confirms the stability of trichloroethylene and perchloroethylene towards water.

TABLE 13-2. HYDROLYSIS RATES OF CHLOROHYDROCARBONS AT 25°C IN THE ABSENCE OF AIR AND METALS

	Quantity Hydrolyzed/Day	
Chlorohydrocarbon	mg/liter	Mole %
Trichloroethylene	0.01	6.8×10^{-7}
Perchloroethylene	0.01	6.2×10^{-7}
Tetrachloroethane	1.5	945×10^{-7}
Pentachloroethane	0.435	260×10^{-7}
Ethylene dichloride	0.35	280×10^{-7}
1,1,2-Trichloroethane	0.052	36×10^{-7}

Tables 13-1 and 13-2 illustrate the importance of oxygen or air in the decomposition of chlorinated hydrocarbons. This was one of the first areas of stability investigated.

The susceptibility of trichloroethylene to oxidation was studied by Erdman who postulated a decomposition mechanism involving the formation of a highly reactive epoxide which readily isomerized to dichloroacetyl chloride.[110]

$$Cl_2C{=}CHCl + \tfrac{1}{2}O_2 \rightarrow Cl_2C\overset{}{\underset{O}{\diagdown\diagup}}CHCl \qquad (13\text{-}7)$$

$$Cl_2C\overset{}{\underset{O}{\diagdown\diagup}}CHCl \rightarrow Cl_2CHCOCl \qquad (13\text{-}8)$$

In the presence of moisture, the acetyl chloride is hydrolyzed to dichloro-acetic acid and hydrogen chloride.

$$Cl_2CHCOCl + H_2O \rightarrow Cl_2CHCOOH + HCl \qquad (13\text{-}9)$$

The epoxide has been verified experimentally by McKinney and co-workers.[211] Perchloroethylene has likewise been shown to form an epoxide which rearranges in a similar manner to trichloroacetyl chloride.[125,173,223]

The decomposition of trichloroethylene is accelerated by ultraviolet light in the presence of oxygen, indicating a free radical mechanism. The breakdown of trichloroethylene by this means leads to a variety of products, such as phosgene, carbon monoxide, dichloroacetic acid, formic acid, glyoxylic acid, HCl, Cl_2, and various polymers.[211,215] As trichloroethylene may yield two free-radical forms, $Cl_2C\!=\!\overset{\cdot}{C}Cl$ and $Cl_2C\!=\!\overset{\cdot}{C}H$, the number of decomposition products are many. The nature of the initiating free radical is not known, but considerable work has been done which indicates a chlorine free radical may be involved.[215] Based on the work of several investigators,[96,110,190,215] the reaction probably involves a successive growth and decomposition of free radical chains, initiated by chlorine-radical attack followed by oxidation. In the case of trichloroethylene the reaction would be as follows:

$$Cl\!\cdot\ +\ Cl_2C\!=\!CHCl\ \rightarrow\ Cl_2CHCCl_2 \qquad (13\text{-}10)$$

$$Cl_2CHCCl_2\ +\ O_2\ \rightarrow\ Cl_2CHCCl_2OO\!\cdot \qquad (13\text{-}11)$$

$$Cl_2CHCCl_2O\text{------}OClCHCCl_2O\text{------}OClCH\overset{\cdot}{C}Cl_2 \rightarrow (13\text{-}12)$$

$$Cl_2CHCOCl + Cl\!\cdot \quad HCl + CO + COCl_2 \quad \underset{\displaystyle O}{ClCHCCl_2}$$

The stabilization of trichloroethylene and perchloroethylene against oxidation and free radical attack was originally with the use of gasoline and unsaturated hydrocarbons. Later, amines such as triethylamine and pyridine were used. The use of amines as stabilizers led to the designation of commercial grades of trichloroethylenes as "Alkaline Tri." While these amines did act as acid neutralizers, it was largely coincidental that for many years the best antioxidants happened to be alkaline. At present, more exotic amines such as N-methylpyrrole are used, as well as such conventional antioxidants as thymol and t-amylphenol.

In the early applications of trichloroethylene as an extraction solvent, thermal decomposition was not a problem, and manufacturers were able to provide a product of adequate quality by adding small amounts of an antioxidant such as triethylamine or thymol to the solvent before and after final distillation to insure solvent stability during storage, transit, and use. Then, during the early 1930's, vapor degreasing became the major outlet for trichloroethylene. Its use spread rapidly, particularly during World War II when heavy production demands showed the advantages of trichloroethylene vapor degreasing for both ferrous and nonferrous metals. The stresses imposed on the solvent and its stabilizers by contact at the boiling point with various metals and a wide variety of contaminants (such as fatty acids, sulfurized cutting oils, etc.) made it necessary to develop specific stabilizer compositions for use in vapor degreasing.

In an atmosphere of nitrogen or of its own vapors, pure trichloroethylene is stable at temperatures up to 130°C.[49] Higher temperatures decompose trichloroethylene by simultaneous homogeneous radical-chain and bimolecular mechanisms.[132] The dimer of trichloroethylene is known to exist,[227] and experience has shown that, especially in the presence of anhydrous metal chlorides, tar-like and coke-like compositions can result from overheating pure trichloroethylene. Thermal breakdown occurring under conditions of normal use presumably involves the following reactions:[74,131]

$$2\,ClCH{=}CCl_2 \rightarrow (ClCH{=}CCl_2)_2 \tag{13-13}$$

$$(ClCH{=}CCl_2)_2 \rightarrow HCl + Cl_2C{=}CHCCl{=}CCl_2 \tag{13-14}$$

Perchloroethylene does not polymerize[40,180] but when heated to 300 to 350°C at 1810 atmospheres, it forms hexachloroethane and hexachlorobutadiene.[130]

The prevention of thermal decomposition of tri- and perchloroethylene is largely accomplished by the introduction of unsaturated hydrocarbons such as cyclohexene, diisobutylene, and amylene, although a wide variety of compounds have been used.

Tables 13-3 and 13-4 summarize a large portion of the patent literature on the stabilization of trichloroethylene and/or perchloroethylene. They show how the art of solvent stabilization has progressed from the simple selection of additives to fulfill specific needs (storage stability, resistance to thermal decomposition) to the blending of synergistically acting components to meet the varied demands imposed by heavy-duty or hypercritical vapor-degreasing operations (resistance to air, heat, acids, metals, metal salts; freedom from chloride ions). To secure satisfactory performance under these demands, the stabilizing systems of degreasing-grade trichloroethylene must

(a) inhibit the action of light and air on the solvent,
(b) inhibit the action of high temperatures on the solvent,

TABLE 13-3. PATENTED STABILIZERS FOR TRICHLOROETHYLENE AND/OR PERCHLOROETHYLENE

Chemical Class and Example	Ref.	Chemical Class and Example	Ref.	Chemical Class and Example	Ref.
Hydrocarbons and Derivatives		Ketones		Nitriles	
Gasoline	81	Monoketone (b.p. 130°C)	158	Propionitrile	219
Liquid petrolatum	325	Essential oils, camphor	99	Acrylonitrile	174
Cyclohexane	201	2-Hydroxy-2-methyl-3-		3-Methoxypropionitrile	289
Hydrogenated naphthalene	257	butanone	64	2-Isopropylaminoethyl	
Amylene	144	Acetyl acetone	186	cyanide	93
Diisobutylene	175				
Cyclohexene	185	Esters and Salts		Alkaloids, Azines,	
3-Pentene-1-yne-3-methyl	269	Butyl tartrate	257	Oximes, etc.	
2,5-Dimethyl-1,5-hexadiene-		Isopropyl acetate	239	Caffeine	98,47
3-yne	44	Propargyl acetate	268	Diethylidene azine	263
3-Chloro-1-propyne	20	Ethyl nitrate	176	Acetalazine	224
Nitromethane	271	Diisopropyl ammonium acetate	244	Tetramethylpiperazine	288
		Calcium stearate	134	1-Isopropyl-2-methylene	
Alcohols				aziridine	105
Isoamyl alcohol	237	Lactones		Acetophenone oxime	238
Allyl alcohol	198	Butyrolactone	282	N-Methyl morpholine	284
Propargyl alcohol	270			2-Methyl-2-oxazoline	184
3-Methyl-1-pentyne-3-ol	108	Amines		Pyrazole	159
3,3-Dimethoxy-2-methyl-		Triethylamine	240,87		
butan-2-ol	65	Diisopropylamine	30	Resins	
3-Amino-2-methyl-butan-2-ol	66	Naphthylamine	221	Gum mastic, sandarac,	
		Diphenylamine	222	rosin	272
Phenols		Guanidine	220		
Phenol	48	Triphenylguanidine	191	Sulfur Compounds	
p-tert-Butyl phenol	192	Hexamethylenetetramine	218	Sulfur dioxide	236
Hexylresorcinol	241	Pyridine	97	Butyl mercaptan	285

(c) protect the solvent from the action of small amounts of strong acid,

(d) protect the solvent from the action of metals and their salts,

(e) correct any incipient localized breakdown of the solvent.

Thus, a properly stabilized system for vapor degreasing may contain several components—all carefully selected to inhibit one or more of the specific actions outlined above. In addition, each component must be

(1) sufficiently volatile to exist in both the liquid and the vapor phases of the solvent and to be recoverable from distillation residues,

(2) unreactive with the many and varied contaminants encountered in the cleaning of metal parts from modern fabricating operations,

(3) compatible with the other stabilizers in the system.

TABLE 13-4. PATENTED STABILIZER COMPOSITIONS FOR TRICHLOROETHYLENE AND/OR PERCHLORETHYLENE

CHEMICAL CLASSES OF COMPONENTS	Ref.
2-Component Compositions	
Alcohol–terpene	116
Acetylenic alcohol–phenol	45,150,267
tert-Alcohol–phenol	276
Phenol–epoxide	82
Alcohol–ammonia	248
prim-Pentanol-pyridine	83
Tetrahydrofuran–1-methyl pyrrole	46
Ester–epoxide	279
Alkali soap–fatty acid	256
Metallic soap–amine	136
Organometallic chelate–pyrrole	278
Amine–epoxide	77
Amine–aldehyde (condensation products)	169
Amine–ion exchange resin	197
Amine (high-boiling)–amine (low-boiling)	196
Aminoborane–pyrrole	206
Amide–epoxide	78
3-Component Compositions	
Hydrocarbon–alcohol–amine	119
Acetylenic alcohol-amine-pyrrole	322
Phenol–amine–pyridine	118
Amine–epoxide–epoxide	133
4-Component Compositions	
Alcohol–phenol–hydrazine–epoxide	235
Acetylenic alcohol–phenol–amine–amine	67
Alcohol–ester–pyridine–epoxide	117

Stabilizers often do not completely prevent a reaction, but reduce the rate; therefore, some decomposition products are usually present. In a vapor degreasing application, if a small amount of hydrochloric acid is produced it can lead to the formation of metal salts. Metal salts such as ferric chloride or aluminum chloride accelerate the breakdown of both tri- and perchloroethylene. To combat this kind of decomposition, two types of compounds are used, acid acceptors and metal deactivators. Those that react with the hydrochloric acid to prevent further reaction are called "acid acceptors." Epoxides may be used for this purpose as illustrated by the following reaction:

$$CH_3CH_2CHCH_2 + Cl^- + H_3O^+ \rightarrow CH_3CH_2CHOHCH_2Cl + H_2O \qquad (13\text{-}15)$$
$$\underset{O}{\diagdown\diagup}$$

A synergistic effect from the combination of certain epoxides and amines increases the effectiveness of the stabilizer system.[77]

The effects of metal salts on the decomposition of trichloroethylene can be largely prevented by alcohols, esters, ethers[172] and sulfur compounds.[277,281] The compounds deactivate the metal salts primarily by complex-formation. In many cases these additives also modify the mechanism of solvent breakdown.[172]

Many applications of trichloroethylene do not impose the severe stresses frequently encountered in vapor-degreasing operations. In others, for example in the preparation of medicines and in hypercritical cold-cleaning uses, extensive stabilization is undesirable. Manufacturers therefore offer several grades of trichloroethylene, stabilized for specific end uses. The labels applied identify these grades as technical (or general-purpose), extraction, C.P., U.S.P., electronic, missile-flushing, freezing-point depressant (for fire-extinguishing fluid); alkaline-stabilized (high-amine) or neutrally stabilized (low-amine) vapor-degreasing grade. Perchloroethylene is available under technical, C.P., U.S.P., dry cleaning, and degreasing-grade labels.

PROPERTIES

Large-volume uses of trichloroethylene and perchloroethylene are dependent mainly on the physical properties of these solvents, e.g., their nonflammability, high solvent power and low water solubility. Their chemical properties chiefly concern solvent stability under use conditions (see under "Stabilization"). The following paragraphs discuss those characteristics which are important to the dominant uses of the two solvents and mention chemical reactions of commercial importance.

Physical Characteristics

Trichloroethylene and perchloroethylene are both colorless, volatile liquids with characteristic odors. Both are miscible with alcohol, ether, and most organic solvents. Both are practically insoluble in water.

Table 13-5 gives selected values for the fundamental physical characteristics of trichloroethylene and perchloroethylene. Figure 13-6 compares the pressure-temperature relationships of each solvent.

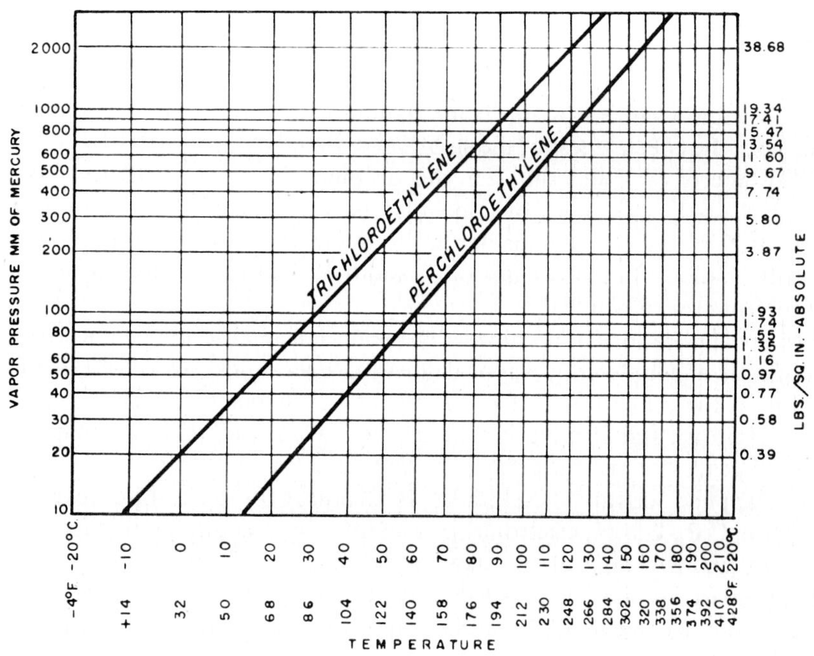

Figure 13-6. Pressure-temperature relationship of trichloroethylene and perchloroethylene.

The high vapor densities, low latent heats, and low viscosities of trichloroethylene and perchloroethylene favor their use in vapor degreasing. Further, the low boiling point of trichloroethylene keeps metal parts from becoming too hot to handle immediately after degreasing. Similarly, the high liquid density, low viscosity, and low solvent capacity for water favor the use of perchloroethylene in dry cleaning. Nonflammability is an important characteristic for each application.

Lecat[189] and Horsley[155] have published extensive compilations of binary azeotropes. They report azeotropes formed by trichloroethylene and by perchloroethylene with various organic acids, alcohols, ethers, halides, and nitro compounds, among others. Both list more azeotropes for perchloroethylene than for trichloroethylene and nearly twice as many non-azeotropic mixtures for trichloroethylene as for perchloroethylene. Azeotropic distillation is a common method for separating the constituents of mixtures and in drying applications. As data on the steam-distillation points show (Table 13-5), perchloroethylene offers a favorable solvent:water ratio in such applications and has the advantage of very low solubility in water.

Chemical Reactions

Under "Stabilization" the oxidation of trichloroethylene and perchloroethylene to dichloro- and trichloroacetyl chloride are mentioned, respectively, as well as the dimerization of trichloroethylene. The oxidative decomposition of both solvents relates to their safe handling, as described under "Safety Factors."

Beilstein[31] and Huntress[157] list numerous references to patents and reports on these and many other reactions of trichloroethylene and perchloroethylene. Four of these reactions having commercial importance are (1) the chlorination of trichloroethylene to pentachloroethane (and subsequent dehydrochlorination to perchloroethylene), mentioned under "Manufacture"; (2) the chlorination of perchloroethylene to hexachloroethane, large volumes of which were produced during World War II for use as a smoke screen ingredient; (3) the fluorination of perchloroethylene to *sym*-difluorotetrachloroethane, "Freon-112" refrigerant; and (4) the hydrofluorination of perchloroethylene to 1,1,2-trifluorotrichloroethane, *sym*-tetrafluorodichloroethane, and pentafluorochloroethane (see Chapter 29, "Freons").

According to Hauptschein and Bigelow, fluorination of trichloroethylene under nitrogen gives an inseparable mixture of chlorofluorohydrocarbons.[145] Recent patent literature describes the preparation of 1,1,1-trifluoro-2-chloroethane[113,254] and of fluoro-1,1,2-trichloroethane[112] by the action of hydrogen fluoride on trichloroethylene in the presence of a metal halide catalyst.

A French process for the oxidation of trichloroethylene[68a] to monochloroacetic acid by 90 percent sulfuric acid at over 190° C reportedly gives a much purer product than that obtainable by conventional chlorination of acetic acid. Although this process has never appeared attractive

TABLE 13-5. PHYSICAL PROPERTIES OF TRI- AND PERCHLOROETHYLENE

	Trichloroethylene	Perchloroethylene
Chemical formula	$CHCl:CCl_2$	$CCl_2:CCl_2$
Molecular weight	131.40	165.85
Boiling point (760 mm), °C	86.9	121.2
°F	188.4	250.2
Freezing point, °C	−86.4	−22.35
°F	−123.5	−8.2
Steam distn. point (1 atm), °C	73.2	87.9
°F	163.8	190.2
Solvent: water ratio, by wt.	13.4 : 1	5.2 : 1
Latent heat of vaporization (B.P.), CHU = cal/g	57.2	50.0
BTU/lb	103.0	90.0
Specific heat, cal/g/°C or BTU/lb/°F, liquid (20°C)	0.225	0.205
vapor (C_p) (1 atm)	0.156 (80°C)	0.154 (100°C)
Critical temperature, °C	271.0	347.1
°F	519.8	656.8
Critical pressure, atm	49.5	46[a]
Thermal conductivity, BTU/hr (sq ft)(°F/ft), liquid (20°C)	0.0685	0.0638
vapor (B.P.)	0.00482	0.00505
Specific gravity (20°/4°C)	1.464	1.623
Pounds per gallon (20°C, 68°F)	12.22	13.55
Coefficient of cubical expansion, Av/°C, liquid	0.00117 (0°–40°C)	0.00103 (0°–25°C)
Vapor Density (B.P., 1 atm), g/liter	4.45	5.13
lb/cu ft	0.278	0.320
Specific gravity (air = 1)	4.54	5.73
Diffusivity in air (25°C, 1 atm), sq cm/sec	0.073	0.067
Evaporation rate (ether = 100)	30	12
Solubility in water (25°C), g/100 g	0.11	0.015
Solubility of water in chlorohydrocarbon (25°C), g water/100 g	0.027	0.0105

Viscosity, centipoises, liquid (20°C)	0.58	0.896
vapor (60°C)	0.0103	0.0099
Refractive index (N_D^{20}), liquid	1.4782	1.5044
vapor (0°C)	1.001784	1.002009
Surface tension (in air), dynes/cm	approx. 29 (30°C)	32.32 (20°C)
Dielectric constant, liquid	3.42 (16°C)	2.30 (25°C)
vapor	—	—
Heat of formation, kg-cal/mole, liquid	+1[a]	+3[a]
vapor	−7[a]	−6[a]
Flammability[b]	nonflammable	nonflammable
Underwriters' Laboratories rating	3	0
Flash point	none	none

[a]Estimated

[b]Trichlorethylene is practically nonflammable and nonexplosive at ordinary temperatures, but at higher temperatures under favorable conditions may form weakly combustible mixtures with air. The rating numbers accord with Underwriters' Laboratories standard of classification in which:

Ether rates	100
Gasoline rates	90–100
Alcohol (ethyl) rates	60–70
Kerosene (100 °F. flash) rates	30–40
Paraffin oil rates	10–20

to U. S. producers, imports of the French product have apparently found a ready market among producers of weed killers, carboxymethylcellulose, and other organic derivatives.[11a]

Also described in the literature is the use of trichloroethylene or perchloroethylene as interpolymer or as reaction medium for polyester and vinyl resins.

USES

Although in many of their applications trichloroethylene and perchloroethylene appear to be interchangeable, the differences in their physical properties have channeled each into specific large-volume uses. Thus, over 90 percent of the trichloroethylene consumed in the United States goes into vapor degreasers for cleaning metals.[5] Nearly 80 percent of the perchloroethylene consumed is for dry-cleaning[11,259] and about 10 percent for vapor degreasing. The following discussion highlights the technical aspects of cleaning and finishing operations for metals and fabrics. Solvent-extraction applications are reviewed and other small-volume uses of trichloroethylene and perchloroethylene are mentioned briefly.

Metal Cleaning

Trichloroethylene and perchloroethylene find wide application in cleaning metal parts between fabricating steps and before finishing or assembly. The physiological properties of the two solvents (see under "Safety Factors") require the exercise of certain precautions in their use, but present-day equipment design and handling techniques make elaborate safety measures unnecessary. Although adaptation of cold-flushing methods is an obvious approach to the use of these solvents in metal cleaning operations, the vapor-degreasing process, because of its inherent versatility and efficiency, has predominated since its introduction in Germany and in England in the late 1920's and in this country in the early 1930's.

Vapor Degreasing. In its simplest form, the vapor-degreasing process consists merely of dipping the oil- and grease-covered work in an atmosphere of trichloroethylene or perchloroethylene vapors. The hot vapors condense immediately on the cold metal, dissolving the oil and grease and flushing them from the work. Within minutes, the work heats to vapor temperature and condensation ceases, leaving the work thoroughly clean, dry, and ready for further treatment.

The only equipment needed for such a process is an open tank with a heating element in the bottom (to boil the trichloroethylene or perchloroethylene) and cooling coils near the top (to contain the vapors within the tank). Because of the high density of the vapors, diffusion into the air

Figure 13-7. Simple vapor degreaser consists of open tank with a heating element in the bottom and cooling coils near the top. The operator lowers work below the coils to clean by condensing vapors.

above the tank is very low in properly designed equipment. Adding the physical action of a liquid spray or dip facilitates the removal of insoluble soil and provides one of the most versatile, rapid, and economical methods available for cleaning metal parts (see Figure 13-8).[262] In special cases where soil is unusually tenacious or deeply imbedded the work is scrubbed by solvent cavitation, induced by ultrasonic waves propagated in the liquid rinse through transducers (usual frequency range, 20 to 50 kc/sec).[39] Regardless of degreasing cycle, the work always receives a final rinse of clean solvent-vapors condensate as it leaves the degreaser.[170]

Embellishments to the vapor degreasing process have been proposed from time to time, e.g., the addition of a glycol or a polyglycol ether to

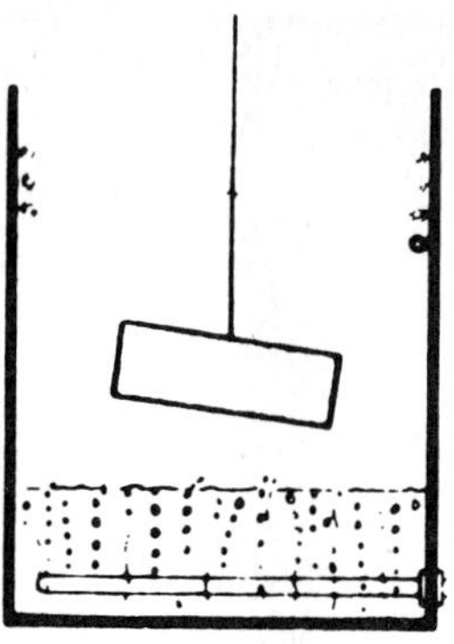

(1) Vapor

The work is suspended in the vapors of boiling solvent. The vapors condense on the cool metal surface. The condensed solvent dissolves the greasy contaminants, and drips back into the degreaser.

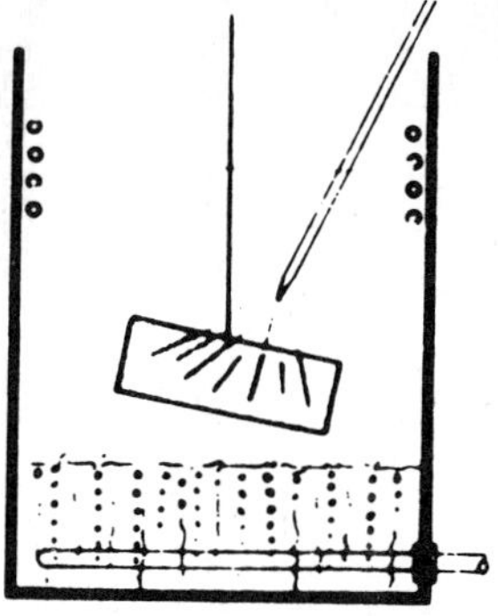

(2) Vapor—Spray—Vapor

The work is suspended in the vapors of boiling solvent. A spray of warm liquid solvent is directed over the work, and after this, the work receives a vapor rinse before removal.

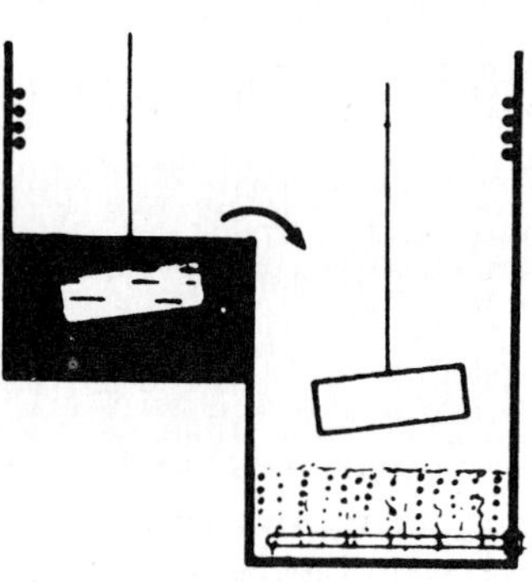

(3) Warm Liquid—Vapor

The work is immersed in warm liquid solvent, then suspended in a bath of solvent vapors.

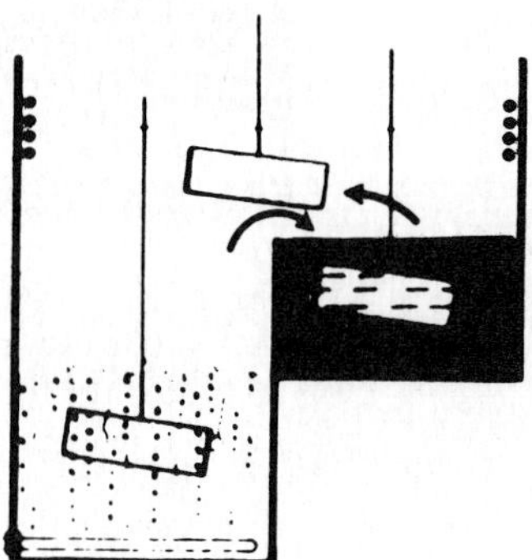

(4) Boiling Liquid—Warm Liquid—Vapor

The work is immersed in boiling solvent to utilize the natural mechanical action. It is then cooled in warm liquid solvent, and finally rinsed with the vapors of the boiling solvent.

Figure 13-8. Vapor degreasing cycles.

the solvent to reduce cleaning time,[225] the addition of cellulose acetate to raise the liquid viscosity,[308] and the suspension of ball-shaped particles of polyvinyl chloride to enhance mechanical action.[188] As a rule, however, the excellent solvent power and comparatively high densities of the fluids themselves ensure adequate cleaning action.

Vapor degreasing competes successfully with alkali- and emulsion-cleaning processes for metals, despite the comparatively high chemical cost of vapor degreasing solvents. Investment and operating costs for vapor degreasing are low[147,170] and recovery of the solvent is normally routine. Losses are through diffusion, drag-out on cleaned work, discard in degreaser and still sludges, and leaks.[14,141] In conveyorized, canopied equipment operating at near-capacity, solvent losses amount to 0.5 to 1 gallon per ton of free-draining work cleaned. Utility requirements are low— 3000 to 5000 BTU and 8 to 10 gal cooling water for each ton of steel cleaned. Solvent and utility requirements of open, manually operated units are about 20 to 45 percent greater.[14]

Figure 13-9. Conveyorized, canopied vapor degreasers use only a gallon or less of solvent, 3000 to 5000 BTU, and 10 gallons or less of cooling water to clean a ton of free-draining steel parts.

With vapor degreasing any common construction metal may be cleaned without danger of etching or staining. Cost of cleaning by vapor degreasing is almost independent of the amount of soil removed. Fragility and complexity of work structure are minor factors, solved by proper racking.[147] Efficiency of soil removal is relatively insensitive to normal variations in operating conditions (work load, changes in type of soil, etc.) so that a major asset of degreasing is the consistent quality of the work processed. This versatility of application has placed vapor degreasing second only to alkali washing in nationwide use for cleaning steel[262] and has made it the preferred process for cleaning aluminum.

Trichloroethylene accounts for over 90 percent of U. S. total vapor-degreasing solvent consumption. Perchloroethylene accounts for most of the balance.[259,262] The higher-boiling solvent is useful in special operations, as in the removal of high-melting waxes or of water from work.[170,262] European reports on "spontaneous decomposition" of trichloroethylene in aluminum-degreasing operations[212,300] do not reflect general U. S. experience where an estimated 75 percent of fabricated-aluminum cleaning operations are by trichloroethylene vapor degreasing. This may be attributable to the types of stabilizers employed in U. S. degreasing grades of trichloroethylene (see under "Manufacture—Stabilization").

Specialty applications of vapor degreasing include the cleaning of electric motors protected by a solvent-resistant insulating varnish or the stripping of electric motor coils which have solvent-sensitive varnishes.[142] Similarly, trichloroethylene vapors can soften and so bond together turns of coil to form a self-supporting electromagnetic coil.[121]

In the cleaning of tank cars and trucks, trichloroethylene vapors from a separate generator piped to the top opening of the car or truck clean the tank of asphalt and heavy oils much more quickly than conventional alkali cleaners.[143] Such a system is also effective in cleaning the interiors of large liquid oxygen tanks for missiles.[9] A variation of this system passes steam through the solvent vaporizer to create mixed vapors of trichloroethylene and water.[321]

Another special use of trichloroethylene vapors is the removal of wax patterns from shells for investment casting.[41,294] A fast photoengraving and photolithographic process[298] uses trichloroethylene vapors to remove unexposed portions of a special photosensitized coating from exposed plates.

In the vapor-drying of metals and glass the use of perchloroethylene, which may contain a surfactant, is preferred.[8,161] Alternatively, a pre-dip in deionized water containing a small amount of surfactant aids water removal without spotting. Canned foods can be sterilized by passing them through vapors of trichloroethylene and/or perchloroethylene.[139,171,286,304]

Cold Cleaning. Trichloroethylene and especially perchloroethylene find use, alone or in compositions, for room-temperature cleaning operations. When used in open tanks, a supernatant layer of water prevents the solvent from evaporating into the atmosphere. The water may also serve as a component of a diphase cleaning system to remove water- and solvent-soluble types of soil. Trichloroethylene-water diphase systems find application in paint-stripping hooks and racks used to carry work through conveyorized painting systems.

If means, such as a ventilated hood or exhausted work-shroud, are available for protecting the worker, cleaning-solvent application can be by liquid flush or aspirated spray. Examples include the cold flushing of rocket motors and the cleaning of electric motors.[142]

The patent literature cites hundreds of cleaning compositions based on trichloroethylene and/or perchloroethylene. These may contain other solvents, water, emulsifying agents, and alkalies. Strong alkalies such as caustic soda (sodium hydroxide) must never be used with trichloroethylene due to their possible reaction to form the highly flammable dichloroacetylene (see under "Safety Factors").

Components selected for quick-drying cleaning mixtures (see Table 13-6) preferably volatilize at approximately equal rates to insure freedom from oily residues and safety in use. Wide variations in volatilities cause composition changes during evaporation that can result in an originally nonflammable mixture becoming gradually more flammable during use (see Figures 13-10 and 13-11). Specialty cleaners, such as carbon removers,[32] usually seek to ensure uniformity of composition through homogenization or emulsification, unless diphase action is desirable.

Metal Finishing

The ability of trichloroethylene to remove oils and greases quickly from metal parts suggests a value as carrier for nonpolar compositions, such as slushing oils and waxes, in fast-drying, rust-preventive formulations. Such formulations have long been applied as special dips which follow the normal degreasing operations. Where the degreaser has a built-in tank, the degreased parts are immersed in the oil or wax dip and withdrawn slowly through the vapor zone to flash off the solvent and so leave the work with a thin uniform layer of rust-preventive, ready for packing or shipping.

Painting. Borushko applied the vapor-dry coating principle to the application of trichloroethylene-soluble resin coatings in 1950 when he patented a coating system resembling a vapor degreaser.[35] Subsequent improvements in equipment design[165,166] and process handling[36,37] have resulted in a nonflammable painting process based on trichloroethylene-thinned paints.

TABLE 13-6. COLD CLEANING COMPOSITIONS—SUGGESTED FORMULAS

Intended Use	Composition				Approx. Extent of Evaporation before Mixture begins to Flash (% by volume)[a]	Mixture No.[b]
	Chlorohydrocarbon	% (vol.)	Flammable Solvent	% (vol.)		
General maintenance (Tools, mechanical equipment, automobile parts, etc.)	Trichloroethylene	100	–	–	100	–
	Trichloroethylene	70	V.M.&P. Nephtha	30	slight	1
	Trichloroethylene	40	Stoddard solvent	60	slight	2
	Perchloroethylene	75	Toluene	25	100	3
	Perchloroethylene	60	Stoddard solvent	40	25	4
Cleaning electrical equipment (Motors, contacts, wiring, etc.)	Methylene chloride and	25	} Stoddard solvent	70	slight - 10	–c
	Perchloroethylene	5				
	Perchloroethylene	60	Stoddard solvent	40	25	4
Textile cleaning ("Spotting")	Trichloroethylene	100	–	–	100	–
	Perchloroethylene	100	–	–	100	–
	Perchloroethylene	60	Stoddard solvent	40	25	4
Paint, varnish, and lacquer removal	Trichloroethylene	90	Methyl ethyl ketone	10	100	5

NOTES: [a]As indicated by flash points of related mixtures, determined by the Cleveland Open Cup method at temperatures up to the boiling point of the mixture.

[b]See Figures 13-10 and 13-11. Nonflammable solvents used in laboratory tests were standard commercial grades. Stoddard solvent: boiling range, $306°-387°F$.; flash point (closed cup), $101°F$. V.M.&P. Naphtha: boiling range, $249°$ to $286°F$; flash point (closed cup), $56°F$. Evaporation rates were determined using solvent-impregnated filter paper; changes in composition using 100-ml samples.

[c]So-called Cleaning Mixture No. 49.

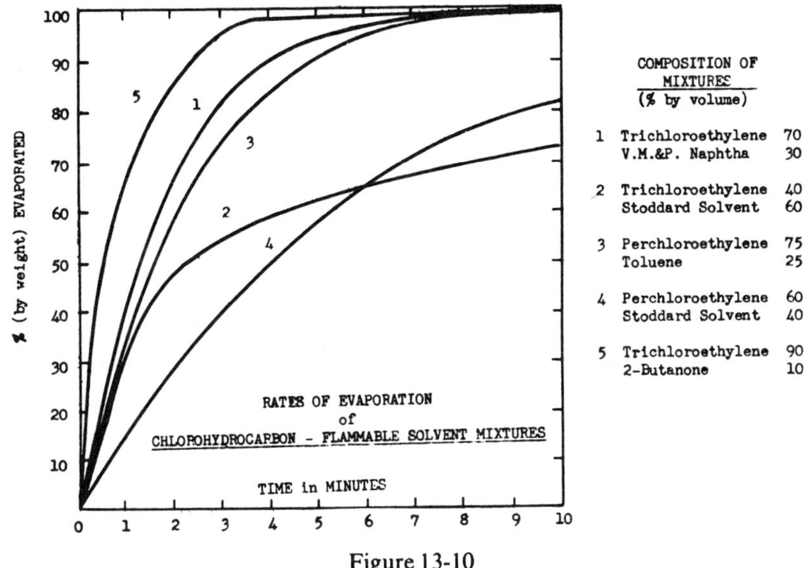

Figure 13-10

In the dip-painting process, the clean work enters a tank of hot paint (temperature generally not lower than 150 to 160° F) where it remains submerged long enough to heat up to the temperature of the paint. As the hot painted work emerges from the liquid, it passes quickly through a concentrated trichloroethylene vapor zone where leveling of surface imperfec-

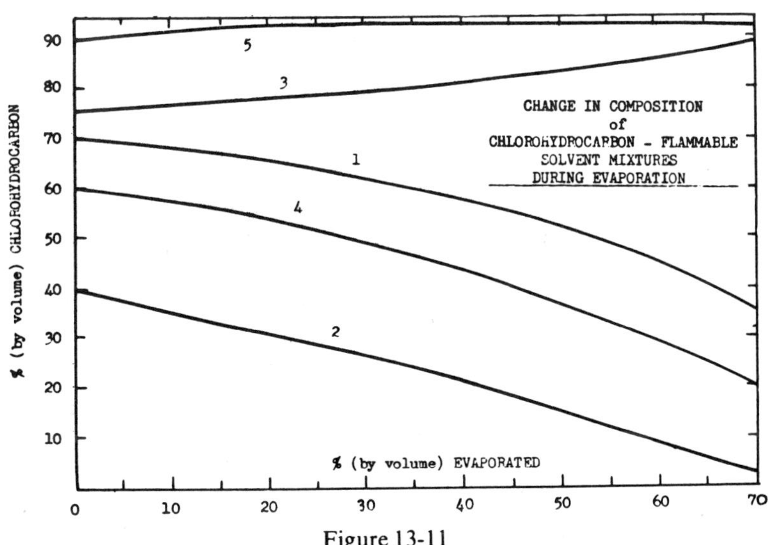

Figure 13-11

tions takes place. The painted work then emerges from the vapor and the trichloroethylene thinner evaporates quickly from the warm work, leaving it dry to the touch. Cooling coils in the paint machine condense the trichloroethylene for reuse.

"Spray-in-vapor" systems using trichloroethylene-thinned paints are also available. The principle applied may be one of either true spraying or of flow-coating paint at a temperature of 190 to 225° F in an air-free atmosphere of trichloroethylene vapor. Work entering the Spray-in-vapor unit passes downward through the controlled trichloroethylene vapor blanket which fills the machine. The vapor condensing on the cooler work continuously heats the part as it advances through the machine. By the time the work reaches the paint-spray section, it is practically at the temperature of the vapor. The paint, which has passed through a preheater, is at a temperature some 10 to 20° F above its normal boiling point as it leaves the spray nozzle. The higher temperature of the paint minimizes solvent-vapor condensation on the work during and immediately after spraying. As the work moves out of the vapor zone, the trichloroethylene thinner evaporates almost instantly, leaving the work dry to the touch.

Paint manufacturers offer trichloroethylene-thinned paints based on most resin types, including alkyd, epoxy ester, phenolic, rosin, acrylic, cellulose ester, and chlorinated rubber, as well as asphaltum. Clear coatings and specialty finishes, such as metalescent types, have also been formulated.

Coupling vapor degreasing with trichloroethylene-thinned painting eliminates the drying step needed before painting alkali-washed work. Since painted work emerges dry from a trichloroethylene system, it need not pass through an oven unless a baked finish is desired. These factors contribute important savings in investment, space, time, operating labor, and utilities costs. Savings in materials are also realized through use of these systems because wedging is avoided, drip loss is eliminated and overspray is recoverable, as is thinner.[258]

Phosphatizing. Phosphatizing processes are used to provide protective coatings for metal products, principally iron and steel. In most applications the phosphate coating serves as a base for paint, lacquer, oils, or waxes. The coatings improve adhesion, restrict moisture penetration, and prevent the spread of underfilm corrosion.

The first important commercial application of phosphatizing stemmed from a patent issued to Coslett covering the treatment of iron and steel parts in an aqueous solution of phosphoric acid and ferrous sulfate.[79] Since then phosphatizing from aqueous solutions has been a recognized part of metal-finishing operations, and numerous proprietary formulations have been devised along similar lines.[253]

The first reported use of an organic solvent for phosphate application is credited to Feidt who cleaned steel with an aqueous alcoholic solution of phosphoric acid.[115] Verner and Wood found that a true phosphate coating could be applied to 'iron and nonferrous metals by treating these with organic solvents containing 1 to 7 percent phosphoric acid. Specific solvents suggested include acetone, methyl alcohol, and combinations thereof with carbon tetrachloride to raise the flash point of the solution.[301] Copelin preferred to dissolve the acid, together with alkyl acid phosphate, in a nonflammable solvent such as trichloroethylene or perchloroethylene to give a solution capable of phosphatizing metals.[76]

A new process, announced by Du Pont in 1959, likewise uses a trichloroethylene solution of phosphatizing chemicals to apply an iron phosphate coat to steel surfaces. Application is by spray or dip in a degreaser-like machine which provides the characteristic trichloroethylene vapor zone over the phosphatizing solution. Coating weights of 40 to 200 or more milligrams per square foot are obtainable in one-half to three minutes. The process is potentially applicable to other active construction metals such as aluminum and zinc.[258]

The availability of phosphatizing and painting processes based on trichloroethylene solutions makes possible fully integrated cleaning-phos-

Figure 13-12. Three-stage system on automobile engine line cleans, phosphatizes, and paints parts in compact equipment, completely hooded to ensure economical operation.

phatizing-painting systems which can clean and finish metal parts in 15 minutes. The principle advantages are reported to stem from the lower initial investment required due to the elimination of rinsing stages, drip zones, and drying ovens, which are necessary parts of conventional finishing systems. The integrated process has a low heat requirement (relative to aqueous cleaning and phosphatizing) and appears to provide special advantages in areas where the local water supply is of variable quality or where the discharge of contaminated waste water is prohibited by local ordinance.

Textile Cleaning

The frequent, disastrous fires which occurred in dry cleaning and textile-processing plants using volatile flammable solvents (benzine) at the turn of the century prompted European investigators to design equipment for the use of carbon tetrachloride and trichloroethylene. Industrial acceptance of the nonflammable chlorohydrocarbons was delayed, however, by their relatively high cost. In this country, so-called synthetic dry-cleaning did not really gain general acceptance until the mid-1930's, after the Du Pont Company introduced perchloroethylene for use in closed dry-cleaning systems which provided for solvent recovery.

In the late 1930's, Roland E. Derby successfully applied synthetic dry cleaning to the processing of open-width woolen piece goods. Several Derby installations are operating in woolen mills today utilizing trichloroethylene and perchloroethylene. Raw-wool scouring with chlorohydrocarbons, on the other hand, has yet to achieve commercial significance, although a number of patents exist—testifying to its potential interest.

The patent literature describes many other textile processing operations using trichloroethylene and perchloroethylene—dry dyeing and special finishing treatments, for example—as well as similar operations for paper and leather products. The techniques employed resemble those described below for textile processing.

Dry Cleaning. Since its introduction in 1933 as a replacement for carbon-tetrachloride and trichloroethylene, perchloroethylene has become the prime chlorinated dry cleaning solvent. It is more stable and less toxic than carbon tetrachloride, and few dyes bleed into it. Trichloroethylene is a solvent for many cellulose acetate dyes, as well as some vat and azoic colors.[33]

A dry cleaning system normally comprises a washer-extractor, a dryer, and a filter and/or still. The filter removes suspended soil by absorption on a coat of diatomaceous earth, over which an active layer of treated clay may serve to remove certain soluble soils such as free fatty acids. The filter also holds small amounts of activated carbon, added to the solvent to absorb coloring matter.

In a system designed for the use of perchloroethylene, the drying tumbler has a condenser and water separator for recovery of the solvent (see Figure 13-13). In early models of perchloroethylene dry cleaning systems the recovery unit was built into the washer-extractor so the operator would not need to transfer solvent-wet garments. With a separate recovery-tumbler, however, a given washer-extractor has a greater hourly capacity,

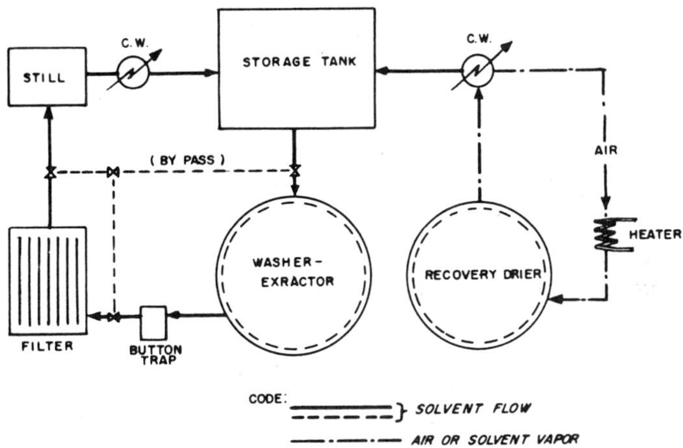

Figure 13-13. Perchloroethylene flow in conventional dry cleaning systems.

and this is now the customary arrangement. Actual air tests in many dry cleaning plants have shown that the transfer of properly extracted garments from a perchloroethylene washer-extractor to an adjacent recovery tumbler creates in the immediate area an atmospheric concentration of only 50 to 60 parts perchloroethylene vapor in a million parts of air.

Table 13-7 illustrates the great savings realized by the recovery-tumbler. With special devices permitting the recovery of solvent from filter sludges (e.g., adaption of a heating jacket and condenser to a sub-filter or provision for steam-stripping sludges and condensing the solvent),[63] further appreciable savings result.

Late in 1957, the Vic Manufacturing Company introduced a carbon adsorptive system for the recovery of vapor losses from around the equipment and from the vents.[302] Several such systems are now on the market. According to claim, they can almost double the "mileage" (that is, the amount of cleaning per drum of perchloroethylene) of even the most efficient standard operation. Experience is not yet sufficiently widespread, however, to establish the over-all economics of such systems.[259]

TABLE 13-7. AVERAGE SOLVENT LOSSES IN PERCHLOROETHYLENE—DRY CLEANING SYSTEMS

(lb solvent/100 lb work)

Source of Loss	Without Recovery— Drying	With Recovery— Drying No Sludge Drier	Sludge Drier
Garments	21.5	3.5	3.5
Filter sludge	5.0	5.0	2.0
Still residue	0.3	0.3	0.3
Still vent	0.2	0.2	0.2
Tank vent and misc.	0.5	0.5	0.5
Leaks	0.5	0.5	0.5
Total	28.0	10.0	7.0
Pounds of garments cleaned/700-lb drum of perchloroethylene	2500	7000	10000

Perchloroethylene readily removes fatty and oily stains from fabrics, but is a poor solvent for sugars, salts, and other water-soluble soil. The low solubility of water in perchloroethylene (0.015 g in 10 cc at 40° C)[207] prohibits the addition of moisture to remove water-soluble soil because free water may cause wrinkling, shrinking, or pilling of fabrics.[10] Removal of water-soluble stains by special handling ("spotting" or "wet-cleaning") is expensive. Palit and Venkateswarlu found that the addition of 1 g of a 1 to 1 mixture of dodecylammonium formate and dodecylammonium oleate solubilized 1.5 g water in 10 cc of perchloroethylene at 40° C.[234] This principle of using 1 to 4 percent concentrations of mixed hydrophilic and lipophilic detergents with moisture in perchloroethylene to facilitate soil removal in the dry-cleaning unit is common practice in the industry.[6,152] Means are available for controlling the detergent and the moisture contents.[10]

Dry cleaning plants also apply water-repellents, e.g., chlorinated wax compositions, and moth-repellents, e.g., dichlorodiphenyltrichloroethane (DDT), from perchloroethylene solutions.[6]

Wool Scouring. Hey emphasized the non-injurious cleansing action (freedom from shrinkage, felting, or alkali damage) of trichloroethylene on wool in his discussion of the advantage of solvent-degreasing woolen fabrics and yarns. He also cited the possibility of recovering both solvent and oils.[151] These advantages are fully realizable in the solvent-scouring of woolen piece goods by the Derby process, but apparently existing raw-wool scouring processes require further refinement in the handling of the wool grease and accompanying impurities.

Woolen Piece Goods. Essentially, the Derby continuous dry-cleaning machine[91,92] consists of a compartmented solvent bowl, a drying chamber, and a solvent-recovery system. Open-width worsted or woolen crepes from the loom, for example, pass counter to the flow of trichloroethylene or perchloroethylene (temperature optional) through the bowl, then over vacuum slots to the drying chamber. The cleaned, dried fabric is ready for dyeing.[15,88,89] Solvent vapors from the bowl and drying chamber are condensed for reuse. The spent solvent is recovered by distillation. Where

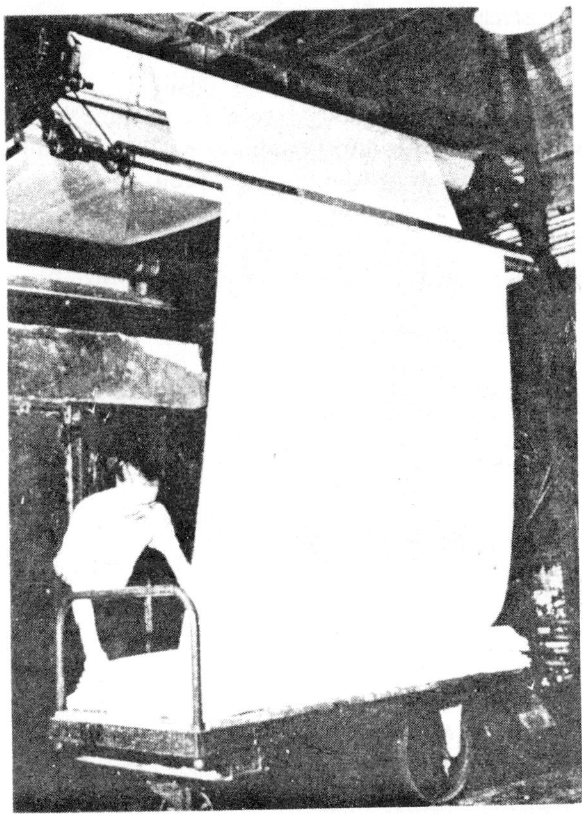

Figure 13-14. Woolen piece goods cleaned and degreased with trichloroethylene or perchloroethylene in the Derby dry cleaner are ready for dyeing. Processing oils removed are recoverable.

necessary, a preliminary treatment with triethanolamine or other alkaline agent will partially saponify the fats and coagulate the dirt and fatty matter. Decanting the clarified liquid and distilling leaves a reusable oil.[90]

A British process degreases the wool at or near the boiling point of trichloroethylene or perchloroethylene, then flashes off the solvent by immersing the fabric in water at the solvent-water steam-distillation point.[312]

The process is also applicable to the removal of waxy and oily matter from cotton and the like.[318,319] The degreased and dewaxed goods can be treated with an aqueous alkaline solution of a peroxide for bleaching and desizing prior to dyeing.[2,313–316] A hypochlorite bleach step can also be incorporated in the treatment.[80,317,320]

Raw Wool. In the early 1920's, Koch[179] and Coghlan[62] obtained patents covering the trichloroethylene-degreasing of raw wool. The Coghlan process progressively degreased the wool in a series of autoclaves in which the solvent was at a temperature of 45°C. The trichloroethylene was subsequently separated from the wool by distillation. This fundamental sequence is used in the Hoffman[154] and Smith, Drum[103] processes which adapt mechanical means for impelling the wool through the scouring and drying chambers to provide continuous operation.

A process patented by Norman seeks to control the grease content of perchloroethylene-scoured wool and to improve the quality of the wool grease obtained. This is accomplished by separating the liquid extract into a perchloroethylene-rich portion, which returns to the extractor, and a grease-rich portion, which is distilled.[232]

Solvent Extraction

An early application of trichloroethylene, the first large-scale use in this country, was as an extraction solvent. Trichloroethylene is an excellent solvent for alkaloids, natural oils and fats, and many waxes and resins. Its moderate boiling point and immiscibility with water make trichloroethylene an ideal extraction solvent. Perchloroethylene is less versatile in such applications, owing to its high boiling point; but it does find application in extracting oil from wet products such as animal wastes.

Three products have dominated the history of trichloroethylene as an extractant—caffeine, olive oil, and soybean oil. These extraction operations are outlined below, following which brief mention is made of the trichloroethylene extraction of waxes.

Caffeine. Trichloroethylene is a unique solvent in processes for the separation of caffeine from coffee beans. The first decaffeinated coffee was produced in Germany about 1903. The process consisted of extracting steamed coffee beans with trichloroethylene until their caffeine content was reduced to 0.03 percent. Residual solvent was removed from the extracted beans by steaming, and the beans were dried and roasted. Steam distillation of the trichloroethylene extract left a mixture containing 60 percent caffeine and 40 percent waste, largely wax.[309]

The patent literature covers numerous modifications of this process. Present day practice is to extract roasted coffee beans in aqueous medium.

Caffeine is then recovered from the water phase by extraction with trichloroethylene.[21,140]

Olive Oil. The trichloroethylene extraction of olive oil from press cakes became fairly widespread in southern Europe in the early 1920's. Trichloroethylene replaced carbon disulfide for this operation, and reduced the problems of fire and corrosion while improving the quality of the oil.[182,213]

The olive press cakes are hot-air dried in rotating drums to a water content of about 15 percent, then charged to extractors where they are treated countercurrently with trichloroethylene at about 40° C. The extract from the final stage contains 40 to 50 percent oil. Solvent is recovered from the extracted cakes and from the oil-trichloroethylene solution by steam distillation. The oil, after treatment with bleaching earth, is satisfactory for use in the manufacture of soap or as fuel oil. Vacuum distillation with superheated steam gives an edible oil.

A similar process for recovering grapeseed oil from pressings is used at some European wineries.[34,202]

Soybean Oil. Trichloroethylene has been used to extract oil from crushed soybeans. However, the resulting meal is unsuitable for use as an animal feed supplement because it can cause a hemorrhagic disease in cattle.[262] Since the disease first appeared in this country in 1951–2, much research has been devoted to identifying the elusive toxic factor. The studies have shown that trichloroethylene is capable of reacting with the sulfhydryl groups of cysteine and reduced glutathione to form S-dichlorovinyl derivatives of L-cysteine and L-glutathione. These derivatives produce in calves the aplastic anemia syndrom typical of that produced by trichloroethylene-extracted soybean oil meal.[209] Dichloroethylenes and perchloroethylene form similar cysteine thioethers, but the trichloroethylene-derived S-(*trans*-dichlorovinyl)-L-cysteine appears to have unique properties.[208]

Trichloroethylene has also been applied to the extraction of cottonseed flakes,[18] fish offal,[19] and meat scrap,[262] but the finding that toxic properties can develop under certain conditions in trichloroethylene-extracted meat scrap[247] has seriously curtailed the commercial usefulness of the trichloroethylene processes.

A recently developed German process uses perchloroethylene to extract raw animal wastes.[1,287] Perchloroethylene-extracted products are apparently less toxic than trichloroethylene-extracted products.[208,210]

Waxes. Trichloroethylene and perchloroethylene are used to a limited extent in wax processing for the recovery of wax values from waste products such as used waxed paper[228] and spent filter cakes. They are also applied in coating operations, particularly the wax coating of paper, leather, and metals.

Miscellaneous

The literature suggests many other uses for trichloroethylene and for perchloroethylene, but few of these are of major commercial importance. Some of the more significant minor applications are cited below.

Trichloroethylene is a good heat-transfer medium in the temperature range –73 to +120° C (–100 to +250° F), and it is used in installations for testing materials under conditions simulating outdoor temperature ranges.

Both trichloroethylene and perchloroethylene are good coolants for cutting tantalum and similar space-age metals.

The excellent solvent properties of trichloroethylene and perchloroethylene for rubber are useful in the compounding of cements. As compared with toluene, trichloroethylene in adhesives based on polyvinyl acetate increases their bite, particularly on waxed paper; raises their viscosity; and improves film coalescence.

Both trichloroethylene and perchloroethylene are useful as silicone mold releases and for cleaning isocyanate resin-forming equipment.

High-pressure lubricants offer a limited use for trichloroethylene as a volatile thinner. Trichloroethylene is a good freezing-point depressant for carbon tetrachloride fire-extinguishing fluid. Numerous other compositions, e.g., metal cleaners and polishes, textile-spotting fluids, sealants, leather dubbings, paint removers, insecticides and fungicides, include trichloroethylene and/or perchloroethylene.

In chemical syntheses, both trichloroethylene and perchloroethylene are reactants in the manufacture of other halohydrocarbons (see under "Properties-Chemical Reactions"). Trichloroethylene and perchloroethylene have also found use in dehydrating and purifying chemical products.

Trichloroethylene of the U. S. P. grade finds use as an anesthetic and analgesic; perchloroethylene, as an anthelmintic. Considerable medical and veterinary literature exists on these applications.

SAFETY FACTORS

Trichloroethylene and perchloroethylene are used extensively in industry because of their safety in handling. Both are classed as nonflammable under normal use conditions. Neither trichloroethylene nor perchloroethylene produce a flash by either the closed-cup or open-cup method. However, trichloroethylene vapors may ignite under specific conditions. For example, if the concentration in the air is allowed to exceed 15 percent, the vapor may burn if exposed to an intense source of ignition (see Underwriters' Laboratories Classification, MH 494).[295]   Therefore, welding operations should never be carried out on equipment containing trichloroethylene liquid or vapor.

The decomposition products of trichloroethylene and perchloroethylene are to be avoided since they contain harmful chlorides; e.g., phosgene and hydrogen chloride.   Thus trichloroethylene and perchloroethylene should not be exposed to sources of high temperatures (above 300° C) such as open flames and electric heaters.   An extensive study of the thermal decomposition of chlorinated solvent vapors in air has been made by Sjöberg.[264]

Toxicity. Laboratory data and industrial experience indicate that trichloroethylene and perchloroethylene are among the least toxic of the chlorohydrocarbon solvents.   However, as is the case with all solvents, inhalation of large quantities should be avoided.   Trichloroethylene is an anesthetic, and prolonged exposure to high concentrations may produce nausea, dizziness, headaches, and ultimately unconsciousness or even death.   In small amounts, however, it is rapidly eliminated from the body, and the usual symptoms of overexposure disappear quickly on access to fresh air, with no permenent effects on the heart, liver or kidneys.   The clinical effects of chlorinated solvents have been summarized by Smyth.[273] The American Industrial Hygiene Association has recommended a maximum atmospheric concentration of trichloroethylene§ or perchloroethylene ‖ of 200 parts (by volume) of solvent per million parts of air for an 8-hour working day.   The average person is usually able to detect these solvents in the atmosphere at 50 ppm.

Contact with the skin should be avoided since absorption through the pores may produce the same physiological effects as inhalation.   Excessive dryness and cracking of the skin will also be produced since the solvents

---

§"RECOMMENDED MAXIMUM ATMOSPHERIC CONCENTRATION (8 hours): 200 parts of vapor per million parts of air, by volume (ppm). This level is currently believed to represent a maximum that should not be exceeded for repeated daily exposure; it is not to be used as an integrated average concentration.

"(An average concentration of 100 ppm is believed, in the United States, to be without any probable risk.) . . ." —AIHA Hygienic Guide Series, Trichloroethylene, March 1956.

‖"RECOMMENDED MAXIMUM ATMOSPHERIC CONCENTRATION (8 hours): 200 parts of vapor per million parts of air, by volume (ppm). This level is believed to represent a maximum, which should not be exceeded in repeated daily exposure; it is not to be used as an integrated average concentration. A level of 100 ppm would be more suitable for the latter." —AIHA Hygienic Guide Series, Tetrachloroethylene (Perchlorethylene), June 1960.

extract the normal skin oils. Particular care should be taken to avoid contact with eyes due to the sensitiveness of these organs.

Safe Handling

The Manufacturing Chemists' Association has published safety data sheets covering the safe handling of trichloroethylene[199] and perchloroethylene.[200] The precautions recommended are as follows:
    (a) Care to avoid leaks and spills:
    (b) Immediate mop-up of spills;
    (c) Out-of-door drying of solvent-wet rags;
    (d) Proper handling of tank cars;
    (e) Care to avoid over-exposure and physical injury during the cleaning and repair of tanks and equipment;
    (f) Care to avoid thermal decomposition of the solvent by hot metal surfaces, heaters, and open flames;
    (g) Care to provide adequate ventilation of the work area.
Strong alkalies, such as caustic soda, may react with trichloroethylene to form explosive mixtures (dichloroacetylene). Weaker alkalies, such as trisodium phosphate or sodium carbonate, may be used with safety.

Control Tests

Proper control of vapor concentrations is of obvious importance. It is possible to estimate concentrations in the range of 0 to 500 parts trichloroethylene and perchloroethylene vapors in a million parts of air with an accuracy of 10 percent, using commercially available photometric equipment. These instruments measure the intensity of the blue lines of a spectrum produced by an arc passing between copper electrodes in a chamber through which passes the atmospheric sample.[214] Halide leak detectors based on the Beilstein test, in which the green-to-blue color of a flame impinging on copper indicates the approximate chlorohydrocarbon vapor content of the air, are useful for examining equipment in areas free from flammable solvents or vapors.

Low-cost gas detector kits are also available. These consist of a carrying case, a number of sealed indicator tubes, and a hand pump or rubber bulb for drawing air through a tube from which the tips have been broken.[129] Gage has described a "home-made" gas detector with two tubes in series, one filled with permanganate-impregnated silica gel to oxidize the chlorohydrocarbon vapors in the aspirated air, the other with *o*-tolidine hydrochloride to measure the chlorine liberated.[127] Calibration charts for each chlorohydrocarbon make direct readings possible.

Many industrial hygienists have recommended routine control tests on workmen handling trichloroethylene. The tests would measure the metabolites formed by the inhaled vapor, specifically trichloroacetic acid and trichloroethanol. The Fujiwara pyridine-alkali reaction is adaptable to such use.[17,22,137,261]

References

1. *Brit. Chem. Eng.,* **2,** 185 (April 1957).
2. *Chem. Age (London),* **81** (2069), 403 (1959).
3. *Chem. Eng.,* **64** (3), 152, 154 (1957).
4. *Chem. Eng.,* **68** (1), 52, 54 (1961).
5. *Chem. Eng. News,* **38** (14), 40 (1960).
6. *Chem. Week,* **83** (9), 61 (1958).
7. *Ind. Marketing,* **45** (13), 56 (1960).
8. *Metal Finishing J.* (London) [N. S.] **4,** 386 (1958).
9. *Metal Progr.,* **77** (2), 88 (1960).
10. *Natl. Cleaner Dyer,* **49** (3), 109 (1958).
11. *Natl. Cleaner Dyer,* **51** (2), 110 (1960).
11a. *Oil, Paint and Drug Reporter,* **179** (3), 3, 38 (1961).
12. *Petrol. Refiner,* **38** (11), 283 (1959).
13. *Petrol. Refiner,* **38** (11), 297 (1959).
14. *Steel,* **138** (25), 184 (1956).
15. *Textile World,* **94** (10), 94 (1944).
16. Aitchison, A. G., and Petering, W. H. (to Westvaco Chlorine Products Corp.), U. S. Patent 2,371,645 (1945).
17. Andersson, A., *Acta Med. Scand.,* **157,** Supp. 323 (1957); *A. M. A. Arch. Ind. Health,* **18,** 335 (1958).
18. Arnold, L. K., *J. Am. Oil Chemists' Soc.,* **32,** 151 (1955).
19. Arnold, L. K., *J. Am. Oil Chemists' Soc.,* **32,** 163 (1955).
20. Baldridge, J. R. (to Diamond Alkali Co.), British Patent 795,891 (1958); *C. A.,* **53,** 290 (1959).
21. Barch, W. E. (to Standard Brands, Inc.), U. S. Patent 2,817,588 (1957).
22. Bardoděj, Z., and Vyskocil, J., *A. M. A. Arch. Ind. Health,* **13,** 581 (1956).
23. Barton, D. H. R., *J. Chem. Soc. (London),* **1949,** 148.
24. Barton, D. H. R., and Howlett, K. E., *J. Chem. Soc. (London),* **1951,** 2033.
25. Basel, G., and Schaeffer, E. (to Alex. Wacker Ges fuer elektrochemische Industrie, G.m.b.H.), U. S. Patent 2,139,219 (1938); *C. A.,* **33,** 2151 (1939).
26. Basel, G., and Schaeffer, E. (to Alex. Wacker Ges. fuer elektrochemische Industrie, G.m.b.H.), U. S. Patent 2,158,213 (1939).
27. Basel, G., and Schaeffer, E. (to Alex. Wacker Ges. fuer elektrochemische Industrie, G.m.b.H.), U. S. Patent 2,178,622 (1939).
28. Basel, G., and Schaeffer, E. (to Alex. Wacker Ges. fuer elektrochemische Industrie, G.m.b.H.), U. S. Patent 2,222,931 (1940).
29. Basel, G., and Schaeffer, E. (to Alex. Wacker Ges. fuer elektrochemische Industrie, G.m.b.H.), U. S. Patent 2,255,752 (1941); *C. A.,* **36,** 100 (1942).
30. Beckers, N. L. (to Diamond Alkali Co.), U. S. Patent 2,903,488 (1959).

31. Beilstein, "Handbuch der Organischen Chemie," 4th Ed., Berlin, Springer, Vol. I (1918) p. 187; 1st Supp. (1928), p. 79; 2nd Supp. (1941), p. 160; 3rd Supp. (1958), pp. 661, 667.

32. Berkeley, B., and Schoenholz, D., *Soap Chem. Specialties,* **32** (9), 47, 173 (1956).

33. Bird, C. L., *J. Soc. Dyers Colourists,* **49,** 379 (1933); *C. A.,* **28,** 6569 (1934).

34. Bonnet, J., *Oil Colour Trades J.,* **64,** 829 (1923).

35. Borushko, M. J. (to Harding Manufacturing Co.), U. S. Patent 2,515,489 (1950).

36. Borushko, M. J. (to Harding Manufacturing Co.), U. S. Patent 2,728,686 (1955).

37. Borushko, M. J. (to Harding Chemical Corp.), U. S. Patent 2,783,165 (1957).

38. Bozel-Maletra (Soc. industrielle de produits chimiques), French Patent 715,421 (1930).

39. Branson, N. G., *Materials in Design Eng.,* **47** (2), 118 (1958).

40. Breitenbach, J. W., Schindler, A., and Pfug, Ch., *Monatsh.,* **81,** 21 (1950); *C. A.,* **44,** 8693 (1950).

41. Broad, E. M., *Steel,* **144** (4), 72 (1959).

42. Brown, D., Colton, J. W., and Landau, R. (to Chempatents, Inc.), U. S. Patent 2,746,998 (1956).

43. Brown, T. E., and Davis, C. W. (to Dow Chemical Co.), U. S. Patent 2,377,669 (1945).

44. Burch, R. J., and Leeds, M. W. (to Air Reduction Co., Inc.), U. S. Patent 2,841,625 (1958); *C. A.,* **52,** 18965 (1958).

45. Burch, R. J., and Leeds, M. W. (to Air Reduction Co., Inc.), U. S. Patent 2,945,895 (1960).

46. Campbell, D. H. (to Hooker Chemical Corp.), Canadian Patent 606,048 (1960).

47. Carlisle, P. J. (to E. I. du Pont de Nemours & Co., Inc.), U. S. Patent 1,996,717 (1935); *C. A.,* **79,** 3353 (1935).

48. Carlisle, P. J., and Harris, C. R. (to E. I. du Pont de Nemours & Co., Inc.), U. S. Patent 2,008,680 (1935).

49. Carlisle, P. J., and Levine, A. A., *Ind. Eng. Chem.,* **24,** 1164 (1932).

50. Carpenter, G. B., "Chlorinated Hydrocarbons from Acetylene," U. S. Dept. Commerce, Office Tech. Services Rept. P.B. 75,813 (1947).

51. Cass, O. W. (to E. I. du Pont de Nemours & Co., Inc.), U. S. Patent 2,308,489 (1943); *C. A.,* **37,** 3767 (1943).

52. Cass, O. W. (to E. I. du Pont de Nemours & Co., Inc.), U. S. Patent 2,327,174 (1943); *C. A.,* **38,** 549 (1944).

53. Cass, O. W. (to E. I. du Pont de Nemours & Co., Inc.), U. S. Patent 2,342,100 (1944).

54. Cass, O. W. (to E. I. du Pont de Nemours & Co., Inc.), U. S. Patent 2,374,923 (1945); *C. A.,* **39,** 3534 (1945).

55. Cass, O. W. (to E. I. du Pont de Nemours & Co., Inc.), U. S. Patent 2,379,414 (1945); *C. A.,* **39,** 4617 (1945).

56. Chemische Fabrik Buckau, German Patent 274,782 (1912); *Chem. Zentr.,* **1914** (II), 95.

57. Chemische Fabrik Griesheim-Elektron, German Patent 263,457 (1912); *Chem. Zentr.*, **1913** (II), 830.

58. Chemische-Werke Huels A.-G. (F. Kruell, R. Baumbach, O. Nitzschke, and W. Krummer, inventors), German Patent 1,074,025 (1960).

59. Chilton, C. H., *Chem. Eng.* **65** (9), 116 (1958).

60. Chloberag Chlorbetrieb Rheinfelden A.-G. (E. Feder, inventor), German Patent 868,294 (1953); *C. A.*, **52**, 18213 (1958).

61. Choporov, Ya. P, Tishchenko, O. A., and Apalenova, V. Ya., U.S.S.R. Patent 107,335 (1957); *C. A.*, **52**, 2885 (1958).

62. Coghlan, T. A., U. S. Patent 1,478,203 (1923); *C. A.*, **18**, 756 (1924).

63. Cohen, M., U. S. Patent 2,702,433 (1955).

64. Cole, G. E., Jr., (to Air Reduction Co., Inc.), U. S. Patent 2,719,181 (1955).

65. Cole, G. E., Jr., (to Air Reduction Co., Inc.), U. S. Patent 2,802,885 (1957).

66. Cole, G. E., Jr., (to Air Reduction Co., Inc.), U. S. Patent 2,802,886 (1957).

67. Columbia-Southern Chemical Corp., British Patent 787,726 (1957); *C. A.*, **52**, 10141 (1958).

68. Columbia-Southern Chemical Corp., British Patent 793,205 (1958); *C. A.*, **53**, 1142 (1959).

68a. Compagnie des Produits Chimiques d'Alais et de la Camargue, Austrian Patent 89,199 (1922); *Chem. Zentr.*, **1923** (IV), 591.

69. Conrad, M.F. (to Ethyl Corp.), U.S. Patent 2,725,412 (1955); *C.A.*, **52**, 2884 (1958).

70. Consortium fuer elektrochemische Industrie Ges., British Patent 302,321 (1929); *C.A.*, **23**, 4231 (1929).

71. Consortium fuer elektrochemische Industrie Ges., German Patent 154,657 (1903); Nieuwland, J. A., and Vogt, R. R., "The Chemistry of Acetylene", (A.C.S. Monograph No. 99), pp. 101–5 (text); 107–9 (bibliography), New York, Reinhold Publishing Corp., 1945.

72. Consortium fuer elektrochemische Industrie Ges., German Patent 171,900 (1905); *Chem. Zentr.*, **1906** (II), 571.

73. Consortium fuer elektrochemische Ind. G.m.b.H. (M. Mugdan and J. Sixt, inventors), German Patent 659,434 (1938); *C. A.*, **32**, 5857 (1938).

74. Consortium fuer elektrochemische Ind. G.m.b.H. (J. Wimmer, inventor), German Patent 896,346 (1953).

75. Cooper, R. S. (to Diamond Alkali Co.), U. S. Patent 2,621,215 (1952); *C. A.*, **47**, 10548 (1953).

76. Copelin, H. B. (to E. I. du Pont de Nemours & Co., Inc.), U. S. Patent 2,789,070 (1957).

77. Copelin, H. B. (to E. I. du Pont de Nemours & Co., Inc.), U. S. Patent 2,797,250 (1957); *C. A.*, **51**, 15109 (1957).

78. Copelin, H. B. (to E. I. du Pont de Nemours & Co., Inc.), U. S. Patent 2,904,600 (1959).

79. Coslett, T. W., U. S. Patent 870,937 (1907).

80. Crowder, N. F., and White, W. A. S. (to Imperial Chemical Industries Ltd.), British Patent 806,536 (1958); *C. A.*, **53**, 8649 (1959).

81. Dangelmajer, C. (to Roessler & Hasslacher Chemical Co.), U. S. Patent 1,816,895 (1931); *C. A.*, **25**, 5436 (1931).

82. Daras, N. (to Solvay et Cie.), U. S. Patent 2,935,537 (1960).

83. Daras, N., and Ryckaert, A. L. (to Solvay et Cie.), U. S. Patent 2,966,524 (1960).

84. Davis, C. W., Dirstine, P. H., and Brown, W. E. (to Dow Chemical Co.), U. S. Patent 2,442,323 (1948); *C. A.*, **42**, 6373 (1948).

85. Decker, R., and Holz, H. (J. Smidt and B. Kellner, inventors), German Patent 806,456 (1951); *Chem. Zentr.*, **1951** (II), 3234.

86. Decker, R., and Holz, H. (W. Fritz and E. Schaeffer, inventors), German Patent 846,847 (1952); *C. A.*, **48**, 1406 (1954).

87. Denny, P. W., and Imperial Chemical Industries Ltd., British Patent 391,156 (1933); *C. A.*, **27**, 4892 (1933).

88. Derby, R. E., *Am. Dyestuff Repr.*, **28** (18), *Proc. Am. Assoc. Textile Chemists Colorists,* P520–9, 537–8 (1939); *C. A.*, **33**, 9003 (1939).

89. Derby, R. E., *Can. Textile J.*, **56** (14), 37 (1939); *C. A.*, **33**, 6605 (1939).

90. Derby, R. E., U. S. Patent 2,607,786 (1952); *C. A.*, **46**, 11700 (1952).

91. Derby, R. E. (to M. T. Stevens & Sons Co.), U. S. Patent 2,176,705 (1939); *C. A.*, **34**, 1196 (1940).

92. Derby, R. E. (to M. T. Stevens & Sons Co.), U. S. Patent 2,176,706 (1939); *C. A.*, **34**, 1195 (1940).

93. Dial, W. R. (to Columbia Southern Chemical Corp.), U. S. Patent 2,781,406 (1957); *C. A.*, **51**, 11676 (1957).

94. Diamond Alkali Co., British Patent 673,565 (1952); *C. A.*, **47**, 3867 (1953).

95. Diamond Alkali Co. (H. S. Weiner, inventor), British Patent 832,425 (1960).

96. Dickinson, R. G., and Leermakers, J. A., *J. Am. Chem. Soc.*, **54**, 3852 (1932).

97. Dinley, C. F. (to Bell, J. H.), U. S. Patent 2,096,735 (1937); *C. A.*, **32**, 107 (1938).

98. Dinley, C. F. (to Bell, J. H.), U. S. Patent 2,096,736 (1937); *C. A.*, **32**, 107 (1938).

99. Dinley, C. F. (to Bell, J. H.), U. S. Patent 2,096,737 (1937); *C. A.*, **32**, 107 (1938).

100. Donau Chemie A.-G., Austrian Patent 162,898 (1949).

101. Donau Chemie A.-G., (O. Fruhwirth, inventor) Austrian Patent 166,909 (1950); *C. A.*, **47**, 141 (1953).

102. Donau Chemie A.-G., Belgian Patent 447,044 (1942).

103. Drum, H. S., and Hopkins, A. (to Smith, Drum and Co.), U. S. Patent 2,593,422 (1952); Norman, D. P., and Johnson, W. W. A., *Am. Dyestuff Repr.*, **44**, 893 (1955).

104. E. I. du Pont de Nemours & Co., Inc., British Patent 575,530 (1946); *C. A.*, **41**, 6891 (1947).

105. Edward, J. W., and Rapp, D. E. (to Dow Chemical Co.), U. S. Patent 2,878,297 (1958); *C. A.*, **53**, 22620 (1959).

106. Eisenlohr, D. H. (to Columbia-Southern Chemical Corp.), U. S. Patent 2,859,254 (1958).

107. Eisenlohr, D. H., and Shelton, R. D. (to Columbia-Southern Chemical Corp.), Canadian Patent 598,192 (1960).

108. Ellis, W. C., Jr., and Leeds, M. W. (to Air Reduction Co., Inc.), Canadian Patent 569,992 (1959).

109. Ellsworth, A. C., and Vancamp, R. M. (to Columbia-Southern Chemical Corp.), U. S. Patent 2,951,103 (1960).

110. Erdman, E., *J. prakt. Chem.*, **85,** 78; **86,** 111 (1912).

111. Faraday, M., *Ann. chim.*, (2), **18,** 48 (1821).

112. Farbenfabriken Bayer (J. Soell, inventor), German Patent 808,832 (1951).

113. Farbwerke Hoechst A.-G. vormals Meister Lucius und Bruening, French Patent 1,074,320 (1954).

114. Feathers, R. E., and Rogerson, R. H. (to Columbia-Southern Chemical Corp.), U. S. Patent 2,914,575 (1959).

115. Feidt, G. D., U. S. Patent 1,109,670 (1914); *C. A.,* **8,** 3550 (1914).

116. Ferri, A., and Patron, G. (to Sicedison, S.p.A.), Canadian Patent 596,215 (1960).

117. Ferri, A., and Patron, G. (to Sicedison, S.p.A.), U. S. Patent 2,887,516 (1959).

118. Ferri, A., and Patron, G. (to Sicedison, S.p.A.), U. S. Patent 2,906,782 (1959).

119. Ferri, A., and Patron, A., (to Sicedison, S.p.A.), U. S. Patent 2,910,512 (1959).

120. Fischer, E., *Jahresber.,* **1864,** 481.

121. Flynn, E. J. (to General Electric Co.), U. S. Patent 2,814,581 (1957).

122. Foerst, W., "Ullmanns Encyclopaedie der technischen Chemie," 3rd Ed., Vol. 5, p. 433, Munich, Urban & Schwarzenberg, 1954.

123. Foster, R. T., and Frankish, S. W. (to Imperial Chemical Industries Ltd.), British Patent 697,482 (1953); *C. A.,* **49,** 1770 (1955).

124. Foster, R. T., Hawkins, P. A., and Imperial Chemical Industries Ltd., British Patent 671,947 (1952); *C. A.,* **47,** 3866 (1953).

125. Frankel, D. M., Johnson, C. E., and Pitt, H. M., *J. Org. Chem.,* **22,** 1119 (1957).

126. Fruhwirth, O., and Waller, H. (to Donau Chemie A.-G.), U. S. Patent 2,538,723 (1951).

127. Gage, J. C., *Analyst,* **84,** 509 (1959).

128. Gelblum, S. (to Dominion Tar & Chemical Co. Ltd.), British Patent 774,125 (1957); *C. A.,* **51,** 15547 (1957).

129. Gisclard, J. B., *A. M. A. Arch. Ind. Health,* **21,** 250 (1960).

130. Gonikberg, M. G., and Zhulin, V. M., *Bull. Acad. Sci. U.S.S.R., Div. Chem. Sci. S.S.R.* (English Translation), **1959,** 592.

131. Gonikberg, M. G., and Zhulin, V. M., *Bull Acad. Sci. U.S.S.R., Div. Chem. Sci. S.S.R.* (English Translation) **1959,** 882.

132. Goodall, A. M., and Howlett, K. E., *J. Chem. Soc. (London),* **1954,** 2599.

133. Graham, J. W. (to Canadian Industries Ltd.), Canadian Patent 603,695 (1960).

134. Graham, J. W., and Lusby, G. R. (to Canadian Industries Ltd.), Canadian Patent 601,668 (1960).

135. Graham, J. W., and Lusby, G. R. (to Canadian Industries Ltd.), Canadian Patent 601,680 (1960).

136. Graham, J. W., and Lusby, G. R. (to Canadian Industries Ltd.), U. S. Patent 2,946,826 (1960).

137. Grandjean, E., Muenchinger, R., Turrian, V., Haas, P. A., Knoepfel, H. K., and Rosenmund, H., *Brit. J. Ind. Med.,* **12,** 131 (1955); *C. A.,* **49,** 11207 (1955).

138. Grebe, J. J., Reilly, J. H., and Wiley, R. M. (to Dow Chemical Co.), U. S. Patent 2,034,292 (1936).

139. Griggs, F. E. P. (to Canadian Hanson & Van Winkle Co., Ltd.), Canadian Patent 463,162 (1950).

140. Hag A.-G. (W. Roselius and G. Petsch, inventors), German Patent 1,017,900 (1957).

141. Hama, G. M. *Nat. Safety News,* **68** (4), 128 (1953).

142. Hargarten, J. J., *Plant Maintenance and Eng.,* **21** (3), 46 (1960).

143. Haring, D. C., *Petrol. and Chem. Transporter,* **21** (12), 8, 32 (1959).

144. Harris, C. R. (to Roessler & Hasslacher Chemical Co.), U. S. Patent 1,904,450 (1933); *C. A.,* **27,** 3481 (1933).

145. Hauptschein, M., and Bigelow, L. A., *J. Am. Chem. Soc.,* **72,** 3423 (1950); *C. A.,* **46,** 1432 (1952).

146. Heitz, R. G., and Brown, W. E. (to Dow Chemical Co.), U. S. Patent 2,442,324 (1948); *C. A.,* **42,** 6373 (1948).

147. Hendrixson, P. R., *Steel,* **135** (11), 124 (1954).

148. Hendrixson, P. R. (to E. I. du Pont de Nemours & Co., Inc.), U. S. Patent 2,861,897 (1958).

149. Hennig, B. (to I. G. Farbenind. A.-G.), German Patent 712,579 (1941); *C. A.,* **37,** 4407 (1943); *Chem. Zentr.,* **1942** (I), 1809.

150. Herman, C. O., and Lehr, W. (to Air Reduction Co., Inc.), U. S. Patent 2,911,449 (1959).

151. Hey, H., *J. Soc. Dyers Colourists,* **36,** 11 (1920); *C. A.,* **14,** 1444 (1920).

152. Hirschhorn, E., *Soap Chem. Specialties,* **36** (7), 51 (1960).

153. Hirschkind, W., *Ind. Eng. Chem.,* **41,** 2752 (1949).

154. Hoffman, M. T., U. S. Patent 2,479,358 (1949).

155. Horsley, L. H., "Azeotropic Data," American Chemical Society, 1952.

156. Houser, T. J., and Bernstein, R. B., *J. Am. Chem. Soc.,* **80,** 4439 (1958).

157. Huntress, E. H., "Organic Chlorine Compounds," pp. 611–15, 695–7, New York, John Wiley & Sons, 1948.

158. Imperial Chemical Industries Ltd., *Australian Application,* 59,612 (1960).

159. Imperial Chemical Industries Ltd., *Australian Application,* 59,786 (1960).

160. Imperial Chemical Industries Ltd., *Australian Application,* 59,788 (1960).

161. Imperial Chemical Industries Ltd., French Patent 1,207,858 (1959).

162. Imperial Chemical Industries of Australia and New Zealand Ltd, British Patent 736,740 (1955); *C. A.,* **50,** 7840 (1956).

163. I. G. Farbenindustrie A.-G., Belgian Patent 449,037 (1943); *C. A.,* **42,** 199 (1948).

164. I. G. Farbenindustrie A.-G., Swiss Patent 213,747 (1941); *Chem. Zentr.,* **1942** (I), 2706.

165. Jones, H. H. (to E. I. du Pont de Nemours & Co., Inc.), U. S. Patent 2,930,349 (1960).

166. Jones, H. H., and Hudoch, R. A. (to Geo. W. Harding), U. S. Patent 2,739,567 (1956).

167. Joseph, W. J., by A. B. Bianchi, administrator for the estate of W. J. Joseph, deceased, (to Dow Chemical Co.), U. S. Patent 2,752,401 (1956); *C. A.,* **51,** 2029 (1957).

168. Kali-chemie A.-G., Italian Patent 383,229 (1940); *Chem. Zentr.,* **1942** (I), 3143.

169. Kauder, O. S. (to Argus Chemical Corp.), U. S. Patent 2,944,088 (1960).

170. Kearney, T. J., and Kircher, C. E., *Metal Progr.*, **77** (4), 87; **77** (5), 93 (1960).

171. Kearney, T. J. (to Detrex Corp.), U. S. Patent 2,771,023 (1956).

172. Kircher, C. E., A.S.T.M. Bull. No. 219, 44–9 (1957).

173. Kirkbride, F. (to Imperial Chemical Industries Ltd.), U. S. Patent 2,321,823 (1943); *C. A.,* **37,** 6676 (1943).

174. Klabunde, W. (to E. I. du Pont de Nemours & Co., Inc.), U. S. Patent 2,422,556 (1947); *C. A.,* **42,** 98 (1948).

175. Klabunde, W. (to E. I. du Pont de Nemours & Co., Inc.), U. S. Patent 2,435,312 (1948); *C. A.,* **42,** 2916 (1948).

176. Klabunde, W. (to E. I. du Pont de Nemours & Co., Inc.), U. S. Patent 2,436,772 (1948); *C. A.,* **42,** 2916 (1948).

177. Klabunde, W. (to E. I. du Pont de Nemours & Co., Inc.), U. S. Patent 2,440,100 (1948); *C. A.,* **42,** 6144 (1948).

178. Klabunde, W. (to E. I. du Pont de Nemours & Co., Inc.), U. S. Patent 2,492,048 (1949); *C. A.,* **44,** 2543 (1950).

179. Koch, F., U.S. Patent 1,358,163 (1920); *C. A.,* **15,** 445 (1921).

180. Korshak, V. V., and Matveeva, N. G., *Doklady Akad. Nauk S.S.S.R.,* **85,** 797 (1952); *C. A.,* **47,** 28 (1953).

181. Korshak, V. V., Strepikheev, Yu. A., and Verlatova, L. F., *J. Gen. Chem.* (*U.S.S.R.*), **17,** 1626 (1947) (in Russian); *C. A.,* **42,** 3721 (1948).

182. Kuckhoff, A., Seifensieder-Ztg. **48,** 310 (1921); *C. A.,* **15,** 2202 (1921).

183. Kuntz, T. F., Disbelger, G. J., and Cocherell, A. L. (to Diamond Alkali Co.), U. S. Patent 2,676,998 (1954); *C. A.,* **49,** 3237 (1955).

184. Larchar, A. W. (to E. I. du Pont de Nemours & Co., Inc.), U. S. Patent 2,517,893 (1950); *C. A.,* **45,** 2242 (1951).

185. Larchar, A. W. (to E. I. du Pont de Nemours & Co., Inc.), U. S. Patent 2,517,894 (1950); *C. A.,* **45,** 2242 (1951).

186. Larchar, A. W. (to E. I. du Pont de Nemours & Co., Inc.), U. S. Patent 2,517,895 (1950); *C. A.,* **45,** 2242 (1951).

187. Laurent, A., *Ann. chim. et phys.,* **68,** 377 (1836).

188. Lebsanft, K., and Schmidt, E. (to Wacker-Chemie G.m.b.H.), U. S. Patent 2,901,383 (1959).

189. Lecat, M., "Tables Azéotropiques," 2nd Ed., Vol. 1, Uccle-Brussels, 1949.

190. Leermakers, J. A., and Dickinson, R. G., *J. Am. Chem. Soc.,* **54,** 4648 (1932).

191. Levine, A. A., and Cass, O. W. (to E. I. du Pont de Nemours & Co., Inc.), U. S. Patent 2,125,381 (1938); *C. A.,* **32,** 7479 (1938).

192. Levine, A. A., and Cass, O. W. (to E. I. du Pont de Nemours & Co., Inc.), U. S. Patent 2,155,723 (1939); *C. A.,* **33,** 5869 (1939).

193. Little, Arthur D., Inc., private report.

194. Longiave, C., and Zanetti, F. (to Montecatini Società Generale per l'Industria Mineraria e Chimica), U. S. Patent 2,847,482 (1958).

195. Lukes, J. J. (to Diamond Alkali Co.), Canadian Patent 536,810 (1957).

196. Lusby, G. R. (to Canadian Industries Ltd.), Canadian Patent 601,666 (1960).

197. Lusby, G. R., and Murdock, J. G. (to Canadian Industries Ltd.), Canadian Patent 601,667 (1960).

198. Maass, G. (to Chemische Werke Huels, A.-G.), German Application C5,702 (1955).

199. Manufacturing Chemists' Assoc., "Trichloroethylene," Chemical Safety Data Sheet SD-14, 1956.

200. Manufacturing Chemists' Assoc., "Perchloroethylene," Chemical Safety Data Sheet SD-24, (1948).

201. Martin, L. F., and Elston, A. A. (to Dow Chemical Co.), U. S. Patent 1,858,022 (1932); *C. A.*, **26**, 3806 (1932); *Chem. Zentr.*, **1932** (II), 1073.

202. Martínez Moreno, J. M., *Quím. e ind. (Bilbao)*, **1**, 127 (1954); *C. A.*, **49**, 7266 (1955).

203. Maude, A. H. (to Hooker Electrochemical Co.), U. S. Patent 2,444,661 (1948); *C. A.*, **42**, 7312 (1948).

204. McBee, E. T., and Devaney, L. W., *Ind. Eng. Chem.*, **41**, 803 (1949); *C. A.*, **43**, 5361 (1949).

205. McDonald, L. S., and Hughes, H. G. (to Dow Chemical Co.), U. S. Patent 2,917,554 (1959).

206. McDonald, L. S., Hughes, H. G., and Crabbs, C. R. (to Dow Chemical Co.), U. S. Patent 2,917,555 (1959).

207. McGovern, E. W., *Ind. Eng. Chem.*, **35**, 1236 (1943).

208. McKinney, L. L., Eldridge, A. C., and Cowan, J. C., *J. Am. Chem. Soc.*, **81**, 1423 (1959); *C. A.*, **53**, 16008 (1959).

209. McKinney, L. L., Picken, J. C., Jr., Weakley, F. B., Eldridge, A. C., Campbell, R. E., Cowan, J. C., and Biester, H. E., *J. Am. Chem. Soc.*, **81**, 909 (1959); *C. A.*, **53**, 11696 (1959).

210. McKinney, L. L., Weakley, F. B., Eldridge, A. C., Campbell, R. E., Cowan, J. C., and Picken, J. C., Jr., *J. Am. Chem. Soc.*, **79**, 3932 (1957).

211. McKinney, L. L., Uhing, E. H., White, J. L., and Picken, J. C., Jr., *J. Agr. Food Chem.*, **3**, 413 (1955).

212. Metz, L., and Roedig, A., *Chemie, Ing., Tech.*, **21**, 191 (1949); *C. A.*, **43**, 6515 (1949).

213. Michaelis, O., *Seifensieder Ztg.*, **50**, 124 (1923); *C. A.*, **17**, 2369 (1923).

214. Midgley, T., Jr. (to General Motors Corp.), U. S. Patent 1,990,706 (1935); "Davis Halide Meter," Davis Emergency Equipment Co., Inc., Leaflet 2M 3–51.

215. Miller, W. T., Jr., and Dittmar, A. L., *J. Am. Chem. Soc.*, **78**, 2793 (1956).

216. Missbach, E. C. (to Stauffer Chemical Co.), U. S. Patent 2,043,257 (1936); *C. A.*, **30**, 5240 (1936).

217. Missbach, E. C. (to Stauffer Chemical Co.), U. S. Patent 2,043,258 (1936); *C. A.*, **30**, 5240 (1936).

218. Missbach, E. C. (to Stauffer Chemical Co.), U. S. Patent 2,043,259 (1936); *C. A.*, **30**, 5240 (1936).

219. Missbach, E. C. (to Stauffer Chemical Co.), U. S. Patent 2,043,260 (1936); *C. A.*, **30**, 5240 (1936).

220. Missbach, E. C. (to Stauffer Chemical Co.), U. S. Patent 2,069,711 (1937); *C. A.*, **31**, 2235 (1937).

221. Missbach, E. C. (to Stauffer Chemical Co.), U. S. Patent 2,094,367 (1937); *C. A.*, **31**, 8549 (1937).

222. Missbach, E. C. (to Stauffer Chemical Co.), U. S. Patent 2,094,368 (1937); *C. A.*, **31**, 8549 (1937).

224. Monroe, R. F., and Rapp, D. E. (to Dow Chemical Co.), U. S. Patent 2,906,783 (1959).

225. Montecatini Societá Generale per l'Industria Mineraria e Chimica, British Patent 819,839 (1959).

226. Mugdan, M., and Barton, D. H. R. (to Distillers Co. Ltd.), U. S. Patent 2,379,372 (1945); *C. A.*, **39**, 4617 (1945).

227. Mugdan, M., and Wimmer, J. (vested in the Alien Property Custodian), U. S. Patent 2,338,297 (1944).

228. Myers, R. E., Canadian Patent 553,264 (1958).

229. Nex, R. W., and Redemann, C. T. (to Dow Chemical Co.), U. S. Patent 2,959,556 (1960).

230. Nordenborg, N. H. (to Uddeholms Aktiebolag), Swedish Patent 132,770 (1951); *C. A.*, **46**, 6666 (1952).

231. Nordenborg, N. H., and Tiganik, L. (to Uddeholms Aktiebolag), Swedish Patent 129,532 (1950); *C. A.*, **45**, 3407 (1951).

232. Norman, D. P. (to Pacific Mills), U. S. Patent 2,717,901 (1955).

233. Obrecht, R. P., and Bender, H. (to Stauffer Chemical Co.), U. S. Patent 2,857,438 (1958).

234. Palit, S. R., and Venkateswarlu, V., *J. Chem. Soc. (London)*, **1954**, 2129; *C. A.*, **48**, 12511 (1954).

235. Patron, G., and Ferri, A. (to Sicedison, S.p.A.), Canadian Patent 596,686 (1960).

236. Péchiney (Cie. de Produits Chimiques et Electrométallurgiques), French Patent 1,192,205 (1959).

237. Petering, W. H., and Aitchison, A. G. (to Westvaco Chlorine Products Corp.), U. S. Patent 2,371,644 (1945); *C. A.*, **39**, 4584 (1945).

238. Petering, W. H., and Aitchison, A. G. (to Westvaco Chlorine Products Corp.), U. S. Patent 2,371,646 (1945); *C. A.*, **39**, 3607 (1945).

239. Petering, W. H., and Aitchison, A. G. (to Westvaco Chlorine Products Co.), U. S. Patent 2,371,647 (1945); *C. A.*, **39**, 3607 (1945).

240. Pitman, A. L. (to E. I. du Pont de Nemours & Co., Inc.), U. S. Patent 1,925,602 (1933).

241. Pitman, A. L. (to Westvaco Chlorine Products, Inc.), U. S. Patent 1,910,962 (1933); *C. A.*, **27**, 3951 (1933).

242. Pitman, A. L. (to Westvaco Chlorine Products Corp.), U. S. Patent 2,319,261 (1943); *C. A.*, **37**, 6372 (1943).

243. Plueddemann, E. P., and Rathman, R. (to Westvaco Chlorine Products Corp.), U. S. Patent 2,423,343 (1947); *C. A.*, **42**, 98 (1948).

244. Pray, B. O., and Chisholm, R. S. (to Columbia-Southern Chemical Corp.), U. S. Patent 2,959,623 (1960).

245. Price, W. (to Westvaco Chlorine Products Corp.), U. S. Patent 2,108,390 (1938); *C. A.*, **32**, 2955 (1938).

246. Pyle, D. J. (to Dow Chemical Co.), U. S. Patent 2,752,402 (1956); *C. A.*, **51**, 2029 (1957).

247. Rehfeld, C. E., Perman, V., Sautter, J. H., and Schultze, M. O., *J. Agr. Food Chem.*, **6**, 227 (1958).
248. Reichsmonopolverwaltung fuer Branntwein (K. R. Dietrich and W. Lohrengel, inventors), German Patent 649,118 (1937); *C. A.*, **32**, 95 (1938).
249. Reilly, J. H. (to Dow Chemical Co.), U. S. Patent 1,947,491 (1934); *C. A.*, **28**, 2371 (1934).
250. Reilly, J. H. (to Dow Chemical Co.), U. S. Patent 2,140,551 (1938); *C. A.*, **33**, 2540 (1939).
251. Reilly, J. H. (to Dow Chemical Co.), U. S. Patent 2,140,548 (1938).
252. Reitlinger, O., U. S. Patent 2,674,633 (1954); *C. A.*, **49**, 4702 (1955).
253. Romig, G. C., in "Phosphoric Acid, Phosphates and Phosphatic Fertilizers," W. H. Waggaman, 2nd Ed., pp. 512-13, New York, Reinhold Publishing Corp., 1952.
254. Ruh, R. P., Davis, R. A., and Broadworth, M. R. (to Dow Chemical Co.), U. S. Patent 2,885,427 (1959).
255. Rumianca Societá per Azioni (C. Paoloni and A. Eusepi, inventors), Italian Patent 513,373 (1955); *C. A.*, **51**, 16512 (1957).
256. Savage, J., Pitter, A. V., and Imperial Chemical Industries Ltd., British Patent 378,084 (1932); *C. A.*, **27**, 3941 (1933).
257. Schacken, J. M. G. de, and Schmidt, M. E. A., French Patent 794,878 (1936); *C. A.*, **30**, 5430 (1936); *Chem. Zentr.*, **1936** (II), 864.
258. Schumacher, F. G., *Metal Finishing*, **58** (1), 68 (1960).
259. Seiler, W. U., *Chem. Eng. News*, **38** (22), 124 (1960).
260. Semeluk, G. P., and Bernstein, R. B., *J. Am. Chem. Soc.*, **76**, 3793 (1954); *C. A.*, **48**, 13375 (1954).
261. Seto, T. A., and Schultze, M. O., *Anal. Chem.*, **28**, 1625 (1956).
262. Shepherd, C. B., *Chem. Eng. News*, **31**, 234 (1953).
263. Sicedison S.p.A., British Patent 841,107 (1960).
264. Sjöberg, B., *Svensk Kemisk Tidskrift* (in English), **64**, 63 (1952).
265. Skeeter, M. J. (to Diamond Alkali Co.), British Patent 765,522 (1957).
266. Skeeters, M. J. (to Diamond Alkali Co.), U. S. Patent 2,868,851 (1959).
267. Skeeters, M. J. (to Diamond Alkali Co.), U. S. Patent 2,947,792 (1960).
268. Skeeters, M. J. (to Diamond Alkali Co.), U. S. Patent 2,958,711 (1960).
269. Skeeters, M. J., and Baldridge, J. R. (to Diamond Alkali Co.), Brit. Patent 773,632 (1957).
270. Skeeters, M. J., and Beckers, N. L. (to Diamond Alkali Co.), U. S. Patent 2,775,624 (1956).
271. Skeeters, M. J., and Esselstyn, W. J. (to Diamond Alkali Co.), U. S. Patent 2,567,621 (1951); *C. A.*, **46**, 3062 (1952).
272. Smith, L. C., and De Pree, L. (to Dow Chemical Co.), U. S. Patent 1,971,318 (1934).
273. Smyth, H. F., *Am. Ind. Hyg. Assoc. Quart.*, **17**, 129 (1956).
274. Société d'électrochimie, d'électrométallurgie et des aciéries électriques d'Ugine, British Patent 749,408 (1956); *C. A.*, **50**, 16058 (1956).
275. Solvay et Cie., Belgian Patent 488,683 (1949); *C. A.*, **49**, 1771 (1955).
276. Solvay et Cie.. French Patent 1,215,238 (1959).

277. Starks, F. W. (to E. I. du Pont de Nemours & Co., Inc.), U. S. Patent 2,742,509 (1956).
278. Starks, F. W. (to E. I. du Pont), U. S. Patent 2,795,623 (1957).
279. Starks, F. W. (to E. I. du Pont), U. S. Patent 2,818,446 (1957).
280. Starks, F. W. (to E. I. du Pont), U. S. Patent 2,919,295 (1959).
281. Starks, F. W. (to E. I. du Pont), U. S. Patent 2,945,070 (1960).
282. Starks, F. W. (to E. I. du Pont), U. S. Patent 2,958,712 (1960).
283. Stauffer, W. O. (to E. I. du Pont), U. S. Patent 2,751,421 (1956); *C. A.* **51,** 1242 (1957).
284. Stevens, H. C. (to Columbia-Southern Chemical Corp.), U. S. Patent 2,721,883 (1955).
285. Stewart, L. C., and DePree, L. (to Dow Chemical Co.), U. S. Patent 1,917,073 (1933); *C. A.,* **27,** 4539 (1933).
286. Stimpson, R. H., and Anderson, J. B. (to H. J. Heinz Co.), U. S. Patent 2,716,609 (1955).
287. Stobe, R. (to Wacker-Chemie G.m.b.H.), U. S. Patent 2,788,275 (1957).
288. Stoesser, S. M., and Alquist, F. N. (to Dow Chemical Co.), U. S. Patent 2,111,253 (1938); *C. A.,* **32,** 3754 (1938).
289. Strain, F., and De Witt, B. J. (to Columbia-Southern Chemical Corp.), U. S. Patent 2,737,532 (1956); *C. A.,* **50,** 8712 (1956).
290. Strosacker, C. J., and Schwegler, C. C. (to Dow Chemical Co.), U. S. Patent 1,930,350 (1933).
291. Thermet, R., and Parvi, L. (to Société d'électrochimie, d'électrometallurgie et des aciéries électriques d'Ugine), U. S. Patent 2,919,296 (1959).
292. Thermet, R., and Parvi, L. (to Société d'électrochimie, d'électrometallurgie et des aciéries électriques d'Ugine), U. S. Patent 2,938,931 (1960).
293. Tompkins, H. K., and The Clayton Aniline Co., Ltd., German Patent 222,622 (1907); *C. A.,* **4,** 2866 (1910); *Chem. Zentr.,* **1910** (II), 121.
294. Turnbull, J. S. (to Metropolitan-Vickers Electrical Co., Ltd.), Canadian Patent 536,499 (1957).
295. Underwriters' Laboratories, Inc., File MH494, December 18, 1946.
296. U. S. Tariff Com., "Summary of Tariff Information, 1921, Relative to the Bill H. R. 7456" p. 60, (1922).
297. Vancamp, R. M., and Muren, A. P., Jr. (to Columbia-Southern Chemical Corp.), U. S. Patent 2,914,576 (1959).
298. Van Deusen, W. P. (to Eastman Kodak Co.), U. S. Patent 2,544,905 (1951).
299. Vanharen, L. (to Solvay et Cie.), U. S. Patent 2,610,215 (1952); *C. A.,* **47,** 8769 (1953).
300. van Hinte, J., *Veiligheid (Amsterdam),* **28,** 121 (1952); *C. A.,* **47,** 10152 (1953).
301. Verner, H. G., and Wood, L. S. (to Du-lite Chemical Corp.), U. S. Patent 2,515,934 (1950); *C. A.,* **44,** 8311 (1950).
302. Victor, I. (to Vic Manufacturing Co. Profit Sharing Trust), U. S. Patent 2,910,137 (1959).
303. Vining, W. H. (to E. I. du Pont de Nemours & Co., Inc.), U. S. Patent 2,361,072 (1944); *C. A.,* **39,** 2080 (1945).
304. Voytilia, J. E. (to E. I. du Pont), U. S. Patent 2,709,139 (1955).

305. Wacker, A., Ges. fuer elektrochemische Ind., G.m.b.H. (W. Koerner and A. Suchy, inventors), German Patent 464,320 (1928); *Chem. Zentr.*, **1929**, (1), 1044.

306. Wacker, Alexander, Ges. fuer elektrochemische Ind., G.m.b.H., German Patent 733,750 (1943); *C. A.*, **38**, 980 (1944).

307. Wacker, A., Ges. fuer elektrochemische Ind., German Patent 763,024 (1952); *C. A.*, **49**, 14022 (1955).

308. Wacker-Chemie G.m.b.H., German Patent 1,020,509 (1957).

309. Walters, R. H., in "Encyclopedia of Chemical Technology," R. E. Kirk and D. F. Othmer, Vol. 2, pp. 741–3, New York, Interscience Encyclopedia, 1948.

310. Warren, G. W. (to Dow Chemical Co.), U. S. Patent 2,577,388 (1951); C. A., **46**, 6138 (1952).

311. Weiser, H. B., and Wightman, G. E., *J. Phys, Chem.*, **23**, 415 (1919).

312. White, W. A. S. (to Imperial Chemical Industries Ltd.), Canadian Patent 597,906 (1960).

313. White, W. A. S. (to Imperial Chemical Industries Ltd.), U. S. Patent 2,790,699 (1957).

314. White, W. A. S. (to Imperial Chemical Industries Ltd.), U. S. Patent 2,803,517 (1957).

315. White, W. A. S., and Crowder, N. F. (to Imperial Chemical Industries Ltd.), British Patent 803,021 (1958); *C. A.*, **53**, 3725 (1959).

316. White, W. A. S., and Crowder, N. F. (to Imperial Chemical Industries Ltd.), British Patent 803,796 (1958).

317. White, W. A. S., and Crowder, N. F. (to Imperial Chemical Industries Ltd.), British Patent 812,893 (1959).

318. White, W. A. S., and Ross, H. J. (to Imperial Chemical Industries Ltd.), British Patent 812,894 (1959); *C. A.*, **53**, 16553 (1959).

319. White, W. A. S., and Ross, H. J. (to Imperial Chemical Industries Ltd.), Canadian Patent 562,706 (1958).

320. White, W. A. S., Ross, H. J., and Crowder, N. F., *J. Textile Inst., Proc.*, **50**, 274, 320 (1959).

321. Wiklundh, L. S.-E., Persson, S. H., and Holm, K. A. (to Uddeholms Aktiebolag), U. S. Patent 2,944,924 (1960).

322. Willis, G. C., and Christian, C. A. (to Dow Chemical Co.), U. S. Patent 2,803,676 (1957).

323. Wimmer, J., and Mugdan, M. (to Consortium fuer elektrochemische Ind., G.m.b.H.), U. S. Patent 2,249,512 (1941).

324. Zanetti, F., Longiave, C., and Forni, A., *Riv. combustibile*, **7**, 649 (1953); *C. A.*, **49**, 2298 (1955).

325. Zuckermandel, E. C. (to Dow Chemical Co.,), U. S. Patent 1,835,682 (1932); *C. A.*, **26**, 999 (1932).

Acknowledgment

The valuable contribution of Janet Searles and Alfred A. D'Addieco of the Electrochemicals Department, E. I. du Pont de Nemours & Company in the preparation and review of this chapter are gratefully acknowledged by the author.

14. CHLORINATED BENZENES—THEIR AMMONOLYSIS AND HYDROLYSIS

WALTER H. PRAHL
Hooker Chemical Company

WILLIAM H. WILLIAMS and ALEX H. WIDIGER
The Dow Chemical Company

CHLORINATED BENZENES

Today, large quantities of chlorine are consumed in the manufacture of chlorinated benzenes, annual capacity being approximately 325,000 tons of chlorine in the United States and Western Europe. This is an increase from about 500 tons in 1920—a more than 600-fold increase in 40 years.

The largest use of chlorine is for the manufacture of about 380,000 tons of monochlorobenzene, requiring about 240,000 tons of chlorine annually. In turn, the major use of the monochlorobenzene is as an intermediate for the manufacture of phenol in the United States, Germany and Great Britain.

In addition to these 325,000 tons of chlorine, which are made by the electrolysis of salt brine, there are additional quantities made by the Raschig process which utilizes the vapor phase catalytic air-oxidation of hydrochloric acid to form chlorine and its immediate conversion to chlorobenzenes. This, in turn, is catalytically hydrolyzed to phenol and the hydrochloric acid is recycled as described by its inventor, Walter H. Prahl.

The production capacity for di-, tri-, and tetrachlorobenzene manufacture probably requires about 80,000 tons of chlorine annually.

The chlorination of benzene to these compounds converts over 50 per cent of the chlorine used to hydrogen chloride and thus produces over 170,000 tons of hydrogen chloride annually. Part of this hydrogen chloride is used in processes as a gas as it is produced. Some is liquified and marketed in cylinders for use as a chemical intermediate, and the remainder is dissolved in water to make hydrochloric acid of from 22 to 37 per cent hydrogen chloride content to satisfy the market for such acid in various industries.

Benzene hexachloride is another product of benzene and chlorine and is discussed in another chapter of this monograph.

These products probably represent over 90 percent of the chlorine used for the chlorination of benzene.

Bourion[1, 2, 3] has written an excellent summary of the fundamental science and art of benzene chlorination before 1920. He adequately described the use of gaseous chlorine in the chlorination of benzene in the liquid phase and the control of the conditions required to produce various proportions of monochloro-, dichloro- and trichlorobenzenes.

There are three principal methods for the chlorination of benzene:

(1) Liquid phase chlorination in the absence of metals, etc., but in the presence of molecules activated by ultraviolet light to produce additive products such as benzene hexachloride (hexachlorocyclohexane).

This chapter is not concerned with this method except to mention that it is an important process for the production of benzene hexachloride, the gamma isomer of which is a valuable insecticide.

(2) Gas phase chlorination with or without catalyst to substitute one or more hydrogens of the benzene ring with chlorine.

No commercial development of this method is known to the authors unless the Raschig process utilization of nascent chlorine can be so considered. However, References (4) and (5) indicate it to be advantageous for the control of isomer distribution, e.g., the production of metadichlorobenzene as the major dichlorobenzene isomer.

(3) The substitution of the hydrogen of the benzene ring in the liquid phase, which dissolves a Friedel-Crafts catalyst, by gaseous chlorine is the major commercial process in use today for the production of chlorinated benzenes.

It is hypothesized that in the chlorination of liquid benzene the chlorine is an electronegative substituent and that substitution is directed principally to the para and ortho positions by the resonance effect produced by interaction of an unshared electron pair on the chlorine with the π-electrons in the atomatic nucleus, i. e.,[6, 7]

The typical results obtained by liquid phase chlorination of individual chlorinated benzenes with anhydrous iron chloride as the catalyst at temperatures in the range of 50 to 150° C are indicated in Figure 14-1.*[21]

*The authors wish to thank Kenneth Bradley, of The Dow Chemical Company's Chemicals and Physical Research Laboratory, for his work in the preparation of and spectrographic analysis of the reaction mixtures.

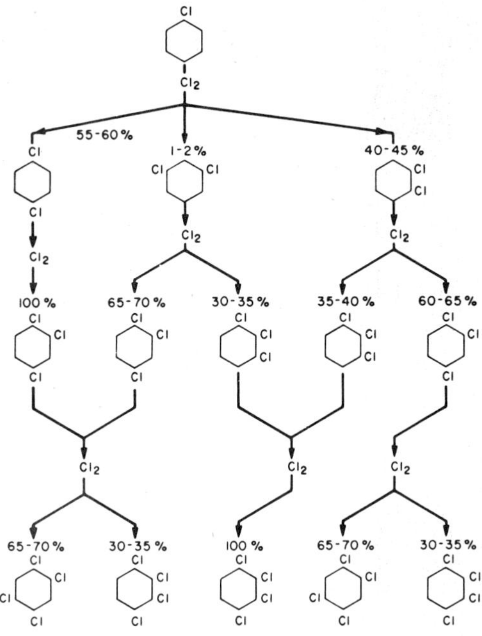

Figure 14-1

The product of the chlorination reaction can be controlled in degree of substitution and isomer ratio by variation of the catalyst, temperature of reaction, extent of chlorination and the isomers in the feed stock used for chlorination.

This can be summarized as follows:

p- Dichlorobenzene Content Varied 60–100%

1,2,4 Trichlorobenzene Content Varied 55–100%

1,2,4,5 Tetrachlorobenzene Content Varied 35–100%

1,2,3,4,5 Pentachlorobenzene 100%

The majority of the chlorobenzenes are produced in the United States by continuous chlorination, and the desired isomers may also be produced by isomerization of unwanted isomers by means of Friedel–Crafts reactions[8] and by dehydrohalogenation with copper-containing catalysts.[9]

Chlorinated benzenes are intermediates in the production of phenols, aromatic amines and chloronitrobenzenes. In 1914, Meyer and Bergius[10] produced phenol in the laboratory from chlorobenzene by hydrolysis with aqueous alkalies. Others[11,12] attempted to develop this reaction into a commercial process in a tubular autoclave but were unsuccessful. Still others also contributed to this art,[13] and the sum of the research resulted in a commercial operation manufacturing phenol from chlorobenzene and aqueous caustic soda in 1924; more phenol probably has been produced by this process than by any other.[14]

Similar research was progressing at about the same time on the ammonolysis of chlorobenzene[15,16,17] which contributed to the commercial process developed in 1926 for the ammonolysis of chlorobenzene to aniline and phenol by the use of aqueous ammonia, with a continuous autoclave development in 1941.[18]

Polychlorobenzenes can be hydrolyzed by alkaline solutions to the corresponding polychlorophenolic alkali salts and thence, by acidification, to the polychlorophenols.[19,20]

This can be summarized as in Table 14-1.

With this system of reactions, it is obvious that there need be few unwanted isomers from the chlorination of benzene since by chlorination to completion or dehydrohalogenation almost any combination of isomer distribution can be utilized.

Over 400 tons of chlorine per day from Michigan brines are used in chlorinating benzene to mono-, di-, tri-, and tetrachlorobenzenes by The Dow Chemical Company at Midland, Michigan.

Its process for manufacturing mono- and dichlorobenzenes consists of contacting chlorine gas and benzene, principally in the liquid phase, with a Friedel–Crafts catalyst by continuous passage through gas-liquid contacting columns at temperatures approaching the boiling point of benzene. (See Figures 14-2, 14-3).[21]

The chlorine is all but completely reacted to hydrogen chloride which is scrubbed substantially free of benzene with refrigerated chlorinated benzenes. It is used as a gas by other processes or absorbed in water to produce 32 percent hydrochloric acid.

Monochlorobenzene, hydrogen chloride and hydrochloric acid are the principal products of this operation.

The sum of the dichlorobenzenes can be reduced or increased in relation to the monochlorobenzene by recycle of more or less benzene through the chlorination system.[1,2,3,22]

TABLE 14-1

Chem. Compound	Minutes Reaction Time	°C Temp.	% Conversion	Major Prod.
Cl (chlorobenzene)	10–20	390	99	OH (phenol)
Cl···Cl (1,4-dichlorobenzene)	60–90	275	88	OH···Cl
Cl, Cl (1,2-dichlorobenzene)	60–90	275	90	OH, Cl
Cl, Cl, Cl (trichlorobenzene)	60–90	250	81	Cl, OH, Cl
Cl, Cl, Cl, Cl (tetrachlorobenzene)	15–30	250	84	OH, Cl, Cl, Cl
Cl, Cl, Cl, Cl, Cl (pentachlorobenzene)	60–90	225	99	Cl, Cl, OH, Cl, Cl
Cl, Cl, Cl, Cl, Cl, Cl (hexachlorobenzene)	60–90	225	98	OH, Cl, Cl, Cl, Cl, Cl

The ration of the ortho, meta and para isomers may be varied through a considerable range by changing the catalyst and temperature of the reaction.[23]

Tri- and tetrachlorobenzenes are produced by chlorination of liquid dichlorobenzenes. The ratios of the various isomers can be controlled within a range of 0 to 30 percent of 1,2,3-trichlorobenzene to 35 to 100 percent of the 1,2,4-trichlorobenzene, and the tetrachlorobenzenes can be controlled through a range of 30 to 60 percent of 1,2,3,4-tetrachlorobenzene and 30 to 60 percent 1,2,4,5-tetrachlorobenzene. These ratios are changed by varying the proportions of the isomers to be chlorinated, the chlorination rate, and the temperature, and by the use of different catalysts and solvents.[24]

Figure 14-2. Continuous benzene chlorinators.

Figure 14-3. Stills for fractionating chlorobenzene products.

Currently unwanted isomers can be continuously dehydrohalogenated with hydrogen over a suitable copper catalyst to reduce them to the tri-, di- or monochlorinated benzenes and hydrogen chloride. The degree of dehydrohalogenation is readily controlled by the temperature, time and catalyst exposure[9] (see Figure 14-4).

Figure 14-4. Unit for dehalogenating polychlorobenzene compounds.

The properties of the chlorinated benzenes, hydrogen chloride and hydrochloric acid are listed in Table 14-2.

PHENOL

Phenol, C_6H_5OH, the main source material of the phenolic plastics industry, is distinguished, compared to other organic chemicals, by the number of different processes by which it both can be and is being, produced.

(1) It is recovered from coal tar by extraction and distillation.

TABLE 14-2. PROPERTIES OF THE CHLORINATED BENZENES AND HYDROCHLORIC ACID

Product	Description	Freezing Point (°C)	Boiling Point °C 760mm Hg	Sp/Gr 25/25°C	Flash Point (°F)	Approx. Sol. g/100g Solvent at 25°C		
						Water	Methanol	Ether
Monochlorobenzene	Colorless liquid	−45.6	131.7	1.105	105	<0.1	−	−
o-Dichlorobenzene	Colorless liquid	−22	180.4	1.303	155	0.008	−	−
p-Dichlorobenzene	White to clear, transparent crystals	53	174	$1.248^{55/40°C}$	155	0.008	34	216
1,2,4-Trichloro-benzene	Colorless liquid	16.5	213.5–214.0	1.454	260	<0.1	−	−
1,2,4,5-Tetrachloro-benzene	Light-colored crystalline solid							
Hydrochloric acid	Colorless to very slightly yellow, fuming liquid					very soluble		

(2) It can be made from benzene by direct oxidation

$$C_6H_6 + \tfrac{1}{2}O_2 \rightarrow C_6H_5OH$$

However this process, investigated and tried several times, does not seem to be in commercial use at this time.

At least four different commercial processes convert benzene to phenol through an intermediate product:

(3) The sulfonation process uses benzenesulfonic acid as intermediary, consuming, in addition to benzene, sulfuric acid and sodium hydroxide. It gives sodium sulfite as a by-product, according to the equations:

$$C_6H_6 + H_2SO_4 \rightarrow C_6H_5SO_3H + H_2O$$

$$C_6H_5SO_3H + \tfrac{1}{2}Na_2SO_3 \rightarrow C_6H_5SO_3Na + \tfrac{1}{2}SO_2 + \tfrac{1}{2}H_2O$$

$$C_6H_5SO_3Na + 2NaOH \rightarrow C_6H_5ONa + Na_2SO_3 + H_2O$$

$$C_6H_5ONa + \tfrac{1}{2}SO_2 + \tfrac{1}{2}H_2O \rightarrow C_6H_5OH + \tfrac{1}{2}Na_2SO_3$$

$$C_6H_6 + H_2SO_4 + 2NaOH \rightarrow C_6H_5OH + Na_2SO_3 + 2H_2O$$

(4) The Dow process uses monochlorobenzene as the intermediate. It consumes chlorine and caustic. This process is treated in detail later in this chapter.

(5) The Cumene process uses cumene (isopropylbenzene) as the intermediary. It consumes propylene and gives acetone as a by-product, according to the equations:

$$C_6H_6 + C_3H_6 \rightarrow C_6H_5CH(CH_3)_2$$

$$C_6H_5CH(CH_3)_2 + O_2 \rightarrow C_6H_5CCOOH(CH_3)_2$$

$$C_6H_5COOH(CH_3)_2 \rightarrow C_6H_5OH + (CH_3)C = O$$

$$C_6H_6 + C_3H_6 + O_2 \rightarrow C_6H_5OH + (CH_3)C = O$$

(6) The Raschig process also uses chlorobenzene as the intermediary, but recovers and recycles the hydrogen chloride. It consumes es-

sentially no chemicals and gives no by-products, according to the equations:

$$C_6H_6 + HCl + \tfrac{1}{2}O_2 \rightarrow C_6H_5Cl + H_2O$$

$$C_6H_5Cl + H_2O \rightarrow C_6H_5OH + HCl$$

$$C_6H_6 + \tfrac{1}{2}O_2 \rightarrow C_6H_5OH$$

(7) Another process using cyclohexane and cyclohexanol as intermediaries and recycling the hydrogen seems to approach commercial production.

(8) A process starting from toluene and using benzoic acid as an intermediary seems to be close to commercial production.

Of these eight processes, the Dow and the Raschig process use chlorobenzene as an intermediary, and thus fall under the scope of this publication.

The Raschig Process to Phenol

After the end of World War I, the rapid increase of the consumption of phenol for phenolics posed the problem of a synthetic production to the Dr. F. Raschig, G.m.b.H., Ludwigshafen, Germany, who at that time produced about one half of the world's supply of cast phenolic resins. In 1928 research work was started in Raschig's laboratory by Walter H. Prahl.

The sulfonation route depended on the availability of sulfuric acid and sodium hydroxide, and on the marketability of sodium sulfite, neither of which were present in Raschig's case. The liquid phase hydrolysis of monochlorobenzene depended upon the availability of chlorine and caustic. Both processes could operate economically only as integral parts of a major chemical operation, and they were thus unsuitable for Raschig as essentially a producer of plastics. The problem was to find a process for producing phenol which would not be a link in a chain of a chemical production, which would not be dependent upon production facilities for chlorine, sodium hydroxide, sulfuric acid, etc., and the economics of which would not be influenced by markets for by-products. Such a process would convert benzene to phenol, without the need of a tie-in with other chemical processes.

A direct oxidation did not appear economically feasible with the knowledge available at that time. (The problem of oxidizing benzene in the presence of the more readily oxidizable phenol does not seem to have been solved yet.)

It was known at that time that catalytic vapor-phase hydrolysis of chlorobenzene produced some phenol and hydrogen chloride. It was further known that hydrogen chloride, when passed with air and benzene over the Deacon catalyst, would produce certain quantities of monochlorobenzene. If these two processes could be perfected to the point where they would give an economically acceptable yield, they could be combined in a phenol process which would come close to the ideal defined above. In this process, in the first stage, chlorobenzene would be chlorinated by hydrochloric acid and air over suitable catalysts in vapor phase to monochlorobenzene according to the equation

$$C_6H_6 + \tfrac{1}{2}O_2 \rightarrow C_6H_5Cl$$

The chlorobenzene would be hydrolized in the second stage according to the equation

$$C_6H_6Cl + H_2O \rightarrow C_6H_5OH + HCl$$

The HCl would be recycled to the first stage, giving the following net equation:

$$C_6H_6 + \tfrac{1}{2}O_2 \rightarrow C_6H_5OH$$

This process was developed, first by W. H. Prahl alone and later, with the assistance of Wilhelm Mathes, in the laboratory of Dr. F. Raschig G.m.b.H., in various larger scale developments including a pilot plant and a plant built by Rhône-Poulenc in France in 1932, and finally in a production plant built by Raschig in 1934. In 1937 the process was transferred to America, where Durez Plastics & Chemicals Inc. acquired a license under it and built a plant at North Tonawanda, New York (Figure 14-5).

In 1950 the Bakelite Corporation followed with their plant in Marietta, Ohio. At the present time plants based on this process are under construction in the United States and in Argentina. Plans for plants in other countries are under way. The process is known under the names: Raschig Phenol Process or Regenerative Phenol Process. Hooker acquired the process by the merger of Durez in 1955. Extensive research work carried out since then under the direction of W. H. Prahl has resulted in decisive improvements, making the process fully competitive with the more recently developed methods. The process incorporating the newly invented features is called "The Hooker Process."

The development of the process from the crude partial processes known in 1927, to the highly economical Hooker process of the present time, required the solution of many problems.

Figure 14-5. North Tonawanda, New York, Raschig phenol plant.

First there were new catalysts to be found. In the chlorination stage, use of the Deacon catalyst required temperatures above 300° C and resulted in large scale decomposition and oxidation of the starting materials and products. By the proper combination of catalytically active metal compounds with surface active carriers, the temperature could be reduced to about 200° C and the yield increased to above 90 percent.

A more difficult problem was to increase the yield of the hydrolysis catalyst. The best yield reported in the literature at that time was about 25 percent phenol. Practically any catalytic surface will cause chlorobenzene to react, but most catalysts will predominantly form benzene rather than phenol. The hydrogen required in the formation of benzene obviously originates in a kind of water-gas reaction of the organic decomposition products with the water vapor at the high reaction temperature. Several catalysts give almost quantitative yields of benzene, hydrogen chloride and carbon dioxide, but catalysts giving high yields of phenol by a straight hydrolysis reaction were hard to find. After testing hundreds of substances, a satisfactory catalyst was found which is rather unique in its ability to convert chlorobenzene to phenol with a yield of about 95 percent, with less than 5 percent going to benzene.

Another problem was posed by the limits in the conversions. While a liquid-phase chlorination of benzene proceeds at a much faster rate than that of monochlorobenzene under the same conditions, and that of monochlorobenzene proceeds much faster than that of dichlorobenzene, etc., the rates of chlorination of benzene, monochlorobenzene, dichlorobenzene, etc. under the conditions of the vapor-phase chlorination of the Raschig process are not very different. The quantities of more highly chlorinated benzenes formed in a given percentage of conversion of benzene by the Raschig process are therefore considerably higher than in a liquid-phase chlorination. In order to keep them within acceptable limits, it is necessary to operate with conversions on the order of 10 percent. At this conversion rate a quantity of dichlorobenzenes is formed which corresponds to roughly 10 percent of the monochlorobenzene. The more highly chlorinated benzenes constituted the only by-product of the Raschig process, until recently a method was found to convert them in the process itself to phenol. Thus, the Hooker process no longer has by-product dichlorobenzenes.

In the hydrolysis stage the conversion is limited by the fact that, with increasing concentration of phenol, side reactions set in, resulting in the formation of tarry materials which at present are worthless. In order to keep their formation on the order of 1 percent or less, a conversion of only about 10 to 15 percent of the chlorobenzene is permissible.

With these low conversions in both stages, the process requires the circulation of large volumes of vapors through the catalysts. An economic method of recovering the products from such large vapor volumes had to be found. In the hydrolysis stage this problem could be solved with comparative ease owing to the large chemical difference between products and starting materials. Several possibilities are available and were used at various times. The present method is to remove phenol as well as hydrochloric acid from the chlorobenzene-water mixture in a scrubbing column, a mixture of water and chlorobenzene acting as scrubbing medium. The small quantities of by-products, mostly benzene and carbon dioxide, are removed by purging a small percentage of the circulating vapors.

In the chlorination stage the problem of recirculation was more difficult to solve. Supplying hydrogen chloride in the form of weak hydrochloric acid, and oxygen in the form of air, introduces into the system large quantities of water and nitrogen which limit the percentage that can be recirculated.

The process requires the circulation of a large quantity of hydrochloric acid which is generated at high temperatures in the hydrolysis reaction, cooled down to near its boiling point, separated from phenol, evaporated again, heated to the proper reaction temperature, passed through the catalyst chambers, etc. It is obvious that this operation requires a careful design of the equipment in order to prevent excessive corrosion. This problem was solved with regard to most of the equipment by carefully controlling conditions in such a way as to stay within the area in which mixtures of hydrogen chloride, water and organic materials are not corrosive to steel.

For equipment where that is not possible, the usual protections against hydrochloric acid corrosion are necessary. Thus the Hooker process uses glass or brick-lined steel where no heat transfer is involved, while for heat-transfer surfaces tantalum, karbate or similar substances are available.

As they come from the reaction, the products of neither of the two stages are pure. In order to make the monochlorobenzene suitable for use in the hydrolysis, and in order to make the phenol suitable for all commercial requirements, certain purifications are required.

The problem of purification of the products is complicated on one side and facilitated on the other by the existence of numerous azeotropic mixtures. The dichlorobenzenes and trichlorobenzenes form azeotropes with phenol, and all organics involved form azeotropes with water and with aqueous hydrochloric acid. In the Hooker process proper use of the various azeotropes existing in these systems, resulted in a comparatively simple system of purification by a combination of plain and azeotropic distillations.

With the low temperatures and the oxidizing atmosphere under which the first catalyst operates, its lifetime depends mostly on the absence of impurities such as sulfur compounds, resin-forming compounds, etc. in the benzene. Under normal conditions it extends over a period of one or several years. The second stage catalyst, operating at temperatures between 400 and 500° C and under neutral or reducing conditions, is covered in the course of a few hours with carbonaceous substances and thereby loses activity. These deposits are removed by periodically passing air over the catalyst. In order to maintain continuous hydrolysis this system requires two or more catalyst chambers, one of which is being burned out, while the other is operating. An arrangement consisting of three or four catalyst systems with burning-out times of two hours, and operating times between four and six hours, was found most economical.

The original U. S. patents covering the Raschig process were taken out by Walter H. Prahl and assigned to Raschig. During World War II they were vested as alien property and have since then expired. The original Raschig process, after it was introduced in 1937 into this country, did not find the acceptance which would be expected. One reason was its reputation for high maintenance cost resulting from corrosion. This reputation, however, originated from an erroneous interpretation of certain facts. The plant in Tonawanda did suffer heavily from corrosion in its first year. However, this corrosion was not caused by the hydrochloric acid or other corrosive substances forming part of the process. The corrosion problems encountered by the Tonawanda plant were caused by the fact that a new well, drilled for the purpose of supplying cooling water to the new phenol plant, gave water with a considerable hydrogen sulfide content. This had not been expected at the time of the design, and no precautions had been taken. Considerable time was required to modify the equipment so that it operated without too much corrosion by hydrogen-sulfide-containing cooling water. Even with these modifications, however, the hydrogen sulfide-caused corrosion remained heavy, until finally a cooling tower was installed and recirculated cooling water was used.

There were, of course, in the original plant a few minor corrosion spots, the occurrence of which could not be foreseen theoretically. These were eliminated in the course of the development over the last twenty years. The maintenance rate of plants using the Raschig process is now considerably below the rate of many other chemical operations.

Operation of the Process

A schematic view of the Hooker process is given in Figure 14-6. The recycled acid enters the acid evaporator 1 in the form of 17 percent hydrochloric acid and is vaporized by low pressure steam. The heat-transfer surfaces are either 2 in. tantalum bayonet heaters or shell and tube-type karbate evaporators (Figure 14-7).

Benzene vapor coming partly from the product purification train, and partly from recirculation passes through a heater where by means of medium-temperature flue gas from the second stage heaters its temperature is raised to about 300° C in order to bring the temperature of the mixture up to the desired reaction temperature.

Air is pulled into the system from the atmosphere through a heater and is mixed with the acid and benzene vapors. The mixture enters the catalyst chamber 2. Here the chlorination takes place at temperatures between about 200 and 330° C, depending on the age of the catalyst and other conditions. The reaction is controlled so as to consume about 99 percent of the hydrogen chloride for chlorination. The mixture passes into a packed column 3, where water and benzene serve to scrub out the products, together with the small quantity of unreacted acid. A mixture of benzene and water vapors with the inerts leaves the top of the column and goes to a condenser 4 where, depending on the temperature of the cooling water, most of the benzene is condensed. The tail gas goes to a scrubbing system in which the benzene content is recovered in the conventional manner.

The bottoms of column 3 consisting of chlorinated benzenes and slightly acidic water, pass through a separator from whence the water goes on to the second stage, while the crude chlorobenzene product goes to the purification system.

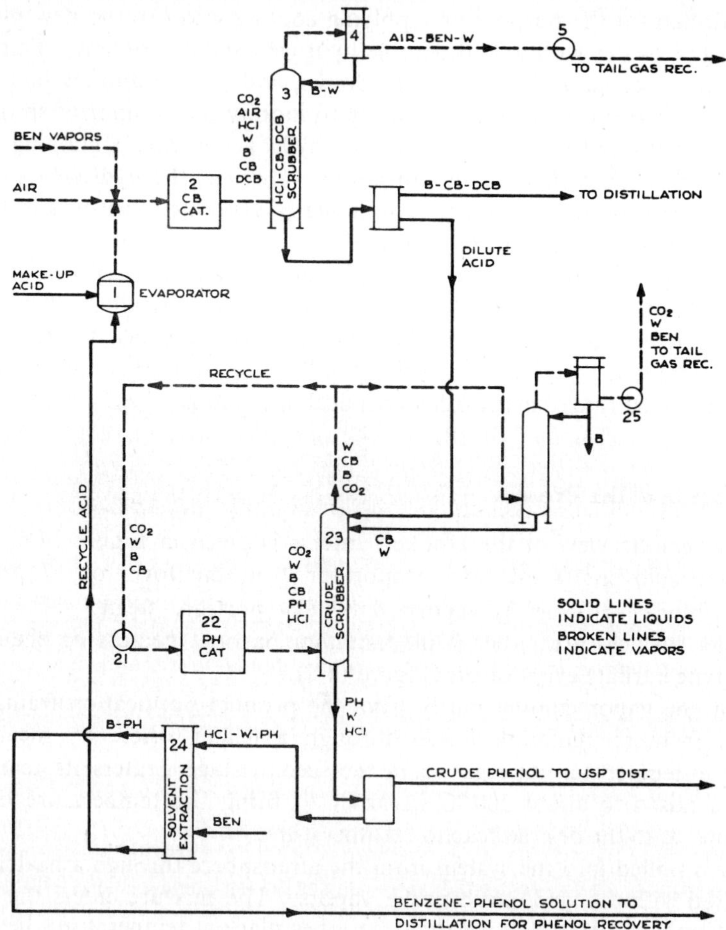

Figure 14-6. Flow sheet Raschig phenol process.

The hydrolysis stage can best be visualized as a circular train of equipment, in which a mixture of chlorobenzene and water vapor is cycled by means of a blower 21. The mixture is heated in a heat interchanger and a heater; it reacts adiabatically in the catalyst chamber 22, is cooled in the heat interchanger, passes the scrubbing column 23 and goes back to the blower 21 for a new cycle. Column 23 scrubs out the reaction products by means of water and chlorobenzene, supplied in sufficient quantity to replace the reacted materials. A small purge stream is taken off between 23, and 22 to remove the products of side reactions, mostly benzene, carbon dioxide and hydrogen. The reaction products from the bottom of 23 form two layers, which are mechanically separated. The acid layer goes through an ex-

Figure 14-7. Raschig phenol plant.

traction 24 in which its phenol content is recovered by extraction. The extracted acid returns to the acid evaporators 1. The extractant phenol mixture goes to the distillation train where it is separated, together with the other process streams, into its constituents.

The hydrolysis train from the heat interchangers through the heaters and the reactors consists of three individual systems, two of which are operating at any given time, while the other is being burned out with air. The change-over is effected by means of valves timed automatically.

The dichlorobenzenes produced in the first stage can either be worked up (by distillation, crystallization, etc.) for sale as ortho- and paradichlorobenzene, or they can, under certain conditions, be injected into the hydrolysis system where they are finally converted to phenol. In the former case the quantity of chlorine taken out of the system in the form of the dichlorobenzenes may be considerable. If the dichlorobenzenes are recycled for conversion to phenol, only the mechanical losses of hydrochloric acid have to be replaced. Either replacement can be effected in the form of chlorine, hydrochloric acid or monochlorobenzene, or in some other suitable form. If the dichlorobenzenes are recirculated, the consumption is approximately 0.93 lb of benzene per lb of phenol. The consumption of chlorine is then about 0.02 lb of HCl per lb of phenol. The consumption of utilities is given in Table 14-3.

TABLE 14-3. CONSUMPTION OF UTILITIES IN THE RASCHIG PHENOL PROCESS

Utilities	per lb of Phenol	
Steam	10.0	lbs
Fuel Oil	0.022	gal
Power	0.10	KWH
Cooling Water	40.0	gal

Discussion of the Operation

The description of the process above indicates that the original goal of creating a process which converts benzene to phenol without by-products and without consumption of chemicals has essentially been reached. However, small quantities of by-products still remain, even in the most modern version of the regenerative process. The chlorination produces, in addition to di- and trichlorobenzenes, which can be recycled, small quantities of more highly chlorinated benzenes all the way up to hexachlorobenzene. The recovery of these is normally not attractive, owing to the small quantity involved. They have to be considered as waste products. The hydrolysis produces further a certain quantity, normally less than 1 percent (phenol basis) of a high molecular-weight tar, which appears as a residue of the phenol distillation.

It consists of a mixture of very many different substances, among which high molecular-weight and chlorinated phenols predominate. It also contains smaller quantities of dibenzofuran homologues, some of which have caused difficulties because of their peculiar properties.

Another unexpected by-product of the hydrolysis reaction is acetylene, which, although formed at the rate of less than 0.1 milligram per pound of phenol, still can lead to considerable difficulties if not taken care of properly.

In a system of this nature, dealing with large volumes of vapors, the idea suggested itself to decrease the size of the equipment by operating under pressure. Considering the nature of the chemicals involved, however, it appeared preferable to take the opposite view and to operate both systems under a pressure slightly below atmospheric. This sub-atmospheric pressure is maintained in the chlorination stage by the location of the blower 5 which, being arranged at the end of the air path, thus pulls rather than blows air through the system. In the second stage the desired pressure is maintained by a small blower 25, which pulls the purge gas out of the system. The advantages of running under a slight minus pressure become apparent in operation, where any leakage is from the outside to the inside; thus neither corrosive vapors of hydrochloric acid, poisonous vapors of benzene, nor other noxious substances can escape, and practically all repair work can be done without shutdown.

In the chlorination stage, the exothermic character of the reaction and, in the hydrolysis stage, the need for oxidative regeneration suggest the use of fluid bed catalysts. In both cases, however, certain technical problems and especially economic questions exist, which do not favor the use of fluid beds.

Although the Raschig-Hooker process, in competition with the many other phenol processes, accounts for less than one fourth of the total phenol production in this country, it is steadily improving its position as the only presently available process which converts benzene to phenol with essentially no consumption of chemicals and no by-products.

The Dow Process To Phenol

The Dow Chemical Company converts monochlorobenzene to phenol and aniline by hydrolysis and ammonolysis, resulting in the formation of valuable by-products. The processes are quite similar and are accomplished with aqueous caustic soda or aqueous ammonia as the hydrolytic or ammonalytic reagents, respectively, as indicated by the flow sheet in Figure 14-8.[12]

The phenol reaction takes place continuously during a period of 10 to 30 minutes in tubular autoclaves at temperatures in the range of 360° to 400° C and pressures above the corresponding vapor pressure of the reactants. The temperature and time of reaction is varied within the range indicated above, and the caustic soda molecular and aqueous concentrations are varied from 2 to 2.5 moles of caustic per mole of chlorobenzene plus phenyl ether and from 12 to 20 percent aqueous caustic, dependent upon the ratio of the products (phenol, phenyl phenols and phenyl ether) that are required for the current market situation.[13,25]

The autoclave reaction products are sodium phenolates and sodium chloride which are in an aqueous solution and an oil layer that contains any unreacted chlorobenzene and phenyl ether which is decanted and recycled

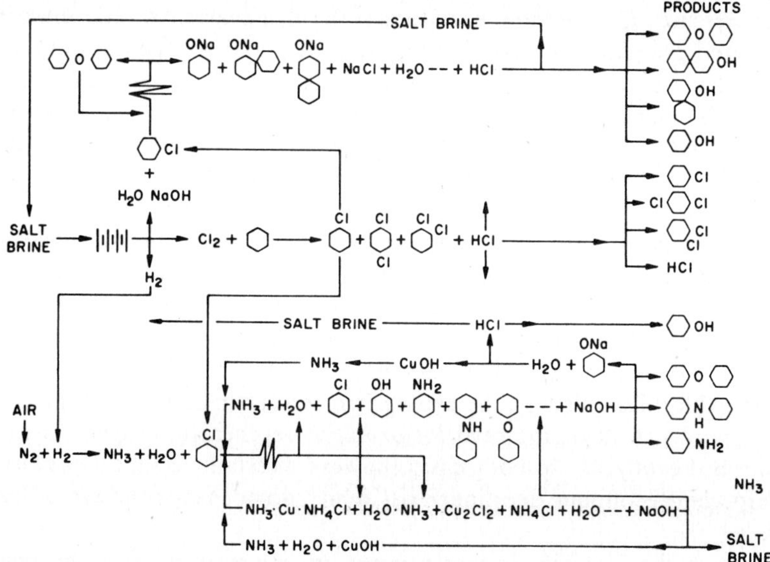

Figure 14-8. Chlorobenzene-phenol-aniline process.

through the autoclaves or refined for sale as the occasion requires (see Figure 14-9).

The aqueous layer from the decanter is extracted to remove phenyl ethers and traces of naphthalenes. It is then neutralized to a pH of 6.5 with hydrochloric acid obtained from the chlorination of benzene. This is settled into an oil and a sodium chloride brine layer which are decanted.

The sodium chloride brine layer is extracted with benzene[26] to contain less than 100 ppm of phenol and may be returned to the electrolytic cells to produce chlorine, caustic and hydrogen.

The oil layer decanted from the neutralization and settling operation contains the phenols contaminated with brine and traces of phenyl ethers and naphthalenes. It is dried azeotropically and filtered free of salt; the filtrate is fractionally distilled to produce phenol, ortho-, meta- and para-phenylphenols and fractions of a percent of alkyl phenols, hydroxy phenyl diphenyls and naphthalenes (see Figure 14-10).

The properties and specifications of the refined products are listed in Table 14-4. For additional properties on some of these products see, "Physical Properties of Chemical Compounds," Advance in Chemistry Series, Volume 15.

The Dow Process To Aniline

In the aniline process, chlorobenzene is reacted continuously during a period of 60 to 100 minutes with 20 to 40 percent aqueous ammonia in

Figure 14-9. High pressure autoclave units for hydrolyzing chlorobenzene with aqueous caustic.

Figure 14-10. Phenol distillation unit.

TABLE 14-4. PHENOLS

Product	Description	Freezing Point (°C)	Boiling Point °C 760 mm Hg	Sp/Gr. 25/25°C	Flash Point °F	Approx. Sol. g/100 g Solvent at 25°C		
						Water	Methanol	Ether
Phenol, U.S.P.	White, crystalline; solid	40.9	181.75	$1.0413^{60/4°C}$	195	9.5		
p-Phenylphenol	White flakes	166.5	321	1.275	330	<0.1	36	27
Diphenyl oxide	Colorless liquid	27	257	$1.070^{27/4°C}$	205	Ins.		
o-Phenylphenol	White to light-buff or pink flakes	56–57 (m.p.)	275	–	–	<0.1	975	
2,4,5-Trichloro-phenol	Gray flakes to a sublimed mass	61–63 (m.p.)	252	–	–	<0.2	615	
Pentachlorophenol	Dark-colored flakes and sublimed needle crystals	188–189 (m.p.)	309	$1.978^{22°}$	–	<0.1	202	148

tubular autoclaves at temperatures within a range of 200 to 300°C and at pressures above the vapor pressure of the reactants to produce aniline, phenol, phenyl ether, diphenylamine and ammonium chloride.

The molecular ratio of chlorobenzene to ammonia and the latter's concentration in the aqueous solution as well as the concentration of the copper catalyst can be varied with the temperature and time to make more or less by-products, dependent upon the market situation. The phenol and phenyl ether are common to both the phenol and aniline processes[17] (see Figure 14-11).

Figure 14-11. High pressure pump and autoclave chamber for converting chlorobenzene to aniline with aqueous ammonia.

The autoclave reaction product is flashed to recover the majority of the unreacted ammonia for recycle to the autoclave. The liquid remaining is settled and separates into two layers. The two layers are decanted and a portion of the aqueous layer containing the excess ammonia, the ammonium chloride and copper catalyst may be recycled to the reactor. The remainder is treated with aqueous caustic soda and boiled, recovering the ammonia for recycle and the organic chemicals which join the oil layer. The cuprous hydroxide is recovered for recycle by filtration from the residual aqueous sodium chloride brine which is discarded.

The oil layer from the decantation contains the aniline, diphenylamine, phenol, phenyl ether and any unreacted chlorobenzene. It is contaminated with some of the aqueous layer containing ammonia, ammonium chloride and copper catalyst. This oil layer is treated with aqueous caustic soda to free the ammonia and hold the phenol as sodium phenate. The ammonia, water and chlorobenzene are recovered by distillation and are recycled.

The aniline, phenyl ether and diphenylamine are separated by distillation from the residual salts which are then dissolved and suspended in water and filtered to recover the cuprous compounds for recycle.

The filtrate is neutralized with hydrogen chloride to free the phenol as an oil which is settled, decanted, dried azeotropically and refined by fractional distillation.

The phenol in the aqueous brine layer is extracted and is combined with the phenol oil previously mentioned for drying and fractional distillation.

After extraction the aqueous brine is stripped of solvent by steaming and is then discarded.

The distillate of aniline, phenyl ether and diphenylamine is separated into its components by fractional distillation to produce the refined products.

Their properties and specifications are listed in Table 14-5.

2,4,5-trichlorophenol is manufactured from 1,2,4,5-tetrachlorobenzene which has been separated from its 1,2,3,4 isomer by continuous crystallization. The 1,2,4,5-tetrachlorobenzene may be hydrolyzed by caustic soda in tubular autoclaves similar to those used for the hydrolysis of monochlorobenzene to phenol at a temperature of between 225° and 300°C during a period of 15 to 30 minutes[19,20] (see Figure 14-12).

It is processed by neutralization of the sodium 2,4,5-trichlorophenate aqueous solution with hydrogen chloride, settled and the oil layer decanted. The brine layer is discarded after extraction to recover dissolved trichlorophenol. The 2,4,5-trichlorophenol is refined by fractional distillation. Its properties and specifications are listed in Table 14-4.

PRODUCT USES

Phenol—phenol intermediates are used for synthetic resins and plastics, agricultural chemicals, pharmaceuticals, dye intermediates, photographic chemicals and synthetic perfumes. Phenol is also used in the manufacture of germicidal and disinfecting paints and soaps, as a preservative in the leather, glue and adhesive industries, and in the refining of low-grade distillate by the petroleum industry.

Orthophenylphenol—used for microbiological control, pharmaceuticals, cosmetics, chemical intermediate, plastisizers, resins and certain adhesives.

Figure 14-12. Pumping units and autoclave chambers for converting 1,2,4,5-tetrachlorobenzene to sodium 2,4,5-tricholorophenate solution.

Paraphenylphenol—used as a chemical intermediate for microbiological control and for manufacture of phenolic resins.

Diphenyl oxide—used as an ingredient in a heat transfer agent and as a chemical intermediate.

2,4,5-trichlorophenol—used as a chemical intermediate in the manufacture of weed killers and some products for the control of bacteria and fungi.

Pentachlorophenol—used as a preservative for wood, microbiological control, etc.

Aniline—used as a chemical intermediate.

Diphenylamine—used as a chemical intermediate, in rubber compounding and in agricultural chemicals.

Benzene—used as a raw material for aromatic chemical products.

Monochlorobenzene—used as a chemical intermediate, in DDT, and in the manufacturing of phenol and aniline.

Orthodichlorobenzene—used as a solvent, as a chemical intermediate and in a heat transfer agent.

Paradichlorobenzene—used for moth proofing, as a chemical intermediate and as a deodorant.

1,2,4-trichlorobenzene—used as a solvent.

1,2,4,5-tetrachlorobenzene—used as a chemical intermediate and in the manufacture of 2,4,5-trichlorophenol.

1,2,3,4-tetrachlorobenzene—used as a chemical intermediate.

TABLE 14-5. ANILINE PROCESS

Product	Description	Freezing Point (°C)	Boiling Point °C 760mm Hg	Sp/Gr 25/25°C	Flash Point (°F)	Approx. Sol. g/100g Solvent at 25°C		
						Water	Methanol	Ether
Aniline	Colorless liquid	−6.0	184.0	1.021	190	4	−	−
Diphenylamine	Yellow flake	51.5						
Ammonia, anhydrous, technical grade	Colorless gas	−77.7	−33.35	0.6819 (gas) −33.4/4°C (liquid)	−	Very sol.	sol.	sol.

Hydrochloric acid (muriatic acid)—has many acid uses such as pickling steel.

Ammonia—used as a chemical intermediate and for fertilizer.

References

1. Bourion, M. F., *Compt. rend.*, **170,** 1319 (1920).
2. Bourion, M. F., *Ann. Chim.*, **14,** (9), 215 (1920).
3. Bourion, M. F., *Ann. Chim.*, **14,** (9), 272 (1920).
4. Engelsma, J. W., *et al, Rec. trav. chim.*, **76,** 325, (1957).
5. U. S. Patents 2,123,857; 2,827,502; 2,880,246.
6. Ingold, C. K., "Structure and Mechanism of Organic Chemistry" p. 163 (1953).
7. Fieser, L., and Fieser, M., "Organic Chemistry," 3rd Ed., 567 New York, Reinhold Publishing Corp., 1956.
8. U. S. Patents 2,666,085; 2,727,075; 2,819,321; 2,866,829.
9. U. S. Patents 2,726,271; 2,826,617; 2,886,605.
10. Meyer, K. H., and Bergius, F., *Ber.*, **47,** 3155 (1914).
11. U. S. Patent 1,213,142; British Patent 103,664.
12. Brown, Kirk, *Ind. Eng. Chem.*, **12,** 279 (1920).
13. Hale, W. J., and Britton, E. C., *Ind. Eng. Chem.*, **20,** 114 (1928).
14. U. S. Patents:

1,602,766	1,868,140
1,607,618	1,882,824
1,737,841	1,882,826
1,756,110	1,921,373
1,806,798	1,924,313
1,814,796	1,925,321
1,821,800	1,986,194
1,824,867	2,137,587
1,833,485	2,275,044
1,856,021	2,391,848

15. U. S. Patents:

1,607,824	1,823,026
1,726,170	1,840,760
1,726,171	1,885,625
1,726,172	1,890,246
1,726,173	1,932,518
1,764,867	2,028,065
1,764,869	2,052,349
1,775,360	2,371,543
1,804,466	2,432,551
1,814,822	2,432,552
1,823,024	
1,823,025	

16. Vorozhtzov and Kobelev, *J. Gen. Chem.* (U.S.S.R.) **4,** 310 (1934).
17. Groggins, P. H., and Stirton, A. J., *Ind. Eng. Chem.*, **25,** 1, 42 (1933).
18. U. S. Patents 2,432,551; 2,432,552.
19. U. S. Patent 2,799,713.

20. U. S. Patent 2,799,714.
21. Lee, J. A., *Chem. Eng.*, **54**, 11 (118) (1947).
    Hatta, Shiroji, and Shoji, Kiyoshi, *Chem. Eng.* (Japan), **19**, 482 (1955), *C.A.* **50**, 6f.
    French Patents 866,787; 963,087.
    German Patent 643,387.
    British Patents 605,693; 691,504.
    U. S. Patents 1,858,521; 2,470,336.
22. Zilberman, G. B., and Slobodnik, *J. Applied Chem.* (U.S.S.R.) **10**, 1080–5 (1937), *C.A.*, **32**, 1664[6].
23. Holleman, H. F., and Van der Linden, T., *R*, **30**, 328 (1911).
    Gupta, S. B., *et al.*, *J. Indian Chem. Soc., Ind. & News Ed.*, **11**, 139–45 (1948).
    Wiegandt, H., *Ind. Eng. Chem.*, **43**, 2167 (1952).
    Kovacic, Peter, and Brace, Neal O., *J. Am. Chem. Soc.*, **76**, 5491 (1954).
24. U. S. Patent 1,934,675.
25. P. B. Reports 79314; 97853; 76032; 63627; 6687.
    B.I.O.S. Report 664.
    Hausman, E., *Erdol u Kohle*, **7**, 496 (1954).
    Lee, J. A., *Chem. Eng.*, **54**, 9 (122) (1947).
26. U. S. Patent 2,137,587.

15. WATER CHLORINATION

Edmund J. Laubusch
The Chlorine Institute, Inc.

DEVELOPMENT

Historical

The first occasion on which chlorine was used for water disinfection has not been definitely ascertained. History records its use as a general disinfectant at about 1800 by De Morveau in France and by Cruikshank in England, but apparently it was not until the beginning of the twentieth century that its role as a water disinfectant was established. By the 1890's, however, after Schwann, Pasteur, Koch and others had made their contributions to the science of bacteriology, the use of chlorine and chlorine bleaches for the disinfection of water and sewage was receiving the attention of various American, English, French and German scientists.

Chlorine gas, electrolytically produced, was experimentally employed in 1896 by William M. Jewell and George W. Fuller at the Louisville Experiment Station. The following year Sims Woodhead used bleach solutions as a temporary measure to sterilize potable water distribution mains at Maidstone, Kent, subsequent to a typhoid fever epidemic. In 1909, Jewell again experimented with chlorine gas, and later chlorinated lime, for temporary treatment of filter effluents at Adrian, Michigan.[2,34]

Chlorinated lime was used[2] primarily with iron ("Ferro-chlor") to aid coagulation as early as 1902 by Dr. Maurice Duyk at Middlekerke, Belgium, and this plant-scale treatment continued until 1921.[15] However, the first continuous, municipal application of chlorine to raw water specifically for disinfection is considered to have occurred in England at the City of Lincoln filter plant, in 1904–05, when Sir Alexander C. Houston and Dr. McGowan employed a sodium hypochlorite solution ("Chlorous," containing 10 to 15 percent available chlorine) in an effort to combat an epidemic of typhoid. For this and other pioneering activities, Houston has been credited as the "Father of Water Chlorination."

In North America, the first commercial application of chlorine to water[24] was made in 1908 by G. A. Johnson who used chloride of lime to chlorinate the Bubbly Creek water supply of the Union Stock Yards Company at Chicago. The overwhelming success of this application led, in the same year, to its introduction[30] by G. A. Johnson and J. L. Leal into the 40 mgd Boonton reservoir supply of the Jersey City, New Jersey water works—the first continuous, municipal application of chlorine for water disinfection in the United States. Its adoption the following year by J. C. Otis at Poughkeepsie, New York, following court recognition and acceptance of the practice as a public health safeguard, paved the way for rapid extension of chloride of lime disinfection of public water supplies.[2]

In these early years of water chlorination, the only commercial sources of chlorine were chlorinated lime (also termed chloride of lime, bleaching powder, etc.) and sodium hypochlorite solutions. The poor stability and variable effective chlorine content of such compounds caused many operating difficulties, and often inadequate disinfection dosages were used. Moreover, equipment feeders then available yielded erratic results. Hypochlorite water chlorination received renewed stimulus with the commercial availability, in 1928, of high-test calcium hypochlorite, $Ca(OCl)_2$. This material, containing a high available chlorine content (70 percent), is relatively stable throughout production, packaging, distribution and storage. Sodium hypochlorite solutions contain a lesser available chlorine content and are much less stable; for this and other reasons, they do not enjoy the popularity of the high-test hypochlorites at water (or waste) treatment plants.

In 1910 liquid chlorine was produced as an article of commerce. The following year Major C. R. Darnall of the U. S. Army Medical Corps experimentally employed it for water disinfection[10] at Fort Myer, Virginia, using a dry gas chlorine feeder of his own design which was later patented. The first plant-scale use of liquid chlorine for water disinfection followed in 1912 at the Niagara Falls, New York, filter plant, with solution-feed equipment developed by Dr. Georg Ornstein. This step was taken by Health officer Dr. E. E. Gillick and eliminated a recurring typhoid condition. In 1913 improved equipment invented by C. F. Wallace and M. F. Tiernan to measure chlorine gas, dissolve it in water and apply the solution was installed at Boonton, New Jersey. These developments paved the way for the rapid adoption of liquid chlorine for water (and wastes) disinfection.[16,39]

Evolution of Chlorination Practices

In its earliest applications chlorine either was used as a unit process of water purification or applied for disinfection as the sole treatment process. Prechlorination (the addition of chlorine prior to filtration) was practiced for disinfection in 1910 at Avalon, Maryland, and in 1914 at Exeter, New

Hampshire, as an aid to coagulation.[43]   During World War I it was practiced by Sir Alexander Houston in London as a substitute for prolonged water storage.   In 1921, largely as a result of the research findings of R. S. Weston and N. J. Howard, the city of Toronto commenced prechlorination of its water supply.[22,23]   However, it was not until about 1927 that the practice received any marked impetus in the United States.[12]

It was again Sir Alexander Houston who led the way in the development of pre-ammoniation treatment to prevent taste production in chlorinated water.[43]   Chlorine-ammonia treatment of water (using ammonia water and bleaching powder) was first begun in the period from 1915 to 1917 by Joseph Race at Ottawa, Ontario, where he demonstrated the ability of the treatment to reduce taste formation.[2,34]   Although the treatment was adopted in the United States in 1926 at Denver, Colorado,[2] the first practical application of the process was successfully made, also in 1926, at Greenville, Tennessee, by J. W. McAmis[32] who used liquid chlorine and anhydrous ammonia to prevent the formation of iodoform tastes.   The process was studied elaborately in 1929 by Braidech[43] at Cleveland, Ohio, and subsequently became widespread in the United States later to decline in popularity largely due to the advent and development of more efficient "free-residual chlorination" processes. (Its principal application currently is as post-ammoniation to facilitate maintenance of long-lasting chlorine residuals in potable water distribution systems.)

Experimental treatment by excess chlorine (and dechlorination with sulphur dioxide) for taste and odor destruction was reported in 1912 and again in 1925 by Sir Alexander Houston.[43]   The research by N. J. Howard and R. E. Thompson (about 1922) brought the concept to ultimate fruition with its practical establishment in 1926 at the 70 mgd Toronto filter plant where it was successfully employed for the destruction of tastes due to chlorophenols.[23]   Later (1926 to 29) C. R. Cox and F. E. Hale reported extensively on its use in New York for destruction of tastes due to algae.[13]

The "breakpoint" concept of water chlorination began to emerge in 1939 through the observation of Faber[14] and Griffin,[21] and was developed more fully in 1940 when H. A. Brown demonstrated plant-scale applicability and C. K. Calvert demonstrated the fundamentals of chlorine-ammonia reactions and the existence of free chlorine in water when all ammonia has been destroyed by chlorine.[20]   Impetus for its adoption was soon provided by P. C. Laux's demonstration[29] that chlorine alone is present when full color develops instantly in the orthotolidine colorometric test.   Various test procedures for distinguishing between free and combined chlorine were subsequently introduced, mostly during the 1940's, and perfected,[20,34] and by 1950 breakpoint treatment was employed at about 500 water treatment plants in the United States.[2]

In 1940 sodium chlorite was developed commercially as an active bleaching agent. This paved the way for the introduction in 1944 at Niagara Falls, New York,[37] of chlorine-dioxide—a more active oxidant than chlorine—applied to improve the destruction of taste-producing chlorophenols and other substances. Although the use of chlorine dioxide for water disinfection had been tried experimentally about 1898, its earlier practical adoption was not feasible because of its instability and the necessity of on-the-spot generation. The reader is referred to Chapter 17 for additional information on chlorine dioxide. (For various reasons, however, the practice does not appear to be widely used for disinfection, but the literature contains several reports of its successful application to the control of difficult taste and odor problems, for iron and manganese oxidation, disinfection and other purposes.)

Undoubtedly improvements in chlorine disinfection practices and equipment feeders that were developed in the early days of water chlorination contributed to its popularity and widespread use. Disinfection dosages were based largely on the application of fixed amounts of chlorine. It soon became abundantly apparent that this practice made no provision for the effects of variations in water quality and fluctuations in chlorine demand. Gradually, however, the concept of varying chlorine dosage on the basis of residual chlorine was established, and iodometric methods for qualitative and quantitative assay of residual chlorine were developed. The use of orthotolidine as a qualitative indicator of residual chlorine was proposed in 1909 by E. G. Phelps,[34,43] subsequently evaluated by Kinnicut and F. W. Jones,[43] and in 1913 extended quantitatively by J. W. Elms and S. J. Houser,[11] who also developed colorometric standards for its use. Chlorine disinfection finally was established on a scientific basis in the period from 1917 to 1919 by the work of Abel Wolman and L. H. Enslow who demonstrated the suitability and reliability of the orthotolidine test for even the smallest water supplies.[42] Since that time, a better understanding of water chlorination processes has occasioned many refinements in the test and suggestions of some other tests. Today, the orthotolidine-arsenite test, developed[18] in 1944 by F. J. Hallinan and F. W. Gilcreas in order to distinguish various forms of residual chlorine in water, is the standard colorometric plant control test for residual chlorine.[1] The amperometric test for residual chlorine suggested in 1942 by H. C. Marks and J. R. Glass[1,31] is more sensitive and more accurate. This test is becoming more widespread, especially in the automatic residual recorders now used in many treatment plants.

Current Status

The past half century of experiment, refinement and understanding of water chlorination processes has crystallized certain policies and practices to ensure effectiveness for disinfection and numerous other applications of chlorine in water treatment technology. Today, no other community water treatment process enjoys such popularity and almost universal adoption in the United States. Certainly not all of the public health protection afforded by current water treatment technology can be attributed to chlorination, but its value, as evidenced by its development, is suggested in Table 15-1.

In 1948, the latest year for which U. S. statistics are available at this writing, more than half of all existing U. S. water treatment facilities employed chlorination as the only treatment. Moreover, about 88 percent (6137) of the existing facilities employed chlorine disinfection alone or in conjunction with other water treatment processes and served more than 80 million persons. These facilities represented more than 96.3 percent of the U. S. population served by *all* community water systems, both with and without treatment, and 96.9 percent of the U. S. population served with treated water.

The writer estimates that in 1960 more than 80,000 tons of liquid chlorine were used in the United States for disinfection and other treatment of community and industrial water supplies. Although the use of liquid chlorine surpasses that of hypochlorites for these purposes, particularly at larger facilities, hypochlorite chlorine forms are by no means obsolete or diminishing in importance. For the period from 1940 to 1948, the number of U. S. plants using liquid chlorine increased by about 40 percent while the number using hypochlorites increased by over 110 percent. This undoubtedly is due to the adoption of chlorine disinfection at smaller treatment plants where hypochlorites are especially appropriate.

BASIC CHEMISTRY

Reactions with Water

The utility of chlorine in water treatment is attributable to its toxicological characteristics and its oxidative capacity. Chlorine is employed to destroy bacteria and other microorganisms, to modify the chemical characteristics of the water being treated, or both (see Figure 15-1).

When chlorine (solution of chlorine gas or hypochlorite) is added to water, a mixture of hypochlorous (HOCl) and hydrochloric (HCl) acids is formed:

$$Cl_2 + H_2O \rightleftharpoons HOCl + H^+ + Cl^- \qquad (15\text{-}1)$$

This reaction is essentially complete within a very few seconds. In dilute solution and at pH levels above about 4, the equilibrium shown in Equation (15-1) is displaced to the right and very little Cl_2 exists in solution. Hypo-

TABLE 15-1. TREND IN DEVELOPMENT OF COMMUNITY WATER CHLORINATION IN THE U. S.[a]

| | Plants Chlorinating | | | Chlorine Forms Employed | | | |
Year	Number	Increase (%)	Population Served $\times 10^6$	Hypochlorite	Gas	Ammonia-Chlorine	Misc. or Unknown
1910[33]	18	–	1.18+	–	–	–	–
1915[33]	340	1790	15.58	–	–	–	–
1930[33,38]	2917	755	–	–	–	–	–
1940[38,41]	4650	60	63.50[b]	728	2563	668	683
1948[41]	6137	32	80.63	1557	3613	702	226

Notes: [a]Data are not strictly comparable due to slight variations in statistical reporting and analytical procedures.
[b]Estimated.

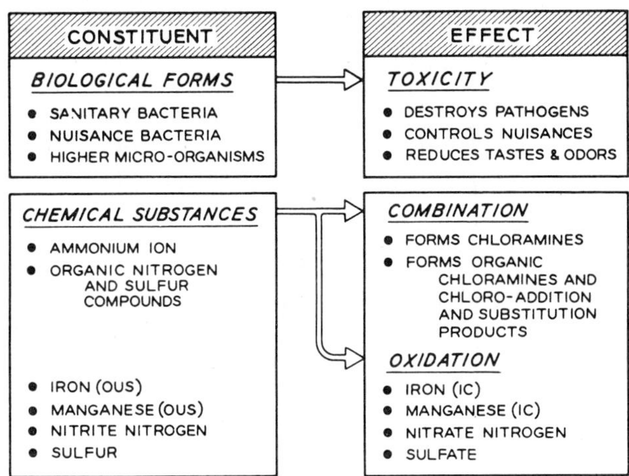

Figure 15-1. Principles of water chlorination.

chlorous acid is weak ($K = 2.8 \times 10^{-8}$) and dissociates poorly at levels of pH below about 6. Between pH 6.0 and 8.5 there occurs a very sharp change from undissociated HOCl to almost complete dissociation:

$$HOCl \rightleftharpoons H^+ + OCl^- \tag{15-2}$$

Thus, chlorine exists predominantly as HOCl at low pH levels. At 20°C above about pH 7.5, and at 0°C above about pH 7.8, hypochlorite ions (OCl⁻) predominate, and they exist almost exclusively at levels of pH around 9 and above (see Figure 15-2). Chlorine existing in water as hypochlorous acid and hypochlorite ions (or both) is defined as *free available chlorine.*[2]

Hypochlorite chlorine forms, i.e, high-test (70 percent available chlorine) calcium hypochlorite, $Ca(OCl)_2$, ionize in water and yield hypochlorite ions:

$$Ca(OCl)_2 + H_2O \rightleftharpoons Ca^{++} + H_2O + 2\,OCl^- \tag{15-3}$$

The hypochlorite ions establish equilibrium with hydrogen ions, depending on pH, as shown in Equation (15-2). Thus, the same equilibria are established in water regardless of whether elemental chlorine or hypochlorites are employed. The significant difference is in the resultant pH and its influence on the *relative* amounts of HOCl and OCl⁻ existing at equilibrium. Chlorine tends to decrease pH while hypochlorite tends to increase it. The pH value of chlorinated water supplies is normally within the range where chlorine may exist both as HOCl and OCl⁻.

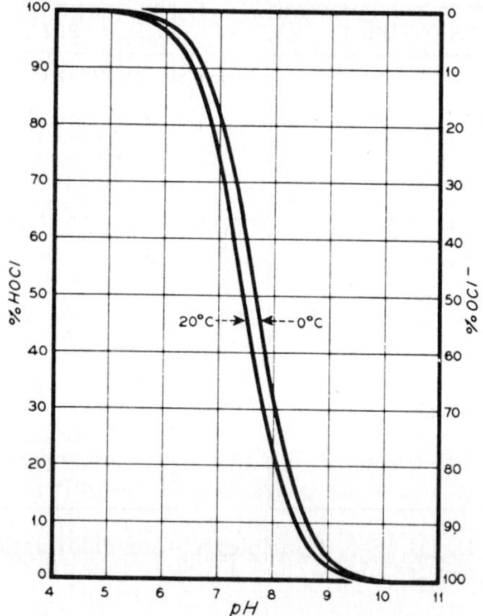

Figure 15-2. Relative amounts of HOCl and OCl⁻ formed at various pH levels.[17]

Reactions with Water Impurities

Chlorine and chlorine-containing compounds used in water treatment are potent oxidizers and often may be dissipated in reactions with a variety of chemical impurities before significant disinfection is accomplished. Some of these reactions are very rapid while others are very slow. The chemical reactions that occur and the rate at which they proceed are dependent upon the oxidation potential of the chlorine and chlorine compounds formed in the reactions. Both the rate and the range of reactions complicate the use of chlorine for water disinfection, especially on a heavily contaminated surface, and industrial and swimming pool water supplies.

Among those variable components of water that often affect water chlorination practices are the following:

(1) Alkalinity
(2) Inorganic Reducing Materials
(3) Ammonia and Amines
(4) Oxidizable Organic Materials
(5) Bacteria and Other Organisms

Alkalinity. As implied earlier, the efficacy of chlorine disinfection depends in part on the maintenance of an optimum pH range. From Equation (15-1) it is evident that hypochlorous and hydrochloric acids produced will tend to lower the pH; each part of chlorine added reduces the alkalinity by up to about 1.4 parts (as $CaCO_3$). Hardness-producing materials, principally the carbonates and bicarbonates of Ca and Mg, tend to buffer water against significant changes in pH that otherwise would accompany the addition of small amounts of acid and alkaline materials. If sufficient amounts of chlorine are applied to deplete natural alkalinity to such an extent that the pH is substantially decreased, pH correction may be indicated as auxiliary treatment. Hypochlorites always contain excess alkali for stability and tend to raise pH levels, but usually by insignificant amounts.

Inorganic Reducing Materials. The reaction between hydrogen sulfide and chlorine is typical of the reaction which occurs with other inorganic reducing agents:

$$H_2S + 4\,Cl_2 + 4\,H_2O \rightleftharpoons H_2SO_4 + 8\,HCl \qquad (15\text{-}4)$$

One part of H_2S theoretically can be oxidized by about 8.5 parts of chlorine, and chlor-oxidation treatment sometimes is practiced with this objective in mind.

Iron and manganese in water also affect chlorination practices. If the pH is high enough for hydroxide formation and sufficient chlorine is added, it will effectively oxidize the metals to the insoluble precipitate. (Chloramines do not enter vigorously into these oxidation reactions. For satisfactory manganese oxidation, free available chlorine, as distinct from combined available chlorine, is required). Nitrites are occasionally found in some waters, particularly those contaminated by sewage and certain industrial wastes. They, too, are readily oxidized to nitrates by chlorine (particularly free available chlorine).

Reactions such as these consume the chlorine of hypochlorous acid solutions. In some instances substantial amounts of chlorine may be consumed by these reactions before disinfection is achieved.

Ammonia and Amines. The reactions of chlorine with ammonia are of great significance in water chlorination processes, especially disinfection. When chlorine is added to water containing natural or added ammonia (ammonium ion exists in equilibrium with ammonia and hydrogen ions), the ammonia reacts with HOCl to form various chloramines depending on the pH of the solution and the initial $Cl_2 : NH_3$ ratio. The following equations suggest the reactions that occur:

$$NH_3 + HOCl \rightarrow NH_2Cl + H_2O \qquad (15\text{-}4)$$
Monochloramine

$$NH_3 + 2\,HOCl \rightarrow NHCl_2 + 2\,H_2O \qquad (15\text{-}5)$$
Dichloramine

$$NH_3 + 3\,HOCl \rightarrow NCl_3 + 3\,H_2O \qquad\qquad (15\text{-}6)$$

Trichloramine or nitrogen trichloride

Whether one of these compounds, or a combination of them, is formed depends upon the pH value of the water and upon an excess of ammonia. In general, low pH levels and high $Cl_2 : NH_3$ ratios favor dichloramine formation. Above a pH of approximately 8.5, monochloramine exists almost exclusively. At a pH between about 8.5 and 5, monochloramine and dichloramine exist simultaneously; at a pH between about 5.5 and 4.5, dichloramine exists almost exclusively; and, below a pH of about 4.4 nitrogen trichloride will be produced (see Figure 15-3). Reactions anal-

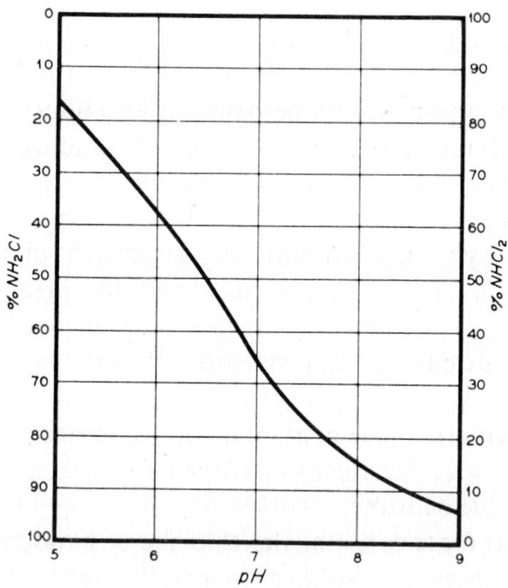

Figure 15-3. Relative amounts of chloramines formed at various pH levels.[3]

ogous to those between chlorine and ammonia also occur in varying degrees with organic amino groups. Chlorine existing in water in chemical combination with ammonia or organic nitrogen compounds is defined as *combined available chlorine.*[2]

Chloramines (both inorganic and organic) dissolved in water are partially hydrolyzed to form hypochlorous acid. The hydrolysis constants for a number of chloramines are given in Chapter 17. The disinfecting capacity of chloramines is a function of the amount and rate of hypochlorous acid formation. The hydrolysis constant of monochloramine is too low for hypochlorous acid to be formed in significant amounts; dichloramine has a higher hydrolysis constant and is, therefore, more effective for disinfection. The organic chloramines also have different hydrolysis constants and, accordingly, different disinfecting capacities; some are more effective than monochloramine and dichloramine, while others may be bacteriostatic rather than bactericidal.

Both inorganic and organic chloramines are important in water chlorination processes from the standpoint of their effectiveness as disinfectants and oxidants, and in the measurement of chlorine residuals.

Oxidizable Organic Materials. There are a variety of organic materials in water that react with chlorine and chloramines at various rates and under various conditions. These may exert a substantial influence on chlorine requirements for disinfection or other purposes, depending on available chlorine concentration and reaction time. Under some conditions, chlor-addition or chlor-substitution products might be formed, while under others the organic material may be completely oxidized. Certain organic chlorine complexes possess some low bactericidal or bacterostatic capacity, but their particular significance in water chlorination practices is due to the chlorine which they consume and the consequent tastes and odors they may impart.

Bacteria and Other Organisms. Although chlorine as a chemical oxidant has considerable utility in water treatment, its primary and principal use by far is for disinfection. In water treatment, the terms *chlorination* and *disinfection* have largely come to mean the same and imply processes designed to destroy harmful or nuisance organisms of sanitary significance. Such processes do not ordinarily accomplish sterilization.

Over the years several theories have been advanced to explain the mechanism of bacterial destruction or inactivation by chlorine. In the early years of water chlorination it was thought that chlorine reacted directly with water to produce nascent oxygen which, in turn, exerted a bactericidal effect on cells. This theory has long since been abandoned. Green and Stump[19] suggested in 1946 that the death of organisms results from a chemical reaction of HOCl with an enzyme system which is essential as a catalyst for glucose utilization and cellular metabolic activity, and that the enzyme probably attacked is triosephosphate dehydrogenase. Not only is the ability of the effective disinfectant to react with the enzyme of concern, but also its ability to gain access to the enzyme by penetrating through the cell membrane. It is apparently the rate of diffusion of the active disinfectant through the cell membrane which largely determines the rate of disinfection and relative efficiency of various disinfecting materials. Thus, it apparently is not the strong oxidizing power of HOCl that makes it a superior disinfectant, but rather its small molecular size and electrical neutrality which allow it to pass readily through the cell membrane. OCl⁻ has little if any bactericidal effect since its negative charge impedes pentration. Spores and cysts exhibit greater resistance to chlorine than do bacteria; Fair *et al.*[17] suggested in 1948 that this condition is due to the thickness and resistance to diffusion of the cyst or spore wall as compared to cell membranes.

The bactericidal capacity of a solution of chlorine, hypochlorites or chloramines is directly proportional to the undissociated HOCl concentration of the solution. As noted earlier, this is a function of the pH of the solution.

BASIC OPERATING PRACTICES

The earliest water chlorination practices (variously termed "plain chlorination," "simple chlorination" and "marginal chlorination)" were applied for the purpose of disinfection. As previously noted, chlorine-ammonia treatment (previously termed "chloramination") was soon thereafter introduced to *limit* the development of objectionable tastes and odors often associated with marginal chlorine disinfection. Subsequently, "superchlorination" (as then practiced) was developed for the additional purpose of *destroying* objectionable taste and odor producing substances often associated with chlorine-containing organic materials. The introduction of "break point chlorination" and the recognition that chlorine residuals can exist in two distinct forms established modern, conventional water chlorination as *combined residual chlorination* or *free residual chlorination*. Either type, properly employed, can yield a bacteriologically safe water; but both types are not equally applicable to all waters for disinfection and other water chlorination objectives (see Figure 15-4).

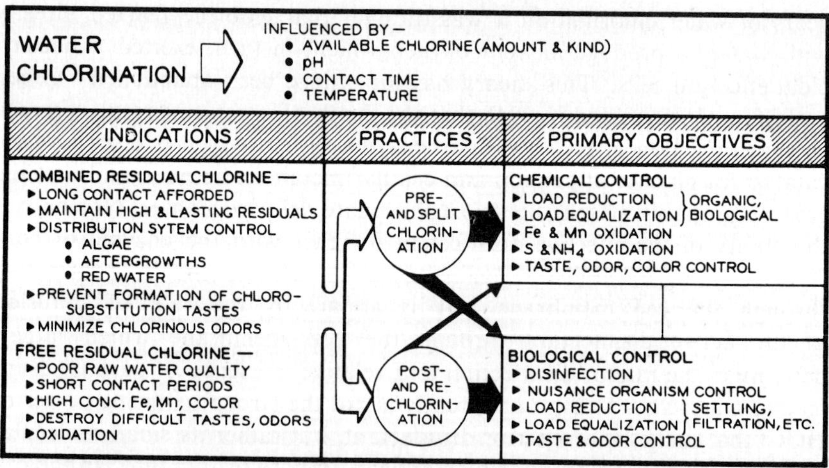

Figure 15-4. Summary of conventional water chlorination practices.[25]

Combined Residual Chlorination

Combined residual chlorination practice involves the application of chlorine to water to produce, with natural or added ammonia, a combined available chlorine residual and to maintain that residual through part or all of a water treatment plant or distribution system. *Combined available residual chlorine* is that residual chlorine existing in water in chemical combination with ammonia or organic nitrogen compounds. Combined chlorine forms have a lower oxidation potential than free chlorine forms, but the chlorine in these forms is still available for chemical reactions.[2] The low hydrolysis constants of chloramines, and in some cases the slow rate of establishing equilibrium, is responsible for the slow bactericidal action of these forms.

If water contains sufficient ammonia to produce with added chlorine a combined available chlorine residual of the desired magnitude, the application of chlorine alone suffices. If water contains too little or no ammonia, the addition of both chlorine and ammonia is required. If water has an existing free available chlorine residual, the addition of ammonia will convert the residual to combined available chlorine.

The optimum $Cl_2:NH_2$ ratio varies with specific conditions but in practice a ratio of 3:1 or more usually is effective when it is desired to insure the presence of an excess of ammonia. The maximum amount of stable chloramine will be formed by a ratio of 5:1 when the water has a pH value of 9.0, and by a ratio of 9:1 when the pH value is 5.0. Ratios of $Cl_2:NH_3$ less than 3:1 are seldom desirable because of the excess of ammonia which remains in the treated water.

Unlike free available chlorine, combined available chlorine does not greatly alter tastes existing in raw water, but often limits or prevents the formation of chlor-addition and chlor-substitution products that may cause serious taste and odor problems. The practice of combined residual chlorination is especially indicated after filtration (post treatment) for controlling certain algae and bacterial aftergrowths and reducing dead-end red water troubles in potable water distribution systems, and for providing and maintaining a stable, tasteless residual throughout the system to the point of consumer use. It frequently is preceded by free residual chlorination to ensure water potability except possibly where very long contact periods are obtainable.

Free Residual Chlorination

Free residual chlorination practice involves the application of chlorine to water to produce, either directly or through the destruction of ammonia, a free available chlorine residual and to maintain that residual through part or all of a water treatment plant or distribution system. *Free available residual chlorine* is that residual chlorine existing in water as hypochlorous acid (HOCl) and hypochlorite ion (OCl⁻).[2]

If water contains no ammonia (or other nitrogenous materials), the application of chlorine will yield a free residual. If water does contain natural or added ammonia, sufficient chlorine must be added until all ammonia is converted to chloramine.

With molar Cl_2:NH_3 concentrations of 1:1, both monochloramine and dichloramine are formed, the relative amounts of each depending on pH. Further increases in the Cl_2:NH_3 ratio result in the formation of some nitrogen trichloride and oxidation of part of the ammonia to nitrogen and other gases. These reactions in practice are essentially complete when two moles of chlorine have been added for each mole of ammonia present. Chloramine residuals generally reach a maximum when one mole of chlorine has been added for each mole of ammonia, and then decline to a minimum value when the Cl_2:NH_3 ratio is 2:1. Further addition of chlorine produces free residual chlorine. The point at which all ammonia is converted to NCl_3 or oxidized to free nitrogen (and possibly nitrous oxide) is referred to as the "breakpoint"—the point of minimum chlorine residual beyond which the residual rises in proportion to the amount of chlorine added (see Figure 15-5). The breakpoint varies with different waters and is primarily dependent on ammonia content; in practice, up to 25 times as much chlorine as the ammonia nitrogen content may be required to accomplish free residual chlorination. After free residual chlorination of water by breakpoint treatment, the resulting residual should consist of at least 90 percent free available chlorine and should contain little combined available chlorine.

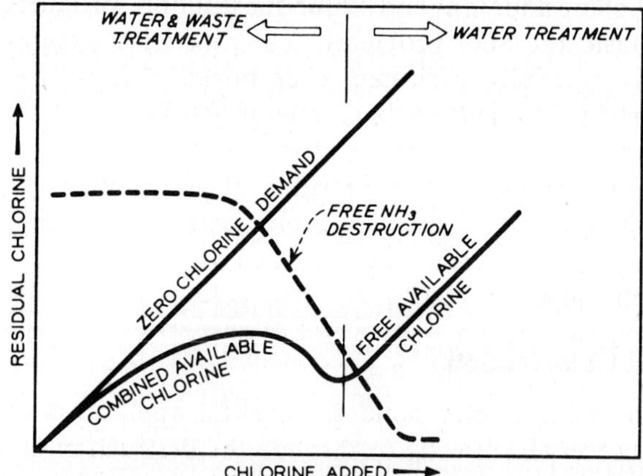

Figure 15-5. Free and combined residual chlorination.[25]

Free residual chlorination is especially indicated in the following cases:

(1) The quality of water is poor.

(2) Oxidation of iron and manganese.

(3) Contact time is insufficient to accomplish disinfection by combined available chlorine forms.

(4) Tastes exist before treatment (also to destroy objectionable tastes and odors produced by some substances that react with chlorine and by certain algae growths).

(5) To minimize biological growths on filters and lengthen fiter runs.

(6) To eliminate chlorine-resistant bacteria and aftergrowths in distribution systems.

The practice is adaptable to pre-, post-, and re-chlorination schemes.

Points of Application

When chlorination of water was first employed for disinfection, terminal treatment of the purification plant effluent was almost invariably practiced. Today, the use of chlorine in various stages of water treatment, and even in the distribution system, is common practice. Split chlorination schemes are increasingly employed for disinfection and other purposes and frequently enhance the efficiency of many unit water treatment processes. Fundamentally, the points at which chlorine is applied depend on the specific objectives of chlorination, technical, practical, safety, economic and other considerations.

Prechlorination. Prechlorination involves the application of chlorine to water prior to any other unit treatment process.[2] Among the benefits claimed are: improved filter operation by reduction and equalization of the bacterial and algae load, and control of slime and mud ball formation; improved coagulation; reduction of taste- ordor- and color-producing materials by oxidation and retardation of decomposition (in settling units); and, importantly, the provision of a safety factor in disinfecting heavily contaminated waters while keeping the chlorine residual in the distribution system at a minimum.

When prechlorination is indicated it is usually desirable to obtain a contact period which is as long as possible, preferably by adding the chlorine to the raw water suction intake to provide contact during the entire purification process. The dosage, of course, depends on the objective. In some cases free available chlorine residual forms are indicated; in others combined chlorine forms may suffice. Care must be exercised to maintain the proper residual necessary to accomplish the desired objective, however, as some of the above benefits can be sacrificed when operating with reduced oxidation potentials of chloramines.

Post-chlorination. Post-chlorination involves the application of chlorine to water subsequent to any other unit treatment process.[2] The most important form of post-chlorination is that following filtration for disinfection and to provide a potent (free) or a stable (combined) chlorine residual in a part or the entire potable water distribution system. The contact period provided to effect disinfection, as discussed elsewhere, is an important consideration; chlorine usually is added to the filter effluent or at the filter clear well. If post-chlorination precedes filtration greater filter efficiency usually is obtained.

Re-chlorination. Re-chlorination involves the application of chlorine to water, following previous chlorination treatment, at one or more points in the distribution system. The practice, which may involve free or combined residual chlorination, is especially common where the distribution system is long and complex and where the plant effluent residual is insufficient to control bacterial and algae regrowths, red-water troubles, etc. The chlorine (and/or ammonia) may be added at the end of a long main line in the distribution system, at a point where a main supplies water to an outlying community, at a reservoir, a standpipe, a pumping station, etc.[44]

Chlorine-ammonia. If a long contact period is available, chlorine-ammonia treatment may be employed as pre-, post-, or re-chlorination for the various purposes for which its use is intended. If a long contact period is not available, the general trend is to obtain free residuals through prechlorination and, by the addition of ammonia, to convert the residual to a combined form through post- or re-chlorination to provide a more persistent residual in the distribution system.

The points of application of ammonia and chlorine should be sufficiently far apart to permit thorough mixing of the ammonia with the water before chlorine is applied, but not so distant as to permit dissipation of ammonia by organic matter before chlorine is added. The earlier addition of chlorine may result in the formation of chloro-phenols and other products unaffected by chlorine-ammonia treatment.

Chlorine Dosages

The chemical reactivity of chlorine is a function of its capacity to combine with or to oxidize organic and inorganic matter. Chemical reactions that occur upon the addition of chlorine to water, and the intensity at which they proceed, are dependent on the oxidation potential of free and/or combined chlorine formed.* The activity of free and combined avail-

---

*The reaction between $HOCl$ and NH_3 is not instantaneous and varies considerably with temperature and pH. The reaction rate of monochloramine formation is at a maximum at a pH of 8.3; above and below this pH level slower rates occur. Thus, for a short time after the addition of chlorine to water containing ammonia, both free chlorine and combined chlorine will coexist.[35]

able chlorine is affected by temperature and pH. The rate and extent of chemical reactions are accelerated by an increase in temperature and a decrease in pH. Even under optimum conditions of temperature and pH the oxidation of some compounds may occur only slowly and require an appreciable amount of time for completion. To achieve the desirable equilibrium of chemical reactions it is necessary to allow sufficient time to elapse and to provide an unreacted residual. This residual may consist of free available and/or combined available chlorine.

The *chlorine demand* of a water is the difference between the amount of chlorine added to water and the amount of chlorine (free available and/or combined available) remaining at the end of a specified contact period. For a particular water, it varies with the amount of chlorine added, temperature, pH and contact time. When a small amount of chlorine is added to water some of it is utilized in initial disinfecting action. The rest combines with organic amines to produce a residual that is predominantly combined available chlorine. As larger amounts are added to obtain a higher residual the chlorine demand will be increased, and the remaining residual may include some free available chlorine. If sufficient amounts are added to produce a residual that is predominantly free available chlorine, the oxidation potential will be such that oxidation reactions prevail. Thus, instead of combining with organic substances which frequently cause undesirable tastes and odors, the chlorine may completely oxidize them, destroying or changing them into less complex substances that do not cause disagreeable tastes and odors.

To ensure freedom from pathogenic bacteria and other organisms of sanitary significance it is necessary to provide and maintain a certain chlorine residual for a certain period of time after chlorination, depending on temperature, pH, the amount and kind of other chlorine-consuming materials present, and the resisitivity to chlorine of the organisms to be destroyed. With long contact times a low concentration of disinfectant suffices, while short contact times require higher concentrations to accomplish equivalent bacterial kills. Moreover, the killing power of chlorine increases with rising temperature in addition to being markedly influenced by pH. Reference has already been made to the reduction of bactericidal power which occurs with increasing pH—in the case of free available chlorine due to the degree of dissociation of HOCl, and in the case of inorganic combined available chlorine due to the swing of equilibrium from di- to monochloramine. Thus, it is important to know both the concentration and *kind* of residual chlorine acting. As an operating guide, to obtain 100 percent kill during the same exposure period requires about 25 times as much chloramine as free chlorine; to obtain the same kill with the same amounts of chlorine, combined forms require approximately 100 times the exposure period necessary for free forms.[4,5,40] Based on the investigations of

Butterfield *et al.*[4,5,6,40] and Chang *et al.*[7,8] minimum free and available chlorine residuals for cysticidal and bactericidal doses have been developed[36] (see Figure 15-6). Such doses do not, of course, apply in the case of viruses, protozoa and other organisms of sanitary significance.[9,26]

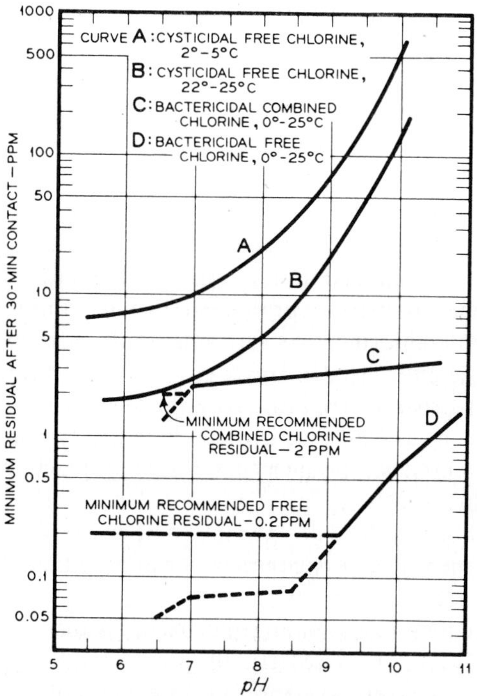

Figure 15-6. Minimum chlorine residuals for cysticidal and bactericidal doses.[36]

OPERATING CONTROL

Chlorine Dosage and Residual Chlorine

Because of variations in water (or waste) quality, effective and economic operating control depends on frequent adjustment of chlorine dosage to compensate for variations in chlorine demand. To facilitate maintenance of suitable chlorine residual levels, chlorine demand determinations are employed. The chlorine dosage represents the sum of the chlorine demand and the amount and type of chlorine residual required at the end of the specific contact period. As previously noted, the nature and amount of residual, as well as the contact time required, necessarily depend on the objectives of chlorine treatment and vary according to local circumstances. The practical basis of control of chlorination is the attainment of a defined residual which, after the contact afforded, has been established by experimentation and experience to accomplish the purposes intended. Although demonstration of bacterial (or other organism) destruction currently is the only positive means of establishing the efficacy of chlorine disinfection, residual chlorine determinations represent the sole practical criterion for demonstrating this with sufficient rapidity.

Quality control by bacteriological testing, coupled with statistical interpretations of bacteriological sampling, has become largely a matter of confirming treatment results to satisfy authoritative requirements.

Conventional methods for determining chlorine residuals depend on their oxidizing power and are based on reactions with reducing agents. Basically three tests are currently employed: volumetric starch iodide (iodometric titration); modified colorimetric orthotolidine; and amperometric titration.

Starch-iodide Test. The starch iodide method served as the only satisfactory basis of chlorination control until about 1913 when the orthotolidine test was developed. It depends on the oxidizing power of free and combined chlorine residuals to convert iodide to free iodine:

$$HCl + HOCl + 2\,KI \rightarrow 2\,KCl + I_2 + H_2O \qquad (15\text{-}7)$$

In the presence of starch, free iodine produces a blue color indicative of the presence of residual chlorine. *Quantitative* measurement of total residual chlorine is achieved by acid titration of released iodine with a standard reducing agent solution, commonly sodium thiosulfate:

$$I_2 + 2\,Na_2S_2O_3 \rightarrow Na_2S_4O_6 + 2\,NaI \qquad (15\text{-}8)$$

This method is better than the orthotolodine test for determining total chlorine residuals when the concentration exceeds 1 mg/l; however, when the concentration is less than 1 mg/l the titration end point is difficult to detect. The test is not widely used because of the relative instability of indicator solutions, the large volumes of sample and the greater complexity and time required, and because currently it is satisfactory only for total available chlorine determinations. Moreover, the interference effects of nitrites, and manganic and ferric salts may be considerable.

Basic Orthotolidine (OT) Test. Orthotolidine is an aromatic compound that is oxidized in acid solution by chlorine, chloramines and other oxidants to produce a yellow colored complex which, at pH levels below

1.8, is proportional to the amount of oxidants present:

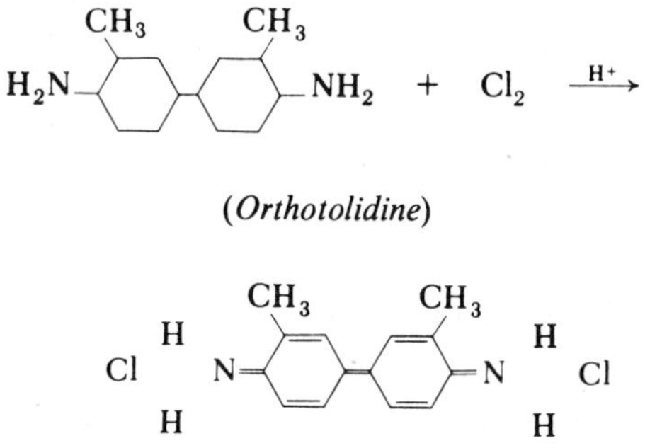

$$H_2N-\underset{CH_3}{\bigcirc}-\underset{CH_3}{\bigcirc}-NH_2 \;+\; Cl_2 \;\xrightarrow{\;H^+\;}$$

(Orthotolidine)

(15-9)

$$Cl^-\,H_2N=\underset{CH_3}{\bigcirc}-\underset{CH_3}{\bigcirc}=NH_2\,Cl^-$$

The test is especially suitable for routine determination of total chlorine residuals where the concentration does not exceed about 10 mg/l. The colorimetric OT test is rapid and simple, but it is somewhat affected by natural color and turbidity, temperature variations, nitrites, ferric and manganic ions, and possibly by organic iron, lignocellulose, algae and other materials. Because of these inherent limitations and difficulties in compensating for them, some investigators believe that the basic test has only limited utility in research applications and for continuous automatic recording of chlorine residuals.

An *orthotolidine flash test* modification, performed near $0°C$ to minimize the effect of chloramines, can be used for the determination of free available chlorine. Although oxidized manganese affects the test results, slow-acting interfering substances, nitrites, and oxidized iron do not have a significant influence.

With the development of current concepts of free and combined available residual chlorination, test methods were designed to permit differentiation and quantitative evaluation of residual chlorine forms. Two methods are commonly employed for this purpose.

Orthotolidine-arsenite (OTA) Test Method. The OTA test measures total free and combined available chlorine forms in the presence of significant concentrations of interferences (see Figure 15-7), and it is based on the fact that free chlorine residuals react instantaneously with orthotolidine whereas combined residuals react slowly. Total chlorine is determined by the OT method (as previously discussed), with readings made after five minutes. A test is made on a second sample adding sodium arsenite reducing agent immediately (within five seconds) after the orthotolidine is added; this permits oxidation of orthotolidine by free chlorine and allows only a small part of combined chlorine to act before the addition of arsenite (which reduces the chloramines instantly and prevents further action with orthotolidine). The resulting color intensity is primarily a

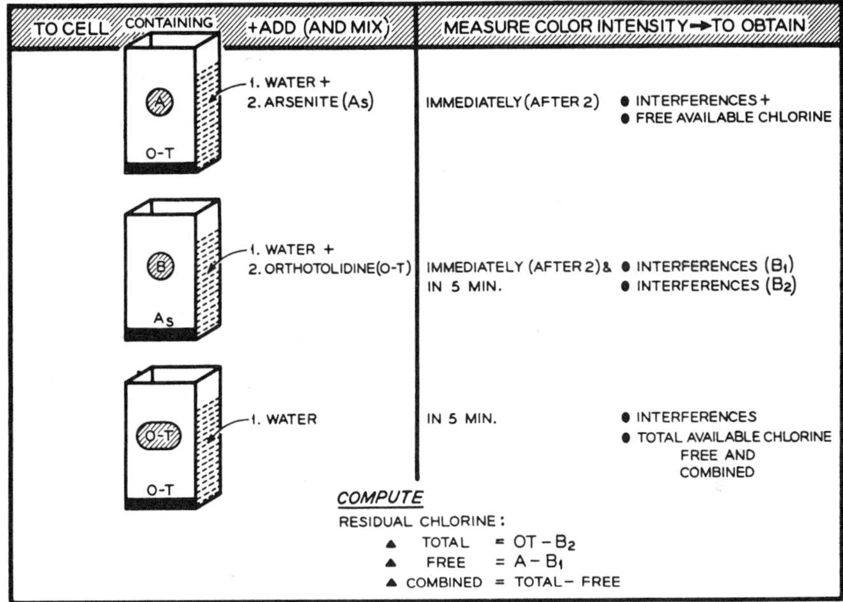

Figure 15-7. Orthotolidine-arsenite test for residual chlorine.

measure of free chlorine. If interferences are present additional test modifications are indicated to correct the values of total and free chlorine as above determined.

Various residual chlorine test kits employing either permanent glass or liquid OT color standards are available and widely used. Most units are designed to compensate for the effects of color and turbidity.

Amperometric Titration Test. This test measures free and combined chlorine residuals through oxidation-reduction titration procedures that require the use of electrometric devices employing a suitable electrode system to indicate completeness of reactions. Accordingly, the test is not suitable for field use. Phenylarseneoxide (C_6H_5AsO), the reducing agent normally used as the titrating agent, reacts with free chlorine residuals quantitatively at a pH from 6.5 to 7.5, and reduces chloramines at pH levels below 6.0 providing iodide ion is present. Presumably, chloramines oxidize iodide ion to free iodine and the phenylarseneoxide reduces the free iodine, thereby measuring the chloramine concentration. By conducting a two-stage titration with pH at about 7 and 4, free and combined chlorine residuals can be measured separately. Interferences from nitrites and oxidized manganese are eliminated by conducting the titrations above pH 3.5.[35] The test accuracy, however, is affected by the presence of nitrogen trichloride, chlorine dioxide, free halogens other than chlorine, and copper.

An adaptation of the amperometric technic is available for differentiation of mono- and dichloramines. The adaptation is subject to the usual interfering influences; in addition, it cannot be determined in which fraction those organic chloramines that might be present will titrate.

Equipment-Instrumentation

Feeders. The use of chlorine in connection with water (or waste) treatment may be indicated for various reasons, the most important of which is to protect the public health. For this purpose, chlorination processes usually must be continuous, and dual, stand-by equipment should ordinarily be provided.

Chlorine dispensing equipment is designed to control and measure chlorine in the gaseous state and to apply it either as a gas or as an aqueous chlorine solution. If the chlorine storage container is emptied in the liquid phase, vaporization or evaporation of the chlorine to gas is indicated before discharge to chlorine feeders. Gaseous feed is predicated on these factors: the practice lends itself to accurate flow regulation, either manual or automatic; gas may be metered with variable area or fixed orifice-type flow meters (with or without recording instruments); and, chlorine gas can be diffused into water.

Basically there are two types of equipment employed for applying chlorine.[27] *Direct-feed* equipment involves metering of dry chlorine gas and conducting it under necessary pressure conditions to the water (or waste) being chlorinated. *Solution-feed* equipment involves metering of dry chlorine gas under a vacuum and pre-dissolving it in a minor flow of water so that a high-strength solution is applied. In either case, the chlorine is dispensed to the feed equipment from an upright cylinder, the top (gas) outlet valve of a horizontal ton container, the gas valve of a single unit tank car, the outlet of a chlorine evaporator, or a header valve in the chlorine gas piping system.

The operation of direct- or solution-feed equipment is based on the common principle of regulating flow by establishing a pressure relationship on the upstream and downstream sides of a constant or variable orifice in the chlorine gas flow line. Means are provided for controlling the feed rate either by varying the pressure differential across a fixed orifice (variable differential units) or by varying the orifice size (constant differential units).

Direct-feed equipment involves gas application directly through a diffuser into the solution being treated. Because chlorine gas is only slightly soluble in water (see Chapter 3), the rate at which it can be fed is limited. Moreover, undissolved gas in a moist environment presents corrosive conditions of serious magnitude. These and other technical limitations and safety reasons have resulted in greater acceptance and wider use of solution-feed, vacuum-type equipment.

Dosage Control. The most elementary form of chlorination control involves manual adjustment of the chlorine feed rate as necessary. The earliest and most simplified form of automatic control provided for automatic maintenance of a fixed dosage rate in proportion to variations in the flow being treated. Later, additional forms of automation emerged which greatly enhanced the efficiency of chlorine treatment and simplified control by operating personnel. Among some schemes developed are the following.

(1) *Semi-automatic control.* The chlorine feeder is automatically actuated by pumps or timers to deliver a fixed, manually preset chlorine feed rate. In multiple pump installation a step-control regimen can be provided, and chlorine feed is automatically synchronized to provide proper dosage for one or a combination of pumps in operation.

(2) *Program-control.* Chlorine feed rate is automatically varied in accord with a pre-established time schedule. (This type of control is utilized more frequently in waste chlorination processes than for water treatment.)

(3) *Automatic proportional control.* The chlorine feeder automatically and continuously adjusts chlorine feed rate in accord with changes of flow to provide a constant, pre-established dosage at all flows. The desired dosage is manually preset by the operator and, by means of various flow-measuring devices, the chlorine feed rate is automatically controlled to yield the desired dosage.

Most proprietary chlorine feeders can be adapted readily to program, semiautomatic or automatic proportional control by responding to a vacuum, pneumatic or electrical transmission signal of primary metering elements. The transmitted signal regulates the chlorine feed in direct proportion to the main volume of flow. Moreover, where chlorine feed rate is more advantageously controlled at a point remote from the feeder, most equipment can be set up in such a way as to provide for immediate adjustment of the chlorine flow as and when needed. At such installations, an indicating chlorine flow transmitter at the feeder location and a flow recorder at the remote station often are provided.

Not one of these schemes, by itself, automatically compensates for variations in chlorine demand. Unfortunately, maintenance of a fixed chlorine dosage often does not provide that a chlorine residual necessary to accomplish the treatment objective will be maintained. Moreover, when chlorine requirements fluctuate due to demand variations, it is difficult to adjust the dosage rate of the feeder to prevent the application of too much or too little chlorine.

Within recent times an approach to fully automatic chlorine feeder control was devised which responds to changes in both flow and chlorine demand.[3] The unit automatically adjusts the feed rate to compensate for changes in chlorine demand and to maintain a constant residual in the water passing through it. The amount of chlorine required to produce this residual is measured and recorded as chlorine demand. This record is then transmitted electrically to the chlorinator which it actuates in proportion to the chlorine demand. In this scheme a signal from the demand recorder, together with a signal from a flow-measuring device, can be fed to the chlorinator which then actuates the chlorinator in accordance with both demand and flow changes.

Residual Indicators and Recorders. Where variations in chlorine demand are erratic and fluctuate within wide limits, a continuous indication and record of residual chlorine is an especially desirable operation adjunct.

Residual chlorine recording facilities provide more effective and economical chlorination practice. They are especially useful in detecting and compensating for sudden or natural changes in chlorine demand and in furnishing a permanent record of the efficiency of chlorine treatment. Remote location of chlorine residual recorders facilitates efficient and prompt adjustments of chlorine feed from central operating stations and minimizes the necessity of on-the-spot surveillance of treatment facilities. A variety of commercial units are available.

Chlorine flow rate recorders also offer many advantages. For example, they provide evidence and a permanent record of operation at the chlorinator or at remote sites as well as a constant check on chlorine consumed. Moreover, these records also are useful in determining correct chlorine dosages, seasonal and other variations in chlorine demand, and in evaluating other valuable operating information. Several commercial units are available.

MISCELLANEOUS APPLICATIONS

The foregoing discussion pertains primarily to chlorination processes and practices applied to potable water purification systems where biological control and/or modifications of chemical characteristics are the objectives of treatment. Chlorination of non-potable waters (e.g., bathing waters) and a variety of industrial process waters also is widely employed for similar purposes.

In many cases, treatment requirements are even more exacting than for potable water supplies.

Swimming Pools. Although bathing waters have not been epidemiologically indicted in the spready of respiratory, dermatological and gastrointestinal disorders often associated with bathing in polluted waters, most health authorities maintain that sanitary conditions exist only when there is provided and maintained at all times an active disinfectant in sufficient concentration to effect prompt and continuous pollution control. By far, chlorine, in its various forms, is the disinfectant most widely recommended and used for this purpose.

Bathing water chlorination does not differ greatly from drinking water chlorination except that in addition to destruction of pathogenic organisms naturally present in the water source, destruction of organisms of sanitary importance that are casually or deliberately introduced by bathers is indicated. The need for continuous chlorination during filtration and recirculation of pool water is generally indicated and intermittent treatment usually is considered makeshift and applicable only at small and private pools. Chlorine-ammonia treatment generally is disfavored and free residual practice is becoming more widespread in the United States.

Demonstration of bacterial destruction remains the sole criterion of adequate disinfection. Customary treatment involves maintenance of a free available chlorine residual of 0.4 to 1.0 mg/l except where breakpoint processes are employed. Combined residuals in the range of 0.7 to 1.0 mg/l, and preferably higher, usually are indicated.

Industrial Waters. Probably the greatest single outlet for chlorine in industrial water treatment is in the control of bacterial, algae, slime and macroscopic biological fouling organisms in fresh and saline condenser cooling waters. Depending on local conditions and treatment objectives, chlorination treatment may be continuous or intermittent. For the control of slimes, tubercles and the spat or larvae stages of larger organisms, free residual chlorine concentrations in the order of 0.5 to 2.0 mg/l often are employed. Once a satisfactory regimen is established, however, considerable economy is effected through chlorination treatment. Bacteria and slime control of process waters in the pulp, paper, beverage, canning and other food processing industries is often effected by chlorination.

Alteration of the chemical characteristics of industrial waters may also be accomplished by chlorination. Among the various objectives of treatment might be sulfur dioxide and ammonia destruction; iron and manganese reduction (brewing, beverage, paper, textile, laundering and photographic processing); color reduction (textile and paper-pulping bleaching operations); and oxidation of organics (chemical and food processing industries). For such purposes chlorination schemes employed usually are tailored to individual needs and objectives to provide safe, economical and efficient treatment.

Acknowledgment

The author acknowledges the helpful assistance and suggestions of Harry A. Faber, formerly of the Chlorine Institute, and currently Engineering Research

Coordinator for the Public Health Service, Department of Health, Education and Welfare, Washington, D. C.

References

1. American Public Health Association, Inc., "Standard Methods for the Examination of Water, Sewage and Industrial Wastes," Ed. 10, Baltimore, Md., Waverly Press, Inc., (1955).
2. American Water Works Association, Inc., "Water Quality and Treatment," Ed. 2, Lancaster, Pa., Lancaster Press Inc., (1955).
3. Baker, Robert J., "Types and Significance of Chlorine Residuals," *J. Am. Water Works Assn.*, **41**(9), 1185 (September 1959).
4. Butterfield, C. T., "Bactericidal Properties of Free and Combined Available Chlorine," *J. Am. Water Works Assoc.*, **40**(12), 1305 (December 1948).
5. Butterfield, C. T., Wattie, Elsie, Mergregian, S., and Chambers, C. W., "Influence of pH and Temperature on the Survival of Coliforms and Enteric Pathogens When Exposed to Free Chlorine," *Public Health Reports*, **58**, 1837 (December 1943).
6. Butterfield, C. T., and Wattie, Elsie, "Influence of pH and Temperature on Survival of Coliforms and Enteric Pathogens when Exposed to Chloramine," *Public Health Reports*, **61**, 157 (February 1946).
7. Chang, S. L., "Studies in *Endamoeba histolytica*," *War Med.*, **5** (46) (1944).
8. Chang, S. L., and Fair, G. M., "Viability and Destruction of the Cysts of *Endamoeba histolytica*," *J. Am. Water Works Assoc.*, **33**(10), 1705 (October 1941).
9. Clarke, Norman A., and Chang, Shih Lu, "Enteric Viruses in Water," *J. Am. Water Works Assoc.*, **51**(10), 1299 (October 1959).
10. Darnall, C. R., "The Purification of Water by Anhydrous Chlorine," *Am. J. Pub. Health*, **1**, 783 (1911).
11. Ellms, J. W., and Hauser, S. J., "Orthotolidine as a Reagent for the Colorometric Estimation of Small Quantities of Free Chlorine," *Ind. Eng. Chem.*, **5**, 915 (1913).
12. Enslow, L. H., "Chlorine in the Fields of Sanitation," *Chem. Markets*, **24**(2), (February 1929).
13. Enslow, L. H., "Progress in Chlorination of Water, 1927–1928," *J. Amer. Water Works Assoc.*, **20**(6), 819 (December 1928).
14. Faber, Harry A., "Super-chlorination Practice in North America," *J. Am. Water Works Assoc.*, **31**, 1539 (1939).
15. Faber, Harry A., "Contemporary Chlorination Practices," *J. Am. Water Works Assoc.*, **39**(3), 200 (March 1947).
16. Faber, Harry A., "How Modern Chlorination Started," *Water & Sewage Works*, **98**, 455 (November 1952).
17. Fair, Gordon M., and Morris, J. Carrell, Chang, Shih Lu, Weil, Ira and Burden, Robert P., "Chlorine as a Water Disinfectant," *J. Am. Water Works Assoc.*, **40**(10), 1051 (October 1948).

18. Gilcreas, F. W., and Hallinan, F. J., "The Practical Use of the Orthotolidine Arsenite Test for Residual Chlorine," *J. Am. Water Works Assoc.,* **36,** 1343 (1944).

19. Green, D. E., and Stumpf, P. K., "The Mode of Action of Chlorine," *J. Am. Water Works Assoc.,* **38,** 1301 (1944).

20. Griffin, A. E., "Chlorination—A Five-Year Review," *J. New England Water Works Assn.,* **58,** 322 (1944).

21. Griffin, A. E., "Reaction of Heavy Doses of Chlorine in Various Waters," *J. Am. Water Works Assoc.,* **31,** 2121 (1939).

22. Howard, N. J., "Modern Aspects of Chlorination of Water," *J. Am. Water Works Assoc.,* **40**(10), 546 (October 1948).

23. Howard, N. J., "Twenty Years of Chlorination of Public Water-Supplies," *The American City,* p. 791 (June 1927).

24. Johnson, George A., "Hypochlorite Treatment of Public Water Supplies," *Am. J. Pub. Health,* **1** (1911).

25. Laubusch, Edmund J., "Chlorination of Water," *Water and Sewage Works,* **105** (October 1958).

26. Laubusch, Edmund J., "How Safe is Your Chlorine Residual?," *Public Works* **106** (March 1959).

27. Laubusch, Edmund J., "Chlorine Feeding Equipment," *Water and Sewage Works,* **106** (June 1959).

28. Laubusch, Edmund J., "Chlorination Control," *Water and Sewage Works,* **106** (August 1959).

29. Laux, P. C., "Break-Point Chlorination at Anderson," *J. Am. Water Works Assoc.,* **32,** 1027 (1940).

30. Leal, J. L., "The Sterilization Plant of the Jersey City Water Supply Company at Boonton, N. J.," *Proc. Am. Water Works Assoc.,* p. 100 (1909).

31. Marks, H. C., and Glass, J. R., "A New Method of Determining Residual Chlorine," *J. Am. Water Works Assoc.,* **34,** 1227 (1942).

32. McAmis, J. W., "Prevention of Chlor-Phenol Taste with Ammonia," *J. Am. Water Works Assoc.,* **17,** 341 (March 1927).

33. Porges, Ralph, "The Background of Municipal Water Treatment," *Public Works,* **89**(2), 172 (February 1958).

34. Race, Joseph, "Chlorination of Water," Ed. 1, New York, John Wiley & Sons, Inc. (1918).

35. Sawyer, Clair N., "Chemistry for Sanitary Engineers," Ed. 1, New York, McGraw-Hill Book Company, Inc. (1960).

36. Snow, Brewster, "Recommended Chlorine Residuals for Military Water Supplies," *J. Am. Water Works Assoc.,* **48**(12), 1510 (December 1956).

37. Synan, John F., MacMahon, J. D., and Vincent, G. P., "Chlorine Dioxide: A New Development in the Treatment of Water," *J. New England Water Works Assoc.,* **458,** 264 (1944).

38. Thoman, John R., "Statistical Summary of Water Supply and Treatment Practices in the United States," Public Health Service Publication No. 301 (1953).

39. Tiernan, Martin F., "Controlling the Green Goddess," *J. Am. Water Works Assoc.,* **40**(10), 1042 (October 1948).

40. Wattie, Elsie, and Butterfield, Chester T., "Relative Resistance of *E. coli* and *E. typhosa* to Chlorine and Chloramines," *Public Health Reports,* **59,** 1661 (December 1944).

41. Weibel, S. R., "A Summary of Census Data on Water Treatment Plants in the U. S., *Public Health Reports,* **57**(45), (November 1942); Public Health Service Reprint No. 2416 (1942).

42. Wolman, Abel, and Enslow, L. H., "Chlorine Absorption and the Chlorination of Water," *Ind. Eng. Chem.,* **11,** 209 (1919).

43. Wolman, Abel, Donaldson, Wellington, and Enslow, L. H., "Recent Progress in the Art of Water Treatment," *J. Am. Water Works Assoc.,* **22**(9) 1161 (September 1930).

44. Value and Limitation of Chlorine Residuals in Distribution Systems, *J. Am. Water Works Assoc.,* **51** (February 1959).

16. WASTE-WATER CHLORINATION

Edmund J. Laubusch

The Chlorine Institute, Inc.

DEVELOPMENT

Historical

The use of chlorine for the treatment of sanitary wastes preceeded that for the treatment of potable water (see Chapter 15). About a half century before it was demonstrated (c. 1880) that certain bacteria are the cause of specific diseases, the application of chloride of lime for deodorizing sanitary wastes had been advocated, and practiced to some extent, on the theory that control of sewer odors would limit the spread of infection.[28]

The earliest large plant-scale adoption of sewage chlorination (using chlorinated lime) is believed to have been in 1854[12] by the Royal Sewage Commission who employed it to deodorize London sewage. Later, in 1859, tests reported by Hofman and Frankland in Germany demonstrated that putrefaction of sewage could be retarded four days by treating it with 400 lb of chlorinated lime per 1 mgd of waste.[16] Elsewhere chlorinated lime was used to minimize or control stream putrefaction in 1872 (Northampton, England) and in 1884 and 1887 (lower Thames River, England).[1]

Application of chlorinated lime for disinfection was practiced in England in 1879 by William Soper[32] prior to discharge of infectuous typhoid wastes into sewers. Experiments on sewage disinfection with hypochlorites followed in 1894 at Brest and Nice, France, and at Worthing, England; these involved hypochlorite addition to partially treated sewage.[1,17] From then until 1907 the action of chlorinated lime on raw and pretreated sewage was extensively investigated in England, France and Germany where the success of limited applications was demonstrated. Thereafter many bacteriological investigations also were undertaken in the United States.[1,12]

Prior to these investigations a U. S. patent was granted in 1887 to Powers[26] covering chemical production of gaseous chlorine (from manganese dioxide, salt and sulfuric acid) for deodorizing and disinfecting sewage; the process was adopted experimentally at a number of treatment plants in

New York State. Two years later, electrolytic processes for hypochlorite manufacture were being proposed for the purpose, chiefly in England. By 1893 the Woolf process,[42] involving electrolytic production of chlorine from brine which subsequently was reacted with caustic soda to form sodium hypochlorite, had been introduced and was adopted at Brewster, New York to treat the sewage from thirty homes in order to protect a portion of the New York City watershed.[1] Similar processes were later developed and applied at a few treatment plants, but all fell into disuse before 1910.

Meanwhile, experimental observations abroad were being evaluated and confirmed in the United States through laboratory studies—in 1906 to 1907 notably by E. B. Phelps and W. T. Carpenter at the Massachusetts Institute of Technology.[24] In 1907, Phelps made plant-scale studies[23] of sewage disinfection at Red Bank, New Jersey, which marked the beginning of effective sewage chlorination practices in the United States. During the next few years additional studies and plant applications involving both raw and partially treated sewage were made.[1,12]

The adoption of chlorination for sewage disinfection and odor control progressed only slowly, however, primarily because of the inconvenience of on-the-spot generation of hypochlorites and the expense and difficulties in handling chloride of lime, the best commercial source of chlorine existing at that time. Although a patent (British Patent 13755) covering the electrolytic production of chlorine had been granted to Charles Watt in 1851, due largely to the lack of commercial electric power, four more decades passed before chlorine became an article of commerce.[27] In 1884, the Elektron Company of Griesheim, Germany, began investigations of electrolytic chlorine production and, in 1890, started the first commercial electrolytic manufacture of chlorine in the world.[3,21]

Impetus was finally given to the adoption of sewage chlorination in the United States with the commercial production and liquefaction of chlorine at Niagara Falls in 1909, and with the production of proprietary chlorinators in 1912 and 1913. With these developments renewed interest in sewage chlorination became widespread due to the greater stability and more uniform composition of gaseous chlorine over chloride of lime and similar products, and to the relative ease and accuracy of measuring and dispensing chlorine into large amounts of waste water. The first municipality in the United States to employ liquid chlorine for disinfection of sewage effluent was Altoona, Pennsylvania in 1914.[12]

The uses of chlorine in sanitary waste treatment processes for purposes other than odor control and disinfection were recognized and adopted only many years afterward, as later noted. Today, disinfection remains the primary and principal use of chlorine in wastes treatment; however its uses for numerous other purposes are well recognized, and new applications are becoming increasingly apparent.

Although the possibilities of chlorination for treatment of industrial wastes was noted about 1916 when it was approved for disinfection of tannery wastes effluents, its development for purposes other than disinfection also progressed only slowly.

Current Status

Progress in the use of chlorine at municipal and industrial waste treatment facilities has been steady, but the development pattern is not nearly comparable with that for water treatment. The development of community wastes chlorination practice in the United States is indicated in Table 16-1.

TABLE 16-1. TREND IN DEVELOPMENT OF COMMUNITY WASTES CHLORINATION IN THE U.S.A.[a]

| | Treatment Plants | | | Est. Population Served $\times 10^6$ | | |
Year	All Plants	Using Chlorine	% Using Chlorine	All Plants	Plants Using Chlorine	% Benefiting Wastes Chlorination
1910[25]	619	22	3.6	4.46	0.11	2.4
1916[25]	846	55	6.5	6.14	0.28	4.6
1934[40]	3697	655	17.7	22.20	–	–
1940[41]	5580	1127	20.2	40.62	14.34	35.3
1945[34]	5786	1262	21.8	46.86	16.03	34.2
1948[35]	6058	1307	21.6	48.70	18.04	37.0
1957[36]	7518	2216	29.5	76.44	37.84	49.5

Note: [a]Data are not strictly comparable due to slight variations in statistical reporting and analytical procedures.

It is noteworthy that the proportion of total plants having chlorination and the number of persons served by such facilities increased significantly between 1948 and 1957. The percentage increase in numbers of plants and population served during this period is even more pronounced, being 69.5 and 109.8 percent respectively. (North Dakota was the only state in 1957 having no sanitary wastes treatment facilities using chlorine). In 1957, chlorine gas was used by at least 71.5 percent of all plants with chlorination facilities and hypochlorites were used by at least 6.4 percent. (The disinfectant used at the remaining plants is unknown). As is the case for water treatment, the utility of hypochlorites is generally limited to plants serving about 5000 or fewer persons.

The distribution of chlorination facilities at community wastes treatment plants in the United States where such are available is indicated in Table 16-2.

The writer estimates that in 1960 more than 60,000 tons of liquid chlorine were used in the United States for disinfection and other treatment of

TABLE 16-2. COMMUNITY WASTES TREATMENT PLANTS IN THE U.S. USING CHLORINE[36]—1957

Plants	Minor Treatment		Primary Treatment		Intermediate Treatment		Secondary Treatment	
	Plants	Est. Pop. served $\times$ 10^6	Plants	Est. Pop. served $\times$ 10^6	Plants	Est. Pop. served $\times$ 10^6	Plants	Est. Pop. served $\times$ 10^6
All	41	1.86	2730	25.67	100	5.59	4647	7.52
Using chlorine	27	0.74	775	14.23	60	3.49	1354	19.38
% Using chlorine	65.9	39.9	28.4	55.4	60.0	62.4	29.1	44.7

community and industrial wastes. Now that there is strong official and public sentiment in favor of water resources conservation (through waste treatment and water reuse) and pollution abatement, established and new uses of chlorine for such purposes may be expected to become more widespread.

BASIC CHEMISTRY

Initial Reactions

The utility of chlorine in wastes treatment is attributable to its toxicological characteristics, its oxidative capacity, and its adaptability as a coagulant. Chlorine is employed primarily to inactivate or destroy bacteria and other organisms, and to modify the chemical or physical characteristics of the waste being treated. The most important application of chlorine in sanitary wastes treatment is for disinfection, while alteration of physical and chemical characteristics is the most general use for chlorine in industrial wastes treatment.

The initial reactions on adding chlorine (a solution of chlorine gas or hypochlorites) to sanitary and industrial wastes usually are considered to be essentially the same as those that occur when chlorine is added to water (see Chapter 15). Depending on the pH of the waste, hypochlorous acid ($HOCl$) and hypochlorite ions (OCl^-) exist in various proportions at equilibrium (see Figure 15-1). At normal pH levels chlorine exists first as free available chlorine and, subsequently, in various combined forms. As in water treatment, but to a much greater degree, free chlorine reacts immediately with a variety of impurities contained in liquid wastes.

Reactions with Impurities

Sanitary and industrial wastes contain a large and complex variety of suspended and dissolved inorganic and organic materials in water, sometimes in addition to large numbers and varieties of bacteria, protozoa and other organisms. Many of these materials have an effect on wastes chlorination practices similar to that which they have on water chlorination practices; moreover, their frequently greater numbers, variety and complexity make them even more significant in the establishment of a chlorination regimen suitable for treatment objectives. (This subject is covered fully in Chapter 15).

Chemical reducing agents such as hydrogen sulfide (a normal decomposition product of unstable components of sanitary wastes), sulfur dioxide and soluble sulfite salts, hydrosulfites, sulfoxylates and ferrous iron salts consume oxygen and may cause septic conditions. The oxygen-consuming characteristics of reducing substances has an analogy in their chlorine demand. When chlorination is included in sanitary and/or industrial wastes treatment processes, these reducing agents markedly increase chlorine requirements. Among strongly reducing industry wastes are: sulfite pulp mills (sulfur dioxide and sulfites), oil fields and petroleum refineries (hydrogen sulfide and other sulfides), textile dyehouses (hydrosulfites and sulfoxylates), tanneries (sulfides) and many chemical manufacturing plants. Treatment of such wastes frequently involve their oxidation, in whole or in part, by chlorine. Where disinfection is an objective (e.g., certain effluents from tanneries and meat packing operations, etc.) the chlorine demand due to reducing agents, organics and other chlorine-consuming materials will have to be supplied.

Chlorine reacts with ammonia to form chloramines that have a lower oxidation potential than free available chlorine forms even though chlorine is still available for chemical reaction. Chlorine also has a distinct affinity for organic nitrogenous material normal to sanitary and some industrial wastes. For example, it readily combines with the nitrogen-containing molecules of protein and amino acids to form organic chloramines. The oxidation potential and disinfecting capacity of organic chloramines, as noted in Chapter 15, is substantially less than that for inorganic chloramines and free available chlorine forms.

Unstable organic substances normal to sanitary and some industrial wastes comprise the bulk of settleable and (especially) non-settleable suspended solids. The removal of settleable organic solids along with stable, settleable inorganic materials (sand, grit etc.) presents no serious operational problems in wastes treatment although the former exert a substantial chlorine demand until their separation is accomplished. The removal of nonsettleable, colloidal solids that are chiefly of organic composition is more challenging, and many such materials exert a very substantial chlorine demand in the process of chlorine absorption and formation of chlorine-addition products. For example, when sufficient chlorine is added to domestic sewage to carry the oxidation reaction to an equilibrium in a 10-minute reaction period, the nonsettleable solids (principally unstable organics, but also including some finely-divided inorganics) consume about 50 per cent of the chlorine added; the settleable and soluble substances (unstable organics, stable inorganics) each consume about 25 percent of the chlorine added; only a negligible amount of chlorine is consumed by bacterial cells and higher biological forms.[12]

Oxidation reactions involving chlorine reduce the putrescibility of the waste and the oxygen required to satisfy its biological oxygen demand (BOD), as later discussed. The amount of chlorine required to carry chlor-oxidation and addition reactions to equilibrium at any given time interval is a function of the concentration and type of oxidizable and chlorine-absorbing materials present. Thus, the chlorine required to reach an equilibrium or completion of reactions in a given time interval is a measure of the concentration of such materials present in the waste.

Chlorine causes physical changes in many wastes, apparently as a result of its effect on the electrical charge of colloidal particles (it is reported to destroy the emulsifying substances that stabilize colloids). Coagulation may be aided, leading to improved settling of solid particles or floating of oily liquids. Coalescence of colloidal droplets aids in the rising of oils and greases to the surface where they can be recovered.[14] The application of chlorine for such purposes has been especially prominent in the treatment of meat-packing industry wastes.

Like other chemical reactions, the activity of chlorine dissolved in liquid wastes is dependent on such factors as temperature and pH. Increase in temperature and hydrogen ion concentration accelerates chemical reactions. The speed and completeness of oxidation also is governed by the law of mass action. Thus, the concentration of chlorine, chloramines and oxidizable substances controls the mechanism of the reactions and the types of end-products formed. With some organic materials, oxidation reactions proceed slowly even under optimum conditions of temperature, pH and concentration of reacting substances, and an appreciable time period may be necessary to establish equilibrium.

TREATMENT OBJECTIVES

Disinfection

Purposes. Not all of the myriad organisms in sanitary wastes are harmful as potential disease-bearing vectors. Some organisms, classified broadly as *saprophytes*, are essential to aerobic and anaerobic biological purification processes. Disinfection of sanitary or industrial wastes by chlorine, where indicated, must be a selective process so that saprophytic organisms can continue their breakdown of putrescible organic constituents of the waste.

In wastes treatment, as in water treatment, the coliform density remaining after chlorine treatment is the accepted index of its effect. Because wastes generally are discharged into receiving waters where dilution and some degree of natural purification are offered, complete coliform destruction is not the usual treatment objective. The degree of bacterial destruction and the effective chlorine dosage are largely determined by the characteristics and subsequent uses of the receiving water body and discharge restrictions that might be imposed by authoritative requirement. Continuity of chlorine treatment might depend on local needs and requirements. For example, if the receiving water is used as a source of potable water, continuous chlorine treatment of the waste water may be required even though complete disinfection may not be necessary. Requirements may also vary on a seasonal basis where potable water supplies are not involved. For example, the Ohio River Valley Water Sanitation Commission, ORSANCO, provides that wastes discharges into the Ohio River be treated to effect an 85 percent reduction in coliform organisms during May through October, and a 65 percent reduction during the balance of the year.[20] The basic criterion used by state public health departments or stream pollution control authorities in establishing the need for sewage disinfection is the effect of the discharged wastes on the receiving water, and its subsequent uses. Chlorination for disinfection may be required in cases where wastes are discharged in close proximity to public water supply intakes, bathing and other recreational areas, and shellfish propagation areas, where crop irrigation is involved (Arizona), or where dairy cattle pastures are involved (Illinois). Thus, chlorine disinfection of wastes is practiced continuously, intermittently, or not at all, depending on local considerations and subject to the requirements of appropriate regulatory authorities involved.

Chlorine Requirements. Only broad generalizations can be made regarding chlorine requirements for disinfection of waste waters. The mechanism of bacterial destruction is considered to be the same as that thought to apply to water chlorination (see Chapter 15). The extent to which biological destruction or attenuation occurs depends on factors such as the amount and form of available chlorine, the contact time and temperature and the composition of the waste being treated. To effect a degree of bacterial kill it is not essential to apply a chlorine dose sufficient to produce a measurable free chlorine residual. Reductions in bacterial concentration can be and are effected through marginal chlorination, the amount varying with the proportion of the chlorine demand satisfied. Obviously, this is subject to wide variations and depends largely on the character and strength of the waste being treated.

The variation in chlorine demand for some wastes may be fairly predictable, but the amount of chlorine needed in the control of disinfection can be assured only by the determination of residual chlorine. The chlorine required to yield a specified residual varies with the flow, composition, age and temperature of the waste and the time of contact. For combined wastes, the amount and kind of industrial wastes exert a substantial influence on chlorine requirements. It is well established, for example, that chemical reducing substances common to all wastes in varying amounts react with chlorine to form compounds having little or no bactericidal capacity. The formation of chloramines by the action of chlorine on nitrogenous constituents of wastes renders these constituents unsuitable as food sources; however, as already noted, they do possess definite disinfecting power. In wastes treatment, in fact, free residual chlorination is rarely practiced.

Regardless of the kind or amount of available chlorine, that which does not penetrate suspended solids is ineffective in the destruction of bacteria and other organisms encased by these solids. Waste effluents containing substantial amounts of suspended solids that are discharged into water bodies may cause serious aftergrowths due to the release of imbedded bacteria by the action of hydraulic turbulence. Furthermore, some biological species that are relatively unaffected by chlorine dosages typically encountered in waste treatment practices may reproduce prolifically in the absence of large numbers of metabolically-competitive, chlorine-susceptible forms. Estimated chlorine requirements for disinfecting normal sanitary sewage are listed in Table 16-3.

TABLE 16-3. ESTIMATED CHLORINE REQUIRED FOR DISINFECTING NORMAL SANITARY SEWAGE

	mg/L Chlorine		
Treatment	"Ten-State Standards"[37]	WPCF[12]	Fair and Geyer[11]
Pre-chlorination			
(mechanically cleaned tanks)	20–25	–	–
Raw sewage, depending on age	–	6–12 (fresh)	6–24
		12–25 (septic)	–
Primary effluent	20	–	3–18
Chemically precipitated sewage	–	–	3–12
Trickling filter effluent	15	3–10	3–9
Activated sludge effluent	8	2–8	3–9
Sand filter effluent	6	1–5	1–6

Effective disinfection requires that there be a chlorine residual after a suitable contact period. Because the waste of each community or industry necessitates a special determination of that chlorine dosage which will yield the desired or required chlorine residual, it is essential that chlorine application be regulated as often as necessary to insure efficient and economical operation. Only by frequent determinations of chlorine residual can the dosage rate, and thus the efficacy of disinfection, be satisfactorily established. Currently, little significance is placed by control authorities on the definition of minimum chlorine residuals; rather, the trend is toward the establishment of stream standards in contrast to waste effluent standards with the provision that wastes be managed prior to stream discharge in such a way that stream criteria are supported. Satisfactory disinfection of secondary sewage effluents generally is obtained when the chlorine residuals after 15 to 30 minute contact are within 0.2 and 1.0 mg/l.   A residual of 0.5 mg/l after 15 minute contact appears to be a safe average.[14]

Point(s) of Application. In employing chlorine in waste treatment processes, care must be exercised to select the point(s) of application and dosages in such a manner as to accomplish the objective as efficiently as possible without interference to the action of saprophytic organisms essential to aerobic or anaerobic purification processes.

Flexibility in operation should be provided so that chlorine can be applied wherever it can best accomplish the desired objective. When that objective is primarily disinfection, pre-chlorination and/or post-chlorination usually is employed. Up-sewer chlorination (i.e., ahead of the treatment plant) should be employed for disinfection only in those instances where subsequent treatment is limited to screening. It is inefficient for such applications because the chlorine demand of raw sewage is great and the chlorine so applied can not fully penetrate bacteria-laden solids. (As will be noted later, however, up-sewer chlorination frequently is applicable for many purposes other than disinfection).

When secondary waste treatment is provided, post-chlorination is practiced where disinfection is indicated. Chlorine is applied at a contact tank, preceding the effluent discharge, or at the influent of the final clarifier. The latter is increasingly employed except where contraindicated by activated sludge and biological filter recirculation practices. Prechlorination in dosages necessary to effect disinfection also is contraindicated where subsequent biological processes are involved. Lesser amounts of chlorine for other purposes, however, can be applied throughout the treatment plant, as later discussed.

When only primary treatment is provided, limited reductions in bacterial density are accomplished and disinfection often is required by local authorities. Either or both pre-chlorination and post-chlorination may be employed; split chlorination practices are becoming increasingly popular. Applied at one or more points in the interceptor, or to raw or screened wastes ahead of the settling units, maximum contact is provided. Moreover, though disinfection is the primary objective, other benefits may accrue such as odor control, septicity control and improved settling.

Terminal application alone is satisfactory, but the practice usually is less desirable since a special tank or other means to provide adequate contact with chlorine is required; also, the chlorine demand of settled waste actually may be greater than that of the raw waste if it is stale or putrescing.

Operational Control. From the foregoing discussion it is apparent that the considerable variations in the chemical and bacteriological character of wastes, and in disinfection objectives, preclude the establishment of rigid, uniform chlorine residual requirements. The value of maintaining a fixed chlorine residual is based principally on (1) local demonstration of the effect of such practices on desired bacterial kills, and (2) conformance to requirements relative to the quality of the plant effluent or of the receiving water at any downstream point.

Operation control of waste chlorination processes creates a somewhat difficult situation. Due to the presence of oxidizable and chlorine-absorbing constituents, residual chlorine in conventional waste treatment exists almost invariably in combined forms while an appreciable, unsatisfied chlorine demand might simultaneously exist. Since wastes chlorination seldom is practiced to the extent that free chlorine forms persist, differentiation of free and combined chlorine is not made.

For the control of disinfection, a sufficient number of bacteriological tests must be made to establish that chlorine residual which is necessary to destroy coliforms or, more generally, to reduce the coliform density to a stipulated amount. The amount of chlorine needed to effect the desired coliform density (as determined by the MPN or other acceptable tests) or to meet stream quality criteria depends on the volume of flow and varies with the composition, age and temperature of the waste, the amount and kind of inorganic and organic components, the contact time, the condition of the receiving water at the time of discharge and several other factors. As in waterworks practice, periodic bacteriological confirmation of the objectives is desirable and may be legally required.

For purposes other than disinfection, chlorination control also can be obtained by frequent and routine chlorine residual determinations, using the same residual testing technics as later described. For some purposes, control is based on a percentage reduction of the total chlorine requirement.

Operating criteria until recently were based largely on determination of the *chlorine demand,* defined as the amount of chlorine required to produce a residual chlorine content of 0.1 ppm after a 15 minute contact. This term is not recognized in the current edition of "Standard Methods"[2] because it is meaningless in those numerous instances where a higher chlorine residual is required to accomplish the results desired. Necessary chlorine dosages now are determined on the basis of *chlorine requirement,* which represents the amount of chlorine (in mg/l) which must be added to a waste to produce a chlorine residual of such strength that, after a definite contact time, the effluent will show a desired coliform density or will meet the requirements of some other treatment objective. This operating index affords more flexibility than the earlier concept of chlorine demand, and it embodies, as does the chlorine demand index for water, quality variations and differences in treatment objectives.

Chlorine requirement tests are adaptable to plant or laboratory use and are similar to chlorine demand tests employed in water works practice. It should be noted that the value of the chlorine requirement test is relative and cannot be used for comparing operating results from time to time or from place to place. Its utility lies in its application to the control of disinfection and in the estimate it affords of chlorine needed to satisfy other treatment objectives.

Tests for chlorine residual serve the same general purpose in water and waste chlorination control, but there are some differences in technic between the two. In part, as previously noted, no distinction is made between free and combined available chlorine residual forms in sanitary and some industrial wastes treatment practices. The same three basic tests for residual chlorine determinations in water (see Chapter 15) are used for residual chlorine determinations of wastes and other highly polluted waters. For combined and some industrial wastes the use of one or another may be precluded because of interferences, or special modifications may have to be adopted.

The *amperometric test* is unaffected by color and turbidity—frequently an important consideration in waste treatment control. An unusually high organic content sometimes causes uncertainty in establishing the titration end point, but the error is still less than that common to other tests. Provision is made in the test for minimizing interferences due to iron, manganese and nitrites.

The *starch-iodide test* is especially suitable where use of the orthotolidine test is precluded due to the presence of nitrites in excess of 2 ppm and in the presence of substantial amounts of manganic manganese. The precision of the test decreases with increasing concentration of color and organic matter. Residuals so determined generally are in good agreement with those obtained by the amperometric method.

The *orthtolidine test* consistently yields lower chlorine residuals than any other test—probably because of the presence of organic matter. The disparity may be in the range of 2 to 5 mg/l for ordinary settled sewage, and it may be of even higher magnitude for sanitary and industrial waste mixtures. A turbidity-compensating comparator is essential, and the use of permanent orthotolidine color standards is preferred. The OTA test modification (see Figure 15-7) is recommended for industrial wastes analyses if iron and manganese are present and the organic nitrogen content is low.

Other Applications

Chlorination is employed for a number of specialized applications. Next to disinfection, perhaps one of the principal uses of chlorine is to control odor and retard putrefaction. Additionally, a number of other uses have been established (see Figure 16-1).

Odor and Septicity. As mentioned earlier, chlorinated lime was used for controlling sewage odors for many decades prior to the realization that certain bacteria are disease vectors. Chlorination specifically for the control of odors in sanitary and industrial (dairy) wastes was first practiced in 1928.[12]

Most sanitary wastes are potential sources of odors because they present an ideal environment for prolific bacterial metabolism and subsequent putrefaction (anaerobic) of organic waste solids. The composition of sanitary and industrial wastes often is such that chemical treatment is indicated to supplement other conventional treatment practices. Chlorine, being easy to apply and quick acting, is one of the more important chemicals used for control of odors, particularly that esthetically objectionable and economically important product of anaerobic decomposition, hydrogen sulfide.

In theory, about 2 parts by weight of chlorine are required to precipitate sulfur for each part by weight of sulfide (as H_2S), and about 8 parts by weight are required to oxidize it to sulfate:

$$HOCl + H_2S \rightarrow S\downarrow + HCl + H_2O \qquad (16\text{-}1)$$

$$4\,HOCl + H_2S \rightarrow H_2SO_4 + 4\,HCl \qquad (16\text{-}2)$$

These reactions take place at pH 5-9, but the first (16-1) occurs optimally at the lower pH level and the second (16-2) at the higher level. Because of the effect of other chlorine-demanding substances, considerably more chlorine actually may be needed to effect the reactions specifically desired for odor and septicity control.

Many factors contribute to and influence hydrogen sulfide formation in waste collection and distribution systems. Where indicated, the up-sewer addition of chlorine provides additional oxidation potential for oxidation of hydrogen sulfide and, through its bacteriostatic or bactericidal action, inactivates or destroys sulfate-reducing organisms.

Within waste treatment plants there are innumerable opportunities for odor development, especially if dissolved oxygen or a chlorine residual is not maintained. These odors are due to the escape of dissolved gases or volatile compounds produced by the biochemical degradation of organic constituents of the waste. The function of chlorine is to check those biochemical changes which result in odor production. Often this is accomplished through partial satisfaction of the chlorine demand (without a measurable chlorine residual), but the indicated dosage usually is greater if odor-producing materials other than hydrogen sulfide are present.

Industrial wastes high in organic constituents sometimes are chlorinated to prevent or retard septicity preliminary to, during, or after other treatment, including natural purification in receiving water where sufficient dilution and other factors preclude the necessity of other treatment.

The utilization and efficiency of chlorine in odor control has permitted substantial savings, as treatment plants may be built within or in close proximity to city limits or at the manufacturing site, thus precluding added expenses inherent to remote locations.

Corrosion Control. Concrete and metal collection or disposal systems conveying septic wastes having hydrogen sulfide in excess of about 1 mg/l deteriorate, especially where the temperature is high and sluggish flows at low pH levels prevail. These conditions enhance accelerated growth of slime-ensheathed, sulfate-reducing organisms that provide additional sulfide. Some of this sulfide may be absorbed by moisture or grease above the liquid surface in the pipeline where it is converted to sulfuric acid which, in turn, destroys concrete and masonry structures by reacting to form calcium sulfate. These conditions also are damaging to metallic structures and equipment; if allowed to proceed unchecked, the economic loss can be great.

Sub-residual, up-sewer chlorination at one or more points, and usually at somewhat higher dosages than those indicated for odor control only, will limit hydrogen sulfide formation to concentrations which prevent escape of the gas to the air. Residual concentrations may be indicated for slime destruction.

BOD Reduction. The use of chlorine for the reduction of the biochemical oxygen demand of sewage was first demonstrated on a plant scale in Germany in 1925. In the United States, the treatment was evaluated experimentally in 1926; it was adopted the following year at Dallas, Texas. Chlorine for BOD reduction of industrial (cannery) wastes was first practiced in 1929.[12]

The literature is replete with evidence that chlorine reduces the BOD of a waste. According to Fair[11] four kinds of reactions are conceivably involved: (1) direct oxidation of BOD-exerting compounds; (2) formation with nitrogen compounds of bactericidal chloramines by substitution of chlorine for hydrogen; (3) formation with carbon compounds of substances that are no longer decomposable (again by substitution of chlorine for hydrogen); and, (4) addition of chlorine to unsaturated compounds to form non-decomposable substances. The addition of chlorine is not advocated as a substitute for properly designed and operated treatment processes; it is recommended, however, for reducing the BOD load on various unit processes by minimizing the effect of overloading caused by unusual flows, by the presence of certain industrial wastes, digester supernatant or activated sludge return to inflow, and for other purposes.

BOD reduction usually is effected where up-sewer chlorination is practiced for odor control, for prechlorination for disinfection or odor control, or wherever else chlorine is applied for any purpose either before or within a waste treatment plant. While free residual chlorination will effect as much as an 80 percent BOD reduction in activated sludge effluents, these and other secondary treatment effluents seldom are chlorinated for the specific purpose of BOD reduction, except possibly where the units are grossly overloaded. Similarly, while free residual chlorination of raw sewage might effect as much as a 50 percent BOD reduction, this treatment is impractical under usual circumstances largely because of the high chlorine requirement due to the concentration of ammonium compounds and other nitrogeneous materials present in sewage.

In practice, and where no other treatment is provided, a reduction of 5-day BOD of sewage of 15 to 30 percent might be anticipated in the presence of 10-minute chlorine residuals of 0.2 to 0.5 mg/l. Recent investigations,[31] however, suggest that the condition of the waste treated is an important factor. Furthermore, significant additional over-all BOD reduction is effected under any condition for which chlorine is employed.

Biological Filter Operation. The utility of chlorine as an adjunct to trickling filter operation was first demonstrated about 1925.[12] The release of odors over trickling filters associated with the treatment of septic wastes usually can be controlled by prechlorination. Where this is not possible, chlorination of the primary tank effluent or at the filter dosing chamber may be indicated. To avoid interfering with the biological nature of the process only a portion of the chlorine demand should be satisfied—usually about 50 to 75 percent—and there should be no residual chlorine at the filter influent.

The psychoda fly nuisance especially associated with filter operation during warm weather oftentimes is amenable to intermittent chlorine treatment. Various successful treatment regimens have been reported, but optimum treatment should be based on local experiences.

Filter ponding, due to the accumulation of excess solids that clog the interstices of the filter, sometimes can be alleviated by chlorine. Intermittent shock treatment to produce a residual of 5 to 10 mg/l at the filter nozzles during periods of low chlorine demand, or continuous chlorine application to produce a residual of 2 to 5 mg/l usually will be effective in loosening excess zoogleal material that can then be flushed from the filter.[12,14] Where clogging has deeply penetrated the bed, or otherwise is due to the accumulation of insoluble mineral particulates, chlorine treatment usually is of little lasting or no benefit.

Activated Sludge Operation. The activated sludge process of wastes treatment is a complicated biological process which requires considerable care in operation and control. The process can become upset easily by a variety of influences. The condition of sludge bulking, manifested by poor settleability of activated sludge in the final clarification unit, and often accompanied by an abnormal concentration of filamentous fungi, *Sphaerotilus natans,* sometimes can be eliminated by chlorine treatment. This was first recognized about 1928.[12] The chlorination regimen must be carefully established to prevent upsetting the delicate biological flora and fauna necessary to the process and to prevent an imbalance between it and the organic matter applied. In instances involving excessive overloading of treatment facilities, prechlorination often is beneficial. Where bulking of activated sludge is attributable to an abnormal concentration of filamentous fungi, the application of chlorine to return activated sludge at a point which allows 2 to 3 minute contact before mixing with the contents of the aeration tank also may be of benefit. The proper chlorine dosage is dependent on the sludge index and the amount of dry solids in, and the rate of, the return sludge. It appears that continuous chlorination yields results superior to those obtained by intermittent practice.

It was first observed about 1933 that the disposal of waste activated sludge can be facilitated by chlorine treatment.[12] In practice, chlorine is added to the incoming sludge of concentrating or thickening units to maintain the stability of the sludge during the holding time necessary for effective thickening, prior to discharge to digestors. The maintenance of a chlorine residual of 1.0 mg/l in the clear supernatant of the thickening unit appears to yield satisfactory results. Excessive chlorine causes peptization or dispersion of sludge, destroying its settleability and decreasing the efficiency of the sludge thickening process.

Chlorine treatment also is of benefit in reducing the immediate oxygen requirements of return activated sludge and of digester liquor returned to the waste treatment process.

Miscellaneous Uses. Among other applications in sanitary and industrial wastes treatment processes for which the utility of chlorine has been established are:

(1) Chemical Processes—to aid coagulation and settling of wastes and to prevent septicity.

(2) Digester Supernatant—chlorination preceeding discharge to primary settling units to reduce the load and minimize odors.

(3) Digester Sludge—chlorination prior to discharge to drying beds for odor control.

(4) Microorganism Control—up-sewer, pre- and post-chlorination to control or destroy algae, slime-forming, sulfate-reducing and other organisms detrimental to the waste collection, treatment or discharge system.

(5) Grease Control—aero-chlorination preliminary to primary settling, to minimize interferences with biological and other treatment processes.

(6) BOD Reduction—to reduce or delay the BOD of waste waters and effluents discharged into receiving waters.

(7) Others—plant improvements designed to reduce the plant load, to overcome odor or septic waste conditions, to alleviate unsightly grease accumulations, etc. (see Figure 16-1).

Industrial Wastes

The treatment and disposal of industrial wastes frequently presents a great challenge for their characteristics often are such that they cannot be combined with the sanitary wastes of the community. For some, only controlled discharge into community sewers or receiving dilution waters may be indicated, while for others partial or complete treatment may be required. Unit treatment processes applicable to industrial wastes are similar to those used for treating sanitary wastes and include various physical, chemical and biological systems, or combinations thereof, designed to alter the characteristics of the waste to permit safe and unobjectionable disposal. The use of chlorine for treating industrial wastes began about 1916, with the disinfection of tannery effluents. Further progress was slow until about 1928, but since then new chlorination applications have increased rapidly. A review[12] of the subject lists chlorination of dairy wastes for odor control in 1928; chlorination of cannery wastes for BOD reduction in 1929; chlorination of meat-packing wastes for BOD reduction and coagulation, also in 1929; chlorination of textile wastes for coagulation and color reduction in 1933; chlor-oxidation of cyanide wastes commencing in 1942; hypochlorination of wool-scouring wastes and recovery of wool grease, industrially applied in 1943; and plant-scale development of chlorine oxidation of phenolic wastes in 1950. In addition, the use of chlorine in paper mills, at power stations and in other industry areas has come into wide practice to control slime growths in recirculation water that otherwise would be unsuitable.

Analagous to sanitary wastes treatment, chlorination alone for industrial waste treatment is practiced rarely; rather, its primary use is as an adjunct to, or component part of other treatment processes, or as a final polishing medium. Some wastes have such a high chlorine demand that chlorine treatment is contraindicated for practical or economic reasons, and certain wastes are unaffected or adversely affected by chlorine. Each waste constitutes a special problem that requires thorough study of the suitability and means of chlorine application. Objectives of chlorine treatment might include one or more of the following: odor and septicity control, chemical oxidation, BOD reduction, color reduction, coagulation and partial or complete disinfection.

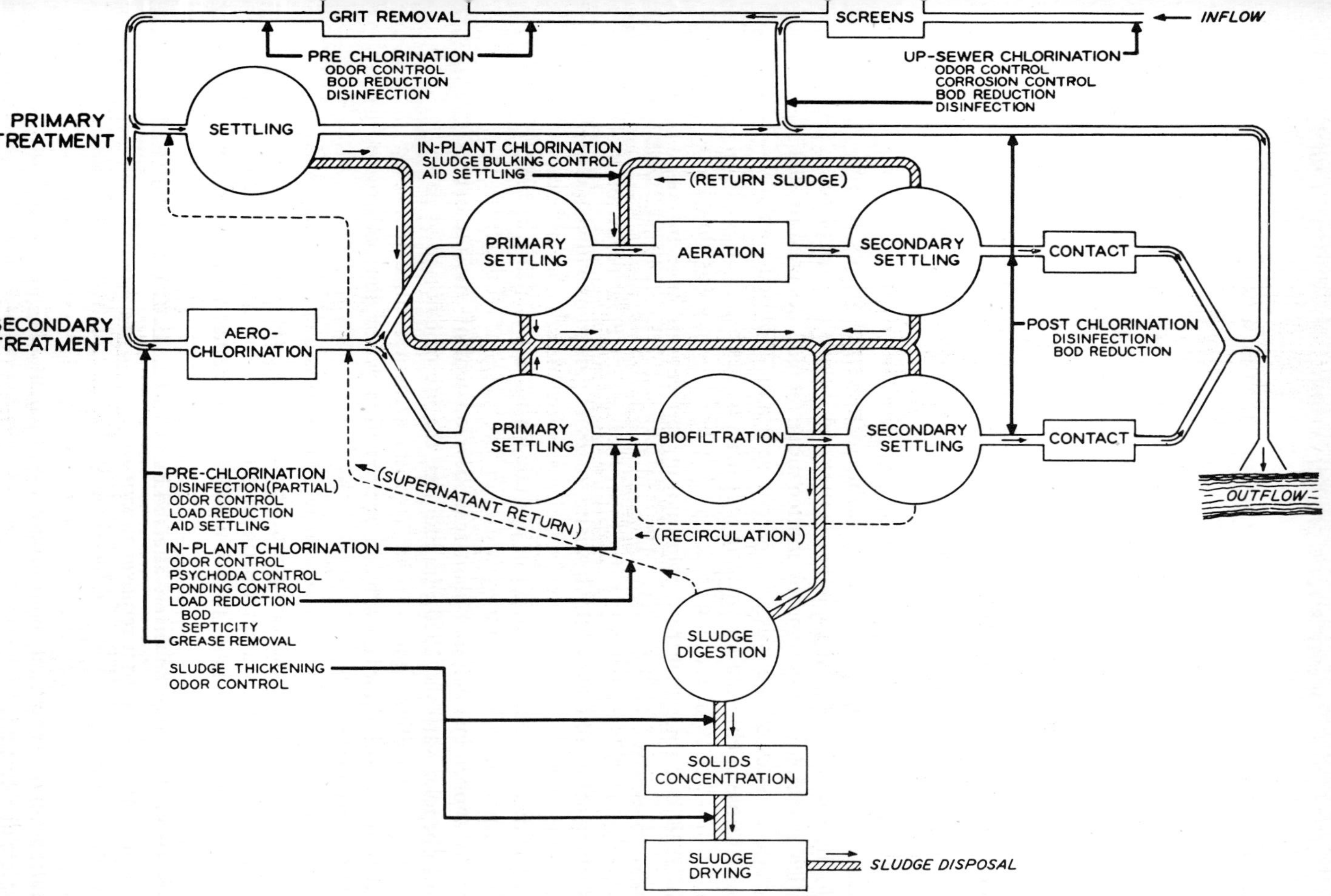

Figure 16-1. Typical applications of chlorine in wastes treatment processes.[19]

The utility of chlorine in industrial wastes treatment is based on its oxidative capacity (e.g., oxidation of cyanides and phenols), its value as a coagulant (e.g., packing house wastes), and its disinfecting action (e.g., slime and bacteria control in paper and food processing industries). Industry wastes for which chlorination is employed, alone or in combination with other processes, are highlighted in Figure 16-2 and in the following discussion.

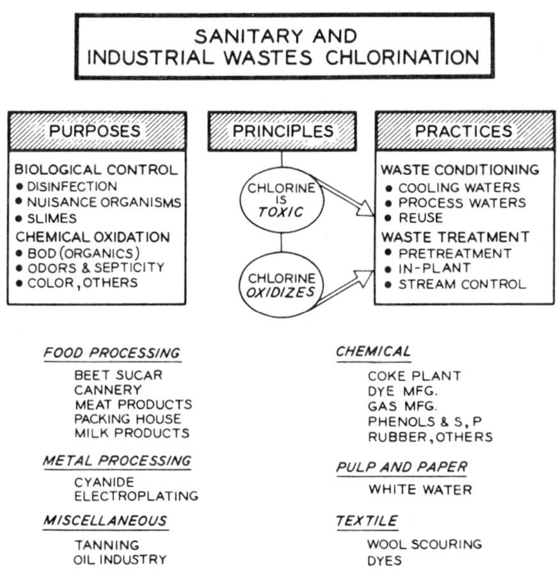

Figure 16-2. Typical applications of chlorine in wastes treatment for resources conservation and pollution abatement.

Chemical. Phenols, the hydroxy derivatives of benzene which include various phenols, cresols and related compounds of higher order, are constituents of many chemical plant effluents, gas manufacturing wastes, coke-plant wastes, petroleum refinery wastes, wood distillation wastes and others. Aside from their toxicity to fish life, minute concentrations in water cause serious taste and odor problems. As little as 0.001 to 0.005 mg/l of phenol on chlorination produces a medicinal taste and odor of chlorophenol.[14] Phenolic or chlorophenolic derivatives in concentrations less than about 200 mg/l can be completely oxidized by chlorine or chlorine dioxide; wastes containing concentrations of phenol in excess of this limit usually require preliminary phenol recovery by various means.[12]

Theoretically, less than 10 parts of chlorine are required to destroy each part of phenol, but due to the usual presence of large amounts of other organic chlorine-consuming materials, as many as 20 parts may be required. Whether chlorine treatment is the only one to be employed depends on the type and concentration of chlorine-consuming materials; if the waste contains large amounts of settleable and colloidal chlorine-consuming materials, preliminary sedimentation and coagulation may be required. The use of chlorine for phenol oxidation and color and odor removal at the Pequanoc Rubber Company (reclaiming) plant at Butler, New Jersey has been reported highly successful.[5,30] In a cooperative study sponsored by the Ohio River Valley Water Sanitation Commission, complete oxidation of ammonia still (coke) wastes by chlorine was demonstrated to destroy phenols and to reduce the BOD and oxygen consumed of the waste by more than 60 percent.[5] Chloroxidation of phenolic wastes must be complete as partial chlorination results in chlorophenol formation.

Elemental phosphorous in waste waters poses a difficult treatment problem because of its insolubility. Phosphorous in contact with air ignites or burns to phosphorous pentoxide which either produces a dense cloud or reacts with water to form phosphoric acid. Successful chlorine treatment at Oldbury Electro-Chemical Company's electric-furnace phosphorous manufacturing plant at Niagara Falls, New York, has been reported.[39] Treatment is based on reaction of chlorine with a dilute phosphorous waste solution to form phosphorous trichloride (PCl_3) which subsequently hydrolyzes to phosphorous acid (H_3PO_3) and hydrochloric acid (HCl):

$$2P + 3\,Cl_2 \rightarrow 2\,PCl_3 \tag{16-3}$$

$$PCl_3 + 3\,H_2O \rightarrow H_3PO_3 + 3\,HCl \tag{16-4}$$

At that plant, condenser wastes containing about 150 mg/l of elemental phosphorous are consumed by about 500 mg/l chlorine during treatment.

Wastes produced in the manufacture of synthetic rubber are objectionable primarily because of their BOD and taste and odor producing properties. A complete abatement method for synthetic rubber industry wastes has not been discovered, but some odor removal reportedly can be accomplished by chlorination.[14]

A variety of other chemical wastes are amenable to chlorine treatment. Hydrogen cyanide, acrylonitrile, vinyl chloride and acetylene reportedly are effectively removed from wastes at Monsanto Chemical Company's Texas City plant by the combination of stripping, chlorination and incineration.[38]

Food Processing. Chlorination does not constitute a method for complete treatment of wastes from food processing industries. These wastes usually have a very high BOD and chlorine demand and can be more economically treated by biological or other means. In conjunction with such treatment and disposal processes, however, chlorination is frequently employed for odor control, BOD reduction, algae control and disinfection.

Pearson and Sawyer[22] have demonstrated that the treatment of beet sugar process waters with 25–60 mg/l chlorine in a pH range of 5.5 to 7.0 materially reduces sugar losses in process waters and permits recovery of about 2 tons/day of sugar from a 1000-ton plant. Chlorine addition results in more efficient precipitation of solids; moreover, it destroys bacteria which otherwise cause decomposition and septic conditions in the reuse of process waters. Chlorination of flume water from sugar beet handling, in addition to other methods of treatment, prevents biological growths and permits reuse of this large volume of water in a closed recirculation system.

In recent times there has been a growing tendency to dispose cannery wastes by spray irrigation systems. The 8-mgd vegetable-processing wastes of Seabrooks Farms (Seabrook, New Jersey) are handled in this manner and are often chlorinated (after pre-settling) prior to irrigation to control algae growths in the distribution system, to preclude biological clogging of the irrigation system, and for odor control.[14]

In general, chlorine requirements for complete treatment of packing house wastes are too high for economical and practical use of this as a complete treatment process. Prechlorination of packing house wastes for preventing septicity, however, is commonly practiced. At some installations larger doses of chlorine are employed as an aid to coagulation and sedimentation. Treatment is based on the ability of chlorine to coagulate and precipitate protein. The use of super-chlorination for this purpose at the Hormel plant in Austin, Minnesota, where the process originated and was employed for several years, reportedly has been abandoned[14, 15, 22, 33] because of cost, the toxicity of the sludge and the necessity of excluding much of the waste from the treatment process. According to Eldridge[9] the super-chlorination method, followed by chemical precipitation, is a superior method to biological treatment for treating wastes of small packinghouses. At Phoenix, Arizona, a packing plant reportedly[7] uses ferric chloride made from the action of chlorine on iron wire. BOD reduction from 1448 to 188 ppm, and suspended solids reduction from 2975 to 167 ppm, have been obtained using 1000 lb of wire and 1150 lb of chlorine per million gallons of waste.

Continuous vacuum filters have found some application in the dewatering of food industry wastes after preliminary concentration. Chlorine sometimes is used as a treatment adjunct. For example, in the present Hormel process for treating meat-packing wastes, mixed primary and secondary sludge is treated with chlorine and ferric chloride as conditioning agents before dewatering on vacuum filters.[14]

Reinke[29,30] has reported that the application of chlorine or sodium hypochlorite at a rate of 20 mg/l is sufficient to destroy odors that often are associated with land irrigation disposal of milk products wastes.

Metal Processing. Cyanide (CN) is a common component of wastes of steel and metal-finishing, electro-plating industries, and a few wastes of chemical manufacturing industries (e.g., certain plastics and synthetic fibers). Cyanides are toxic to aquatic life, interfere with normal biological processes of stream self-purification and are a menace to agricultural, bathing and potable water sources. The alkaline chlorination process[6,8] is perhaps the most widely used means of cyanide destruction by oxidation. When chlorine is added to a waste alkalized to a pH of 8.5 to 10.0, free cyanide is rapidly oxidized to cyanate (CNO). Cyanogen chloride (CNCl) frequently is formed as an intermediate product of the reaction:

$$Cl_2 + NaCN \rightarrow CNCl + NaCl \tag{16-5}$$

$$CNCl + 2\,NaOH \rightarrow NaCNO + NaCl + H_2O \tag{16-6}$$

The complete reaction theoretically requires 2.73 parts of chlorine to oxidize each part of cyanide, but due to the presence of other chlorine-consuming materials equipment usually is sized for a Cl_2: CN ratio of about 4:1.

In the presence of chlorine cyanates are decomposed as shown in Equation (16-7) within 1 to 1.5 hours, and ammonium carbonate is rapidly oxidized by chlorine as shown in Equation (16-8):

$$3Cl_2 + 4H_2O + 2NaCNO \rightarrow 3Cl_2 + (NH_4)_2CO_3 + Na_2CO_3 \tag{16-7}$$

$$3Cl_2 + 6NaOH + (NH_4)_2CO_3 + Na_2CO_3 \rightarrow$$
$$2NaHCO_3 + N_2 + 6NaCl + 6H_2O \tag{16-8}$$

These reactions also occur at pH 8.5 to 10.0. The destruction of cyanates theoretically requires 4.09 parts of chlorine for each part of cyanate (as CN), but, again, due to the presence of other oxidizable matter and the formation of small amounts of nitrous oxide and nitrogen trichloride (not shown), the chlorine requirements are somewhat higher. The over-all reaction is shown in Equation (16-9).

$$5\,Cl_2 + 10\,NaOH + 2\,NaCN \rightarrow$$
$$2\,NaHCO_3 + N_2 + 10\,NaCl + 4\,H_2O \tag{16-9}$$

In theory 6.82 parts of chlorine are required to completely oxidize each part of cyanide but in practice equipment usually is sized for a Cl_2:CN ratio of about 8:1.

The alkaline chlorination process can be accomplished in either a batch-type or flow-through type (continuous or intermittent) chlorination plant. The former normally is used for complete oxidation of cyanides and cyanates, whereas the latter normally is used for oxidation of cyanides to cyanates. With adequate control, the flow-through type plant can also be designed to accomplish complete oxidation of cyanides and cyanates.

Paper Mills. The use of chlorine in the treatment of paper mill wastes is almost entirely confined to the chlorination of white water for slime control. The maintenance of a chlorine residual of 0.2 to 1.0 mg/l reportedly eliminates organic growths and permits a substantial reduction in water consumption and fiber losses.[14] Higher residuals are likely to be harmful to the paper or paper machines and residuals of less than 0.2 mg/l are ineffective. The quantity of chlorine used varies from 0.33 to 5.0 lb/ton of product, where in the same mills the chlorine requirement for fresh water ranges from 0.11 to 6.0 lb/ton of product.[12]

Textile. There are several applications of chlorine in the treatment or pretreatment of wastes from the processing of cotton, wool and other textile fibers. Chlorine is sometimes employed as a coagulant or coagulant aid and in the preparation of chlorinated copperas used in the coagulation of textile wastes. It also oxidizes or bleaches many dyes, at least partially, especially those containing certain sulfur and nitrogen derivatives. In some cases, i.e., for sulfur dyes, it has been used in place of chemical coagulation. In such instances chlorination is effective in removing the major portion of BOD, but with most textile wastes the BOD reduction with chlorine amounts to only 15 to 25 percent. Generally speaking chemical precipitation must be relied upon to remove most of the BOD, with chlorine being simultaneously applied, or separately as a polishing medium.[4]

Wastes from wool-scouring plants are among the strongest and most objectionable of industrial wastes and contain a substantial amount of recoverable wool grease. The treatment of such wastes with chlorine was suggested by De Raeve as far back as 1918. A process was patented in the United States about 1925, but it has never been used commercially.[30] A process using calcium hypochlorite was developed and used by Fields Point Manufacturing Corp., Providence, Rhode Island, about 1946. Reported by Faber,[10] it consists essentially of adding to warm wastes sufficient calcium hypochlorite to reduce the pH to about 7.5, causing coagulation, separation and recovery of wool grease and some other colloidal or suspended solids. Primary treatment of chlorinated wastes showed BOD removals of 80 to 90 percent and suspended solids and grease removals of about 95 percent.[14] Chlorine required for treatment varies for different wastes in proportion to the alkali used in the scouring operation, but 30 to 40 lb of chlorine/1000 gal of waste frequently is indicated.[12]

Others. Chlorine does not appear to have been used to any great extent in the treatment of tannery wastes because of their high chlorine demand and high organic composition. It is, however, sometimes employed as a polishing treatment in the final effluent of more economical treatment processes.[14] Howard[18] has reported the successful emergency application of 300 mg/l of chlorine which caused good effluent coagulation and marked odor reduction in the wastes receiving stream.

Oil industry wastes generally are of large volume and high chlorine demand and disinfection is not a problem. Accordingly, chlorine has been little used except for controlling slime growths in recirculated refinery wash water.

Granstrom[13] has reported significant BOD, suspended solids and color reduction when chlorine was used with alum in treating rendering plant wastes.

The reuse of treated sewage and some industrial waste effluents and process waters is made possible by chlorine treatment. Chlorine serves as a disinfectant to prevent disease transmission and to control slimes, marine and other biological nuisance organisms, in addition to oxidizing objectionable waste products. Resources conservation and water pollution abatement are thereby enhanced—a factor of growing importance and necessity in the increasingly competitive, technological era of current times.

Acknowledgement

The author wishes to acknowledge the helpful assistance and suggestions of Harry A. Faber, formerly of the Chlorine Institute, and currently Engineering Research Coordinator for the Public Health Service, Department of Health, Education and Welfare, Washington, D. C.

References

1. American Public Health Association, Inc., "Chlorination in Sewage Treatment," New York, 1934.
2. American Public Health Association, Inc., "Standard Methods for the Examination of Water, Sewage and Industrial Wastes," Ed. 10, Baltimore, Md. Waverly Press, Inc., 1955.
3. Baldwin, Robert T., "The One-Ton Liquefied Chlorine Gas Container," *J. Am. Water Works Assoc.,* **18** (October 1927).
4. Chamberlin, N. S., "Application of Chlorine and Treatment of Textile Wastes," *Am. Dyestuff Reptr.,* **43**(13), 389 (June 21, 1954).
5. Chamberlin, N. S., and Griffin, A. E., "Chemical Oxidation of Phenolic Wastes With Chlorine," *Sewage and Industrial Wastes,* **24**(6), 750 (June 1952).
6. Chamberlin, N. S., and Snyder, H. B., "Treatment of Cyanide and Chromium Wastes," Proceedings, Regional Conference on Industrial Health, Houston, Texas (September 27–29, 1952).

7. Dept. of Health, Education and Welfare, U. S. Public Health Service, "An Industrial Waste Guide to the Meat Industry," Public Health Service Publ. No. 386 (1954).

8. Dodge, B. F., and Reams, D. C., "Disposal of Plating Room Wastes, Part II: A Critical Review of the Literature Pertaining to the Disposal of Waste Cyanide Solutions," Am. Electroplating Soc. Research Report No. 14 (1949).

9. Eldridge, E. F., "Industrial Waste Treatment Practice," Ed. 1, New York, McGraw-Hill Book Company, Inc., 1942.

10. Faber, H. A., "The Hypochlorite Process for Treatment of Wool Scouring Wastes and for Recovery of Wool Grease," *Sewage Works J.*, **19**, 248 (1947).

11. Fair, Gordon Maskew, and Geyer, John Charles, "Water Supply and Waste Water Disposal," Ed. 1, New York, John Wiley & Sons, Inc., 1954.

12. Federation of Sewage and Industrial Wastes Assn., Manual of Practice No. 4, "Chlorination of Sewage and Industrial Wastes," Washington, D. C. (1951).

13. Granstrom, Marvin L., "Rendering Plant Waste Treatment Studies," *Sewage and Industrial Wastes*, **24**(12), 1478 (December 1952).

14. Gurnham, C. Fred, "Principles of Industrial Waste Treatment," Ed. 1, New York, John Wiley & Sons, Inc., 1955.

15. Halvorson, H. O., Cade, A. R., and Fullen, W. J., "Recovery of Proteins from Packinghouse Waste by Superchlorination, *Sewage Works J.*, **3**, 488 (1931).

16. Hofman, A. W., and Frankland, E., Report to Metropolitan Board of Works, (August 12, 1859).

17. Hooker, Albert H., "Chloride of Lime in Sanitation," Ed. 1, New York, John Wiley & Sons, Inc., 1913.

18. Howard, N. J., "Chlorination Practice in Sewage and Trade Waste Disposal," *Water and Sewage* (Canada), **85**(11), 21 (1947).

19. Laubusch, Edmund J., Chlorination of Waste Water, *Water & Sewage Works*, **105** (December 1958).

20. Laubusch, Edmund J., "State Practices in Sewage Disinfection," *Sewage and Industrial Wastes*, **30**(10), 1233 (October 1958).

21. Ornstein, George, "Liquid Chlorine," *Trans. Am. Electrochemical Soc.*, **29** (1916).

22. Pearson, Erman A., and Sawyer, Clair N., "Beet Sugar Process Waters— Treatment and Utilization," *Chem. Eng. Prog.*, **46**(8), 380 (August 1950).

23. Phelps, Earle B., Chlorination of Water & Sewage, *J. Boston Soc. Civil Engrs.*, **13** (4–5) (April–May, 1926).

24. Phelps, Earle B., and Carpenter, W. T., "Sterilization of Sewage Filter Effluents," *Munic. J. and Eng.*, **22**, 444 (1907).

25. Porges, R., "United States Sewage Treatment Practices During the Early Twentieth Century," *Sewage and Ind. Wastes*, **29**(12), 1321 (December 1957).

26. Powers, J. J., "Apparatus for Disinfecting Sewage," U. S. Patent 362,657 (May 10, 1887).

27. Pritchard, D. A., "Economics of Chlorine," Paper presented at the American Electro-chemical Society Meeting, (April 22–24, 1926).

28. Race, Joseph, "Chlorination of Water," Ed. 1, New York, John Wiley & Sons, Inc., 1918.

29. Reinke, E. A., "Prevention of Odor Nuisance in the Disposal of Creamery Wastes with Chlorine Preliminary to Broad Irrigation," *Calif. Sewage Works J.,* **1**(3); (1928).

30. Rudolfs, Willem, "Industrial Wastes: Their Disposal and Treatment," ACS Monograph 118, Ed. 1, New York, Reinhold Publishing Corp., 1953.

31. Snow, W. B., "Biochemical Oxygen Demand of Chlorinated Sewage," *Sewage and Industrial Wastes,* **24,** 689 (1952).

32. Soper, William, *J. Royal Sanit. Inst.,* 255 (1879).

33. Southgate, B. A., "Treatment and Disposal of Industrial Waste Waters," Ed. 1, Britain, HMS Stationery Office, 1948.

34. Thoman, J. R., "Statistical Summary of Sewage Works in the United States," *Public Health Reports,* (April, 1950).

35. Thoman, J. R., "Statistical Summary of Sewage Chlorination Practice in the the United States," *Sewage and Industrial Wastes,* **22**(9), 1128 (September 1950).

36. Thoman, John R., and Jenkins, Kenneth H., "Statistical Summary of Sewage Chlorination Practice in the United States," *Sewage and Industrial Wastes,* **30**(12) 1461 (December 1958).

37. Upper Mississippi River Board of Public Health Engineers and Great Lakes Board of Public Health Engineers. Standards for Sewage Works—Illinois, Indiana, Iowa, Michigan, Minnesota, Missouri, New York, Ohio, Pennsylvania, Wisconsin, (May 1952).

38. Weyermuller, Gordon, and Morris, H. E., "Monsanto Controls Chemical Waste Disposal," *Chem. Processing,* (October 1955).

39. Weyermuller, Gordon, and Morris, H. E., "Oldbury Electro-Chemical's Phosphorous Plant Solves Problem of Waste Disposal by Installing Unit for Treatment of Effluent With Chlorine," *Chem. Eng. News,* **30,** 466 (1952).

40. "Sewage Treatment Facilities in the United States," *Eng. News-Rec.,* **115**(7), 224 (August 15, 1935).

41. "A Summary of Census Data on Sewerage Systems in the United States," *Public Health Reports,* **57,** 12 (March 20, 1942).

42. "Woolf Process at Brewster, N. Y." *Engr. News,* **30,** 41 (1893).

Additional Selected References

1. Cecil, Lawrence K., "Underground Disposal of Process Waste Waters," *Ind. and Eng. Chem.,* **42**(14), 594 (April 1950).

2. Connell, C. H., Dreyer, Donald A., and Berg, E. J. M., "Reduction of Coliform Bacteria in Sewage Sludge by Halogens," *Sewage and Industrial Wastes,* **30**(5), 634 (May 1958).

3. Crosby, Edwin S., Rudolfs, Willem, and Heukelekian, Houhaness, "Biological Growths in Refinery Waste Waters," *Ind. and Eng. Chem.,* **46**(2), 296 (February 1954).

4. Eliassen, Rolf, Heller, Austin N., and Krieger, Herman L., "A Statistical Approach to Sewage Chlorination," Sewage Works J., **20**(6) 1008 (November 1948).

5. Hess, Seth G., Diachishin, Alex N., and De Falco, Paul, "Bactericidal Effectiveness of Sewage Chlorination," Part I, *Sewage and Ind. Wastes,* **25**(7), 751 (July 1953); Part II, **25**(8), 909 (August 1953).

6. Ingols, Robert S., and Jacobs, George M., "B.O.D. Reduction by Chlorination of Phenol and Amino Acids," *Sewage and Ind. Wastes,* **29**(3), 258 (March 1957).

7. Katz, S., and Heukelekian, H., "Chlorine Determinations in Waste Waters," *Sewage and Ind. Wastes,* **31**(9), 1022 (September 1959).

8. Moore, Edward W., "Fundamentals of Chlorination of Sewage and Industrial Wastes," *Water and Sewage Works,* **1005**(5) R-197(1953).

9. Pomeroy, Richard and Bowlus, Fred D., "Progress Report on Sulfide Control Research," *Sewage Works J.,* **18**(4), 597 (July 1946).

10. Van Kleeck, Leroy W., "How To Disinfect Sewage by Pre- and Post-Chlorination," *Wastes Eng.,* 371 (August 1956).

17. CHLORINATED BLEACHES AND SANITIZING AGENTS

W. H. SHELTMIRE
Olin Mathieson Chemical Corporation

HISTORY

Prior to the development of chlorinated bleaches, the process of bleaching textiles was a long and laborious one which required several months at a minimum and did not produce the high degree of whiteness to which we are accustomed today. Goods were spread on grass fields and bleached by the sun, a process known as "crofting". A large expense in land rent and maintenance of the bleach fields thus were required, and only a few could afford the luxury of cloth bleached in this manner. In the early and middle eighteenth century, attempts were made to improve the bleaching process by steeping the goods in an alkali (bowking) prior to crofting. The practice of souring was also introduced in this period. Initially, it consisted of steeping the partially bleached goods in sour milk or buttermilk for a week and then rinsing before final crofting. About 1750, Home published his discovery that a dilute sulfuric acid sour gave better results with less than 24 hours steeping time. Even with these improvements, bleaching of linen goods required at least three months.

Late in the eighteenth century, a combination of enormous land rental increases and improvements in spinning and weaving brought about a pressing need for improved bleaching processes. Following the discovery of chlorine, the development of bleaching with chlorine completely revolutionized the textile industry.

Although Scheele had discovered chlorine and observed its bleaching property in 1774, it remained for Berthollet to develop the first application of chlorine as a commercial bleaching agent[11] about 1886. His first efforts consisted of applying water solutions of chlorine to the goods. Since this process was difficult to control and resulted in excess tendering, Berthollet experimented on the development of a more satisfactory bleach. He succeeded by dissolving chlorine gas in a dilute solution of potash of lye. The resulting solution was called "Eau de Javelle," a term which has been retained and is frequently applied to sodium hypochlorite solutions. The tendering of cloth which resulted from initial techniques was reduced with experience. Labarraque replaced the expensive potassium hydroxide with caustic soda, obtained from soda ash. Although soda ash was expensive around 1800, its price fell with the introduction of the LeBlanc process around 1810 and dropped sharply when the Solvay process was generally established toward the close of the century.

Labarraque's development resulted in what was probably the first use of sodium hypochlorite as a bleach. It completely replaced Berthollet's solution, and within a short time the bleach fields of Ireland and Holland entirely disappeared. Other methods for producing sodium hypochlorite solutions were later developed, but the chlorination of caustic soda remains the most widely used.

Tennant, in 1798, patented a successful liquid bleach (called bleach liquor today) by passing chlorine through milk of lime. Since textile bleachers had been accustomed to boiling goods in lime water and solutions of soda ash or caustic soda, Lancashire bleachers successfully contested the patent. Tennant continued his experimentation and developed the method of passing chlorine over slaked lime to produce Tennant's Bleach, which soon became known as bleaching powder. This process was patented in 1799.

The development of bleaching powder was of tremendous importance to the textile industry and, in fact, to the economy and living standards of the entire world. It provided, for the first time, a solid form of chlorine bleach which could be transported easily and which needed only to be dissolved in water to be available for use. Bleaching powder became a cornerstone of the early chemical industry. Much of our chemical engineering heritage dates to early efforts to make soda ash, caustic soda, chlorine and bleaching powder for use in the bleaching of cloth.

True bleaching powder, freshly prepared, contains about 36 percent available chlorine. Its storage life is short, especially in warm climates. Because of the instability of bleaching powder at higher temperatures, a more stable bleaching compound was sought. The addition of quick lime to bleaching powder produced what is known as tropical bleach which is fairly stable at tropical temperatures and contains 25 to 30 percent available chlorine. This product was developed about 1920 and is still used in significant tonnages in the underdeveloped areas of the world.

In 1921, Schmidt discovered that cellulose could be bleached by the use of chlorine dioxide with no appreciable degradation. Chlorine dioxide bleaching has gained considerable recognition in the pulp and textiles industries, and a more widespread use is developing in the continuous bleaching of synthetic fibers and cotton-synthetic blends.

In 1928, the first dry calcium hypochlorite produced in the United States entered the market. This effective bleaching agent, containing approximately 70 percent available chlorine (HTH, Perchloron, Pittchlor, etc.), largely replaced bleaching powder in the United States. Dry mixtures of calcium hypochlorite and other chemicals are marketed for easy preparation of sodium hypochlorite solutions.

Chlorinated trisodium phosphate was marketed in 1928. A strongly alkaline compound containing about 3.5 percent available chlorine, it has become widely used in cleansers and some cleaning formulations.

Liquid bleach came into widespread use about 1930 for laundry, household and general disinfecting uses. Its preparation is a modification of Labarraque's method, the trend being toward lower residual alkali than was originally employed. This simplifies purification and sedimentation, while maintaining a pH of around 10.5 to 11 for stability.

Many organic chlorine compounds have been proposed as bleaching agents from time to time. Chlorinated hydantoins and cyanuric acid derivatives have probably been most successful. Dichlorodimethylhydantoin won an appreciable market from 1950 to about 1957. Trichloroisocyanuric acid was used for military purposes during World War II, and since 1955 it and its homologs, dichloroisocyanuric acid and the sodium and potassium salts, have been produced in increasing volume for use as ingredients in bleach and cleanser formulations.

Chlorine-based bleaches should not be confused with the weaker oxidizing bleaches such as the perborates, whose mode of action is entirely different from that of the chlorine-based bleaches.

Sanitizing

According to Megele and Wiseman,[23] the first recorded instance for employing chlorine as a disinfectant was the use of bleaching powder to flush out polluted mains following a typhoid outbreak in England in 1897. Chlorine was also used for this purpose in the United States in 1912 to control a typhoid outbreak in Niagara Falls. However, as early as 1908, hypochlorite solution was being applied to Chicago City water as a general practice for purification. Today, the chlorination of water supplies is nearly universal. Liquid chlorine and dry calcium hypochlorite products (70 percent available chlorine) are the two principal chemicals used for this purpose. Chloramination is also employed (see Chapter 15).

Since about 1915, with the introduction of Dakin's solution, dilute solutions of hypochlorite from both inorganic and organic "available chlorine" compounds have gained importance in general sanitation. Dakin's solution contained chlorinated lime, sodium carbonate and sodium bicarbonate, and had an available chlorine content of about 0.4 percent. It was widely used during World War I to irrigate wounds by the Carrel-Dakin technique. Chloramine-T entered the sanitation field in 1916 and was used in the treatment of infected wounds during the same war. Halazone was introduced by Dakin and Dunham in 1917. Being one of the more palatable chloramines, it was carried by the soldiers of both world wars as an emergency means of water purification. One tablet containing four milligrams of Halazone was used to disinfect one liter of water. A large number of other chloramines, including Dichloramine-T, chloramine-B, succinchloramide, and a series of hydantoins, have since been used for general sanitizing.[9] The germicidal property of monochloramine was first observed by Rideal in 1910. Race treated the water supply of Ottawa with chloramines by adding ammonia and chlorine (as bleaching powder) in a ratio of 1:2, and published his work in 1918.[30]

When chloramine-T came into widespread use, the U. S. Public Health Service set minimum standards of two minutes contact time and 50 parts per million available chlorine for sanitizing dairy equipment. This regulation is still in effect today and covers all "available chlorine" compounds used to sanitize equipment from which the public is served. Although these standards are much higher than necessary with faster acting sanitizing agents such as hypochlorite solutions, they are sufficiently high to assure effective sanitation with the slower acting chloramines.

Dry calcium hypochlorite products have been widely used for sanitizing since their introduction into the market in the late 1920s. Formulations containing approximately 70,50,35 and 15 percent available chlorine are marketed for use in industrial plants, on farms, in swimming pools and in general household sanitizing. The 70 percent products are also used for emergency chlorination of water supplies.

Chlorine dioxide has been used as a disinfectant since 1944.[41] Its primary use in this connection is in food processing and water purification. It is also used in industrial plants as a slimicide.

The chlorinated derivatives of cyanuric acid have received much attention in recent years. They have excellent bleaching and disinfecting properties. They are recommended for sanitizing family pools and in formulations for laundry bleaches, scouring powders and dish washing compounds.

THE "AVAILABLE CHLORINE" FAMILY

The bleaching and disinfecting power of chlorinated bleaches is dependant upon the hypochlorous acid concentration of their solutions. When chlorine is added to water, hypochlorous acid is formed according to the equation:

$$Cl_2 + H_2O \rightleftharpoons HOCl + HCl \qquad (17\text{-}1)$$

The chlorine atom of the hypochlorous acid may be considered positive (valence of plus 1) and is the "active" chlorine referred to in bleaching and sanitation literature. It should be observed that only half of the chlorine is utilized in the formation of hypochlorous acid, and subsequently in bleaching and sanitizing action.

The hypochlorites and the chloramines, both organic and inorganic, also form hypochlorous acid in water solution.

$$Ca(OCl)_2 + 2\,H_2O \rightleftharpoons Ca(OH)_2 + 2\,HOCl \qquad (17\text{-}2)$$

$$NH_2Cl + H_2O \rightleftharpoons NH_3 + HOCl \qquad (17\text{-}3)$$

It is seen from Equations (17-2) and (17-3) that all of the chlorine in the hypochlorite or the chloramine is replacable and can be utilized in bleaching and sanitizing application. Therefore, one hundred pounds of a hypochlorite or chloramine formulation containing fifty percent chlorine would produce the same amount of hypochlorous acid as would one hundred pounds of gaseous or liquid chlorine.

In the early days of sanitation studies, it was believed that liquid or gaseous chlorine was one hundred percent utilized, or "available," in sanitation and bleaching reactions. The term "available chlorine content" was established as the basis for comparing the potential bleaching or disinfecting power of chlorine compounds with that of commercial chlorine. The available chlorine content of a solution is determined by titrating the iodine which the solution will liberate from an acidified iodide solution. The calculated weight of elemental chlorine (Cl_2) required to liberate the same amount of iodine is the "available chlorine content" of the solution. Since only half of the chlorine of the Cl_2 molecule is positive in solution, the available chlorine content of hypochlorites and chloramines is just twice the "active" (positive) chlorine content. For example, the percentage of chlorine in monochloramine is 68.9, but the available chlorine content is 137.9 percent.

It is unfortunate that the term "active chlorine compounds" has reappeared in recent literature after having been out of favor for thirty years. It leads to the misuse of "active" chlorine content in place of available chlorine content by those who believe the two terms are synonymous, with confusion resulting.

Although it is convenient that the term "available chlorine" compares the strength of bleach solutions to that of chlorine solution, much confusion has resulted from its use. In actual fact, the titration described is a measure of the oxidizing power of a solution, and the "available chlorine content" of solutions containing no chlorine at all (anhydrous hydrogen peroxide—208 percent available chlorine) can readily be determined by the same method. It is preferable to consider such compounds as having a certain "equivalent" available chlorine content (see footnote page 534).

The *available chlorine family* is comprised of the group of chemicals which, when dissolved in water, yield solutions of hypochlorous acid. These compounds may be further subdivided into those which contain free available chlorine and those which contain combined available chlorine.

"Free available chlorine" describes the oxidizing power attributable to chlorine in solution as hypochlorous acid, as hypochlorite ion or, in strong acid solutions, as free chlorine. The term "combined available chlorine" is associated with the organic chloramines. The small hydrolysis constants of the organic chloramines permit only a small amount of hypochlorous acid to be formed upon initial solution. The rest of the chlorine remains "combined." As hypochlorous acid is depleted through bleaching or disinfecting action, more is supplied with continuing hydrolysis of the chloramines.

With free available chlorine, the proportion which is present as hypochlorous acid is controlled by the pH of the solution (see Table 17-3). With combined available chlorine, the concentration of hypochlorous acid is mainly controlled by the hydrolysis constant of the chloramine.

Reducing substances react with the chlorine of hypochlorous acid solutions, consuming it. Nitrogenous substances convert a part of the free available chlorine to combined available chlorine, reducing the balance to chlorides.

Chlorine dioxide does not belong to the available chlorine family. It does not form hypochlorous acid in solution, nor is its bleaching action in any way similar to that of the hypochlorites. Chlorine dioxide bleaching is discussed separately at the end of the chapter.

The Mechanism of Bleaching with Chlorine

All of the bleaching agents in the available chlorine family form water solutions of hypochlorous acid. The bleaching action of hypochlorous acid solution may follow any of three types of reactions: oxidation, chlorination or chlorhydrination. At pH levels below 4, an appreciable portion of the available chlorine is present as free chlorine:

$$HOCl + H^+ + Cl^- \rightleftharpoons H_2O + Cl_2 \qquad (17\text{-}4)$$

Consequently, chlorination is more likely to occur in acid solutions.

TABLE 17-1. FREE AVAILABLE CHLORINE CONTENT

	Bleaching Agent	% Available Chlorine
	Cl_2, chlorine	100.0[a]
Pure Compounds	$Ca(OCl)_2$, calcium hypochlorite	99.2
	NaOCl, sodium hypochlorite (unstable)	95.2
	LiOCl, lithium hypochlorite	121.6
	"HTH," "Pittchlor," "Perchloron"	70–74
	Bleaching powder	35–37
Commercial preparations	Tropical bleach	25–30
	"Lo-Bax," B-K (dairy sanitizers)	50–55
	Liquid bleach	12–15
	Household bleach	3–5¼
	Bleach liquor (lime bleach liquor)	3–10

[a] by definition

At higher pH levels, oxidation and chlorhydrination are the predominant reactions. Saturated compounds are oxidized, whereas unsaturated compounds are chlorhydrinated, thereby destroying the resonating structure of the molecule with a resulting loss of color:

$$R-C=C-R' + HOCl \rightarrow R-\overset{\overset{\displaystyle Cl}{|}}{C}-\overset{\overset{\displaystyle OH}{|}}{C}-R' \tag{17-5}$$

Bleaching with hypochlorite solutions is seldom a reversible reaction.

TABLE 17-2. COMBINED AVAILABLE CHLORINE CONTENT

Bleaching Agent	% Available Chlorine
NH_2Cl, Monochloramine	137.9
$NHCl_2$, Dichloramine	165.0
NCl_3, Nitrogen trichloride (explosive)	176.7
$CH_3C_6H_4SO_2NClNa \cdot 3H_2O$, Chloramine-T	25.2
$C_6H_5NClNa \cdot 1½ H_2O$, Chloramine-B	29.4
$HOOCC_6H_4SO_2NCl_2$, Halazone	52.4
$NClCONClCOC(CH_3)_2$, Dactin (Halane)	77.6
NClCONClCONCl CO, Trichloroisocyanuric acid	91.5
CONClCONClCONH, Dichloroisocyanuric acid	71.7
CONClCONClCON Na, Sodium dichloroisocyanurate	64.5
CONClCONClCONK, Potassium dichloroisocyanurate	60.1

The Mechanism of Disinfecting with Chlorine

Hypochlorous acid has outstanding bactericidal power. This is generally attributed to its ability to diffuse through cell walls and thereby reach the vital parts of the bacterial cell. A widely accepted theory credits the death of the cell to a reaction between hypochlorous acid and an enzyme.[14] The hypochlorite ion has little if any bactericidal effect since its negative charge impedes penetration of the cell wall.

The bactericidal power of a solution of chlorine, a hypochlorite, or a chloramine is directly proportional to the hypochlorous acid concentration of the solution. The percent available chlorine as undissociated hypochlorous acid is therefore the true measure of the bactericidal effectiveness of a solution containing one of the chemicals of the available chlorine family.

The relationship between pH and the degree of dissociation of hypochlorous acid is shown in Table 17-3[32] (see Figure 15-1 also). Hydrolysis increases rapidly as the pH rises above neutrality. With rapidly hydrolizing compounds, such as hypochlorites, a pH near 9 is necessary for stability but is satisfactory since they are rapid acting. With compounds that hydrolize slowly, such as chloramines, the pH must be kept lower to provide a larger proportion of undissociated HOCl since they are less rapid acting.

TABLE 17-3. DISSOCIATION OF HYPOCHLOROUS ACID
AS A FUNCTION OF pH AT 25°C

pH	% HOCl Undissociated
5.0	99.6
6.0	96.5
7.0	73.0
7.4	50.0
8.0	21.0
9.0	2.7
10.0	.3

LIQUID BLEACH
(Sodium Hypochlorite Solution)

Liquid bleach, called soda bleach liquor in the paper and textile industries, is the most widely used of all chlorinated bleaches. Over 150 tons of available chlorine as liquid bleach is used in the United States every day for household and laundry bleaching. Other uses include chemical processing (e.g. chlorhydrination), textile bleaching, water treatment and general disinfecting.

Commercial liquid bleach usually contains 12 to 15 percent available chlorine. It is sold in carboys and in rubber-lined drums and trucks. Liquid bleach solution of 3 to 5¼ percent available chlorine, the majority of the household bleach trade, is packaged in brown or amber glass bottles for household, laundry and sanitizing use. The recent trend here is toward the 5 or 5¼ percent product.

The stability of hypochlorite solutions is greatly affected by heat, light, pH and the presence of heavy metal cations. Greatest stability is attained with a pH close to 11 and with the absence of heavy metal cations. Storing in a colored bottle, as opposed to a clear bottle, will increase the half-life six fold. Storage temperatures should not exceed 85° F, above which the rate of decomposition (which nearly doubles with each 10° F increase in temperature) becomes too high and the available chlorine content is rapidly depleted.

For the production of relatively large quantities of liquid bleach, the modern trend is to use high purity liquid caustic soda shipped in lined tank cars. Chlorine is passed through a dilute caustic soda solution in either a batch or a continuous operation. Chlorination of caustic soda solutions is accompanied by a heat rise which must be carefully considered. When liquid chlorine is used, the heat rise is approximately 0.6° F for each gram per liter of available chlorine; with gaseous chlorine, the heat rise is approximately 0.7° F for each gram per liter of available chlorine.

At temperatures above about 100°F, sodium chlorate formation becomes appreciable with resulting loss of sodium hypochlorite. Consequently, precautions must be taken to assure that this temperature is not reached. Cooling the caustic soda solution is one method frequently used. Another is to prepare the bleach solution at a sufficiently low concentration to avoid overheating (e.g., 25 g/l available chlorine).

Figure 17-1 shows a typical arrangement for producing sodium hypochlorite in moderate quantities. Water is added to a predetermined mark and then caustic soda is added to another mark, giving the required caustic soda concentration.* Chlorine is passed into the caustic soda solution through a sparger tube near the bottom of the make-up tank at a rate at which chlorination of the batch is completed in three to five hours. Thorough agitation of the contents is effected by means of an air stream since a metallic stirrer would contaminate the solution. A barometric loop is preferred to a vacuum breaker for preventing suck back of hypochlorite solution which would result in severe corrosion of lines and contamination or "poisoning" of subsequent batches.

Selection of proper materials for the equipment is important since sodium hypochlorite solution is very corrosive. The make-up tank may be

*The caustic soda solution is usually received as a 73 or 50 percent solution. It is diluted to about 20 percent and cooled before it is piped to the mixing tank.

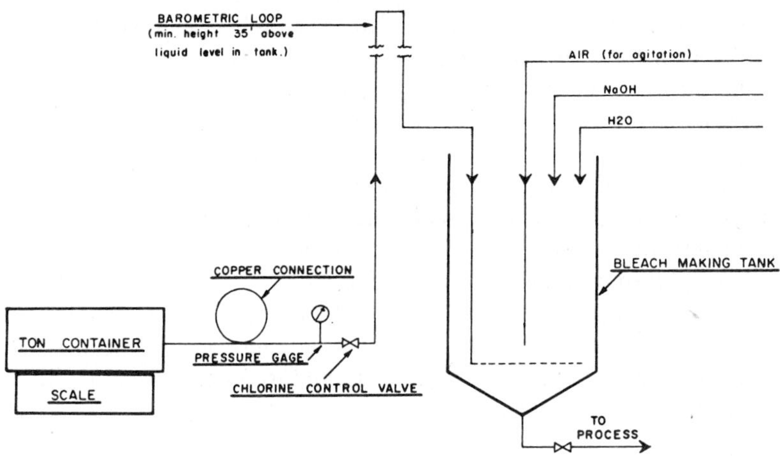

Figure 17-1. Preparation of NaOCl using ton containers of chlorine.

cement or it may be steel lined with rubber or vinyl chloride sheet lining. Lines conveying chlorine should be of vinyl chloride, but silver is sometimes used for the sparger tube. Where conditions require that the bleach solution be cooled as it is made, the current trend is to use titanium heat exchangers to recirculate the solution. Another method is direct cooling with ice, allowing for the weight of the ice in diluting the caustic soda. About three pounds of ice is required for each pound of liquid chlorine added; about 3.6 pounds of ice is required for each pound of gaseous chlorine. The amounts vary with the starting and finishing temperatures.

If liquid bleach is to be stored, it should contain sufficient excess caustic soda to maintain stability of the solution. The amount of excess caustic soda will vary with the degree of stability desired, and may vary up to 10 percent of the available chlorine content. The amount of excess caustic soda may be calculated and the chemicals mixed in the proper proportions, or the residual alkali may be titrated at suitable intervals.

In some cases, such as textile bleaching, the liquor should have little or no caustic alkalinity. In such an instance, the caustic soda excess may be brought low or completely consumed; or the chlorination may be stopped when a small residual of caustic is present, and the caustic may then be removed by adding CO_2 or sodium bicarbonate. Alternately, sodium carbonate may be added to the solution and the chlorination stopped when the free caustic has been consumed, but the carbonate has not. The carbonate provides a buffer alkalinity, preventing local over-chlorination which would give some decomposition (carbonate may enter the solution from the caustic supply, from the diluting water and from the air used for agitation).

For commercial sale, the finished liquor is settled to remove most of the metallic impurities and thus improve the stability. The clear liquor is pumped out with rubber-lined pumps, and is packed promptly in carboys or bottles.

For large-scale bleach production, chlorine is supplied directly from a tank car. Where the liquid bleach solution is captive and is used in large quanity, the chlorination may be conducted continuously with potentiometric control to regulate the flow of chlorine. Continuous production is usually employed where no settling is required.

A consumer may prepare small lots of liquid bleach, e.g., the 1 percent solutions used in commercial laundries, by adding soda ash to a solution or slurry of calcium hypochlorite and decanting off the clear liquor after the sludge has settled. While this method has the disadvantage of giving an insoluble precipitate of calcium, it is convenient for preparing small quantities of bleach.

If caustic soda solution of over 20.5 percent is chlorinated, solid sodium chloride is formed. This limits the maximum concentration of liquid bleach to approximately 15 percent available chlorine. Most commercial bleach makers maintain production in the range of 12 to 14 percent available chlorine to avoid salting difficulties. Household bleach is usually produced with an available chlorine content of 5.25 percent.

BLEACH LIQUOR
(Lime Bleach Liquor)

"Bleach liquor" is a water solution of calcium hypochlorite and calcium chloride, containing less than 100 g/l available chlorine. At concentrations above 100 g/l, the dibasic salt of calcium hypochlorite is precipitated from solution with a corresponding loss of available chlorine.

The method of preparation dates back to 1798 when Tennant patented the process of passing chlorine through a milk of lime solution. In present day adaptations of this method, lead pipe cooling coils are employed to keep the temperature below 80° F, thereby avoiding decomposition and at the same time causing better settling of excess lime. Following chlorination, the solution is allowed to settle and the clear liquor is siphoned or decanted.

The principal use of bleach liquor is in the bleaching of pulp where large tonnages are consumed annually. Pulp bleaching is a stepwise process (see Chapter 10) in which the hypochlorite bleach step usually follows the first extraction. A typical concentration for the bleach liquor is 30 to 35 g/l available chlorine, although concentrations as high as 85 g/l may be used.

NOTE: The terms "bleach liquor," "lime bleach liquor," "soda bleach liquor" and "liquid bleach" grew in industry after 1930. Prior to that, "bleach" meant bleaching powder. The term "chemic" in the textile industry refers to hypochlorite bleach solution.

CALCIUM HYPOCHLORITE

Calcium hypochlorite is the dominant form of dry bleach in the United States, Canada, England and most of the Continent. It is marketed at 70 to 74 percent available chlorine content for use in textile bleaching, for preparation of laundry bleach, and as a universal sanitizing agent. The first calcium hypochlorite marketed was German "Perchloron" which consisted of hemibasic calcium hypochlorite and impurities and contained about 65 percent available chlorine.

Mathieson workers found that when sodium hypochlorite and calcium chloride are mixed in equivalent amounts, calcium hypochlorite and sodium chloride are formed as a stable pair.[13,20,34,35] In the Mathieson process, a slurry of lime and caustic soda is chlorinated and then cooled to minus 10° F. The crystals which form are centrifuged to remove the mother liquor and the insoluble impurities. The crystals are then added to a slurry of chlorinated lime which contains calcium chloride in an amount equivalent to the sodium hypochlorite content of the crystals. Upon being warmed, the mixture precipitates calcium hypochlorite dihydrate and the sodium chloride remains in solution. The slurry is then filtered and the resulting cake is granulated, dried and sized. The product contains over 70 percent available chlorine and less than 3 percent lime. It leaves little insolubles when dissolved.

The American "Perchloron" process, developed by Pennsylvania Salt Manufacturing Company, represents an improvement over the German process. By carrying the chlorination step past the hemibasic point, the American process forms calcium hypochlorite dihydrate crystals and at the same time retains a sufficient amount of the large hemibasic crystals to permit filtration of the material. This product also contains over 70 percent available chlorine. The original product had 12 to 16 percent lime content, but this has recently been reduced to 3 to 5 percent, thereby reducing the amount of sludge remaining in its solution.

The Columbia (Columbia-Southern Chemical, Division of Pittsburgh Plate Glass Co.) process depends upon production of hypochlorous acid solution from chlorine monoxide and water and subsequent neutralization with a lime slurry to produce a calcium hypochlorite solution.[7,8] The clear liquor is first spray dried and then granulated and vacuum dried to yield a produce containing 70 percent available chlorine and 4 to 6 percent lime.

All commercial calcium hypochlorites contain some calcium carbonate and other insolubles. In hard water an additional amount of calcium carbonate is formed. There is, thus, some sediment which may be removed from solution, but this is very much less than the amount of precipitate formed when bleaching powder is dissolved to make a solution of equal available chlorine content.

The decomposition of calcium hypochlorite is exothermic and will proceed rapidly if any part of the material is heated to 350° F. The decomposition releases oxygen and chlorine monoxide. Calcium hypochlorite should never be stored where it is subject to heating, nor should it be allowed to contact any organic material of an easily oxidized nature.

Under normal storage conditions calcium hypochlorite preparations will lose 3 to 5 percent of their available chlorine content in a year. Contact with water or the atmosphere induces a pronounced increase in the rate of decomposition.

The principal use of calcium hypochlorite is for disinfecting purposes. Water treatment provides the largest single outlet for this chemical. It is also used for sewage treatment and in industrial plant sanitation. Its use to prepare sodium hypochlorite bleaching solutions by adding a sodium alkali to a calcium hypochlorite solution has been discussed previously. A stable dry mixture containing 15 percent available chlorine as calcium hypochlorite, together with soda ash, has been on the market since 1930 as "HTH-15." This product permits ready preparation of sodium hypochlorite bleach in small amounts without mixing several ingredients. An alternative to reacting calcium hypochlorite with soda ash is to sequester the calcium with sodium tripolyphosphate or other agent. The commercial product "Ad-Dri" is a laundry preparation containing calcium hypochlorite and sufficient sodium tripolyphosphate to yield a nearly complete conversion to sodium hypochlorite when it is dissolved for use.

BLEACHING POWDER

Bleaching powder as known today was patented by Tennant in 1799. His method of passing chlorine gas over finely divided lime is still used. From the time of its introduction, bleaching powder was the only commercial form in which chlorine was transported until the advent of shipping liquid chlorine about 1912. In fact, prior to World War I, it was substantially the only form in which chlorine was marketed.

Bleaching powder contains about 37 percent available chlorine. It is produced by chlorinating lime by one of several methods. In any of these methods the temperature is kept below 95° F to minimize the formation of chlorate. If the temperature rises over 105° F, some hemibasic calcium hypochlorite will form and the product will become deliquescent. If the chlorination step were carried to completion, the theoretical available chlorine content would be 49 percent. However, at available chlorine levels above 37 percent the product becomes deliquescent and loses its stability.

The two basic methods for producing bleaching powder are the chamber process[26] and the mechanical process.[36] In the chamber process slaked lime is placed on a cooled floor in a lead-lined room, and diluted chlorine gas is passed over the lime until chlorination is complete. In the mechanical process chlorine and slaked lime are passed countercurrently through inclined tubes, with careful temperature control.

Although the composition of bleaching powder has been the subject of many studies and some controversy over 150 years, its actual composition has not been settled. It is not a stable phase in contact with any solution, and on heating to somewhat over 105° F it becomes deliquescent, changing to a mixture of hemibasic calcium hypochlorite, basic chlorides and sometimes calcium chloride hydrates. Bunn, Clark and Clifford[6] showed that the first phase recognized by X-ray diffraction is the basic chloride, but most of their preparations were made above the 105° F limit, which is never exceeded in commercial preparation of bleaching powder. Other studies have shown that the chlorine adds to the hydrated lime to form an indefinite compound, plus the basic chloride. Common bleaching powder is probably a mixture of the basic chloride and a mixed hypochlorite crystal of questionable structure.

With the advent of shipping liquid chlorine, it became cheaper to purchase liquid chlorine and prepare bleach liquor at point of use than to ship bleaching powder. This method replaced bleaching powder where large quantities of bleach were required. Where small quantities of bleach are needed and stability is desirable, calcium hypochlorite products have almost completely replaced bleaching powder in the United States. Production of bleach powder has therefore steadily declined since 1925 when 115,000 tons were produced in this country.

TROPICAL BLEACH

The instability of bleaching powder in warm climates lead to the development of "tropical bleach" which is prepared by adding quick lime (CaO) to fairly dry bleaching powder. When this is done, the free water in bleaching powder becomes bound as calcium hydroxide. This reaction generates a large quantity of heat, and the method of preparing tropical bleach must provide for sufficient removal of heat to keep the temperature below 140° F. Overheating will cause decomposition with loss of available chlorine.

Tropical bleach mixtures contain hemibasic and dibasic calcium hypochlorite and the basic chloride in addition to lime, and they usually contain 25 to 30 percent available chlorine. Because of the high lime content of tropical bleach, it is difficult to dissolve and a large amount of sludge is formed. It has therefore been largely replaced by the more stable calcium hypochlorite products. However, large tonnages are still used in less developed countries as it is the cheapest and simplest chemical to use for sanitation purposes.

LITHIUM HYPOCHLORITE

Lithium hypochlorite (LiOCl) contains about 12 percent available chlorine and has excellent stability; however, the high and unstable price of lithium has prevented commercial production of this chemical. It should come into production if the price of lithium is ever stabilized at a competitive level.

Lithium hypochlorite solutions are excellent bleaches and they have the advantage over calcium hypochlorite solutions of being clear when prepared, thereby eliminating the need for settling and discarding of any sludge. In addition, lithium hypochlorite will not precipitate soap from warm laundry solutions, whereas calcium hypochlorite will.

Lithium hypochlorite may be made by mixing a strong solution of lithium chloride with a strong solution of sodium hypochlorite. Sodium chloride precipitates and the 20 to 25 percent solution of lithium hypochlorite is evaporated and drum dried.

CHLORINATED TRISODIUM PHOSPHATE

Chlorinated trisodium phosphate (TSP) is produced in substantial tonnages for use in domestic and industrial cleaning, disinfecting and sterilizing formulations. It is marketed in crystalline form with an available chlorine content of about 3.5 percent. Chemically, the material is one of the hydrated trisodium phosphate complexes[5] having the formula $4(Na_3PO_4 \cdot 11 H_2O)NaOCl$. Mathias [22] first isolated and produced the salt by crystallization and separation from an aqueous solution containing Na_3PO_4, Na_2HPO_4 and $NaOCl$. A later process by Adler[1] involving the direct crystallization on solidification of a molten mass of these materials approximates the present manufacturing technology. For example, sodium acid phosphate and caustic soda are melted at 220° F with sufficient water to form a solution having a specific gravity of 1.63. Strong sodium hypochlorite solution is added rapidly to the phosphate melt and the whole mass is cooled under vigorous agitation until it crystallizes (95° F). This procedure yields a dry, crystalline solid.

Use of chlorinated TSP involves not only the oxidizing property of the hypochlorite ion but also the general detergent properties of the alkaline phosphate. Several detergent applications in the dishwashing, cleaning and sanitation field have been patented.[2,3,25]

CHLORAMINES

The term "Chloramine" (or "chloroamine") is a broad term which includes all compounds containing one or more chlorine atoms attached to nitrogen, e.g., chloramines, chlorimines, chloramides and chlorimides. The chlorine which is attached to the nitrogen in these compounds is the source of the "combined available chlorine" of the chloramines. Although the bond between nitrogen and chlorine is covalent, some of its properties differ from those of the carbon-chlorine bond. For the purpose of considering the disinfecting and bleaching properties of the chloramines, it is convenient to regard the chlorine bonded to nitrogen as "positive" chlorine, i.e., chlorine with a valence of plus 1.

When chloramines are dissolved in water, the N—Cl bond is partially hydrolysed.

$$RR'NCl + H_2O \rightleftharpoons RR'NH + HOCl \qquad (17\text{-}6)$$

The hypochlorous acid thus released is responsible for the bleaching and disinfecting properties of the chloramine solution. The hydrolysis constant may be written as follows:

$$\frac{(RR'NH) \times (HOCl)}{(RR'NCl)} = K$$

Soper[39] gives the hydrolysis constant of a number of chloramines, as follows:

Chloramine-T	4.9×10^{-8}	N-chloro-*o*-acetotoluide	3.3×10^{-7}
Dichloramine-T	8.0×10^{-7}	N-chloro-*p*-acetotoluide	2.2×10^{-6}
N-chloroformanilide	2.4×10^{-7}	N, *o*-dichloroacetanilide	6.9×10^{-7}
N-chloroacetanilide	7.3×10^{-7}	N, *p*-dichloroacetanilide	1.5×10^{-6}

Corbett, Metcalf and Soper,[40] give the hydrolysis constant for monochloramine as 2.8×10^{-10}.

Useful chloramines have hydrolysis constants ranging from 10^{-4} to 10^{-8}. In general, chloramines should have constants higher than 10^{-6} to give useful bleaching, while those with constants lower than 10^{-7} may give slow disinfection, but will usually avoid reaction with nitrogenous matter. Since hypochlorite would be rapidly consumed by nitrogenous and other reducing substances present in milk, dairy equipment should be cleaned before it is disinfected with hypochlorite solutions. Chloramines are less affected by such contaminants. Several chloramines have been formulated into cleanser-sanitizer combinations for use in the dairy industry.

In hypochlorite sanitation, a given concentration of hypochlorite at a given pH has a fixed hypochlorous acid concentration, and hence a fixed bactericidal effectiveness, which is not affected by the amount of hypochlorite previously consumed. With chloramines, however, the residual amine tends to repress the release of hypochlorous acid. Repeated treatment of the same solution with chloramines and consumption of the available chlorine by reducing materials will allow the free amine to build up to a point where the hypochlorous acid concentration may be only one tenth of that formed when chloramine was first added. A chloramine can also react with an amine to form another chloramine of lower hydrolysis constant. Chloramines of low hydrolysis constant may be bacteriostatic rather than bactericidal.

The ability of a chloramine to bleach or disinfect is primarily dependent upon the degree of "activity" of the nitrogen-bonded chlorine. In monochloramine (NH_2Cl), the chlorine is not sufficiently active for this compound to be used as a bleaching agent, nor is it a satisfactory germicide against many organisms. Its hydrolysis constant is too low for hypochlorous acid to be formed in sufficient amount. Attempts have been made to increase the "activity" of the chlorine by replacing the hydrogens. It was discovered that by replacing the hydrogens with organic radicals the "activity" of the chlorine could be increased. Continuing work showed that replacing one of the hydrogens with a carbonyl group further increased the "activity" of the chlorine. The superior ability of the carbonyl group to increase the "activity" of the chlorine lead to the development of such compounds as succinchloramide, the chlorinated hydantoins and the cyanuric acid derivatives. The bactericidal success of the N-chloro sulfonamides (Chloramine-T, dichloramine-T, chloramine-B and halazone) depends upon the chlorine "activation" by the sulfonyl group, which also prevents the migration of the chlorine to the ring structure.

Another aspect of chloramines which makes them of particular interest is the safety of strong solutions. In very dilute solution (measured in parts per million) the chloramines are quite highly hydrolyzed and an appreciable portion of the available chlorine is present as hypochlorous acid, or "free" available chlorine. Examination of the hydrolysis constant shows that in a solution containing several percent of a chloramine, only a very small portion of the available chlorine is present as hypochlorous acid. With a hypochlorite solution of comparable available chlorine strength, the concentration of hypochlorous acid is very high. For example, a 15 percent solution of calcium hypochlorite contains approximately 30,000 ppm available chlorine as hypochlorous acid, a strength injurious to fabrics; but the strongest known solution of chloramine only contains about 3,000 ppm available chlorine as hypochlorous acid. This means that accidental contact of strong solutions of chloramine with cloth will not cause serious damage, whereas strong solutions of the hypochlorite may cause rapid degradation of fibers.

Another advantage of the commercial organic chloramines is the fact that they can be marketed as dry powders and granules which are conveniently stored and which only lose two or three percent of their available chlorine over the course of a year under normal storage conditions. Inorganic chloramines do not offer this advantage. The half life of monochloramine is measured in terms of hours; that of dichloramine, in terms of minutes; and nitrogen trichloride is well known for its explosive characteristics. Shipment of inorganic chloramines is therefore out of the question. They are formed at the point of application as required for use. N-chlorimidophosphates[42] and N-chloroimidosulfates are the exception, but have never won a market.

The principal uses of chloramines today are for sanitizing and disinfecting purposes. However, they are receiving greater notice in the constant effort to develop a satisfactory chloramine-based dry bleach to replace liquid bleach in the household and small laundry bleach market.

Inorganic Chloramines

Inorganic chloramines are formed by the reaction of ammonia with dilute hypochlorous acid. The nature of the chloramine depends upon the pH of the solution (see Figure 15-3). The simplest inorganic chloramines are monochloramine, dichloramine and nitrogen trichloride. Much work has been done on the formation of mono- and dichloramine in water treatment, and the instability of dichloramine is important in "breakpoint" chlorination today (see Chapter 15).

Inorganic chloramine treatment of water supplies as a method of providing available chlorine saw its widest use in the 1930's and early 1940's. It was primarily used in waters which contained phenols and other organic compounds which form objectionable tastes and odors when water is chlorinated. However, chloramine does not have sufficient bactericidal properties in the presence of certain organisms. The work of Butterfield and Wattie[44] shows that to obtain a 100 percent kill, the amount of combined available chlorine required is 25 times that of free chlorine in a given time; and the exposure time with chloramines is 100 times that required with equivalent free available chlorine levels. Chloramine treatment has been generally discontinued in the United States, with some states even forbidding its use in the treatment of water supplies. Nitrogen trichloride was used in bleaching and maturing of flour from 1920 through the middle 1940's.[12]

Organic Chloramines

Several dozen organic chloramines (including trichloroisocyanuric acid, chloramine-B and others) were prepared and studied by Chattaway and his co-workers around 1900.[10] However, none of these were brought into commercial production at that time. With the onset of World War I, the British showed renewed interest in the chloramines in an effort to find one suitable for disinfecting use in field hospitals. Since then several thousand organic chloramines have been synthesized. Of these, only a few have the combination of properties necessary for a successful bleaching or sanitizing agent. They must be stable under normal storage conditions; they must be readily soluble, and they must have a suitable hydrolysis constant. They must also be easily prepared and isolated.

Forty or fifty organic chloramines have received some commercial attention and then been abandoned. There are many other chloramines which have no readily separated derivatives and therefore have not been exploited. Only a few of the chloramines are stable.

The important organic chloramines are the N-Chloro derivatives of the following four groups of compounds:
(1) Sulfonamides.
(2) Heterocyclic compounds with nitrogen in the ring (e.g., hydantoin).
(3) Condensed amines from cyanamide derivatives.
(4) Anilides.

Chloramines are formed by the reaction of hypochlorous acid on an amine, amide, imine or imide. While the preparation and properties of a large number of chloramines have been published, only those of some commercial interest are discussed here.

Chloramine-T. The chemical formula for this compound, a sulfonamide derivative, is:

$$CH_3C_6H_4SO_2NClNa \cdot 3 H_2O$$

It is a white crystalline material which is soluble in water to the extent of 12 percent at 77° F. The available chlorine content is about 25 percent. This chloramine was the first to be successfully marketed. First prepared by Chattaway in 1905 and later introduced by Dakin in 1916, it has been widely used as a germicide, especially during World War I in the treatment of wounds. It has also been used as a surgical disinfectant and as a mouthwash.

Chloramine-T may be prepared by hypochlorination of *p*-toluene-sulfonamide which is prepared from ammonia and *p*-toluenesulfonyl chloride, a by-product in the manufacture of saccharin.

Dichloramine-T. Dichloramine-T ($CH_3C_6H_4SO_2NCl_2$) is prepared by chlorination of chloramine-T or by dissolving *p*-toluenesulfonamide in chlorinated lime solution. Unlike most other chloramines, it decomposes in acid solution to liberate chlorine.

It is a greenish-yellow crystal or crystalline powder which melts at 160 to 167° F and decomposes at 320° F. Its available chlorine content is about 58 percent. It is soluble in chlorinated paraffins, petroleum benzene and carbon tetrochloride, but it is nearly insoluble in water. In solution of chlorinated paraffins it has been used for surgical disinfection and as an antivesicant salve.

Chloramine-B. This sulfonamide derivative ($C_6H_5SO_2NClNa \cdot 1\frac{1}{2} H_2O$) is an odorless, white, crystalline material which decomposes at about 338° F. It is stable in air and light, and has an available chlorine content of 29.5 percent. Chloramine-B received increasing attention as a disinfectant in the 1940's and is still used for this purpose.

Halazone. Another sulfonamide derivative, halazone ($HOOCC_6H_4SO_2NCl_2$) is a white, crystalline powder which has the odor of chlorine. Its available chlorine content is about 50 percent.

Halazone is prepared by oxidizing *p*-toluenesulfonamide to convert the methyl group to a carboxyl group and then reacting this compound with hypochlorite.

The major use of halazone was for emergency purification of water supplies during both world wars.

Chlorinated Hydantoins. The structure of hydantoin, a heterocyclic amine, is shown below (I). A large number of mono- and dichlorinated derivatives have been prepared and studied.[45] Chief among these is 1, 3-dichloro-5, 5-dimethyl hydantoin (II), having the trade names "Dactin" and "Halane." In commercial laundry use, it will bleach to within 5 degrees of the whiteness obtainable with hypochlorite. In this compound, the chlorine in the (3) position is the more "active" since it has a carbonyl group on each side.

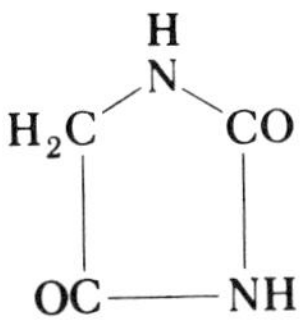

(I) 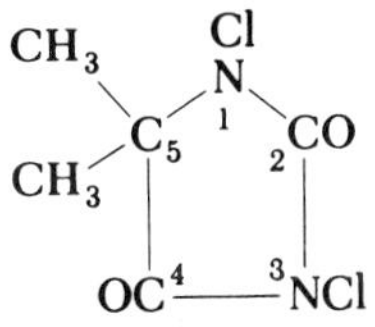   (II) 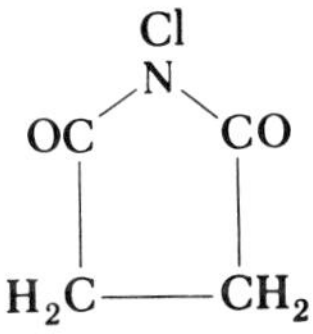   (III)

The chlorinated hydantoins have very limited solubility in water. This has limited their use to disinfecting and antivesicant purposes, although some of the derivatives have been patented as bleaching agents.[18,29,38]

Hydantoins can be prepared by treating an alcoholic solution of a ketone with aqueous ammonium carbonate solution, cooling, and acidifying.[37] An alternate method of preparation is to treat a ketone with hydrogen cyanide, forming the cyanohydrin which is then treated with aqueous ammonium carbonate solution.[21] In the preparation of chlorinated hydantoins, the parent hydantoins are dissolved in sodium carbonate solution and chlorinated to neutrality. This yields dichlorimines which are filetered off. Monochlorinated hydantoins can be prepared by mixing equivalent amounts of unchlorinated hydantoins and dichloro hydantoins in a slurry.[21] The monochlorinated hydantoins have lower oxidation potentials than the dichlorinated hydantoins, but they are usually more soluble.

Succinchloramide. Succinchloramide (III) is another heterocyclic chloramine. It is a white, crystalline powder which has the odor of chlorine. The available chlorine content is about 52 percent. The solubility of succinchloramide in water is 1.4 percent at 77° F. It is prepared by chlorination of succinamide, which is made either by heating the ammonium salt of succinic acid or by the reaction of ammonia upon succinic anhydride.

Succinchloramide has been proposed as a disinfectant for water supplies[9] and has found some use in treatment of water for livestock.

Cyanuric Acid Derivatives. The four most important derivatives of cyanuric acid are trichloroisocyanuric acid (91.5 percent available chlorine) dichloroisocyanuric acid (71.7 percent available chlorine), sodium dichloroisocyanurate (64.5 percent available chlorine) and potassium dichloroisocyanurate (60 percent available chlorine). All are useful as bleaching agents and disinfectants.

The reaction of hypochlorous acid with cyanuric acid is the basis of the preparation of chlorinated cyanurics. Cyanuric acid is first solubilized with caustic soda, and then chlorine is passed through the solution until the desired degree of chlorination is achieved. Precipitation of trichloroisocyanuric acid is essentially completely at a pH of 3.5.

Trichloroisocyanuric acid may be purified by dissolving in cold concentrated sulfuric acid, decanting, and adding an equal volume of ice water. The crystals of TCCA which then precipitate can be separated easily with ceramic filters.[18]

Dichloroisocyanuric acid is the starting material used to make the sodium and potassium salts. While solutions of salts can be processed by spray drying, slow drying in air is preferred. In this method, a temperature exceeding 110° F is avoided where the mother liquor is present. Hydrated crystals of the potassium salt are about 1 millimeter across and are easily dried. The hydrate of the sodium salt forms in needles which produce a much wetter cake than does the hydrate of the potassium salt.

The chloroisocyanurics as a class are dusty and electrostatic (dichloroisocyanuric being most so, then the sodium salt, followed by the potassium salt and trichloroisocyanuric acid), and they require some modification for marketing. Their high available chlorine content and good solubility, combined with long storage life, make them attractive as dry packaged bleaches for laundry and household use.

The chloroisocyanurics are diluted for marketing as bleaches and sanitizers (5 to 20 percent available chlorine), scouring powders (0.2 to 1.0 percent available chlorine), and dishwashing compounds (0.5 to 2.0 percent available chlorine.)[15] For swimming pool sanitation the cyanurics are used at a 55 to 60 percent available chlorine level.

Synthetic detergents and bleaches are usually blended with diluent materials (builders) having a synergistic nature. Special care must be taken in the selection of builders for chloroisocyanurate formulations. They must be anhydrous to prevent decomposition of the chloramine by water. Sodium chloride and sodium sulfate were the first builders used in preparing laundry bleaches. Sodium chloride has declined in use since it has no beneficial effect. It has been replaced by sodium tripolyphosphate, sodium metasilicate and, in some cases, sodium carbonate. Sodium and potassium silicate are also used in scouring cleanser formulations. Detergents should be used in small amounts. Anionic sodium dodecylbenzenesulfonate and sodium alcohol sulfate are generally preferred.

The bleaching action of the chloroisocyanurates is rapid and comparable to that of the hypochlorites. However, with the hydrolysis constant of these chloramines on the order of 1×10^{-4}, the danger of fiber degradation is negligible as compared with the hypochlorites.

The first application of chloroisocyanurates to swimming pool water gives bactericidal action nearly as strong as the hypochlorites. As cyanuric acid builds up in the pool water, the "activity" of the chlorine is repressed, giving a longer life to the chloroisocyanurate. While these chloramines are regarded as sufficient for the sanitation of family pools, they are not as effective algaecides as the hypochlorites and cannot replace them in this respect.

CHLORINE DIOXIDE

In 1921, E. Schmidt discovered that cellulosic fibers can be purified by chlorine dioxide with no appreciable degradation. This discovery led to extensive research to develop economical methods for bleaching with chlorine dioxide. Being extremely explosive at high concentrations, chlorine dioxide gas could not be transported as such. It was necessary to develop a stable chemical which could be safely transported and which could be easily reacted to form chlorine dioxide at the bleaching site. Mathieson Chemical Corporation successfully developed sodium chlorite as the economical commercial chemical for this purpose. The sodium chlorite bleaching of textiles was introduced in 1939 and soon became an established process.

Chlorine dioxide can also be generated by the reduction of sodium chlorate solution. This method requires a sizeable capital outlay for generation equipment and is only economical where tonnage quantities of chlorine dioxide are required. Thus, its use is restricted to the bleaching of pulp.

Chlorine dioxide is the one bleaching agent discussed in this chapter which does not belong to the family of "available chlorine" compounds (those compounds which form hypochlorous acid when dissolved in the water). Nevertheless, the oxidizing power of chlorine dioxide is referred to as "available chlorine content," and is determined by titration as described of page 516. It has an equivalent* available chlorine content of 263 percent.

Chlorine dioxide undergoes a valence change of 5 in going to chloride, and is most effective in reactions which require five or six valence changes. Neither chlorine dioxide nor chlorous acid passes through a hypochlorous acid stage during reduction. Chlorine dioxide cannot chlorhydrinate a double bond, a reaction which requires two valence changes. It does not attack most amines. It is effective in removing phenols from water, a reaction where six valence changes are involved, and tenths of a part per million of chlorine dioxide can reduce phenols to trace amount. It thus undergoes reactions which are entirely different from those of hypochlorous acid. The end result is an oxidation, although the mechanism of bleaching with chlorine dioxide has not been undisputedly established. According to one theory, the bleaching action is dependant upon the formation of chlorous acid. However, recent experimentation has shown that excellent bleaching is obtained by exposing dry greige goods to chlorine dioxide gas.

Chlorine Dioxide From Sodium Chlorite

Dilute solutions of chlorine dioxide may be produced by adding sodium chlorite directly to the bleach saturator. Normal practice is to prepare a 10 percent solution of sodium chlorite which is then added to the solution in the saturator as required.

*In the case of bleaching compounds which do not bleach through the medium of hypochlorous acid, it is preferable to speak of the oxidizing power as *equivalent* available chlorine content to distinguish such bleaching compounds from those which belong to the available chlorine family.

When sodium chlorite is dissolved in water, chlorous acid is formed:

$$H_2O + NaClO_2 \rightleftharpoons HClO_2 + Na^+ + OH^- \qquad (17\text{-}6)$$

If the solution is made acid, the chlorous acid is converted to chlorine dioxide according to the equation:[46]

$$5\,HClO_2 \xrightarrow{\quad H^+ \quad} 4\,ClO_2 + HCl + 2\,H_2O \qquad (17\text{-}7)$$

Some side reactions may lead to the formation of chlorate, but under proper operating conditions this is negligible. Earlier methods of activating sodium chlorite solutions for bleaching have, in general, been replaced by the acid chlorite bleach. Chlorine activation[17] is still used in water purification, however.

Potentiometric studies have shown that the oxidizing power of chlorite solutions is less than that of hypochlorite, but greater than that of peroxide. The oxidation potential of chlorite solutions increases rapidly as the pH is lowered (see Figure 17-2). Sodium chlorite solutions give excellent bleaching without degradation with all fibers except wool and silk.

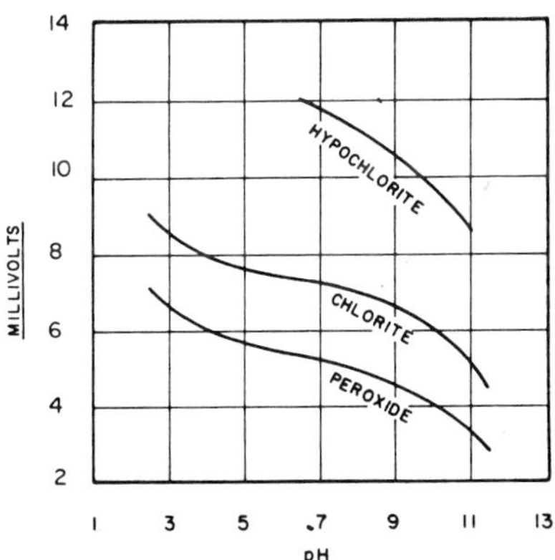

Figure 17-2. Oxidation potentials referred to the normal hydrogen electrode.

Very few metals are completely resistant to chlorine dioxide. Titanium is the most resistant metal used for this purpose today, and its use is restricted to pulp and paper mills which bleach with chlorine dioxide on a large scale. In the textile industry molybdenum alloy stainless steels are used for acid chlorite bleach apparatus. Type 316 stainless steel is preferred. For best operation, the stainless steel sheet should be cold rolled and polished, and all welds should be ground smooth and polished. The addition of sodium nitrate to the chlorite solution is general practice to inhibit pitting of stainless steels, and, with the operating conditions that have been developed recently, this has proved quite successful. A double salt of sodium chlorite and sodium nitrate has also received some commercial acceptance.

The odor problem which had discouraged the use of acid chlorite bleaching has been overcome by careful control of chlorite concentration, temperature and pH. In present practice, the temperature in the saturator is kept below 100° F and the pH is maintained between 3 and 4. Chlorite concentrations range between 0.10 and 0.75 percent. The goods are held in a bleach J-chute for 30 to 60 minutes at about 205° F and then washed.

Continuous sodium chlorite bleaching has several advantages over peroxide bleaching. With sodium chlorite, all synthetics and synthetic-cotton blends are readily bleached. In the case of cotton, sodium chlorite bleaching produces a fabric with softer hand, superior dyeing and printing properties, and freedom from localized tendering caused by metallic contaminants. In addition, there is less degradation of fibers with chlorite bleaching. This becomes evident with the greater tensile and tear strengths which are achieved with sodium chlorite bleaching to a given whiteness and which are even maintained after resination.

Chlorine dioxide was first used for water purification in 1944. The unique ability of chlorine dioxide to break down phenolic compounds and completely remove phenolic taste and odors from water has lead to its use in the purification of municipal water supplies. Many plants are equipped for emergency treatment of phenolic contamination of water. Some other tastes and odors encountered in water treatment can also be controlled by chlorine dioxide.

Some municipal facilities employ chlorine dioxide treatment of water as one of the pretreatment steps in the regular water purification program. For this application, sodium chlorite is first activated with a water solution of chlorine to form a dilute solution of chlorine dioxide according to the equation:[17]

$$2\,NaClO_2 + Cl_2 \rightarrow 2\,ClO_2 + 2\,NaCl \qquad (17\text{-}8)$$

The chlorine dioxide solution is then mixed with the water at a level of several parts per million. At this low concentration, the activation would take place much too slowly if the chemicals were added directly to the water being treated.

For large municipal water works a solution of sodium chlorite is supplied in tank cars or trucks in place of the drummed flake material at a considerable economic advantage.

Chlorine dioxide has also been used to improve the quality of low-grade fats and oils.[28,48]   In this application, chlorine dioxide is generated externally and is then passed into the fat or oil to be upgraded.   Other uses of chlorine dioxide include the bleaching and maturing of flour,[12] bleaching of sugar,[43] and blacking certain metal surfaces.[19,24]

Sodium chlorite may be prepared by passing chlorine dioxide through a moderately strong solution of sodium hydroxide.   However, sodium chlorate is formed in an equimolar quantity and is difficult to separate.

$$2\,ClO_2 + 2\,NaOH \rightarrow NaClO_2 + NaClO_3 + H_2O \qquad (17\text{-}9)$$

A means of reducing the chlorate must therefore be provided.   This may be done with hydrogen peroxide, sodium peroxide or amalgam.   In an early method, carbonaceous matter was used for the reduction.   Following chlorate reduction, the sodium chlorite solution is evaporated and drum-dried.

The quivalent available chlorine content of pure sodium chlorite is about 157 percent.   Commercial preparations (Textone, C2, etc.) contain about 81 percent sodium chlorite along with some sodium chloride, sodium chlorate and water.   The equivalent available chlorine content of these products averages 125 percent.

Chlorine Dioxide From Sodium Chlorate

Chlorine dioxide is produced from sodium chlorate at a rate far exceeding 100 tons per day in pulp bleaching mills in the United States alone. Since the first chlorine dioxide bleaching plant in the U. S. A. began operation at the Natchez mill of the International Paper Co. in 1950, this bleaching process has rapidly gained widespread adoption.   In 1959 there were over 46 mills using chlorine dioxide in this country.

Bleaching with chlorine dioxide yields an exceedingly white pulp which gives a paper of high brightness, great strength and excellent color stability.   Prior to the development of chlorine dioxide bleaching, it was impossible to achieve this combination of desirable properties (whiteness could be improved at the expense of strength, and vice versa), and to even approach the present standards which have been established by chlorine dioxide bleaching was economically out of the question.

For this application, the chlorine dioxide is produced in a special generating plant and then is piped to the bleach plant.   It is generated as a dilute gas (10 percent or less in air) which is dissolved in water in an absorption tower to a concentration of about 0.5 to 1.0 percent. This solution is mixed with high density pulp (10 to 12 percent) which is then steam heated and pumped into a retention tower where it is held from 3 to 5 hours, allowing the pH to drop to near 4.5.

There are six methods which have been used to generate chlorine dioxide for pulp bleaching. All are based upon the reduction of sodium chlorate. Three of these originated and are operated in Europe. They are the Holst, the Persson and the Kesting processes. First to be operated on this continent was the C.I.P. process in Canada, followed by the Solvay process in 1952 and the Mathieson process in 1954 in the United States. All of these methods were successful and are in operation today.[31]

The Holst process is a batch operation in which sodium chlorate in solution is reduced by gaseous sulfur dioxide in the presence of sulfuric acid. Two reaction vessels are used alternately. The sulfur dioxide is produced from a sulfur burner. The sodium acid sulfate produced can be used in Kraft mills.

In the Persson process, sulfur dioxide is used to reduce chromic acid to chromic sulfate which is reacted with the chlorate in the presence of sulfuric acid to form chlorine dioxide and chromic acid. The acid is recycled. This avoids the side reactions which accompany sulfur dioxide reduction of the chlorate. The only disadvantage is high capital investment and a little more complex equipment than other systems.

In the Kesting process, hydrochloric acid is the reducing agent employed. The reduction is carried out at relatively high temperatures. High chlorate concentrations with low acid concentration are maintained to prevent the reaction forming chlorine gas. This method is made practical through integration of the generator with a sodium chlorate plant.

The C.I.P. process is based upon the countercurrent flow of dilute sodium chlorate solution and dilute sulfur dioxide in a reaction tower. The sulfurous acid formed is oxidized to sulfuric acid while the chlorate is reduced to chlorine dioxide. This provides continuous operation without purchase, storage or handling of sulfuric acid. However, difficulties in large-scale operation and lower yields than other methods discouraged its general adoption.

The Solvay process is based upon the use of methanol for the reduction of sodium chlorate to chlorine dioxide. The reaction of methanol with chloric acid is relatively slow and is therefore carried out in two stages with two reaction vessels. This process yields chlorine dioxide of about the same purity as the sulfur dioxide processes, but it does not have any advantages over sulfur dioxide reduction.[31] The cost per pound of chlorine dioxide via methanol reduction is about two cents greater than via sulfur dioxide reduction when a sulfur burner is used. If sulfur dioxide is purchased as a liquid, the cost of chlorine dioxide is about the same for both processes. Having been the first to operate in this country however, the Solvay process has received wide acceptance, with about twenty mills using it today.

The Mathieson process is the most widely used in this country with twenty-nine mills operating it at present. It is a continuous process (see Figure 17-3) in which sulfur dioxide diluted with air is reacted with sodium chlorate solution in the presence of sulfuric acid. The air used to dilute the sulfur dioxide becomes the diluent for the chlorine dioxide until it is absorbed in water. With high acid concentration and low chlorate concentration the chlorine dioxide produced has a negligible chlorine content. Sulfur dioxide, either purchased as a liquid or produced by burning sulfur, is introduced into the bottom of a large reaction vessel at one or more points. The contents of the vessel are kept in continuous turbulence to maintain the contents at a constant composition. Units producing up to

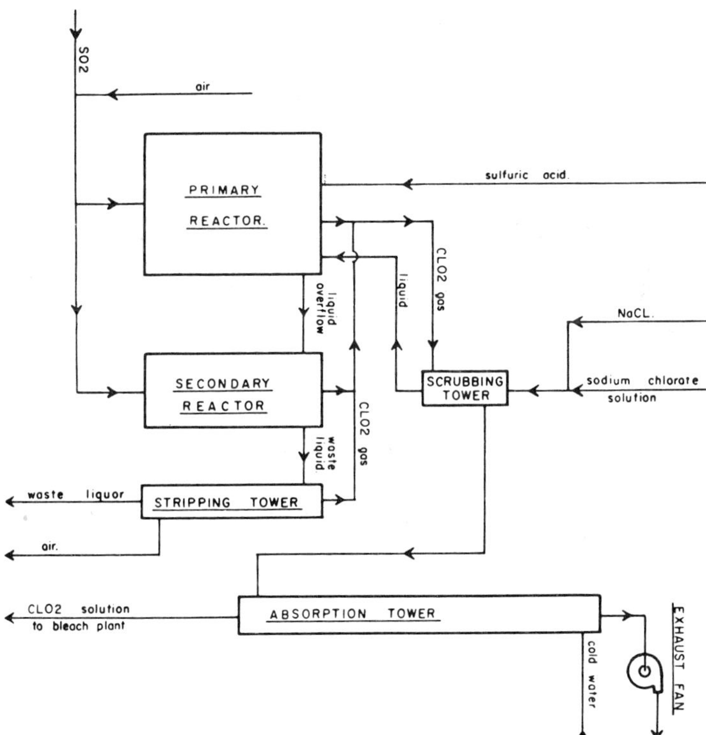

Figure 17-3. Mathieson ClO$_2$ flow diagram.

6 tons per day are currently in operation and show a yield of 90 percent or higher of the theoretical conversion of sodium chlorate.

This process has the advantages of high efficiency, low chlorine content, easy start-ups and shutdowns and low capital investment. It produces the cheapest chlorine dioxide of any of the present processes when a sulfur burner is used to provide the sulfur dioxide.

The major hazard in chlorine dioxide handling is the possibility of explosion. This is overcome by keeping the chlorine dioxide gas diluted with an inert gas or air to a partial pressure less than one seventh of an atmosphere. As an added precaution, vessels in which safe concentrations may be exceeded are provided with explosion lids.

The bleaching of paper is a multistage process (see Chapter 10). Bleaching with chlorine dioxide may be accomplished in either one or two stages of the bleaching sequence. When chlorine dioxide is used in conjunction with the hypochlorite bleaching, it is usually applied in one stage following the hypochlorite bleach. A modern trend is to eliminate hypochlorite bleaching and to employ the chlorine dioxide bleach in two stages. A typical sequence is as follows: (1) chlorination, (2) hot alkali extraction, (3) chlorine dioxide, (4) hot alkali extraction, and (5) chlorine dioxide. A wide variety of other sequences are also employed.

Sodium chlorate is shipped in tank cars as a dry crystalline material. When the car is received, water is pumped in at a temperature usually between 130 and 140° F to make a solution of approximately 45 percent concentration. The solution is pumped out at between 100 and 120° F and stored.

Chlorine dioxide can also be transported as a frozen hydrate. This method is rather expensive, but it is convenient where small amounts of chlorine dioxide are used. Chlorine dioxide is passed into cold water until the octahydrate crystallizes.[47] The hydrate is separated and is shipped and stored under refrigeration.

References

1. Adler, H., U. S. Patent 1,965,304 (1934).
2. Albrecht, K., U. S. Patent 2,756,214 (1956).
3. Anderson, D. E., *et al.*, U. S. Patent 2,589,225 (1954).
4. Barton, L. J. (to Chlorox Chemical Co.), U. S. Patent 2,170,108 and U. S. Patent 2,918,351.
5. Bell, R. W., *Ind. Eng. Chem.*, **41**, 2901 (1949).
6. Bunn, C. W., Clark, L. M., and Clifford, I. L., *Proc. Roy. Soc. (London)*, **A151**, 141 (1935).
7. Cady, G. H. (to Pittsburg Plate Glass Co.), U. S. Patents 2,157,524; 2,157,525 (1939).

8. Cady, G. H., and Muskat, I. E. (to Pittsburg Plate Glass Co.), U. S. Patents 2,157,558 and 2,157,559 (1939); U. S. Patent 2,225,923 (1940).

9. Chang, S. L., "Destruction of Micro-Organisms," *J. Am. Water Works Assoc.,* **36,** 1192 (1944).

10. Chattaway, F. D., *J. Chem. Soc.,* **57–91** (1890 to 1907).

11. Edelstein, S. M., "Origins of Chlorine bleaching in America," *Am. Dyestuff Reptr.,* **49,** (1960).

12. Ferrari, C. G., and Hutchinson, W. S. (to General Mills, Inc.), U. S. Patent 2,324,203 (1943).

13. George, A., and MacMullin, R. B. (to Mathieson Alkali Works), U. S. Patent 1,713,650 (1929).

14. Green, D. E., and Stumpf, P. K., "The Mode of Action of Chlorine," *J. Am. Water Works Assoc.,* **38,** 1301 (1944).

15. Hardy, E. A. (to Monsanto Chemical Co), U. S. Patent 2,607,738 (1952).

16. Hinshaw, J. O. (to Purex), U. S. Patent 2,610,905.

17. Logan, J. O. (to Mathieson Alkali Works), U. S. Patent 2,131,447 (1938).

18. Lorenz, W. K. (to Purex), U. S. Patent 2,828,308 (1955).

19. MacMahon, J. D. (to Enthone, Inc.), U. S. Patent 2,481,854 (1949).

20. MacMullin, R. B., and Taylor, M. C. (to Mathieson Alkali Works), U. S. Patent 1,787,048 (1930).

21. Magill, P. L. (to Du Pont), U. S. Patent 2,430,233 (1947).

22. Mathias, L. D., U. S. Patent 1,555,474 (1925).

23. Megele and Wiseman, *Water Treatment,* London, 1958.

24. Meyer, W. R. (to Enthone, Inc.), U. S. Patent 2,460,897 (1949).

25. Miller, D. E., U. S. Patent 2,536,456 (1951).

26. Montgomery, G. L., *Chem. & Met. Eng.,* **26,** 1038 (1922).

27. Muskat, I. E. (to Pittsburg Plate Glass Co.), U. S. Patent 2,240,342 (1941).

28. Olin Mathieson Chemical Corp., "Upgrading Inedible Fats with Chlorine Dioxide," (1956).

29. Quinn, J. A. (to Theobald Industries), U. S. Patent 2,795,556 (1957).

30. Race, Joseph, "Chlorination of Water," 1st Ed., New York, John Wiley & Sons, Inc., 1918.

31. Rapson, W. H., *TAPPI,* **37,** (4) 129 (1954).

32. Ridge, B. P., and Little, A. H., "The Control of pH in Hypochlorite Solutions," *J. Textile Inst.,* **33,** (4) (1942).

33. Ritter, B. H. (to Hooker Chemical Co.), U. S. Patent 2,288,841.

34. Robson, H. L. (to Mathieson Alkali Works), U. S. Patent 2,368,042 (1942).

35. Robson, H. L., and Petroe, G. A. (to Mathieson Alkali Works), U. S. Patent 2,195,775 (1940).

36. Robson, H. L. (to Olin Mathieson Chemical Corp.) and Stoneman, A. C. (to Purex Corp.), U. S. Patent 2,806,765 (1957).

37. Rogers, A. O. (to Du Pont), U. S. Patents 2,398,598 and 2,398,599 (1946).

38. Scheer, W. E., and Levy, M. F. (to Argus Chemical Corp.), U. S. Patent 2,798,875 (1956).

39. Soper, F. G., *J. Chem. Soc.,* **125,** 1899 (1924).

40. Soper, F. G., Metcalf, W. S., and Corbett, R. E., *J. Chem. Soc.,* Part II, 1927 (1953).
41. Synan, J. F., MacMahon, J. D., and Vincent, G. P., "A New Development in the Treatment of Water," *J. New England Water Works Assoc.,* **58,** 264 (1944).
42. Taylor, M. C., U. S. Patent 2,796,321 (1956).
43. Vincent, G. P., and Fenrich, E. G., *Sugar,* **37,** (10) 26 (1942).
44. Wattie, E., and Butterfield, C. T., "Relative Resistance of *E. Coli* and *E. Typhosa* to Chlorine and Chloramines," *Public Health Reports,* **59,** 1661 (1944); (Reprint No. 2593).
45. Wave, E., "The Chemistry of the Hydantoins," *Chem. Review,* **46,** 403 (1950).
46. White, J. F., Taylor, M. C., and Vincent, G. P., *Ind. Eng. Chem.,* **34,** 782 (1942).
47. Williamson, H. V., and Hampel, C. A. (to Cardox Corp.), U. S. Patent 2,768,877 (1956).
48. Woodward, E. R., and Vincent, G. P., *Soap Sanit. Chemicals,* **22,** (9) 40, 137 (1946).

18. ETHYL CHLORIDE AND ETHYLENE DICHLORIDE

PETER W. SHERWOOD

Ethyl Corporation

ETHYL CHLORIDE

History

The known history of ethyl chloride dates back to the early fifteenth century when alchemist Basil Valentine described a sweet product obtained by the combination of common salt with spirit of wine, followed by distillation. This "sweet spirit of salt" was well known to later chemists; Glauber refers to it in 1648 and Ludolff describes, in 1749, a method of preparation from alcohol, sulfuric acid and sodium chloride. The product obtained by the action of muriatic gas on alcohol was sold in 1801 by a German apothecary under the name "Basse's hydrochloric ether."[13]

The anesthetic properties of ethyl chloride were discovered in 1847, but, because of objectionable symptoms caused by the early product, adoption of ethyl chloride for anesthetic use did not become widespread until after 1895.

Various anesthetic applications remained the only commercial outlet for ethyl chloride, as late as 1913. The first large-scale use of the product came with the development of Kraus' tetraethyl lead synthesis[70] which had its initial commercial application in 1923. To this date, tetraethyl lead remains by far the principal consumer of ethyl chloride.

Commercial Aspects

The only generally known very large use for ethyl chloride is as an intermediate in the manufacture of tetraethyl lead (1958 U. S. consumption was 455 million pounds).[113] Other reported areas of consumption are principally in the production of ethyl cellulose, as a catalyst in the manufacture of ethyl benzene, and as an anesthetic and refrigerant. The latter four uses are minor, amounting in total to no more than ten percent of ethyl chloride consumption.[39] Several additional uses, mentioned in the subsequent discussion of ethyl chloride reactions, are of minor economic significance.

Comparison of these consumer products with reported ethyl chloride production shows that at least one major use not identified by the literature must exist. One such outlet for ethyl chloride may be its conversion to higher chlorinated compounds.

United States production of ethyl chloride is reported by the U. S. Tariff Commission as follows:[73]

Year	Production, Million Pounds
1954	547.1
1955	541.6
1956	645.5
1957	604.4
1958	535.7
1959	550.8
1960	545.4

As in the past, future consumption of ethyl chloride will hinge principally on requirements for the antiknock compound, tetraethyl lead (TEL). Among new uses for ethyl chloride, the brightest outlook is in the manufacture of ethyl aluminum sesquichloride which is the leader in a growing market for Ziegler-type polymerization catalyst systems.

Commercial Methods of Manufacture

Three types of process are of commercial importance for the manufacture of ethyl chloride: reaction between ethanol and hydrogen chloride (Frei reaction, Equation 18-1), the direct addition of hydrogen chloride and ethylene (Equation 18-2) and the thermal chlorination of ethane (Equation 18-4). Of these, the Frei reaction was the predominant method of manufacture until the early 1940's but is today nearly obsolete (for 1960, Messing[77a] still reports U. S. ethyl chloride capacity of 80 million pounds per year, based on ethyl alcohol vs. a total ethyl chloride capacity of 770 million pounds). The analogous route to methyl chloride continues to be significant.

The bulk of ethyl chloride output in the United States is derived from ethylene and hydrogen chloride. At least five American producers employ this route today.

The chlorination of ethane as a source of ethyl chloride made its commercial entrance during the early 1950's and is known to be practiced by one manufacturer in the United States and by one British firm. Its main economic advantage is the lower cost of feed hydrocarbon (1960 average ethane price is reported at 1.0 cents per pound vs. 5.0 cents for ethylene.[73]

On the debit side of ethane chlorination is the more extensive formation of higher chlorinated C_2-by-products and the formation of co-product HCl which constitutes a difficult disposal problem in many instances.

This question of HCl balance is a key to the relative economics of ethylene hydrochlorination and ethane chlorination as routes to ethyl chloride. Equations (18-2) and (18-4) show that ethylene hydrochlorination is a *consumer* of HCl while ethane chlorination is a net *producer* of HCl. In technology today HCl is in most locations a surplus commodity, the only large-scale end uses being in muriatic acid and in the production of vinyl chloride (from acetylene), chloroprene, ethyl chloride (from ethylene or ethanol), and methyl chloride (from methanol). On the other hand, HCl is generated as a by-product in any number of chlorination processes, including the production of allyl chloride, vinyl chloride (from 1,2-dichloroethane), perchloroethylene, chlorinated methanes, etc.

Thus, the manufacture of ethyl chloride may be integrated with some other consumer or producer of HCl. The types of integration listed in Table 18-1 are typical:

TABLE 18-1. MANUFACTURE OF ETHYL CHLORIDE BY INTEGRATION

Source of Hydrocarbon Radical in Ethyl Chloride Process	Producer or Consumer of HCl
Ethylene or ethanol	HCl Producer: Chlorination of methane, ethane, propylene (allyl chloride synthesis), propane, etc.
Ethane	HCl Consumer: Ethyl chloride synthesis from ethylene, vinyl chloride synthesis from acetylene, manufacture of chloroprene, methyl chloride production from methanol.

It should be noted that the possibility of such integration in ethyl chloride synthesis is not yet fully realized, i.e., significant quantities of HCl for reaction with ethylene are still obtained by the reaction of hydrogen and chlorine. However, there is a very strong trend toward obtaining overall HCl balance in ethyl chloride synthesis wherever factors of location permit.

Thus, aspects of plant location and size, raw materials position and the possibility of HCl balance with other syntheses are the determining factors in the selection of the most economic route to ethyl chloride synthesis. A consideration of the technical aspects of the three commercially significant methods of production in their historical sequence is undertaken in the following pages.

Hydrochlorination of Ethanol (Frei Reaction)

This process is carried out in accordance with the following over-all reaction:

$$C_2H_5OH + HCl \rightleftharpoons C_2H_5Cl + H_2O \qquad (18\text{-}1)$$

Equation (18-1) represents a reversible reaction. Kilpi[67] has shown that the following equilibrium coefficient

$$K = \frac{C_2}{C_1} = \frac{k_2(C_2H_5Cl)\,(H_2O)}{k_1(C_2H_5OH)\,(HCl)}$$

increases slightly as the electrolyte concentration is raised (in the range 0.05 to 0.8N). This effect is due almost entirely to a reduction in the velocity coefficient k_1 as electrolyte concentration rises. Similarly, the velocity coefficient of ethyl chloride formation declines with increasing alcohol concentration.

Since the decrease in velocity coefficient k_1 is slow with increasing electrolyte content, the *over-all* rate of reaction rises with increasing HCl concentration. This is confirmed by Walvekar *et al.*[120] who have shown that the initial rate of reaction of ethyl alcohol with hydrogen chloride increases with HCl concentration, as shown in Table 18-2.

TABLE 18-2. INITIAL REACTION RATE OF ETHYL ALCOHOL WITH HYDROGEN CHLORIDE

HCl Concentration (moles/liter)	Initial Reaction Rate Percent per min., $r \times 10^{-4}$
.1979	36.67
.420	57.99
.6099	67.50
.690	70.0
1.013	77.50

The reaction rate is significantly influenced by the presence of solvents. For example, addition of benzene increases initial reaction rate, presumably because a differential solubility effect causes more extensive association of hydrogen chloride with the alcohol molecule.

The mechanism of the reaction between ethanol and HCl has not been fully explored. Most recently, Gerrard *et al.*[44] have shown evidence that the conversion entails two steps—the rapid formation of an oxonium halide, $ROH_2^+X^-$, followed by its rate-determining dehydration to ethyl chloride.

Operating aspects of the Frei reaction, as covered by the patent literature, include catalytic conversion in the vapor phase over zinc chloride,[102,79] alumina,[32] magnesium chloride,[77] or phosphoric acid[38] and non-catalytic conversion in the liquid phase.[22,36,37,1,83,21] However, the overwhelming interest is given to execution of the reaction in liquid aqueous phase and in the presence of a catalyst.

Most of the catalysts proposed for the reaction exhibit predominantly dehydrating character. Of particular interest in the liquid phase are zinc chloride,[53,41,106,24] calcium chloride,[122] ferric chloride,[41] cadmium chloride, tin chloride, antimony chloride[40] and sulfuric acid.[60] The pseudo-catalytic effect of benzene has already been referred to above.

Billitzer[17] has investigated the continuous reaction of HCl with ethyl alcohol in the presence of aqueous calcium chloride solution at $100°C$. Rate of ethyl chloride evolution reached a maximum after two hours, after which time it dropped off—presumably due to excessive dilution of the reaction medium by water formed in the reaction. Several runs are reported in which yield of ethyl chloride is essentially quantitative (on ethanol).

This decline in catalyst activity due to water-dilution is recognized as a major problem and several approaches have been proposed to maintain constant catalyst concentration.

Carter[24] calls for operation of the reactor at temperature-pressure conditions which will maintain zinc chloride concentration between 45 and 50 percent. Temperature ranges from 138 to $160°C$, and a corresponding pressure between 20 and 50 psig is maintained. As an example, 95 percent ethanol is fed continuously into 45 percent aqueous zinc chloride at the rate of 0.17 parts per hour per part $ZnCl_2$ solution. Anhydrous hydrogen chloride is fed concurrently at 0.13 parts per hour per part $ZnCl_2$ solution. The reaction medium is held at $145°C$ and 30 psig to maintain constant catalyst concentration. The reactor overhead is dephlegmated at $80°C$ for removal of aqueous HCl and is then condensed at $20°C$ and 30 psig. Product ethyl chloride is separated from water by decantation. Yield is reported at 98 percent (on ethanol).

Also to remove water of reaction from the catalyst system, McCurdy[27] proposes the use of entraining agents, e.g., carbon tetrachloride or trichloroethylene, which form ternary azeotropes with water and alcohol. Azeotropic composition at atmospheric pressure is high in concentration of entraining agent (e.g. 86.3 percent CCl_4 or 69.0 percent trichloroethylene).

Still another approach,[60] proposed for ethyl chloride synthesis in the presence of sulfuric acid, involves continuous withdrawal of liquid from the reactor, distillation for removal of contained alcohol and ethyl chloride, refortification to 45 percent H_2SO_4 content by addition of strong acid and recirculation to the reactor. Obviously, this approach necessitates a corresponding bleed stream of spent catalyst solution.

In general, it is preferred to use hydrogen chloride in about 10 to 50 percent excess above the stoichiometric equivalent of alcohol. Scott[98] points out that this leads to serious corrosion problems downstream of the converter. For this reason, and to avoid formation of by-product ethers and tarry materials which occur in the presence of $ZnCl_2$ and $FeCl_3$ catalysts, he recommends the use of aliphatic tertiary amine hydrochlorides as catalysts. Catalyst concentration is in excess of 50 percent. Reaction is carried out at 115 to 140°C with HCl and C_2H_5OH fed in stoichiometric balance or with excess alcohol. Conversion is reported in the range 35 to 70 percent.

Hydrochlorination of Ethylene. The commercially important synthesis of ethyl chloride from ethylene proceeds in accordance with Equation (18-2):

$$C_2H_4 + HCl \rightleftharpoons C_2H_5Cl \tag{18-2}$$

This is an equilibrium reaction with the following relationship.[92]

for the pressure equilibrium constant $K_p = \dfrac{p_{C_2H_5Cl}}{p_{C_2H_4} \cdot p_{HCl}}$

$$\log K_p = \frac{2925}{T} - 4.96 \tag{18-3}$$

where temperature (T) is in °K and partial pressure (p) in atmospheres. A graph of this relationship is published by Thodos and Stutzman[109] who have also investigated rates of this reaction over silica gel-supported zirconium oxychloride catalyst.

From the practical point of view, hydrochlorination of ethylene in both the vapor-phase and the mixed liquid-vapor phase is of interest. In both types of process Friedel-Crafts and related catalysts are employed.

Vapor-phase. For vapor-phase processes, Arnold[7] specifies the catalytic use of supported chlorides of iron, aluminum, antimony, arsenic and bismuth, as well as of beryllium, magnesium, zinc, cadmium and mercury at 100 to 150°C. Bond[19] covers the use of supported thorium halides and oxyhalides at 180 to 260°C. Cherniavsky[26] employs supported zinc chloride, promoted by the halides of copper, lithium, antimony, magnesium, calcium or bismuth, at 150 to 250°C and 75 to 300 psig.

The usual types of catalyst carrier have been proposed, but interest centers on porous alumina, e.g. calcined bauxite,[26,10] characterized by low iron and titanium content.[107]

Slotterbeck[104] points out that chlorides of metals in the fifth group of the Periodic Table are at first highly active but that they suffer from rapid loss in activity. Such degradation can reportedly be retarded by adding 10 to 15 percent moisture to the mixed reactor feed gases.

Arnold[7] cites as a chief drawback of vapor-phase conversion, the rapid decrease in equilibrium conversion as the temperature is raised above 100° C. Furthermore, he points out that catalysts of the fifth metal group are volatile and are therefore carried from the reaction chamber and that further catalyst deterioration occurs due to formation of stable complexes with olefins. The latter objections are reportedly less pronounced for catalysts consisting of supported chlorides of metals in the second group of the periodic table, and particularly in the sub-group consisting of Be, Mg, Zn, Cd and Hg.

In most instances of vapor-phase operation, the reaction gases are passed through a fixed bed of catalyst and the gases leaving the converter are condensed and fractionated. Axe[10] proposes combination of conversion and crude stripping in a single unit. Here, the reaction is carried out in the upper half of a fractionation column. Product ethyl chloride is condensed and refluxed; contained ethylene and hydrogen chloride are stripped from the product in the lower half of the column and are returned directly to the reaction zone.

Cherniavsky[26] describes the hydrochlorination of ethylene at 175° C and 250 psig over copper chloride-promoted zinc chloride on porous alumina. At a rate of 1 lb mixed feed per hour per 1.8 lb catalyst, ethylene conversion is 90 percent per pass, and ethyl chloride yield is 99.5 percent on ethylene converted. Arnold[7] reports sustained conversion of only 84 percent over zinc chloride on charcoal.

Nekrasov[81] has investigated the vapor-phase reaction between HCl and ethylene over various catalysts. His best result (93.5 percent ethyl chloride yield) was obtained over ferric chloride-impregnated clay briquets at 200° C and 32.2 hour^{-1}. The catalyst was reported to be short-lived.

Liquid-phase. "Liquid-phase" operation here refers to the use of a liquid-phase catalyst-containing reaction medium through which the ethylene and HCl are passed. According to Axe,[10] operation in the liquid phase permits easier control of catalyst concentration and activity. Catalysts employed for the liquid-phase hydrochlorination of ethylene are of the Friedel-Crafts type.

Axe[10] employs an equimolar boron trifluoride-phosphoric acid combination as catalyst. Operating in the presence of an inert hydrocarbon diluent (to facilitate temperature control) at 38° C and 500 to 600 psi, and using an HCl: ethylene ratio between 2:1 and 5:1, about 95 percent of the olefin is reportedly converted to ethyl chloride and five percent is consumed by the formation of polymerization and hydrochlorinated polymerization products.

Schwegler[97] describes the use of complexes containing aluminum or ferric halides. To suppress a foaming tendency, he proposes the addition of 1 to 2 weight percent sulfur to such liquid-phase catalysts.

Kobler[69] and Neher[80] state that the use of aluminum chloride in a liquid catalyst solution has long been practiced on a commercial scale. In commercial practice, gaseous streams containing HCl and ethylene are mixed and introduced into the bottom of a so-called flooded reactor. This reactor is filled or partially filled with a liquid reaction medium which consists essentially of ethyl chloride plus less than one percent aluminum chloride. Preferred aluminum chloride concentration, for minimum catalyst waste, is 0.2 to 0.3 weight percent.[80]

Addition between HCl and ethylene is extremely rapid. The product may be discharged entirely in the vapor phase and subsequently recovered by condensation of the overhead stream. Alternatively, a side stream of liquid may be withdrawn, stripped of part of its ethyl chloride content, and recycled to the reactor.[84]

The ethylene feedstream may be dilute; concentration as low as 25 percent has been reported.[82] Higher olefins should be essentially absent because of their strong tendency to undergo polymerization reactions and thus cause operating difficulties and rapid catalyst deactivation. Also objectionable are moisture and aqueous-reacting substances. Saturated hydrocarbons are permissible diluents.

The HCl:ethylene ration in the feedstream is essentially stoichiometric, but the introduction of some excess hydrogen chloride is conducive to improved ethylene utilization and a low rate of catalyst deactivation.[82] To operate with excess HCl in the reactor, while avoiding the occurrence of this corrosive acid in the recovery system, O'Connell[82] proposes execution of the reaction in two stages in series. The first stage receives all of the HCl but only 50 to 95 percent of the ethylene. The residual 5 to 50 percent of the ethylene feed is admixed to the primary reactor effluent at a point preceding the secondary converter.

Considerable attention has been given to means for decreasing the rate of catalyst deterioration. To prevent build-up of deactivated catalyst, some of the reactor liquid is bled off, contained ethyl chloride is recovered by distillation, and the resulting polymer bottoms, which still contain some active catalyst, must be removed from the system.

Rapid loss of catalyst life has been traced to the following causes:[69] high temperature, presence of higher olefins, presence of aqueous-reacting substances and high aluminum chloride concentration.

When make-up catalyst is introduced into the reactors intermittently, high aluminum chloride concentration will occur unavoidably at times when such make-up is provided, and during these periods the rate of catalyst loss is particularly great. It is possible to cope with this problem by the frequent addition of small batches of aluminum chloride or by continuous addition of catalyst. The latter approach, however, introduces serious mechanical problems; a method for continuous introduction of an aluminum chloride spray with the reactor feed gas is described by Neher.[80]

A catalyst system, claimed by Blue[42] to be less susceptible to deterioration, is prepared by reacting aluminum metal with an organic halide (such as propylene chloride or ethyl chloride) in the presence of a small amount of bromine.

Kobler[69] claims that addition of acetylene or carbon dioxide in a concentration range of 0.5 to 2 volume percent (on ethylene) results in substantial improvement, both in catalyst and raw materials consumption. Thus, operation with HCl: ethylene feed ratio of 1.06:1 at 52°C and 125 psi gave the following results:

TABLE 18-3. EFFECT OF ACETYLENE OR CO_2 ON
ETHYL CHLORIDE CONSUMPTION

Consumption, lb/100 lb Ethyl Chloride	Without added Acetylene or CO_2	With Added Acetylene and CO_2
Ethylene	47.1	45.4
HCl	64.9	62.6
$AlCl_3$	1.0	0.6

Neher[80] reports a typical commercial $AlCl_3$ consumption of 0.5 to 0.8 lb/ 100 lbs of ethyl chloride at an $AlCl_3$ concentration of 0.2 to 0.3 weight percent.

Keyl[66] describes a method for regenerating partially spent $AlCl_3$ complex by passing through it a direct electric current from an aluminum anode to an inert cathode. In the process, aluminum is dissolved in the catalyst solution.

Regardless of method of production the resulting ethyl chloride must be freed of residual acid. It is then purified by fractionation. Here, the presence of ethane presents a problem because its volatility is close to that of ethyl chloride. According to Wall,[119] this separation can be facilitated by step-wise condensation at successively lower temperatures followed by fractionation of the liquid formed in each condensation stage.

Chlorination of Ethane. The chlorination of ethane to yield ethyl chloride plus some higher chlorinated ethanes, is part of the more general subject of paraffin chlorination. Key considerations in this technology are covered in other chapters of this book (cf. Chapters 12 and 23), and elsewhere (e.g., Reference 52). The present discussion will be confined to aspects of the process which are specific to the monochlorination of ethane.

The following stoichiometric relation is involved:

$$C_2H_6 + Cl_2 \rightarrow C_2H_5Cl + HCl \qquad (18\text{-}4)$$

This highly exothermic (26.8 Kcal/mole at 25°C) reaction proceeds thermally at satisfactory reaction rates at temperatures above 230°C. Operation at lower temperatures calls for use of catalysts or the influence of actinic light.

The mechanism of thermal ethane chlorination has been investigated by Vaughn and Rust.[114] At intermediate temperatures, free radical chain mechanism appears to predominate while thermal bimolecular processes are important at higher temperatures.

Above 230°C, chlorine will react much more readily with ethane than with ethylene or with ethyl chloride (see Figure 18-1). The practical result of this finding is that (a) a mixture of ethane and ethylene will undergo only little addition reaction in the high temperature range (the extent of such chlorine addition to the double bond declines to nil at 293°C in the presence of ethane), and (b) product ethyl chloride is quite stable toward further chlorination in the presence of ethane.

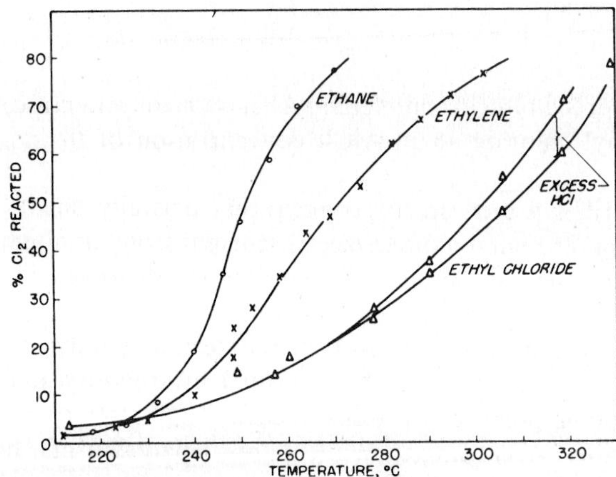

Flow (cc/min.): 50 Cl_2; 100 C_2H_4, C_2H_6 or C_2H_5Cl; 150 N_2 or CO_2.

Figure 18-1. Chlorination of ethane, ethylene and ethyl chloride (temperature profiles).

Despite this resistance of ethyl chloride to chlorination, polychlorinated products are obtained in the ethane chlorination reaction. These are presumably formed by the interaction of energy-rich, newly-formed monohalide with chlorine. Here, chlorination of ethane provides an exception to Hass' rule that "in vapor phase chlorination the presence of a chlorine atom on a carbon atom tends to hinder further reaction upon that carbon atom during the second substitution." Thus, it is found that, at 320° C, 1,1-dichloroethane predominates over the 1,2-compound in the ratio of 4:1, while a ratio of 7.7:1 is observed at 415°. Even the amount of 1,1,1-trichloroethane exceeds the 1,2-dichloroethane content by a substantial margin.

Another side reaction is observed in the formation of vinyl chloride, which apparently is derived from ethylene formed by pyrolysis of ethyl chloride at high temperature.

Below 350° C, where a free radical chain mechanism predominates, oxygen is a powerful inhibitor of ethane chlorination. This effect is not observed in the high temperature range. According to Vaughan,[116] the presence of oxygen merely elevates the reaction onset temperature. This rise in initiation temperature, which is due to oxygen, can be reduced by adding unsaturated hydrocarbons to the ethane-chlorine mixture.

McBee *et al.*[75] have studied the effect of the ratio of reactants in the thermal and photochemical chlorination of ethane. These authors find that formation of over-chlorinated products can be held at a minimum by maintaining a low chlorine concentration at all points of the reactor, including the critical mix point. To attain this end and to permit a relatively high over-all ratio of chlorine: ethane, they introduce chlorine into the hydrocarbon stream through a series of jets. The effect of this type of operation on product distribution is shown in Figure 18-2.

Catalytic Methods. An incubation period which is observed in the chlorination of ethane may be reduced by addition of initiators or catalysts. Numerous materials have been proposed for this function in the monochlorination of ethane. There has been much work on the catalytic use of activated charcoal in various forms, both in fixed-bed and fluidized-bed arrangement (see Reference 56), but Herzog[57] points out that the effectiveness of this material is probably due to the increased surface which it presents and to its excellent heat exchange properties rather than to its specific catalytic action. Other proposed heterogeneous catalysts include pumice,[46] copper salts on various carriers,[91] chlorides of aluminum, sodium, copper or bismuth in liquid phase.[88]

Use has also been made of oxides and chlorides of titanium, zirconium, cerium, copper and silver on a porous support.[90] Among gas-phase catalysts, work includes the use of TEL (at 135° C in .002 mol percent concen-

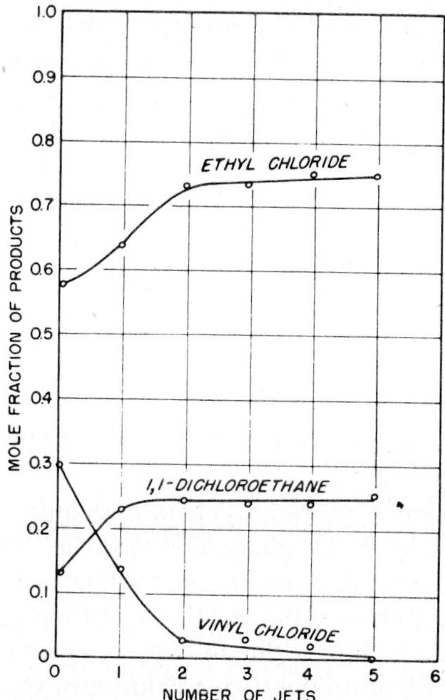

Figure 18-2. Effect of number of jets on organic chlorides in thermal chlorination. (Nickel tubing, 0.75 in.; bath temperature, 380°C; ratio of chlorine to ethane, 0.976 to 1.08; chlorine flow, 2000 to 2100 grams per hr.)

tration),[114] iodine,[74] benzoyl peroxide, dimethyl percarbonate,[105] etc. In this connection, the catalytic effect of TEL has been ascribed to its ability to undergo decomposition to free radicals which will initiate the chain reaction of ethane with chlorine.[114]

Photochemical Methods. The advantage of photochemical chlorination is that, at its lower feasible temperature, it is possible to avoid pyrolysis and isomerization reactions which occur in thermal chlorination. Much work has been concerned with photochemical initiation of the reaction between chlorine and ethane.

Generally, a light source with a radiation of 3,000 to 4,000 A is employed.[57] Typical light requirements are 400 watts per 5 to 15 kg chlorine reactant.[8] As in thermal chlorination, high chlorine concentration must be avoided to prevent formation of polychlorinated ethanes.

Photochemical chlorination has been reported both in vapor and liquid phase. An example of the former approach is given by Hill[58] who finds that admixture of methane to the ethane-chlorine system raises ethyl chloride yield substantially, without formation of any chlorinated methanes. The optimum proportion of feed gases is stated to be methane, chlorine and ethane in equal volumes. Preferred temperature range is 100 to 200° C, and residence time is 1 to 1.5 seconds. At these conditions, the results in Table 18-4 are reported in the presence and absence of methane.

TABLE 18-4

Gas Comp.		Ratio	% C_2	Product, mol %			% Yield
C_2H_6	CH_4	Cl_2/C_2H_6	Conv.	EtCl	$EtCl_2$	CH_3Cl	$EtCl/C_2H_6$
100	–	1.03	74.5	77.4	22.6	–	57.6
50	50	1.03	88.9	86.9	13.1		77.2

Drawbacks observed in photochemical gas-phase chlorination of ethane include fouling of the light transmitting surfaces by carbonization products and the difficulty of removing heat through the poorly conducting surfaces constructed of light-transmissive material.

Governale *et al.*[51] propose to overcome these difficulties by dispersing a fine fog of hydrochloric acid through the gas-phase reactor zone in an amount sufficient to remove the heat of reaction but insufficient to shield the gaseous chlorine and ethane from the actinic light rays.

A different approach is taken by Archibald[6] who effects ethane chlorination in liquid medium, preferably in 30 percent aqueous HCl, through which the reaction gases are finely dispersed. Operating in such medium with a hydrocarbon:chlorine ratio of 2:1 at 20 to 75° C, he obtained 75 to 80 percent ethyl chloride yield. Decrease in feed ratio led to corresponding drop in useful conversion.

Thermal Methods. It is apparent from the preceding discussions that catalytic chlorination of ethane is of questionable effectiveness and that photochemical methods raise severe engineering problems in reactor design.

Considerable effort has therefore been directed toward thermal chlorination of ethane. It is well recognized that ethane undergoes chlorination much more readily than ethylene at elevated temperatures and Vaughan[115] points out that chlorine will attack ethane almost exclusively in ethane-ethylene mixtures at 225 to 275° C.

Practical difficulties which must be overcome in carrying out thermal chlorination of ethane include, above all, the formation of higher chlorinated products, occurrence of carbonization reactions and pyrolysis of ethyl chloride to yield ethylene and hydrogen chloride. Of these three side reactions, the first is caused particularly by high localized chlorine:hydrocarbon ratios, while carbon formation and product pyrolysis are largely high-temperature phenomena.

Thus, a successful thermal chlorination process must provide means to maintain a low chlorine : ethane ratio at all points of the converter since, at thermal conditions, substitution chlorination is extremely rapid and all chlorine is consumed in a fraction of a second. Furthermore, it is essential that adequate means be provided for removing the highly exothermic heat of reaction.

Conrad[29a] describes a fluidized-bed thermal chlorination reactor which is claimed to solve both of these key problems. The fluidized solids are non-catalytic materials such as graphite, alundum or sand, preferably in a size distribution between 60 and 140 mesh. Chlorine- and ethane-containing feed streams are premixed at essentially room temperature and are injected into the bottom of the fluidized-bed reactor through a number of feed jets, each of which has an opening between 1/16 and 1/4 inch.

At a given pressure and temperature, the chlorine : ethane ratio determines product distribution. Some typical results are listed in Table 18-5.

TABLE 18-5. DISTRIBUTION OF PRODUCTS IN THERMAL CHLORINATION IN A FLUIDIZED-BED REACTOR

	Average product distribution, mole per cent (hydrogen chloride- and excess ethane-free basis)		
Mole Ratio Cl_2/C_2H_6	Ethyl Chloride, C_2H_5Cl	Dichloroethanes, Chiefly $1,1\text{-}C_2H_4Cl_2$	Trichloroethanes, Chiefly $1,1,1\text{-}C_2H_3Cl_3$
0.2	95.5	4	0.5
0.4	90	7	3.0
0.6	84	10.5	5.5

At higher mole ratios significant formation of ethylene and trichloroethanes as well as of some 1,2-dichloroethane is observed.

Conrad's data[29a] indicate best ethyl chloride formation in the range of 350° to 450° C. Furthermore, carbon build-up reaches a minimum between 400 to 450° C, and this is the preferred temperature range.

Operating *pressure* should be above atmospheric in order to increase converter capacity and to facilitate product recovery. The typical range is 35 to 40 psig. *Contact time* is not very critical; it is reported to be between one and two seconds.

The Integrated Process. The two main routes to ethyl chloride—ethane chlorination and ethylene hydrochlorination—may be combined into an integrated process in which by-product HCl from the chlorination stage (Equation 18-4) serves as raw material in the hydrochlorination step (Equation 18-2).

Each of the reactions may be carried out in accordance with any of the several approaches outlined above. However, there are a number of ways in which the process components may be combined. Most particularly, the following choices must be made:

Hydrocarbon flow. Make-up ethylene may be introduced preceding the chlorination or the hydrochlorination reactor, and the same choice applies to the point of ethane feed.

Ethyl chloride removal. Product ethyl chloride may be removed after each reaction stage, or following the hydrochlorination step alone. In the latter case, the product of ethane chlorination is passed as ballast through the hydrochlorination converters.

Two opposite approaches to the problem of ethyl chloride process integration are described by Cherniavsky[26] and by Neher,[80a] and may be outlined briefly as follows:

(1) In Cherniavsky's process (Figure 18-3), a tubular empty chlorination reactor is employed and ethylene hydrochlorination is carried out in

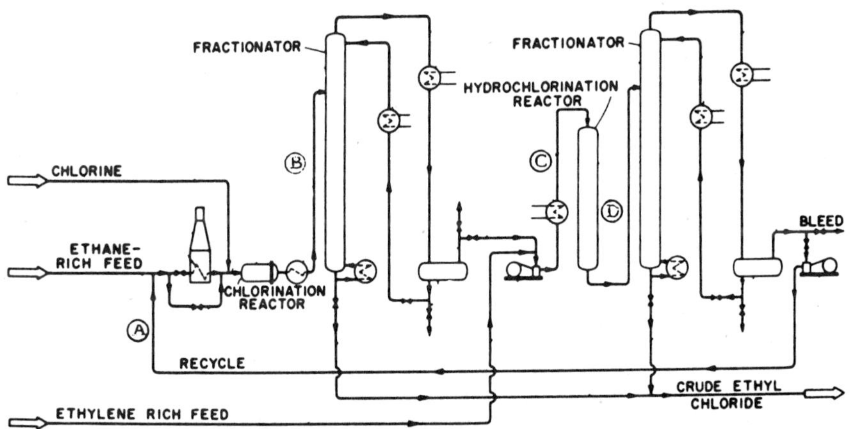

Figure 18-3. One version of the integrated chlorination-hydrochlorination process for ethyl chloride production.

the vapor phase. Each of the feed hydrocarbons is introduced separately, and product ethyl chloride is removed following each of the reaction stages.

Typical conditions are as follows:

Chlorination. 420° C, 80 psig, 3 seconds contact time.

Hydrochlorination. Vapor phase, copper chloride-promoted zinc chloride on porous alumina ("Celite"), 193°C, 250 psig.

Composition of the various process streams is shown in Table 18-6.

TABLE 18-6. COMPOSITION OF THE VARIOUS PROCESS STREAMS IN CHERNIAVSKY'S PROCESS

(Composition in parts by weight)

	A	B	C	D
Ethylene	232	498	2447	244
Ethane	5550	5665	5930	5930
Hydrocarbons other than C_2's	251	256	291	265
Ethyl chloride	59	4059	812	5853
Chlorinated by-products	97	816	134	250
HCl	240	3181	3181	253
Total	6429	14,475	12,704	12,704

(2) In Neher's process,[80a] chlorination is carried out in a fluidized-bed, non-catalytic system. Hydrochlorination is effected in the liquid phase, using aluminum chloride in ethyl chloride solution as catalyst. Ethane and ethylene are introduced separately; all ethyl chloride product is removed after the hydrochlorination stage. The following conditions are typical:

Chlorination. 390°C, 110 psig, 185 lb sand per square foot of cross-sectional area, 2 ft. depth of fluidized solids, superficial velocity of 1.5 ft/second.

Hydrochlorination. Aluminum chloride dissolved in ethyl chloride. 50 to 55° C, 120 to 130 psig.

Process stream composition obtained under these conditions is shown in Table 18-7.

An alternative approach to process flow, also described by Neher,[80a] provides for introduction of an ethane-ethylene mixture to the hydrochlorination stage and separate removal of ethyl chloride product after each of the conversion steps.

TABLE 18-7. PROCESS STREAM COMPOSITION UNDER SPECIFIC CONDITIONS IN NEHER'S PROCESS

Component	Hydrochlorination			Chlorination	
	Feed		Product		
	Fresh	Recycle		Fresh feed	Recycle
Ethane	7.1	424.0	431.1	203.9	388.7
Ethylene	183.1	5.5	0.9	1.0	0.8
Methane	0.9	22.3	22.8	3.1	19.4
HCl		204.2	27.0		22.3
Chlorine				177.7	
Ethyl chloride		160.4	377.8		4.1
Other Cl_2 products		8.2	1.1		

Miscellaneous Methods of Synthesis

Aside from the three commercially significant methods of production which have now been described, ethyl chloride can be synthesized by a host of reactions of lesser importance. Included among these secondary routes to ethyl chloride are the following:

(a) From diethyl sulfate and sodium chloride:

$$(C_2H_5)_2SO_4 + 2\,NaCl \rightarrow 2\,C_2H_5Cl + Na_2SO_4 \tag{18-5}$$

This reaction is of potential interest because it employs a low-cost source of chlorine. It has been investigated in detail by Giraitis[47,48] who obtained yields as high as 90 percent on diethyl sulfate.

(b) A related reaction obtains ethyl chloride from ethane, sodium chloride and sulfur trioxide.[47]

$$2NaCl + 2SO_3 + C_2H_6 \rightarrow C_2H_5Cl + Na_2SO_4 + HCl + SO_2 \tag{18-6}$$

Fairly good yield is reportedly obtained by passing ethane and sulfur trioxide through a bed of coarse granular salt at temperatures between 200 and 400° C.

(c) A sequence of reactions yields ethyl chloride from ethane, hydrogen chloride and oxygen, using copper chloride as oxygen carrier (e.g. 85):

$$2\,CuCl_2 + C_2H_6 \rightarrow C_2H_5Cl + Cu_2Cl_2 + HCl \tag{18-7a}$$

$$Cu_2Cl_2 + \tfrac{1}{2}\,O_2 \rightarrow CuCl_2 \cdot CuO \tag{18-7b}$$

$$CuCl_2 \cdot CuO + 2HCl \rightarrow 2\,CuCl_2 + H_2O \tag{18-7c}$$

The method has the obvious advantage of being a consumer of HCl and of low-cost ethane. However, it shares the major technical and high investment problems which have so far prevented successful commercial realization of chlorine production by oxidation of hydrogen chloride (cf. Chapter 9).

(d) Three phosphorus compounds—PCl_3, PCl_5, and $POCl_3$—have been used as chlorinating agents to convert ethyl alcohol to ethyl chloride, e.g.:

$$3\,C_2H_5OH + PCl_3 \rightarrow C_2H_5OP(OH)_2 + 2\,C_2H_5Cl + HCl \tag{18-8}$$

The frequently published reaction of ethyl alcohol with PCl_3 to yield only ethyl chloride and $P(OH)_3$ is, according to Gerrard *et al.*,[45] not in accord with the experimental facts. Yield rarely exceeds 50 percent of such an assumed conversion, but it can reportedly[72] be raised to 60 to 80 percent by addition of zinc chloride to the reaction system. The same catalyst serves usefully in raising the mediocre yield in the reaction between ethyl alcohol and PCl_5. Use of $POCl_3$ in reaction with ethanol results in very low ethyl chloride yield.

(e) Darzens[31] has worked out a method for producing alkyl chlorides from thionyl chloride and alcohol in the presence of a tertiary base such as diethylamine:

$$ROH + SOCl_2 \xrightarrow{\text{Base}} RCl + SO_2 + HCl \qquad (18\text{-}9)$$

(f) Some recent work has been concerned with the production of organic chlorides from ammonium chloride and acetylene or organic oxygen- sulfur- and nitrogen-bearing compounds.[110]

(g) In a conversion related to the Frei reaction, ethyl chloride has been prepared by cleavage of diethyl ether with HCl in the presence of catalysts (e.g. zinc chloride).[20]

(h) Vinyl chloride has been hydrogenated to ethyl chloride.[14]

(i) A non-catalytic hydrochlorination of ethylene has been carried out in a Siemens ozonizer at short reaction time and low reaction temperature.[33] At these conditions, ethylene and HCl were converted to a mixture of ethyl chloride and 1,4 dichlorobutane plus a small amount of 1-chlorobutane.

For various other methods of ethyl chloride synthesis, reference is made to Huntress "Organic Chlorine Compounds."[59]

Reactions of Ethyl Chloride

The commercial synthesis of tetraethyl lead involves reaction of ethyl chloride with monosodium-lead alloy:

$$4\,NaPb + 4\,C_2H_5Cl \rightarrow Pb(C_2H_5)_4 + 4\,NaCl + 3\,Pb \qquad (18\text{-}10)$$

It is immediately apparent that this reaction makes it necessary to recover and recycle three-quarters of the initial lead feed. Considerable work has been carried out to minimize the costly lead recovery operation by feeding an alloy of higher initial sodium content. However, a comparative study by Shapiro[99] indicates that none of these alloys permits TEL formation in yields which come even close to the 85 percent reportedly possible with the monosodium alloy.

Shapiro also cites evidence that the rate of the sodium-lead-ethyl chloride reaction may be increased significantly by the use of .005 to 4 percent of an accelerator such as a ketone, aldehyde, acetal, anhydride, ester or amide. Of further importance to reaction rate and yield is the physical structure of the sodium-lead alloy. The reactivity can, for example, be improved by wet-grinding the alloy immersed in ethyl chloride at low temperature or by quick-cooling such as can be achieved by means of drum-casting or by feeding the alloy in molten state to the ethyl chloride.

Some details of tetraethyl lead plants are reported by Schaefer[96] and Sittig.[103] As described by Thomas,[111] the commercial synthesis of TEL is carried out by the reaction of sodium lead alloy and liquid ethyl chloride, the latter in stoichiometric excess. Upon termination of reaction, the batch reaction mixture contains over three-fourths of the original lead charged, tetraethyl lead product, sodium chloride, some excess ethyl chloride and other minor components.

This stiff, pasty reaction mass, after venting off excess ethyl chloride, is charged to a pool of water and the tatraethyl lead is removed in large part by steam distillation. Concurrently, the bulk of the sodium chloride content is dissolved in the aqueous phase. The residue from this operation, after free draining of the liquid phase, is a predominantly lead solids system, wet with a dilute salt brine and having only minor proportions of entrapped liquid tetraethyl lead. This material is dried in steam-heated dryers, vaporizing most of the liquid in the free drained sludge. The tetraethyl lead vaporized is later recovered by stratifying from the aqueous condensate and drawing off separately. The dried sludge is then fed to smelting furnaces for remelting and recovery of the lead.

Industrially, the reaction of Equation (18-10) is carried out both on a batch and on a continuous basis, with batch technique accounting for the overwhelming bulk of production.

Alternative routes to TEL which involve use of ethyl chloride, include the following reactions:

$$Pb + 4\,C_2H_5Cl + 2\,Mg \rightarrow Pb(C_2H_5)_4 + 2\,MgCl_2 \qquad (18\text{-}11)$$

$$PbCl_2 + 3\,C_2H_5Li + C_2H_5I \rightarrow Pb(C_2H_5)_4 + 2\,LiCl + LiI \qquad (18\text{-}12)$$

$$Mg_2Pb + 4\,C_2H_5Cl \rightarrow Pb(C_2H_5)_4 + 2\,MgCl_2 \qquad (18\text{-}13)$$

$$2\,CaPb + 4\,C_2H_5Cl \rightarrow 2\,CaCl_2 + Pb(C_2H_5)_4 + Pb \qquad (18\text{-}14)$$

Details concerning execution of these reactions are contained in various references.[23,49,100,71]

A commercially important outlet for ethyl chloride is its use as an intermediate in the production of ethyl cellulose, a plastic material which is used as such and in protective coatings. A typical process for production of ethyl cellulose[52] calls for treating purified cellulose with sodium hydroxide solution and reacting the resulting sodium cellulose with ethyl chloride in an autoclave at a temperature up to 205° C and under pressure to maintain liquid phase.

Of growing importance is the reaction between aluminum and ethyl chloride to yield a mixture of ethyl aluminum chlorides referred to as ethyl aluminum sesquichloride.[28]

$$4\,Al\ +\ 6\,C_2H_5Cl\ \rightarrow\ (C_2H_5)_4Al_2Cl_2\ +\ (C_2H_5)_2Al_2Cl_4 \qquad (18\text{-}15)$$

The product is used, as such, in the preparation of Ziegler-type catalysts. It also serves as intermediate in one route to triethyl aluminum, another source of stereospecific catalysts.

A variety of other organometallic compounds may be prepared by analogous syntheses. Of special importance in this field are the Grignard reagents.

The reaction of ethyl chloride with certain amines yields compounds of interest as intermediates. Illustrative is the reaction with toluidine.[112]

$$Me(C_6H_4)NH_2\ +\ C_2H_5Cl\ \rightarrow\ Me(C_6H_4)NHC_2H_5 \cdot HCl\ and \qquad (18\text{-}16)$$

$$Me(C_6H_4)N(C_2H_5)_2 \cdot HCl$$

Chlorination of ethyl chloride under thermal or photochemical conditions will proceed readily and leads predominantly to 1,1-dichloroethane and 1,1,1-trichloroethane.

Reaction between ethyl chloride and benzene in the presence of Friedel-Crafts catalysts yields ethyl benzene:

$$C_6H_6\ +\ C_2H_5Cl \rightarrow (C_6H_5)C_2H_5\ +\ HCl \qquad (18\text{-}17)$$

The reaction is not commercially important. However, in the synthesis of ethyl benzene from ethylene and benzene, which is practiced on a very large scale, ethyl chloride (in lieu of anhydrous HCl) serves as promoter for aluminum chloride catalyst.

Other reactions of ethyl chloride which deserve mention in this place include: synthesis of ethyl mercaptan,[115] alkylation of primary nitriles,[16] alkylation of phenols,[52] esterification reactions, production of alkyl silicon halides by reaction with silicon or silicon alloy.[12] Finally, ethyl chloride will undergo a series of downgrading conversions such as the Wurtz synthesis of butane, hydrolysis to ethyl alcohol, conversion to ethyl sulfate, etc.

ETHYLENE DICHLORIDE

Commercial Aspects

1,2-Dichloroethane, generally referred to as "ethylene dichloride," is today the largest-volume organic derivative of chlorine. U. S. production rose from 510 million pounds in 1955 to 1.3 billion pounds in 1960.

Most of the recent growth rate stems from demand for ethylene dichloride as intermediate for vinyl chloride (q.v.). Vinyl chloride polymers and copolymers make up 75 to 80 percent of the total group of vinyl resins which are expected to grow at an average annual rate of eight percent.[108,4] To be sure, only part of U. S. vinyl chloride output is derived from ethylene dichloride, with the remainder obtained by hydrochlorination of acetylene. The ratio between these two routes to vinyl chloride fluctuates from year to year. Typically, ethylene dichloride conversion accounts for 45 to 55 percent of vinyl chloride output.[3,5]

The second-most important consumer of ethylene dichloride are tetraethyl lead and tetramethyl lead-containing antiknock fluids for motor (as distinct from aviation) use, in which ethylene dichloride and ethylene dibromide serve as scavenging agents. Typical is "Ethyl" Motor Mix which contains 30.6 pounds of ethylene dichloride per one hundred pounds of TEL.

Over-all, the following 1959–60 use pattern in the United States has been estimated[77a] for ethylene dichloride:

	(%)
Vinyl chloride production	77
TEL	15
Miscellaneous	8

Included in the "miscellaneous" category are uses of EDC as a solvent for azeotropic dehydration of acetic acid, for degreasing metals and textiles, for plastics (notably polyvinyl chloride), for lube oil refining, as a grain fumigant and as intermediate in the manufacture of Thiokol rubbers and of ethylene diamine.

Commercial Methods of Manufacture

More than 85 percent of ethylene dichloride output is produced by the reaction of ethylene and chlorine. An alternative synthesis which appears to have commercial potential is the reaction of ethylene and hydrogen chloride in the presence of air (oxychlorination). Both of these techniques will be discussed below.

The remaining ethylene dichloride is obtained as by-product of ethylene oxide by the chlorohydrin route (see Chapter 11). Here, too, ethylene dichloride is formed by the reaction of ethylene with elemental chlorine, but the emphasis is on minimizing this reaction which detracts from the yield of ethylene chlorohydrin—the desired product. The extent of EDC formation in the ethylene chlorohydrin process depends largely on reactor design, ethylene:chlorine ratio, and the rate of chlorine feed. Typically, ethylene dichloride is formed at the rate of 8.5 lb/100 lb ethylene chlorohydrin (16 lb ethylene dichloride per 100 lb ethylene oxide).[101]

Direct Chlorination of Ethylene. Production of ethylene dichloride from ethylene and elemental chlorine proceeds in accordance with Equation (18-18).

$$C_2H_4 + Cl_2 \rightarrow Cl-CH_2-CH_2-Cl \qquad (18\text{-}18)$$

At fairly low temperatures (25° C), this addition reaction proceeds to the almost complete exclusion of substitution reactions. As the temperature is raised, 1,1,2-trichloroethane formation becomes significant.[50]

Chlorine addition to ethylene can be carried out in both the liquid and the vapor phase, with and without catalyst. According to Golev,[50] the liquid-phase reaction is catalyzed by water, ethylene dichloride and by the walls of the reactor.

Other suggested liquid-phase catalysts include calcium chloride,[93] tetraethyl lead,[94] the chlorides of elements of the fourth and seventh groups, e.g., antimony chloride,[11] ferric chloride[43] and ferric oxide.[87] These catalysts serve in part to suppress substitution reaction above 25° C and in part to accelerate chlorine addition at low temperature.

Williams[121] has investigated the effect of solvents on the addition of chlorine to ethylene. In dissociating solvents (such as acetic acid or nitrobenzene), halogen addition is definitely a two-stage process which is homogeneous and insensitive to stray catalysts. In non-dissociating solvents such as CCl_4 and $CHCl_3$, halogen addition tends to have less reproducible rates and an induction period is observed.

Ethylene dichloride itself has been repeatedly cited as a suitable solvent for carrying out the addition chlorination.[50,87,2,65] According to Anantakrishnan,[2] the presence of ethylene dichloride enhances the reaction rate and reduces the extent of substitution. By contrast, pronounced substitution reaction is observed when the conversion is carried out in liquid pentane; its extent depends on the chlorine:olefin ratio.

To suppress occurrence of substitution reactions, even at low chlorination temperature, Reese[87] and Benedict[15] call for operation with olefin in 5 to 25 percent excess above chlorine equivalent. In addition, Benedict withdraws EDC from the reactor zone in vapor form (operation in this instance is at 80 to 120°C) and, following rectification, refluxes the higher-boiling products.

Golev[50] describes two approaches to execution of addition chlorination in the liquid phase: injection of liquid into the gas stream, and bubbling the gaseous reactants through the solvent. In both approaches, pure as well as dilute ethylene is a suitable feedstock. Golev prefers use of the bubbling method because it employs simpler apparatus and permits higher yield. Also, according to Golev, the product ethylene dichloride is washed with caustic, followed by a sulfuric acid wash, and finally is rectified.

Illustrative of liquid-phase chlorination is the process piloted by Shell at Imujden, Holland, which serve as basis for the commercial ethylene dichloride-vinyl chloride plant at Rotterdam.[117]

At this plant, EDC is produced by direct chlorination of ethylene in the presence of $FeCl_3$ catalyst suspended in liquid ethylene dichloride. The reaction is carried out at 75 psig. Reactants enter the converter coils at 30°C and leave at 46°C.

Due to the exothermic nature of the reaction (70,000 Btu/pound-mole), intimate mixing must be provided in the reactor to prevent local overheating. By refrigerating the ethylene dichloride, which serves as catalyst suspension medium, en route to the reactor, temperature control during conversion is facilitated.

The reaction is complicated, and chlorine consumption is uselessly raised, by the presence of higher olefins; it is rendered hazardous by acetylene. The feed hydrocarbon is therefore a close C_2-fraction, freed of acetylene. At Imujden, the stream fed to the reactor contained only 40 percent ethylene. Associated ethane, for all practical purposes, is inert to chlorination at conditions prevailing in the converter. The hydrocarbon stream is joined by the recirculating liquid reaction medium (ethylene dichloride) at a point preceding the reactor, and chlorine is injected into the system at a point closely downstream of the reactor entrance.

By effecting the conversion at elevated pressure (75 psi) considerably faster reaction rates are achieved than would otherwise be possible. Furthermore, useful components may be recovered from the reactor effluent with minimum recourse to refrigeration.

The reaction is carried out on a once-through basis. Vlugter[117] reports that, at optimum conditions, 98 percent conversion of ethylene is obtained to yield a chlorination product that contains 96 percent 1,2-dichloroethane. The product is freed of permanent gases (mainly ethane and methane), residual ethylene and chlorine (and some by-product HCl), in a short Raschig ring-packed column. The off-gases are cooled to –10°C in a refrigerated dephlegmator.

The bottoms product leaves the column at 46°C. It is released to atmospheric pressure and is then water-washed for the extraction of iron chloride (catalyst) and contained HCl. To prevent corrosion in storage and subsequent process equipment for conversion to vinyl chloride, Imujden found it necessary to free its EDC of water by means of azeotropic distillation.

A number of processes have been described for the catalytic addition chlorination of ethylene in the *vapor phase*. Here, too, fairly low temperature must be maintained to avoid substitution chlorination. According to Ramage,[86] the temperature should not exceed 40 to 45° C.

Catalysts proposed for such vapor-phase processes include finely divided lead[9] or, more specifically, a mixture of 63 percent lead, 16 percent silver and 21 percent antimony on asbestos fiber as carrier.[86] Harding[54] effects the reaction in the presence of a large number of surfaces, including sand, coke, glass, Fe_2O_3, etc., in the absence of moisture; Dow[35] describes the catalytic effect of ethylene dibromide.

Vapor-phase chlorination of ethylene has been proposed[25] at about 200° C over bauxite catalyst, in the presence of at least one volume percent oxygen to suppress substitution reaction. 82 percent of the feed ethylene is reportedly converted to chlorinated products which contain 86 percent 1,2-dichloroethane.

According to Doroganevskaya,[34] the gas-phase reaction between chlorine and ethylene is catalyzed by the walls of the reaction vessel, but the gases react best when dissolved in previously prepared ethylene dichloride (i.e. by liquid-phase technique).

Ethylene Dichloride from Ethylene and Hydrogen Chloride. In the above discussion of ethyl chloride, consideration has been given to the crucial role played by hydrogen chloride in the over-all manufacturing economics of hydrocarbon chlorination plants. Many such installations find themselves in the position where by-product HCl has little or no value, and must even be neutralized at a cost.

This situation exists, for example, in plants for the production of vinyl chloride from ethylene dichloride. Several attempts have been made to convert by-product HCl to chlorine by partial combustion or electrolysis. However, such methods are not commercial for reasons of technical and corrosion difficulties and high capital cost. (The largest known, presently operating plant for conversion of HCl to chlorine has a daily capacity of 25 tons).

An alternative, but related approach involves the reaction of HCl with oxygen and ethylene to yield ethylene dichloride, which may then serve as part of the feed to the vinyl chloride plant. Success in such a synthesis would conserve up to one-half of the chlorine value charged to the EDC-vinyl chloride unit, and processes of this type appear to have commercial potential.

The process approach parallels closely some of the methods proposed for partial oxidation of HCl to chlorine—with the modification that ethylene is admixed to the feed stream to serve as chlorine acceptor, according to the over-all reaction scheme.

$$2\,C_2H_4 + 4HCl + O_2 \rightarrow 2\,C_2H_4Cl_2 + 2\,H_2O \qquad (18\text{-}19)$$

Available information on this process calls for the use of copper chloride as catalyst and chlorine carrier. Illustrative is a process due to Johnson and Cherniavsky[62] who employ the catalyst in fluid-bed form in order to cope with the large heat evolution.

In one example, a gaseous mixture containing oxygen, hydrogen chloride and ethylene in the mole ratio $1:3.3:1.65$ is contacted with a fluidized, finely divided catalyst consisting of alumina impregnated with copper chloride, at the rate of 0.5 lb of the charge mixture per lb catalyst per hour in a reactor operated at 285°C and atmospheric pressure. The gaseous effluent from the reactor contains 16.2 mol percent organic chlorides consisting essentially of EDC, plus 72 mol percent water and 8.8 percent HCl. Over-all EDC yield is 90.7 percent on ethylene and 97.7 percent on HCl.

In a process due to Joseph[64] and Pye,[85] the supported copper chloride catalyst is employed in a moving-bed system in which the reaction zones are separated as follows: (1) oxidation of cuprous chloride by air, (2) reaction between HCl and cuprous oxychloride to yield cupric chloride; (3) reaction between cupric chloride and ethylene at 350 to 360°C. Pye[85] reports an organic product containing 66.9 percent EDC at 82 percent conversion of ethylene to chlorinated hydrocarbons.

Particularly high EDC yield is claimed by Reynolds[89] for catalysts consisting of 2 to 10 weight percent copper silicate on a porous adsorptive siliceous base. Over such catalysts, ethylene chlorination by HCl plus air at 310°C is reported to achieve 96 percent conversion of HCl and 52 percent conversion of the hydrocarbon. The organic product contains about 80 percent EDC.

Alternative Routes to Ethylene Dichloride

In addition to the commercial production methods for EDC, described above and in Chapter 11, several syntheses for this compound are available. Of special interest is the *chlorination of ethane* which is practiced commercially to produce ethyl chloride (q.v.). In such processes, higher chlorinated hydrocarbons are formed inevitably, among them dichloroethanes. The bulk of this fraction is 1,1-dichloroethane with only minor formation of 1,2-dichloroethane (EDC). Typically, the mole ratio of 1,1-dichloroethane for the 1,2-isomer is between $9:1$ and $14:1$.[29a]

Similarly, the chlorination of ethyl chloride leads to a dichloroethane product in which the 1,1-isomer is predominant.

Chlorination of ethylene glycol has been effected by means of PCl_3 or PCl_5 in the presence of $ZnCl_2$ catalyst[27] and by $SOCl_2$ in pyridine solvent.[118]

Other methods of synthesis, cited by Beilstein and by Huntress, include the following reactions: NOCl with ethylene; bis-(β-chloroethyl) sulfate with alkali chloride, bis-(β-chloroethyl) sulfite with chlorine; S_2Cl_2 with ethylene oxide; HCl with acetylene in the presence of NO_2; chlorine with acetylene in the presence of CCl_4 and $AlCl_3/NaCl/FeCl_3$; breakdown of diazomethane; disproportionation of ethylene chlorohydrin; and silent electric discharge through methyl chloride.

Reactions of Ethylene Dichloride

(1) *Dehydrohalogenation to vinyl chloride* is the largest commercial consumer of ethylene dichloride. This reaction is discussed in some detail in Chapter 26.

(2) *Reaction with sodium polysulfide* is of commercial significance in the production of one type of "Thiokol" rubber. The reaction is carried out at 70° C in aqueous medium containing a dispersion agent (magnesium hydroxide) and a wetting agent. The product is a latex which is washed and then coagulated and vulcanized into a rubber characterized by outstanding resistance to weathering, oxygen, ozone, and swelling action by oils and solvents.[63]

Of the three types of "Thiokol" polymer offered commercially as rubber, type A is formed from ethylene dichloride and sodium polysulfide and type FA is produced by the reaction of sodium polysulfide with a mixture of EDC and dichloroethyl formal. Type FA is the high-volume crude among "Thiokols" (aircraft applications, liner for paint spray hose, printing rolls, etc.) while "Thiokol" A is essentially confined to use as plasticizer for sulfur in acid-resistant bricks.[63] Other types of "Thiokol," some of which consume EDC in the course of formation, are available as water dispersions and as liquid polymers.

(3) *Reaction with ammonia* is also of commercial importance as a route to ethylene diamine. The reaction is carried out in aqueous emulsion at 180° C and under pressure.

$$2\,NH_3 + Cl—CH_2—CH_2—Cl \rightarrow NH_2—CH_2—CH_2—NH_2 + 2\,HCl \quad (18\text{-}20)$$

The reaction is accompanied by formation of diethylene triamine and triethylene tetramine.

(4) Miscellaneous Conversions of EDC:

(a) Reaction with sodium cyanide and methanol permits formation of succinonitrile in 90 percent yield.[29]

(b) Reaction with aromatics in the presence of Friedel-Crafts catalysts leads to polymers.[68] EDC and benzene will reportedly copolymerize under the influence of air and in the catalytic presence of manganese and bromine to yield a polymer which is akin to polystyrene in its basic chemical structure.[95]

(c) Other conversions, reported by Huntress,[59] include reactions with alcohols, mercaptans, phenols, salts of organic acids, amines, carbon disulfide and ethylene dibromide. Also, reactions with chlorine, sodium, sulfuric acid, and various metal salts have been reported.

References

1. Aickelin, H., U. S. Patent 2,007,322 to General Aniline Works (July 9, 1935).
2. Anantakrishnan, S. V., *et al., Chem. Rev.,* **33,** 27 (1943).
3. Anon., *Chem. Eng.,* 88 January 27, 1958.
4. Anon., *Chem. Week,* 71 November 16, 1957.
5. Anon., *Chem. Week,* 79 May 9, 1959.
6. Archibald, F. M., and Mottern, H. O., U. S. Patent 2,393,509 to Standard Oil Development Co. (January 22, 1946).
7. Arnold, H. R., and Lessig, E. T., U. S. Patent 2,097,750 to E. I. du Pont de Nemours & Co. (November 2, 1957).
8. Asinger, F., "Chemie U. Technologie der Paraffin-Kohlenwasserstoffe," Berlin, Akademie-verlag, 1956.
9. Askenasy, P., and Heller, A., U. S. Patent 1,851,970 (April 5, 1932).
10. Axe, W. N., U. S. Patent 2,434,094 to Phillips Petroleum Co. (January 6, 1948).
11. Bahr, H., and Fried, F., German Patent 640,827 to I. G. Farbenindustrie (January 15, 1937).
12. Barry, A. J., and DePrie, L., U. S. Patent 2,488,487 to Dow Chemical Co. (November 15, 1949).
13. Baskerville, C., and Hamor, W. A., *Ind. Eng. Chem.,* **5,** 828 (October 1913).
14. Baumann, *et al.,* U. S. Patent 2,118,662 to I. G. Farben Industrie (May 24, 1938).
15. Benedict, D. B., U. S. Patent 2,929,852 to Union Carbide Corp. (March 22, 1960); cf. also British Patent 779,938 to Union Carbide and Carbon Corp. (July 24, 1957).
16. Bergstrom, F. F., and Agostinko, R. J., *Am. Chem. Soc.,* **67,** 2152 (1945).
17. Billitzer, A. W., *J. Austr. Chem. Inst.,* **15,** 261 (1948).
18. Blue, R. D., U. S. Patent 2,180,345 to Dow Chemical Company (November 21, 1939).

19. Bond, D. C., and Savoy, M., U. S. Patent 2,453,779 to Pure Oil Co. (November 16, 1948).
20. Brooks, U. S. Patent 2,015,706 to Standard Alcohol Co. (October 1, 1935).
21. Buc, H. E., and Gleason, A. H., U. S. Patent 2,153,170 to Standard Oil Development Co. (April 4, 1939).
22. Buc, H. E., and Johns, C. O., U. S. Patent 1,440,683 to Standard Oil Co. of New Jersey (Jan. 2, 1923).
23. Calingaert, G., and Shapiro, H., U. S. Patent 2,535,190 to Ethyl Corp. (December 26, 1950).
24. Carter, R. P., U. S. Patent 2,396,639 to Hercules Powder Co. (March 19, 1946).
25. Celanese Corporation, British Patent 781,414 (August 21, 1957).
26. Cherniavsky, A. J., U. S. Patent 2,807,656 to Shell Development Co. (September 24, 1957).
27. Clark, Streight, *Trans. Roy. Soc. Can.,* **23**(3), 80 (1929).
28. Coates, G. E., "Organo-Metallic Compounds," London, Methuen & Co., Ltd., 1956.
29. Codignola, F., *et al.,* Italian Patent 431,407 (February 27, 1948).
29a. Conrad, F., *et al.,* U. S. Patent 2,838,579 to Ethyl Corp. (June 10, 1958).
30. Cook, S., *et al.,* U. S. Patent 2,838,577 to Ethyl Corp. (June 10, 1958).
31. Darzens, *Compt. rend.,* **152,** 1314, 1601 (1911).
32. Daudt, H. W., U. S. Patent 1,920,246 to E. I. du Pont de Nemours & Co. (August 1, 1933).
33. den Hertog, H. J., *et al., Petroleum* (*London*), **21,** 27 (January 1958).
34. Doroganevskaya, E. A., *J. Chem. Ind. (U.S.S.R.),* **8,** 857 (1931) Chem. Abstr. **26,** 789.
35. Dow, H. H., U. S. Patent 1,841,279 to Dow Chemical Co. (Jan. 12, 1932).
36. Ernst, O., and Berndt, W., U. S. Patent 1,918,371 to I. G. Farbenindustrie (July 18, 1933).
37. Ernst, O., and Berndt, W., German Patent 467,185 (October 27, 1938).
38. Ernst, P., German Patent 541,566 to Dr. Alexander Wacker Ges. (Jan. 13, 1932).
39. Faith, W. L., *et al.,* "Industrial Chemicals," 2nd Ed., New York, John Wiley & Sons, Inc., 1957.
40. Frei, J., U. S. Patent 1,784,423 to E. I. du Pont de Nemours and Co. (December 9, 1930).
41. Frei, J., U. S. Patent 1,824,951 to E. I. du Pont de Nemours and Co. (September 29, 1931).
42. Furr, J., *et al.,* U. S. Patent 2,644,016 to Ethyl Corp. (June 30, 1953).
43. Galitzenstein, E. G., and Woolf, C., British Patent 553,959 to Distillers Co., Ltd. (June 11, 1953).
44. Gerrard, W., *et al., J. Appl. Chem.,* **5,** 28 (1955).
45. Gerrard, W., *et al., J. Chem. Soc.,* **1953,** 1920.
46. Giordani, M., *Ann. chim. applicata,* **25,** 163 (1935).
47. Giraitis, A. P., Paper presented at National Meeting of the ACS. Dallas, Texas, April 8–13, 1956, Preprints of the Division of Petroleum Chemistry, **1**(2) (April 1956).

48. Giraitis, A. P., U. S. Patent 2,688,641 to Ethyl Corp. (September 7, 1954).
49. Gittins, T. W., and Mattison, E. L., U. S. Patent 2,763,673 to E. I. du Pont de Nemours and Co. (September 18, 1956).
50. Golev, A. S., *Khim. Referat. Zhur.*, (3), 87 (1940); **36**, 25231.
51. Governale, L. J., U. S. Patent 2,589,689 to Ethyl Corporation (March 18, 1952).
52. Groggins, P. H., "Unit Processes in Organic Synthesis," 5th Ed., New York, McGraw-Hill Book Co., Inc., 1958.
53. Groves, J., *Chem. Soc.*, **27**, 639 (1874).
54. Harding, V. J., British Patent 126,511 (June 11, 1918).
55. Harvey, C. C., *et al.*, U. S. Patent 2,543,575 to Ethyl Corporation (February 27, 1951).
56. Herold, P., *et al.*, German Patent 720,079 (1942).
57. Herzog, R., *et al.*, "Ethyl Chloride" "Encyclopedia of Chemical Technology," Vol. 3, pp. 751–9, Interscience Encyclopedia, Inc., New York 1949.
58. Hill, H. W., and Dance, E. L., U. S. Patent 2,453,691 (November 16, 1948).
59. Huntress, E. H., "Organic Chlorine Compounds," Section 3:7015, New York, J. Wiley & Sons, Inc., 1948.
60. I. G. Farbenindustrie A. G., British Patent 486,453 (June 3, 1938).
61. Imperial Chemical Industries, Ltd., British Patent 639,435 (June 28, 1950).
62. Johnson, A. A., and Cherniavsky, A. J., U. S. Patent 2,644,846 to Shell Development Co. (July 7, 1953).
63. Jorczak, J. C., in "Introduction to Rubber Technology," ed. by Morton, M., New York, Reinhold Publishing Corp., 1959.
64. Joseph, W. J., and Bianchi, A. B., U. S. Patent 2,752,401 to Dow Chemical Co. (June 26, 1956).
65. Kazantsev, M. P., *Khim. Prom.*, (11) 16 (1946).
66. Keyl, A. C., and Blue, R. D., U. S. Patent 2,209,981 to Dow Chemical Co. (August 6, 1940).
67. Kilpi, S. Z., *phys. Chem.*, **141**, 424 (1929); **142**, 195 (1929).
68. Klebanskii, J., *et al.*, *Zhur. priklad. Khim.*, **14**, 618 (1941).
69. Kobler, J. F., and Mercier, G. T., U. S. Patent 2,786,875 to Ethyl Corp. (March 26, 1957).
70. Kraus, C. A., and Callis, C. C., U. S. Patent 1,696,245 to Standard Oil Development Co. (January 1, 1929).
71. Krohn, I. T., and Shapiro, H., U. S. Patent 2,555,891 to Ethyl Corp. (June 5, 1951).
72. Machell, G., *Mfg. Chemist*, 70, February 1957, 70–87.
73. Manufacturing Chemists' Association, "Chemical Statistics Handbook," 5th Ed., p. 48, 1960. Also, U. S. Tariff Commission Reports "Synthetic Organic Chemicals" 1959 and 1960.
74. Martin, L. F., and Lux, A. P., U. S. Patent 1,801,873 to Dow Chemical Co. (April 21, 1931).
75. McBee, E. T., *et al.*, *Ind. Eng. Chem.*, **41**, 799 (1949).
76. McCurdy, J. L., U. S. Patent 2,516,638 to Dow Chemical Co. (July 25, 1950).
77. McKee, R., and Burke, S. P., U. S. Patent 1,738,193 (December 3, 1929).

77a. Messing, R. F., *et al., Chem. Eng.,* 66, (August 22, 1960).

78. Millard, U. S. Patent 2,779,805 to Olin-Mathieson Chemical Corp. (January 29, 1957).

79. N. V. Chemische Fabriek "Naarden," British Patent 610,843 (October 21, 1948).

80. Neher, C. M., U. S. Patent 2,818,447 to Ethyl Corp. (December 31, 1957).

80a. Neher, C. M., and O'Connell, H. E., U. S. Patent 2,905,727 to Ethyl Corp. (September 22, 1959).

81. Nekrasov, A. S., and Karicheva, V. N., *Trudy Inst. Nefti, Akad. Nauk. S.S.S.R.,* 12, 276 (1958).

82. O'Connell, H. E., and Huguet, J. H., U. S. Patent 2,818,448 to Ethyl Corp. (December 31, 1957).

83. Olin, J. F., and Hinds, G. E., U. S. Patent 2,122,110 to Sharples Solvents Corp. (June 28, 1938).

84. Pierce, J. E., U. S. Patent 2,140,927 to Dow Chemical Co. (December 20, (1938).

85. Pye, D. J., U. S. Patent 2,752,402 to Dow Chemical Co. (June 26, 1956).

86. Ramage, A. S., U. S. Patent 2,441,287 to A.A.F. Maxwell (May 11, 1948).

87. Reese, R., U. S. Patent 2,601,322 to Jefferson Chemical Co. (June 24, 1952).

88. Reilly, J. H., U. S. Patent 2,140,547 to Dow Chemical Co. (December 20, 1939).

89. Reynolds, M. B., U. S. Patent 2,783,286 to Olin Mathieson Chemical Corp. (February 26, 1957).

90. Riblett, E. W., U. S. Patent 2,334,033 to Process Management Co. (November 9, 1943).

91. Roka, Koloman, U. S. Patent 1,723,442 to Holzverkohlungs-Industrie Aktien-Gesellschaft (August 6, 1929).

92. Rudkovskii, D. M., *et al., Ukrain. Khim. Zhur.,* 10, 277 (1935).

93. Ruip, J. D., and Edwards, J. W., U. S. Patent 2,099,231 to Shell Development Co.

94. Rust, F. F., and Vaughan, W. E., U. S. Patent 2,284,479 to Shell Development Co. (May 26, 1942).

95. Saffer, A., and Barker, R. S., Belgium Patent 546,191 to Mid-Century Corp. (March 17, 1956).

96. Schaefer, J. H., *Chem. Eng.,* 57 (8) 102, 164 (1950).

97. Schwegler, C. C., and Tennant, F. M., U. S. Patent 2,469,702 to Dow Chemical Co. (May 10, 1949).

98. Scott, N. D., U. S. Patent 2,570,495 to E. I. du Pont de Nemours and Co. (October 9, 1951).

99. Shapiro, H., chapter in "Advances in Chemistry Series," No. 23, pp. 290–8. Copyright 1959 by the American Chemical Society.

100. Shapiro, H., U. S. Patent 2,535,236 to Ethyl Corp. (December 26, 1950).

101. Sherwood, P. W., *Petroleum Refiner,* 28 (7) 120 (July 1949).

102. Shinosaki, H., *Japan,* 156, 749 (May 28, 1943).

103. Sittig, M., "Sodium, Its Manufacture, Properties and Uses," New York, Reinhold Publishing Corp., 1956.

104. Slotterbeck, O. C., and Rosen, R., U. S. Patent 2,174,278 to Standard Oil Development Co. (September 26, 1939).

105. Smolyan, E. S., cited in *Uspekhi Khim.,* **29** (1), 52 (1960).

106. Soc. Anon. des Matieres Colorantes et produits chimiques de Saint-Denis, French Patent 858,724 (December 2, 1940).

107. Strange, E. H., and Kane, T., British Patent 500,880 (February 16, 1939).

108. Sward, A. F., *Chem. Eng. News,* 30, (July 6, 1959).

109. Thodos, G., and Stutzman, L. F., *Ind. Eng. Chem.,* **50,** 413 (1958).

110. Thomas *et al.,* U. S. Patents 2,755,310—2,755,312 to Olin Mathieson Chem. Corp. (July 17, 1956).

111. Thomas, H. A., U. S. Patent 2,904,835 to Ethyl Corp. (Sept. 22, 1959).

112. Tolmachev, N. A., *Trans Leningrad Chem. Technol. Inst.,* **1,** 119 (1934); Chem. Abstr. **29,** 2930 (1934).

113. U. S. Department of Health, Education and Welfare, "Public Health Aspects of Increasing Tetraethyl Lead Content in Motor Fuel," Public Health Service Publication No. 712, Washington, 1959.

114. Vaughan, W. E., and Rust, F. F., *J. Org. Chem.,* **5,** 449 (1940).

115. Vaughan, W. E., and Rust, F. F., U. S. Patent 2,246,082 to Shell Development Co. (June 17, 1941).

116. Vaughan, W. E., and Rust, F. F., U. S. Patent 2,284,482 to Shell Development Co. (May 26, 1942).

117. Vlugter, J. C., *et al., Chimie et Industrie,* **67** (2) pp. 87ff. (1952).

118. Vogel, *J. Chem. Soc. (London),* **48,** 647 (1948).

119. Wall, H. H., U. S. Patent 2,556,833 to Ethyl Corp. (June.12, 1951).

120. Walvekar, S. P., *et al., J. Ind. Chem. Soc.,* **19,** 409 (1942).

121. Williams, G., *Trans. Faraday Soc.,* **37,** 749 (1941).

122. Wirth, W. V., U. S. Patent 2,013,722 to E. I. du Pont de Nemours & Co. (September 10, 1935).

19. MAGNESIUM CHLORIDE

Charles L. Mantell

Consulting Chemical Engineer
Newark College of Engineering

Magnesium chloride is a widely spread constituent of chemical brines, salt lakes, salt beds and sea water. In effect, however, the only chemical manufacture of magnesium chloride is involved in its separation and regeneration from natural sources. Other than having a minor use as a raw material for magnesium oxychloride cement, magnesium chloride is used as the raw material for the electrolytes or "bath" for the electrochemical production of magnesium metal. All of the electrolytic processes involve the preparation of magnesium chloride, and these will be reviewed later in this chapter.

Oxychloride or sorel cement is made by the exothermic reaction of a 20 percent $MgCl_2$ solution on magnesia. In the reaction below the product is hard but is attacked by water.

$$3\,MgO + MgCl_2 + 11\,H_2O \rightarrow 3\,MgOMgCl_2 \cdot 11\,H_2O$$

It is a constituent of flooring cements, serving as a base for tile and terrazzo, soundproofing and sparkproof surfaces. Copper powder or $MgSO_4 \cdot 7H_2O$ may be added to correct excessive expansion and increase water resistance.

Sea water contains approximately 4.2 g/l of magnesium chloride which is readily converted to almost totally insoluble oxide or hydroxide by lime. The hydroxide may be purified, dried and easily converted to chloride by reaction with hydrochloric acid made from chlorine electrolyzed from the chloride, thus allowing a cyclic operation.

Magnesium was first prepared electrochemically by Bunsen in 1852 by the electrolysis of fused anhydrous $MgCl_2$.[1]

Intermittent Chloride Process

Harvey[3] has described the intermittent chloride process. Before the First World War the metal was made in Germany by the electrolysis of fused anhydrous carnallite (KClMgCl$_2 \cdot$ 6H$_2$O) which occurs naturally, has a lower melting point than MgCl$_2$, is less volatile and can be dehydrated more readily. The fused, anhydrous material was fed into the electrolytic cells[9] and electrolyzed until the magnesium chloride content had been lowered to 10 or 15 percent.

Unfortunately, however, an electrolytic bath consisting of 85 percent potassium chloride and 15 percent magnesium chloride has a specific gravity at 700°C so close to that of magnesium metal floating in it that it is impossible to control the position or level of the metal in the bath. As MgCl$_2 \cdot$ 6H$_2$O, the raw material tends to decompose during dehydration and fusion into MgO or oxychloride. The resulting residue contains MgO.

In practice approximately equal parts of NaCl were added to prevent this decomposition. (Potassium chloride is more effective, but its greater cost—at least in the United States—precluded its use.) To retard the decomposition of MgCl$_2$ small amounts of NH$_4$Cl were added. The mixed salts were dehydrated by careful heating in iron pots over a slow fire until five of the six molecules of water were removed, after which the salts were quickly transferred to a hot quick fire for finishing. The double salts melt to a fairly clear fusion at 620° C and contain about 10 percent MgO.

In the furnace or cell, a cast-steel pot formed the electrolyte chamber and also served as the cathode. A graphite anode was suspended centrally in the fused salt, means being provided for varying its height so that the bath level which was lowered during electrolysis could be followed and a constant voltage on the cell maintained. The furnace cycle was a 24-hour one; at the end of this cycle the residual bath containing about 10 percent MgCl$_2$ and 90 percent NaCl was removed, and a new molten charge was added. Sodium displaces magnesium from its fused salts, and within the limits of concentration stated, practically no sodium is electrolyzed. However, the intermittent nature of the operation and an expensive dehydration process are disadvantages of this method of reduction.

The temperature of the electrolyte was maintained, partly by the resistance of the bath to the electric current and partly by a coal fire under the steel pot.

The reduced magnesium metal was lighter than the electrolyte and rose to the surface, collecting in pools near the edge of the cathode pot. The molten chloride salt formed a protecting film covering the floating metal. At a temperature range between 675 and 725° C, although no separating diaphragm was used in the cell, recombination between magnesium and chlorine was not excessive. In practice it was found best to maintain the supply of external heat as uniformly as possible and to obtain temperature regulation almost entirely by electric methods.

Continuous Chloride Process

The Dow Chemical Company developed a continuous chloride process which is described by Gann.[2] The basic raw material used was a salt brine. This process permitted the separation of hundreds of tons per day of a mixture of the chemically allied salts ($MgCl_2$, $CaCl_2$ and $NaCl$) in addition to the liberation of bromine. These brine constituents, when separated from the $MgCl_2$, became the raw materials for other branches of the local chemistry industry. However, the success of the sea-water process, described later, as well as changes in the utilization of brine, has made the process of historic interest only.

Natural brine pumped from wells 1,200 to 1,400 feet deep contained approximately 14 percent $NaCl$, 9 percent $CaCl_2$, 3 percent $MgCl_2$ and 0.15 percent bromine. After the bromine was removed, the brine was treated with a magnesium hydrate slurry to precipitate the iron and other impurities which were separated in continuous thickeners and sedimentation tanks. The decanted liquor was evaporated until the $NaCl$ had crystallized. The salt was removed on rotary filters.

The $MgCl_2$ and $CaCl_2$ in the rotary-filter mother liquor were separated from each other by fractional crystallization. Under controlled composition and temperature conditions, the crystals separating from a complex salt solution may have a different composition from the solids remaining in the mother liquor. This was accomplished by concentrating a solution with a $1:3$ weight ratio of $MgCl_2$ to $CaCl_2$, whereupon crystals of the double salt "tachydrite" ($2\ MgCl_2 \cdot CaCl_2 \cdot 12\ H_2O$) were formed. It has a $2:1$ ratio of $MgCl_2$ to $CaCl_2$, but its crystals are in equilibrium with a mother liquor having a $MgCl_2$ to $CaCl_2$ ratio of $1:10$. The crystals and liquor were separated in false-bottom tanks. The tachydrite crystals were then dissolved in hot water and the solution transferred to a series of crystallizers where substantially pure $MgCl_2 \cdot 6H_2O$ separated, leaving a mother liquor containing $MgCl_2$ and $CaCl_2$ in a ratio of approximately $1:1$. This solution, together with the $MgCl_2 \cdot 6H_2O$ wash water, contained approximately half of the $MgCl_2$ originally present in the brine and was returned to the process for reworking. The $MgCl_2 \cdot 6\ H_2O$ crystals were melted in their water of crystallization, and the fused mass was flaked on rotating steel drums.

Feed for the electrolytic cells was prepared by dehydration operations to remove the six molecules of water of crystallization. Air drying is practical to a composition corresponding approximately to $MgCl_2 \cdot 2\ H_2O$ provided the temperature is controlled to prevent incipient fusion. In earlier practice the last two molecules of water were removed by heating to still higher temperature in an HCl atmosphere which is necessary to prevent hydrolysis and the formation of MgO.

The cell feed, as described by Hunter,[4] is $MgCl_2 \cdot 1.25\ H_2O$ (75 percent $MgCl_2$) in granular form. The commercial electrolysis of a pure or even approximately 100 percent $MgCl_2$ bath has probably never been tried. Its melting point is high ($712°C$). The electric conductivity is low, and the fact that additions of other chlorides—notably of the alkali and alkaline earth groups, all of which have higher decomposition potentials than magnesium chloride—lower the melting point of the mixture and improve the conductivity suggests the use of mixtures for electrolysis. As a result, most fused baths contain a minor quantity of magnesium chloride and consist in the main of sodium or potassium chlorides. It must be borne in mind that the composition of the molten electrolyte is dictated by the impurities. The exact composition of the fused bath used in an electrolytic magnesium process is dependent on the raw materials available in the locality.

Michigan brines used for the production of magnesium contain magnesium chloride, sodium chloride, calcium chloride and, present in very small amounts, strontium chloride. These Michigan brines contain little potassium chloride and practically no sulfates. In the electrolysis of magnesium chloride, sodium chloride is preferred to potassium chloride. Its conductivity is greater, and it has the advantage of yielding a bath with higher specific gravity, thereby facilitating satisfactory separation of the magnesium metal. Since the Midland raw material contained calcium chloride, it was practically impossible to operate a process without the effects of calcium. All impurities added to the cell leave in one of the following ways: by removal in the magnesium metal produced; by removal in the chlorine-rich anode gases; by removal in the cell sludge; or by the intentional "dipping" of the bath for the control of composition.

Magnesium produced in the cell will remove those impurities which alloy with it. These include aluminum, copper, nickel, zinc, silicon and to a small degree manganese—all of which deposit and alloy with the magnesium produced at the cathode or are reduced by chemical reactions in the bath adjacent to the cathode. As these metals are undesirable in the virgin magnesium metal, however, care must be taken to prevent their being included in the cell feed. Fortunately, most of the metals are not present in the raw materials and may be kept out of the cell feed by judicious elimination of them in the chemical processing preceding electrolysis.

Other metals such as iron, boron and, to a lesser extent, manganese are either liberated on the cathode or their compounds are reduced by the magnesium floating in the cell bath. Since these metals do not appreciably alloy with the magnesium, they fall to the bottom of the cell. If cell operating conditions are not correct, these metal impurities may form shells around small particles of magnesium, resulting in globules whose net specific gravity is greater than that of the bath; these globules will therefore fall to the bottom of the cell and the magnesium in them will be lost. Proper operation should permit the formation of larger globules which tend to rise to the top[6] where shells of heavy metal are sloughed off and fall to the bottom.

The magnesium oxide (from hydrolysis of $MgCl_2$ in the feed) introduced into the cell follows several courses. A limited amount dissolves in the bath and may be electrolyzed; a considerable portion is rechlorinated to magnesium chloride, probably owing to the presence of chlorine, carbon and carbon monoxide; and the remainder, amounting to less than half of that added, falls to the bottom of the cell. With the heavy metals and shell-encased magnesium, MgO forms a spongy layer disposed throughout the bottom of the cell. This "sludge" is dipped out periodically. A typical sludge is 18 percent MgO, several percent Mg and less than 1 percent heavy metals, and the balance consists of bath which is interspersed through the sponge. The bath thus enclosed in the sponge is the principal means of escape for the soluble impurities of alkali and alkaline earth groups which are not affected by the electrolysis and which would otherwise accumulate in the bath. Any other adjustment of materials in or out must be accomplished by the addition of the desired salt and ladling of the bath.

Sulfates combined with the alkali or alkaline earth metals are detrimental. The possible reactions which may be written for the behavior of sulfates in the magnesium cell include (1) the reduction[7] by magnesium of the SO_4 radical near the cathode to elementary sulfur or SO_2, (2) the appearance in the sludge of iron sulfides or (3) the liberation of SO_2 and oxygen on the anode.

The unusual difficulties encountered in attempts to separate calcium completely from magnesium (either as oxides, carbonates, or chlorides) mean that when dolomites are used as a raw material some calcium chloride will be present in the feed delivered to the cells. This makes it necessary to adopt a bath composition in which calcium chloride is a major component. The electrochemistry of magnesium permits operation in most of the systems involving calcium, sodium, potassium and magnesium.

The flow sheet for the production of magnesium from sea water is shown in Figure 19-1.[5] The process consists of the following steps:

(1) Magnesium hydrate is precipitated from sea water using milk of lime made from oyster shells.

(2) After filtration, the hydrate is converted into magnesium chloride using 10 percent solution of hydrochloric acid.

(3) The magnesium chloride solution is concentrated, first in direct-fired evaporators, then on shelf driers, and finally in a rotary drier.

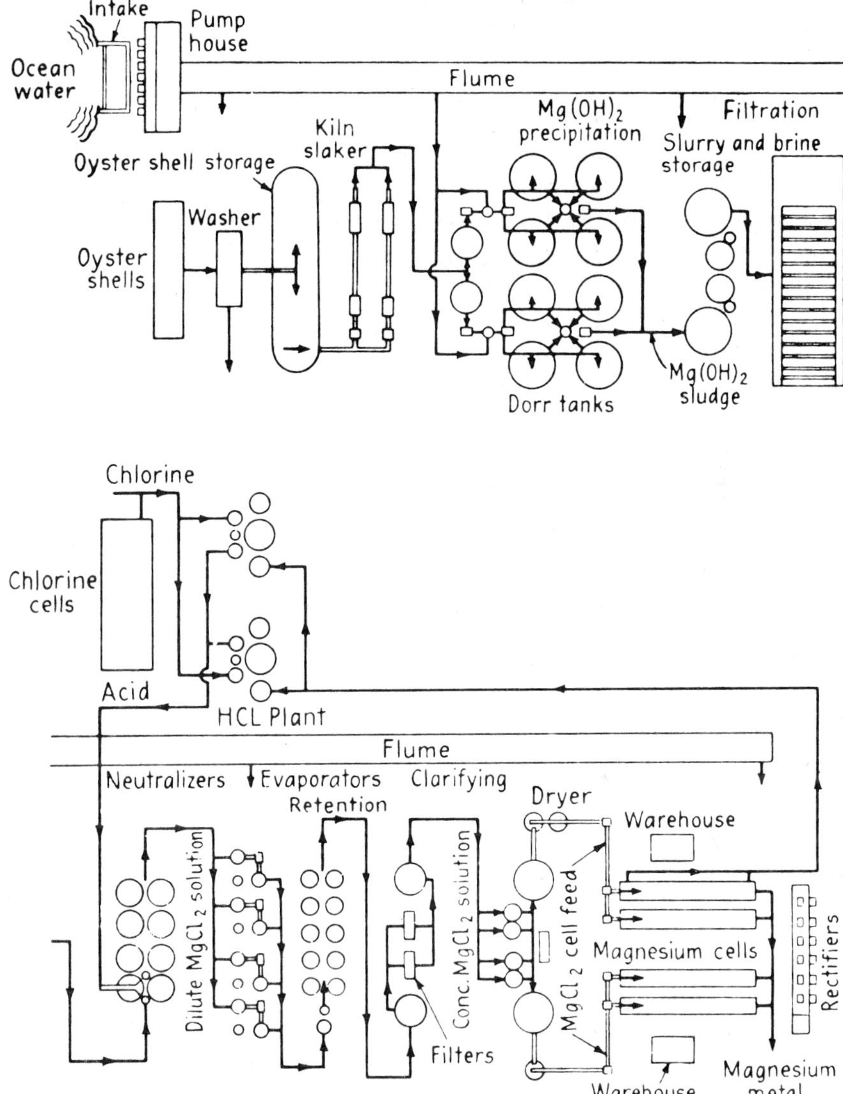

Figure 19-1. Flow sheet of magnesium production from sea water.

(4) Flaked magnesium chloride, in practically anhydrous condition, is fed into the cells where it is electrolyzed to produce metallic magnesium of average purity between 99.9 and 99.95 percent, along with by-product chlorine which is used to make HCl.

The chief impurities to be found in the cell feed made from sea water are sodium chloride, calcium chloride and small amounts of potassium chloride. Sulfate is present, but it may be kept low. Other impurities are present in very small quantities.

An element which caused considerable trouble in cell feed made from ocean water is boron. The vapor pressure of halides of boron, particularly the chloride, is high so that more than 70 percent of the boron is vaporized as boron chloride in the anode gases and later condensed in the hydrochloric acid solution which is formed. Boron may accumulate in the cell feed produced from precipitated ocean-water magnesium hydrate reacted with the recycling hydrochloric acid.

In a pure salt bath operating on boron-free feed, the deposited magnesium leaves the cathode in small spheres which do not coalesce even when strenuously agitated or puddled. Oettel[7] reported this phenomenon as early as 1895 and suggested its remedy by the addition and maintenance of small quantities of calcium fluoride, which is soluble in the bath and probably dissolves films which may form on the surface of the magnesium spheres.

The presence of boron in cell feed results in its liberation at the cathode, where it is believed to react with the magnesium to form metallic borides. This metal-like deposit collects between crystals and exerts a dispersing action on the metal which is impossible to control with calcium fluoride. Less than 10 ppm of boron in cell feed is not particularly troublesome, but if 100 ppm are present things begin to happen. The result is a mass of uncollectible metal, some of which is sufficiently weighted down by the presence of boride to sink and accumulate in the sludge at the bottom of the cell. Sludge formations many times normal result; cell operation deteriorates; the magnesium metal will not coalesce, and production efficiency falls. Boron is best controlled by keeping it out of the cell feed, which is accomplished by chemical treatment of the weak magnesium chloride liquors.

With slow and uniform addition, cell feed containing 10 to 20 percent water may be introduced as most of the water flashes off immediately.

During the Second World War, a large magnesium plant was built by Basic Magnesium, Inc., at Las Vegas, Nevada,[8] starting with dolomite containing 40 percent or more MgO, according to the flow sheet in Figure 19-2.

The chlorinators consisted of refractory-lined steel shells electrically heated by inserted carbon electrodes to 850° C. The MgO was reduced by the coal of the charge to metal, which in turn reacted with chlorine from electrolytic cells to form $MgCl_2$. This collected at the bottom of the furnace and was tapped off for cell feed.

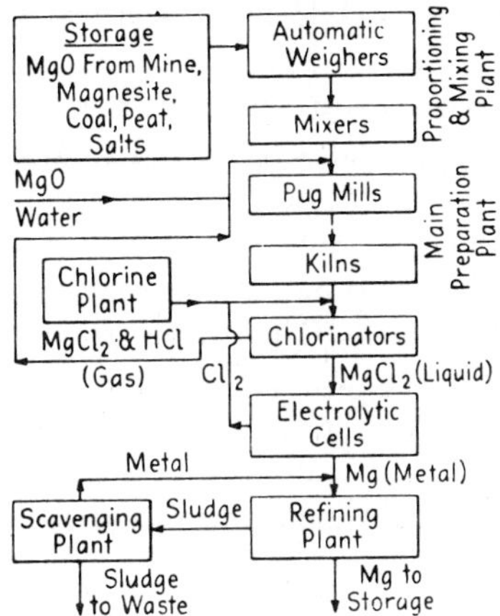

Figure 19-2. Flow sheet of main reduction works of Basic Magnesium, Inc.

German Practice

Procedures at the Bitterfeld and Stassfurt works are typical of German methods. There the raw materials for the operation were carnallite from Stassfurt and high magnesium content dolomite (stated to be 40 percent MgO). The dolomite was calcined and slaked in 26 percent $MgCl_2$ brine, an end product of the potassium chloride operations from carnallite. The precipitated $Mg(OH)_2$ was filtered on an Oliver filter, washed with water and cake dropped with 50 to 60 percent H_2O. This cake was dried at 500° C in a shelf drier with rabble arms on a vertical shaft to move the material across the shelves. It was then calcined at 900°C with "braunkohle" producer gas. The dried MgO was mixed thoroughly with crystal $MgCl_2 \cdot 6 H_2O$ obtained by evaporation of the $MgCl_2$ liquors from carnallite. The product, a mixture with analysis $MgO \cdot MgCl_2$ plus 30 percent H_2O of crystallization, was called "oxychloride" and does not absorb much water. At the magnesium plants it was mixed with braunkohle, which was ground to 25 percent through 250 mesh. At Bitterfeld the mix was 88 percent oxychloride, plus coal to give 6 percent carbon and peat to yield 6 percent carbon.

At Stassfurt 12 percent carbon was obtained from braunkohle; no peat was added. The oxychloride, coal and peat were mixed in dry mixers, a little $MgCl_2$ liquor having been added to aid in briquetting; and the moist mixture was extruded through a screw briquettor to a $1\frac{1}{2}$-in. cylinder of low strength which broke into 2- to 4-in. lengths as it was extruded. These were dried at 400° C to harden and then were calcined to coke the coal and peat. The product contained about 10 percent H_2O. The chlorinators at Bitterfeld are each rated at 10 tons of anhydrous $MgCl_2$ per day. They are vertical steel cylinders acid brick lined and are 20 ft high, having about a 9 ft inside diameter and 13 ft outside diameter. The briquettes were fed in at the top. Chlorine from the cells (90 percent concentration), to which was added make-up chlorine from liquid, was introduced just above the bottom. The temperature was maintained by carbon resistors operating three phase at 2,500 amperes and 22 volts (about 0.6 kva per lb of magnesium). Each chlorinator served about 12 cells. Molten magnesium chloride flowed out of the chlorinators and was transferred into crucibles with tilting mechanism and fed to the cells once per 8-hour shift.

The chlorine losses in this operation result from stack loss of the gas not absorbed in the briquettes, the conversion of CaO impurities to $CaCl_2$ (present in the feed up to 6 percent) and the conversion of a large portion of the 10 percent of H_2O to HCl due to the water-chlorine reaction at the elevated temperature. The $CaCl_2$ and the HCl loss are estimated at 0.2 lb of chlorine per pound of metal for the $CaCl_2$ and 1.5 lb of chlorine per pound of metal for the HCl.

Liquid chlorine was supplied to the process at the rate of 0.5 to 1.0 lb of chlorine per pound of metal. The chlorine balance, therefore, would appear to be 1.48 lb of chlorine per pound of metal in the $MgCl_2 \cdot 6\ H_2O$ added and 0.5 to 1.0 new chlorine, making a total addition of 1.98 to 2.48 and a known loss from HCl and $CaCl_2$ of 1.7.

The chlorine gas outlets are one per dome, three per cell. They have an inside diameter of about 2 in. and are composed of a section of ceramic pipe delivering into a cast-iron line. The line suction was 1 in. of water and the cell suction about 0.1 in. Before compression, the chlorine was dry-filtered through thimbles (baghouse) of asbestos cloth for removal of the dust and was then compressed to about 20 in. of water pressure with a blower and returned to the chlorinators. The four cathode chambers are ceramic-covered, with lids provided in the front for metal and sludge removal, and they are vented at the back through lines similar to those used for chlorine, to remove HCl. These gases were scrubbed and thrown away.

Composition of the cell bath at Stassfurt was given as 30 to 40 percent $CaCl_2$, 20 to 25 percent KCl, 20 percent NaCl and 15 percent $MgCl_2$, with about 100 lb of NaF added per cell per month.

References

1. Bunsen, R., *Lieb. Ann.*, **82**, 137 (1852).
2. Gann, John A., *Trans. Am. Inst. Chem. Engrs.*, **24**, 206 (1930).

3. Harvey, W. G., *Trans. Am. Electrochem. Soc.*, **47,** 327 (1925).
4. Hunter, Ralph M., *Trans. Electrochem Soc.*, **86,** 21 (1944).
5. Kirkpatrick, S. D., *Chem. Met. Eng.*, **48**(11), 76, 130 (1944).
6. Oettel, F., *Chem. Ztg.*, **61,** 156 (1937).
7. Oettel, F., *Z. Elektrochem.*, **2,** 394 (1895).
8. Ramsey, Robert H., *Chem. Met. Eng.*, **50,** 98, 115 (October 1943).
9. Shcherbakov, I. G., *et al.*, *Kaliĭ (U. S. S. R.)*, **5,** 10, 19 (1936).

20. DDT

David J. Porter

Research Department
Diamond Alkali Company

INTRODUCTION

"DDT" is the term applied to the product 2,2-bis-(*p*-chlorophenyl)-1,1,1-trichloroethane obtained by condensing chloral (or its alcoholate or hydrate) with chlorobenzene in the presence of sulfuric acid. During its development into an important article of commerce in the early 1940's, it became known by the mnemonic abbreviation of its generic name—dichloro-diphenyl-trichloroethane.

DDT was first synthesized in 1873 during an investigation of the reactions of aldehydes with substituted aromatic hydrocarbons[88] and for 66 years remained only a Beilstein entry.[5] Its outstanding merit as an insecticide was discovered in 1939 by Dr. Paul Müller of J. R. Geigy-A. G. in Switzerland during a systematic search for a water-insoluble moth-proofing agent for wool which would act as a contact poison, similar to pyrethrum and rotenone.[37,44]

DDT soon proved to be very effective not only as a contact poison but also as a stomach poison, with a remarkable residual effect, when tested against such pests as flies, aphids, mosquito larvae, the Colorado potato beetle, ants, roaches, bedbugs, silverfish, and the body louse.[57,50,43]

In August 1942, Geigy informed the United States Military Attache in Berne of the effectiveness of a DDT dust formulation in the control of body lice. At that time, imports of derris into the United States had been shut off completely by the fall of Singapore, and the supply of pyrethrum was insufficient to meet war time demands. The Bureau of Entomology and Plant Quarantine of the U. S. Department of Agriculture was actively engaged in a search for a substitute for these contact agents, and had screened several thousand compounds with indifferent success when DDT was brought to its attention in October. Two formulations, a dust and a wettable powder, were shipped from Switzerland to New York for testing; undiluted DDT was imported later.[71] Upon confirmation of its reported remarkable performance, an intensive program of testing and evaluation was started, which led to the adoption of a military DDT louse powder formulation in May 1943. A spectacular demonstration of its effectiveness and lasting power was the checking of a typhus epidemic in Naples that winter when its resident and refugee population was deloused in a few months; no U. S. soldier contracted typhus in Italy.

Continued testing demonstrated the effectiveness of DDT against a wide variety of other insect pests including, for example, the codling moth, Japanese beetle, thrips, tomato fruit worm, plant lice, cabbage worm, Oriental fruit moth, apple maggot, potato leafhopper, corn earworm, European corn borer, and pink bollworm.

This remarkable activity against such a broad spectrum of insects, the remarkable success on the Italian front, and the fact that its insecticidal use was discovered and developed under war time conditions created so much public curiosity and interest that when DDT was released in the United States for limited civilian use in 1945, it received enthusiastic acceptance.

In many ways DDT has been and is a unique substance. It was the first synthetic organic pesticide with a real impact, and other than sodium fluoride it was the first synthetic acting both as a contact and stomach poison. It was discovered, developed, and proved useful during a war; it stopped a typhus epidemic in the winter, and for the first time during war; it has a remarkably broad sphere of activity. The chemistry of DDT is relatively simple, its synthesis easy, its raw materials relatively low in cost, and its high insecticidal activity is combined with a relatively low mammalian toxicity. Further, it has a pleasant odor, little color, and is readily dissolved and formulated; its resistance to oxidation and photolysis is great, and its vapor pressure is low, which renders it highly persistent. Its first use was for the protection of health—against the typhus louse in the European Theatre and the malarial mosquito in the Pacific Theatre of war. Subsequently it has found wide application in plant protection, where it is practically harmless to the foliage of all plants with the exception of cucurbits. DDT does not taint plant juices, nor are its deposits unsightly. If applied according to recommendations, any residues at harvest will meet acceptable tolerances and will present no hazard to the consumer.

Within five months of the Swiss reports showing its effectiveness in the control of the body louse, DDT was being produced in the United States for the Armed Forces by the Cincinnati Chemical Works at the rate of 1000 lb per month (April 1943).[57] Under highest war time priorities, added capacity was constructed so that the production rate increased from 200,000 lb per month in January 1944 to 2,000,000 lb per month in December of that year,[1] by installations of Merck, Hercules, Du Pont, General, Baker, Monsanto, Elko, and Sherwin-Williams.[50]

During this rapid build-up of production capacity, the market price for DDT declined from a reported $1.60 per pound in October 1943, to one dollar, and from one dollar to 75¢ in October 1944.[50,10] These figures do not agree with the 53¢ reported by the U. S. Tariff Commission for 1943 DDT sales, as shown in Table 20-1.[77] If $1.47 is used as the price for 1943, this value lies on a straight-line "learning curve"[3] plot of cumulative average price per pound against cumulative production (Figure 20-1) which represents quite satisfactorily the subsequent sales history of this important product. The break in the line from 1950 to 1953 shows the results of an abnormal pricing situation which developed as demand surged during the Korean conflict.

TABLE 20-1. DDT PRODUCTION AND SALES STATISTICS[77]

Year	U. S. Production mm lb	U. S. Sales $mm	Price ¢/lb	Cumulative Sales mm lb	Price ¢/lb
1943	5.5	4.3	.53	4.3	147
4	9.6	9.0	.84	13.4	104.2
5	33.2	31.6	.54	45.0	68.9
6	45.7	43.5	.43	88.5	56.1
7	49.6	36.1	.41	124.5	51.8
8	20.2	30.0	.31	154.5	47.7
9	37.9	37.4	.32	191.9	44.7
1950	78.1	71.0	.29	262.9	40.3
1	106.1	94.6	.42	357.5	40.7
2	99.9	94.4	.35	451.9	39.5
3	84.4	91.5	.25	537.6	37.1
4	97.2	91.5	.24	629.2	35.2
5	129.7	124.7	.21	753.8	32.8
6	137.7	107.9	.20	861.7	31.2
7	124.5	120.3	.21	982.1	29.9
8	145.3	145.2	.18	1127	28.5
9*	156.7	148.7	.20	1276	27.5

*1959 data preliminary

Properties

DDT, also called gesarol, neocid, chlorophenothane (U. S. Pharmacopoeia XVI), zerdane (France), and dicophane (British Pharmacopoeia), has the empirical formula $C_{14}H_9Cl_5$, a molecular weight of 354.51 and a Cl content of 50.01 percent. The pure p,p'-isomer is a white, crystalline compound melting at 108.5 to 109.0° C with a density of 1.556.[32,31] Its vapor pressure is 1.5×10^{-7} mm Hg at 20° C and is represented over the

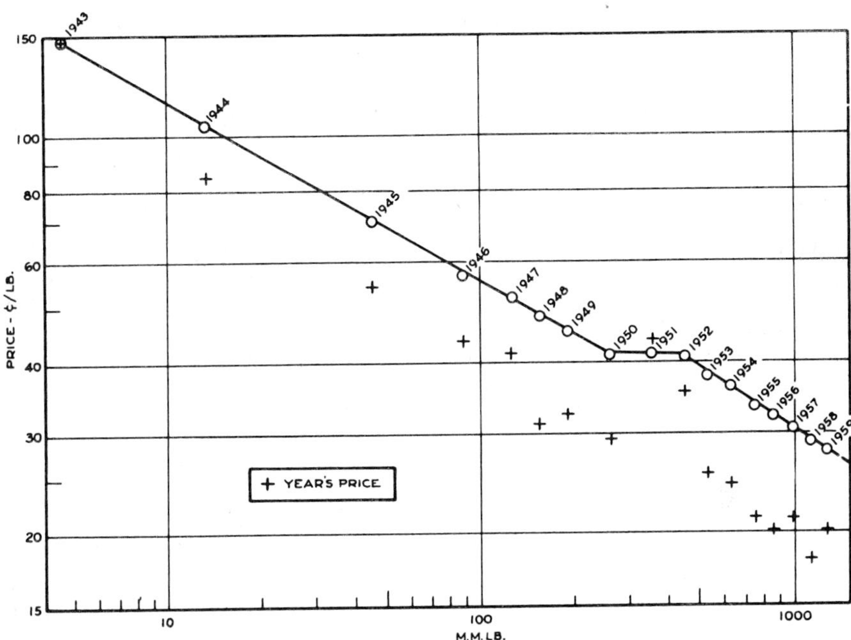

Figure 20-1. Cumulative U.S. sales of DDT.

range 40 to 90° C by the equation $\log_{10}p = 9.60-4370/(273 + t)$.[35] The second order rate constant for its hydrolysis by NaOH in 92.6 percent ethanol is 0.0248 liter/mol sec at 20.1, 0.0711 at 30.4° C; corresponding values for the *o,p'*-isomer are about 1/67th as large.[19] The solubility of DDT in water lies in the range 0.01 to 0.1 ppm.[60] Solubility of the pure p,p'-isomer in various solvents increases with temperature, as shown by Table 20-2.[31] The solubility of technical DDT in various solvents useful for insecticidal formulation has been determined, Table 20-3.[34] Pure DDT is relatively stable, but Friedel-Craft catalysts, alkalies, and HCl-acceptors accelerate its decomposition at 115 to 120° C.[23]

MANUFACTURE

Basic Chemistry

The basic reaction for the DDT synthesis is the Baeyer condensation of two mols of chlorobenzene with chloral in the presence of strong sulfuric acid or oleum:

$$CCl_3CHO + 2C_6H_5Cl \xrightarrow{H_2SO_4} (p\text{-}C_6H_4Cl)_2CHCCl_3 + H_2O \qquad (20\text{-}1)$$

TABLE 20-2. SOLUBILITY OF p,p'-DDT[31]

Temperature, °C	Weight %				
	0	7.2	24.0	45.0	48.0
Solvent					
Acetone	21.2	27.3	40.3		59.0
Benzene	6.8	27.1	44.0		57.8
Carbon tetrachloride	9.0	10.5	18.0	34.8	
Chloroform	18.2	21.9	31.0	47.4	
Dioxane	8	29	46		61
Ethyl ether	15.0	18.9	27.5		
Ethanol-95%	0.8	1.0	2.2		3.9
Petroleum ether 30–60°	1.7	2.4	4.8		
Pyridine	21	36	51		62

Since the chlorobenzene does not all react in the *para* position, the condensation is accompanied by side reactions yielding the o,o'-isomers and o,p'-isomers. For the DDT reaction Equation (20-1) ΔH may be approximated, by considering the heats of formation of benzene, formaldehyde, and diphenylmethane, at ca -38 Kcal/gm mol. For the reaction of SO_3 with water to yield 100 percent H_2SO_4, $\Delta H = -37.3$. Hence the total heat release in the synthesis of DDT using oleum as the condensing agent should approximate 210 cal/g. The addition to chloral of one mole of chlorobenzene, yielding either the *ortho-* or the *para*-isomer of 2-chlorophenyl-1,1,1-trichloroethanol-2, which have been identified in technical DDT, has

TABLE 20-3. SOLUBILITY OF TECHNICAL DDT[34] 27–30°C

	Grams per 100 ml Solvent	Grams per 100 gm Solvent
Acetone	58	74
Benzene	78	89
Chlorobenzene	74	67
Cyclohexanone	116	122
o-Dichlorobenzene	59	45
1,4-Dioxane	92	89
Isophorone	74	80
Methylene chloride	88	66
o-Xylene	57	66
Gasoline	9–10	12–13
Stoddard solvent	9	12
Kerosene	8–10	10–12
Kerosene refined	4	5
Aromatic petroleum naphtha	33–47	38–59
Methylated naphthalenes	48–59	47–61
Xylene $-10°$	53	61

been postulated as the first step in the condensation:

$$CCl_3CHO + C_6H_5Cl \rightarrow CCl_3-CHOH-C_6H_4Cl \cdot \qquad (20\text{-}2)$$

As the result of incomplete chlorination, commercial chloral may contain measurable amounts of dichloroacetaldehyde which may react to yield five corresponding 1,1-dichloro compounds.

The intermediate alcohols, $CCl_3-CHOH-C_6H_4Cl$, may react with HCl to yield the corresponding chlorinated phenylethanes $CCl_3-CHCl-C_6H_4Cl$ (*o*- or *p*-), or may lose HCl and rearrange to give the acid chlorides, $ClC_6H_4-CHCl-COCl$, which, in turn, may yield amides if neutralized with ammonia.

When oleum or concentrated sulfuric acid is used as the condensing agent, an important competing reaction is that of sulfonation. *p*-Chlorobenzenesulfonic acid may be present in incompletely neutralized DDT, or in the form of its Na or NH_4 salt if the DDT is neutralized but washed insufficiently. Condensation of *p*-chlorobenzenesulfonic acid with chlorobenzene yields bis-(*p*-chlorophenyl) sulfone. Chlorobenzene and its normal contaminant, *p*-dichlorobenzene, may remain dissolved in the DDT as the result of incomplete stripping.

Haller *et al.*, working with the products of four of the first United States DDT producers, were able to separate, characterize, and identify fourteen of the large number of possible by-products and contaminants of DDT, of which seven had not been previously described (Table 20-4).[32]

Technical DDT is therefore composed of 70-odd percent of 2,2-bis-(*p*-chlorophenyl)-1,1,1-trichloroethane (*p,p'*-DDT) and about 20 percent of 2-*o*-chlorophenyl-2-*p*-chlorophenyl-1,1,1-trichloroethane (*o,p'*-DDT, melting point 74.0 to 74.5° C). Smaller amounts of 2,2-bis-(*p*-chlorophenyl)-1,1-dichloroethane (*p,p'*-DDD, melting point 109.5 to 110° C) and 2-*o*-chlorophenyl-2-*p*-chlorophenyl-1,1,-dichloroethane (*o,p'*-DDD, melting point 74.5 to 75.5° C) may be present in proportion to the dichloroacetaldehyde content of the chloral. From the large list of other compounds which may be present, including possible intermediates, isomers, and reaction products of impurities, eight others have been identified in technical DDT, for example, bis-(*p*-chlorophenyl)-sulfone (melting point 148 to 149° C).[32]

The commercial production of DDT takes place in five fairly well-defined steps:

(1) Condensation of chloral with chlorobenzene, usually through the action of strong sulfuric acid or oleum.

(2) Separation of the resultant crude DDT from the spent condensing agent.

(3) Neutralization, washing, and stripping of the crude DDT to bring its impurity concentration to within specification limits.

TABLE 20-4. COMPOUNDS IDENTIFIED IN TECHNICAL DDT[32]

Compound	Formula	Melting Point ($^\circ$C)	
2,2,-bis-(p-chlorophenyl)-1,1,1-trichloroethane	(p-ClC$_6$H$_4$)$_2$CH-CCl$_3$	108.5–109.0	p,p'-DDT
2-o-chlorophenyl-2-p-chlorophenyl-1,1,1-trichloroethane	(o-ClC$_6$H$_4$) (p-ClC$_6$H$_4$) CH-CCl$_3$	74.0–74.5	o,p'-DDT
2,2-bis-(p-chlorophenyl)-1,1-dichloroethane	(p-ClC$_6$H$_4$)$_2$CH-CHCl$_2$	109.5–110	p,p'-DDD
2-o-chlorophenyl-2-p-chlorophenyl-1,1,1-dichloroethane	(o-ClC$_6$H$_4$) (p-ClC$_6$H$_4$) CHCHCl$_2$	74.5–75.5	o,p'-DDD
bis-(p-chlorophenyl)-sulfone	(p-ClC$_6$H$_4$)$_2$SO$_2$	148–149	
2-trichloro-1-o-chlorophenylethyl p-chlorobenzene sulfonate	(o-ClC$_6$H$_4$)CH(CCl$_3$)OSO$_2$ (p-C$_6$H$_4$Cl)	106.9–107.4	
α-chloro-α-p-chlorophenyl acetamide	p-ClC$_6$H$_4$-CHCl-CONH$_2$	134–135	
α-chloro-α-o-chlorophenyl acetamide	o-ClC$_6$H$_4$-CHCl-CONH$_2$	103–105	
2-p-chlorophenyl-1,1,1,2-tetrachloroethane	(p-ClC$_6$H$_4$)CHCl-CCl$_3$		b. 112–117/1 mm
2-trichloro-1-p-chlorophenylethanol	(p-ClC$_6$H$_4$)CHOH-CCl$_3$		b. 123–125/1 mm
chlorobenzene	C$_6$H$_5$Cl		
p-dichlorobenzene	p-ClC$_6$H$_4$Cl		
sodium p-chlorobenzenesulfonate	p-ClC$_6$H$_4$OSO$_2$ONa		
ammonium-chlorobenzenesulfonate	p-ClC$_6$H$_4$OSO$_2$NH$_4$		

(4) Solidification and mechanical processing of the purified DDT.

(5) Treatment and recovery of materials from the spent condensing agent.

Condensation

In this step DDT is formed as oleum is added to a chilled mixture of chloral and chlorobenzene, with agitation to promote interphase contact and heat transfer.

The condensation is conveniently carried out as a batch reaction in a closed, agitated vessel provided with a jacket, or internal cooling coils, or both; a continuous system has been proposed.[10] Heat-transfer capacity may be augmented by circulating the reaction mixture through an external heat exchanger. Early condensation vessels were glass-lined[57] but steel has since been found to be satisfactory, thus decreasing equipment cost and permitting greater latitude in equipment design.

Measured amounts of chloral and chlorobenzene are charged to the reaction vessel and cooled to the desired initial temperature, usually in the range 0 to 30°C; the chlorobenzene-chloral mol ratio may lie within the range from about 2.2 to perhaps as much as 4 to 5. To this cooled mixture, oleum is added at a rate which is adjusted to hold the temperature of the reaction mass within the desired range, usually 0 to 30°C. The first DDT dissolves in the chloral-chlorobenzene phase, but the amount of solvent decreases and the amount of DDT increases as the reaction proceeds, until the solution finally discharges its supersaturation. The appearance of the third, solid phase in the system is accompanied by a noticeable temperature increase. Continued addition of oleum leads to almost complete utilization of chloral. The time required for oleum addition depends upon the size of the batch, the temperature at which the reaction is being conducted, the strength of the oleum, and the heat-transfer characteristics of the cooling system utilized, and may vary from 2 to more than 8 hours.

The strength of the condensation acid may vary from 98 percent H_2SO_4 to 20 percent or stronger oleum; 20 percent oleum is customarily used. The amount of water removed from the system per unit acid weight increases rapidly with acid strength, as does the rate of sulfonation of chlorobenzene.[16,42] The amount of oleum is chosen so that after taking up any water which may have been present initially in the chloral and chlorobenzene and that which was liberated in the simultaneous reactions of condensation (yielding DDT) and sulfonation (yielding chlorobenzenesulfonic acid), the final acid strength has a chosen value, generally in the range of 88 to 95 percent on an organic-free basis. Rapid addition of oleum to the mixture of chloral and chlorobenzene at a temperature of 0 to 5°C tends to increase the *p,p′*-isomer content, while raising the system temperature to 25 to 30°C following the oleum addition promotes more rapid crystallization of the DDT.[41]

The quantities, conditions, and reaction time adopted are those which for the particular situation optimize the economics of DDT production. Included are such variables as the raw material costs of chloral, chlorobenzene and oleum; the processing costs of operating labor, refrigeration and utilities, and maintenance; the applicable fixed costs associated with investment and operation; the costs or credits associated with the treatment of spent acid for chlorobenzene recovery and the sale or fortification of this acid; and, finally, the somewhat less quantifiable factors of product quality and processing characteristics.

Separation and Neutralization

When mixtures of chloral with about 2 moles of chlorobenzene are condensed with sulfuric acid or oleum, the DDT separates in the form of soft agglomerates of fine crystals. The acid is quite viscous at the temperature and concentration of the reaction. Consequently, separation of the two phases by filtration or centrifugation is slow. A fine fabric is required for complete retention but blinds quickly, and the resulting DDT cake contains a very high proportion of sulfuric acid.

Techniques of DDT recovery based on the separation of two liquid phases have proved simpler and more efficient, and generally utilize one or more of these principles: (1) adding a solvent for DDT (at the start of, during, or following the condensation), usually an excess of chlorobenzene for convenience; (2) raising the temperature of the system to melt the DDT or to increase its solubility and to decrease the viscosity of the sulfuric acid; and (3) diluting the acid, especially when heating is part of the separation technique, primarily to reduce its reactivity as a sulfonating agent, thus avoiding losses of DDT or residual chlorobenzene, and also to decrease its viscosity, thereby speeding and making more efficient the resulting separation. After a settling period of 30 minutes to an hour, the acid layer is withdrawn by decantation until the relatively colorless organic phase replaces the dark-acid phase in a sight glass.

The supernatant organic layer which remains in the condensation vessel or separation vessel after the acid layer has been withdrawn consists primarily of a solution of DDT isomers in chlorobenzene, contaminated with entrained sulfuric acid, dissolved and emulsified *p*-chlorobenzenesulfonic acid, and traces of other acidic materials. The acidic constituents are extracted by several washes (batchwise or continuous countercurrent) with hot water. One or two washes with dilute alkali (1 to 2 percent sodium carbonate or bicarbonate) neutralize any residual *p*-chlorobenzenesulfonic acid, and subsequent water washes (to specified pH) remove traces of the suspended alkaline wash, and thus reduce the ash content of the product to specification level.

Following neutralization and washing, it is necessary to free the DDT from water and chlorobenzene. This is accomplished by vacuum drying or distillation, steam stripping, or a combination of both. For steam economy, completeness of stripping, and minimum equipment requirement, a continuous stripping column is most efficient. If the column bottoms temperature is held at 105° C or above, the effluent DDT is dry as well as free of chlorobenzene, and can be fed directly to the flaker.

Solidification

The conversion of molten DDT into a usable solid form is not simply a problem of heat transfer. Technical DDT supercools readily; the subsequent crystallization of the supercooled melt may be slow or incomplete, and the resulting solid may range from soft and waxy to hard.

In addition to the two isomers, minor amounts of twelve other compounds have been identified in technical DDT,[32] and the presence of as many as seventeen has been shown by vapor-phase chromatography.[78] The temperature-composition diagram for the two-component system composed of the *o,p'*- and *p,p'*-isomers may be used to approximate the characteristics of the technical mixture. This diagram consists of a liquidus of two straight lines intersecting at the eutectic of 25 percent *p,p'*-DDT, 62° C, and a solidus curve showing the precipitation of relatively rich solid solutions. The melting points of the pure *o,p'*- and *p,p'*-isomers are lowered 0.46 and 0.62° C respectively, per percent of the other isomer.[42] The melting of technical DDT begins at about 55° C and is complete at about 95° C.[76]

In early commercial production, metal pans were filled to depths of two or three inches with molten DDT and allowed to cool, either by natural convection or aided by the forced circulation of air. Various techniques including raking, agitation, and seeding were employed to induce crystallization. After cooling for about a day, the finished slabs of DDT were turned out of the pans and passed through an ice crusher for reduction to an acceptable size.

To eliminate the expensive labor, space, and equipment requirements of this casting and crushing technique, a considerable amount of ingenuity has been devoted to the development of continuous solidification processes. Rapid chilling of molten DDT when it is cast in a thin layer on the surface of a drum or belt flaker is likely to result in supercooling. The appearance of crystals may be deferred and completion of the crystallization may require days, with the solid remaining soft and waxy for this period. If provision is made for the combination of a slow cooling rate, by the use of a relatively warm surface, perhaps 25 to 40° C, and a correspondingly longer cooling period, an acceptably friable product may result. If a rotary drum flaker produces a gummy or semiliquid mass, this can be transferred to a belt conveyor for continued slow cooling and crystallization,[29] but it would, in general, be more attractive to cast directly on the belt.

A device with considerably more flexibility is represented by a continuous metal belt flaker, with which separate zones of specific cooling rates may be maintained, and which provides more readily for the combination of extended cooling time, high productivity, and low investment.[20] Several modifications of the operation of a single belt flaker have been developed, involving casting a layer of molten DDT on the belt, agitating it, providing contact with crystalline DDT for seeding, and controlling the cooling rate by zoning of the cooling sections of the belt.[33]

It is possible to cast DDT in water by cooling an agitated suspension, with care in the range 80 to 85° C, to about 50° C, then filtering, yielding a cake of material ranging in size from 50 to 200 mesh[65,51] which can then be dried with warm air.

If DDT is to be ground, as in the preparation of a dust concentrate or wettable powder, it is commonly first cured for a few days after flaking. This provides additional time for crystal development and improves greatly the grinding characteristics of the product.

Since crystallization of technical DDT is difficult to initiate, and since the temperature at which this occurs cannot be accurately measured or duplicated, despite the theoretical value established by p,p'-isomer content, some other related method of characterization was sought. Such an index, the set point, has been found to be reproducible and related to the p,p'-isomer content.[24] The basis for commercial specifications,[22] it is defined as the highest temperature reached by a stirred cyrstallizing mass of previously supercooled DDT.

Another empirical constant, the dough point, is defined as that temperature at which the particles of a stirred body of granulated DDT, being heated slowly under specified conditions, first begin to agglomerate. It is related to the incipient melting point and thus has been found to be of value in predicting the grinding characteristics of DDT, especially such high-DDT mixtures encountered in the preparation of an air-milled 75 percent powder.

Treatment of Spent Acid

The acid phase separated from the crude DDT-solvent mixture at the end of a condensation consists primarily of a solution of p-chlorobenzene-sulfonic acid in strong sulfuric acid, contaminated with minor quantities of chloral, HCl, sulfonated DDT and degradation products which give it a dark color. The weight of the acid phase can be as low as about 1.2 times the weight of the DDT produced if 20 percent oleum is used as the condensing agent, the chloral is anhydrous, and a minimum amount of sulfonation of chlorobenzene takes place.[78] The spent acid-DDT weight ratio increases if a weaker condensing agent is used, if the chloral or chlorobenzene contains water, or if the amount of sulfonation increases. Consider, for example, a condensation in which the conversions of chloral and chlorobenzene to DDT are both 90 percent, and 2.5 lb of 20 percent oleum and 3.0 moles of chlorobenzene are charged per mole of chloral. If 1.0 mole of chlorobenzene is recovered by stripping from the molten DDT, after adding 0.04 lb of dilution water per lb of oleum charged, the calculated weight of spent acid is 1.43 lb per lb of DDT. The acid layer composition is 80.1 percent H_2SO_4, 8.3 percent H_2O, 8.4 percent *p*-chlorobenzenesulfonic acid, and 3.2 percent chloral. On an organic-free basis, this corresponds to a sulfuric acid strength of 90.7 percent. The *p*-chlorobenzenesulfonic acid content of the crude acid mixture has been reported to be as high as 50 percent.[86,58]

Such a highly contaminated acid rarely can be utilized as is and normally requires some sort of clean-up treatment to render it acceptable industrially. The chlorobenzene content, in the form of *p*-chlorobenzenesulfonic acid, can be profitably recovered. Three methods have been used: the first is based on separating the *p*-chlorobenzenesulfonic acid from the spent acid; in the second chlorobenzene is released by hydrolysis; and the third combines the hydrolysis of *p*-chlorobenzenesulfonic acid with the simultaneous utilization of the spent acid by reaction, for example, with phosphate rock.

The separation of *p*-chlorobenzenesulfonic acid from spent acid is based upon its insolubility in 70 to 74 percent sulfuric acid. A practical requirement is a ready use for this diluted acid, which cannot be transported in steel. If the spent acid is diluted directly to 70 to 75 percent acid content (organic-free basis), the resulting precipitate of *p*-chlorobenzenesulfonic acid may be quite fine and difficult to separate, yielding a centrifuge or filter cake with an impracticably high acid content. By adding spent acid and water simultaneously to an agitated, cooled batch of crystallizing *p*-chlorobenzenesulfonic acid, a solid slurry results which is considerably more filterable and yields a drier cake.[86] The dilute sulfuric acid from such a separation contains chloral, HCl, and a minor amount of *p*-chlorobenzenesulfonic acid in solution.

One technique which has been proposed to facilitate the separation of the *p*-chlorobenzenesulfonic acid from the dilute sulfuric acid makes use of the preferential wetting of the precipitated *p*-chlorobenzenesulfonic acid with an aromatic hydrocarbon[83] followed by separation of the acid crystals from the slurry in the hydrocarbon by dissolving them in water.

The *p*-chlorobenzenesulfonic acid in the spent acid may be converted directly to chlorobenzene without separation by hydrolysis at a temperature in the range of 190 to 240° C, by passing a current of superheated steam through the acid,[68] or spraying an acid-superheated steam mixture into a separating zone,[58] condensing the resulting vapor mixture and separating chlorobenzene from the water phase. This high temperature treatment results in the destruction of some of the dissolved organic impurities with the creation of tars and coke and the reduction of H_2SO_4 to SO_2. The chloride content of the acid is sharply reduced, with the appearance in the condensate of HCl. The resulting stripped acid, after cooling and filtration, is suitable for use directly (e.g., in fertilizer manufacture) or for reconcentration.

If *p*-chlorobenzenesulfonic acid has been separated from the spent acid, it is possible to convert this material to chlorobenzene by high-temperature hydrolysis as outlined above,[68] or by steaming it in the presence of sodium acid sulfate generated by the simultaneous addition of caustic soda, *p*-chlorobenzenesulfonic acid, and superheated steam to a hydrolysis vessel from which it is possible to withdraw a molten, relatively organic-free sodium bisulfate.[39]

In a proposed method of treating spent acid, the high temperature hydrolysis of the *p*-chlorobenzenesulfonic acid is combined with the simultaneous reaction of the acid content and of the acid liberated in the hydrolysis, with phosphate rock, the process yielding superphosphate and chlorobenzene.[25]

Before any of these methods of treatment for the recovery of its chlorobenzene content is employed, the significant residual dehydration value of spent acid may be utilized in the dehydration of chloral[70] prior to its introduction to the condensation reaction.

The quantity of spent acid to be disposed of may be reduced by recycling a portion and bringing it up to the desired starting strength by the addition of SO_3. In addition to that water introduced into the system directly or in the reactants, one mol of H_2O, as H_2SO_4, must still be removed from the system for each mol of DDT made and for each mol of chlorobenzene which is sulfonated.

Alternate Syntheses

The burst of enthusiasm which greeted the DDT announcement resulted in the examination of a wide variety of alternate methods aimed at increasing yields, purity, production rate, *p,p'*-isomer content, utilizing other starting materials, or achieving a more economical method of synthesis.

A laboratory and pilot study of the use of chlorosulfonic acid as the condensing agent gave a DDT yield of 77 percent on chloral hydrate, 70 percent on chlorobenzene, using 2.16 mols of $ClSO_3H$ per mol of chloral hydrate. The use of anhydrous chloral would presumably have cut this requirement by 50 percent. Reactions were conducted in the presence of inert solvents at temperatures ranging from –5 to –40° C; the optimum conditions were no solvent, 14 hours, and 20° C.[64,63,62]

The use of fluosulfonic acid has been studied in the DDT condensation,[63,80] and means were developed for the recovery and utilization of some of the resulting H_2SO_4—FSO_3H—HF spent acid mixture.[36,27] The condensation has also been effected by use of large quantities of (initially) anhydrous HF,[67] or, in low yields by use of 98 percent H_2SO_4 containing 1 to 10 percent P_2O_5.[75]

Chloral

Chloral, trichloroacetaldehyde, 2,2,2-trichloroethanal (CCl_3CHO, molecular weight 147.40) is a colorless liquid, with a melting point of –57.5° C, a boiling point of 97.5° C, a d_4^{20} of 1.512, a n_D^{20} of 1.5468[21] and a characteristic sweetish-sharp odor. First prepared by Liebig in 1832 by chlorination of alcohol, its structure was established in 1848 by Dumas.[79] It dissolves readily in ethanol with the formation of chloral alcoholate $CCl_3CH(OH)OC_2H_5$ (melting point 46° C). With water it forms chloral hydrate, $CCl_3CH(OH)_2$ (melting point 57° C, boiling point 96–7° C with dissociation, sp. gr. 1.908, $pK_A = 11.2^{21}$), one of the few compounds with two hydroxyl groups on a single carbon atom.

Chloral reacts as an aldehyde, reducing ammoniacal silver nitrate solution and reacting with Schiff's reagent. The hydrate, alcoholate, and the ammonia addition compound are stabilized by the trichloromethyl group. Chloral is converted very readily by alkali into chloroform and formate, a reaction which forms the basis for a convenient method of analysis.

Liebriech discovered in 1839 the soporific activity of chloral, and, for a century this was almost its sole use, with minor amounts used as a hypnotic or reduced to trichloroethanol,[79] although numerous reactions and derivatives have been listed.[4,21,55,2] During this period chloral was prepared by slowly chlorinating absolute ethyl alcohol at temperatures gradually increasing during the three-day operation from 0 to 95° C, distilling the resulting chloral alcoholate from concentrated sulfuric acid, and removing HCl from the crude chloral by distilling it, in turn, over $CaCO_3$:[79]

$$2\,C_2H_5OH + 4Cl_2 \rightarrow CCl_3{-}CHOH(OC_2H_5) + 5HCl \qquad (20\text{-}3)$$

$$CCl_3{-}CHOH(OC_2H_5) \xrightarrow{\ H_2SO_4\ } CCl_3CHO + C_2H_5OH \qquad (20\text{-}4)$$

Essentially this technique, either batchwise or continuous, using anhydrous or aqueous alcohol[6] served the entirely modest requirements of the pharmaceutical and specialty organic chemical industry for chloral. The first step in the chlorination of alcohol is presumably its oxidation to acetaldehyde, with the successive reactions of enolization to an unsaturated alcohol, addition of chlorine to the double bond, and splitting out of HCl with the generation of a chlorinated aldehyde, repeated three times, accounting for the chlorination. The resulting chloral, combined with a molecule of ethanol as chloral alcoholate, is released by treatment with concentrated sulfuric acid.[79]

As DDT requirements spurred demand for chloral, the long batch chlorination time, low initial-reaction temperatures, relatively low efficiencies resulting from the simultaneous conversion of alcohol to ethyl chloride and ethyl hydrogen sulfate, initial hazards of ethyl hypochlorite formation, and the possibility of explosive vapor-phase reactions of chlorine with alcohol were recognized as significant shortcomings of the classic chloral synthesis, and a period of rapid development of the technology for its manufacture ensued. These innovations were variously concerned with improved methods for chlorinating ethanol, with the preparation of chloral from acetaldehyde, with the dehydration and purification of chloral, and with its stabilization.

Cass[12] found that by adding ethanol to the products of a previous chlorination the initial explosion hazards were decreased, higher initial reaction temperatures were possible, and the initial chlorination rate was increased, and thus introduced the concept of the continuous stagewise chlorination of alcohol. The reaction between ethanol and HCl yielding ethyl chloride was suppressed effectively by decreasing the concentration of HCl and the time of contact of HCl and ethanol by conducting the high-temperature portion of the chlorination with countercurrent flow in a packed column. A chloral yield of 84 percent on alcohol converted was reported.[46] The supplementary use of light to improve batch or continuous chlorination has been suggested.[13] Brothman proposed an elaborate operation based on the multistage, continuous, countercurrent, $FeCl_3$-catalyzed chlorination of alcohol with addition of water in the final stage to release alcohol by converting the alcoholate to hydrate.[10,7] Sonia chlorinated aqueous ethanol under moderate pressure to yield the hydrate.[69] Rosin improved the yield of chloral in the chlorination of 95 percent alcohol by adding water at the dichloroacetaldehyde stage, giving conditions favoring the subsequent formation of hydrate rather than alcoholate.[61] Coster,[18] by adding water at two successive stages, achieved yields of chloral hydrate of 84 percent based on alcohol.

Production of chloral by vapor-phase combination of formaldehyde and and carbon tetrachloride at 200 to 500° C has been claimed.[26] Yields of 50 to 60 percent have resulted from the anhydrous or aqueous[81] chlorination of ethylene chlorohydrin, and from the aqueous photochlorination of *B,B'*-dichlorodiethylether.[17] Yields of as high as 73 percent were obtained by reaction of trichloroethylene with aqueous hypochlorous acid at 5 to 20° C, with pentachloroethane being a significant by-product.[74]

Another approach to the manufacture of chloral was based on the idea that acetaldehyde, as a postulated intermediate in the chlorination of alcohol, might be utilized as a raw material, thus eliminating some of the problems associated with alcohol taxation and the consumption of chlorine as an oxidant, and, depending on relative prices and efficiencies, perhaps introducing raw material or processing economies.

Pinner first attempted the chlorination of paraldehyde, obtaining butyl chloral as the major product in an anhydrous system, and found that oxidation to acetic acid predominated in the neutral aqueous chlorination of acetaldehyde.[56]

Butyl chloral ($CH_3CHCl—CCl_2—CHO$) results from the condensation of a molecule of monochloroacetaldehyde with a molecule of acetaldehyde, followed by elimination of water and saturation by chlorine of the resulting double bond. To minimize the production of butyl chloral, and thus permit profitable use of acetaldehyde or paraldehyde as a raw material, systems have been developed to minimize the product of the concentrations or activities of the two reactants. Continuous stagewise chlorination provides a particularly effective means if the composition of the first stage is maintained at a chlorine-acetaldehyde ratio such that the concentrations of monochloroacetaldehyde and aldehyde are small. This continuous chlorination may be accomplished in the high aqueous dilution of 4 mols H_2O per mol of acetaldehyde,[15] in the presence of chlorinated acetaldehyde and acetic acid,[48,30] or in an anhydrous system using $SbCl_3$ as a catalyst and chlorinated acetaldehyde as the diluent.[54]

In the two-stage anhydrous chlorination of paraldehyde, about 8 mols of chlorine per mol of paraldehyde is fed to a first-stage reactor provided with a gas dispersing system and reflux condenser. Operation is at atmospheric reflux (*ca.* 82° C) with a holdup time in each stage of two to four days. Reaction is promoted by the periodic introduction of $SbCl_3$ at the rate of about 2 percent based on the paraldehyde fed. The first-stage chlorinator overflows to a second, similar stage, into which about 1 mol of Cl_2 is fed per mol of paraldehyde, the overflow from which is distilled at 100 mm for a 70 to 80 percent yield of chloral containing 1 percent dichloroacetaldehyde.[53,52]

In an aqueous chlorination, continued until the dichloroacetaldehyde level has been sufficiently reduced and accomplished without the production of undue quantities of acetic acid of butyl chloral, no purification step other than dehydration is required to yield chloral suitable for the production of acceptable DDT. It may be convenient, however, to dehydrate the chloral, especially if it is to be stored or shipped.

Crude chloral, as the hydrate alone or mixed with the alcoholate, is most conveniently dehydrated for DDT synthesis by treatment with sulfuric acid. Several procedures in use involve decantation, distillation, or both. Acid stronger than 85 percent H_2SO_4 decomposes chloral on prolonged refluxing, while drying is incomplete upon distillation from acid weaker than 70 percent.[11]

Treatment of crude chloral in solution in chlorobenzene with H_2SO_4 of final strength in the range of 70 to 85 percent removes most water-soluble impurities, while a second wash with acid of strength 93 to 100 percent renders it essentially anhydrous, thus avoiding distillation; this treatment can utilize some of the spent acid from the DDT synthesis.[85]

Distillation of chloral alcoholate-hydrate from mixture with an equal weight of concentrated sulfuric acid[13,61] to a pot temperature of 135° C yields anhydrous chloral.

The dehydration may be accomplished as part of the DDT condensation simply by charging hydrate or alcoholate to the DDT reactor[18] and providing an appropriately greater amount of dehydrating acid at that step, preferably in the form of a preliminary decantation wash with spent acid.

To avoid the problems associated with the use of sulfuric acid as a dehydrating agent several distillation processes have been devised. The application of azeotropy, using benzene or hydrocarbon fractions, has been described by Cave[14] and others,[38,28] in processes which depend upon an initial separation into a column overhead of the immiscible system entrainer plus aqueous HCl, and an underflow of anhydrous crude chloral and entrainer. A final distillation step separates the entrainer, as a column overhead from the chloral, which is then suitable for use, shipment, or storage.

Partial dehydration of chloral can be accomplished by rectification in a stream of anhydrous HCl.[49] Recovery of chloral from the alcoholate by reaction with acetic acid to yield ethyl acetate has been proposed.[45]

Anhydrous chloral is unstable on standing, exposure to light, or in contact with iron. On standing at ordinary temperatures, it is slowly converted to one of several solid polymeric forms, a tendency which may be countered by addition of a small amount of hydroquinone[11] or a long-chain aliphatic amine.[84] Suitably stabilized, anhydrous chloral is stored and shipped in unlined steel drums and tank cars.

APPLICATIONS

Specifications

Three grades of DDT were initially recognized by the War Production Board: (1) technical DDT, the current specifications for which are presented in Table 20-5, the commercial product; (2) purified or aerosol grade DDT, a partially refined product containing greater amounts of p,p'-DDT than the technical, and having a melting point of not less than 103° C; and (3) pure DDT, the highly purified p,p'-isomer, useful as a chemical, biological, or pharmacological standard.[32]

The technical DDT requirements of the Federal Government and of the World Health Organization are procured under specifications which are essentially equivalent in calling for a neutral product containing 70 percent p,p'-DDT and limited amounts of impurities. Technical DDT is sold as flake, lump, or powder, with the added specification of a white-to-cream color and freedom from extraneous matter or added modifying agents.

TABLE 20-5. SPECIFICATIONS FOR TECHNICAL DDT

		Federal[22]		World Health Organization[37]	
		Min.	Max.	Min.	Max.
Setting point	C	89	–	89	
Organic Cl	wt %	49	51	49.0	51.0
Hydrolyzable Cl	wt %	–	–	9.5	11.5
p,p'-Isomer content	wt %	–	–	70.0	–
Melting point separated p,p'-Isomer	C	–	–	104.0	–
pH by extraction		5.0	8.0	–	–
Acidity as H_2SO_4	wt %	–	–	–	0.3
Cyclohexanone insoluble	ml	–	0.2	–	–
$CFCl_3$ insoluble	ml	–	0.01	–	–
Acetone insoluble	wt %	–	–	–	1.0
Chloral hydrate	wt %	–	0.025	–	0.025
Water soluble	wt %	–	0.25	–	–
Water	wt %	–	–	–	1.0
Ash	wt %	–	0.5	–	–

Formulations

Technical DDT is not in a form suitable for direct utilization in insect control. Proper formulation and application can bring forth its versatility, its effectiveness, and its remarkable spectrum of activity, and enable full realization of its potential as an economic poison. DDT formulations may be in the form of solutions, emulsifiable concentrates, wettable powders, dusts, granules, aerosols, or insecticidal smokes.

The use of concentrates—solutions, wettable powders, and emulsifiable concentrates—results in savings on containers and transportation. These may be diluted locally to adaptable formulations which the grower can use by further dilution in a spray. Because of the equipment required for and problems associated with their dilution, mixing, and grinding, dust formulations for the grower are usually in final form for direct application.

DDT solutions can be made in concentrated form and diluted, for household applications and the like to about 5 percent strength in deodorized kerosene, with the addition of auxiliary solvents as required for low-temperature stability.

Emulsifiable concentrates contain from 25 to 50 percent DDT dissolved in xylene, alkylated naphthalenes, or other effective solvents. Mutually soluble wetting agents enable ready preparation of sprays.

Aerosol formulations combine the DDT solution with propellants. Other toxicants are frequently added to provide quick knockdown or kill in combination with DDT's long residual activity.

The low melting point and plastic nature of DDT effectively prevent its being ground alone finely enough to be effective as a dust. Even if this were possible, the required rates of application are so low that good coverage could not be obtained without dilution. At an application of 2 pounds per acre of 325-mesh DDT there would be, on an average, one particle per 0.4 square centimeter if the foliage area were equal to ground area. DDT is customarily ground with clay, talc, or other inert diluents in preparing dusts (1 to 10 percent) or dust concentrates (75 to 90 percent).

If a wettable powder is desired for preparation of sprays, small amounts of wetting agents and adjuvants are incorporated.

Granular formulations are prepared by spraying DDT solutions on sized particles of clay, corncobs, vermiculite, "Milorganite" or other inert substance. The solvent may be removed by drying or evaporation. Granular formulations have the desirable properties of low drift in aerial application, ability to work down through foliage, as in mosquito control, and to reach specific portions of the plant, as in corn for borer control.

The choice of a formulation is governed by the pest or pests to be controlled, the crops or areas to be treated, the equipment to be used, and the method, conditions, and practices of application, as well as by such economic factors as convenience, effectiveness, safety, and the costs of applying the formulation.

Insects Controlled

The following list of insects is presented to illustrate the broad spectrum of economic usefulness of DDT in its various commercial formulations. DDT poisons the nervous system of an insect on contact, as do pyrethrum, rotenone, and nicotine, and thus it is effective against sucking insects. Its activity as a stomach poison enables it to control chewing insects formerly controlled by such inorganic poisons as lead arsenate or Paris green. Though it has no direct ovicidal action, the residual action of DDT is so long-lasting that insects are killed by contact as they hatch. For each species cited, an abundant discussion of crop, season, climate, variety, formulation, dosage, application, and auxiliary information is required for guidance in safe, effective and economic control. The art and science of selecting, formulating, and applying insecticides for the control of the many plant pests that trouble mankind has resulted in a voluminous literature.

Around the home, DDT is useful for the control of flies, lice, roaches, bedbugs, mosquitoes, ticks, carpet beetle, clothes moth, ants, termites, silverfish, flour beetle, and weevil. In home and truck gardens it is effective against the Japanese beetle, cutworm, wireworm, thrips, pea aphids, pepper weevil, cabbage looper, rose chafer, army worm, Colorado potato beetle, corn ear worm, and many caterpillars, leafhoppers and other plant bugs. Fruit crops are protected by DDT against codling moth, tent caterpillar, apple maggot, oriental fruit moth, plum curculio, and casebearers. On farms, DDT is effective against the bollworm, fleahopper, pink bollworm, and, with BHC, the boll weevil on cotton; fall armyworm, corn borer, and earworm on corn; and the Lygus bug and weevil on alfalfa. Forest and ornamental trees are protected by DDT against the spruce budworm, elm leaf beetle, and gypsy moth. DDT has been used in spraying granaries for control of weevil, though residue problems are involved, in dusting seed grain for protection against beetle and weevil, and in control of mosquitoes in rice fields and swamps.[73,47]

The median lethal dose (LD_{50}) for DDT for insects which it controls lies in the range of 0.2 to 5.0 micrograms per gram[40]—considerably less than the corresponding oral dosages for various laboratory animals but of the same order of magnitude as the intravenous values. Normal dusting and spraying application of 0.2 to 2 pounds of DDT per acre provide, nevertheless, great excesses over the theoretical minima required based on the existing insect population. Most field crops receive from 1 or 2 to as much as 5 lb of DDT per acre per year, although for cotton and for European corn borer control as much as 10 lb may be applied. Truck crops are treated with 2 to 5 lb of DDT per acre per crop. Some orchards receive as much as 50 to 60 lb per acre each year.[66]

Limitations. Two factors have tended to limit the applicability of DDT for the control of insect pests: the development in some species of hereditary resistance to its action, and its toxicity to other forms of life.

The development of insect resistance to DDT was first noted in flies in Italy about four years after its first widespread use in Naples; resistant strains of mosquitoes were observed at about the same time. Highly resistant strains of flies have since been grown in the laboratory by continued exposure of both larvae and adults to sublethal dosages for 9 to 18 generations.[72] Similar resistance has been observed among other species of insects to DDT as well as to other chlorinated insecticides. This is combated by control programs involving application of a toxicant with an entirely different mode of action frequently enough to prevent the successful evolution of resistant strains.

Technical DDT retards seriously the growth of squash, a highly sensitive crop, when applied as a dust, although the pure p,p'-isomer is essentially without harmful effect. Oil-based formulations are mildly toxic to young tomato foliage.[66] DDT in the soil has an effect on beans and squash which is strongly dependent upon the soil type, the tolerance being much greater in muck than in sandy soils.[73] DDT persists in soils for several years, decomposing at about 5 percent per year; hence, in orchards or other places where it is used regularly and at heavy dosage rates, a sufficient residue may accumulate in the soil in 3 to 5 years to be injurious to such cover crops as Abruzzi rye, tomatoes, squash, and snap beans.[73,47]

The toxicity of DDT to warm-blooded animals imposes even more significant limitations upon its general applicability for insect control. DDT has a remarkable tendency to accumulate in the fat of animals ingesting it, and to appear in their milk. Consequently, a residue tolerance of 7 ppm has been established on many fruits and vegetables,[47] and on the fat of meat from cattle, sheep and hogs. It is not recommended for fly control in dairies or for treatment of forage or livestock feed crops which could result in its presence in milk and dairy products.

Insecticidal Toxicity

DDT may enter an insect through the cuticle, foot structure, or gut. The rates of absorption into and penetration of the cuticle are high, in most species, with the result that the topical LD_{50} is of the same magnitude as that for injection. The pronounced insecticidal contact action of DDT results from absorption through the feet as insects come into contact with treated surfaces. Hairs and pad structures on the feet of some may increase their effectiveness in absorbing DDT. Although DDT is an effective stomach poison for many insects, resistance in certain grasshoppers has been shown to be associated with lack of absorption through the walls of the gut.[40]

DDT is a nerve poison, and its killing action is comparatively slow. The period from exposure to death can sometimes be as long as 30 hours. The nature, timing, and sequence of symptoms vary with the species affected, but the first noticeable manifestation, following a latent period, is usually a general tremulousness of the body and appendages, the "DDT jitters," followed by uncoordinated walking movements, extreme reactions to insignificant stimuli, repeated rolling over or falling on the back, and continued spasmodic muscular activity diminishing with advancing paralysis until death.[8]

Metcalf[40] has reviewed a number of the theories which have been advanced in explanation of the action of DDT upon insects. Its effects may be ascribable to structural characteristics such as the di-*p*-chlorophenyl-methylene group, or the trichloromethyl group, or the combination of the two. The combination of penetration of insect tissues followed by dehydrochlorination has been suggested. Perhaps there is some relation between the shape and dimensions of the molecule and the dimensions of enzymes important in the insects' nervous systems.

Mammalian Toxicity

DDT can enter man or animals by absorption through the skin, the lungs, or the walls of the digestive system. Since it is almost insoluble in water, the rate at which absorption takes place is increased if the DDT is in solution, or, in the digestive tract, if accompanied by fats. Apparently the first step in its metabolism is dehydrochlorination to bis-(*p*-chlorophenyl)-2,2-dichloroethylene (DDE). This is subsequently converted to di-(*p*-chlorophenyl)-acetic acid (DDA), in which form DDT is slowly eliminated from the system. Addition of DDT to the diet of lactating animals results in its appearance in the milk. It also appears in eggs when added to poultry feed. Since DDT entering the system accumulates rapidly in the body fat, largely in the form of DDE, it is problematical whether that found in milk and eggs is unchanged DDT or DDE. This complicates the appraisal of the toxicity of such contaminated foods.[82]

DDT poisoning in laboratory animals shows up chiefly in its effect on the nervous system; following a latent period of several hours they become apprehensive, excited, develop locomotor disturbances and tremors, and finally convulsions and death. DDT is not a skin irritant, though solvents with which it is associated may sometimes be, nor has it shown carcinogenic activity. The effects on the nerve structure, circulation, blood, internal organs, metabolism, and enzymes have been studied.[82]

The single-dose oral toxicity of DDT, as measured by the LD_{50}, ranges from 150 to 800 mg/kg for various laboratory animals, with values in the lower part of the range being associated with preparations in edible oils. For cutaneous application in various solvents the range is 300 to 3000. For intravenous injection observations range from an LD_{100} of 40 to 60 for small animals to an LD_{50} of 50 to 60 for dogs and monkeys.[82]

TABLE 20-6. STRUCTURES OF SOME COMMERCIAL DDT ANALOGUES

$$X-\text{C}_6\text{H}_4-\underset{\underset{Y}{|}}{\overset{\overset{Z}{|}}{C}}-\text{C}_6\text{H}_4-X$$

Common Names	Chemical Name	Substituent X	Y	Z
Methoxychlor DMDT "Marlate"	2,2-bis-(p-methoxyphenyl)-1,1,1-trichloroethane	CH_3O—	—CCl_3	—H
DDD TDE "Rhothane"	2,2,-bis-(p-chlorophenyl)-1,1-dichloroethane	Cl—	—$CHCl_2$	—H
DFDT "Gix"	2,2-bis-(p-fluorophenyl)-1,1,1-trichloroethane	F—	—CCl_3	—H
DMC "Dimite"	1,1-bis-(p-chlorophenyl)-ethanol-2	Cl—	—CH_3	—OH
"Kelthane"	1,1-bis-(p-chlorophenyl)-2,2,2-trichloroethanol	Cl—	—CCl_3	—OH
"Perthane" Q-137	2,2-bis-(p-ethylphenyl)-1,1-dichloroethane	C_2H_5—	—$CHCl_2$	—H
"Bulan" "CS-674A"	1,1-bis-(p-chlorophenyl)-2-nitrobutane	Cl—	NO_2 $\|$ —CH—C_2H_5	—H
"Prolan" "CS-645A"	1,1-bis-(p-chlorophenyl)-2-nitropropane	Cl—	NO_2 $\|$ —CH—CH_3	—H
"Dilan" "CS-708"	Mixture of 1 part "Prolan" and 2 parts Bulan			

For repeated oral doses, the toxicity is influenced by the absorption. Only a small fraction of dry DDT in capsules is absorbed, while incorporation in the diet or with fats or oils increases absorption. The toxic effects of DDT with continued administration in food increase with its concentration in the diet and with the duration of the exposure. Reported data on the effects of various dosage levels in the diet, but without direct relation of DDT intake and animal weight, range from deaths to mice in less than a week at 125 to 500 ppm to toxic effects and some fatalities in rats at 800 and 12,000 ppm.[82]

DDT is quite toxic to fish; the various species differ in their susceptibility, with LD_{50} concentrations ranging from 0.01 to 0.2 ppm.

The hazard from inhalation of DDT in the form of mists, dusts or sprays is small. Repeated cutaneous application of DDT in solution can give rise to toxic symptoms.[82] Users are normally warned to avoid inhalation and skin contact with solutions.

Cases have been reported of human poisoning from eating food mistakenly prepared with DDT, from ingestion of insecticide solutions, and from application of DDT formulations. When allowances are made for the toxicity of the solvent and the psychic and environmental conditions of the human subjects, it will be realized that the toxicity hazard of DDT by inhalation or skin absorption is low indeed.[9]

Analogues

In their original publication, Lauger, Martin and Müller reported the performance of a large number of compounds, the structures of several of which were analogous to that of DDT.[37] Their subsequent work, and that of others, has led to the introduction and commercialization of the related insecticides illustrated in Table 20-6.

Bibliography

1. Anon., *Chem. Ind.,* **57,** (2), 346 (1945).
2. Anon., Chloral Technical Bulletin No. 206, p. 3, Westvaco Chemical Division, Food Machinery & Chemical Corporation, New York, 1953.
3. Bartels, C. R., *Chem. Eng. Prog.,* **54** (3), 59–61 (1958).
4. Beilstein, F. K., "Handbuch der Organischen Chemie," 4th ed., Vol. 1, pp. 616–24; 328–30, Vol. 2, pp. 677–81, Vol. 3, pp. 2663–72, Berlin, Springer, 1959.
5. *Ibid.,* Vol. 5, p. 606.
6. Besson, J. A., U. S. Patent 774,151 (November 8, 1904).
7. Brothman, A., U. S. Patent 2,478,741 (August 9, 1949).
8. Brown, A. W. A., "Insect Control by Chemicals," pp. 312–318, New York, John Wiley & Sons, Inc., 1951.
9. Brown, A. W. A., *ibid.,* p. 517.
10. Callaham, J. R., *Chem. Met. Eng.,* **51** (10), 109–114 (1944).
11. Cannon, M. R., and Mosher, H. S., DDT Report 44–9, August 2, 1944, British Patent 103,633.
12. Cass, O. W., U. S. Patent 2,443,183 (June 15, 1948).
13. Cass, O. W., U. S. Patent 2,478,152 (August 2, 1949).
14. Cave, W. T., Alexander, J. H., and McCoubrey, J. A., U. S. Patent 2,606,864 (August 12, 1952).
15. Cave, W. T., and Haddeland, G. E., U. S. Patent 2,553,934 (May 15, 1951).
16. Chin-Yee, H. R., and Winkler, C. A., *Can. J. Research,* **26B,** 379-86 (1948).
17. Churchill, J. W., and Schaeffer, B. B., U. S. Patent 2,680,092 (June 1, 1954).
18. Coster, L. R., U. S. Patent 2,669,585 (February 16, 1954).
19. Cristol, S. J., *J. Am. Chem. Soc.,* **67,** 1494–98 (1945).
20. Dickinson, C. L., U. S. Patent 2,808,236 (October 1, 1957).

21. "Encyclopedia of Chemical Technology," ed. by Kirk & Othmer, Vol. 3, pp. 660–4, New York, The Interscience Encyclopedia, Inc., 1949.

22. Federal Specification G-I-514, U. S. Government Printing Office, (July 10, 1957).

23. Fleck, E. E., and Haller, H. L., *Ind. Eng. Chem.*, **37,** 403 (1945).

24. Fleck, E. E., and Preston, R. K., *Soap Sanit. Chemicals,* **21** (5), 111–3 (1945).

25. Ford, D. L., Australian Patent 155,276 (February 17, 1954).

26. Frolich, P. K., and Wiezevich, P. J., U. S. Patent 2,042,303 (May 26, 1936).

27. Gates, C. W., and Woods, W. P., U. S. Patent 2,447,326 (August 17, 1948).

28. Goebel, M. T., Langston, J. W., Loder, D. J., and Siefen, H. T., U. S. Patent 2,559,247 (July 3, 1951).

29. Geigy, J. R., A. G., Swiss Patent 308,565 (October 1, 1955).

30. Gilbert, E. E., U. S. Patent 2,697,120 (December 14, 1954).

31. Gunther, F. A., *J. Am. Chem. Soc.,* **67,** 189 (1945).

32. Haller, H. L., Bartlett, P. D., Drake, N. L., Newman, M. S., Cristol, S. J., Eaker, C. M., Hayes, R. A., Kilmer, G. W., Magerlein, B., Mueller, G. P., Schneider, A., and Wheatley, W., *J. Am. Chem. Soc.,* **67,** 1591 (1945).

33. Hindes, L. W., Curtis, H. S., and Siemoneit, J. R., U. S. Patent 2,902,719 (September 8, 1959).

34. Jones, H. A., Fluno, H. J., and McCollough, G. T., *Soap Sanit. Chemicals,* **21** (11), 110 (1945).

35. Kuhn, W., and Massini, P., *Helv. Chim. Acta.,* **32,** 1530 (1949); *Chem. Abs.* **44,** 888b (1950).

36. Kulka, M., U. S. Patent 2,447,476 (August 17, 1948).

37. Lauger, P., Martin, H., and Müller, P., *Helv. Chim. Acta,* **27,** 892 (1944).

38. Mahoney, J. F., and Pierson, E., U. S. Patent 2,584,036 (January 29, 1952).

39. Manske, R. H. F., Trickey, E. G., and Myers, G. S., U. S. Patent 2,511,166 (June 13, 1950).

40. Metcalf, R. L., "Organic Insecticides," pp. 127–189, Interscience Publishers, New York (1955).

41. Miller, G. A., and Lasco, R. H., U. S. Patent 2,932,672 (April 12, 1960).

42. Mosher, H. S., Cannon, M. R., Conroy, E. A., Van Strien, R. E., and Spalding, D. P., *Ind. Eng. Chem.,* **38,** 916 (1946).

43. Müller, P., *Helv. Chim. Acta,* **29,** 1560 (1946).

44. Müller, P., U. S. Patent 2,329,074 (September 7, 1943); Reissue, U. S. Patent 22,700 (December 4, 1945).

45. N. V. de Bataafsche Petroleum Maatschappij, British Patent 629,699 (September 26, 1949).

46. N. V. de Bataafsche Petroleum Maatschappij, British Patent 644,916 (October 18, 1950).

47. De Ong, Elmer Ralph, "Chemistry and Uses of Pesticides," Ed. 2, pp. 195–201, New York, Reinhold Publication Corporation, 1956.

48. Otto, J. A., and Veldhuis, B., U. S. Patent 2,702,303 (February 15, 1955).

49. Park, R. S., and Stair, R. H., U. S. Patent 2,746,912 (May 22, 1956).

50. Peaker, P., *Soap Sanit. Chemicals,* **20** (6), 133; (7), 105; (10), 127 (1944).

51. Pennsylvania Salt Mfg. Company: Australian Patent Application 6108 (February 27, 1948).

52. Pianfetti, J. A., and Porter, D. J., U. S. Patent 2,615,048 (October 21, 1952).

53. Pianfetti, J. A., and Porter, D. J., U. S. Patent 2,615,049 (October 21, 1952).

54. Pianfetti, J. A., and Porter, D. J., U. S. Patent 2,863,924 (December 9, 1958).

55. Piganiol, P., "Traité de Chimie Organique,"Vol. 7, pp. 272–283, Paris, Masson, 1950.

56. Pinner, *Ann. Chem.*, **179**, 21, 36 (1875).

57. Plowden, J. G.: "History of DDT Insecticides," Preprint 8, 42nd National AIChE Meeting, Atlanta, Georgia, February 1960.

58. Rathmell, R. K., and Zevnik, F. C., U. S. Patent 2,689,871 (September 21, 1954).

59. Riemschneider, R., and Cohnen, W., *Chem. Ber.*, **89**, 2702 (1956).

60. Roeder, K., and Weiant, E., *Science,* **103**, 304 (1946).

61. Rosin, J., U. S. Patent 2,616,929 (November 4, 1952).

62. Rueggeberg, W. H. C., and Cook, W. A., U. S. Patent 2,500,961 (March 21, 1950).

63. Rueggeberg, W. H. C., Catlin, W. E., and Cook, W. A., U. S. Patent 2,554,269 (May 22, 1951).

64. Rueggeberg, W. H. C., and Torrans, D. J., *Ind. Eng. Chem.*, **38**, 211 (1946).

65. Searle, N. E., and Flenner, A. L., U. S. Patent 2,518,191 (August 8, 1950).

66. Shepard, Harold Henry, "The Chemistry & Action of Insecticides," pp. 303–314, New York, McGraw-Hill, 1951.

67. Smith, A. C., and Staubly, J. L., U. S. Patent 2,455,388 (December 7, 1948).

68. Societe d'Electro-Chimie, d'Electro-Metallurgie et des Acieries Electriques d'Ugine, French Patent 1,023,611 (March 20, 1953); *Chem. Abs.,* **52**, 5469e (1958).

69. Sonia, J. A., U. S. Patent 2,558,319 (June 26, 1951).

70. Stange, H., U. S. Patent 2,763,698 (September 18, 1956).

71. Stefferud, Alfred, ed., "Insects, The Yearbook of Agriculture," p. 210, United States Department of Agriculture, U. S. Government Printing Office, Washington, D. C. (1952).

72. Stefferud, Alfred, ed., *Ibid.,* pp. 320–330.

73. Stefferud, Alfred, ed., *Ibid.,* pp. 634–637.

74. Stevens, H. C., Kaman, A. J., and Sellers, J. W., U. S. Patent 2,759,978 (August 21, 1956).

75. Matsuo, N., Suga, K., and Kato, T., Japanese Patent 176704 (September 20, 1948); *Chem Abs.*, **45**, 6661b (1951).

76. Sveda, M., U. S. Patent 2,463,653 (March 8, 1949).

77. Synthetic Organic Chemicals, U. S. Tariff Commission, U. S. Production and Sales (1943–1958).

78. Trocki, J. R., Diamond Alkali Company, Private Communication, April 20, 1961.

79. Ullmann, Fritz, "Enzyklopadie der technischen Chemie," 2nd Ed., Vol. 3, pp. 232–235, Berlin, Urban and Schwarzenberg, 1929.

80. U. S. Rubber Company, Australian Patent 144,623 (January 4, 1952).

81. Vaughn, T. H., U. S. Patent 2,456,545 (December 14, 1948).

82. Von Oettingen, W. F., "The Halogenated Hydrocarbons, Toxicity and Potential Dangers," pp. 335–379, Publication No. 414, U. S. Department of Health, Education and Welfare, U. S. Government Printing Office (1955).

83. Whichello, W. E., Australian Patent 166,780 (February 6, 1956); *Chem. Abs.*, **52**, 17604g (1958).

84. Williams, D., Thomas, R. M., and Haines, G. S., U. S. Patent 2,504,953 (April 25, 1950).

85. Wohlers, H. C., and Lentz, C. W., U. S. Patent 2,768,173 (October 23, 1956).

86. Wohlers, H. C., Pauling, L. E., and Grover, F. N., Jr., U. S. Patent 2,733,264 (January 31, 1956).

87. Specifications for Pesticides, World Health Organization, Genèva, Spec. No. (WHO/SIT/1.R1) September 11, 1954.

88. Zeidler, O., *Ber. deut. chem. Ges.*, **7**, 1180 (1874).

21. PREPARATION AND PROPERTIES OF BENZENE HEXACHLORIDE

DAVID S. ROSENBERG

Hooker Chemical Corporation

HISTORY

Benzene hexachloride (BHC) or 1,2,3,4,5,6-hexachlorocyclohexane (HCH) is the fully saturated product formed by the addition reaction of benzene with chlorine. It was first prepared in 1825 by Michael Faraday.[53,54] When Faraday isolated benzene by compression of oil gas he then proceeded to observe its reaction with chlorine in sunlight. The white crystalline alcohol-soluble solid which he obtained was undoubtedly benzene hexachloride, and his method of preparation, with some refinements, is the same as that used in the principal commercial processes of 1960.

In 1833 Mitscherlich[142] distilled benzene from a mixture of lime with benzoic acid, calling his distillate "benzin." The confusion in terms which he thereby initiated still persists, with "benzin" or "benzine," now referring to an aliphatic petroleum distillate and "benzol" to a commercial product which is predominantly the pure compound, benzene. Minor impurities in benzol are responsible for some of the undesirable properties of crude BHC. Mitscherlich also chlorinated his distillate and showed further that reaction of the solid product with alkalies produced trichlorobenzenes. Today the waste isomers of benzene hexachloride are frequently converted to trichlorobenzenes by the primary producers.

The composition of benzene hexachloride was first established by Pelicot[157] and Lauret[107] in 1836, but the first concept of the structural forms had to wait for Kekulé to propose the ring structure of benzene in 1865 and for Sachse, in 1890, to propose the spatial modifications of cyclohexane based on the tetrahedral carbon atom. In 1884 Meunier[129] isolated the alpha and beta isomers from the crude mixture prepared by feeding chlorine into boiling benzene. In 1891 Matthews[122,123] prepared benzene hexachloride by chlorination of benzene in contact with sodium hypochlorite solution and separated the alpha from the beta isomer by steam distillation. Matthews assumed a planar configuration for benzene hexachloride, since it was derived from benzene. He conceived of the structure as hexachloro-hexamethylene and thus considered that only two configurations could result, a *trans* form, which he called alpha, and a *cis* form, which he called beta. Although his deductions were erroneous, they set a pattern for equally positive but fallacious conclusions drawn from subsequent studies in this field. The status of current knowledge of isomer structure is discussed later in this chapter.

The next major contribution was made by Van der Linden[113] in 1912. He prepared crude benzene hexachloride by a procedure similar to that of Matthews and observed that, in addition to a solid product, he obtained a dense, oily material which was quite resistant to further chlorination. This he combined with the viscous residues from the alpha and beta isomer separation steps and extracted them with ether. From this extract a solid phase slowly formed. These crystals were digested with alcohol and recrystallized from ether to obtain rhombic crystals of the gamma isomer. The work of Van der Linden was the basis for the first commercial process of preparing pure gamma isomer. Using a mixture of carbon tetrachloride and alcohol for extraction and recrystallization of the material remaining from separation of the gamma isomer Van der Linden also isolated the delta isomer. He concluded that the isomers resulted from distribution of the hydrogen and chlorine atoms either above or below the plane of the surface containing the six carbon atoms. With this structure eight isomers are possible, rather than the two of Matthews.

Since the four isomers identified by Van der Linden represent about 95 percent of the total product normally obtained by addition chlorination of benzene, his failure to isolate other isomers is understandable. After a lapse of 35 years Kauer,[91] Du Vall, and Alquist in 1947 first detected the epsilon isomer by infrared spectroscopy. By use of this analytical technique they were able to isolate the pure isomer by a series of extractions and crystallizations. The isomer normally represents from 2 to 3 percent of the chlorination product.

In 1953 Riemschneider[167,168] reported the isolation of the zeta isomer by partition chromatography. He obtained 1.5 grams from 1050 grams of a delta-rich concentrate. The isomer normally represents less than 0.1 percent of the chlorinated product. Kolka[96,98] and his co-workers isolated the same isomer during this period. However, since an isomer of 1,1,2,4,4,5-hexachlorocyclohexane had earlier been reported erroneously as the zeta isomer, Kolka called his isomer the eta isomer, and this terminology is now used by other workers. The theta isomer was also prepared by Kolka and his associates, but the iota isomer has not yet been isolated.

Studies of the preparation and properties of benzene hexachloride were mainly of academic interest until the discovery of the insecticidal properties of the gamma isomer. Credit has been claimed by United States, French and British investigators. In a process patent[219] issued to Harry Bender of Great Western Electrochemical Company in 1935 the use of BHC as an insecticide is indicated. In November 1943, The French investigator Lebas[165] described the effectiveness of BHC in killing larvae of clothes moths. The notice attracted little attention, however. The first major publication was the Hurter memorial lecture delivered by R. E. Slade,[183] the research director of Imperial Chemical Industries, on March 8, 1945. In 1942 investigators of the ICI entomological laboratories discovered that benzene hexachloride would kill the turnip flea beetle. The four principal isomers were isolated and tested for insecticidal activity. The gamma isomer was extremely toxic to weevils; the other three isomers were relatively inactive.

Slade's comprehensive and authoritative report was a major milestone in the history of benzene hexachloride. In the preceding ten year period there was only a handful of issued patents pertaining to this material, starting with the patent[219] granted to Harry Bender in 1935 for reaction of benzene with liquid chlorine. In the subsequent ten-year period the number of patents mounted annually to reach a peak in 1956, with 47 patents issued in the United States alone. The number of patents has since declined rapidly, with only three U. S. patents awarded in the first half of 1960.

Commercial production was begun by Hooker Chemical Corporation in the United States in early 1946. New producers were announced almost monthly until, by 1951, sixteen U. S. manufacturers released a flood of 117 million pounds gross of benzene hexachloride in various forms, equivalent to about 18 million pounds of gamma isomer. The market was able to absorb only 60 percent of this output and production in subsequent years declined steadily to reach a fairly stable level of about $4\frac{1}{2}$ million pounds a year of the gamma isomer by 1960. This output is supplied by five domestic producers with a reported productive capacity about twice that of current market requirements.

The factors responsible for the decline of BHC from its preeminent position in the early fifties are difficult to assess. Most important is probably the cloud that has hung over the entire group of chlorinated insecticides —insect resistance. Many users of chlorinated hydrocarbons, such as cotton growers, have switched to phosphate insecticides when they received reports of resistance of cotton pests to chlorinated materials. Once the picture can be clarified and order restored, there will undoubtedly remain a firm and permanent position to be filled by benzene hexachloride. The factor of insect resistance now appears to have been overstressed, and the unique qualities of the gamma isomer will undoubtedly insure its use for many years in the future.

STRUCTURE AND MECHANISM OF ADDITION CHLORINATION

Up to the time of Slade's study only the structure of the beta isomer had been determined with any certainty, although three other isomers had been isolated and described. In 1928 Dickinson and Bilicke[50] published results of an X-ray examination showing the beta isomer to be centro-symmetrical, the valency bonds being disposed in tetrahedral fashion. The only possible structure was the one assigned (Figure 21-1) which later workers have verified. Although the cyclohexane ring can be written in two conformations which were postulated by Sachse in 1890 as the "chair" and "boat" forms, electron-diffraction studies of cyclohexane and its derivatives have established that the predominant form is always the chair form. All isomers of hexachlorocyclohexane prepared by chlorination of benzene may be considered to have this configuration, with three carbon atoms in one plane and three in a second, parallel plane.

Substituents on the cyclohexane ring may be of two types. In one type the valences are parallel to the axis of symmetry of the molecule. Three valences are directed up from one of the two parallel planes and three are directed down. These bonds are generally termed "axial" or "a" bonds. The term "polar" bonds has also been widely used. The other six valences are nearly in the planes of the two paralled planes, although they alternate up and down around the ring. These bonds are known as equatorial, or "e" bonds. Each carbon atom in the ring contains one axial and one equatorial bond. Each of the six chlorine atoms in hexachloro-cyclohexane may be in either the equatorial or axial position. Steric interaction of a large substituent such as chlorine favors the equatorial position. There are sixteen possible configurations of benzene hexachloride. The ability for a ring to flip,[24,70,72,213] all carbon atoms shifting from one plane to the other, reduces the possibilities to nine—two of which are enantiomorphs or mirror images.

Slade tentatively assigned structures to the four isomers known in 1945.[183] His assumption that the insecticidal value of the gamma isomer is due to its structural similarity to meso-inositol led to interchanging of the gamma and delta structures. Later studies with Raman spectra, electron and X-ray diffraction, chemical reaction, and polar moments have established the structures of six of the isomers that have been isolated.[18,21,36,112,144,174] The alpha isomer has the two enantiomorphic forms. Seven isomers, representing eight of the nine possible configurations have been prepared. The structural configuration of the remaining isomer requiring completely *cis* addition to the ring makes its preparation extremely unlikely.

There is a confusion in terminology of the isomers isolated subsequent to the epsilon isomer. A compound initially reported as the zeta isomer was

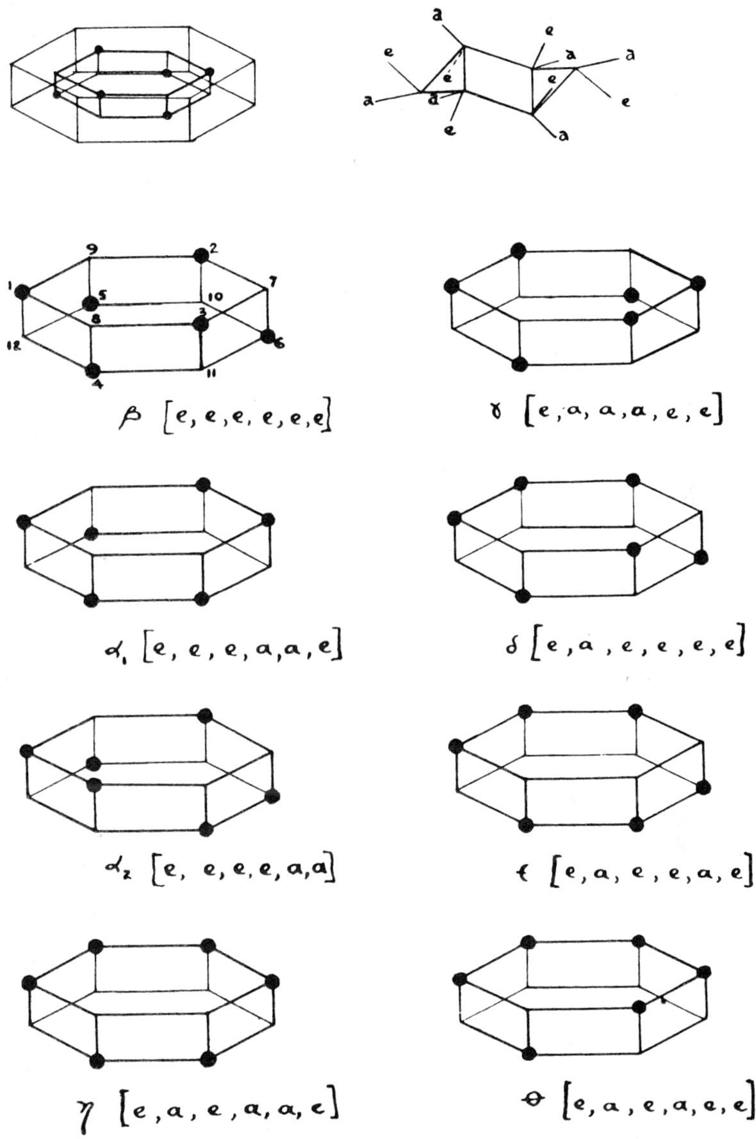

Figure 21-1. Configurations of the isomers of 1,2,3,4,5,6-hexachloro-
cyclohexane.

later identified as an isomer of 1,1,2,4,4,5-hexachlorocyclohexane. Riemschneider,[167, 168] then isolated the zeta isomer from an extract of crude benzene hexachloride. In this same period Kolka[36,96,97,98] and his co-workers
prepared the same isomer by chlorination of benzene tetrachloride and

designated it as the eta isomer.   No studies are reported for the theta isomer, but the probable configuration is indicated in Figure 21-1 which represents all seven isomers in a form more pictorial than the standard chair or star representation.   The numbering on the beta isomer is from the star representation.

Determination of isomer structure is now fairly well completed but the mechanism of chlorine addition is still a matter of considerable speculation.   Chlorine can react with benzene (or partially substituted benzene) in one of two ways:

(1) Electrophilic substitution can take place in the nucleus proceeding by a polar mechanism via positive chlorine ions or chlorine molecules. The reaction is promoted by electrophilic reagents, or Lewis acid catalysts, such as $FeCl_3$.

(2) Chlorine addition can take place by a free radical mechanism. The reaction normally involves the addition of six chlorine atoms to the aromatic reactant, forming a cyclohexane derivative.

By selection of the proper reaction iniator, one or the other type of reaction can be promoted almost exclusively.   Under certain conditions both reactions take place, so that benzene hexachloride prepared by a free-radical mechanism generally contains as well some substituted reaction products, such as heptachloro- and octachlorocyclohexanes.   This factor has been overlooked by some investigators and has greatly complicated the task of isolating and identifying the minor components in the crude product.

The first reported study of the mechanism of the addition reaction was made by Hubert N. Alyea[6] in 1930.   His method was to put a known quantity of Radon into a solution of benzene saturated with chlorine.   He then set the solution in the dark with a blank.   He calculated the number of molecules of $C_6H_6Cl_6$ produced per ion formed (alpha particles emitted x number of ions produced per alpha particle) and obtained chain lengths of the order of 720.   The effects of gamma radiation were neglected and his assumption that the chlorine had to be ionized was incorrect, but his work represented a positive contribution to this little explored field.   When the addition reaction is initiated by production of free radicals of chlorine, the reaction normally proceeds to addition of six atoms of chlorine. However, Kolka and his. co-workers managed to stop the reaction at the tetrachloride stage by adding iodine to the reaction solution.   They separated the benzene tetrachloride isomers and chlorinated them separately to the hexachlorocyclohexanes.   Their results are presented in Tables 21-1 and 21-2.

Using this information, Yosheija Kanda[90] subsequently derived probability ratios for *cis* and *trans* approaches by chlorine in chlorination of

TABLE 21-1. THEORETICAL RELATIONSHIP BETWEEN BTC AND HCH

| | BHC Isomers Expected | | |
| | *Trans* Addition | | *Cis* Addition |
BTC Isomer	ee	aa	ea or ae
α (eeaa)	α	$1/\alpha \longrightarrow \alpha$	η γ
β (aeea and eaae)	ε or α	$1/\alpha \longrightarrow \alpha$	η
		$1/\varepsilon \longrightarrow \varepsilon$	
γ (eeea)	δ	γ	θ α
δ (eeee)	β	α	δ
ε (eeae)	δ	η	ε θ
ζ* (eaea)	θ	$1/\theta \longrightarrow \theta$	η i*

*Never isolated

the benzene ring, after assuming that the reaction proceeds stepwise with formation of benzene dichloride, benzene tetrachloride and finally benzene hexachloride. Formation of the six possible isomers of benzene tetrachloride by this mechanism is shown in Figure 21-2. Assuming bi-radical addition first of Cl^+ and then of Cl^- to the reacting species and certain empirical ratios for *cis-trans* addition in the formation of the intermediate, he obtains results that are in reasonable agreement with reported isomer distributions. The major discrepancy is the 5.4 percent eta isomer that should be present by his mechanism.

It would appear that other mechanisms suit equally well or better the limited data available. Other more plausible mechanisms are suggested and shown schematically in Figures 21-3 and 21-4. The reaction is initiated by addition of a chlorine atom to a benzene molecule to form a free radical.

Since the free-radical concentration is very low as compared to the concentration of molecular chlorine, bi-radical addition as postulated by Kanda is very unlikely. The second reaction step is most probably the reaction of $C_6H_6Cl \cdot$ with molecular chlorine. This reaction may proceed in either of two ways:

$$\text{(a)} \quad C_6H_6Cl° + Cl_2 \rightarrow C_6H_6Cl_2 + Cl° \qquad (21\text{-}1)$$

TABLE 21-2. PRODUCT DISTRIBUTION FROM BTC CHLORINATION

BTC Isomers	α	β	γ	δ	ε	Other
α	89		11			
β	81				19	1
γ	39		40	11		1
δ	70	13		17		
ε				33	40	27

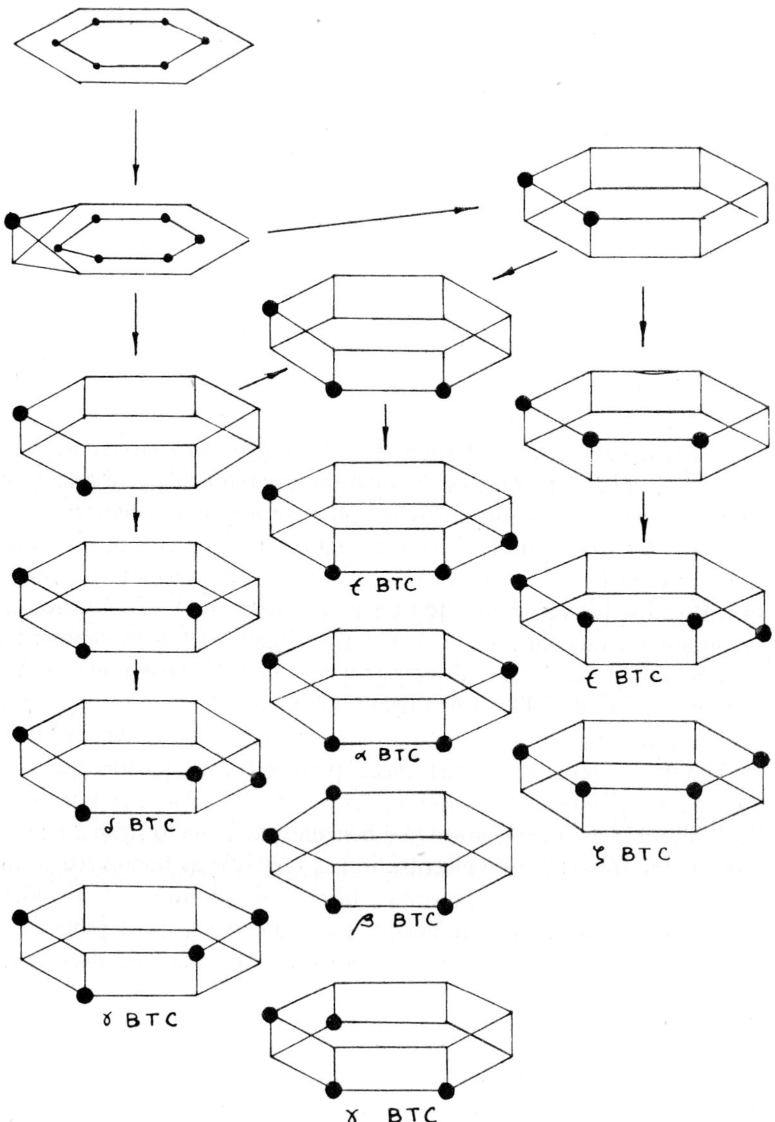

Figure 21-2. Formation of the tetrachloride isomers by ion mechanism.

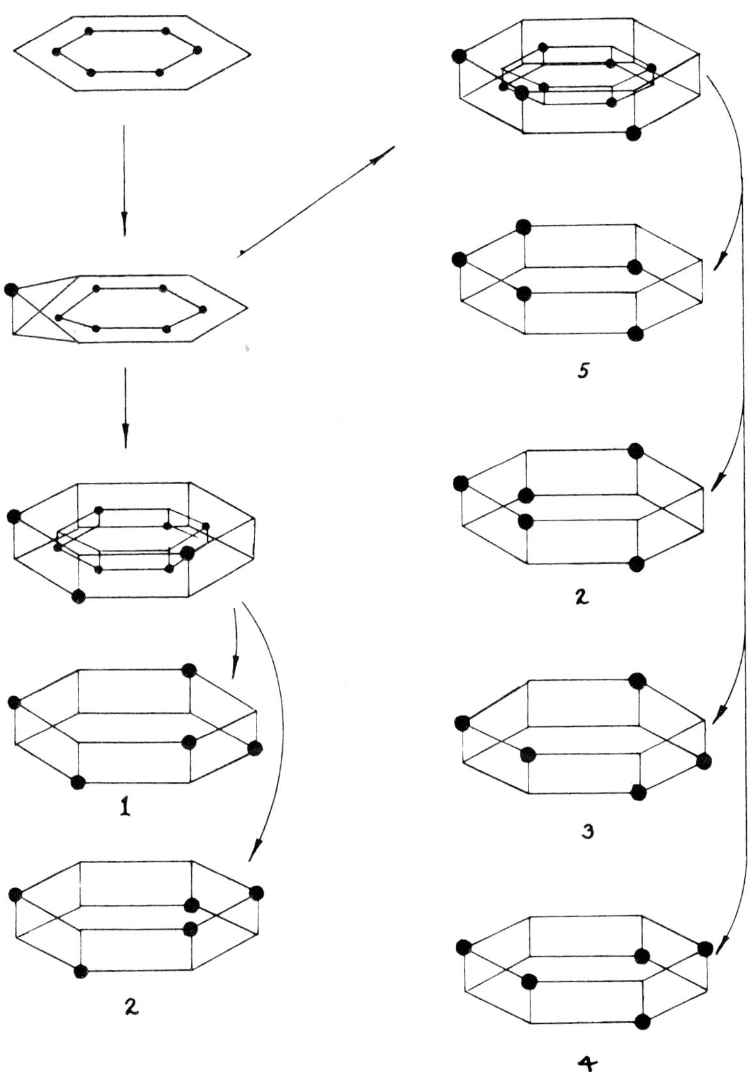

Figure 21-3. Mechanism of molecular addition to the 2,3- positions of
the initial free radical.

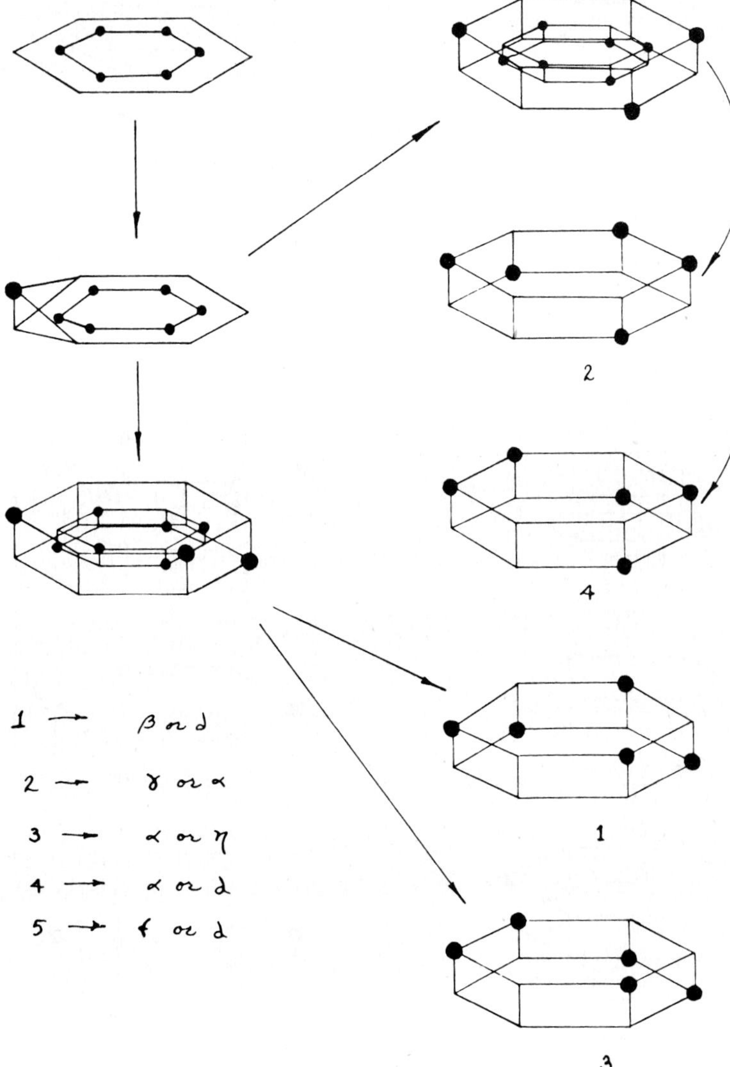

Figure 21-4. Mechanisms of the molecular addition to the 3,4 position of the initial free radical.

$$\text{(a)} \quad C_6H_6Cl° + Cl_2 \rightarrow C_6H_6Cl_3° \qquad\qquad (21\text{-}2)$$

The first mechanism provides chain transfer with formation of a relatively stable, although reactive, intermediate. The reaction proceeds to completion thus:

$$\text{(b)} \quad C_6H_6Cl_2 + Cl° \rightarrow C_6H_6Cl_3° \qquad\qquad (21\text{-}1)$$

$$\text{(c)} \quad C_6H_6Cl_3° + Cl_2 \rightarrow C_6H_6Cl_4 + Cl°$$

$$\text{(d)} \quad C_6H_6Cl_4 + Cl° \rightarrow C_6H_6Cl_5°$$

$$\text{(e)} \quad C_6H_6Cl_5° + Cl_2 \rightarrow C_6H_6Cl_6 + Cl°$$

With this reaction mechanism the reaction product should contain detectable amounts of the stable intermediates, $C_6H_6Cl_2$ and $C_6H_6Cl_4$, particularly if the reaction is effected in a flow system. In fact, no such intermediates have been isolated unless a reactive site is blocked with iodine, as in studies of Orloff. This special condition does not exist in normal chlorination procedures.

The most plausible mechanism would thus seem to require the presence of free radicals in all intermediate steps, with Reaction (21-2) proceeding to completion thus:

$$\text{(b)} \quad C_6H_6Cl_3° + Cl_2 \rightarrow C_6H_6Cl_5° \qquad\qquad (21\text{-}2)$$

$$\text{(c)} \quad C_6H_6Cl_5° + Cl_2 \rightarrow C_6H_6Cl_6 + Cl°$$

It should be noted further that Kolka's tetrachlorides had to be photo-chlorinated to the hexachlorides in a separate step and therefore might not be an actual intermediate in the one-step chlorination to HCH.

The mechanism involving molecular addition of chlorine to the initial free radical is shown in Figures 21-3 and 21-4. This mechanism results in a plausible distribution of isomers of benzene hexachloride in conformity with observed data. Blocking of one of the double bonds by formation of an unstable iodine addition compound will also permit formation of the isomers of benzene tetrachloride prepared by Kolka and his co-workers. It should be emphasized, however, that available data are too fragmentary to establish any mechanism with certainty. The applications of deductions derived from a consideration of reaction mechanisms to commercial preparation of the gamma isomer are discussed in more detail under manufacture and preparation.

Applying this mechanism of molecular addition of chlorine to the free radicals, we arrive at the following kinetic equations:

$$Cl_2 + I_0 \xrightarrow[k_2]{k_1} 2\,Cl \qquad\qquad I_0 = \text{the light intensity}$$

$$I = k_1(I_0)(Cl_2) \qquad\qquad II = k_2(Cl°)^2$$

$$Cl\cdot \ + \ \bigcirc \qquad \bigcirc Cl\cdot$$

$$III = k_3\!\left(\!\left\langle\bigcirc\right\rangle\!\right)(Cl\cdot)$$

$$\bigcirc Cl\cdot \ + \ Cl_2 \ \longrightarrow \ \overset{Cl\ Cl}{\bigcirc}Cl\cdot$$

$$IV = k_4\!\left(\!\left\langle\bigcirc\right\rangle Cl\cdot\right)(Cl_2)$$

$$\overset{Cl\ Cl}{\bigcirc}Cl\cdot \ + \ Cl_2 \ \xrightarrow{k_5} \ \overset{Cl\ Cl}{\underset{Cl\ Cl}{\bigcirc}}Cl\cdot$$

$$V = k_5\!\left(\!\left\langle\overset{Cl\ Cl}{\bigcirc}\right\rangle Cl\cdot\right)(Cl_2)$$

$$\overset{Cl\ Cl}{\underset{Cl\ Cl}{\bigcirc}}Cl\cdot \ + \ Cl_2 \ \xrightarrow{k_6} \ BHC \ + \ Cl\cdot$$

$$VI = k_6\!\left(\!\left\langle\overset{Cl\ Cl}{\underset{Cl\ Cl}{\bigcirc}}\right\rangle Cl\cdot\right)(Cl_2)$$

From stationary state theory:

$$III = IV = V = VI$$

$$d(BHC) = k_3\!\left(\!\left\langle\bigcirc\right\rangle\!\right)(Cl\cdot)$$

$$d(Cl\cdot) = 2I + VI - 2II - III = O$$

$$\therefore \quad I = II$$

$$k_1(Cl_2)I = k_2(Cl\cdot)^2$$

$$d(BHC) = k_3(k_1/k_2)^{1/2}(Cl_2)^{1/2}(I)^{1/2}\!\left(\!\left\langle\bigcirc\right\rangle\!\right)$$

The only reported kinetic studies of the photochemical chlorination of benzene pertains to the vapor phase reaction which was studied by Smith,

Noyes, and Hart.[184]   The data correspond roughly to the following equation:

$$-\frac{dP}{dt} = aI_o^{1/2}(1-e^{-kPA})^{1/2}P_AP_M$$

$$P_A \quad = \text{initial chlorine pressure}$$

$$P_M \quad = \text{initial benzene pressure}$$

$$I_o \quad = \text{absolute intensity of light}$$

They suggested the possibility of the existence of benzene dichloride as an intermediate of the vapor phase reaction but their data are far from conclusive.

MANUFACTURE AND PREPARATION

Economics

The only raw materials for making benzene hexachloride are benzene and chlorine. Since chlorine represents 73 percent by weight of the product, all present (1960) BHC producers in the United States are also chlorine producers. Raw materials represent the major manufacturing cost. Using typical 1960 values, if chlorine is charged at $60 a ton and benzene at $90 a ton, the theoretical minimum raw-material cost for crude benzene hexachloride is $68 a ton. Normal operating factors will raise this to $75 to $80 a ton.

The various forms of crude benzene hexachloride are sold on the basis of gamma isomer content. Price stabilization at $12 to $15 a unit ton can be expected at 1960 price levels. The margin for profitable operation is slim, and only a resourceful and efficient producer can survive in this field today.

There are two ways in which a BHC producer can improve his position in BHC manufacture: First, he can operate so as to produce BHC of a higher gamma content than the "standard value" of 12 percent. Second, he can find some use for the insecticidally inactive isomers and separate part or all of them from the gamma isomer. Both procedures have been the subject of intensive study. The results of these numerous investigations are summarized in the following sections.

Methods of Preparing Benzene Hexachloride

The factor of prime interest to commercial producers of benzene hexachloride is the gamma isomer content. An almost endless variety of procedures is suggested in the patent and scientific literature. A total of 186 U. S. patents in this field had been issued by June 1960. Only a few of the proposals have any major commercial significance. For the sake of clarity the various techniques are collated in four sections: (1) methods of reaction initiation, (2) control of reaction conditions, (3) special additives, and (4) use of solvents. These factors are discussed in sequence.

(1) Reaction Initiation of Benzene with Chlorine. *Disclosed Procedures.* The additive chlorination of benzene is a free-radical reaction. Virtually any method of generating free radicals can be used to initiate the reaction. Much of the electromagnetic spectrum has been investigated for initiating chlorination reactions. However, high-energy radiation such as gamma rays and X-rays, also forms ions and excited molecules as well as free radicals. It is thus possible that radiation-catalyzed reactions might follow a somewhat different course from photochemically catalyzed reactions. Studies carried out thus far show that reaction products of benzene chlorination are the same with high-energy radiation as with energy from mercury-arc lamps.[68] A process employing gamma irradiation has been patented.[396]

Initiation of the reaction by catalysts normally used for polymerization, such as organic and inorganic peroxidic compounds[246,269,312,314,405] and azo compounds[256] has been claimed in the literature. Use of olefins such as ethylene and propylene,[240] and of catalysts such as oxychlorides[311] is disclosed. Chlorinating agents other than elemental chlorine such as $SOCl_2$,[311] SO_2Cl_2,[311] $NaOCl$,[280] and $HOCl$[280] have been used. The reaction has also been effected in an electrical discharge.[148] None of these procedures has yet shown any advantage over the conventional photochlorination techniques with elemental chlorine.

Since these other methods are of academic interest only at this time, the subsequent discussion is based on data derived from photochemical chlorination. Many of the theoretical considerations however apply to other techniques of free radical formation.

Photochemical Chlorination. The procedure used by Michael Faraday in 1825 to prepare benzene hexachloride is still the only widely-used commercial method.[53] Chlorine molecules are dissociated to form free radicals by irradiation with actinic light. Once initiated, the reaction proceeds rapidly to form the fully-chlorinated end-product. No reaction intermediates can be detected in product prepared by standard chlorination techniques. The probable reaction mechanisms have been discussed in some detail earlier in this chapter.

Theoretical Considerations. The elemental unit of radiant energy is a photon or a quantum. Each quantum of energy absorbed by a molecule can either be retained as potential energy or can break a bond to dissociate the molecule into free radicals. If dissociation results, the absorption spectrum for the region of the electromagnetic spectrum under consideration will be continuous. The continuous absorption spectrum for chlorine gas begins at a wave length of 4920 Angstrom units, corresponding to a dissociation energy of 58.1 kcal/g mole. Any radiant energy of wave length less than 4920 Angstrom units will have sufficient energy content to dissociate chlorine molecules.

Various solvents may interact with the free radicals in solution, changing their mobility and modifying the absorption spectrum. In BHC preparation this effect is significant in relation to isomer distribution, and is discussed later in this chapter (Use of Solvents).

In practical applications, photochemical energy is supplied by mercury arc lamps having envelopes of quartz, Vycor, or Corex glass. The lamps are either mounted in Pyrex light wells immersed in the liquid or the irradiated liquid is contained in glass pipes or vessels. The effective radiation is bounded at the lower limit by the transmission characteristics of Pyrex glass, and thus normally falls within the region of 3000–4900 Angstrom units. Commercial mercury arc lamps usually provide about 5.5 percent of their total energy output in this region of the spectrum. A 1000 watt lamp will provide sufficient energy to activate about 0.125 pounds an hour of chlorine.

The actual free radical concentration during the reaction is extremely low and cannot be measured. However, since quantum efficiencies up to several thousand can be obtained by careful exclusion of inhibitors, the reaction must proceed by a chain mechanism once it is initiated by a chlorine free radical. There is little dispute as to the initial reaction step (Reaction 21-2):

$$C_6H_6 + Cl^\circ \rightarrow C_6H_6Cl^\circ$$

The subsequent reactions are a matter of considerable speculation.

A more fruitful approach to control of isomer distribution is the selection and control of the reaction conditions which may affect isomer distribution. The speculative reaction mechanisms are helpful mainly in defining these variables.

(2) Reaction Conditions. *Free-radical Concentration.* A major disadvantage in using light as a free radical initiator for chlorination reactions lies in the high absorption coefficient of chlorine in the near ultraviolet region. The absorption coefficients of chlorine gas for the region 3000–4800 Angstrom units are shown in Figure 21-5.

The absorption coefficient will vary in different solutions, but the order of magnitude can be assessed from data obtained with chlorine gas. Similar

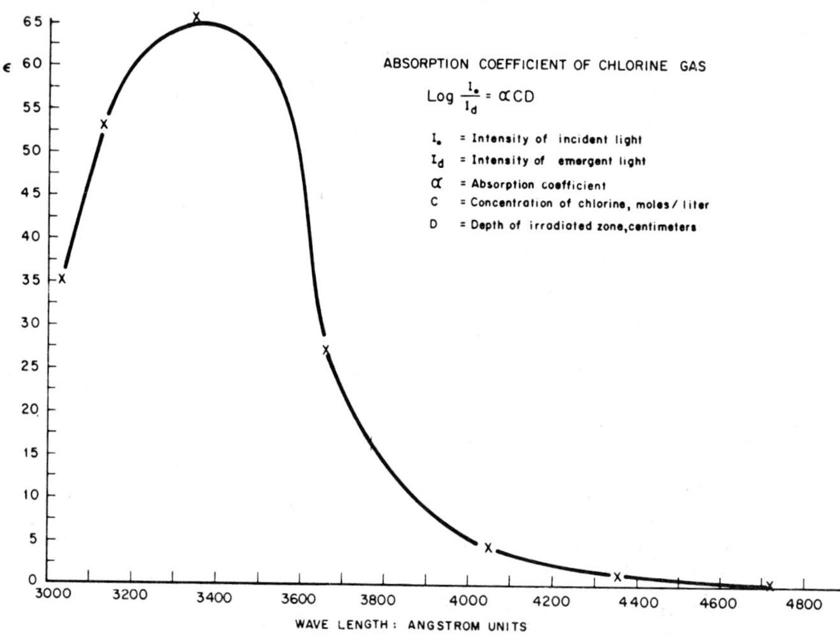

Figure 21-5. Absorption coefficient of chlorine gas.

values have been observed in benzene solution. The depth of a solution containing 1 percent dissolved chlorine which is needed to absorb 90 percent of the incident light is shown in Figure 21-6. The energy output of commercial lamps in this region of the spectrum is shown in Figure 21-7. It will be observed that with commercial mercury arc lamps a large potion of the radiant energy will be absorbed at a distance of less than one inch from the point of incidence to a 1 percent solution of chlorine. Greater penetration and more uniform activation can be obtained by use of lamps providing radiant energy mainly in the blue region of the spectrum, but industrial lamps of this type have a low total energy output.

Thus, for commercial applications employing mercury arc lamps, three choices are offered to approach uniform free-radical concentration:

(1) Maintain a very low concentration of chlorine in solution.
(2) Use a very small depth of solution, such as a falling film.
(3) Provide a high degree of turbulence in the reacting solution.

The first procedure usually results in less efficient chlorine utilization and increased formation of undesirable by-products. Modifications of the second and third procedures provide the best practical methods of process control.

Other methods of free radical initiation, such as decomposition of peroxides, can provide a high, instantaneous degree of uniformity with respect to free-radical concentration. However, since proper design of photochemical chlorination systems can provide equivalent results, the greater expense and complexity of other systems has not thus far justified their commercial application.

Various investigators have claimed improved results by using a narrow band of the electromagnetic spectrum.[411] Most of these claims have little justification in fact, and result from a failure to understand the nature of this particular free-radical reaction.

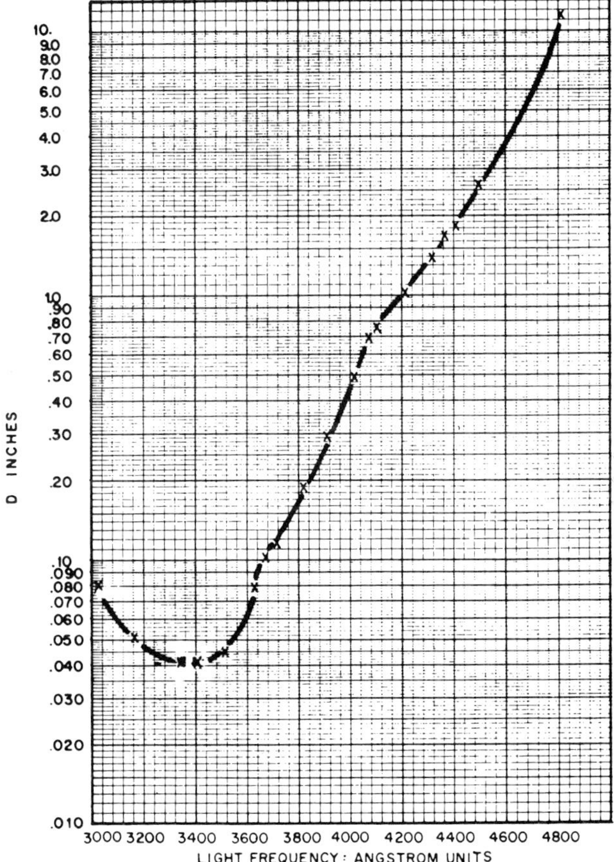

Figure 21-6. Absorption of light by a 1% solution of chlorine in a solution of density = 1.0(0.141 molar) depth of solution for 90% absorption.

Chlorine Concentration. Several investigators of this process have indicated the importance of relating the chlorine feed rate to the rate of free-radical initiation. Chlorine concentration is an important variable in the control of isomer distribution, as shown by the postulated reaction mechanisms. Failure to recognize this factor and its relation to other variables has led to many erroneous conclusions reported in the literature. Although the over-all rate of reaction is determined by the rate of reaction initiation, the rate of conversion of these various free radicals to benzene hexachloride depends on the concentration of molecular chlorine. These intermediate steps are the isomer-determining steps, assuming the reaction

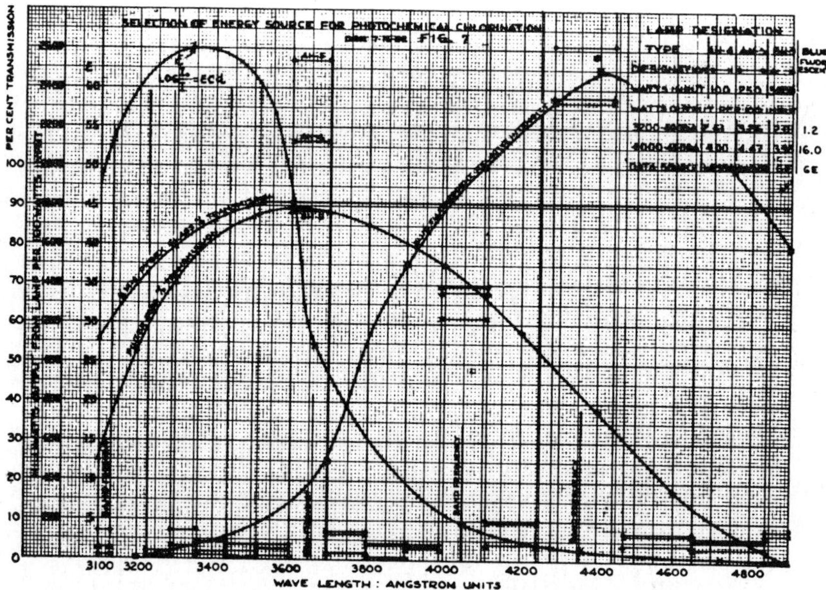

Figure 21-7. Selection of energy source for photochemical chlorination.

mechanism discussed under Structure and Mechanism. If the intermediate steps involve only chlorine free radicals, as proposed by Kanda and other investigators, the free-radical concentration would be the major factor in isomer distribution. Published data indicate that molecular chlorine concentration is the dominant factor, and several patents claim specific concentrations of chlorine in the reaction zone.

A surprisingly large variation in isomer distribution can be obtained by variation of chlorine concentration over a range of .01 to 10 percent. If all other conditions are fixed in a given system the isomer distribution can be related to chlorine concentration. The alpha isomer content will be inversely proportional to chlorine concentration and the gamma isomer content will show a maximum depending on other operating conditions. The concentrations of beta and delta isomers will be directly proportional to chlorine concentration. This factor accounts for many of the variations in isomer distribution reported in the literature which have been attributed to other variables, such as frequency of incident light or type of reaction initiator.

Temperature. Reaction temperature is a secondary variable in control of BHC isomer distribution. However, a lack of understanding of the importance of controlling energy input and chlorine concentration has confused interpretation of much of the reported data concerning temperature as a reaction variable. Higher reaction temperatures also favor secondary reactions, such as substitution chlorination. Temperature, *per se*, has only a secondary effect on the isomer-determining mechanism. At temperatures near the freezing point of benzene (5 to 10° C) gamma contents of 18 to 20 percent are reported in the literature.[230,237] At temperatures near the boiling point of benzene (60 to 70° C) gamma contents of at least 15 percent can be obtained. Lack of control of reaction variables gives the more "normal" gamma content of 11 to 12 percent. By lowering the chlorination temperature the ratio of beta and delta isomers to gamma isomer can be reduced. Thus, by variation in reaction temperature from 10 to 70° C, of chlorine concentration from 0.01 to 10 percent, and of actinic light intensity from 5 to 5000 foot candles, the following variation in isomer distribution can be obtained in bulk chlorination of benzene.

	Percent
Alpha	50–75
Beta	4–15
Gamma	6–20
Delta	5–15
Epsilon	1–4
By-products	1–10

The by-products are mainly hepta and octachlorocyclohexanes formed by substitution chlorination. Trace amounts of the eta and theta isomers may also be present.

Numerous patents have been issued relating to control of reaction conditions and design of the chlorination system for such control.[220,225,231,264,267,274,284,305,322,379] Of these various designs, some modification of a recirculating reaction system, with external cooling and addition of chlorine to the circulating stream, is best adapted to control of reaction conditions within the desired limits.

(3) Special Additives. The addition of minor quantities of various substances is reported in the literature and published patents to promote the formation of gamma isomer. Nitrotoluene,[250] organic mercury compounds,[408] sulfur compounds,[311,410,412] azo compounds,[256] and phosphorus pentoxide[320] are among the diverse substances which are claimed to have a beneficial effect. There appears to be little substance to most of these claims. Any influence which these substances may have is probably due to their effect on the primary reaction variables already discussed under structure and mechanism. For instance, some of these materials may act as free radical inhibitors, thus reducing the free-radical concentration. Others, such as azo compounds, may be free-radical initiators. An objectionable impurity from the viewpoint of inhibition is oxygen, and control of oxygen content at minimum levels is desirable from an economic viewpoint as well as for providing uniform reaction conditions. Continuous purging of the reaction medium with some inert gas such as nitrogen to remove inhibitors such as oxygen is suggested in several patents.

(4) **Use of Solvents.** In contrast to minor additives, solvents can have a major effect on isomer distribution. For this discussion, use of the term "solvent" implies that the "solvent" component is present in a concentration at least equal to that of the benzene.

In April 1947 the Belgian firm of Technique Chimique Belge obtained a Belgian patent[407] for use of polyhalogenated aliphatic derivatives as solvents to enhance the gamma isomer content of benzene hexachloride. This firm did not pursue their studies, and patents of other companies dominate the art in other countries. In May 1947 the idea of diluting benzene with a solvent was also disclosed by Gonze of Solvay and Cie in a Belgian patent.[212,406] Although his examples are confined to carbon tetrachloride, he claims the use of other aliphatic organic solvents. The advantage cited for the solvent is reduction of by-product formation.

The choice of solvents for this reaction is limited in practice to compounds which are not all, or only slowly chlorinated by chlorine free radicals, although solvents having a high specific heat, such as hydroaromatic compounds, are also claimed in issued patents. Of the organic compounds, only highly chlorinated halohydrocarbons and the lower carboxylic acids, anhydrides, or their chlorides can be considered. Extensive studies have been carried out using these solvents.[228,244,252,295,321,327,328,335,352,375,376,387] Of the inorganic compounds, only the liquid metal chlorides, such as titanium tetrachloride, or such compounds as sulfur dioxide or sulfuryl chloride, can be considered. Very little study has been made of such solvents.

The use of solvents permits chlorination at temperatures below the freezing point of benzene (5.5° C). Data are reported in the patent literature at chlorination temperatures as low as −50° C. By judicious selection of solvent and reaction conditions a gamma content as high as 30 percent can be obtained in the crude benzene hexachloride, with a corresponding reduction in concentration of alpha isomer and comparatively small change in concentration of beta and delta isomers. This shift in isomer distribution is consistent with the reaction mechanism postulated earlier employing molecular chlorine in the intermediate reaction steps.

Preparation of the Market Forms of the Gamma Isomer

Concentrates of the Gamma Isomer. *Benzene Hexachloride of 12 to 50 Percent Gamma Content.* The simplest procedure for isolating benzene hexachloride from unreacted benzene is to evaporate the benzene. Under these conditions all of the isomers will stay in the liquid phase except some portion of the beta isomer. If the terminal temperature is raised to 130° C, residual benzene can be stripped from the molten product with steam. The melt can then be cast or crystallized to obtain a solid benzene hexachloride suitable for grinding and blending to produce the desired insecticidal formulation.

Great care must be taken in removal of benzene at atmospheric pressure to exclude materials which promote degradation of the product. For this reason, some producers prefer to effect solvent removal at reduced pressures. Under these conditions the product will crystallize as evaporation proceeds, so that equipment designed for this particular combination of process steps must be used.

In the early days of commercial production a large portion of the benzene hexachloride was marketed as the crude chlorinated product containing about 12 percent gamma isomer. At the present time (1960) some portion of these inactive ingredients is removed by the commercial producers to obtain a concentrate containing 25 to 50 percent gamma isomer. Since the gamma isomer is more soluble than the alpha or beta isomers in most common solvents including benzene, enrichment of gamma isomer is obtained by extraction of the crude benzene hexachloride with the desired solvent in the region of gamma saturation. The solid phase of alpha and beta isomers can then be separated by filtration or centrifugation. The concentration of the gamma isomer in the solution phase will depend on the solvent used. For most producers the solvent used is benzene, giving a concentrate containing 40 to 45 percent of the gamma isomer. With other solvents, such as chlorinated hydrocarbons, a gamma content as high as 60 percent may be obtained by extraction. Various techniques for concentrating the gamma isomer by crystallization are disclosed in issued patents.[221,227,251,255,277,278,285,288,292,293,297,298,299,308,318,332,336,341,342,347,360,361,365,366]

This concentrate, known commercially as fortified benzene hexachloride (FBHC) contains all the gamma, delta, and epsilon isomers present in the original crude benzene hexachloride, as well as minor amounts of hepta and octachlorocyclohexanes and residual alpha and beta isomers. This mixture has a low and indistinct crystallization temperature. The melt can be supercooled to room temperature without inducing crystallization. Once solidified, the last crystal point is about 70 to 80° C upon remelting.

Various procedures have been suggested for inducing or accelerating crystallization of the melt. In one patent, a suggested procedure is casting a sheet in a thin film so that crystallization can be effected in 20 to 30 minutes to obtain a solid form which can be processed readily.[409]

Pure Gamma Isomer ("Lindane"). The name "lindane" was proposed in 1949 for the commercial form containing at least 99 percent of the gamma isomer. The name is now employed universally, and will be used in subsequent discussions.

No commercial process has yet been found which permits recovery of all of the gamma isomer present in crude benzene hexachloride. Because of the inevitable losses and the attendant processing costs, most of the gamma isomer is marketed as "fortified" BHC, as described in the preceding sections. However, there are many applications for which the undesirable properties of this crude material, such as the musty, acrid odor and phytotoxicity, are sufficiently objectionable to warrant the premium price of the purified material. Household and home garden formulations are generally made with lindane.

Numerous procedures have been described and patented for isolation of the gamma isomer. Of these, the most important are:

(1) Crystallization of the gamma isomer from a solution supersaturated with respect to other components.

(2) Fluid classification, based on differences in crystal size of the various isomers.

(3) Crystallization from a concentrate prepared by distillation or solvent extraction.

(4) Fractional crystallization using two or more solvents.

(5) Solvent extraction, using a system employing two or more solvents.

Two of these procedures are known to be used commercially for lindane preparation—the supersaturation process,[347,391] and the fluid classification process.[286,337] Intermediate steps in concentrating the gamma isomer may employ one or more of the other procedures before carrying out the primary stage of isolating the gamma isomer.[393,397,402,403,404] The supersaturation procedure was first employed by Van der Linden to isolate the gamma isomer. In preparing his concentrate he also used solvent extraction, fractional crystallization, and steam distillation procedures and laid the basis for the commercial supersaturation processes which were subsequently developed between 1945 and 1950.

(1) *Supersaturation Process.* Many modifications of the supersaturation process are described in the patent literature. In most cases, the gamma

isomer is crystallized from a solution of a gamma concentrate containing all of the minor impurities present in the crude BHC, such as the delta isomer and the heptachlorocyclohexanes. These impurities greatly promote supersaturation of the alpha and beta isomers, as noted in several issued patents, and therefore assist this process.

The various components in crude benzene hexachloride have a pronounced mutual solubilization effect. Data reported in Table 21-3A on the solubility of the individual isomers in various solvents cannot be applied to solutions of the mixed isomers. Typical data for a particular sample of benzene hexachloride are shown in Figure 21-8. The addition of delta isomer to a methanol solution together with minor impurities associated with the delta isomer, increases the solubility of both gamma and alpha isomers proportionately. There is also a pronounced effect of temperature on solubility of the isomers.

The usual solvent for the supersaturation process is a lower primary alcohol, generally methanol.[347,391] Alcohol-water systems are also described in several patents.[360,361,365,366] An appreciable portion of the gamma isomer is left in the concentrate when the limit of supersaturation of alpha isomer is reached. Further recovery of gamma isomer can be obtained by removal of the alcohol solvent and crystallization of a gamma-rich concentrate from

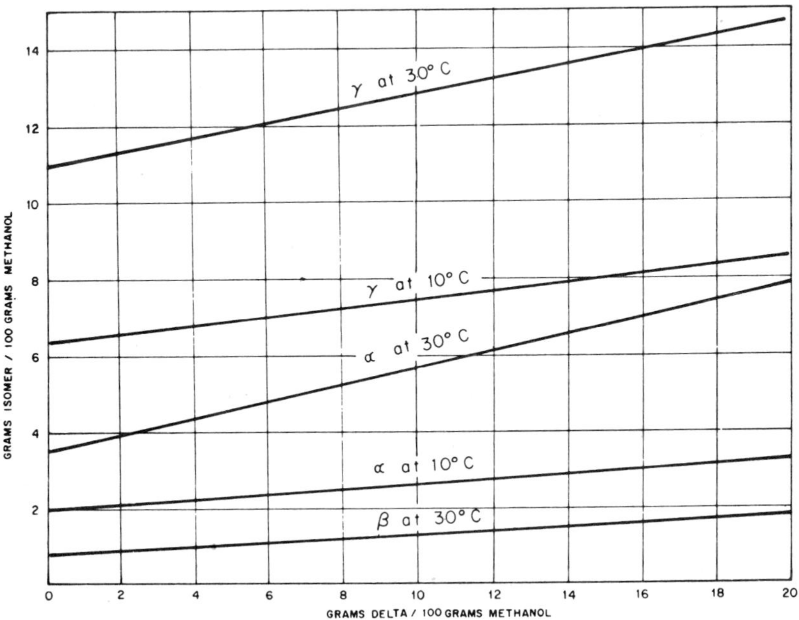

Figure 21-8. Mutual solubility characteristics of HCH isomers in methanol.

TABLE 21-3A. SOLUBILITY OF BHC ISOMERS AT 20°C
GRAMS/100 GRAMS SOLVENT[91, 183]

Solvent	α	β	γ	δ	ε
Acetic acid, glacial	4.4	1.1	14.7	34.4	
Acetone	14.1	7.9	56.0	85.0	33.2
Acetylene tetrachloride	5.22	0.51	18.82	13.64	
Benzene	11.3	1.12	33.7	46.2	14.8
n-Butyl acetate	12.0	7.6	46	119	
n-Butyl alcohol	1.6	0.7	4.6	24	2.4
iso-Butyl alcohol	0.9	0.4	3.1	15.6	
Carbon tetrachloride	1.8	0.3	7.2	3.7	0.5
Chloroform	4.8	0.17	25.2	14.3	2.0
Cyclohexane	1.4	0.8	4.8	2.8	0.6
Cyclohexene	5.8	1.0	21.1	17.1	
Cyclohexanol	1.9	0.6	4.8	20.9	
Cyclohexanone	20.9	13.8	58.0	99.6	
Decahydronaphthalene	2.5	0.4	0.5	11.6	
Diacetone alcohol	5.7	3.0	26.6	43.9	
Diethyl carbonate	11.3	4.3	39.7	86.2	
Diesel oil	1.5	0.3	4.3	10.1	
Dimethyl acetal	16.7	3.5	63.1	120.7	
Dioxane	50.6	8.5	45.8	143.3	21.6
Ethyl alcohol (abs.)	2.5	0.93	6.7	31.2	4.2
Ethyl acetate	12.5	5.9	46.3	75.5	24.5
Ethyl ether (anh.)	5.56	0.36	19.2	31.0	3.0
Ethylene dichloride	4.8	0.17	16.5	11.4	2.4
Ethylene glycol	0.3	0.1	0.6	4.3	
Ethylidene chloride	6.0	0.7	25.3	24.2	
Glycerine	0.02	nil	0.06	0.2	
n-Heptane	1.3	0.1	3.5	1.8	0.4
Methanol	2.5	0.36	8.0	37.6	3.7
Methyl acetate	15.7	7.2	44.5	165.7	24.3
Methyl propionate	14.9	8.6	60.8	156.4	
Monochlorobenzene	8.0	0.4	30.5	27.2	
Naptha, heavy (230–270°)	6.2	1.5	22.1	43.7	
Odourless distillate (198–257°)	0.8	0.02	2.0	1.1	
Paraffin (138–212°)	1.2	0.05	3.3	4.8	
Perchloroethylene	2.9	0.09	7.9	3.9	0.6
Pentane	0.9	0.1	2.2	1.6	0.2
Petroleum ether (50–70°)	0.87	0.10	2.5	1.7	0.32
n-Propyl alcohol	1.6	1.1	5.5	26.7	
iso-propyl alcohol	0.6	0.4	2.9	19.6	2.0
iso-Propyl ether	2.6	0.17	6.5	10.4	1.3
Propylene dichloride	7.1	0.32	20.0	22.2	3.34
Toluene	9.9	2.1	38.1	71.2	15.8
Trichloroethylene	3.8	0.3	17.2	8.2	1.3
Distilled water (ppm)	(10)	(5)	(10)	(10)	
White oil	0.7	0.02	1.9	1.1	
Xylene	9.3	3.4	32.8	72.7	

a solvent having very different solubility characteristics for the various isomers, such as carbon tetrachloride or a paraffinic hydrocarbon.

Preparation of lindane by the supersaturation process requires recovery of the gamma isomer from a solution containing a high concentration of impurities. Final purification of the resultant lindane is troublesome, and trace impurities are difficult to eliminate.

(2) *Fluid Classification Process.* In the late forties R. H. Kimball and G. W. Darling of Hooker Chemical Corporation observed that crystals of the mixed isomers frequently had very different particle sizes, with the gamma isomer having the largest particles. Thus, when mixed crystals are obtained by crystallization from solution of alpha, beta, and gamma isomers, the very small alpha and beta crystals can be separated from the gamma crystals by decanting a suspension of the smaller particles.

The delta isomer and minor soluble impurities are first separated from the alpha, beta, and gamma isomers by preparation of a saturated solution of the gamma isomer in a solvent having a high solubility ratio for the delta and gamma isomers. Alcohols, such as methanol or isopropanol, are very effective for this separation and also yield excellent crystal forms for subsequent crystal classification. An extremely pure lindane is obtained by this process.

(3) *Other Processes.* Multiple solvent extraction processes are proposed in several published patents.[393,397,402,403,404] Several modifications employ countercurrent liquid-liquid extraction using two immiscible solvents. A typical process is disclosed by the Ethyl Corporation in U. S. Patent 2,918,500.[402] One solvent stream uses an alcohol, glycol, dioxane or a nitrated alkyl compound, and the other uses a hydrocarbon, chlorohydrocarbon, or β,β'-dichloroethyl ether. The recovery and purity of the gamma isomer are good; however, extremely large volumes of very dilute solutions must be handled and concentrated. The process has apparently not been operated on a commercial scale.

At temperatures below the melting point of the alpha isomer, the gamma isomer is the most volatile of the isomers of BHC. Therefore, a concentrate can be obtained by sublimation, steam distillation or vacuum evaporation. The procedure has been used to obtain a gamma concentrate. Vacuum-fractional distillation is not feasible, however, since at temperatures above the melting point of the alpha isomer the vapor pressure of the alpha isomer is greater than that of the gamma isomer (see data on Physical Properties.)

Purification of Benzene Hexachloride. Crude commercial benzene hexachloride is a grey or grey-brown solid containing a variety of contaminants which impart a characteristic acrid, musty odor. The identity of these impurities has not been well established. Of the isomers of benzene hexachloride, only the delta isomer has a distinct and sharp odor. This isomer is also quite lachrymatory.

The major portion of the contaminants, other than isomers of benzene hexachloride, consists of heptachlorides and octachlorides, formed by substitutive chlorination of benzene hexachloride. Several of these isomers have been isolated and identified. The pure forms of these materials are also relatively odorless, crystalline solids. The malodorous compounds are apparently minor components which impart their objectionable qualities even when present in very low concentration. In preparing the pure gamma isomer, removal of the last traces of these contaminants is particularly troublesome.

Numerous purification procedures for improving odor and/or color of crude BHC are suggested in the patent literature.[241,247,255,258,261,279,283,366] Most of these proposals involve use of acids or oxidizing agents, such as HNO_3, H_2SO_4, P_2O_5, and alkaline or acid hypochlorite solutions. Steam stripping and treatment with hydrogen are claimed to improve odor. While some benefit may accrue from these various treatments, most of the objectionable contaminants are not affected. The presence of these impurities in crude benzene hexachloride has limited applications in which residual odor may be objectionable.

Disposition of Waste Isomers

In the preparations of the commercial concentrates of the gamma isomer or of lindane, residues consisting of the insecticidally-inactive isomers are obtained. Efforts have been made to find uses for these other isomers, but no significant commercial application has yet been developed. The delta isomer is reported to have bactericidal and fungicidal properties, but the attendant phytotoxic and irritant effects have limited such applications. The alpha isomer has been used on a limited scale as an inert, fire-retardant filler for various plastics, but its poor compatability has restricted these applications.

The most general use of these waste isomers involves conversion to polychlorobenzenes. Dehydrochlorination yields mixed isomers of trichlorobenzene, with a predominance of 1,2,4-trichlorobenzene. The trichlorobenzenes may be used for various commercial products such as transformer fluids, or they can be chlorinated to the higher polychlorobenzenes. The resultant 1,2,4,5-tetrachlorobenzene is used in production of 2,4,5-trichlorophenol, and the hexachlorobenzene is used to make pentachlorophenol.

All of the isomers of benzene hexachloride except the beta isomer can be dehydrochlorinated readily by an aqueous or alcoholic solution of sodium or potassium hydroxide. This procedure is used in analytical determinations. For commercial production, thermal dehydrochlorination is preferred, since no caustic is required and the hydrogen chloride evolved can be recovered. Various catalysts, such as aluminum chloride and ferric chloride can be used to promote dehydrochlorination, and a great variety of procedures is covered in issued patents.[248, 275, 301, 302, 303, 307, 309, 317, 323, 326, 343, 354, 358, 362, 373, 377, 383, 385, 389, 400]

PHYSICAL PROPERTIES

This section on physical properties is a compilation of data from published literature and private files of the Hooker Research laboratories.

Crystalline Forms. The polymorphic crystalline forms of the alpha, beta, gamma, delta, and epsilon isomers were studied in some detail by Adolf Kofler.[95] He reports that the gamma isomer appears in three modifications, one rhombic, and two monoclinic, which are enantiotropic to the rhombic form. The melting points of the two stable forms are very close together.

The beta isomer has two monotropic modifications, a stable cubic form and an unstable monoclinic form. Octahedric crystals are also quite common.

The delta isomer also has two enantiotropic modifications with a conversion point at 132° C. The form stable at room temperature is monoclinic. The high temperature form crystallizes hexagonally and has negative double refraction.

TABLE 21-3B.   REFRACTIVE INDEX

Solid ($5893\,\text{Å}$, 25°C)		Liquid
Alpha 1.630	1.630 ± 0.002	1.5101 (160°C)
Beta (cubic)	1.633 ± 0.004 (calc.)	
Gamma	1.644 ± 0.002	1.5101 (171°C)
Delta		1.5101 (157°C)

TABLE 21-3C.   ABSOLUTE DENSITY AT 25°C

Alpha	1.851
Beta	1.860
Gamma	1.855

TABLE 21-3D.   POLAR MOMENTS (112)

	α	β	γ	δ	ε	η
μ_0 Experimental (benzene and CCl_4)	2.14	0	2.83	2.22	0.26	1.7
μ Calculated	3.2	0	4.6	3.2	0	1.9

TABLE 21-3E. HEAT OF CRYSTALLIZATION

Alpha	20.0 cal/g
Beta	26.0 cal/g
Gamma	17.2 cal/g

TABLE 21-3F. MELTING POINTS

Alpha	157.5–158.0 °C	Delta	138.0–138.4 °C
Beta	309.8–310.7 °C	Epsilon	218.5–219.3 °C
Gamma	112.86 °C	Eta	89.8–90.6 °C
		Theta	124–125 °C

TABLE 21-3G. VAPOR PRESSURE OF SOLID FORMS

$$\mathrm{Log}_{10}\, p = A - \frac{B}{T}$$

Isomer	A	B	Range of Temperature (°C)
Alpha	11.950	4850	51–71
Beta	11.790	5375	95–117
Gamma	15.515	6020	60–92
Delta	12.635	5100	55–75

The alpha isomer has only one form, and produces symmetrical mono-clinic prisms. The crystalline form of the epsilon isomer is also mono-clinic.

Vapor pressure data for the solid forms are shown in Figure 21-9. Data are extrapolated to the melting points for the alpha, gamma, and delta isomers and the limited available vapor pressure data for the liquid forms are included.

Infrared Absorption Spectra

Infrared absorption spectrometry is widely used for analysis of the isomers of benzene hexachloride in laboratories which are equipped for this technique (see Analytical Procedures). Complete absorption spectra[46, 47] have been published. The absorption bands used for analysis are shown in Table 21-4. Carbon disulfide is used as a solvent for alpha, gamma, delta, and epsilon isomers. Acetone is used as a solvent for the beta isomer.

ANALYTICAL PROCEDURES

Mixed Isomers

Of the analytical procedures developed since the commercial introduction of benzene hexachloride in 1945, only three have found general use.

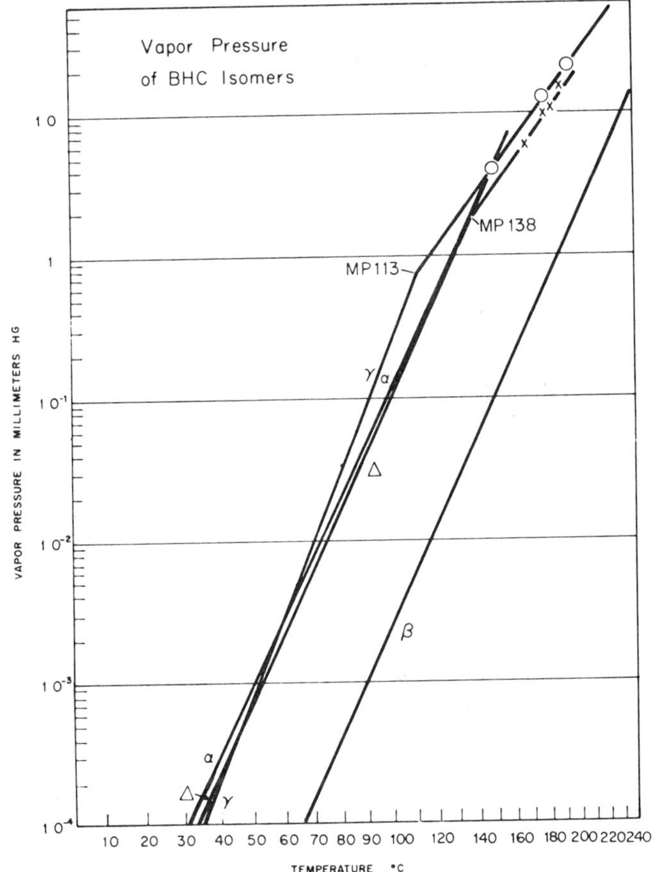

Figure 21-9. Vapor pressure of BHC isomers.

These are discussed briefly below. Data applicable to these techniques are reported in the previous section.

Dehydrochlorination.[85, 105, 163, 92]. Probably the first practical method was that of measuring the rate of dehydrochlorination with potassium hydroxide. As shown in the literature noted the different isomers have different reaction rates with base. This method can distinguish the pure isomers in a mixture, but impurities such as the heptachlorides cause erroneous results. A similar method is the dehydrochlorination with monoethanolamine.

Partition Chromatography.[5, 164] In 1946 Ramsey and Patterson devised a partition chromatography method. Using nitro-methane and *n*-hexane as mutually insoluble solvents and a silica gel column they were able to separate not only the four isomers of HCH but a heptachloride and an octachloride as well. A modification of this procedure was the official method

TABLE 21-4. INFRARED ABSORPTION BANDS FOR ANALYSIS OF ISOMERS OF BENZENE HEXACHLORIDE

A. Sodium chloride prism, range (2–15.5 microns)

Isomer	Band (microns)
Alpha	12.58
Beta	13.46
Gamma	14.53
Delta	13.22
Epsilon	13.96

B. Potassium bromide prism, range (12–25 microns)

Isomer	Band (microns)
Alpha	15.92
Beta	13.46
Gamma	20.7
Delta	22.2
Epsilon	13.96 or 18.30

of the Association of Official Agricultural Chemists[79,80] until 1958, when a paper chromatography method devised by L. L. Mitchell[132–141] was adopted. The paper chromatography gives better results with smaller samples.

Infrared Sprectrometry.[47,102,103] The chromatographic procedures are long and tedious, and a great number of other methods have been tried at various times for specific purposes. Of these, only infrared spectrometry has found widespread application. Infrared analysis is simple and rapid, but the equipment is expensive and complex. This method has not been adopted as the standard because small laboratories usually are not equiped with infrared analysers. A good set of standards for the pure materials and possible contaminents are needed. Care in solvent removal must be exercised. Results from this method are in good agreement with the results of the much more laborious methods. Standards for the BHC isomers as well as the heptachlorides and octachlorides have been published (see Physical Properties).

Gamma Isomer

Four procedures have been used for analysis of the gamma isomer. A fifth, differential refractometry[76] has been studied, but commercial equipment is not available.

Polarography. The polarographic method for determining gamma[87] in BHC is capable of rapid and accurate results in mixtures of BHC isomers with normal impurities. Materials which are reduced in the same voltage as the gamma interfere. The other isomers are not determined.

Cryoscopy. The cryoscopic method[30, 186, 202] could be applied to the determination of all the isomers. It is, however, a laborious method requiring considerable amounts of the pure isomers. It is used most effectively to determine traces of impurities in an isolated isomer. It is the only method which specifies purity in terms of gamma content to an accuracy of ± 0.1 percent when the sample approaches 100 percent purity.

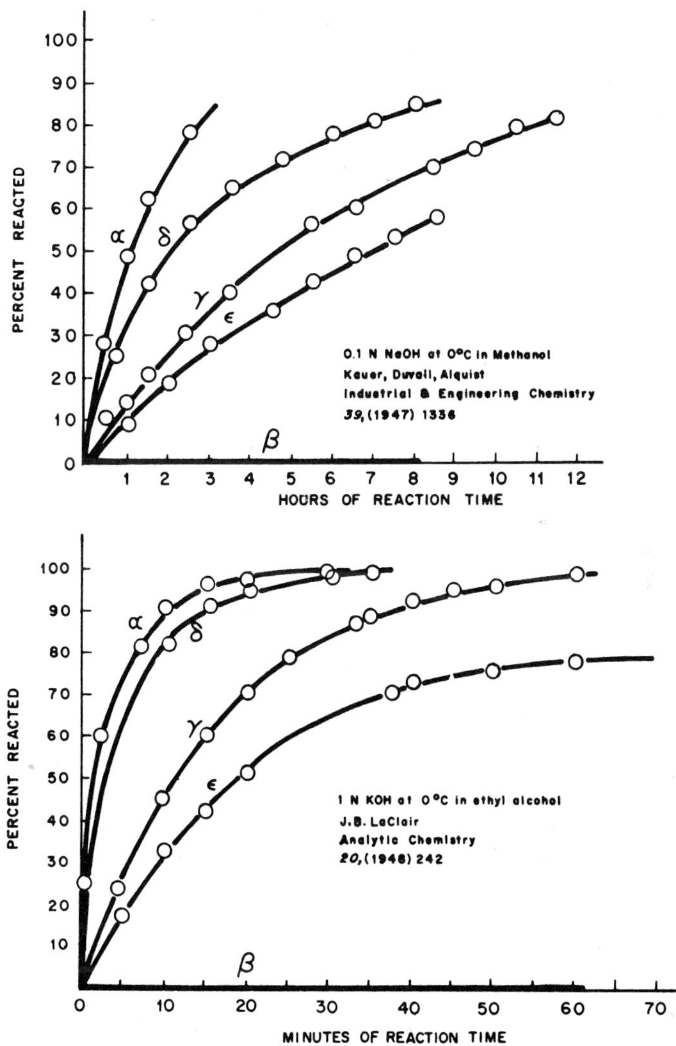

Figure 21-10. Dehydrochlorination of HCH isomers.

Radioisotopic Analyses. The method of Cl^{36} radioisotope dilution[45,81,203] has been applied only to the determination of gamma. However, it could be equally applicable to other isomers. Pure standard radioactive gamma HCH containing Cl^{36} is diluted with ordinary gamma HCH contained in the sample to be analyzed. A sample of pure gamma is separated from the mixture. The standard radio active gamma will be diluted with gamma in proportion to the gamma content of the unknown sample. By decrease in radioactivity of pure gamma, recrystallized from the sample, the amount of gamma in the sample can be determined. It is a laborious method but offers excellent results.

Bioassay. Mention should be made of the bioassay methods of determination as a specialized technique.[149,159,173] After the L/D_{50} of an insect has been accurately determined minute quantities of the gamma isomer can be estimated by exposing the insect to a small sample. Using a number of insects, effects of gamma HCH can be separated from effects of DDT and other insecticides. This method, of course, is useless for large samples but has application for small samples such as residues on plant leaves. The biological properties of the gamma isomer are more fully covered in the next two sections.

INSECTICIDAL PROPERTIES

Dr. Thomas discovered the insecticidal properties of HCH when he was testing various chlorinated ring compounds.[183] He later found the chief insecticidal agent was the gamma isomer. After the similarity between the inositols and the BHC isomers was noted, Slade suggested that the gamma isomer[183] inhibited the metabolism of meso-inositol, which is a growth-promoting vitamin of the B complex. It was later found that it was the delta isomer rather than the gamma isomer which resembled meso inositol. Experiments in metabolism on insects, which were given sublethal doses of gamma BHC and then supplied with inositol,[200] indicated there was another mode of action by the insecticide.

Since that time researchers have studied such factors as the increase of free nucleotides in the brain[37] after application of insecticides. In other studies the gamma isomer was also found to completely inhibit cytochrome Coxidase[145] from the American cockroach and alkaline phosphatase in Aedes aegypti.[48] Its action is, therefore, something more subtle than at first proposed.

Table 21-5 is a listing of some of the insects and pests that are killed by gamma-BHC. This is by no means a complete listing and the mean lethal dose varies greatly from insect to insect. It should be noted also that certain species of flies and mosquitoes have developed resistant[197,115] strains. Studies on the nature of this resistance showed that these resistant strains were able to metabolize the gamma isomer much more readily than nonresistant strains.

Other investigators have also studied effects of climatic conditions on toxicity. Temperature,[124] humidity,[15] amount of sunlight,[131] and the time delay before a rain have been considered. The results of these tests indicate that the best conditions for application would be a hot, humid day with a rain fall within a couple of days. These conditions are most likely to be met in late July or early August and a number of field reports show better results when application is made during this time.

The comparative toxicity of different insecticides to Melanoplus differentialis is shown in Table 21-6[149] and of Thrips[128] in Table 21-7. The value of gamma-BHC can be seen in the extremely small dose needed for kill. Numerous patents have been issued for insecticidal compositions containing gamma BHC.[222,226,229,232,234,235,236,242,243,253,254,257,259,260,262,268,276,287,304,333]

TABLE 21-5. SOME OF THE INSECTS AND OTHER PESTS KILLED BY GAMMA HCH.

A. INSECTA
    1. Siphonaptera (Aphaniptera)[1,183]
        Various fleas, e.g. Ctenocephalus canis
    2. Hemiptera[183]
        Bed bug, Cimex Lectularious L.
    3. Mallophaga[183,44]
        Chicken lice, e.g. Menopon pallidum
    4. Heteroptera[183]
        Squash bug, Anasa tristis
    5. Isoptera[183,74,75]
        Subterranean termite, Reticulitermes flavipes K.
    6. Siphunculata (Anopleura)[183,32,78]
        Head louse, Pediculus humanus L.
        Body louse, Pediculus vestimenti
    7. Coleoptera[183,31,23,27,26,100,180,29,146,194,188,155,154,175,176,109,161]
        Cotton boll weevil, Anthonomus grandis
        Apple blossom weevil, Anthonomus pomerum L.
        Wireworm, Cononderus lividus
        Plum curculio, Conotrachelus nenuphur
        Larder beetle, Dermestes lardarios
        Hide beetle, Dermestes vulpinus F.
        Striped cucumber beetle, Diabrotica vittata
        Spotted cucumber beetle, Diabrotica punctata
        Alpha weevil, Hypera postica
        Cigarette beetle, Lasioderma sericorne F.
        Red legged ham beetle, Necrobia rufipes D.
        Mustard beetle, Phaedon cochleariae F.
        Japanese beetle, Popilla japonica

TABLE 21.5.—Continued.

Corn weevil, sitophilus Cryzae L.
Grain weevil, sitophilus granarius L.
Flour beetles, Tribolium spp.
Rhinocerous beetle, Xyloryctes satyrus

8. Orthoptera[181,86,56,33,183]

Oriental cockroach or Black beetle, Blatta orientalis L.
German cockroach or Croton bug, Blatella germanica L.
Field cricket, Gryllus assimilis F.
House cricket, Gryllulus domesticus L.
African migratory locust, Locusta migratoria migratrioides
Lesser migratory grasshopper, Sphenarius purpuracens

9. Hymenoptera[3,183,199]

Wasps, Vespidae spp. and Philanthus triangulum
Ants, Lasius spp.
Argentine ant, Iridomyrex pruinosus
Road or meat ant, Iridomyrex rufoniger
European apple sawfly, Hoplocampa testudinea

10. Lepidoptera[59,89,183]

Cabbage caterpillars, Pieris spp.
Winter moth caterpillars, Cheimatobia brumata L.
Clothes moth caterpillars, Tineola bisselliella
Mellon worm, Diaphania hyaliniata L.
Southern army worm, Laphygma eridana
Gypsy moth, Porthetria dispar
Angoumois grain moth, Sitotroga cerealella
Tobaco leaf miner, Gnorimoshema operculella Zell.
Cotton boll worm, Heliothis obsoleta armigera

11. Homoptera[8,23,42,95,130,158,160,183,193,214]

Black aphid, Alphis fabae Scop
Ezo-spruce gall aphid, Adelges japonicus
Pine spittle bug, Cercopidae
Taxus mealy bug, Pseudococcus cuspidatae
Aphis betae

12. Diptera[2,34,35,115,104,52,4,62,66,67,204,182,150,218,196,111,118,125,183]

Mosquitoes, Aedes aegypti L. and Anopheles spp., and Theo-
baldia spp.
Houseflies, Musca domestica L. and Musca vacina
Oriental fruit fly, Dacus dorsalis
Mangold fly, Pegomyia betae
Asparagus fly, Platyparea poecilopera
Button top midge, Rhabdophaga heterobia

B. ARACHNIDA

1. Acarina[10,14,40,153,183,189,190,192,212]

Poultry red mite, Dermanyssus gallinae D.
Red legged earth mite, Halotydeus destructor
Trombicula akamushi
Cattle fever tick, Boophilus annulatus
Cattle tick, Hyalomma scunpense
Onithodorus brasiliensis and moubata

C. CRUSTACEA[183]

Woodlice, Oniscidae spp.

TABLE 21-6. COMPARATIVE TOXICITY OF INSECTICIDES TO MALANOPLUS DIFFERENTIALIS[149]

Material	Dose in Mg	% in Emulsion	% Dead in 24 Hours
DDT	100	2	30
	50	1	20
	25	0.5	0
	10	0.5	0
	5	0.5	0
Chlordane	50	1	70
	25	0.5	30
	10	0.5	10
	5	0.5	0
δ BHC	25	0.5	100
	10	0.5	90
	5	0.5	10
	2	0.5	0
Control			0

EFFECTS ON MAMMALS

One of the major considerations of any insecticide is its effect on animals and especially human beings. Numerous studies were made on animals under various conditions. The results of dipping mice in an emulsion containing gamma HCH are presented in Table 22-8. If the mice[57] were restrained from licking their fur no mice were killed by a 5 percent dip. When the gamma isomer was introduced into the stomach it took 0.19 g/ kilogram body weight for a 50 percent kill of rats.

In tests on dogs,[22] with 50 mg/kilogram, there were no ill effects shown by the animals. Dipping tests on cows showed no ill effects, except when killing the insects resulted in increased bacterial counts. In these cases the addition of a bactericidal agent was necessary to prevent disease.

Some ill effects have been reported by workers who were exposed to large quantities of the vapors.[28] Over exposure to the fumes can cause

TABLE 21-7. COMPARITIVE TOXICITY OF INSECTICIDES TO HELIOTHRIP HAEMORRHOIDALIS

100% Mortality of Heliothrip Haemorrhoidalis

Insecticide	Concentration in Solution
γ BHC	0.001%
2,2- bis (*p*-chlorophenyl) 1,1,1-trichloroethane	0.005%
Chlordane	0.05%
Toxaphene	0.1%
Hexaethyltetraphosphate	0.1%

TABLE 21-8. TOXICITY OF γ-HCH TO MICE[57]

Dosage in g/Kg	Treated	Number of Symptoms	Symptom, but Recovered	Died
0.3	14	13	0	1
0.6	19	6	6	7
1.2	15	1	0	14

skin irritation or acne-like sores. For this reason people handling large quantities of the material must exercise the necessary precautions. For the consumer who handles only small quantities there is little danger. The estimated lethal dose for an adult is 13 grams of the gamma isomer.

A second necessary consideration is the result of spraying food crops. Tests for residues on plants have been designed and maximum tolerances defined by the FDA. This area presents no major problem to the crop grower. One aspect that has limited the use of HCH is the report of bad taste from plants that have been exposed to either the crude or purified product. Even minute quantities of the gamma isomer have been detected in taste tests of certain foods. Table 21-9 is the result of triangular taste tests on various crops. If the food was cooked or canned the results were much worse, with bad taste resulting in almost all cases.

The insecticide can be used to great benefit on flowers and wood trees, against household pests, and as a dip for animals. Among the specialized uses are such applications as impregnation of clothing in North Borneo[211] to protect against the terrestial leaches.

TABLE 21-9.   TRIANGULAR FOOD TEST[25]

Food	Application of Lindane	Results
Beets	Foliage and soil	Flavor tainted, both samples
Carrots	Foliage and soil	Flavor not tainted, both samples
Cabbage	Foliage and soil	Flavor not tainted, both samples
Cucumber	Foliage and soil	Foliage applied, sample not tainted
Rutabagas	Foliage and soil	Flavor tainted, both samples
Radishes	Foliage and soil	Flavor tainted, both samples
Squash	Foliage and soil	Soil applied, sample not tainted
Sauerkraut	Foliage and soil	Flavor tainted, both samples
Potatoes	Soil	Flavor tainted, one sample
Onions	Foliage and soil	Foliage applied, sample not tainted
Pumpkin	Foliage and soil	Flavor not tainted, both samples

References

1. Anon., *Chem. Wk.*, **69** (11), 37 (1951).
2. Anon., *J. Econ. Entomol.*, **40**, 588 (1947).
3. Anon., *Agr. Gaz, N.S. Wales,* **58**, 638 (1947).

4. Acher and Levinson, *Ruista Parissitol,* **14,** 57 (1953).

5. Aepli, Munter, Gall, *Anal. Chem.,* **20,** 610 (1948).

6. Alyea, *J. Am. Chem. Soc.,* **52,** 2743 (1930).

7. Anderson, Bray, Martin, *Proc. Intern. Cong. Peaceful Uses of Atomic Energy,* **15,** 235 (1956).

8. Anderson and Brooks, *J. Econ. Entomol.,* **40,** 508 (1947).

9. Armour Res. Found., *Anal. Chem.,* **21,** 882 (1949).

10. Avet, *Nature* (London), **157,** 699, (1946).

11. Baccaredo and Beati, *Ricerca Sci.,* **20,** 13.3 (1950).

12. Baker and Schoof, *J. Econ. Entomol ,* **48,** 181 (1955).

13. Balson, *Faraday Soc. Trans.,* **43,** 54 (1947).

14. Barber and Bath, *J. Dept. Agr., Victoria,* **53,** 228 (1955).

15. Barlow and Hadaway, *Nature,* **178,** 1299 (1956).

16. Barton, Hassel, Pitzer, Prelog, *Science,* **119,** 49 (1954).

17. Barton, Hassel, Pitzer, Prelog, *Nature,* **172,** 1096 (1953).

18. Bastiansen, *Acta Chem. Scand.,* **6,** 875 (1952).

19. Bastiansen and Hassel, *Acta Chem. Scand.,* **1,** 683 (1947).

20. Bastiansen, Ellefson, Hassel, *Research* (London), **2,** 248 (1949).

21. Bastiansen, and Markali, *Acta Chem. Scand.,* **6,** 442 (1952).

22. Batte and Turk, *J. Econ. Entomol.,* **41,** 102 (1948).

23. Becnel, *J. Econ. Entomol.,* **40,** 508 (1947).

24. Bijvoet, *Rec. trav. chim.,* **67,** 777 (1948).

25. Birdsall, Weckel, Chapman, *J. Agr. Food Chem.,* **5,** 523 (1957).

26. Block, Va. *Fruit,* **36,** (1) 56 (1948).

27. Bobb, Va. *Fruit,* **35,** (1) 47 (1947).

28. Bonnazzi, Mucci, Troisi, *Folia Med.* (Naples), **35,** 382 (1952).

29. Bovington & Stock, *Internatl. Soc. Leather Trades Chem., J.,* **31,** 115 (1947).

30. Bowen and Pogorelskin, *Anal. Chem.,* **20,** 346 (1948).

31. Brett and Rhoades, (Sci. Note) *J. Econ. Entomol.,* **46,** 572 (1947).

32. Brown and Gamlin, *Med. Off.,* **79,** 121 (1948).

33. Brown and Hurtig (Sci. Note) *J. Econ. Entomol.,* **40,** 276 (1947).

34. Burnett, *Nature,* **177,** 663 (1956).

35. Busvine, *Nature,* **177,** 533 (1956).

36. Calingaert, Griffing, Kerr, Kolka, Orloff, *J. Am. Chem. Soc.,* **73,** 5224 (1951).

37. Caper, *Arch. Exptl. Pathol. Pharmokol,* **232,** 265 (1957).

38. Caper, *et al., Klinisch Wochenschrift,* 264, April 1951.

39. Collie (London), *Chem. Soc. Proc.,* **13,** 143 (1897).

40. Corbett, *American J. Vet. Res.,* **8,** 280 (1947).

41. Coutin and Hennequin, *Phytiat Phytopharm.,* **6,** 93 (1957).

42. Cox, *J. Econ. Entomol,* **40,** 195 (1947).

43. Craig, Tryon, Brown, *Analy. Chem.,* **25,** 1661 (1953).

44. Creighton, Hatrick, Hunt, Duncun, *Poultry Sci.,* **26,** 674 (1947).

45. Cristol, *J. Am. Chem. Soc.,* **69,** 338 (1947).

46. Cupples, *Anal. Chem.,* **21,** 630 (1949).

47. Daasch, *Anal. Chem.,* **19,** 779 (1947).

48. Dasgupta and Roy, *Bull. Calcutta School Trop. Med.,* **3,** 169 (1955).

49. Dawson and Escritt, *Nature* (London), **158**, 448 (1946).
50. Dickinson and Bilicke, *J. Amer. Chem. Soc.*, **50**, 764 (1928).
51. Dupire, Raucourt, *Acad d' Agr. de France*, **29**, 470, *Comp. Rend.*
52. Du Tort and Fiedler, *Onderstepoort J. Vet. Research 25*, —53 (1952).
53. Faraday, *Phil Trans, 1825*, 440.
54. Faraday, *Ann. de Chim et de Phys. 30*, 269 (1825).
55. Fiedler and Du Tort, *Onderstepoort J. Vet. Research*, **25**, 409 (1954).
56. Flores and Delgado, *Proc. First Latin-Am. Congr. Phytoparasitol. Mex.* (1950).
57. Furman, *J. Econ. Entomol.*, **40**, 518 (1947).
58. Gannon and Decker, *J. Econ. Entomol.*, **48**, 240 (1955).
59. Geisthardt and Madel, *Z. angew Zool.*, **1955**, 441.
60. Gjullin, *Proc. New Jersey Mosquito Enterm. Assoc.*, **41**, 127 (1954).
61. Gonzales, Bunyi, Ramos, Araneta, *J. Agr.*, **4** (1), 84 (1957).
62. Guile, *Plant Pathol.*, **7**, 50 (1958).
63. Gunther, *Chem. and Ind.*, **44**, 399.
64. Gunther and Blinn, *J. Am. Chem. Soc.*, **69**, 1215 (1947).
65. Halban, Siedentopf, *Z. physik. Chem.*, **103**, 71 (1922).
66. Harbour, *Scot. Farmer*, **54**, 583 (1946).
67. Harbour and Watt, *Vet. Rec.*, **57**, 685 (1945).
68. Harmer, *U. S. Atomic Eng. Com.*, *Publ.* No. 12, 582 (1955).
69. Harms, *Farm Chemicals*, **117** (11), 24 (1954).
70. Hassel, *Tids Kjemi Bregresen Met.*, **3**, 32 (1943).
71. Hendricks, *Chem. Rev.*, **7**, 431 (1930).
72. Henriquez, *Proc. Acad. Sci. Amsterdam*, **37**, 532 (1934).
73. Herken, *Klinische Wochenschrift*, 582 (1950).
74. Hetrick, *J. Econ. Entomol*, **43**, 57 (1950).
75. Hetrick, *J. Econ. Entomol.*, **45**, 235 (1952).
76. Hill, Jones, Palin, *Chem. and Ind.*, p. 12, Feb. 6, 1954.
77. Hinreiner and Simone, *Hilgardia*, **26**, 76 (1956).
78. Hoffman, *J. Econ. Entomol.*, **49**, 347 (1956).
79. Hornstein, *J. Ass. Off. Agr. Chemist.*, **38**, 290 (1955).
80. Hornstein, *J. Ass. Off. Agr. Chemist*, **39**, 373 (1956).
81. Hornstein, *J. Ass. Off. Agr. Chemist*, **40**, 737 (1957).
82. Hornstein, *Science*, **21**, 206 (1955).
83. Horton, Karel, Chadwick, *Science*, **107**, 246 (1948).
84. Hough and Hill, *J. Econ. Entomol.*, **49**, 217 (1956).
85. Howard, *Analyst*, **72**, 427 (1947).
86. Hynes, *Nature* (London), **159**, 200 (1947).
87. Ingram and Southern, *Nature* (London) **161**, 437 (1948).
88. Jatkur and Kulkarsi, *Science & Culture*, **14**, 482 (1949).
89. Kaempf, *Pharmazie*, **8**, 575 (1953).
90. Kanda, *J. Am. Chem. Soc.*, **82**, 3085 (1960).
91. Kauer, Du Vall, Alquist, *Ind. Eng. Chem.*, **39**, 1335 (1947).
92. Kimball and Tufts, *Ind. Eng. Chem.*, (1938) *Anal. Ed.*, **10**, 530.
93. Klosa, *Arch. Pharm.*, **286**, 216 (1953).

95. Kofler, *Chem. Ber.*, **84,** 376 (1951).

96. Kolka, *et al.*, *J. Am. Chem. Soc.*, **76,** 3940 (1954).

97. Kolka, *et al.*, ACS 121st meeting 1952, p. 5k.

98. Kolka, *et al.*, *J. Am. Chem. Soc.*, **75,** 4243 (1953).

99. Kramer and Ditman, *Food Technol.*, **10,** 155 (1956).

100. Kulash, *Euclides* (Madrid), **15,** (173) 199 (1955).

101. Kulkaim, *J. Ind. Chem.*, **27,** 273 (1950).

102. Kuratani, *Report Inst. Sci. Technol.*, **4,** 236 (1950).

103. Kuratani, Sakaskita, *Reports Radiation Chem. Res. Inst.*, **5,** 8 (1950).

104. Labrecque, Noc, Guham, *Mesquito News,* **16,** 1 (1956).

105. La Clair, *Analyt. Chem.*, **20,** 241 (1948).

106. Lakewing and Vogeliorle, *Naturwissenschuflis* **37,** 211 (1950).

107. Lauret, *Ann. Chim et Phys.*, **63,** 27 (1836).

108. Leland, *Chem. Specialties Mfrs. Assoc. Proc.* p. 110 (June 1952).

109. Le Pelley and Kockum, *Bull Entomol Research,* **45,** 295 (1954).

110. Letard and de Sacy, *Soc. de Biol.* (Paris), *Compt. Rend.* **139,** 353 (1945).

111. Light and Gould, *Plant Pathol.*, **4,** 58 (1955).

112. Lind, Hobbs, Gross, *J. Am. Chem. Soc.*, **72,** 4474 (1950).

113. Linden, Van der, *Deut. Chem. Gesell Ber.*, **45,** 231 (1912).

114. Lindsley, *Hilgardia,* **26,** 1 (1956).

115. Livadas, *Mesquito News,* **15,** 67 (1955).

116. Ludicke and Pranter, *Z. Naturforsch.*, **13b,** 111 (1958).

117. Luther, Goldberg, *Z. physik Chem.*, **56,** 43 (1906).

118. Machegan, *West. Scot. Agr. Coll. Research Bull No. 12,* 1 (1955).

119. Mackie, Stewart, Cutler, Misra, *Brit. J. Pharmacol.*, **10,** 7 (1955).

120. Maercks, *Z. Angew Zool.*, **1955,** 375.

121. Malbrunot and Richard, *Prog. agr. vit.*, **75,** 175 (1958).

122. Mathews (London) *Chem. Soc. Proc.*, **14,** 52 (1897).

123. Mathews (London) *Chem. Soc. Jour.*, **59,** 165 (1891).

124. McIntosh, *Chem. & Ind.* (London), **1957,** 2.

125. McLeod, *J. Econ. Entomol.*, **39,** 631 (1946).

126. McNamara and Krop, *J. Pharmocol. Expt. Ther.*, **92,** 140 (1948).

127. Menary, *Acta Cryst,* **8,** 840 (1955).

128. Metcalf, *J. Econ. Entomol.*, **40,** 522 (1941).

129. Meunier, (Paris) *Acad des Sci. Compt. Rend.*, **98,** 436 (1884).

130. Missonier, *Compt. rend. acad. agr. France,* **41,** 669 (1955).

131. Mistric, Martin, *J. Econ. Entomol.*, **49,** 757 (1956).

132. Mitchell, *J. Assoc. Off. Agr. Chemist,* **35,** 920 (1952).

133. Mitchell, *J. Assoc. Off. Agr. Chemist,* **36,** 553 (1953).

134. Mitchell, *J. Assoc. Off. Agr. Chemist,* **36,** 1183 (1953).

135. Mitchell, *J. Assoc. Off. Agr. Chemist,* **37,** 216 (1954).

136. Mitchell, *J. Assoc. Off. Agr. Chemist,* **37,** 530 (1954).

137. Mitchell, *J. Assoc. Off. Agr. Chemist,* **37,** 996 (1954).

138. Mitchell, *J. Assoc. Off. Agr. Chemist,* **39,** 484 (1956).

139. Mitchell, *J. Assoc. Off. Agr. Chemist,* **39,** 985 (1956).

140. Mitchell, *J. Assoc. Off. Agr. Chemist,* **40,** 294 (1957).

141. Mitchell, *J. Assoc. Off. Agr. Chemist*, **41**, 31 (1958).

142. Mitscherlich, *Ann der Phys u Chem.*, **35**, 370 (1835).

143. Morino, Miyagawa, Oiwa, *Batyu-Kagaku*, **15**, 181 (1950).

144. Morino, Miyagawa, Chiba, Schimozawa, *J. Chem. Phys.*, **25**, 185 (1956).

145. Morrison, Brown, *J. Econ. Entomol.*, **47**, 723 (1954).

146. Muka, *J. Econ. Entomol.*, **50**, 216 (1957).

147. Nakajima and Oiwa, *Botyu-Kagaku*, **15**, 114 (1950).

148. Nakajima and Nagano, *Botyu-Kagaku*, **16**, 183 (1951).

149. Nakajima, *Mem. Coll. Agr. Kyoto Univ.*, **65**, 1 (1952).

150. Newton, Satchell, Shaw, *Nature* (London), **158**, 417 (1946).

151. Nieman, *Angeu Botan.*, **30**, 1 (1956).

152. Norman, *Acta Chem. Scand.*, **4**, 251 (1950).

153. Norris, *Austral. Inst. Agr. Sci. Jour.*, **12**, 51 (1946).

154. Ochoa, *Acta agron* (Columbia), **5**, 29 (1955).

155. O'Connor, *Agri. J. Colony Figi Dept. Agr.*, **25**, 84 (1954).

156. Orloff, *Chem. Review.* **54**, 347 (1954).

157. Pelicot, *Ann. Chim et Phys.*, **2**, 59 (1834).

158. Petrosyan, *Vinodelic i Vinogradarstro S.S.S.R.*, **12**, 39 (1952).

159. Phillipot and Dallemagne, *Experientia*, **3**, 118 (1947).

160. Pierce, *Proc. 33rd ann. meeting Texas Pecan Growers Assoc.*, **1954**, 46.

161. Prodham and Rangarao, *Bull. Entomol. Research*, **48**, 261 (1961).

162. Radeleef and Woodard, *Proc. Am. Vet. Med. Assoc.*, (92) 109 (1955).

163. Raffaelli, *Ann. chim. applicata*, **38**, 552 (1948).

164. Ramsey and Patterson, *J. Ass. Off. Agr. Chem.*, **29**, 237 (1946).

165. Raucourt and Bouchet, *Chemie & industrie*, **56**, 449 (1946).

166. Riemschneider, *Z. Naturforsh*, **6b**, 48 (1951).

167. Riemschneider, Spat, Rausch, Bottger, *Monatsh*, **84**, 1068 (1953).

168. Riemschneider, Spat, Rausch, Bottger, *Z. Naturforsch*, **8b**, 70 (1952).

169. Riemschneider, *et al.*, *Z. Naturforsh*, **106**, 605 (1955).

170. Riemschneider, *et al.*, *Z. Naturforsh*, **116**, 675 (1956).

171. Riemschneider, *et al.*, *Anz. Schadlingskunde*, **23**, 62 (1950).

172. Riemschneider, *et al.*, *Chem. Zentr.*, **II**, 2515 (1951).

173. Rosin and Radan, *Anal. Chem.*, **25**, 817 (1953).

174. Schneider, Heuer, *Z. Naturforsch*, **8b**, 695 (1953).

175. Schopp, Brindley, Hinman, *J. Econ. Entomol.*, **46**, 860 (1953).

176. Schread, *Conn. Agr. Expt. Sta. Circ.*, **184**, (1953).

177. Schwabe, and Rammelt, *Z. physik Chem. (Liepzig)*, **204**, 310 (1955).

178. Schwabe, and Seiten, *Chem. Tech.* (Berlin), **3**, 3 (1951).

179. Schwabe, *Z. Electrochem.*, **60**, 151 (1956).

180. Schwerdtfeger, *Chem. Zentr.*, **1951**, (II), 2942.

181. Seal and Eden, *J. Econ. Entomol.*, **49**, 262 (1956).

182. Shaw, *Plant Panthol.*, **4**, 11 (1955).

183. Slade, *Chem. and Ind.*, p. 314, October 13 (1945).

184. Smith, Noyes, Hart, *J. Am. Chem. Soc.*, **55**, 4444 (1933).

185. Sobotka, *Research* (London), **2**, Suppl. 393 (1949).

186. Sperone, Jerine, Perini, *Chemica e Industria* (Milan), **234**, 4 (1952).

187. Sternberg and Kearns, *J. Econ. Entomol.*, **49**, 548 (1956).

188. Stone and Foley, *J. Econ. Entomol.*, **46**, 1075 (1953).
189. Suzuki, *Japan J. Exptl. Med.*, **23**, 487 (1953).
190. Suzuki, *Japan J. Exptl. Med.*, **24**, 181 (1954).
191. Suzuki, *Japan J. Exptl. Med.*, **23**, 487 (1953).
192. Tagiltsev, *Med. Parazitoli Parazitar Bolezui*, **1953**, 450.
193. Taki, *Spec. Rept. Forestry Expt. Sta., Hakkaido*, (5) 31 (1956).
194. Tenket and Bone, *J. Econ. Entomol.*, **45**, 218 (1952).
195. Thomas and Bevan, *Plant Pathol.*, **3**, 124 (1954).
196. Thomas and Bevan, *Plant Pathol.*, **5**, 115 (1956).
197. Thorp, and de Meillon, *Nature* (London) **160**, 264 (1947).
198. Thurston, Stark, Boush, *J. Econ. Entomol.*, **49**, 828 (1956).
199. Tidman, *Plant Protect. Overs. Res.*, **1949**, 1, 42.
200. Tirunarayama and Sarma, *J. Sci. Ind. Research (Ind.) BB*, **488**, (1954).
201. Todzdowski and Moose, *J. Econ. Entomol.*, **44**, 1016 (1951).
202. Toops, and Riddick, *Anal. Chem.*, **23**, 1106 (1951).
203. Trenner, Walker, Arison, Buks, *Anal. Chem.*, **21**, 285 (1949).
204. Van Dinther, *Tydschi Phanteriziekten*, **59**, 217 (1953).
205. Van Vloten, *et al.*, *Nature*, **162**, 771 (1948).
206. Van Vloten, *et al.*, *Acta Cryst.*, **3**, 139 (1950).
207. Vashkov, Serebiyakova, *Parasitic Dis. (U.S.S.R.)*, **16**, 41 (1947).
208. Verma, *J. Econ. Entomol.*, **48**, 205 (1955).
209. Wakefield, *Chem. and Ind.*, **46**, 367 (1945).
210. Wallace, *Austr., Commonwealth Sci. Ind. Res. Org. Dir. Ent. Paper 1* (1951).
211. Walton, Traub, Newsom, *Am. J. Trop. Med. Hyg.*, **5**, 190 (1957).
212. Weintraub, *Proc. Entomol. Soc. Brit. Columbia*, **48**, 51 (1952).
213. Weightman, *Chem. Ind.*, **58**, 604 (1937).
214. Wilcox, and Howland, *J. Econ. Entomol.*, **48**, 581 (1955).
215. Winteringham, and Barnes, *Physical Revs.*, **35**, 701 (1955).
216. Woodard, Davidow, Lehman, *Eng. Chem.*, **40**, 711 (1948).
217. Woodside, and Turner, *J. Econ. Entomol.*, **49**, 640 (1956).
218. Wright, *Natl. Vegetable Research Sta.*, 2nd Report, 10, 1950–51.

United States Patents

	Year	Patent No.	Holder	Company Assigned
219.	1936	2,010,841	Bender	Great Western Electro-Chem. Co.
220.	1940	2,218,148	Hardy	Imperial Chemical Industries, Ltd.
221.	1948	2,438,900	Cooke & Smart	Imperial Chemical Industries, Ltd.
222.	1948	2,440,082	Flanders & Jones	Imperial Chemical Industries, Ltd.
223.	1949	2,474,590	Morey	Commercial Solvents Corp.
224.	1949	2,486,688	Thomas, Stager	Shell Development Company
225.	1950	2,499,120	D. B. Storman	Hooker Electrochemical Co.
226.	1950	2,499,396	Lynn	Dow Chemical Co.
227.	1950	2,502,258	Hay & Webster	Imperial Chemical Industries, Ltd.
228.	1950	2,513,092	Gonze	Solvay & Cie
229.	1950	2,523,420	Burrage	Imperial Chemical Industries, Ltd.
230.	1950	2,524,970	Gonze	Solvay & Cie
231.	1950	2,529,803	Gonze	Solvay & Cie

United States Patents—Continued

	Year	Patent No.	Holder	Company Assigned
232.	1950	2,532,349	Taylor & Holm	Imperial Chemical Industries, Ltd.
233.	1950	2,534,485	Towle	Pittsburgh Plate Glass Co.
234.	1950	2,534,926	Rea	
235.	1951	2,538,595	Sharp	Imperial Chemical Industries, Ltd.
236.	1951	2,539,269	Parr	Penn. Salt Manufacturing Co.
237.	1951	2,542,602	Christopher & Spear	American Aerovap, Inc.
238.	1951	2,543,955	Boyd	Standard Oil Co. of Indiana
239.	1951	2,550,046	H. L. de Wall	Agricura Laboratoria, Ltd.
240.	1951	2,552,562	Kauer, Alquist, Britton	Dow Chemical Company
241.	1951	2,553,956	Burrage & Smart	Imperial Chemical Industries, Ltd.
242.	1951	2,554,274	Smith, Hill, Cantsell	Gulf Oil Corp.
243.	1951	2,557,814	Dinsdale, Holmes, Martin	Waeco, Ltd.
244.	1951	2,558,363	Kolka & Orloff	Ethyl Corp.
245.	1951	2,559,569	Orloff	Ethyl Corp.
246.	1951	2,564,406	Neher and Hall Kolka & Orloff	Ethyl Corp.
247.	1951	2,567,034	Sconic	Dow Chemical Co.
248.	1951	2,569,441	Alquist, Wasco, Kauer	Dow Chemical Co.
249.	1951	2,569,677	LaLande, Aeugler, Molyneux	Pennsylvania Salt Manfg. Co.
250.	1951	2,572,002	Berl	Tennessee Products & Chem. Co.
251.	1951	2,573,676	Campbell	Stauffer Chemical Co.
252.	1951	2,574,165	Bender & Pitt	Stauffer Chemical Co.
253.	1951	2,575,098	Crawford & Isquier	Phillips Petroleum Co.
254.	1951	2,578,858	Taylor, Holm, Hutchinson	Imperial Chemical Industries, Ltd.
255.	1952	2,584,376	Williams	Dow Chemical Company
256.	1952	2,584,992	Dykstra	E. I. du Pont de Nemours & Co.
257.	1952	2,585,755	Duspera	Pest Control Ltd.
258.	1952	2,585,898	Kauer	Dow Chemical Co.
259.	1952	2,586,519	Collins	Allied Chemical & Dye Corp.
260.	1952	2,590,529	Gillies & Rowe	Imperial Chemical Industries, Ltd.
261.	1952	2,603,664	Burrage	Imperial Chemical Industries, Ltd.
262.	1952	2,606,858	Gillies, Cunningham	Imperial Chemical Industries, Ltd.
263.	1952	2,606,868	Alquist, Kauer	
264.	1952	2,607,723	Pranfetti, Sexton, Williams	Food Machinery & Chemical Corp.
265.	1952	2,610,940	Endicot	Gaspray Corp.
266.	1952	2,622,105	Miller, Dunn, Neher, Hall	Ethyl Corp.
267.	1952	2,622,205	Miller, Dunn, Neher	Ethyl Corp.

United States Patents—Continued

	Year	Patent No.	Holder	Company Assigned
268.	1953	2,627,488	Zakheim	Pennsylvania Salt Manfg. Co.
269.	1953	2,628,260	Britton, Alquist, Kauer	Dow Chemical Co.
270.	1953	2,633,444	Marke	Imperial Chemical Industries, Ltd.
271.	1953	2,642,351	Swezey	Dow Chemical Co.
272.	1953	2,656,313	Miller, Dunn, Weber	Ethyl Corp.
273.	1953	2,658,017	Marhofer	Phillips Petroleum Co.
274.	1953	2,662,186	Governale, Horn, O'Connell	Ethyl Corp.
275.	1953	2,662,924	Humphreys	Ethyl Corp.
276.	1954	2,667,438	Gardner	California Spray-Chemical Corp.
277.	1954	2,673,857	Tryon	Commercial Solvents Corp.
278.	1954	2,673,883	Tryon	Commercial Solvents Corp.
279.	1954	2,674,569	Burrage & Beverage	Imperial Chemical Industries, Ltd.
280.	1954	2,691,050	Goenee & Stedeharder	Koninkligh Industrieele maatschapij voorhun noury & van der Lande N.V.
281.	1954	2,691,051	Orloff, Worrel	Ethyl Corp.
282.	1954	2,691,625	Clark	Ethyl Corp.
283.	1954	2,692,900	Bissinger	Columbia Southern Chemical Corp.
284.	1955	2,696,509	LaLande, Molyneux Aeugle	Pennsylvania Salt Manf. Co.
285.	1955	2,699,455	Smith & Sconce	Hooker Electrochemical Co.
286.	1955	2,699,456	Kimball & Smith	Hooker Electrochemical Co.
287.	1955	2,700,011	Taylor	Imperial Chemical Industries, Ltd.
288.	1955	2,705,731	Dunn	Ethyl Corp.
289.	1955	2,706,172	Dunn, Hall, Miller, Neher	Ethyl Corp.
290.	1955	2,705,731	Dunn	Ethyl Corp.
291.	1955	2,707,695	Courtier	Societé anon. Des manufactures des glaces et produits chemiques de Saint Gobain Chauny & Arey
292.	1955	2,708,681	Vossen	Solvay & Cie
293.	1955	2,713,076	Ellsworth & Hersehorn	Columbia Southern Corp.
294.	1955	2,714,572	Hansen	California Spray-Chemical Corp.
295a	1955	2,717,233	Trotter	Ethyl Corp.
295b	1955	2,717,238	Neubauer, Strain, Kung	Columbia Southern Chem. Corp.
296.	1955	2,717,272	Trotter	Ethyl Corp.
297.	1955	2,718,531	Berl, Smith, Gricon	Tennessee Products & Chem. Corp.
298.	1955	2,719,869	Tryon	Commercial Solvents Corp.

United States Patents—Continued

	Year	Patent No.	Holder	Company Assigned
299.	1955	2,719,870	Tryon	Commercial Solvents Corp.
300.	1955	2,724,002	Viriot	Solvay & Cie
301.	1955	2,724,003	Piester	Columbia Southern Chem. Corp.
302.	1955	2,725,404	Montes	Ethyl Corp.
303.	1955	2,725,406	Merritt	Ethyl Corp.
304.	1955	2,727,824	P. ter Horst	Olin Mathieson Chemical Corp.
305.	1955	2,728,799	Hetrick & Donaldson	Ethyl Corp.
306.	1956	2,729,603	Clark & Hall	Ethyl Corp.
307.	1956	2,729,683	Kolka & Orloff	Ethyl Corp.
308.	1956	2,729,684	Donaldson, Hetrick	Ethyl Corp.
309.	1956	2,729,686	Humpreys	Ethyl Corp.
310.	1956	2,731,338	Fike & Morrell	Monsanto Chemical Corp.
311.	1956	2,731,409	Nicolaisen	Olin Mathieson Chemical Corp.
312.	1956	2,733,276	Bartlett	Columbia Southern Chemical Corp.
313.	1956	2,739,175	Brodbacker	Commercial Solvents Corp.
314.	1956	2,739,988	Kung	Columbia Southern Chemical Corp.
315.	1956	2,740,818	Nicolaisen	Olin Mathieson Chemical Corp.
316.	1956	2,742,342	Dew & Sliepcevich	American Cyanamid Co.
317.	1956	2,742,508	Johnson	Commercial Solvents Corp.
318.	1956	2,743,282	Thonnessen	Dr. F. Roschig, G.M.B.H.
319.	1956	2,743,548	Christopher & Spear	Americal Aerovap, Inc.
320.	1956	2,744,145	Twiehans & Prindle	Columbia Southern Chemical Corp.
321.	1956	2,744,146	Dehn	Columbia Southern Chemical Corp.
322.	1956	2,744,862	Nicolaisen	Olin Mathieson Chemical Corp.
323.	1956	2,745,883	Jenny	Olin Mathieson Chemical Corp.
324.	1956	2,749,373	Meyer	Dow Chemical Corp.
325.	1956	2,755,235	Governale	Ethyl Corp.
326.	1956	2,757,211	Giraitis, Humpreys, Leeper	Ethyl Corp.
327.	1956	2,758,077	Lande, Knorr, Aeugle	Pennsylvania Salt Mfg. Co.
328.	1956	2,758,078	Lande, Knorr, Aeugle	Pennsylvania Salt Mfg. Co.
329.	1956	2,759,903	Epstein, Falck	Aaron S. Epstein
330.	1956	2,759,982	Pianfetti, Seaton, Williams	Food Machinery
331.	1956	2,760,900	Glenn, Dowling	United States Rubber Co.
332.	1956	2,760,995	Cragg, Dunn	Ethyl Corp.
333.	1956	2,761,805	Huidobro, Escobar	

United States Patents—Continued

	Year	Patent No.	Holder	Company Assigned
334.	1956	2,765,255	Swarbrick	Shell Development Co.
335.	1956	2,765,272	Neubauer, Strain, Kung, Bissinger	Columbia Southern Chemical Corp.
336.	1956	2,767,223	Donaldson, Park	Allied Chemical & Dye Corp.
337.	1956	2,767,224	Kimball	Hooker Electrochemical Co.
338.	1956	2,768,111	Butler, Harvey	Ethyl Corp.
339.	1956	2,768,216	Schmitz	Commercial Solvents Corp.
340.	1956	2,770,658	Lowdermilk, Bruce	Pennsylvania Salt Manufg. Co.
341.	1956	2,773,102	Clark, Cragg	Ethyl Corp.
342.	1956	2,773,103	Calingaert	Ethyl Corp.
343.	1956	2,773,105	Kolka	Ethyl Corp.
344.	1957	2,777,795	Le Veaux	Food Machinery & Chemical Corp.
345.	1957	2,778,860	McCoy, Inman	Pennsylvania Salt Mfg. Co.
346.	1957	2,786,012	McHan	Calcium Carbonate Co.
347.	1957	2,787,645	Danzker, Wood	Olin Mathieson Chemical Corp.
348.	1957	2,788,307	Torrent, Fitzwater	
349.	1957	2,789,077	Zarheim	Pennsylvania Salt Mfg. Co.
350.	1957	2,790,012	Dunn, Grandjean, Padgitt	Ethyl Corp.
351.	1957	2,793,154	Shillitoe, Harford	British Petroleum Co., Ltd.
352.	1957	2,797,195	Neubauer, Strain Jung, Dehn	Columbia Southern Chemical Corp.
353.	1957	2,801,745	Piester	Columbia Southern Chemical Corp.
354.	1957	2,777,882	Humpreys	Ethyl Corp.
355.	1957	2,788,320	Bracey	
356.	1957	2,792,434	Becke, Durkheim, Sperber	Badische Anilin-u Soda-Fabrik Akt.-Ges.
357.	1957	2,803,673	Johnson	Commercial Solvents Corp.
358.	1957	2,806,070	Mattner	Schering A-G
359.	1957	2,806,523	Nicolaisen	Olin Mathieson Chemical Corp.
360.	1957	2,812,368	Gruber, Filliettaz	Dr. R. Maag A-G Chemische Fabrik Dielsdorf
361.	1957	2,812,369	Evans	Columbia Southern Chemical Corp.
362.	1957	2,812,370	Bell, Haines	Pennsalt Chemicals Corp.
363.	1958	2,819,198	Goodhue	Phillips Petroleum Co.
364.	1958	2,821,500	Jackson, Mayeux Head	Wilson, Toomer Fertilizer Co.
365.	1958	2,826,616	Hetrick	Ethyl Corp.
366.	1958	2,827,501	Walach, Kudszns	C. H., Boehringer Sohn
367.	1958	2,834,817	Degeorges, Lehureau	Societe Progil
368.	1958	2,842,476	Schreiber	McLaughlin Gormley King Co.

United States Patents—Continued

	Year	Patent No.	Holder	Company Assigned
369.	1958	2,845,380	Mayhew, Numm	General Aniline & Film Corp.
370.	1958	2,846,011	Miller	Pan American Petroleum Corp.
371.	1958	2,849,498	Danzker	Olin Mathieson Chemical Corp.
372.	1958	2,852,465	Jensen	Esso Research & Engineering Co.
373.	1958	2,852,571	Frejacques, Guinst	Pechiney Compagnie
374.	1958	2,854,491	Kung	Columbia Southern Chemical Corp.
375.	1958	2,857,437	Pitt, Bender	Stauffer Chemical Co.
376.	1958	2,858,260	Neubauer, Strain, Kung, Dehn	Columbia Southern Chemical Corp.
377.	1958	2,859,252	Lawlor, Miville	Pennsalt Chemicals Corp.
378.	1958	2,859,822	Wright	Pan American Petroleum
379.	1959	2,870,335	Pitt	Stauffer Chemical Co.
380.	1959	2,872,368	Sanders, Knaggs, Nussbaum	
381.	1959	2,874,769	Bell, Stafford, Lowdermilk	Pennsalt Chemicals Corp.
382.	1959	2,875,128	Kirkpatrick, Land, Seale	Visco Products Co.
383.	1959	2,880,246	Walach, Kottler Scheffler	C. H. Boehringer Sohn
384.	1959	2,884,128	Witman	Columbia Southern Chemical Corp.
385.	1959	2,886,605	McClure, Melbert, Hoblit	Dow Chemical Co.
386.	1959	2,886,606	Mattner	Schering ATK
387.	1959	2,889,376	McConnell, Jones	Imperial Chemical Industries, Ltd.
388.	1959	2,893,913	Wiedow	Monsanto Chemical Company
389.	1959	2,895,998	Crowder, Gilbert	Allied Chemical Corp.
390.	1959	2,899,771	Burris	E. I. du Pont de Nemours & Co.
391.	1959	2,902,520	Chuffort	Imperial Chemical Industries, Ltd.
392.	1959	2,904,475	Bell	Pennsalt Chemical Corp.
393.	1959	2,904,597	Aepli	Pennsalt Chemical Corp.
394.	1959	2,905,609	Bermano	Solvay & Cie
395.	1959	2,906,744	Jancosek, Brown	Standard Oil (Indiana)
396.	1959	2,911,342	Kung	Columbia Southern Chemical Corp.
397.	1959	2,911,446	Nicolaisen, Wood	Olin Mathieson Chemical Corp.
398.	1959	2,913,372	Velde, Linke	Atlas Powder Company
399.	1959	2,914,573	McCoy, Inman, Kyker	Pennsalt Chemicals Corp.
400.	1959	2,914,574	Zinn, Luces	Diamond Alkali Company
401.	1959	2,916,414	Raecke, Dreners	Henkel & Cie
402.	1959	2,918,500	Cragg	Ethyl Corp.

United States Patents—Continued

	Year	Patent No.	Holder	Company Assigned
403.	1959	2,926,197	Kung	Columbia Southern Chemical Corp.
404.	1960	2,926,198	Kung, Stevens	Columbia Southern Chemical Corp.
405.	1960	2,942,035	Strain, Bissinger	Columbia Southern Chemical Corp.

Selected Foreign Patents with no U. S. Counterpart

	Year	Patent No.	Company Assigned
406.	1947	Belgian 469,299	Solvay & Cie
407.	1947	Belgian 471,772	Technique Chemique Belge, soc. anon.
408.	1947	Brit. 646,917	Dupine
409.	1958	Brit. 792,610	Columbia Southern Chemical Corp.
410.	1951	Can. 476,480	Burrage, Williams
411.		Ger. PA 114239-8351	
412.	1957	U.S.S.R. 105,898	Strongin

22. MISCELLANEOUS CHLORINATED ORGANIC PESTICIDES

EDWARD D. WEIL

Hooker Chemical Corporation

INTRODUCTION

The large production rates of DDT and benzene hexachloride have warranted the detailed discussion of these insecticides in separate chapters of the present volume. The majority of the other pesticides currently in use contain chlorine or are made using chlorine-containing intermediates. It is estimated from 1959 Tariff Commission production figures[119] that well over 300 million pounds of chlorine per year are used in the production of chlorinated pesticides (exclusive of DDT and BHC).

This survey, therefore, will present the history, the principal uses, and the known or probable methods of manufacture of these pesticides. Reliable production figures can be given on only a few of these products since most of them are manufactured by only one or two companies.

Trade names of products discussed herein are placed in quotation marks; common names are uncapitalized. No attempt has been made to list exhaustively the frequently-numerous alternative common names and trade names which have been employed for these pesticides.

The reader interested in the current approved uses and recommended rates of application for these pesticides is advised to consult the labels and brochures put out by the respective manufacturers as well as the recommendation leaflets of the appropriate government agricultural agencies, since revision is frequent and local variations are numerous.

INSECTICIDES AND MITICIDES

Toxaphene (Chlorinated Camphene)

Toxaphene was invented by Buntin of Hercules Powder Company.[18] In 1947 it was introduced as an insecticide by Hercules, who remains the sole manufacturer. In 1956 toxaphene was reported to have reached sales of well over 40 million pounds per year.

Toxaphene is used in agricultural and household insect control, the former use being by far the more important. Toxaphene has broad effectiveness against many of the common insect pests of cotton, and usually it gives control lasting for several weeks. Toxaphene also is used for insect control on cereal crops, forage crops, fruits, potatoes, truck crops, tobacco, and livestock. In household formulations such as aerosols, toxaphene is effective against houseflies and cockroaches. In bait formulations, it is effective against grasshoppers.

Toxaphene is a mixture of chlorination products of camphene, containing 67 to 69 percent chlorine and thus having the empirical formula $C_{10}H_{10}Cl_8$. The structural chemistry of the end product is obscure. It appears that the components may not differ greatly from one another, at least in solubility properties, since only a single spot is obtained by use of a paper chromatographic technique which is capable of resolving BHC isomers and the components of chlordane.[85] The starting material, camphene, is made by isomerization of pinene, a major constituent of turpentine.

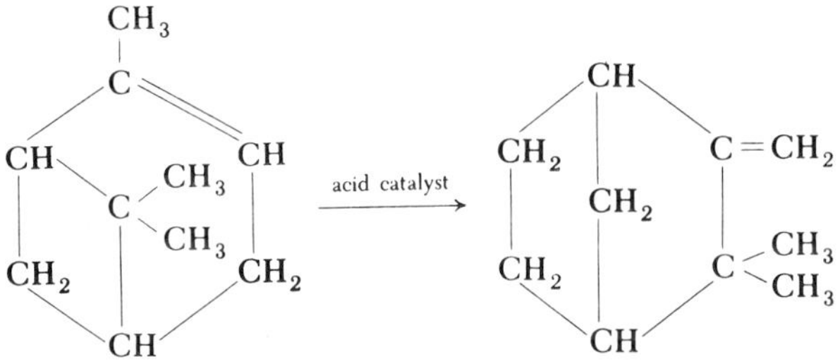

acid catalyst

Pinene   *Camphene*

The chlorination, according to the patent of Buntin,[18] is conveniently carried out in a solvent such as carbon tetrachloride using photochemical catalysis. The reaction is highly exothermic at first, but toward the end, heat must be supplied. Finally the solvent is evaporated off and acidic impurities are removed by washing the residue. The end product is a hard, yellowish-amber wax with a melting range of 70 to 95° C and a specific gravity of 1.66 (27° C).

The structure of the end product may or may not contain the camphene skeleton. Tischenko showed[118] that on monochlorination of camphene, some products retain the camphene ring, but also certain skeletal rearrangements take place yielding at least one product with the tricyclic structure:

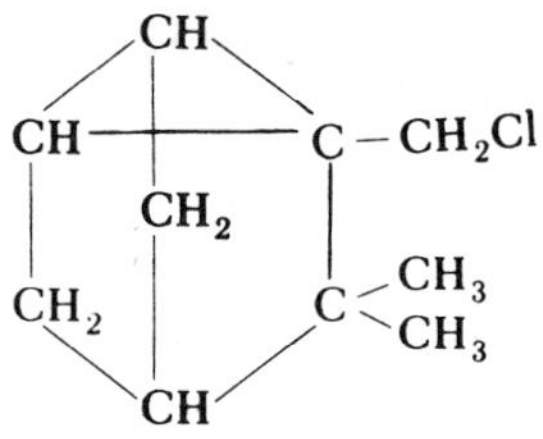

$$
\begin{array}{c}
\text{CH} \\
\text{CH} \text{---} \quad \text{C} - \text{CH}_2\text{Cl} \\
\text{CH}_2 \\
\text{CH}_2 \qquad \text{C} \underset{\text{CH}_3}{\overset{\text{CH}_3}{<}} \\
\text{CH}
\end{array}
$$

A related chlorinated terpene containing 65–66 percent chlorine is available commercially under the trade name "Strobane." It has a pattern of activity similar to toxaphene.

Relatives of DDT

Since the recognition of DDT as an insecticide of major importance, a large number of chlorine substituted 1,1-diarylalkanes have been evaluated for insecticidal activity. The published work on experimental DDT analogs has been surveyed by Frear.[38]

The DDT analogs of commercial significance are the following:

2,2 - Bis(*p* - methoxyphenyl) - 1,1,1 - trichloroethane (methoxychlor). The insecticidal properties of methoxychlor were discovered by Geigy researchers and reported in 1944 by Läuger, Martin, and Müller.[75] The insecticidal use of methoxychlor was developed and patented in the United States by du Pont.[14]

The principal advantages of methoxychlor over DDT are lower mammalian and avian toxicity, and lesser tendency to accumulate in mammalian fatty tissues. Methoxychlor is, however, more costly than DDT. Because of its low toxicity, it may be used for insect control in grain storage bins, dairy barns, and on certain crops such as potatoes, forage crops, fruits, and vegetables up to 3 to 21 days before harvest. Methoxychlor is also widely used in household aerosol insecticides.

The manufacture of methoxychlor on a pilot plant scale has been described by Scheller and Smith.[94] These workers prepared methoxychlor by the reaction of anisole and chloral, using a process analogous to that for DDT (*q.v.*), but commercial concentrated sulfuric acid was added in place of the 100 percent acid or oleum employed for DDT synthesis. Reaction times of one hour were required at 25 to 30° C. Carbon tetrachloride was added after a time to permit the thick mixture to be stirred. The mixture was quenched in cold water, the product slurry washed and stripped of carbon tetrachloride.

The technical product contains about 88 to 89 percent of the active *p,p'*-isomer, the rest being mostly the relatively inactive *o,p'*-isomer.

2,2-Bis(p-chlorophenyl)-1,1-dichloroethane (TDE or DDD). TDE was first described as an insecticide in 1944 by Läuger *et al.* (Geigy) in Switzerland.[75] It was also investigated by Schering and I. G. Farbenindustrie in Germany during World War II. It is manufactured and sold in the United States by Rohm and Haas.

TDE is particularly effective on the red-banded leaf roller (a pest of apples), and on the various lepidopterous larvae (hornworm, fruitworm, etc.) which attack tomatoes and tobacco. It has lower mammalian toxicity than DDT.

Although TDE can be synthesized in a manner analogous to DDT from chlorobenzene and dichloroacetaldehyde, the latter is not readily made in good yield. Meitzner and Hester (Rohm and Haas)[81] have described a more practical process wherein ethanol is chlorinated at below 35° C until a layer with a specific gravity of about 1.29 and a chlorine content of 50 to 60 percent is formed. The principal component is a trichloroethyl ether, $CHCl_2—CHCl—O—C_2H_5$. This layer, without purification, is then condensed with chlorobenzene using 98 percent sulfuric acid as the condensation medium at between 0 and 75°. More recently, Rohm and Haas workers[78] have found that by separating the trichloroethyl ether layer and refluxing it at atmospheric pressure, dehydrochlorination occurs and 2,2-dichlorovinyl ethyl ether can be fractionated off. Condensing the latter with chlorobenzene at 10 to 25° C in 99 percent sulfuric acid or oleum gives TDE of better purity and grindability than that made directly from the ethanol chlorination product.

$$C_2H_5OH + Cl_2 \rightarrow CHCl_2CHCl—OC_2H_5 \xrightarrow{-HCl} CCl_2{=}CH—OC_2H_5$$

$$\xrightarrow[H_2SO_4]{ClC_6H_5} CHCl_2—CH(C_6H_4Cl)_2$$

1,1-Bis(p-ethylphenyl)-2,2-dichloroethane ("Perthane"). "Perthane" was introduced in 1950 by Rohm and Haas Company and is also the outcome of research based on DDT analogs. It has a very low mammalian toxicity and thus may be used on forage and food crops (especially cole crops) close to harvest. In regard to this characteristic, it competes with Du Pont's methoxychlor. "Perthane" also finds commercial use in mothproofing of wool.

"Perthane" may be manufactured by a process analogous to that used for TDE.

1,1-Bis(p-chlorophenyl)-2,2,2-trichloroethanol ("Kelthane"). "Kelthane" was discovered as a result of an extended investigation of *alpha-*

substituted benzhydrols by the Rohm and Haas Company[122] and was introduced by Rohm and Haas in 1955.

It is exclusively an acaricide (miticide) having relatively low insecticidal activity.[6] It is used principally as a summer miticide, being especially toxic on motile forms of mites, and it gives an extended residual control. The principal uses are on deciduous fruits, citrus, cotton, irrigated vegetables, and field crops.

"Kelthane" may be produced by chlorination of bis(*p*-chlorophenyl)-methyl carbinol.[92] However, the most probable commercial process is that described by Wilson and Wolff.[123] In one example of this process, 500 parts of 1,1-bis(*p*-chlorophenyl)-1,2,2,2-tetrachloroethane (produced presumably by chlorination of DDT), 29 parts of water, 12 parts of 96 percent sulfuric acid, and 98.5 parts of *p*-toluenesulfonic acid are refluxed at 140° C until the theoretical amount of HCl is evolved.

Hexachlorocyclopentadiene-derived Insecticides

This important group of insecticides is based on the Diels-Alder reaction products of certain olefins with hexachlorocyclopentadiene, as shown in the following general equation:

$$\text{hexachlorocyclopentadiene} + \text{olefin} \longrightarrow \text{Diels-Alder adduct}$$

While a large number of Diels-Alder adducts of hexachlorocyclopentadiene are known,[118c] relatively few of these adducts have useful insecticidal properties, and the majority of such adducts are non-insecticidal.

The preparation of hexachlorocyclopentadiene has been reviewed by Ungnade and McBee.[118c] A variety of processes, many of only academic interest, have been described. Two processes of commercial importance are: (a) chlorination of pentane photochemically in the liquid phase to polychloropentane followed by vapor phase catalytic chlorination of polychloropentane to hexachlorocyclopentadiene and/or octachlorocyclopentene (the latter being thermally cracked to hexachlorocyclopentadiene),[72b,80b,82a,92e] and (b) the chlorination of cyclopentadiene by alkaline hypochlorite solution.[72c,92b,108b]

Chlordane. Early work on the synthesis and insecticidal properties of chlordane was done by Hyman at the Velsicol Corporation in the mid-1940's.[60a]

Chlordane has broad-spectrum activity as a contact and ingestion insecticide, with especially notable activity on ants, grasshoppers, termites, houseflies, roaches, and many species of beetle adults and grubs. Consequently, it is employed on fruits, vegetables, livestock, turf, and for household insect control. At high rates of application to turf, it also exerts selective pre-emergence herbicidal action on crabgrass, and has been used in important amounts for this purpose.

Chemically, chlordane is 2,3,4,5,6,7,8,8-octachloro-2,3,3a,4,5,5a-hexahydro-4,7-methanoindene.

The technical product is a yellowish viscous fluid with a camphoraceous odor. Chromatographic separation of technical chlordane yielded heptachlor (see below), "chlordene" (the hexachlorocyclopentadiene-cyclopentadiene adduct), and two isomeric chlordanes, (α and β-chlordane) believed to differ by having the chlorines at the 2 and 3 positions in a cis and trans relationship.[78b] Nonachlorinated analogs have also been isolated from technical chlordane.[120b]

Chlordane is manufactured by the Diels-Alder addition of hexachlorocyclopentadiene to cyclopentadiene (the former functioning as a diene, the latter as a dienophile) followed by subsequent addition of chlorine to the adduct ("chlordene").

"Chlordene"

The Diels-Alder addition reaction proceeds under relatively mild conditions (self-initiating at room temperature) and in good yield.[51c,92a,92c]

The chlorination to chlordane is complicated by the fact that substitutive as well as additive chlorination can occur. However, chlordane is the principal product when one mole of chlorine is introduced into chlordene in the liquid phase below 200° (Hyman, *loc. cit*) The use of sulfuryl chloride in the presence of a Lewis acid catalyst (AlCl$_3$, SnCl$_4$, FeCl$_3$) has been claimed.[72d]

Heptachlor. Chromatographic separation of chlordane showed the presence of a highly insecticidal heptachlorinated compound containing two double bonds.[78b,109a,120a] Methods of direct synthesis from chlordene were developed by workers at Velsicol.[60a]

Heptachlor, like chlordane, is active as a contact and ingestion insecticide versus a wide variety of insects. Major applications include control of cutworms, boll weevil, grasshoppers, alfalfa weevil, spittlebug, onion thrips, armyworms, and root weevils. Major crops treated with heptachlor include corn, wheat and other grains, cotton, alfalfa, sugar cane, potatoes, cabbage, and onions.

Heptachlor, chemically, is 1,4,5,6,7,8,8-heptachloro-3a, 4, 7, 7a-tetrahydro-4,7-*endo*methanoindene.

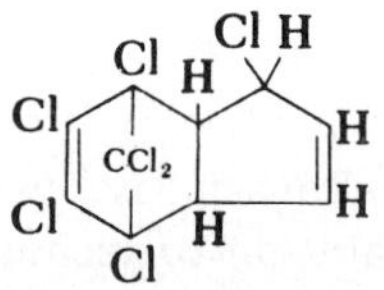

The technical product is said to consist of 72% of this compound accompanied by related compounds.[79a]

Heptachlor is manufactured by the substitutive chlorination of "chlordene" (see under chlordane), under conditions whereby the additive chlorination is suppressed.

Several variations of the basic process have been described: the chlorination of chlordene by sulfuryl chloride in the presence of catalytic amounts of benzoyl peroxide (Hyman, *loc. cit.*) and the chlorination of chlordene in the dark in the presence of fullers' earth.

A Russian group has made a study of the process variables which favor substitution of chlorine into chlordene. In the presence of infusorial earth as a catalyst, the ratio of substitution to addition appears to increase as the temperature is lowered, and the best yields (70%) were at 5°.[120c]

Aldrin. The insecticidal utility of Aldrin was first described by Kearns in 1949.[65a]

Aldrin has been patented and developed as a commercial insecticide by Shell.[75a]

Aldrin possesses broad contact insecticidal properties, with rapid knockdown properties. Because of its stability in the soil and resistance to breakdown by alkaline fertilizers, aldrin can be used effectively in soil and turf insecticidal treatments, such as control of wireworms and rootworms in corn, cabbage root maggot, onion maggot, ants, root weevils, potato tuber beetles, and various grubs. Other uses for aldrin include control of grasshoppers, and various insect pests of forage, fruits, vegetables, grains and cotton.

Aldrin, chemically, is 1,2,3,4,10,10-hexachloro-1,4,4a,5,8,8a-hexahydro-1,4-*endo,exo*-5,8-dimethanonaphthalene, and has the following steric structure:

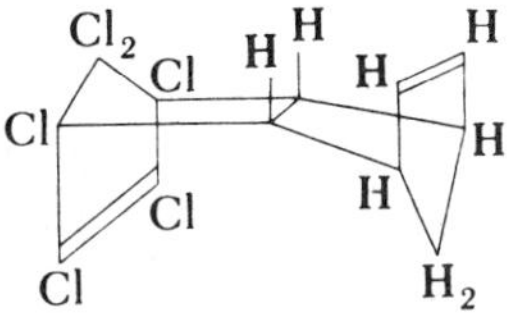

Aldrin is produced by the Diels-Alder addition of hexachlorocyclopentadiene with bicyclo [2.2.1]heptadiene-2,5. The latter is produced by the Diels-Alder addition of acetylene to cyclopentadiene.[60b]

An extensive study of the process variables has been published by Russian authors.[10a]

Bicyclo[2.2.1]heptadiene-2,5 is an extremely reactive dienophile toward hexachlorocyclopentadiene, and the Diels-Alder addition reaction proceeds readily at 90–110° C (Belikova, et al., *loc. cit.*).

Dieldrin. Dieldrin was developed by Julius Hyman & Co. and first described as an insecticide in 1949.[65a]

Dieldrin has been developed commercially by the Shell Chemical Corporation who hold a patent on the composition and insecticidal use.[102a]

It is a highly active contact insecticide of broad utility with pronounced residual properties. Having a high degree of chemical stability, dieldrin persists for many months in the soil and is effective for control of soil insects such as wireworms, corn seed maggots, and various soil-dwelling larvae. Dieldrin is used in control of many cotton insects, (including boll weevil), grasshoppers, alfalfa weevils, thrips, ants, termites, and insect pests of fruit. Dieldrin may be used in seed treatment to protect against seedcorn maggots, wireworms, and other soil insects.

Dieldrin is the 6,7-epoxide of aldrin, and has the steric structure

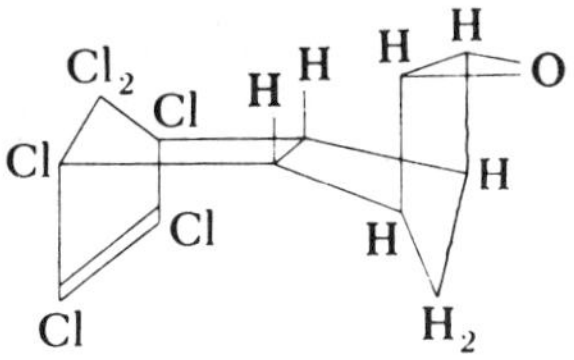

expressed by the systematic name 1,2,3,4,10,10-hexachloro-*exo*-6,7-epoxy-1,4,4a,5,6,7,8,8a-octahydro-1,4-*endo, exo*-4,8-dimethanonaphthalene.

Dieldrin may be manufactured by epoxidation of aldrin (*q.v.*), using as the oxidant a peroxyacid such as peracetic (Soloway, *loc. cit.*), or using hydrogen peroxide in the presence of pertungstic acid (or WO_3).[101b] Apparently hydrogen peroxide can also be used without a catalyst.[10a]

Endrin. Endrin is another member of the group of insecticides discovered by J. Hyman and coworkers. It has been commercialized in the United States by the Shell Corp. who hold a patent on the composition and insecticidal use.[12b]

Endrin has broad insecticidal activity of a high order and is especially effective on lepidopterous larvae (such as armyworms, cutworms, hornworms, cornborers, loopers, etc.) which are major pests on cotton, potatoes, tobacco, grains, and cabbage. It is highly effective against red-banded leaf rollers on apples, armyworms on cabbage and forage crops, fleabeetles, aphids, and cutworms. Endrin has rather substantial mammalian toxicity, and for this reason is a useful contact rodenticide.

Endrin is a stereoisomer of dieldrin, and has the following structure:

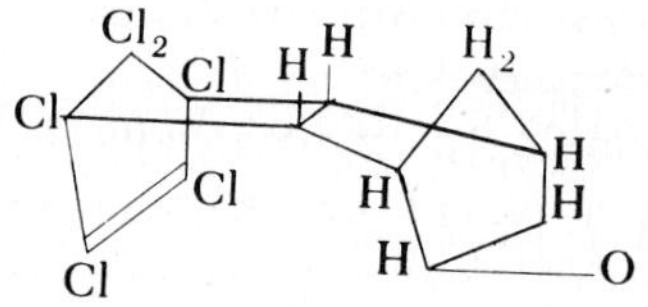

The technical material is a light tan powder of not less than 85 percent purity.

The manufacture of endrin requires first the preparation of "isodrin," itself an insecticidal compound, having a double bond at the site of the epoxy group of endrin.[75b] Isodrin is the cyclopentadiene adduct of 1,2,3,4,7,7-hexachlorobicyclo[2.2.1]heptadiene-2,5, which in turn may be produced by the Diels-Alder condensation of hexachlorocyclopentadiene with vinyl chloride, followed by dehydrochlorination by a strong base such as potassium hydroxide in alcohol.[4a,12b]

An alternative process for 1,2,3,4,7,7-hexachlorobicyclo[2.2.1]heptadiene-2,5 having the advantage of lower chemical cost and fewer steps, is the direct Diels-Alder reaction of acetylene with hexachlorocyclopentadiene.[56a,60b] According to a patent example, the reaction is conducted by bubbling acetylene thru the diene in a tubular nickel alloy reactor under pressure. A run at 300 psi, 155° C, and 6 hours time afforded a 95% yield at 59% conversion.

The Diels-Alder addition of the resultant hexachlorobicycloheptadiene with cyclopentadiene proceeds under mild conditions to yield "isodrin"

(Lidov, *loc. cit.*).   Isodrin is then epoxidized by processes using either peracetic acid or hydrogen peroxide plus pertungstic acid catalyst, as described under "dieldrin." The overall process is as follows:

Isodrin

Thiodan. Thiodan is a relatively new member of the hexachlorocyclopentadiene insecticide group.   The discovery of Thiodan at Farbwerke Hoechst (Germany) is said to have resulted from attempts to convert the corresponding diol to the dichloride using thionyl chloride.[38a,38b] Thiodan was commercialized as an insecticide in Germany by Hoechst and in the U. S. by Niagara Chemical Division of Food Machinery Corporation.

Thiodan is a contact and ingestion insecticide of broad activity spectrum. It is especially active on aphids, various lepidopterous larvae, flea beetles, Colorado potato beetle, boll weevil, and Mexican bean beetle. Major uses are on potatoes, tobacco, cotton, fruit, flowers, nursery crops, cabbage, and other vegetables.

Thiodan is 6,7,8,9,10,10-hexachloro-1,5,5a,6,9,9a-hexahydro-6,9-methano-2,4,3-benzodioxathiepin-3-oxide, a cyclic sulfite.

It is a mixture of two isomers both of which yield the same diol on mild hydrolysis.   Therefore, the isomerism is believed to result from the two possible orientations of the S=O bond relative to the CCl_2 bridge.[38a,76a,92d] Both isomers are insecticidal, apparently to a similar degree.

Processes described for the manufacture of Thiodan involve preparation of "HET diol," the hexachlorocyclopentadiene adduct of cis-2-butene-1,4-diol. The adduct may be made directly by the Diels-Alder condensation of cis-2-butene-1,4-diol with hexachlorocyclopentadiene. It has been found that a large enhancement of the yield in this reaction results from slow addition of the butenediol to the hexachlorocyclopentadiene at the reaction temperature, in the presence of an acid scavenger such as calcium carbonate or an epoxide.[40a]

Alternatively, "Het diol" may be made by condensing hexachlorocyclopentadiene with cis-2-butene-1,4-diol diacetate followed by hydrolysis of the adduct to the diol (Frensch, *loc. cit.*).

The "Het diol" is then treated with thionyl chloride between 50°ι and the reflux temperature. Hydrogen chloride is evolved, and the cyclic sulfite ring is formed.

Phosphorus Insecticides

Parathion and Methyl Parathion. Although neither of these compounds contain chlorine, their inclusion in this review is warranted because of their manufacture through chlorine-containing intermediates, the phosphorochloridothioates. The history and development of this group of insecticides has been reviewed up to 1949 by Cassaday *et al.* and by Hall in a previous ACS publication.[2]

Parathion is widely used for control of insects and mites on a large number of fruits, forage crops, and vegetables. Production of parathion was about 7,434,000 lb in 1960.[119a]

Methyl parathion has only a fraction of the toxicity of parathion. It is used on a large scale for cotton insect control. Methyl paratnion production was about 11,800,000 lb in 1960.[119a]

O,O-Diethyl phosphorochloridothioate, $(EtO)_2PSCl$, and its dimethyl homolog are key intermediates not only for parathion and methyl parathion but also for synthesis of a number of other phosphorus insecticides (see Table 22-1 below). There are three principal routes to these intermediates.

$$ROH \cdot + PSCl_3 + base \rightarrow (RO)_2PSCl + 2HCl \text{ (combined with base)} \quad (22\text{-}1)$$

$$ROH + P_2S_5 \rightarrow (RO)_2PSSH + H_2S$$

$$(RO)_2PSSH \text{ (or salt)} + Cl_2 \text{ (or chlorinating agent)} \rightarrow (RO)_2PSCl + S_2Cl_2 \quad (22\text{-}2)$$

$$ROH + PSCl_3 \rightarrow ROPSCl_2 + HCl$$

$$ROPSCl_2 + ROH + NaOH \rightarrow (RO)_2PSCl + NaCl \quad (22\text{-}3)$$

The first route (22-1) has been discussed by Cassaday *et al.*[2] More recently, the reaction conditions were reviewed by Mel'nikov *et al.*[82] in a Soviet symposium on phosphorus insecticides. It would appear that a practical method is to add the calculated amount of methanol or ethanol to

$PSCl_3$ at about $-8°$ C. Then about 10 percent more than the theoretical amount of NaOH is added as an aqueous or alcoholic solution, or as a dry powder. The mixture is then held at $20°$ C until the reaction is complete. By this method, 89 percent yields may be obtained.

The thiophosphoryl chloride required for this first route is produced, usually, in two steps from the elements.[120] First, PCl_3 is prepared by adding chlorine and liquid phosphorus to a refluxing vessel of already-formed PCl_3. The heat of reaction is largely removed by the reflux condenser. PCl_3 is continually distilled off, treated with excess chlorine to remove traces of dissolved phosphorus, and fractionated. The yield in this step is typically 95 percent. The conversion of PCl_3 to $PSCl_3$ is accomplished by passing the PCl_3 through excess molten sulfur or a solution of sulfur in an inert solvent. The reaction can be carried out at $180°$ C without catalyst, at 150 to $160°$ C with alkali or alkali earth sulfides as catalyst, or at $115°$ C with aluminum chloride as catalyst. The thiophosphoryl chloride is distilled out of the reaction mixture. Yields of 98 percent or better are obtained. A stainless steel autoclave is ordinarily required.

The second route (22-2) is believed to be the preferred commercial method for manufacture of O,O-dialkyl phosphorochloridothioates. Laboratory operation of the process has been described by Fletcher and co-workers[36a] of American Cyanamid Company. Gaseous chlorine was passed into a benzene solution of the O,O-dialkyl phosphorodithiolate at 25 to $30°$ C for $1\frac{1}{4}$ to $1\frac{1}{2}$ hours. The reaction mixture was then added to water to hydrolyze the sulfur chlorides, and the sulfur thus precipitated was removed by filtration. Over-all yields of diethyl and dimethyl phosphorochloridothioate were 59 and 53 percent respectively, based on P_2S_5. Bayer[64a] has patented a process for removing the by-product sulfur by means of sodium sulfite, thus saving a filtration, and a yield of 90 percent of diethyl phosphorochloridothioate (based on the dithiolate) is said to be obtained.

Several variations on this basic process have been patented. Hechenbleikner[51b] has described the use of chlorine in carbon tetrachloride at below $20°$ C for chlorination of the sodium salt as well as for chlorination of the free acid, and he also indicated that sulfur dichloride and phosphorus pentachloride could be used as chlorinating agents. Malatesta[78a] used chlorine diluted with nitrogen, and separated the organic product from the S_2Cl_2 by vacuum distillation. He also employed chlorine in methylene chloride and PCl_5 in carbon tetrachloride. Toy and McDonald[118b] have employed PCl_5 diluted with $PSCl_3$. Hu *et al.*[57] have claimed that sulfur monochloride may be employed as chlorinating agent. In a closely related process, Hechenbleikner[51a] employed sulfur monochloride or dichloride to chlorinate compounds of the type $(RO)_2PS—S_n—PS(OR)_2$ to $(RO)_2PSCl$.

A serious explosion has occurred in at least one plant which has utilized the direct chlorination route for the manufacture of the methyl parathion intermediate.[20]

The third route (22-3) is described by Bland and Young in a Monsanto patent.[12] Dry ethanol is added to $PSCl_3$ at 10°C; the mixture warmed to 47°C to drive off the HCl; then the crude $EtOPSCl_2$ treated at 15°C with NaOH in ethanol to form $(EtO)_2PSCl$ in 80 percent yield.

A novel route to O,O-dialkyl phosphorochloriodothioates has recently been described[80a] wherein an O,O-dialkyl phosphorodithioate is treated with hydrogen chloride in the presence of an H_2S acceptor such as a nitrile.

The reaction of diethyl or dimethyl phosphorochloridothioates with phenolic or phenol-like compounds was reviewed in 1949 by Cassaday.[2] Several interesting variations have been described since 1949. Toy[118a] found trialkylamines to be catalysts for the reaction with sodium *p*-nitrophenoxide in chlorobenzene. The catalytic activity of copper in this reaction was discovered by Schrader at Farbenfabriken Bayer.[98] Schrader also recommended the addition of a small amount of KBr as a co-catalyst.

Orochena (Pittsburgh Coke)[87] has described the reaction of a NaOEt suspension in toluene with $PSCl_3$ to form a solution of $(EtO)_2PSCl$, this being reacted with *p*-nitrophenol and Na_2CO_3 in toluene, using a copper catalyst at 85 to 90°C.

Young and Hensel (Pittsburgh Coke)[124] describe the reaction with sodium *p*-nitrophenoxide in a basic aqueous medium using powdered copper or $CuSO_4$ catalyst. A Monsanto patent[30] illustrates the carrying out of the reaction in alcohol at –10 to 10°C, forming the $(EtO)_2PSCl$ from sodium ethoxide and $PSCl_3$, and the sodium *p*-nitrophenoxide from sodium ethoxide and *p*-nitrophenol. The avoidance of excess *p*-nitrophenol is said to be essential. American Cyanamid[4] claims the use of Na_2CO_3 or Na_3PO_4 to form the salt of *p*-nitrophenol in acetone, no catalyst being necessary for the subsequent reaction with the phosphorochloridothioate in this solvent.

Other Phosphorus Insecticides from Chlorinated Intermediates. (See Table 22-1).

Sulfur-containing Miticides

2,4,4′,5-Tetrachlorodiphenyl Sulfone ("Tedion"). This miticide resulted from an extensive study of substituted diaryl sulfones by researchers of N. V. Philips-Roxane in the Netherlands.[58,59] "Tedion" was introduced experimentally into the United States in 1958 by Niagara Chemical Division of Food Machinery Corporation, licensees of Philips-Roxane, and is now a major commercial miticide.

TABLE 22-1. PHOSPHORUS INSECTICIDES FROM CHLORINATED INTERMEDIATES

Name	Structure	Manufacturer	Process	Principal Uses	Process Reference
"Trithion"	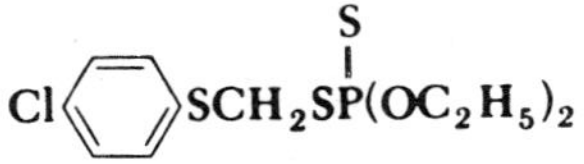 $Cl\langle\rangle SCH_2SP(OC_2H_5)_2$ (with =S)	Stauffer	condensation of p-chlorothiophenol with HCHO and HCl, then with salt of $(EtO)_2PSSH$	broad-range insecticide-miticide on fruit, vegetables, cotton; relatively residual	34
EPN	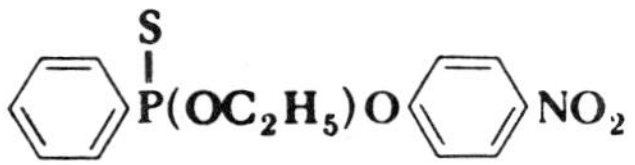 $\langle\rangle P(OC_2H_5)O\langle\rangle NO_2$ (with =S)	Du Pont	reaction of $C_6H_5PSCl_2$ (from benzene and $PSCl_3$) with $NaOC_2H_5$, then with p-$NO_2C_6H_4$-ONa	miticide on various fruits and vegetables, insecticide vs. corn borer	61
"Diazinon"	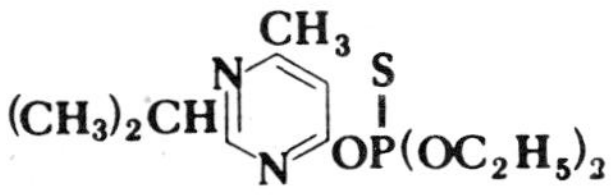 $(CH_3)_2CH$ pyrimidine with CH_3 and $OP(OC_2H_5)_2$ (with =S)	Geigy	reaction of $(C_2H_5O)_2$-PSCl with condensation product of ethyl acetoacetate and isobutyramidine	fly control in dairy barns, etc.	45
Ronnel ("Korlan")	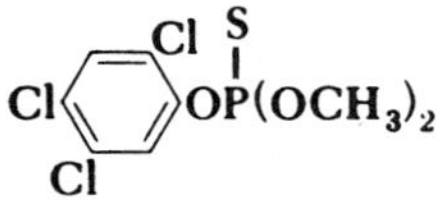 Cl-benzene(Cl)(Cl)$OP(OCH_3)_2$ (with =S)	Dow	reaction of 2,4,5-Cl_3C_6-H_2OH with $PSCl_3$, then with CH_3ONa or $CH_3OH + NaOH$	systemic control of grubs in cattle, fly control around livestock	84, 15
Demeton	$C_2H_5SCH_2CH_2OPS(OC_2H_5)_2$	Chemagro	reaction of $(C_2H_5O)_2$-PSCl with $C_2H_5SCH_2$-CH_2OH	systemic insecticide-miticide for fruit, vegetables and cotton	97
"Phosdrin"	$(CH_3O)_2POOC$=$CHCO_2CH_3$ with CH_3	Shell	reaction of trimethyl phosphite (from PCl_3, CH_3OH and a base) with chlorinated methyl acetoacetate	non-residual systemic, control of aphids, mites, leaf miners, etc. close to harvest	109

TABLE 22-1 (continued)

DDVP	$CCl_2{=}CHOPO(OCH_3)_2$	Shell	reaction of trimethyl phosphite with chloral	fly control around livestock, stored product insect control, household use	9, 120a
"Dibrom"	$CCl_2BrCHBrOPO(OCH_3)_2$	Cal-Spray	addition of Br_2 to DDVP	broad spectrum crop use, especially close to harvest	18a
"Delnav"	(dithiane bis-phosphorothiolate structure) $CHSP(OC_2H_5)_2$	Hercules	dichlorination of dioxane, followed by reaction with $(C_2H_5O)_2$ PSSH salt or $(C_2H_5O)_2$ PSSH and metal chloride catalyst	miticide-insecticide for citrus, other fruits, cotton	27, 106
"Guthion"	(benzotriazinone structure) $NCH_2SP(OCH_3)_2$	Chemagro	condensation of benzotriazinone with formaldehyde, reaction with $SOCl_2$, then with $(CH_3O)_2$ PSSH salt	broad-spectrum insecticide-miticide for fruit, cotton, other crops	76b

"Tedion" is strictly a miticide, having low mammalian toxicity and excellent residual action, and it is particularly effective on egg and nymph stages.

"Tedion" may be manufactured by chlorosulfonating 1,2,4-trichlorobenzene using an excess of chlorosulfonic acid at about 100° C to obtain 2,4,5-trichlorobenzenesulfonyl chloride, which is then reacted with chlorobenzene at 110° C in the presence of about 1.1 molar equivalent of aluminum chloride to produce the sulfone. The yield of the desired isomer is reported to be 63 percent.[58]

2,4-Dichlorophenyl Benzenesulfonate ("Genite" EM-923). "Genite" was introduced in the late 1940's by Allied Chemical and Dye Corporation, who hold a patent on its use.[41]

"Genite" is a miticide having low mammalian toxicity and good residual properties. Its principal use is for early season control of mites on apples and other fruit trees.

It may be manufactured by the reaction of sodium 2,4-dichlorophenate with benzenesulfonyl chloride.

p-Chlorophenyl p-Chlorobenzenesulfonate ("Ovex"). "Ovex" was invented by Hummer and Kenaga by Dow Chemical Company.[60] It has pronounced activity against the eggs of phytophagous mites. The mode of manufacture is analogous to that of "Genite."[101]

2-(p-tert-Butylphenoxy)isopropyl 2-Chloroethyl Sulfite ("Aramite"). "Aramite" was discovered as the result of an extensive investigation of sulfites as insecticides by Harris, Tate, and Zukel of U. S. Rubber Company.[49, 50] It was introduced as a miticide by the Naugatuck Division of U. S. Rubber in 1949.

"Aramite" is a miticide of moderately low residual action, and has found use on fruits, vegetables, and ornamentals. In 1958, the Food and Drug Administration set a zero tolerance for "Aramite," based on new toxicological data, with the result that its usage has been reduced. "Aramite" is manufactured by the following reaction sequence:[49]

$$ClCH_2CH_2OH + SOCl_2 \xrightarrow{\text{low temp.}} ClCH_2CH_2O-SO-Cl \quad (I)$$

$$(I) + (CH_3)_3C\langle\bigcirc\rangle O-CH_2\overset{\overset{\displaystyle CH_3}{\displaystyle |}}{C}HOH \rightarrow$$

$$ClCH_2CH_2O-SO-O\overset{\overset{\displaystyle CH_3}{\displaystyle |}}{C}H-CH_2O\langle\bigcirc\rangle C(CH_3)_3$$

Carbamate Insecticides—"Sevin"

The use of carbamates for insecticidal purposes was first explored by Geigy researchers[39,44] who brought several of these compounds to commercialization. The first carbamate, however, to show promise of major commercial use in the United States is 1-naphthyl N-methylcarbamate.

This compound was invented by Lambrech of Union Carbide and has been patented by this company.[73] "Sevin" was introduced on an experimental basis in 1956. Its pattern of utility was first described by Haynes *et al.*[51] It is a wide-spectrum contact insecticide having good residual activity and slight systemic properties. "Sevin" has a sufficiently low level of mammalian toxicity to be used close to harvest. The principal uses are in fruit, cotton, and vegetables. Consumption in 1960 is estimated at over 20 million pounds.[20a]

"Sevin," like most carbamates, is a phosgene derivative. It may be manufactured (Lambrech, *loc. cit.*) by agitating together at 25° C an aqueous solution of sodium 1-naphthoxide with a toluene solution of phosgene. The intermediate 1-naphthyl chloroformate thus formed goes into the toluene layer, and it is isolated by separating this layer and distilling. The 1-naphthyl chloroformate is then stirred at 25° C with aqueous methylamine, causing "Sevin" to separate as a solid. Alternatively, 1-naphthol may be reacted with methyl isocyanate (from phosgene and methylamine).

The production of phosgene required for "Sevin" (and for several herbicides discussed in a later section) is accomplished commercially by the reaction of carbon monoxide with chlorine.[67] Carbon dioxide from flue gas, lime kilns, or the burning of fuel is reacted with coke to produce carbon monoxide. The latter must be freed of water, hydrogen and noncondensibles. The chlorine must also be dry. The two gases are carefully metered and blended in the stoichiometric ratio, or with CO in slight excess, and passed over an activated charcoal catalyst—preferably in tubes of small diameter so as to permit efficient removal of the heat of reaction.

FUNGICIDES

Derivatives of Trichloromethanesulfenyl Chloride

N-(trichloromethylthio)tetrahydrophthalimide (captan). Captan was invented by Kittleson of Standard Oil Development Co.[68] Kittleson gives data on twenty other imides and nitrogen heterocycles bearing the $-NSCCl_3$ group, which would appear to be conducive to fungitoxicity. Captan was described in the non-patent literature by Kittleson in 1952.[69]

Captan is at present the major organic fungicide not containing a metal cation. It is a safe and effective preventative fungicide for foliar application on apples (against scab), stone fruits (against brown rot), ornamentals, turf, and grapevines. Where powdery mildew is a problem, captan must be supplemented with sulfur or "Karathane." Captan is also useful for seed treatment and potato seed piece treatment for protection against damping off and root rot. It is useful as a soil fungicide in seedbeds and greenhouse benches.

The synthesis of captan is shown by the following equations:

$$CS_2 + Cl_2 \longrightarrow CCl_3SCl + S_2Cl_2$$

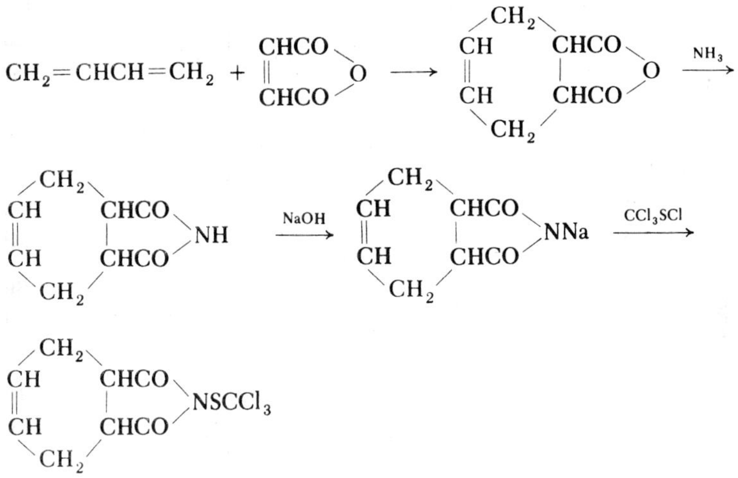

The manufacture and chemistry of trichloromethanesulfenyl chloride (perchloromethyl mercaptan) has recently been reviewed by Sosnovsky.[105] The commercial production of trichloromethanesulfenyl chloride is at present used almost entirely for captan manufacture. An imposing number of recent patents have been issued on new fungicides from CCl_3SCl, but the only new example to become commercial is Cal-Spray's "Phaltan" (see below) which was covered by Kittleson's original patent. Trichloromethanesulfenyl chloride is made commercially by passing chlorine into carbon disulfide at 15 to 30° C. The presence of iron must be avoided since it causes chlorinolysis of the C—S bond with resultant formation of carbon tetrachloride. The product may be isolated by distillation to remove most of the sulfur chloride, and is then washed with water or with various mild oxidizing or reducing solutions to which CCl_3SCl is unreactive. A patented process which can be run in a continuous fashion involves heating the chlorination mixture with more CS_2 plus a catalytic amount of iodine, thereby converting the sulfur chlorides to CCl_3SCl and sulfur.[24,86]

Tetrahydrophthalimide (from the butadiene-maleic anhydride adduct plus ammonia) is converted to the sodium salt by reaction with aqueous caustic at a low temperature, then CCl_3SCl is added with agitation at 6 to 14°.[68,70] A continuous process has been developed wherein the reaction is carried out at 10 to 30°C, and a pH of 10 to 10.5, the crude product being continuously fed to a holding zone where the pH is allowed to drop to 8 to 8.5, and the product is continuously filtered off.[72]　The yields appear to be improved by the addition of sodium chloride prior to the addition of CCl_3SCl.[71]　Some loss of yield occurs because of hydrolysis of reactants or of the product in the alkaline reaction mixture which results when caustic is employed; improved yields have been claimed using a tertiary amine as the base.[72a]

"Phaltan."　N-(Trichloromethylthio)phthalimide　("Phaltan")　has recently been commercially introduced by Cal-Spray.　It was also included in Kittleson's original patent[68] and is made similarly to captan, using phthalimide instead of tetrahydrophthalimide.　Phaltan appears to be a superior fungicide for control of mildew and black spot of roses and other ornamentals, and shows promise for control of powdery mildew of grapes and melons, late blight of potatoes and tomatoes, and various fruit diseases.

Chlorinated Aromatic Compounds

Hexachlorobenzene.　Hexachlorobenzene (HCB) was first reported to be effective for fungicidal seed treatment of cereals by Steindorff *et al.*[108]　It was shown to be particularly effective against soil-borne and seedborne bunt (stinking smut).[74,90]　HCB is used for wheat seed treatment in the Pacific Northwest, where bunt is a significant problem.[54]

The chlorination of benzene to hexachlorobenzene is easily accomplished by iron-catalyzed liquid phase chlorination.　The bulk of the hexachlorobenzene produced is used to make pentachlorophenol (see below).

Pentachloronitrobenzene (PCNB).　This compound was developed by I. G. Farben in Germany in the late 1930's.　It has been commercialized since 1950 in the United States by Olin-Mathieson Chemical Corporation as "Terrachlor."　Principal uses are as a soil fungicide or seed treatment for control of soil-borne fungus diseases of potato (against scab), lettuce (against *botrytis*), cotton (against damping off), legumes (against *sclerotinia* rot), crucifers (against *pythium*), and wheat (against bunt).　PCNB may remain active in the soil for as long as twelve months.

PCNB may be made by direct chlorination of nitrobenzene in chlorosulfonic acid at 60 to 70°C using iodine as a catalyst,[79] or by nitration of pentachlorobenzene.

Pentachlorophenol.　Pentachlorophenol is a broad spectrum biotoxicant.　It is too phytotoxic for fungicidal use on crops and, in fact, finds commercial use in desiccation of the foliage of crops such as cotton to facilitate harvesting, and for nonselective foliar destruction of weeds.　As a fungicide and insecticide, its major field of use is in wood preservation, where it competes with creosote.　The use of pentachlorophenol for this purpose began in the late 1930's; and at the present time, about 39,336,000 lb/year are produced,[119a] mostly for wood preservation. The use of pentachlorophenol and its salts in wood treatment has recently been reviewed by Sproule.[107a]

Pentachlorophenol may be manufactured by two principal routes, chlorination of phenol and hydrolysis of hexachlorobenzene.

The direct chlorination of phenol is advantageously carried out in the liquid phase, in the presence of a small amount of a catalyst such as $AlCl_3$.[108a] Chlorination is usually stopped when the reaction product reaches 88 to 90 percent pentachlorophenol, the rest being largely tetrachlorophenol. Further chlorination tends to introduce strongly-colored less soluble by-products. The product is usually flaked or neutralized with a 0.3 to 1 percent excess of caustic soda to produce the sodium salt, sold either in solution or as a dried powder or dry pellets.[107a]

The reaction of hexachlorobenzene (*q.v.*) with sodium hydroxide can be conducted using 5 to 15 percent sodium hydroxide in methanol at 130 to 140° C.[101a]

Chlorinated Quinones

Chloranil. Although quinones had been patented in 1930 as fungicides by Broderson and Ext,[16] the particular utility of chloranil (tetrachloro-*p*-benzoquinone) for crop protection was first recognized by Ter Horst in 1938.[115] The early history of chloranil as a fungicide has been described in detail by Horsfall.[56] Chloranil was introduced commercially in 1939 to 1940 by U. S. Rubber under the name "Spergon," as a seed protectant for peas. It is now used as an anti-fungal seed protectant on a large number of crops and ornamentals. It has limited value for foliar applications, apparently because of photochemical breakdown.[56]

The manufacture of chloranil by oxidative chlorination of various phenols and anilines has long been known. There are a number of patented processes, e.g., the chlorination of polychlorophenols in the presence of strong sulfuric acid;[1,116] the chlorination of sulfuric acid solutions of aniline;[35] and the chlorination of phenols in a fatty acid solution.[36] An interesting process recently patented is the vapor phase oxidative chlorination of cyclohexane by HCl and O_2 over an alumina catalyst impregnated with Cu, Co, and Fe chlorides.[37]

2,3-Dichloro-1,4-naphthoquinone ("Phygon," dichlone). Research done by Ter Horst and Felix at U. S. Rubber[113-5] on quinone fungicides led to the discovery that 2,3-dichloro-1,4-naphthoquinone is sufficiently stable in sunlight to permit foliar use, such as for prevention of apple scab and brown rot of stone fruits. "Phygon" is also used widely as a seed protectant.

Several routes of manufacture are described by Fletcher (U. S. Rubber). In one process chlorine is passed at 80° C through a 10 percent solution of 1,4-aminonaphthalenesulfonic acid in 50 percent aqueous sulfuric acid with iron as a catalyst.[35]　Alternatively, "Phygon" may be made by chlorination of 1-naphthol in aqueous acetic-sulfuric acid mixture using an iron sulfate catalyst.　A yield of 86 percent is reported.[36]

HERBICIDES

Chlorophenol Derivatives

2,4-Dichlorophenoxyacetic acid (2,4-D) and its derivatives.　The powerful plant growth regulatory activity of 2,4-D was discovered at Boyce Thompson Institute by Zimmerman and Hitchcock.[125]　Its use as a herbicide was covered in a broad group of halogenated arylaliphatic acid herbicides patented by Jones (American Chemical Paint Company) in 1945.[63]　The commercial introduction of 2,4-D occurred in about 1946.

2,4-D in various forms is the major herbicide of the present day, and in its main field of utility it has no close competitors in the United States. In Europe, 4-chloro-2-methylphenoxyacetic acid (see below) competes successfully with 2,4-D.

U. S. production of 2,4-D is now about 36,185,000 pounds annually (see Figure 22-1).[119a]　Assuming the phenol used to make 2,4-D is produced *via* chlorobenzene (Dow), then a total of four molecules of chlorine

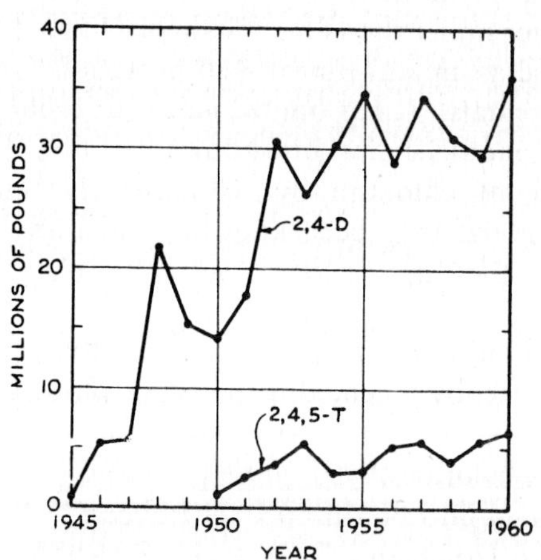

Source—U. S. Tariff Commission

Figure 22-1. Production of 2,4-D and
2,4,5-T (total acid in all formulations).

are stoichiometrically involved in making one molecule of 2,4-D. This would mean that at least 50 tons of chlorine per day are used to make 2,4-D.

The usefulness of 2,4-D depends mainly on its remarkable ability to kill broadleaf plants by systemic action (involving transport throughout the plant) at low rates of application, 1/4 to 1/16 pound/acre often sufficing. It is selective and can be used to kill broadleaf weeds in grassy (monocotyledonous) crops without damage to the crops. Major uses are in wheat, rye, corn, rice, and sugar cane. In turf, it may be safely and economically used to remove dandelions, plantain, and other broadleaf weeds. It has enough residual life in soil to be useful for pre-emergence broadleaf weed control, for example in corn. 2,4-D is used to destroy broadleaf weeds along rail lines, highways, power lines, and ditches. It is often supplemented by 2,4,5-T (*q.v.*) to control resistant brush, or by sodium TCA or dalapon to control grasses.

2,4-D is not commonly used as the free acid. The major formulations and their advantages and limitations are tabulated in Table 22-2.

It is possible to produce 2,4-D by chlorination of phenoxyacetic acid in a melt,[80] but yields are 70 percent or lower and purification of the product to remove isomers is difficult; therefore, this route does not appear to be useful commercially. The commercial process is to react 2,4-dichlorophenol with chloroacetic acid plus sodium hydroxide.

The Dow plant may be considered representative of current practice and has been described in detail by Sanders and Prescott.[93] Production is done batchwise. A 50 percent molar excess of molten 2,4-dichlorophenol (produced by chlorination of phenol) is agitated with 50 percent aqueous caustic to form the sodium phenoxide; then monochloroacetic acid (produced by chlorination of acetic acid as described below) is added, and the mixture is held at 60 to 80°C for 6 to 8 hours, maintaining the pH at 7.0 or above. Aqueous hydrochloric acid is added to a pH of 5, decomposing the excess sodium 2,4-dichlorophenoxide. The 2,4-dichlorophenol is recovered by steam distillation for recycling. The sodium 2,4-dichlorophenoxyacetate solution is then acidified by aqueous hydrochloric acid to a pH of 1, causing precipitation of the 2,4-D acid which is filtered out, washed, and dried.

Esterification is carried out by a conventional azeotropic procedure. It would appear that a solid filtrable acid catalyst (possibly a sulfonated resin) is used in the Dow esterification process.

The requisite 2,4-dichlorophenol is produced by direct chlorination of phenol. In a typical process,[36b] phenol at 80 to 100°C is treated with gaseous chlorine until the freezing point of the reaction mixture reaches 34 to 36°C. The product is isolated by distillation and may be further

TABLE 22-2. 2,4-D DERIVATIVES.

Derivative	Amount Produced[a]	Advantages	Disadvantages	Main Uses
Sodium salt	no data, mostly made by user from purchased free acid and caustic soda	inexpensive where large scale mixing facilities are available	low water solubility; cannot prepare liquid concentrate	railroad line spraying; weed control on Hawaiian plantations
Amine salts	dimethylamine salt, 2,749,000 lbs; di-ethanolamine and triethanolamine salts also used	inexpensive, relative to esters; water soluble; less volatility hazard than esters: easily rinsed from spray equipment.	less powerful than esters, especially on brush; tendency to be washed off plants and to leach from soil.	weed control in vicinity of 2,4-D sensitive crops; weed control in grains
Lower alkyl esters	Isopropyl ester, 5,059,000 lbs; butyl ester, 7,895,000 lbs.	inexpensive relative to low-volatile esters; resist washing off plants by rain. Readily soluble in inexpensive hydrocarbon solvents	volatility hazard, require solvents and emulsifiers.	brush control; control of "hard to kill" weeds, in pastures, hayfields, hedgerows; weed control in grains and other 2,4-D resistant crops.
Low-volatile esters	Isooctyl ester, 1,677,000 lbs.; others (butoxyethyl, butoxy-ethoxypropyl, other alkoxyalkyl), below but probably near 8,784,000 lbs.	less volatility hazard than with lower alkyl esters; resist washing off plants by rain.	relatively expensive; require solvents and emulsifiers; may require coupling agent to permit use of aliphatic hydrocarbon diluents.	brush control, pre-emergence weed control in corn. Control of weeds in turf, control of weeds in grain where nearby susceptible crops make volatility a factor.

[a] U.S. Tariff Commission Report No. 206, 2nd Series, 1959.

purified by recrystallization. Yields may reach 90 percent. The mono-chloroacetic acid required for 2,4-D and 2,4,5-T is manufactured by direct chlorination of acetic acid. Catalysis is required; a typical commercial process[103] employs 5 to 12 percent of PCl_3. To avoid the formation of substantial amounts of by-product dichloroacetic acid (the removal of which requires recrystallization), it is found best to employ moderate pressure (10 to 100 psi) and to start the chlorination at 25° C, gradually raising the temperature to 90 to 130° C as the chlorination approaches 50 percent conversion, at which point the chlorination is stopped and the mixture fractionally distilled, A recent patent[65] claims molybdenum pentachloride to be advantageous over PCl_3 in supressing dichloroacetic acid formation.

Recently, the hydrolysis of trichloroethylene to chloroacetic acid has been practiced commercially in Europe and imports of this product have attained importance in North America.

2,4,5-Trichlorophenoxyacetic Acid (2,4,5-T). The special activity of 2,4,5-T on 2,4-D resistant weeds was first reported by Hanner and Tukey.[48] It was introduced by the American Chemical Paint Company in 1944. A large amount of 2,4,5-T is used for brush control, 2,4,5-T being particularly advantageous on brambles, certain oak species, and woody vines. Combinations of 2,4-D and 2,4,5-T esters are commonly used for control of mixed populations of weeds and brush along rail lines, roadways, around industrial installations, and for control of weeds in turf.

Total U. S. production of 2,4,5-T is about 6.3 million pounds per year, most of which is used in the form of esters.[119a] Amine salts are also used.

2,4,5-T is produced from 2,4,5-trichlorophenol, chloroacetic acid and caustic soda by a process analogous to that used for 2,4-D (*q.v.*). The manufacture of the requisite 2,4,5-trichlorophenol is performed by the reaction of caustic soda on 1,2,4,5-tetrachlorobenzene in an alcohol solvent. In methanol at 150° C, this reaction is substantially complete in several hours.

2-Methyl-4-chlorophenoxyacetic Acid (MCPA). MCPA was developed concurrently with 2,4-D in the 1940's.[100] Its use has been more extensive in Europe than in the United States; perhaps this is because of the lower cost of *o*-cresol in Europe, and the relative safety of MCPA as compared to 2,4-D in grains and certain legumes. Its principal use in North America is for weed control in grains, particularly in oats underseeded with legumes, in flax, rice, and in established stands of certain legumes.

MCPA is made by a process analogous to that used for 2,4-D. The requisite 4-chloro-*o*-cresol is made by chlorination of *o*-cresol at low temperature without a catalyst, and a mixture of about 60 to 65 percent 4-chloro-*o*-cresol and 35 to 40 percent 6-chloro-*o*-cresol results. Commonly the isomers are not separated, and the MCPA formulation (usually an amine salt) contains the 6-chloro isomer as an inert ingredient.

alpha-(2,4,5-Trichlorophenoxy)propionic Acid (2,4,5-TP, "Silvex"). 2,4,5-TP was introduced in 1953 by Dow for the control of 2,4-D/2,4,5-T resistant brush such as oak and maple, and resistant weeds such as bedstraw.

It is also useful in turf since it not only kills dandelions and plantain but also chickweed which is resistant to 2,4-D. Recent reports indicate that 2,4,5-TP has promise as an aquatic weed killer.

2,4,5-TP is produced from 2,4,5-trichlorophenol and *alpha*-chloropropionic acid by a process analogous to that used in making 2,4-D. The introduction of chlorine into the *alpha* position of propionic acid requires that the chlorination be done in the presence of a catalyst such as PCl_3.[10,28]

Derivatives of Anilines and Phosgene

This group may be subdivided into the carbamates (urethanes) and the ureas, both being derivatives of the hypothetical carbanilic acid $C_6H_5NHCOOH$, which is not stable as the free acid.

Carbamates. Templeman and Sexton[112] investigated the N-arylurethanes and reported that isopropyl carbanilate was outstandingly active as a plant growth regulator and as a herbicide against grasses. Isopropyl phenylcarbamate (IPC, or isopropyl carbanilate) attained commercial usage on a moderate scale for controlling wild oats in peas and sugar beets in the Western United States.

In the late 1940's it was found by researchers at Columbia-Southern[121] that isopropyl *m*-chlorophenylcarbamate (CIPC or isopropyl 3-chlorocarbanilate) is a generally more potent pre-emergence grass-killing herbicide, with more soil persistance than IPC. CIPC is used commercially at present for pre-emergence weed control in cotton but is under severe competition from diuron (see below) which performs at lower rates and is reputedly less costly to use. The other principal uses of CIPC are for weed control in onions, spinach, and other vegetables.

A recent estimate (1959) would place CIPC and IPC volume as 1 to 2 million pounds per year.

A new carbamate with a single potential use, albeit on one of the world's major weed problems, is 4-chloro-2-butynyl *m*-chlorocarbanilate (barbane, "Carbyne") introduced in 1959 by Spencer Chemical Company for post-emergence control of wild oats.[55] The carbamate herbicides may be made by several alternative routes, illustrated below for CIPC:

$$m\text{-}ClC_6H_4NH_2 + COCl_2 \rightarrow m\text{-}ClC_6H_4NHCOCl + HCl$$

$$m\text{-}ClC_6H_4NHCOCl + (CH_3)_2CHOH \rightarrow m\text{-}ClC_6H_4NHCOOCH(CH_3)_2 + HCl$$

$$(22\text{-}5)$$

$$m\text{-}ClC_6H_4NHCOCl \xrightarrow{\text{heat}} m\text{-}ClC_6H_4N{=}C{=}O + HCl$$

$$m\text{-}ClC_6H_4N{=}C{=}O + (CH_3)_2CHOH \rightarrow m\text{-}ClC_6H_4NHCOOCH(CH_3)_2 \quad (22\text{-}5a)$$

$$(CH_3)_2CHOH + COCl_2 \rightarrow (CH_3)_2CHOCOCl + HCl$$

$$(CH_3)_2CHOCOCl + m\text{-}ClC_6H_4NH_2 \rightarrow m\text{-}ClC_6H_4NHCOOCH(CH_3)_2 + HCl$$
$$(22\text{-}6)$$

Reaction (22-5a) is frequently the most convenient for laboratory synthesis, but Reaction (22-6) is preferred commercially.

The reaction of phosgene with alcohols to form chloroformates must be run so as to minimize the formation of dialkyl carbonates as by-products. A process patented by Columbia-Southern[117] utilizes a packed column into which phosgene and the alcohol are fed countercurrently. Even with a large excess of isopropanol, a good yield of isopropyl chloroformate is obtained. A stripping tower for removing phosgene from isopropyl chloroformate has been described by Columbia-Southern.[53]

Arylureas. Four arylureas were introduced on an experimental basis in the period from 1951 to 1955 by du Pont: fenuron (3-phenyl-1,1-dimethylurea), monuron [3-(*p*-chlorophenyl)-1,1-dimethylurea], diuron [3-(3,4-dichlorophenyl)-1,1-dimethylurea], and neburon [3-(3,4-dichlorophenyl)-1-methyl-1-*n*-butylurea]. These are now sold commercially.

Fenuron is a nonselective weed and brush killer, useful for fencerows, rail lines and roadsides. Because of its rather high solubility in water (2900 ppm), it is capable of penetrating to the root system of brush rather quickly. Monuron is less water-soluble (230 ppm) and is an excellent non-selective soil sterilant with long persistance. Some species are resistant, e.g., plantain and some deeply rooted perennial vines. Since established asparagus plants are resistant to low rates of monuron, it is effective for control of weeds and undesired asparagus seedlings in established beds. Diuron is still less water soluble (42 ppm) and therefore may be used for weed control in certain deeply-seeded or deeply-rooted crops (*e.g.*, cotton, grapes). At high rates diuron is a soil sterilant with long persistance, but, as in the case of monuron, it is not effective on many deeply rooted perennials. Neburon, having very low solubility (4.8 ppm) is useful for control of chickweed, crabgrass, and other annual weeds in nursery plantings, (by use of directed sprays) and has shown some promise for use in turf.

The manufacture of the aryl urea herbicides is rather analogous to the carbamates, according to patents, and involves the use of phosgene and an aniline in either of the two routes as follows (illustrated for monuron):

$$p\text{-}ClC_6H_4NH_2 + COCl_2 \rightarrow p\text{-}ClC_6H_4NHCOCl + HCl$$

$$p\text{-}ClC_6H_4NHCOCl \xrightarrow{\text{heat}} p\text{-}ClC_6H_4N{=}C{=}O \quad (22\text{-}7)$$

$$p\text{-ClC}_6\text{H}_4\text{N}{=}\text{C}{=}\text{O} + (\text{CH}_3)_2\text{NH} \rightarrow p\text{-ClC}_6\text{H}_4\text{NHCON}(\text{CH}_3)_2$$

$$(\text{CH}_3)_2\text{NH} + \text{COCl}_2 \rightarrow (\text{CH}_3)_2\text{NCOCl}$$

$$p\text{-ClC}_6\text{H}_4\text{NH}_2 + (\text{CH}_3)_2\text{NCOCl} \rightarrow p\text{-ClC}_6\text{H}_4\text{NHCON}(\text{CH}_3)_2 \qquad (22\text{-}8)$$

Chlorinated Aliphatic Acids and Derivatives

Sodium trichloroacetate. Trichloroacetic acid has long been known to have biotoxicant properties, particularly at high concentrations as a protein precipitant. The use of its salts as grassy-weed herbicides was reported in 1947.[31] No patent covers the use of the sodium salt *per se,* although the ammonium salt (now rarely used) or a mixture of the ammonium salt plus a hygroscopic substance such as calcium chloride were patented by Du Pont.[13]

Sodium trichloroacetate has both industrial and agricultural uses. As a water solution, usually in combination with other herbicides such as sodium chlorate, it is sprayed along railroad rights-of-way at around 40 to 100 lb per acre. This combination, patented in 1952 by Taylor of R. H. Bogle Co.[111] has had widespread use and is still popular. The TCA gives top-kill of grasses and seasonal retardation of regrowth. It works to a large degree by root uptake.[7]

In agricultural practice, sodium TCA is used at low rates (below 15 pounds per acre) for pre-emergence weed control in sugar beets and red beets, flax, cabbage, direct-seeded tomatoes, established alfalfa, and sugar cane, and at higher rates (around 100 pounds per acre) for spot eradication of perennial grasses such as Johnson grass, Bermuda grass, and quack grass. The compound is nonselective at these higher rates, but it rarely persists in the soil for more than one season.

At its peak usage in the mid-1950's sodium trichloroacetate was manufactured in the United States in the several-million-pound per year range. Imports from Germany and the promotion of dalapon by Dow (see below) as a substitute for sodium TCA have probably reduced U. S. production.

The manufacture of sodium trichloroacetate is mostly by the direct chlorination of acetic acid. A workable commercial process has been described by Sonia and Scremin.[104] Glacial acetic acid and about 3 percent of PCl_3 as a catalyst are placed in a pressure vessel fitted with light wells containing mercury vapor lamps. Chlorination is begun at about 30°C and is conducted at about 24 to 45 psi. The temperature is gradually raised to 190 to 220°C and held in this range during the last half of the reaction period. The crude trichloroacetic acid thus produced contains a few percent of trichloroacetyl chloride and anhydride. Small increments of water are added until a maximum freezing point is obtained, the product then being trichloroacetic acid suitable for neutralization to sodium TCA. The salt is dried at moderate temperature and sold in the substantially anhydrous state.

Sodium 2,2-dichloropropionate.   Sodium 2,2-dichloropropionate (dalapon) was invented by Barrons of the Dow Chemical Company.[8] The chemical was introduced on an experimental basis in 1953 and was offered commercially shortly thereafter. Several million pounds per year are now sold in the United States.

Like sodium TCA, it is principally a grass killer, but is several times as active on a weight basis on most grass species. In terms of cost, dalapon competes with TCA for uses such as railroad right-of-way spraying. In comparison with TCA, dalapon has a markedly greater ability to translocate from grass foliage down to the root system, and may therefore give more complete kill of perennial grasses. It has been reported to be somewhat less persistant in the soil than TCA.

Nonagricultural uses include grass control along rail lines, highway guard rails, and around oil storage areas and lumber yards.

Agricultural uses include spot eradication of established perennial grasses such as Johnson grass and Bermuda grass where it competes with herbicidal oils, sodium TCA, and sodium chlorate. Dalapon is also used for grass control in orchards, for pasture renovation to get rid of sod containing unwanted grasses, and for control of cattails in drainage ditches and marshes.

The manufacture of dalapon is described in a Dow patent.[28] The chlorination requires a phosphorus catalyst, which may be a chloride, oxychloride, oxide, or oxy-acid, and is carried out in two stages; the first chlorine atom is introduced at 50 to 140°C, the second at 150 to 225°C. A typical large run utilizes 190 lb of phosphorus trichloride catalyst per 4500 lb of propionic acid for the first stage which was carried out over 32.5 hours at 120 to 125°C. At the end of this stage, the isomer distribution was 98 percent *alpha* and 2 percent *beta*. Another 100 lb of phosphorus trichloride was required for the second stage which requires two days at 176 to 180°C. The product consisted of 10 percent *alpha*-monochloro-, 88 percent *alpha, alpha*-dichloro-, 1 percent *alpha, alpha, beta*-trichloropropionic acid, and 1 percent *alpha, alpha*-dichloropropionic acid anhydride. The acid is neutralized to form the sodium salt which is dried and sold as the solid salt.

An alternative dalapon process recently reported in a Russian patent[32] involves catalytic chlorination of methyl ethyl ketone or 2-butanol to pentachlorobutanone, followed by cleavage by cold aqueous caustic, presumably as follows:

$$CH_3CCl_2COCCl_3 + NaOH \rightarrow CH_3CCl_2COONa + CHCl_3$$

Monochloroacetic acid derivatives.   Sodium monochloroacetate has some contact herbicidal activity and defoliant activity but is too weak a herbicide to be of commercial importance.   Its main use is as an intermediate for 2,4-D and 2,4,5-T (*q.v.*).

Surprisingly, certain amides of chloroacetic acid have quite pronounced pre-emergence grass-controlling activity. An intensive study by Hamm and Speziale of Monsanto[47] showed that N,N-disubstituted chloroacetamides where the substituents have three carbons in a straight chain are outstandingly active and from amongst these, N,N-diallyl-*alpha*-chloroacetamide (CDAA, "Randox") was selected by Monsanto for commercialization.

Although not a large tonnage chemical, CDAA finds significant commercial use for pre-emergence annual grass control in corn and soybeans, particularly to control foxtail and crabgrass.   Recently, a combination of CDAA dissolved in trichlorobenzyl chloride (which serves as solvent and as

a pre-emergence broadleaf weed killer) was introduced commercially by Monsanto for broad-spectrum annual weed control pre-emergence in corn.[43]

A convenient preparation of chloroacetamides, requiring only one mole of amine, is described by Speziale and Hamm.[107]   Chloroacetyl chloride is added at –10°C to a stirred two-phase mixture of the amine, aqueous sodium hydroxide, and an inert solvent such as ethylene dichloride.   The product is recovered from the ethylene dichloride layer by distillation.

Chlorinated Aromatic Acids

Chlorobenzoic Acids.   The plant growth regulating activity of certain halogenated benzoic acids was first noted by Zimmerman and Hitchcock,[125] and the use of dichlorobenzoic acids as herbicides was described by Jones.[63]   The exceptional activity of 2,3,6-trichlorobenzoic acid was first noted by Bentley in 1951.[11]   Trichlorobenzoic acid (TBA) consisting of about 68 to 70 percent of the 2,3,6-isomer and 20 to 22 percent of the 2,4,5-isomer was introduced experimentally as a herbicide by Heyden Chemical Company in 1954 and is protected as a synergistic mixture by a U. S. patent.[42]   Polychlorobenzoic acid (PBA) consisting of about 11 percent 2,5-di-, 16 percent 2,3,5-tri-, 8 percent 2,3,6-tri-, 6 percent 2,4,5-tri, 26 percent 2,3,5,6-tetra, 17 percent 2,3,4,5-tetra-, and 4 percent penta- was introduced experimentally by the Hooker Electrochemical Company in 1955. Both of these products are now sold commercially for eradication of bindweed (which is legally required in certain Great Plains states). TBA and PBA are persistant soil sterilants with marked ability to penetrate deep root systems. The principal competitive product is sodium chlorate. PBA and TBA are also effective against leafy spurge, Russian knapweed, and various species of woody vines and brush resistant to 2,4-D and 2,4,5-T.

The manufacture of TBA involves chlorination of *o*-chlorotoluene in the presence of ferric chloride at 0 to 7° C to introduce a total of three chlorine atoms into the ring. This trichlorotoluene may then be oxidized directly to TBA using nitrogen dioxide at 170 to 210° C and a bismuth or titanium oxide catalyst,[26] but milder oxidation conditions may be employed (refluxing 60 percent HNO_3 at atmospheric pressure) if the side chain is first chlorinated to introduce one or two chlorine atoms.[52] The yield is further improved by adding some Fullers' earth to the oxidation mixture.

PBA may be made by direct chlorination of benzoyl chloride in the presence of ferric chloride or other Lewis acid catalyst at 90 to 200° C.[17,25,66] Hydrolysis of the chloride to the acid may then be effected by the passage of steam into the chloride using an acid catalyst.[17]

Dimethyl Tetrachloroterephthalate ("Dacthal"). This compound was discovered to be herbicidal in screening conducted by Diamond Alkali Company at Boyce Thompson Institute. "Dacthal" does not resemble the chlorobenzoic herbicides in its pattern of activity; it appears to be a pre-emergence inhibitor of annual weed germination, especially grasses,[76] and is being commercially developed for crabgrass control in turf. "Dacthal" may be made by exhaustively chlorinating terephthaloyl chloride at 175° C in the presence of iron, and then refluxing the tetrachloroterephthaloyl chloride with methanol.[91]

Cyanuric Chloride Derivatives (Triazines)

The herbicidal activity of compounds containing the symmetrical triazine ring (numbering of the positions is shown below) was discovered by researchers at the Geigy Laboratories and was first reported in 1955.[40]

$$\begin{array}{c} 1 \\ \text{N} \\ 6\,\text{C} \qquad \text{C}\,2 \\ 5\,\text{N} \qquad \text{N}\,3 \\ \text{C} \\ 4 \end{array}$$

Although 2,4-bis(diethylamino)-6-chloro-*s*-triazine was the first example of this series released experimentally in the United States, it soon became evident that this compound is relatively weak compared to 2,4-bis(ethylamino)-6-chloro-*s*-triazine (simazine). Simazine is now sold commercially for pre-emergence weed control in corn, ornamentals such as taxus and juniper, and for soil sterilization around industrial installations, railroad yards, and highway guard rails.

Occasionally simazine fails because of insufficient rainfall; therefore the somewhat more water soluble homolog, 2-chloro-4-ethylamino-6-isopropylamino-*s*-triazine (atrazine) was introduced especially for use in areas of lower precipitation. Atrazine, because of its deeper penetration of the soil is also more active on perennial grasses, e.g., quackgrass, than is simazine. Numerous new triazine herbicides are currently in the experimental stage. This group of compounds has been reviewed recently by Gysin.[46]

Simazine and atrazine are manufactured respectively by the reaction of two equivalents of ethylamine or one equivalent of ethylamine and one equivalent of isopropylamine with cyanuric chloride. Gysin (*loc. cit.*) has reviewed the chemistry of this reaction. A single chlorine atom of cyanuric chloride can be replaced by the alkylamino group at –15 to 0° C using two moles of amine, or one mole of amine plus equivalent quantities of sodium hydroxide, sodium carbonate, or sodium bicarbonate as HCl acceptor. To replace the second chlorine atom, the temperature generally must be raised to 20 to 50° C and another mole of amine and acid acceptor employed. Because of the clean-cut stepwise course of the reaction, it is easily possible to replace two chlorines by two different amino groups, as in the synthesis of atrazine.[89]

The manufacture of cyanuric chloride has recently been reviewed by Smolin and Rapoport.[102] It appears that a commercial method of production is the reaction of hydrogen cyanide with chlorine in the vapor phase to continuously form cyanogen chloride, which is then trimerized by passage through a bed of activated charcoal.[3] A fluidized bed may be used to permit continuous exchange of fresh catalyst for partially spent catalyst. A 90 to 100 percent conversion and 95 percent yield, based on cyanogen chloride, is reported.[99] The chlorination and trimerization may be done as a single step.[29a]

Cyanogen chloride may also be trimerized by passage into aluminum chloride in nitromethane at 40 to 60° C but yields are said to be only 60 percent.[102]

NEMATOCIDES

Allyl Chloride Derivatives

Dichloropropene and "D-D Mixture." "D-D mixture" is a liquid consisting mainly of 1,3-dichloropropene (60 to 66 percent) and 1,2-dichloropropane (30 to 35 percent) and some trichloride (5 percent).[23] Its use as a soil fumigant was first reported by Carter[19] in 1943, and it was placed on the market by Shell in 1945. D-D mixture has attained major use as a nematocidal fumigant for soils to be used for the growing of tobacco, pineapple, citrus, and certain vegetables. In the tobacco market, it is competitive with ethylene dibromide.

"D-D mixture" is produced as a by-product of the manufacture of allyl chloride.[83] In the crude allyl chloride made by propylene chlorination, 1,3-dichloropropenes and 1,2-dichloropropane amount to about 15 percent.[33]

The 1,3-dichloropropene component may be made by vapor phase chlorination of allyl chloride-propylene mixtures, using an olefin-to-chlorine ratio of between 2 : 1 and 3 : 1 according to a Shell patent.[21] According to a Dow patent, dichloropropenes can be made in 83 percent yield by passing a 1.5 : 1 mixture of chlorine and propylene through a heated zone at 435° C.[88]

The principal component of D-D, namely, 1,3-dichloropropene, was made available in a purer form as "Telone" by Dow in 1956. A 1:5 by volume formulation of ethylene dibromide and "Telone" was recently introduced by Dow under the name "Dorlone," for use against a mixed nematode infestation, such as the stylet and lesion nematodes of tobacco, by row treatment.[29]

1,2-Dibromo-3-chloropropane. This compound ("Nemagon," "Fumazone") was first described as a nematocide in 1955 by McBeth and Bergeson.[77] A patent was issued in 1960 to Schmidt (Pineapple Research Institute of Hawaii)[95] covering soil fumigation with 1,2-dibromo-3-chloropropane. It has a remarkable level of activity against root knot nematodes, 8-16 times that of D-D and 2-5 times that of ethylene dibromide. It is promising for use in cotton, vegetables, grapes, citrus, and other trees. Recent tests show that it can be applied by emulsification in irrigation water. "Nemagon" may be used to fumigate the root systems of some living plants (grapes, ornamentals, various fruits). It has not found usage in the large tobacco soil fumigation market because of phytotoxicity. "Nemagon" is made by addition of bromine to the double bond of allyl chloride.

OTHER FUMIGANTS

Chlorocarbons: Ethylene Dichloride, Carbon Tetrachloride

Carbon tetrachloride and ethylene dichloride have been used for several decades for destruction of insects in stored grain. Carbon tetrachloride is commonly used in admixture with the more toxic but dangerously flammable carbon disulfide. Ethylene dichloride by itself also poses a fire hazard and therefore is commonly admixed 3 : 1 by volume with carbon tetrachloride. Over 30 million pounds of carbon tetrachloride and 10 million pounds of ethylene dichloride are used per year in fumigation. The manufacture of these chemicals is discussed elsewhere in this volume.

Chloropicrin

Chloropicrin has long been used as a fumigant for stored grain and for soil, principally to control insects. It appears to be a general biotoxicant, and is effective against nematodes[62] and most soil-borne fungi. Chloropicrin is also used for fumigation of mills and milling equipment.

The use of chloropicrin for large-scale field fumigation is generally not economical, but it is well suited to greenhouse and seedbed fumigation. It is also used as a warning odorant in admixture with fumigants having mild odors, such as methyl bromide.

The oldest process for manufacture of chloropicrin is by treatment of calcium picrate with calcium hypochlorite and steam or warm water. It was reported (1940) that chloropicrin could be produced at a chemical cost of 24 cents/lb by this route.[110] Some present manufacturers employ an alternative route wherein nitromethane is chlorinated by a hypochlorite. According to Cheylan,[22] the reaction may be conducted in 90 percent yield by adding nitromethane to 30 to 35 percent aqueous sodium hypochlorite at 15 to 20° C.

References

1. Alquist, Groom, and Haney (to Dow Chemical Co.), U. S. Patent 2,414,008 (1947).
2. American Chemical Society, "Agricultural Control Chemicals," Advances in Chemistry Series, No. 1, pp. 143–159, 1949.
3. American Cyanamid Co., British Patent 602,816 (1948).
4. American Cyanamid Co., British Patent 644,616 (1950).
4a. Arvey Corporation, British Patent 714,830 (1951).
5. Asquith, *J. Econ. Entomol.,* **48,** 329 (1955).
6. Barker and Maughan, *J. Econ. Entomol.,* **49,** 458 (1958).
7. Barrons and Hummer, *Agric. Chem.,* **6** (6), 48 (1951).
8. Barrons (to Dow Chemical Co.), U. S. Patent 2,642,354 (1953).
9. Barthel, Alexander, Giang, and Hall, *J. Am. Chem. Soc.,* **77,** 2424 (1955).
10. Bass (to Dow Chemical Co.), U. S. Patent 2,010,685 (1935).
10a. Belikova, Vol'fson, Kuznetsova, Mel'nikov, Person, Platé, and Pryonishnikova, *Zhur. Priklad. Khim.,* **33,** 454 (1960).
11. Bentley, *Nature,* **165,** 449 (1951).
12. Bland and Young (to Monsanto Chemical Co.), U. S. Patent 2,663,723 (1953).
12a. Bluestone, Lidov, Knaus, and Howerton (to J. Hyman & Co.), U. S. Patent 2,576,666 (1951).
12b. Bluestone (to Shell Development Co.), U. S. Patent 2,676,132 (1954); Canadian Patents 556,391–2 (1958).
13. Bousquet (to E. I. du Pont de Nemours & Co.), U. S. Patent 2,393,086 (1946).
14. Bousquet and Goddin (to E. I. du Pont de Nemours & Co.), U. S. Patent 2,420,928 (1947).

15. Britton and Blair (to Dow Chemical Co.), U. S. Patent 2,922,811 (1960).

16. Broderson and Ext (to Winthrop Chemical Co.), U. S. Patent 1,752,293 (1930).

17. Brown, and Gobeil (to E. I. du Pont de Nemours & Co.), U. S. Patent 2,890,243 (1959).

18. Buntin (to Hercules Powder Co.), U. S. Patent 2,565,471 (1945).

18a. California Spray-Chemical Corp., Belgian Patent 585,182 (1960).

19. Carter, *Science,* **97,** 383 (1943).

20. Anon., *Chemical Week,* 78, (November 3, 1956).

20a. Anon., *Chemical Week,* 117, (March 24, 1962).

21. Cheney and McAllister (to Shell Development Co.), U. S. Patent 2,430,326 (1947).

22. Cheylan, *Mem. poudres,* **32,** 417 (1940); *Chem. Abs.,* **47,** 8960 (1953).

23. Chisholm, U. S. Dept. of Agriculture, Yearbook of Agriculture *1952,* 331.

24. Churchill (to Mathieson Chemical Corp.), U. S. Patent 2,666,081 (1954).

25. Cocker Chemical Co., British Patent 827,422 (1960).

26. DiBella (to Heyden Chemical Co.), Canadian Patent 565,899 (1958).

27. Diveley and Lohr (to Hercules Powder Co.), U. S. Patent 2,725,328 (1955).

28. Dow Chemical Co., British Patent 752,761 (1956).

29. Dow Chemical Co., *Down to Earth,* **12** (2), 7 (1956).

29a. Drott and Oliver (to Monsanto Chemical Co.), U. S. Pat. 2,965,642 (1960).

30. Dvornikoff, and Young (to Monsanto Chemical Co.), U. S. Patent 2,663,721 (1953).

31. Elder, Timmons, Willard, and Neville (3 papers), *Proc. North Central Weed Control Conf.,* 284 (1947).

32. Englin, USSR Patent 112,346 (1958); *Chem. Abs.,* **52,** 2093f (1958).

33. Faith, Keyes, and Clark, "Industrial Chemicals," 2nd Ed., New York, John Wiley and Sons, 1957, p. 412.

34. Fancher (to Stauffer Chemical Co.), U. S. Patents 2,793,224 (1957), 2,827,492 (1958).

35. Fletcher (to U. S. Rubber Corp.), U. S. Patent 2,422,089 (1947).

36. Fletcher (to U. S. Rubber Corp.), U. S. Patent 2,422,229 (1947).

36a. Fletcher, Hamilton, Hechenbleikner, Hoegberg, Sertl, and Cassaday, *J. Am. Chem. Soc.,* **72,** 2462 (1950).

36b. Foster, and Bennett (to Imperial Chemical Industries, Ltd.), U. S. Patent 2,440,602 (1948).

37. Fox (to Monsanto Chemical Co.), U. S. Patent 2,722,537 (1955).

38. Frear, "Chemistry of the Pesticides," 3rd Ed., New York, Van Nostrand, 1955, pp. 31–39.

38a. Frensch, *Med. u. Chem.,* **6,** 556–572 (1958).

38b. Frensch, Goebel, Staudermann, and Finkenbrink (to Farbwerke Hoechst AG), U. S. Patent 2,799,685 (1957).

39. Gasser, *Z. Pflanzenkrankh. Pflanzenschutz,* **58,** 467 (1951).

40. Gast, Gysin, and Knüsli, *Experientia,* **11,** 107 (1955).

40a. Geering and Nelson (to Hooker Chemical Corp.), U. S. Patent 2,983,732 (1961).

41. Gilbert (to Allied Chemical and Dye Corp.), U. S. Patent 2,618,583 (1952).
42. Girard, Di Bella, and Sidi (to Heyden Chemical Co.), U. S. Patent 2,848,470 (1958).
43. Gleason, *Proc. North Central Weed Control Conf.*, 54 (1959).
44. Gysin, *Chimia (Zürich)*, **8**, 205 (1954).
45. Gysin and Margot (to J. R. Geigy A.-G.), U. S. Patent 2,754,243 (1956).
46. Gysin, *Advances in Pest Control Research*, **3**, 289 (1960).
47. Hamm and Speziale, *J. Agr. Food Chem.*, **4**, 518 (1956).
48. Hamner and Tukey, *Science*, **100**, 154 (1944).
49. Harris, Tate, and Zukel (to U. S. Rubber Corp.), U. S. Patent 2,529,494 (1950).
50. Harris and Zukel, *J. Agr. Food Chem.*, **2**, 140 (1954).
50a. Harrison, Peters, and Rowe, *J. Chem. Soc.*, 235-7 (1943).
51. Haynes, Lambrech, and Moorefield, *Contrib. Boyce Thompson Inst.*, **18**, 507 (1957).
51a. Hechenbleikner (to American Cyanamid Co.), U. S. Patent 2,482,063 (1949).
51b. Hechenbleikner (to American Cyanamid Co.), U. S. Patent 2,692,893 (1954).
51c. Herzfeld, Lidov, and Bluestone (to Velsicol Corp.), U. S. Patent 2,606,910 (1952).
52. Heyden-Newport Chemical Co., British Patent 796,361 (1958).
53. Holden (to Columbia-Southern Chemical Corp.), U. S. Patent 2,732,914 (1956).
54. Holton, *Agric. Chem.*, **12** (7), 42-3, 95 (1957).
55. Hopkins and Pullen (to Spencer Chemical Co.), U. S. Patent 2,906,614 (1959).
56. Horsfall, "Principles of Fungicidal Action," Chronica Botanica Co., Waltham, Mass., 1956, p. 187.
56a. Howald and Marshall (Shell Development Co.), U. S. Patent 2,813,915 (1957).
57. Hu, Li, and Cheng, *Hua Hsüeh Hsüeh Pao*, **22**, 49 (1956); *Chem. Abs.*, **52**, 6156 (1958).
58. Huisman, Uhlenbroek, and Meltzer, *Rec. trav. chim.*, **77**, 103 (1958).
59. Huisman, van der Veen, and Meltzer, *Nature*, **176**, 515 (1955).
60. Hummer and Kenaga (to Dow Chemical Co.), U. S. Patent 2,528,310 (1950).
60a. Hyman (to Velsicol Corp.), U. S. Patent 2,519,190 (1950).
60b. Hyman, Belg. Patent 498,176 (1951); *C.A.*, **49**, 372 (1955).
61. Jelinek (to E. I. du Pont de Nemours & Co.), U. S. Patent 2,503,390 (1950).
62. Johnson and Godfrey, *Ind. Eng. Chem.*, **24**, 311 (1932).
63. Jones (to American Chemical Paint Co.), U. S. Patent 2,390,941 (1945).
64. Jonas (to Farbenfabriken Bayer), German Patent 951,718 (1956).
65. Jordan (to Frontier Chemical Co.), U. S. Patent 2,917,542 (1959).
65a. Kearns, Weinman, and Decker, *J. Econ. Ent.*, **42**, 127 (1949).
66. Kimball and Loverde (to Hooker Electrochemical Co.), U. S. Patent 2,211,466 (1940).
67. Kirk and Othmer, "Encyclopedia of Chemical Technology," Vol. 10, pp 394-6, New York, Interscience, 1953.
68. Kittleson (to Standard Oil Development Co.), U. S. Patent 2,553,770 (1951).

69. Kittleson, *Science,* **115,** 84 (1952).

70. Kittleson, and Yowell (to Standard Oil Development Co.), U. S. Patent 2,553,771 (1951).

71. Kittleson (to Standard Oil Development Co.), U. S. Patent 2,553,776 (1951).

72. Kittleson (to Standard Oil Development Co.), U. S. Patent 2,653,155 (1953).

72a. Kittleson, and Nelson (to Esso Research and Engineering Co.), U. S. Patent 2,856,410 (1958).

72b. Kogan, Burmakin, and Cherniak, *J. Gen. Chem. USSR,* **28,** 27 (1957) (Engl. transl.).

72c. Kleiman (to Arvey Corp.), U. S. Patent 2,658,085 (1953).

72d. Kleiman (to Velsicol Corp.), U. S. Patent 2,598,561 (1952).

73. Lambrech (to Union Carbide Corp.), U. S. Patent 2,903,478 (1959).

74. Lansade, Parasitica, **5,** 1 (1949).

75. Läuger, Martin, and Müller, *Helv. chim. Acta,* **27,** 892 (1944).

75a. Lidov (to Shell Development Co.), U. S. Patent 2,635,977 (1953).

75b. Lidov (to Shell Development Co.), U. S. Patent 2,717,851 (1955).

76. Lindemann (to Diamond Alkali Co.), U. S. Patent 2,923,634 (1960).

76a. Lindquist and Dahm, *J. Econ. Ent.*, **50,** 483 (1957).

76b. Lorenz (to Farbenfabriken Bayer), German Patent 927,270 (1955), *Chem. Abs.* **52,** 2908e (1958).

77. McBeth and Bergeson, *Plant Disease Reptr.,* **39,** 223 (1955).

78. McKeever and Nemec (to Rohm and Haas Co.), U. S. Patent 2,917,553 (1959).

78a. Malatesta, Italian Patent 458,770 (1950).

78b. March, *J. Econ. Ent.,* **45,** 452 (1952).

79. Martin, "Guide to the Chemicals Used in Crop Protection," Canada Dept. of Agriculture, London, Ont., 1957, p. 234.

79a. *ibid,* p. 173

80. Manske (to U. S. Rubber Corp.), U. S. Patent 2,471,575 (1949).

80a. Meinhardt, Cardon, and Vogel, *J. Org. Chem.,* **25,** 1991 (1960).

80b. Maude and Rosenberg (to Hooker Electrochemical Co.), U. S. Patents 2,650,942 (1953); 2,714,124 (1955).

81. Meitzner and Hester (to Rohm and Haas Co.), U. S. Patent 2,464,600 (1949).

82. Mel'nikov and Mandel'baum, Khim. i Primenie Fosfororgan. Soedinenii Akad. Nauk SSSR, Trudy 1-oĭ Konferents., *1955,* 185–93 (pub. 1957); *Chem. Abs.,* **52,** 237a (1958).

82a. Mel'nikov, Volodkovich, and Kogan, USSR Patent 108,590 (1957).

83. Miner and Dalton, "Glycerol," ACS Monograph No. 117, Reinhold Publishing Corp., New York, 1953, pp. 80–3.

84. Moyle (to Dow Chemical Co.), U. S. Patents 2,559,515 and 2,559,516 (1952).

85. O'Colla, *J. Sci. Food Agric.,* **3,** 130 (1952).

86. Ohsol, Carlson, and Hudson (to Standard Oil Development Co.), U. S. Patent 2,575,290 (1951).

87. Orochena (to Pittsburgh Coke and Chemical Co.), U. S. Patent 2,657,229 (1953).

88. Partansky (to Dow Chemical Co.), U. S. Patent 2,688,642 (1954).

89. Pearlman and Banks, *J. Am. Chem. Soc.,* **70,** 3726 (1948).

90. Purdy and Holton, *Phytopath.,* **46,** 385 (1956).

91. Rabjohn, *J. Am. Chem. Soc.,* **70,** 3518 (1948).

92. Reuter and Ascher, Experientia, **12** (8), 316 (1956).

92a. Riemschneider, *Z. Naturforsch.,* **66,** 395 (1951).

92b. Riemschneider, *La Chemica e L'Industria,* **34,** 266 (1952).

92c. Riemschneider, *Monatsh.,* **83,** 802 (1952).

92d. Riemschneider, *Naturwissenschaften,* **48,** 130 (1961).

92e. Rosenberg (to Hooker Chemical Corp.), U. S. Patent 2,899,370 (1959).

93. Sanders and Prescott, *Ind. Eng. Chem.,* **51,** 974 (1959).

94. Scheller and Smith, *Ind. Eng. Chem.,* **41,** 1027 (1949).

95. Schmidt (to Pineapple Research Institute of Hawaii), U. S. Patent 2,937,936 (1960).

96. Schoene, Tate, and Brasfield, *Agr. Chem.,* 24, (November 1959).

97. Schrader (to Farbenfabriken Bayer A.-G.), U. S. Patent 2,571,989 (1951).

98. Schrader (to Farbenfabriken Bayer A.-G.), U. S. Patent 2,624,745 (1953).

99. Schulz and Huemer, *Research and Production,* 454 (1953).

100. Slade, Templeman, and Sexton, *Nature,* **155,** 498 (1945).

101. Slagh and Britton, *J. Am. Chem. Soc.,* **72,** 2808 (1950).

101a. Smith and Livak (to Dow Chemical Co.), U. S. Patent 2,107,650 (1938).

101b. Smith and Payne (to Shell Development Co.), U. S. Patent 2,786,854 (1957).

102. Smolin and Rapoport, "*s*-Triazines and Derivatives," Interscience, New York, 1959, pp. 50–2.

102a. Soloway (to Shell Development Co.), U. S. Patent 2,676,131 (1954).

103. Sonia and Lisman (to Hooker Electrochemical Co.), U. S. Patent 2,595,899 (1952).

104. Sonia and Scremin (to Hooker Electrochemical Co.), U. S. Patent 2,674,620 (1954).

105. Sosnovsky, *Chem. Reviews,* **58,** 509 (1958).

106. Speck (to Hercules Powder Co.), U. S. Patent 2,815,350 (1957).

107. Speziale and Hamm, *J. Am. Chem. Soc.,* **78,** 256 (1956).

107a. Sproule, *Pest Technology,* 41, (November 1960).

108. Steindorff, Pfaff, and Erlenbach (to Winthrop Chemical Co.), U. S. Patent 1,947,926 (1934).

108a. Stoesser (to Dow Chemical Co.), U. S. Patent 2,131,259 (1938).

108b. Strauss, Koflek, and Heyn, *Ber.,* **63B,** 1868 (1930).

109. Stiles (to Shell Development Co.), U. S. Patent 2,685,552 (1954).

109a. Sun, *J. Econ. Ent.,* **43,** 45 (1950).

110. Sweeney, Arnold, Dix, and Stelzer, *Proc. Iowa Acad. Sci.,* **47,** 197 (1940); *Chem. Abs.,* **35,** 7055 (1941).

111. Taylor (to R. H. Bogle and Co.), U. S. Patent 2,589,162 (1952).

112. Templeman, and Sexton, *Nature,* **156,** 630 (1945).

113. Ter Horst and Felix, *Ind. Eng. Chem.,* **35,** 1255 (1943).

114. Ter Horst (to U. S. Rubber Corp.), U. S. Patent 2,302,384 (1943).

115. Ter Horst (to U. S. Rubber Corp.), U. S. Patents 2,349,771 and 2,349,772 (1944)

116. Ter Horst, Stiteler, and Smith (to Dominion Rubber Co.), Canadian Patent 448,130 (1948).

117. Theis and Richardson (to Columbia-Southern Chemical Corp.), U. S. Patent 2,778,846 (1957).

118. Tischenko, *Zhur. Obshcheĭ Khim.,* **23,**1002 (1953).

118a. Toy (to Victor Chemical Works), U. S. Patent 2,471,464 (1949).

118b. Toy and McDonald (to Victor Chemical Works), U. S. Patent 2,715,136 (1955).

118c. Ungnade and McBee, *Chem. Revs.,* **58,** 249 (1958).

119. U. S. Tariff Commission Report No. 206, 2nd Series, 1959.

119a. U. S. Tariff Commission Report, T. C. Publ. 34, 1960.

120. Van Wazer, in Kirk and Othmer, "Encyclopedia of Chemical Technology," Vol. 10, New York, Interscience, 1953, pp. 478–480.

120a. Whetstone and Harman (to Shell Development Co.), U. S. Patent 2,956,073 (1960).

120b. Vogelbach, *Angew. Chem.,* **63,** 378 (1951).

120c. Volodkovich, Vol'fson, Kogan, Mel'nikov, and Sapozhkov, *Zhur. Priklad. Khim.,* **33,** 227 (224 in Engl. translation) (1960).

121. Whitman and Newton, *Proc. Northwest Weed Control Conf.,* 45 (1951).

122. Wilson, and Barker, Abstracts of 126th Am. Chem. Soc. Meeting, 1954, p. 29A.

123. Wilson and Wolff (to Rohm and Haas Co.), U. S. Patents 2,812,280 (1957) and 2,812,362 (1957).

124. Young, and Hensel (to Pittsburgh Coke and Chemical Co.), U. S. Patent 2,693,891 (1954).

125. Zimmerman and Hitchcock, *Contrib. Boyce Thompson Inst.,* **12,** 321 (1942).

23. ALLYL CHLORIDE AND DERIVATIVES

B. H. Pilorz

Shell Chemical Company

INTRODUCTION

General

The development just before World War II of an economic and commercially feasible process for manufacture of allyl chloride was an event of significance to the chemical industry. Allyl chloride had been known for over eighty years, but it remained a laboratory chemical until research in the field of hydrocarbon halogenation revealed a previously unknown reaction—one that permitted synthesis of allyl chloride from basic raw materials.

Exploitation of the development was delayed until after the war; first commercial scale production of allyl chloride was achieved in 1945. However, within five years thereafter almost one-quarter of the glycerol marketed in the United States was derived from allyl chloride, and developments in plastics and resins involving other allyl chloride derivatives were exerting strong pressure on allyl chloride production capacity. The last ten years have seen continuing growth; indications are that allyl chloride production has been between 100 and 200 million pounds a year during the past several years. Thus, in two decades, a chemical originally not available industrially had become an almost commonplace basic intermediate and one of significant commercial importance to the petrochemical industry.

Discovery and Early Preparation

The allyl group, $CH_2\!=\!CH\!-\!CH_2\!-$, was first defined in 1844 by Wertheim,[91] who isolated what he took to be allyl sulfide (later shown to be principally the disulfide)[65] from oil of garlic, or *alium sativum*. Berthelot and De Luca prepared allyl iodide in 1854, treating glycerol with phosphorus and iodine.[5] About a year later Cahours and Hoffman converted the iodide to allyl alcohol via the oxylate, and subsequently they prepared the chloride for the first time by reacting allyl alcohol with phosphorus chloride.[10]

In 1870 Tollens and associates reported a more direct synthesis of allyl alcohol employing oxalic acid,[83] and preparation of allyl halides from the alcohol and appropriate hydrogen halide became the preferred procedure. Double decomposition preparations of the halides, one from another using a halide salt, were also known and utilized.[52] There is an extensive literature covering preparation of allyl halides by these classic routes; Reference 67 contains a comprehensive bibliography.

Development of Propylene Chlorination

While the classic preparation of allyl chloride made it available for experiment and investigation, the procedures could not be considered for commercial application. Indeed, commercial interest in allyl chloride was based upon the realization that it could be used in the synthesis of glycerol, allyl alcohol, etc.—the very raw materials required for its classic preparation.[92] Hence, attention was directed toward production from basic raw materials, i.e., chlorine and hydrocarbons.

Prior to the 1930's, chlorination of propylene had been studied extensively; under the conditions employed, the only significant reaction known was the additive one producing 1,2-dichloropropane. Preparation of allyl chloride from the dichloride was studied by several investigators. It had been shown as early as 1850 that reaction of dichloropropane with alkali led principally to 1-chloropropene,[9,59] an isomer of allyl chloride and less interesting because of the lack of a functional group on the third carbon. Subsequently, it was found that passing dichloropropane over calcium chloride at about 350° C led to the formation of some allyl chloride, but the reaction was slow and the principal unsaturated monochloride again was 1-chloropropene; tars and hydrocarbons were also formed, apparently poisoning the catalyst in a short while.[22,42,93]

Direct pyrolysis of 1,2-dichloropropane at 600 to 700° C has been found fairly satisfactory.[28,93] With proper control, some 30 percent of the dichloride is decomposed per pass with yields of about 50 percent allyl chloride and 35 percent 1-chloropropene on the basis of consumed dichloride. Allyl chloride can be recovered by several techniques, but the over-all process is not industrially attractive.

During the mid-1930's, research directed at developing an understanding of hydrocarbon halogenation mechanisms led to the discovery of a means for obtaining substitutive halogenation of olefins. It was found that, at temperatures elevated considerably above those previously investigated, halo-substitution for hydrogen at saturated carbons occurred rather than the well-known addition at the double bond.[30] In the case of propylene chlorination, yield of allyl chloride and distribution of reaction products were favorable, thereby making economic commercial production of allyl chloride feasible.[92]

A very recent development is the discovery that allyl chloride can be obtained by reaction of propylene, hydrogen chloride and oxygen in the presence of a supported lithium chloride catalyst.[77] The reaction has been known for some time—but over metal oxide catalysts which are not so selective to allyl chloride as lithium chloride apparently is. No commercial application of the process is yet known.

General Significance of Allyl Chloride

The interest that promoted the research effort resulting in development of a means for direct synthesis of allyl chloride stemmed from consideration of its chemical properties. Allyl chloride is reactive both as an organic halide and as an olefin. Additions can be made at the double bond, replacements at the chlorinated carbon, or both; hence, modifications can be directed to any point of the structure by proper selection of reagents and environment. Furthermore, the allyl group when introduced into other molecules is usually reactive, making a number of interesting syntheses possible. Many simple derivatives of allyl chloride preserve the activity of two functional groups and are also important starting points for further syntheses.

The first commercially produced derivative was allyl alcohol, formed by dilute alkaline hydrolysis.[24] The other major commercial derivatives of allyl chloride are epichlorohydrin and glycerol.[96] Epichlorohydrin is formed by chlorohydrination of the double bond with subsequent elimination of hydrogen chloride, while the synthesis of glycerol involves both attack on the double bond and hydrolysis of the substituted chlorine.

ALLYL CHLORIDE

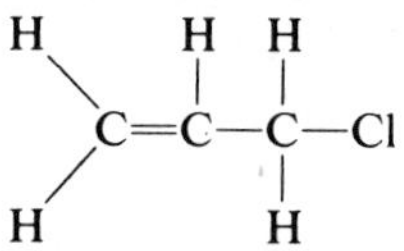

$$\text{H}_2\text{C}=\text{CH}-\text{CH}_2-\text{Cl}$$

3-Chloropropene

Manufacture

The Allyl Chloride Reaction.* *Mechanism.* Allyl chloride is synthesized directly by chlorinating propylene under conditions favoring sub-

---

*Although other reactions leading to allyl chloride are known, the high temperature substitutive chlorination of propylene is the reaction referred to; it is believed to be the only route used commercially at this time.

stitution of a chlorine atom for a hydrogen atom on the saturated carbon;[47] the double bond is preserved, and hydrogen chloride is produced as a by-product:

$$CH_2=CH-CH_3 + Cl_2 \rightarrow CH_2=CH-CH_2Cl + HCl \qquad (23\text{-}1)$$

Propylene   *Chlorine*   *Allyl chloride*   *Hydrogen Chloride*

The reaction occurs only at elevated temperature, the threshold being about 300° C. Exothermic reaction heat has been calculated as -26.7 kg. cal/mole at 355° K.[24]

This high temperature allylic substitution reaction was unknown until approximately twenty-five years ago. Prior to its discovery, work at ambient to moderate temperatures (below 200° C) had indicated that addition at the double bond was the principal result of halogenating straight-chain olefins.[31] Sheshukov had discovered in 1884 that ambient temperature liquid phase chlorination of isobutylene occurred largely by substitution to form unsaturated chlorides,[8,70] but the mechanism of this low temperature reaction differs from that of high temperature chlorination. In 1936 Stewart and Weidenbaum reported small amounts of pentenyl chlorides in the chlorination product of 2-pentene.[78] Shortly thereafter, Groll, Hearne, Burgin and LaFrance reported that allylic substitution of halogens into straight-chain olefins occurred if high temperature was employed, and they clearly demonstrated that temperature level was the critical factor in promoting substitution at the expense of addition.[29,30]

It is believed that high temperature substitutive halogenation of olefins proceeds by a free radical mechanism, the postulated chain being:[18]

$$Cl_2 \xrightarrow{\Delta} 2Cl\bullet \qquad (23\text{-}2)$$

$$Cl\bullet + CH_2=CH-CH_3 \rightarrow HCl + CH_2\text{-----}CH\text{-----}CH_2 \qquad (23\text{-}3)$$

$$CH_2\text{-----}CH\text{-----}CH_2 + Cl_2 \rightarrow CH_2=CH-CH_2Cl + Cl\bullet \qquad (23\text{-}4)$$

It is probable that the substitution is not, strictly speaking, a direct replacement; a hydrogen atom is removed from the propylene molecule, but because of mobility of the resultant radical, the chlorine may enter at a position different from the one vacated.[33]

Evidence supporting this view of the reaction is ample. Rust and Vaughan have shown that the general characteristics of the reaction, e.g., its response to presence of oxygen or tetraethyl lead, are consistent with those of a radical chain mechanism.[60] Groll and Hearne observed that bromine substitution occurred more readily than chlorine substitution at a given temperature; they attributed the effect to the higher degree of thermal dissociation of bromine.[29] Hearne and others observed that the high temperature chlorination of either allyl chloride or of 1-chloropropene leads to the same mixture of dichlorides, namely 90 percent 1,3-dichloropropene and 10 percent 3,3-dichloropropene.[33] Some allylic rearrangement from the 1,3-form to the 3,3-form is conceivable, but the reverse rearrangement is considered unlikely. It is believed that this identical product distribution can best be explained by a free radical mechanism consistent with that postulated for the formation of allyl chloride:

$$Cl_2 \rightarrow 2Cl^{\bullet} \tag{23-5}$$

$$\left.\begin{array}{c} CH_2{=}CH{-}CH_2Cl \\ \text{or} \\ CHCl{=}CH{-}CH_3 \end{array}\right\} + Cl^{\bullet} \rightarrow HCl + \left\{\begin{array}{c} CH_2{=}CH{-}\overset{\bullet}{C}HCl \\ 90\% \downarrow \uparrow 10\% \\ CHCl{=}CH{-}\overset{\bullet}{C}H_2 \end{array}\right\} \tag{23-6}$$

$$\left\{\begin{array}{c} CH_2{=}CH{-}\overset{\bullet}{C}HCl \\ 90\% \downarrow \uparrow 10\% \\ CHCl{=}CH{-}\overset{\bullet}{C}H_2 \end{array}\right\} + Cl_2 \rightarrow Cl^{\bullet} + \left\{\begin{array}{c} CH_2{=}CH{-}CHCl_2\ 10\% \\ \text{and} \\ CHCl{=}CH{-}CH_2Cl\ 90\% \end{array}\right\} \tag{23-7}$$

Finally, it is observed that the distribution of unsaturated monochloride isomers obtained by high temperature chlorination of propylene differs from the distribution resulting from pyrolysis of 1,2-dichloropropane. Compositions of unsaturated monochloride fractions (given in Reference 29) are listed in Table 23-1. The distribution of products of high temperature chlorination is affected by mixing conditions, and there is some evidence that at least 2-chloropropene is formed by cracking 1,2-dichloropropane.[88] The discrepancy in monochloride distribution is so marked, however, that it is concluded that 1,2-dichloropropane is not an intermediate in the principal high temperature allyl chloride reaction.

TABLE 23-1. COMPOSITIONS OF UNSATURATED MONOCHLORIDE FRACTIONS

Unsaturated Monochlorides	High Temperature Chlorination (%)	Pyrolysis of $CH_2Cl{-}CHCl{-}CH_3$ (%)
$CH_2{=}CH{-}CH_2Cl$	96	55 to 70
Cis- and trans-$CHCl{=}CH{-}CH_3$	1	30 to 40
$CH_2{=}CCl{-}CH_3$	3	5

The reaction is not peculiar to propylene. A wide variety of unsaturated hydrocarbons behave in similar fashion,[30,47] provided that appropriate temperatures are employed. Generally, olefins with tertiary carbons at the double bond substitute more readily than those with secondary carbons, and increased chain length promotes ease of substitution; ethylene, having only primary carbons, is most difficult to substitute.[29] Isobutylene, incidentally, chlorinates by substitution quite readily at low temperature in the liquid phase, with very little addition; this reaction is not inhibited by the presence of oxygen and so a free radical chain mechanism is considered unlikely. At 200 to 240°C in the gas phase, however, both addition and substitution are encountered; as temperature is raised, substitution is favored as in the case of straight-chain olefins, and this substitution is inhibited by oxygen.[60] It has been concluded that the low temperature isobutylene reaction may be unique, while the high temperature reaction is of the free radical type.

Reaction Variables. Temperature is the only variable that affects the monochloride substitution reaction itself, at least within the ranges of environmental conditions usually encountered. Pressure, mole ratio, residence time, impurities and degree of mixing affect side reactions, and hence they are also of importance.[24]

Chlorine and propylene in contact with each other at temperatures below approximately 200°C react largely by addition. As temperature is raised above that level, substitution may be observed. The substitution reaction predominates as 300°C is reached, and further temperature increases lead to greater degrees of substitution with corresponding reductions in degree of addition. At extremely high temperatures (about 600°C and above), degradation reactions set in and yield of allyl chloride decreases, while production of heavier boiling products such as benzene increases.

Observations on pilot plant and commercial reactors demonstrate general temperature dependence of the reaction and of over-all chlorination. Chlorine conversions of 99.3 percent at 510°C, 99.98 percent at 560°C are reported, other conditions being the same, as well as conversions of 99.95 percent in 1.8 seconds residence time and 99.997 percent in twice that time, both at 510°C.

The other unsaturated monochloride isomers are formed to some extent as co-products, as noted in the preceding discussion of mechanism. The chief secondary products are, however, the unsaturated dichlorides formed by further chlorination of allyl chloride. Some dichloropropane is always found, while degradation products include benzene, tars and carbon. More than two dozen components are known or suspected in the reaction product.

Production of unsaturated dichlorides can be limited to some extent by using a molar excess of propylene. Thorough mixing to avoid zones of localized improper reactant concentration and localized overheating is also necessary.[88] Pressure level does not affect the main reaction, but carbon deposition may increase slightly with increasing pressure.

The usual feed impurities encountered are water and propane. Water may react with chlorinated hydrocarbons to form undesirable by-products and may also provide a corrosive environment, while propane and other hydrocarbons are chlorinated to undesirable saturated chlorides.

Commercial Practice in Allyl Chloride Production. Industrial scale allyl chloride facilities perform three major processing operations: Feed Preparation, Reaction, and Product Recovery. Variations in design and operation are common because each step of the processing usually must mesh with other operations or adapt to local conditions. The following discussion therefore is generalized, with emphasis upon principles rather than upon practice at any one location. Figure 23-1 presents a schematic flow diagram. Principal references are 24 and 34.

Feed Preparation. Requirements are that both reactor feeds be dry and reasonably pure to limit yield losses. Measures to insure chlorine purity are discussed elsewhere in this monograph. Propylene purification meas-

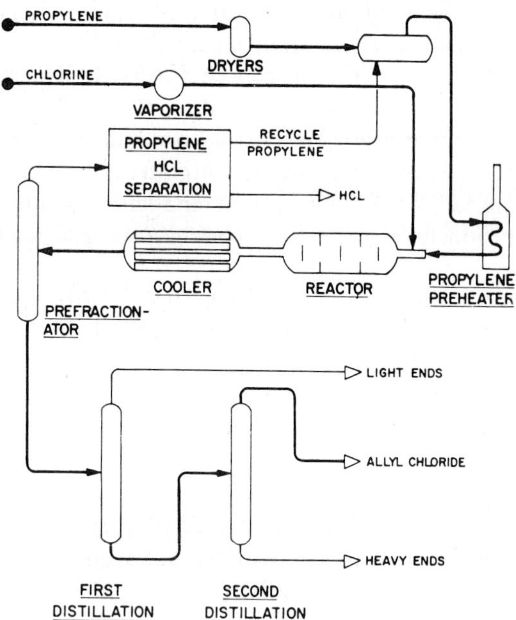

Figure 23-1. Simplified flow scheme—allyl chloride production.

ures vary with the nature of the feedstock; fractionation is employed if necessary, as are standard drying techniques.

The reaction liberates enough heat to permit adiabatic reactor operation if some preheat is supplied. To limit the addition reaction that proceeds readily at low temperature reactor feed is preheated before reactants are mixed, but it is usually not possible to preheat to a level that immediately precludes addition chlorination without subsequently exceeding optimum reactor temperature. Either stream, or both, may be preheated, but it is advantageous to preheat propylene alone because it is present in excess, has a more favorable heat capacity, and is more easily handled while hot.

The degree of preheat required to maintain a given reaction temperature varies with the propylene/chlorine mole ratio. With chlorine feed fixed and hence with heat of reaction fixed, the heat necessary to bring an increasing excess of propylene to the given reaction temperature must be supplied as increased preheat. A reaction temperature of about 500°C is maintained by preheating propylene to about 110°C at a mole ratio of 2.0, while preheating to about 340°C is necessary at a ratio of 4.0.

Chlorine is fed as a vapor. Any suitable vaporizing facilities may be used, exchange with hot water being adequate.

Reaction System. The reaction is carried out in an adiabatic reactor designed to provide rapid and intimate mixing. Failure to mix rapidly and well results in excessively high localized temperatures, by-product formation, and even severe carbon deposition. Slight changes in flow pattern may cause radical deviations in product composition.

Feed is introduced under conventional instrument control through a suitable mixing nozzle. Reaction temperature is controlled by balancing feed mole ratio and propylene preheat temperature. The usual reaction temperature range is 500 to 510°C; pressures in the order of 15 psig are used.

General reactor conditions are illustrated by pilot plant data presented in Table 23-2. Details of the heat balance calculations will be found in Reference 24.

Chlorine utilization in excess of 99.9 percent is obtained in commercial reactors, and that which is not converted in the reactor is completely consumed by reaction with olefinic materials as product is cooled.

Reactor product is cooled to about 50°C in conventional exchangers before routing to further processing. Gradual fouling in the reactor system is experienced, and parallel reactor-cooler trains manifolded for alternate operation are often provided. An average of two weeks between cleanouts is reported.[34] Steel construction may be used throughout the reaction system except for chloride-resistant materials in valve trim, etc.

Product Recovery. Allyl chloride is separated from other components of

TABLE 23-2. ALLYL CHLORIDE PILOT PLANT REACTOR DATA[a,b]

	Boiling Point (°C)	Gram Moles	Sensible Heat (kg.-cal)	Heat Due to Reaction, (kg.-cal)
Reactants				
Propylene		406	2683	
Chlorine		107	9	
Total heat			2692	
Products				
HCl	−83.7	106	383	
Propylene	−47.0	306	3460	
2-Chloropropene	22.5	2.3	33	67
Isopropyl chloride	36.5	0.2	3	3
Allyl chloride	44.9	85.1	1214	2280
2,3-dichloropropene	93.8	0.8	14	44
1,2-Dichloropropane	96.8	2.6	46	107
Cis 1,3-Dichloropropene	103.8	} 6.5	} 103	} 332
Trans 1,3-Dichloropropene	112.1			
Higher boiling	−	2.6	33	120
Total heat			5289	
Heat losses from reactor			240	
Net heat due to reaction			2830	2950

[a]Basis: 100 gram moles propylene reacted
[b]Temperatures:

Base for heat calculations	0°C
Propylene feed	212°C
Chlorine feed	12°C
Reaction zone and outlet	510°C

the reactor product stream by a series of fractionations. After initial removal of hydrogen chloride and propylene, the organic chloride fraction is separated in a conventional two-step distillation. Other schemes are possible; for example, it is reported that, because of preferential absorption, kerosene can be used to extract allyl chloride from other reaction gases.[97]

The initial removal of hydrogen chloride and propylene is accomplished in a fractionating column reboiled to about 40° C and refluxed with liquid propylene at –40° C. The overhead stream is subsequently split, propylene being recycled and hydrogen chloride taken off to other use or to disposal. The choice of means of splitting depends to some extent upon ultimate utilization of hydrogen chloride. Fractional distillation may be employed if anhydrous acid is desired, or hydrogen chloride can be selectively absorbed into water to produce 20° Be hydrochloric acid.[51]

In the distillation of the organic chloride fraction, low boiling constituents are taken overhead in the first column and allyl chloride in the second. The heavy boiling fraction taken off as a bottom product in the second column is made up largely of unsaturated dichlorides. This stream, after further processing, has been found useful as a soil fumigant for nematode control.

Chemical Properties of Allyl Chloride

Allyl chloride exhibits reactivity as an olefin and as an organic halide. Activity as a chloride is enhanced by the presence of the double bond, but activity as an olefin is somewhat less than that of propylene. In general, allyl chloride participates in most types of reactions characteristic of either functional group. Reactions can be directed by control of environment, selection of reagents, and provision of suitable catalysts.

TABLE 23-3. PHYSICAL PROPERTIES OF ALLYL CHLORIDE[a]

General: Colorless liquid, pungent odor, hazardous chemical	
Molecular weight	76.53
Specific gravity of liquid	0.9392^{20}_{4}
Solubility in water at 20°C	0.33%w
Viscosity at 20°C	3.36 c p
Freezing point	−134.5°C
Boiling point at 760 mm	44.96°C
Vapor pressure (T°K, P mm Hg)	$\log P = 20.0151 - 2098.0/T - 4.2114 \log T$
Latent heat of vaporization at b.p.	6,940 cal/gram mole
Specific heat, vapor, 100°C	0.230 cal/gram/°C
liquid, 40°C (calc.)	0.395 cal/gram/°C
Heat of combustion, vapor	440.8 K cal/gram mole
Inflammability limits, air lower limit	3.28%v
upper limit	11.15%v
Flash point, closed cup	−16°F

[a]References 20 and 67 present literature sources and further data.

The chemistry of the allyl halides is the subject of an abundant literature dating from the early preparations a century ago. Extensive discussions and bibliographies can be found in References 20, 37, 67, 90; this review presents only a general view of the subject.

Typical Additions to the Double Bond. Chlorine, bromine and iodine chloride add to allyl chloride at temperatures below the inception of the substitution reaction, producing the 1,2,3-trihalides. (High temperature halogenation follows the free radical mechanism, leading to unsaturated dihalides; see discussion of allyl chloride reaction above). Hypochlorous and hypobromous acids add to form glycerol dihalohydrins, principally the 2,3-dihalo isomer. A more extensive discussion of this reaction will be found below under the heading "Manufacture of Epichlorohydrin;" the chlorohydrins are intermediates in that process as well as in the production of glycerol from either allyl chloride or allyl alcohol.

Hydrogen halides normally add to form 1,2-dihalides, although an abnormal addition of hydrogen bromide is known, leading to 1-chloro-3-bromopropane. The latter reaction is believed to proceed by a free radical mechanism, for it occurs under conditions favorable to formation of free radicals. 1-Chloro-3-bromopropane is converted to cyclopropane, an anesthetic, by treatment with metallic zinc or sodium.

Water can be added by treatment with sulfuric acid at ambient or lower temperatures, followed by dilution with water. The product is 1-chloro-2-propanol.

Simple Replacements of Chlorine. The alkaline hydrolysis of allyl chloride to allyl alcohol is discussed below under the heading "Manufacture of Allyl Alcohol." Passing allyl chloride and steam over calcium or potassium chloride is also reported to lead to the alcohol.[20] Acidic hydrolysis in the presence of cuprous chloride is also known.[87]

Several other replacement reactions are summarized in Table 23-4, although the list is by no means complete.

TABLE 23-4. A SUMMARY OF SIMPLE REPLACEMENT REACTIONS.

Product	Reagent	Remarks
C_3H_5I	CaI_2	
C_3H_5CN	$CuCN$	
C_3H_5SCN	$NaNCS$	cold
C_3H_5NCS	$NaNCS$	hot
$C_3H_5SSO_2ONa$	$Na_2S_2O_3$	
$C_3H_5SC_3H_5$	KSH	major product
C_3H_5SH		minor product
$C_3H_5AsO(OH)_2$	Na_3AsO_3 to salt, then HCL	

In some cases it is preferable to use one of the other halides; they may be prepared from allyl chloride by either direct replacement or action of the phosphorus halide on allyl alcohol. For example, 3-nitropropene and allyl nitrite are prepared from silver nitrite and allyl bromide or iodide.

Allyl Esters. Allyl esters are formed by reaction of allyl chloride with sodium salts of appropriate acids under conditions of controlled pH. Esters of the lower alkanoic, alkenoic, alkandioic, cycloalkanoic, benzene carboxylic, alkyl substituted benzene carboxylic, and aromatic dicarboxylic acids may be prepared in this manner; see also preparation from allyl alcohol.

Nitrogen Compounds. Mono-, di-, and triallyl amines are prepared by reaction with ammonia, the ratio of reagents determining product distribution; with sufficient time and excess of allyl chloride, tetra-allylammonium chloride and triallyl amine predominate. The reactions proceed in steps with formation of substituted ammonium chlorides. Mixed amines are prepared in similar fashion, using a substituted amine in place of ammonia; they may also be prepared with allyl amine and a suitable organic chloride. Amines can be modified by various means to a number of end products; e.g., N-allyl derivatives of heterocyclic bases.

Allyl amides, imides, hydrazines, tetrazines, nitramines and azides have been prepared.

Syntheses of Complex Molecules. A number of syntheses of complex substances take advantage of the reactivity of allyl chloride, which facilitates introduction of allyl groups into other structures, and subsequently of the activity of the introduced allyl group. Some of the initial allyl compounds are of interest in their own right.

Mixed allyl ethers may be prepared from allyl chloride and the appropriate alcoholate or alcohol-alkali mixture. Polyol ethers, especially those having more than one allyl group, form resinous polymers. Allyl aryl ethers undergo the Claisen rearrangement to allyl-substituted phenols and are used as starting points for several syntheses, e.g.,

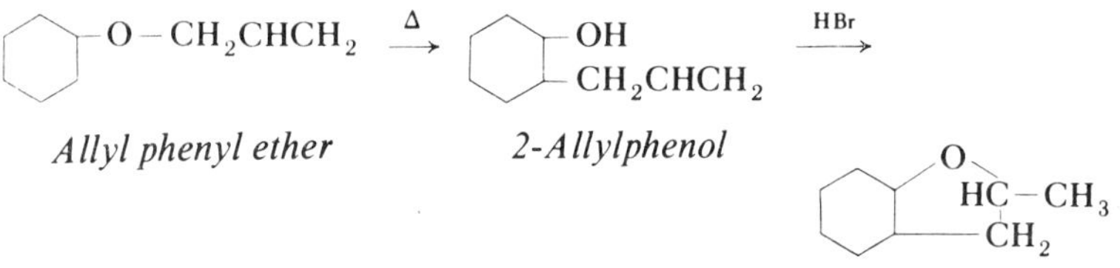

$$\text{Allyl phenyl ether} \quad \xrightarrow{\Delta} \quad \text{2-Allylphenol} \quad \xrightarrow{HBr} \quad \text{2-Methylcoumaran} \qquad (23\text{-}8)$$

Allyl Grignard reagent is readily prepared from allyl chloride by the usual procedures and is used as a means of substituting the allyl group to prepare olefins and diolefins, substituted methanols, allyl-chlorosilanes, etc.

A number of alkylation-type reactions are known; reactivity of either functional group may be involved. The reactions of allyl chloride with benzene are typical of reactions involving the double bond. In the presence of ferric or zinc chloride, the products are 2-chloropropylbenzene and 1,2-diphenylpropane:

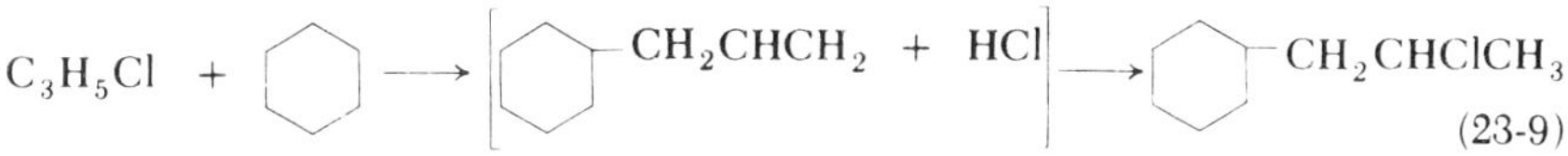

$$C_3H_5Cl \; + \; \bigcirc \; \longrightarrow \; \bigcirc\text{-}CH_2CHCH_2 \; + \; HCl \; \longrightarrow \; \bigcirc\text{-}CH_2CHClCH_3 \qquad (23\text{-}9)$$

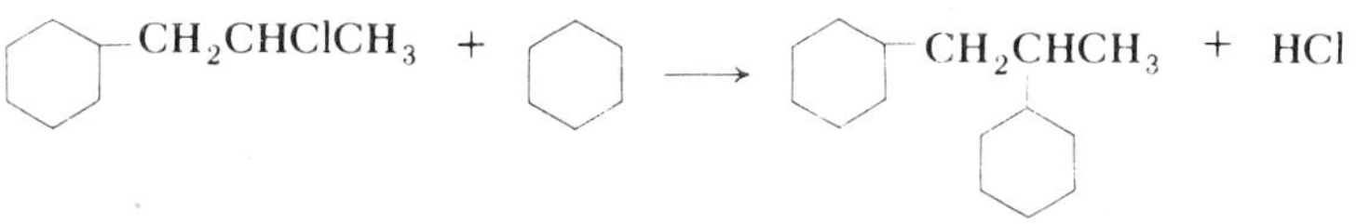

$$\bigcirc\text{-}CH_2CHClCH_3 \; + \; \bigcirc \; \longrightarrow \; \bigcirc\text{-}CH_2CHCH_3 \; + \; HCl \qquad (23\text{-}10)$$

When aluminum chloride containing traces of water is used as a catalyst, further reaction occurs and products include 9,10-diethylanthracene and *n*-propylbenzene. The general ferric chloride-catalyzed reaction can be carried out with a number of benzene derivatives. The 1-aryl-2-chloropropanes may be converted to 2-amino derivatives, i.e., benzedrine and its analogues.

A number of allylation reactions are known, frequently involving the use of a metallo-organic derivative of the compound being allylated or the use of a strongly electropositive metal in conjunction with the reactants; Grignard reactions are in this group.

Illustratively, allyl chloride reacts with sodamide in liquid ammonia to produce 1,3,5-hexatriene; when sodamide is in excess, hexadiene dimer is the principal product, with some trimer and tetramer (C_{24}, 6 double bonds). Allylation at carbon atoms alpha to polar groups is made use of in preparation of α-allyl-substituted ketones and nitriles. Preparation of β-diketone derivatives, methionic acid derivatives and malonic ester, cyanoacetic ester, and β-keto-ester derivatives, etc., involving substitution on a carbon atom between two polar groups, is particularly easy. Typically:

$$C_3H_5Cl + R-\underset{\underset{\displaystyle COOR'}{|}}{\overset{\overset{\displaystyle C=O}{|}}{C}}-Na \rightarrow R-\underset{\underset{\displaystyle COOR'}{|}}{\overset{\overset{\displaystyle C=O}{|}}{C}}-C_3H_5 + NaCl \qquad (23\text{-}11)$$

General Usage Pattern of Allyl Chloride

In commercial application, allyl chloride is an intermediate. It is used almost exclusively in the synthesis of derivatives, many of which, in turn, do not themselves reach an end-use market, but are part of further syntheses. Broadly speaking, the largest end-use market to which allyl chloride can be traced is that of polymers—various plastics and resins that are used *per se* or incorporated into surface coatings, adhesives, etc.

The principal "first generation" derivatives of allyl chloride are glycerol, epichlorohydrin and allyl alcohol. It is estimated that about 99 percent of United States allyl chloride production is converted to one or another of these three; estimates are based upon 1958 figures, the latest available at the time of writing. As can be seen in Figure 23-2, glycerol production absorbs almost three-quarters and epichlorohydrin production most of the remaining available allyl chloride. Principal end-use markets are indicated; more detailed discussion of these applications will be found below under appropriate derivative headings.

Medicinal products collected under miscellaneous intermediate derivatives include allyl-substituted barbiturates and mercury diuretics derived

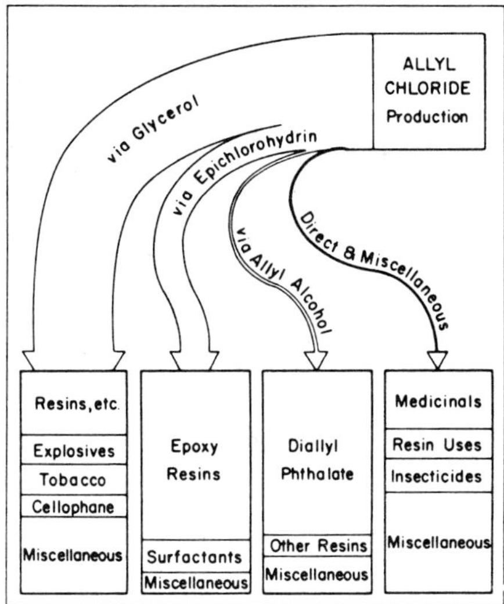

Figure 23-2. Approximate commercial distribution pattern of allyl chloride to various end uses.

from allylamine,[2] as well as the anesthetic, cyclopropane. A number of useful specialty resins are derived from allyl esters and polyesters that may be made directly from allyl chloride.[17] Resin uses also include a number of copolymers and interpolymers of allyl chloride with, e.g., acrylonitrile,[11] vinylidene cyanide,[27] styrene and diallyl esters[80] developed to provide special properties, while allyl chloride also serves as a catalyst[56] or modifier in production of other resins. Alkenylhalosilanes are useful textile water-repellants.[82]

As noted in the discussion of manufacture, important byproducts of the allyl chloride reaction include 1,3- and 2,3-dichloropropene, which result from the further high-temperature chlorination of allyl chloride. These dichlorides are important nematocides.[13] Other insecticidal derivatives of allyl chloride include 1,2-dibromo-3-chloropropane, also used in nematode control.[62]

ALLYL ALCOHOL

$$H_2C=CH-CH_2-OH$$

2-Propen-1-ol

Manufacture

Allyl alcohol is prepared from allyl chloride by alkaline hydrolysis:[3,24,43,57,79,86,93]

$$CH_2{=}CH{-}CH_2Cl + OH^- \rightarrow CH_2{=}CH{-}CH_2OH + Cl^- \qquad (23\text{-}12)$$

Allyl chloride   *Allyl alcohol*

Under hydrolyzing conditions, ether formation can be significant:

$$C_3H_5Cl + C_3H_5OH + OH^- \rightarrow C_3H_5{-}O{-}C_3H_5 + Cl^- + H_2O \quad (23\text{-}13)$$

Allyl   *Allyl*   *Diallyl*
chloride   *alcohol*   *ether*

The extent of ether formation depends upon reaction conditions. Heavy materials are also produced in small amounts; these presumably arise from polymerization of the alcohol itself or of propionaldehyde formed by rearrangement of the alcohol.

In commercial practice the concerns are to attain maximum conversion of allyl chloride and to limit ether formation. Aside from the obvious yield considerations, maximum conversion is important in limiting corrosion by allyl chloride during the subsequent processing of aqueous alcohol streams. Allyl chloride conversion is favored by long contact or residence time and to some extent by elevated temperature, while ether formation is suppressed by high dilution and by avoiding high alkalinity; as a result, a compromise between residence time provisions and degree of dilution is necessary at any given production rate. Good dispersion of the insoluble allyl chloride phase is essential to efficient hydrolysis.

Typical hydrolysis conditions are summarized: Allyl chloride concentration, 0.04 to 0.08 lb/lb H_2O; caustic concentration, 2.5 to 5 weight percent; temperature, 150 to 160°C; residence time, 10 to 30 minutes; ether formation, <10 percent. The interplay of reaction variables is discussed quantitatively in References 24 and 93, and kinetic effects in Reference 41; the reaction is of first order with respect to allyl chloride concentration.

Production facilities typically consist of a reaction or hydrolyzing system and alcohol recovery columns. Generally, the hydrolyzer vessel is equipped for recirculation; allyl chloride and caustic solution feeds are dispersed in the recirculating streams. Dilute crude alcohol is drawn from the hydrolyzer and concentrated by stripping; alcohol, diallyl ether and azeotroped water are taken overhead, while heavy products, salt and water are rejected to waste. Subsequently, allyl alcohol may be recovered by either batch or continuous distillation. Advantage is taken of ternary alcohol-ether-water azeotrope formation for dehydration; upon condensation, the low-boiling azeotrope separates into an upper ether phase and a lower aqueous phase; the ether phase is water washed to remove the alcohol, and the washings plus the water phase are recycled to the stripper for alcohol recovery.

TABLE 23-5. PHYSICAL PROPERTIES OF ALLYL ALCOHOL[a]

General: Colorless liquid, pungent odor, hazardous chemical

Molecular weight	58.078
Specific gravity of liquid	0.8535^{20}_{20}
Miscible all proportions, 25°C, in water, acetone, benzene, carbon tetrachloride, methanol	
Viscosity at 20°C	1.23 cp
Freezing point	unknown, glass at $-190°$C
Vapor–Liquid data	
Boiling point, 760 mm	96.95°C
Vapor pressure (T°K, P mm Hg)	$\log P = 32.62580 - 3451.8/T - 7.94975 \log T$
Latent heat of vaporization at b.p.	9550 cal./gram mole
Specific heat, vapor	$Cp = 0.3114 + 7.447 \times 10^{-4}t$ cal./gram/°C
liquid	$Cp = 0.665$ cal./gram/°C avg., $20.5 - 95.5$°C
Heat of combustion, vapor	442.4 K cal./gram mole
Inflammability limits,	
Air at 100°C—lower limit	2.50%v
—upper limit	18.00%v
Flash point, open cup	90°F
closed cup	72°F

[a]References 19 and 66 present literature sources and further data.

Chemical Properties of Allyl Alcohol

Allyl alcohol has two functional groups, the olefinic bond and the hydroxyl group in the alpha position relative to the bond. Additions to the double bond, typical alcohol reactions, or both can be carried out. Reactivity as an alcohol in replacement reactions is considerably greater than that of saturated primary alcohols. In some instances, reactivity as an olefin is likewise greater than that of simple olefins; a number of addition derivatives are readily obtainable which are formed with difficulty from the olefins, if at all.

As in the case of allyl chloride, complete treatment of allyl alcohol chemistry or even of literature references is not practical. A few of the more significant types of reaction are illustrated; the literature should be consulted for detailed discussion of these and other reactions. References 19, 66, and 89 cover the field in detail and contain bibliographies.

Esterification and Polymerization. Allyl esters of saturated monobasic acids are usually prepared in the presence of acid catalysts or dehydrating agents or under conditions in which water is distilled off; combinations are frequently used. Acid anhydrides and acid chlorides react more rapidly with allyl alcohol and may be used if provision for dehydration is inconvenient.

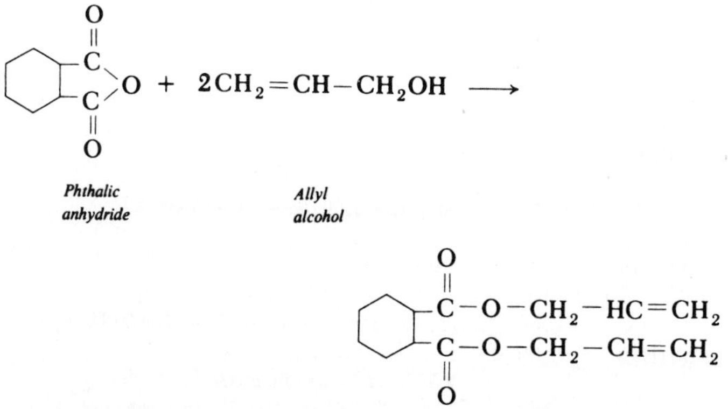

$$\text{Phthalic anhydride} + 2\,CH_2{=}CH{-}CH_2OH \longrightarrow$$

Phthalic
anhydride

Allyl
alcohol

$$\text{Diallyl phthalate} + H_2O \quad (23\text{-}14)$$

Diallyl phthalate

The esters derived from dibasic acids are particularly important in that their heat-induced polymerization occurs by branching and cross-linking to produce strong, hard polymers. Monobasic acid esters generally polymerize to light, straight-chain polymers.

Miscellaneous Additions to the Double Bond. In general, halogens and hypohalous acids add to the double bond:

$$CH_2{=}CH{-}CH_2OH + X_2 \rightarrow CH_2X{-}CHX{-}CHOH \quad (23\text{-}15)$$

$$CH_2{=}CH{-}CH_2OH + HOX \rightarrow CH_2OH{-}CHX{-}CH_2OH \text{ and} \quad (23\text{-}16)$$

$$CH_2X{-}CHOH{-}CH_2OH$$

Dry chlorine acts to some extent as an oxidizing agent, producing some acrolein. Distribution of products from these reactions can be controlled by varying the conditions. (Hydrogen halides generally react to form allyl halides).

Primary and secondary amines add in the presence of sodium to give 3-hydroxypropylamines. Thiocyanogen adds in organic solvents, giving 2,3-dithiocyanopropanol-1 (cyanogen reacts with the hydroxyl group).

Oxidation and Reduction. Reduction to *n*-propyl alcohol occurs over a number of catalysts.

Oxidation products vary, depending upon agents and conditions employed. Some typical reactions are summarized in Table 23-6.

TABLE 23-6. OXIDATIONS OF ALLYL ALCOHOL

Principal Product	Oxidizing Agent	Conditions
$CH_2{=}CH{-}CHO$ Acrolein	O_2	Cu, Ag, etc. catalysts
$\overset{\displaystyle O_3}{\underset{\displaystyle \text{Ozonide}}{CH_2{-}CH{-}CH_2OH}}$	O_3	cold CCl_4 or HAc solution
$CH_2OH{-}CHOH{-}CH_2OH$ Glycerol	H_2O_2	OsO_4 catalyst or ultraviolet light
HCOOH Formic acid	MnO_4	acid solution
$CH_2{=}CH{-}COOH$ Acrylic acid		Electrolytic; Pb anode, chromic and sulfuric acids

Formation of Ethers. Allyl ethers are used in preparation of a number of copolymers and resins. Generally, synthesis of ethers from allyl alcohol is possible, although formation of the more complex ethers from an allyl halide may be more convenient. The enhancement of activity of the hydroxyl group in its alpha position relative to the double bond is utilized in many preparations; e.g., *n*-propyl alcohol shows little reactivity under conditions that lead readily to formation of allyl methyl ether from allyl alcohol and dimethyl sulfate or diazomethane.

Miscellaneous. Without attempting to provide any description or to be comprehensive, it is noted that allyl alcohol undergoes reactions of the following classifications in addition to those touched upon already:

(1) Rearrangement, decomposition, polymerization
(2) Formation of allylates
(3) Reactions with aldehydes and ketones
(4) Reactions with inorganic acids, anhydrides, sulfur and phosphorus compounds
(5) Alkylations and condensations
(6) Formation of complex metal compounds and coordination complexes.

Commercial Uses of Allyl Alcohol

The fields of plastics, resins, plasticizers, varnish ingredients and other surface coating applications comprise the greater part of allyl alcohol end uses. Manufacture of certain agricultural chemicals, medicinals, perfumes, flavorings and other fine chemicals accounts for the remainder.[19,66,89]

As noted in the preceding section, diallyl esters of dibasic acids (or monoesters containing two or more unsaturated groups) are polymerized to thermosetting materials which generally combine good solvent resistance, transparency, toughness, hardness and dimensional stability. These properties suit them for manufacture of cast or molded objects, surface coatings, etc. Polymerization proceeds in stages upon application of heat.[44] Typical of resins in this group are those made from diallyl phthalate.

Naturally occurring allyl compounds are frequently responsible for characteristic tastes and flavors of foods, and synthesized allyl compounds are incorporated in a number of artificial flavorings. For example, artificial oil of garlic is now prepared with allyl sulfide. Other synthetic flavorings include allyl isothiocyanate, isovalerate and caproate in mustard oil, apple flavoring and pineapple flavoring respectively.[64,85]

The alcohol itself is reported to be a useful blood powder preservative,[61] and N-allylmorphine is used as an antidote for poisoning by morphine and some of its derivatives.

EPICHLOROHYDRIN

$$\begin{array}{ccc} & O & H \\ & /\backslash & | \\ H-C & -C-C-Cl \\ & | & | & | \\ & H & H & H \end{array}$$

1-Chloro-2,3-
Epoxypropane

Manufacture

Epichlorohydrin can be made from allyl chloride in a two-step process. A mixture of glycerol dichlorohydrins (2,3-dichloro-1-propanol and 1,3-dichloro-2-propanol) is first prepared by chlorohydrinating allyl chloride, i.e., reacting it with chlorine in aqueous phase.[23,26,35,36,38,39,74,93] These dichlorohydrins are converted to epichlorohydrin by heating under alkaline conditions.[6,14,93]

The chlorohydrination is thought to proceed through formation of a positively charged intermediate complex; trichloropropane may also be formed, as may chloroethers:

$$CH_2{=}CH{-}CH_2Cl + Cl_2 \rightarrow CH_2Cl{-}\overset{+}{C}H{-}CH_2Cl + Cl^- \qquad (23\text{-}17)$$

Allyl chloride   *Complex*

$$Complex + H_2O \rightarrow CH_2Cl-CHOH-CH_2Cl + H^+ \qquad (23\text{-}18)$$

1,3-Dichloro-2-propanol

$$Complex + Cl^- \rightarrow CH_2Cl-CHCl-CH_2Cl \qquad (23\text{-}19)$$

1,2,3-Trichloropropane

$$Complex + CH_2OH-CHCl-CH_2Cl \rightarrow$$

Dichlorohydrin

$$(CH_2Cl)_2CH-O-CH_2-CHCl-CH_2Cl + H^+ \qquad (23\text{-}20)$$

Chloroether

This same general result can be accomplished by reacting allyl chloride with sodium or potassium hypochlorite in acid medium:[20,23,74,93]

$$CH_2=CH-CH_2Cl + NaOCl + H^+ \rightarrow CH_2OH-CHCl-CH_2Cl + Na^+$$

$$(23\text{-}21)$$

High dilution and thorough dispersion of allyl chloride are employed since presence of an organic phase adversely affects yields.[40] Chlorine and allyl chloride dissolve preferentially in the organic phase and react there to form trichloropropane.

When crude, dilute dichlorohydrins are heated under alkaline conditions, epichlorohydrin is formed. The reaction can be carried out, e.g., in a stripping column; epichlorohydrin is distilled overhead as it is formed.

$$(23\text{-}22)$$

$$CH_2Cl-CHOH-CH_2Cl + NaOH \rightarrow CH_2\overset{\displaystyle O}{\overset{\diagup\diagdown}{-}}CH-CH_2Cl + NaCl + H_2O$$

1,3-Dichloro-2-Propanol   Epichlorohydrin

Chemical Properties

Like allyl chloride and allyl alcohol, epichlorohydrin is a highly reactive chemical intermediate. The epoxide ring takes precedence in most reactions, with the result that most reaction products are derivatives of glycerol α-monochlorohydrin. Such substances as nitrogen bases, metal alkoxides and organic acid salts react in this manner, rather than by their usual direct metathetical replacement of the active organic chlorine atom. The chlorine atom may be eliminated as hydrogen chloride in a subsequent step of the same reaction or in a later independent reaction to form glycidol derivatives; these, in turn, undergo the additive reactions typical of epoxy compounds to form further derivatives.

TABLE 23-7. PHYSICAL PROPERTIES OF EPICHLOROHYDRIN[a]

General: Colorless liquid, chloroform-like odor, hazardous chemical

Molecular weight	92.53
Density of liquid	1.1828^{20}_{20}
Solubility in water at 20°C	6.58%w
Viscosity at 25°C	1.03 cp
Freezing point	−57.2°C
Boiling point at 760 mm	116.1°C
dt/dp at boiling point	0.044°C/mm
Latent heat of vaporization (calc.) at b.p.	9060 cal./gram mole
Heat of combustion	4,524.4 cal./gram
Flash point, open cup	105°F

[a]References 21 and 68 present literature sources and further data.

A brief survey of illustrative reactions follows. More thorough coverage will be found in references.[21,37,54,55,68,94,95] Original papers, patents, etc., concerning reactions of epichlorohydrin are counted in the hundreds.

Formation of Ethers. A number of ether syntheses based upon epichlorohydrin are summarized in the accompanying diagram. The general pattern is applicable to reactions other than ether formations since all reactions of epichlorohydrin may generally be classified as additions at the epoxide ring or replacements, by one means or another, of the chlorine atom. The ether reactions are, however, of principal importance because they are involved in most of the commercial condensations and polymerizations of epichlorohydrin derivatives. The following paragraphs refer by numbers to designated reactions.

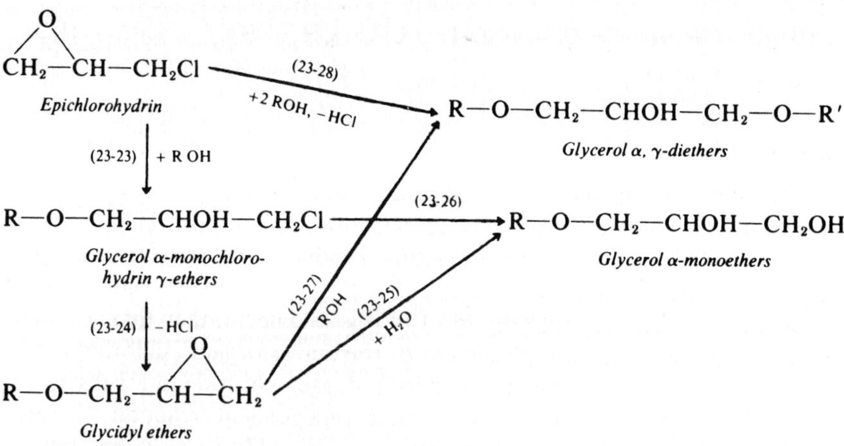

(23-23) Lower primary and secondary alcohols combine with epichlorohydrin at moderate temperature in the presence of several acid catalysts; stannic chloride is used with tertiary alcohols. Elevated temperature is required with aralkyl and higher aliphatic alcohols. Phenols also form the α-monochlorohydrin γ-ethers under acid catalysis; under basic conditions, the reaction may proceed further to glycidyl ethers and glycerol mono- and diethers. (With R defined as a hydrogen atom, the description applies to addition of water to form glycerol α-monochlorohydrin).

(23-24) Elimination of hydrogen chloride from the monochlorohydrin ethers is generally accomplished by treatment with strong caustic. As noted above, glycidyl aryl ethers may be formed directly from epichlorohydrin, essentially by combining Reactions (23-23) and (23-24). The glycidyl ethers are used principally in the preparation of mixed diethers (see Reaction 23-27).

(23-25) Hydrolysis of glycidyl ethers under acid catalysis leads to glycerol α-monoethers.

(23-26) However, glycerol α-monoethers are more usually prepared from a monochlorohydrin derivative. The appropriate α-monochlorohydrin γ-ether can be hydrolyzed in hot, dilute sodium bicarbonate solution; or glycerol α-monochlorohydrin, the appropriate alcohol, and caustic are reacted together.

(23-27) Acidic or basic catalysis may be used to promote formation of mixed diethers, wherein a second alcohol is added to the epoxide ring of a glycidyl ether. The hydroxyl group of the diether is then available for combination with, for example, a glycidyl ether, forming a diglycerol triether. As can be seen, formation of higher condensation products is possible by the same mechanism.

(23-28) Glycerol diethers of lower primary alcohols (R = R′) may be prepared by reacting epichlorohydrin, the alcohol and caustic under anhydrous conditions. Glycerol diaryl ethers of phenol and its near homologues are prepared in this manner, using powdered caustic and working in dioxane solvent.

This summary is not exclusive; for example, glycerol diaryl ethers may in some cases be prepared by treating α-monochlorohydrin γ-monoethers with alkali phenoxides. Obviously, Reactions (23-26) and (23-28), while not carried out in sequential steps, may still proceed through formation of intermediates. And, as noted above, the general pattern of these ether reactions is descriptive of a number of the reactions discussed below if free definition is given to the R groups.

Other Reactions. The reactions of epichlorohydrin with mercaptans are analogous to those with alcohols; γ-thioethers, glycidyl thioethers, and glycerol α, γ-dithioethers may be prepared. Reactions with H_2S in alkaline solution give various products (see Table 23-8), depending upon temperature and H_2S excess.

TABLE 23-8. REACTIONS OF EPICHLOROHYDRIN WITH H$_2$S IN ALKALINE SOLUTION.

Product	Temperature (°C)	Remarks
HSCH$_2$—CHOH—CH$_2$Cl 1-Chloro-3-mercapto-2-propanol	0	excess H$_2$S
┌────S────┐ CH$_2$—CHOH—CH$_2$ 3-Hydroxytrimethylene sulfide	50	excess H$_2$S
S(CH$_2$—CHOH—CH$_2$Cl)$_2$ Bis(γ-chloro-β-hydroxypropyl) sulfide	125	no excess H$_2$S

Amines behave in similar fashion, initially opening the epoxide ring and then eliminating HCl. For example, with a secondary amine:

$$ECH + R_2NH \rightarrow R_2NCH_2—CHOH—CH_2Cl \qquad (23\text{-}29)$$

$$R_2NCH_2—CHOH—CH_2Cl + R_2NH \rightarrow R_2NCH_2—\overset{\displaystyle O}{\overset{\displaystyle /\backslash}{CH—CH_2}} + R_2NH \cdot HCl \qquad (23\text{-}30)$$

There are also further combinations.

Monoesters (predominantly γ) and diesters of α-monochlorohydrin are formed by reaction with organic acids and acid anhydrides:

$$ECH + (RCO)_2O \rightarrow R\overset{\displaystyle O}{\overset{\|}{C}}—O—CH_2—CH—CH_2Cl \qquad (23\text{-}31)$$
$$\underset{\displaystyle RC=O}{\underset{|}{\underset{\displaystyle O}{|}}}$$

In reactions with aldehydes and ketones, epichlorohydrin forms cyclic acetals and ketals, 4-chloromethyl-1,3-dioxolanes substituted in the 2-position:

$$ECH + R—\overset{\displaystyle O}{\overset{\|}{C}}—R' \rightarrow \begin{matrix} CH_2Cl \\ | \\ CH—O \\ | \quad\quad C \\ CH_2—O \end{matrix} \begin{matrix} R \\ \diagup \\ \diagdown \\ R' \end{matrix} \qquad (23\text{-}32)$$

Grignard reactions follow an unusual course. While some yield of the expected 3-alkyl- or 3-aryl-1-chloro-2-propanol is experienced (the 2-substituted-3-propanol apparently is not formed), the major product from the acidified reaction mixture is a glycerol α-chloro-γ-halohydrin (1-chloro-3-halo-2-propanol), particularly with lower alkylmagnesium halides.

Commercial Usage of Epichlorohydrin

Epichlorohydrin is an intermediate in the manufacture of a variety of glycerol and glycidol derivatives and of various resin products. The largest use is for the preparation of epoxy resins.

The most important epoxy resins are condensation products of epichlorohydrin and bisphenol A—2,2-bis(p-hydroxyphenyl)propane.[16,32,45,46,54,63,73] They were first prepared during the late 1930's. A variety of mechanical and chemical properties meeting specific end-use requirements can be obtained by appropriate selection of preparation conditions and of curing agents.

The initial condensation proceeds by formation of ether linkages between epichlorohydrin and bisphenol, leading to a molecule of this general structure:

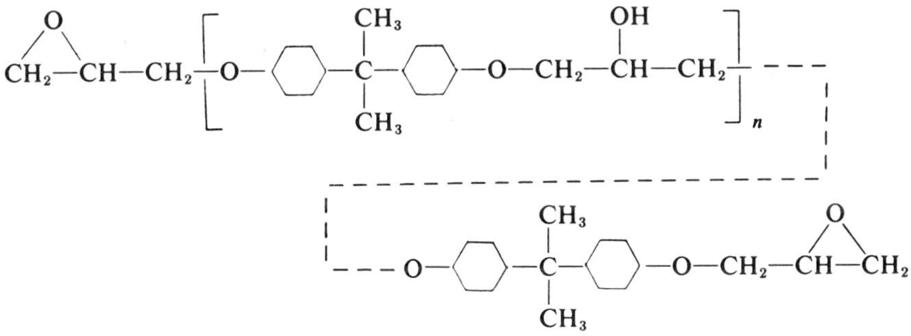

$$CH_2\text{—}CH\text{—}CH_2\text{—}\left[O\text{—}\bigcirc\text{—}\underset{CH_3}{\overset{CH_3}{C}}\text{—}\bigcirc\text{—}O\text{—}CH_2\text{—}\underset{}{\overset{OH}{CH}}\text{—}CH_2\right]_n\text{—}O\text{—}\bigcirc\text{—}\underset{CH_3}{\overset{CH_3}{C}}\text{—}\bigcirc\text{—}O\text{—}CH_2\text{—}CH\text{—}CH_2$$

The resin is generally produced in this form, n varying from grade to grade. Subsequently, these molecular groups are cross-linked to form the final thermosetting resin. Several cross-linking mechanisms are available: for example, homopolymerization of epoxy molecules can be induced; epoxy molecules can be linked together by interconnecting additives; side chains can be added which link either to other epoxy molecules or to other side chains. Important curing agents used to obtain various types of cross-linking are amines, anhydrides, fatty acids and Lewis acids.

General characteristics of epoxy resins are good resistance to chemical attack because of the high proportion of stable ether linkages and a delay of gelation until polymerization is far advanced, so that residual stresses in cured solid resins are minimum. The amine linked resins are almost unique in their ability to cure in the presence of alkali solutions.

Currently, the widest application is in surface coatings, where toughness, durability and chemical resistance are important; coatings include appliance enamels, paints for corrosive atmospheres, and wear-resistant floor coverings. Epoxy resins are used specifically to resist corrosion in chemical facilities, either as coatings to protect metal equipment or as laminating agents with other materials to make nonmetallic equipment and piping. They are the base of a series of industrial and structural adhesives; they are used to insulate or encapsulate electrical and electronic components, and they make excellent, economic foundry cores, forming die tools, etc. because of their ease of casting, strength, toughness and dimensional stability. Some resins have useful ion exchange properties.

A number of surface-active agents are made from epichlorohydrin, either by condensation with a polyamine and a fatty acid or by reaction with sodium acid sulfite and reacting the product with a sodium carboxylate. These products are used as detergents, demulsifiers, etc., and they form the second most important group of epichlorohydrin derivatives.

A number of pharmaceuticals are prepared from epichlorohydrin; other derivatives are used in the manufacture of textile conditioners and dyes, paper sizing agents, etc. Epichlorohydrin is used as a solvent (e.g., for cellulose acetate), as a stabilizer for polyvinyl chloride, chlorinated rubbers and several chlorine-containing insecticides, and as an additive to lubricating oils and greases. A complete catalogue of end-uses would be almost impossible to assemble; some idea of the multiplicity of epichlorohydrin applications can be gained from References 21, 68, and 94.

GLYCEROL

$$\begin{array}{ccc} OH & OH & OH \\ | & | & | \\ H-C- & C- & C-H \\ | & | & | \\ H & H & H \end{array}$$

1,2,3-Propanetriol

Manufacture

Two commercially feasible routes from allyl chloride to glycerol are available, both of which involve intermediate production of glycerol chlorohydrins. Historically, it was recognition of the possibility of producing these chlorohydrins from allyl chloride that promoted development of the allyl chloride synthesis, for it was known that the chlorohydrins could be hydrolyzed to glycerol.

The chlorohydrins are prepared from allyl alcohol in one process and directly from allyl chloride in the other. Another route—hydrolysis of

1,2,3-trichloropropane, which is prepared by low-temperature chlorination of allyl chloride—has been proposed and studied, but the hydrolysis step is difficult and not adaptable to commercial practice.[92,93]

The production of allyl alcohol from allyl chloride and the direct chlorohydrination of allyl chloride have both been discussed in preceding sections of this chapter. Chlorohydrination of allyl alcohol is similar to that of allyl chloride, although somewhat simpler because the alcohol is readily miscible in water. Treatment of the alcohol with chlorine water or treatment of the aqueous alcohol with chlorine produces a mixture of mono- and dichlorohydrins. The chlorohydrination of aqueous alcohol with chlorine is catalyzed by the chloride and hydrogen ions which are formed, together with hypochlorous acid, on reaction of chlorine and water, and thus it is faster than chlorohydrination with simple hypochlorous acid. The reaction is thought to involve formation of a positively charged intermediate complex similar to that postulated for the chlorohydrination of allyl chloride:

$$CH_2{=}CH{-}CHOH + Cl_2 \rightarrow CH_2Cl{-}\overset{+}{C}H{-}CH_2OH + Cl^- \qquad (23\text{-}33)$$

Allyl alcohol　　　　　　　　*Complex*

The complex combines with chloride ions to form a dichlorohydrin and with water to form monochlorohydrins. Chloroether by-products are formed by reaction of the complex with allyl alcohol so that increased dilution favors high yields of the desired chlorohydrins.[7,58,81,93]

Hydrolysis of the dilute solution of chlorohydrins obtained from either process may proceed along several paths, depending upon temperature, residence time, mixing and pH. One or more intermediates may be involved. Possible dichlorohydrin reactions may be summarized thus:

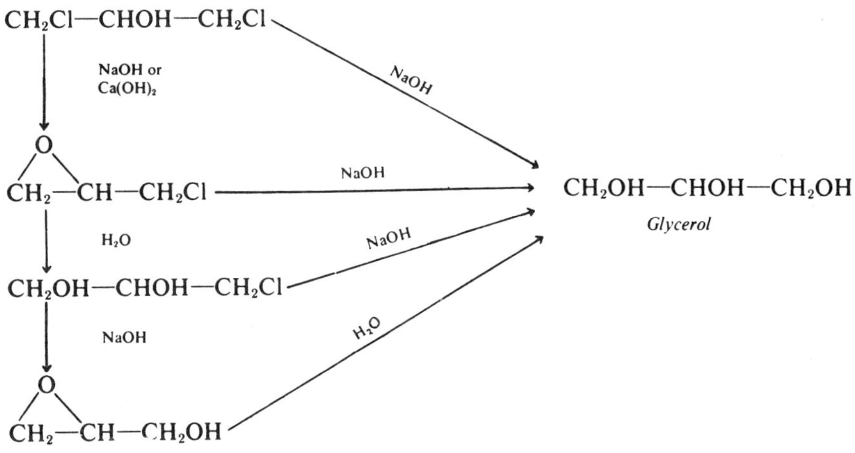

The product of the hydrolysis step is a dilute glycerol stream containing sodium chloride. Concentration and finishing operations are largely conventional; successive vacuum evaporation, centrifuging, and vacuum-steam distillation steps may be followed by hydrocarbon extraction and further vacuum-steam distillation.[49,93] The principal impurity in the finished product is water, usually amounting to considerably less than one percent. Nature and concentration of other trace impurities depends upon processing variations employed.

Physical and Chemical Properties

The physical and chemical properties are amply treated in the literature. Because glycerol is so well known, no attempt is made to present even a digest. Reference 49 treats the subject of glycerol exhaustively.

Commercial Uses of Glycerol

Glycerol is used principally as a raw material in the manufacture of alkyd resins and ester gums and in the production of nitroglycerine and allied explosives, as a humectant in tobacco processing, and as a plasticizer in cellophane manufacture; these uses account for approximately two-thirds of United States consumption. Manufacture of cosmetics, drugs, etc. also consumes appreciable quantities. (See Reference 49 for a comprehensive detailing of glycerol utilization).

Production of glycerol is the chief *raison d'etre* of the allyl chloride process. When the high temperature chlorination reaction was discovered, glycerol was the only allyl chloride derivative for which a large-volume market existed or could be predicted with any assurance, and glycerol production still absorbs more than 70 percent of domestic allyl chloride production.

The effect of synthetic glycerol on the market has been instructive. In 1938, some ten years before commercial-scale glycerol synthesis was achieved, E. C. Williams discussed the implications of the allyl chloride development, particularly in regard to glycerol.[92] He cited glycerol price fluctuations through the range from 13 to 60 cents a pound (see also Reference 49), and he stated that the chemical industry had achieved the potential of supplying the entire world glycerol market with a stably priced commodity of uniform quality.

In 1959, some ten years after synthetic glycerol had become available, almost half of the glycerol produced in the United States was synthesized; total glycerol production was almost exactly twice that of 1937;[84] and in recent years, as synthetic glycerol's fraction of the market has become increasingly significant, prices have remained stable.[50] There have been other changes in the structure of the glycerol market during the intervening 20 years, so that changes in nature of the market cannot be attributed entirely to the availability of synthetic glycerol. For example, synthetic detergents (some of them based on allyl chloride derivatives) have affected production patterns in the soap industry, which formerly supplied the bulk of the glycerol; and a number of polyols have become commercially available to compete with glycerol. Nonetheless, synthetic glycerol production capacity sufficient to have supplied the entire pre-war domestic demand now exists. The synthesized portion of total glycerol production is made as a primary product whose volume is not influenced by market conditions of co-products. Williams' predictions of 1938 were remarkably accurate.

SAFETY AND HANDLING

General

This section summarizes information that is believed to be correct, but should not be considered to be a complete treatment. It is essential that persons contemplating handling allyl chloride and its derivatives first consult available detailed publications dealing with toxicity, handling precautions and, particularly, treatments for exposures as, in general, these are potentially hazardous chemicals.

Toxicity

Hazard. Allyl chloride is a toxic chemical potentially capable of causing severe tissue damage and death. Local skin contact causes irritation and burns; eye damage results from liquid or vapor exposure; absorption through the skin is rapid, with subsequent distribution throughout the system causing damage to internal tissues; inhalation of vapors results in irritation and lung damage, and ingestion is quite dangerous. Allyl alcohol and epichlorohydrin display the same characteristics, as do a number of other allyl chloride derivatives; cumulative effects from chronic exposure of experimental animals to epichlorohydrin at levels above 5 ppm have been reported. Several references discussing toxicity are: allyl chloride, 1, 72, 73; allyl alcohol, 4, 15, 48; epichlorohydrin, 12, 25, 76. Glycerol is considered non-hazardous (see Reference 49). In addition, manufacturers of these chemicals publish information on toxicity and safety measures (e.g. Reference 69).

Prevention. The fundamental precautionary measure is prevention of contact. Suitable protective clothing, impervious gloves, goggles, etc. should be used, and this gear should always be washed before reuse after any contact; because rubber is slowly penetrated by these materials, thorough and immediate cleansing of rubber gear is necessary to remove residual traces.

If there is any possibility that vapors can enter working areas, good exhaust ventillation is imperative. It is considered that a chronic hazardous vapor concentration has been reached when odor can be detected, i.e., the detection level is the warning level. Gas masks equipped with "Organic Vapor" cannisters or, preferably, fresh air masks must be worn whenever entry into a contaminated area is necessary.

Remedial. Because effects of severe exposure may not be evident at once, every contact should be treated immediately and thoroughly. The first treatment in case of skin contact is immediate and extremely thorough flooding and washing with soap and water; a physician should be called if there is the slightest question as to severity. Speed in washing is especially critical for eye exposure, and treatment by a physician is almost always necessary.

Persons exposed to significant vapor concentrations should be referred to a physician and alerted to seek medical attention at any sign of respiratory effects. In the case of allyl alcohol, severe eye irritation may also be expected. Both responses may be delayed as much as twenty-four hours. Generally, severity of damage is inversely proportional to time elapsed before the onset of symptoms.

Entry of liquid into the mouth should be considered an emergency. First aid consists of thorough oral flushing, administration of an emollient, if available, and of a mild emetic. In every case immediate medical attention is essential, and a stretcher should be used to prevent exertion if it is necessary to move the patient.

General Handling

Moist allyl chloride is corrosive. Nickel, "Monel," and other chloride resistant materials are usually recommended for its handling. If not excessively hot, dry allyl chloride can be handled in mild steel, cast iron, stoneware and red brass; aluminum is not considered satisfactory.[67] The alcohol is considered non-corrosive. Uncontaminated, dry epichlorohydrin can generally be handled in steel equipment.

A precaution must be noted in the handling of epichlorohydrin and other derivatives that contain an epoxide ring. Rapid polymerization may occur if traces of acidic, alkaline, or other catalytic substances are present, and extreme evolution of heat and explosions may result.

Allyl chloride and its derivatives present typical hydrocarbon fire hazards; the low flash point of allyl chloride in particular (-16°F) warrants extreme care.

References

1. Adams, E. M., Spencer, H. C., and Irish, D. D., *J. Ind. Hyg. Toxicol.*, **22**, 79 (1940).

2. A.M.A. Council on Pharmacy and Chemistry, "New and Unofficial Remedies," Philadelphia, Lippincot, 1956.

3. Andrews, L. J., and Kepner, R. E., *J. Am. Chem. Soc.,* **70,** 3456 (1948).

4. Atkinson, H. V., *J. Pharmacol. Exptl. Therap., Proc.,* **25,** 144 (1925).

5. Berthelot, M., and de Luca, S., *Ann.,* **92,** 306 (1854); *Ann. chim. phys.,* (3) **43,** 257 (1855).

6. Braun, Geza, *Org. Syntheses,* **16,** 30 (1936); Coll. Vol. II, 256 (1943).

7. Brooks, B. T. (to Standard Alcohol Co.) U. S. Patent 2,311,023 (February 16, 1943).

8. Burgin, J., Engs. W., Groll, H. P. A., and Hearne, G., *Ind. Eng. Chem.,* **31,** (11), 1413 (1939).

9. Cahours, A., *Compt. rend.,* **31,** 291 (1850).

10. Cahours, A., and Hoffman, A. W., *Ann.,* **100,** 356 (1856); *Compt. rend.,* **42,** 217 (1856); *Ann. chim. phys.,* (3) **48,** 286 (1856); *Ann.,* **102,** 285 (1857); *Proc. Roy. Soc.,* **8,** 32, 353 (1868).

11. Caldwell, J. R. (to Eastman Kodak Co.), U. S. Patent 2,656,337 (October 20, 1953).

12. Carpenter, C. P., Smyth, H. F. Jr., and Pozzani, U. C., *J. Ind. Hyg. Toxicol.,* **31,** 343 (1949).

13. Carter, Walter, *Science,* **97** (2521) 383 (April 23, 1943); *J. Econ. Entom.,* **37** (1), 117 (1944).

14. Clarke, H. T., and Hartman, W. W., *Org. Syntheses,* Coll. Vol. I, 233–4 (1941).

15. Clerc, A., Stern, J., and Paris, R., *Compt. rend. soc. biol.,* **116,** 864 (1934).

16. Coderre, R. A., "Epoxy Resins," Ref. 53, Supp. Vol. I, p. 312 (1957).

17. De Benedictis, A. (to Shell Oil Co.), U. S. Patent 2,939,879 (June 7, 1960).

18. De la Mare, H. E., and Vaughan, W. E., *J. Chem. Educat.,* **34,** (1), 10 and **34** (2), 64 (1957).

19. Dow Chemical Co., "Allyl Alcohol," Bulletin Code 164-75, Midland, Michigan, April 1958.

20. Dow Chemical Co., "Allyl Chloride," Bulletin Code 164–78, Midland, Michigan, July 1958.

21. Dow Chemical Co., "Epichlorohydrin," Bulletin, Midland, Michigan, 1958.

22. Essex, H., and Ward, A. L. (to E. I. du Pont de Nemours & Co.), U. S. Patent 1,477,047 (December 11, 1923).

23. Essex, H., and Ward, A. L. (to E. I. du Pont de Nemours & Co.) U. S. Patent 1,477,113 (December 11, 1923).

24. Fairbairn, A. W., Cheney, H. A., and Cherniavsky, A. J., *Chem. Eng. Progress,* **43** (6) 280 (1947).

25. Freuder, E., and Leake, C. D., *U. of Calif. Publ. Pharmacol.,* **2** (5) 69 (1941).

26. Geyerfelt, H. von, *Ann.,* **154,** 247 (1870); *Ber.,* **6,** 720 (1873).

27. Gilbert, H., Miller, F. F., and Falt, L., (to B. F. Goodrich Co.), U. S. Patent 2,650,911 (September 1, 1953).

28. Groll, H. P. A. (to Shell Development Co.), U. S. Patent 2,207,193 (July 9, 1940)

29. Groll, H.P.A., and Hearne, G., *Ind. Eng. Chem.,* **31,** (12) 1530 (1939).

30. Groll, H. P. A., Hearne, G., Burgin, J., and LaFrance, D. S. (to Shell Development Co.), U. S. Patent 2,130,084 (September 13, 1938).

31. Groll, H. P. A., Hearne, G., Rust, F. F., and Vaughan, W. E., *Ind. Eng. Chem.,* **31** (10) 1239 (1939).

32. Harvard Univ. Grad. School of Business Admin., "Epoxy Resins: Market Survey and User's Reference," (July 1959).

33. Hearne, G., Evans, T. W., Yale, H. L., and Hoff, M. C., *J. Am. Chem. Soc.,* **75** (6) 1392–4 (1953).

34. Henderson, J. B., and McKay, N. H., Paper presented at the ACS Southwest Regional Meeting, Little Rock, Ark., December 5, 1952.

35. Henry, L., *Ber.,* **3,** 347 (1870); **7,** 409, 757 (1874); *J. prakt. chem.,* (2) **10** 185 (1874).

36. Henry, L., and Massalske, *Bul. Acad. roy. Belg.,* **1906,** 523.

37. Huntress, E. H., "Organic Chlorine Compounds," Allyl Chloride, pp. 948–53; Epichlorohydrin, pp. 665–78; New York, John Wiley, 1948.

38. Ioffe, I. I., USSR Patent 64,931 (1945).

39. Ioffe, I. I., and Yampol'skaya, E. S., *J. App. Chem.* (*USSR*), **18,** 50 (1945).

40. Johnson, G. F. (to Shell Development Co.), U. S. Patent 2,714,123 (1955).

41. Kirrmann, A., Pouratt, H., Schmitz, R., and Saito, E., *Bull. soc. chim. France,* **1952,** 502, 515.

42. Klebanskii, A. L., and Vol'kenshtein, A. S., *J. Appl. Chem.* (*USSR*), **8,** 106 (1935).

43. Köhler, F. (to Röhm and Haas Co.), U. S. Patent 2,323,781 (July 6, 1943).

44. Kronstein, A., *Ber.,* **46,** 1812 (1913).

45. Lee, H. L., and Neville, K. O., "Epoxy Resins, A Practical Review of their Chemistry and Technology," New York, McGraw-Hill, 1957.

46. Leininger, R. I., "Epoxy Resins," article in "Encyclopedia of Chemistry," (p. 363), ed. by Clark, G. L., and Hawley, G. G., New York, Reinhold Publishing Corp., 1957.

47. McBee, E. T., and Pierce, O. R., "Halogenation," Ref. 53, Vol. 7, pp. 344–51, esp. 347-Olefins.

48. McCord, C. P., *J. Am. Med. Assn.,* **98,** 2269.

49. Miner, C. S., and Dalton, N. N., ed., "Glycerol," ACS Monograph No. 117, New York, Reinhold Publishing Corp., 1953.

50. *Oil, Paint and Drug Reptr,* Hi-Lo Index (February 1960).

51. Oldershaw, C. F., Simenson, L., Brown, T., and Radcliffe, F., *Chem. Eng. Progress,* **43** (7) 371 (1947).

52. Oppenheim, A., *Ann.,* **140,** 204 (1866) and *Ann. Supp.,* **6,** 353 (1868).

53. Othmer, D. F., and Kirk, R. E., ed., "Encyclopedia of Chemical Technology," New York, Interscience Publishers, 1947–1956, with supps. 1957, 1960; see under authors for individual articles

54. Paquin, A. M., "Epoxydverbindungen und Epoxydharze," Berlin, Springer-Verlag, 1958.

55. Parker, R. E., and Isaacs, N. S., *Chem. Reviews,* **59** (4), 737 (1959).

56. Pines, H., and Ipatieff, V. N. (to Universal Oil Products Co.), U. S. Patent 2,721,887 (October 25, 1955).

57. Pollack, M. A., and Chenicek, A. G. (to Pittsburgh Plate Glass Co.), U. S. Patent 2,313,767 (March 16, 1943).
58. Read, J., and Hurst, E., *J. Chem. Soc., 1922,* 989.
59. Reynolds, J. W., *J. Chem. Soc., 1851,* 11.
60. Rust, F. F., and Vaughan, W. E., *J. Org. Chem.,* **5** (5) 472 (1940).
61. Salkowski, E., *Biochem. Z.,* **108,** 244 (1920).
62. Schmidt, C. T. (to Pineapple Research Institute of Hawaii), U. S. Patent 2,937,936 (May 24, 1960).
63. Schrade, J., "Les Resines Epoxy," Paris, Dunod, 1957.
64. Seldner, A., *Am. Perf. & Essential Oil Rev.,* **54,** 295 (1959).
65. Semmler, F. W., *Arch. Pharm.,* **230,** 434 (1892).
66. Shell Chemical Corp., "Allyl Alcohol," Technical Pub. SC:46-32, San Francisco (1946).
67. Shell Chemical Corp., "Allyl Chloride," Technical Pub. SC:49-8, San Francisco (1949).
68. Shell Chemical Corp., "Epichlorohydrin," Technical Booklet SC:49-35 (1949).
69. Shell Chemical Co., Safety and Toxicity Data Sheets
    Allyl Alcohol   SDS SC:57-77, TDS SC:57-78
    Allyl Chloride   SDS SC:57-79, TDS SC:57-80
    Epichlorohydrin SDS SC:57-85, TDS SC:57-86.
70. Sheshukov, J., *Russ. Phys. Chem. Soc.,* **16,** 478 (1884).
71. Silverman, M., and Abreu, B. E., *U. of Calif. Publ. Pharmacol.,* **1,** 119 (1938).
72. Silverman, M., and Leake, C. D., *J. Pharmacol.,* **60,** 118 (1937).
73. Skeist, I., "Epoxy Resins," New York, Reinhold Publishing Corp., 1958.
74. Smith, L., *Z. Physik. chem.,* **92,** 717 (1918).
75. Smith, L., *Acta. Chem. Scand.,* **5,** 1415 (1941).
76. Smyth, H. F., and Carpenter, C. P., *J. Ind. Hyg. Toxicol.,* **30,** 63 (1948).
77. Steen, D. E. (to Monsanto Chemical Co.), U. S. Patent 2,966,525 (December 27, 1960).
78. Stewart, T. D., and Weidenbaum, B., *J. Am. Chem. Soc.,* **58,** 98 (1936).
79. Tamele, M. W., and Groll, H. P. A., (to Shell Development Co.), U. S. Patent 2,072,016 (February 23, 1937).
80. Tawney, P. O. (to U. S. Rubber Co.), U. S. Patents 2,569,959 (August 2, 1951); 2,569,960 (August 2, 1951); 2,592,211 (April 8, 1952); 2,597,202 (May 20, 1952).
81. Taylor, K. A., Maass, O., and Hibbert, H., *Can. J. Research,* **4,** 110 (1931).
82. Thompson-Houston Co., Ltd., British Patent 618,608 (February 24, 1949).
83. Tollens, B., Henninger, A., Weber, R., and Kempf, T., *Ann.,* **156,** 129, 151 (1870).
84. U. S. Department of Commerce, Current Ind. Reports, Fats and Oils; Series M20K(60)-8.
85. "The Use of Chemical Additives in Food Processing," Pub. 398, Nat. Acad. Sci., Nat. Res. Council, 63-4, 1956.
86. Van de Griendt, G. H., Marple, K. E., and Peters, L. M. (to Shell Development Co.), U. S. Patent 2,318,033 (May 4, 1943).

87. Van de Griendt, G. H., and Peters, L. M. (to Shell Development Co.), U. S. Patent 2,475,364 (July 5, 1949).

88. Van Dijk, C. P., and van der Plas, F. J. F. (to Shell Development Co.), U. S. Patent 2,763,699 (September 18, 1956); (to N. V. Bataafsche Petroleum Maatschappij) British Patent 762,153 (1956) and Netherlands Patent 84,704 (1957).

89. Vesper, H. G., "Allyl Alcohol," Ref. 53, Vol. 1, pp. 584–9.

90. Vesper, H. G., "Allyl Chloride," Ref. 53, Vol. 3, pp. 800–7.

91. Wertheim, T., *Ann.,* **51,** 289 (1844).

92. Williams, E. C., "Modern Petroleum Research," paper API Meeting, November 16, 1938; *Ind. Eng. Chem.,* News Ed., **16,** 630 (December 10, 1938).

93. Williams, E. C., *Trans. AIChE,* **37** (1) 157 (1941); *Chem. & Met.,* **47,** 834 (1940).

94. Williams, P. H., "Epichlorohydrin," Ref. 53, Vol. 3, pp. 865–9.

95. Winstein, S., and Henderson, R. B., Chapter on Ethylene and Trimethylene Oxides; "Heterocyclic Compounds," Vol. I, ed. by Elderfield, R. C., New York, John Wiley, 1950.

96. Yabroff, D. L., and Anderson, J., *Third World Petroleum Congress Proceed.* Sect. V, 22–30 (1951).

97. Zielinski, A. Zb., and Suknarowska, S., *Pozemysl Chem.,* **13,** 279 (1957).

24. VINYLIDENE CHLORIDE

E. D. Serdynsky

Dow Chemical Company

Historical

Commercial production of vinylidene chloride monomer and of the homo- and copolymers derived from the monomer occurred in 1939 as a result of extensive research and development work on chlorinated aliphatic compounds in the laboratories of The Dow Chemical Company. The existence of vinylidene chloride had been known for almost a hundred years prior to this time, since the French chemist Regnault[36,37] reported in 1838 the formation of a white noncrystalline precipitate from a liquid boiling between 35 and 40°C. The liquid was apparently an impure 1,1-dichloroethylene, prepared from trichloroethane by reaction with alcoholic potassium hydroxide—a method used by Kraemer[32] and Bauman.[2] In 1872 the latter described the formation of a white substance when dichloroethylene was exposed to sunlight. Ostromislenski[35] found that vinylidene chloride polymerized in light to a white amorphous mass insoluble in many solvents.

Later, Feisst,[23] together with Staudinger[45] and Brunner,[46] investigated polydichloroethylene in greater detail. A portion of the polymer studied was prepared from an apparently impure unsymmetrical dichloroethylene fraction from commercial trichloroethylene. Staudinger and Feisst[45] indicated that the liquid polymerizes quickly in light or slowly when kept in the dark. Other studies were made on polydichloroethylene formed as a by-product in the Rheinfelden works of I. G. Farbenindustrie. The polymer could be separated into molecular weight fractions as it was completely soluble in warm tetra-chloroethane, Decalin, and Tetralin; partially soluble in benzene, chloroform and carbon disulfide; it was insoluble in ethers and alcohols. The major part of the studies involved reactions of the polymer with aniline, resorcinol, quinoline, pyridine, trimethyl amine and hydrazine; there were also studies of the products formed. The polymeric material was completely saturated, and its structure was represented as thread-like molecules. As evidence, it was shown that high-molecular weight hydrocarbons were formed when the polymer was reduced with hydrogen iodide and phosphorus. Feisst reported that the polymer was crystalline, which was later confirmed by G. Natta and R. Rigamonti.[34]

These studies followed shortly after B. T. Brooks[11] indicated in 1922 that halogenated ethylenes other than vinyl chloride and vinyl bromide show a tendency toward polymerization. The relationship of vinylidene chloride to other polymerizable compounds was becoming established during this same period of time. The effect of unsymmetrical substitution in ethylene compounds is well illustrated by comparing vinylidene chloride with its two isomers, *cis-* and *trans-*1,2-dichloroethylene, which polymerize only with great difficulty. Ellis[18] and Brooks[11] mention the unsymmetrically substituted ethylenes prepared by early investigators. Sawitsch[43] has observed that 1,1-dibromoethylene polymerized to a white solid. Denzel[17] and Biltz[4] had made 1-chloro-1-bromoethylene, the polymer of which was also found to be insoluble in many solvents. In general, 1,1-disubstituted ethylenes appear to polymerize more easily and form less soluble polymers than do the vinyl halide compounds.

Based upon the extensive study of vinylidene chloride, its homopolymer and its copolymers, a family of thermoplastics based on vinylidene chloride was introduced to the plastics industry in 1939. This family of thermoplastics is now known generally as "saran," although in many foreign countries "Saran" is a registered trademark of The Dow Chemical Company. The term "saran" refers to polymers that contain 50 or more percent of vinylidene chloride. Vinylidene chloride has commercial use primarily when it is copolymerized with other monomers such as acrylonitrile, vinyl chloride, styrene, vinyl acetate, etc. Saran copolymers are noted for their outstanding resistance to chemicals and solvents, for their extremely low water-absorption and water-vapor transmission and for their excellent dimensional stability. They tend to be odorless, tasteless and non-toxic. Their toughness and abrasion resistance are of a high order. The diversity of copolymers, with their many different properties, has led to the use of these copolymers in a large number of applications.

Vinylidene Chloride

Vinylidene chloride monomer (molecular weight 96.95) is a colorless liquid which has a sweet, pleasant odor although it develops a sharp disagreeable odor after contact with air. It polymerizes slowly in the absence of oxygen. Table 24-1 lists its important properties.[48]
As supplied, because of the manufacturing process, vinylidene chloride monomer contains a small amount of impurities and additives. A typical analysis is listed in Table 24-2.

TABLE 24-1. PHYSICAL PROPERTIES AND CONSTANTS OF VINYLIDENE CHLORIDE

1. Molecular weight — 96.95
2. Odor — Pleasant, sweet
3. Appearance — Clear liquid
4. Color — 10–15 APHA
5. Liquid density at $T/4°C$

Temperature °C	Density (gm/cc)
−20	1.2902
0	1.2517
20	1.2132

6. Boiling point (760 mm Hg) — 31.56°C
7. Freezing point — −122.5°C
8. Vapor pressure

Between 50 and 785 mm Hg;

$$\log P_{mm} = 6.98200 - 1104.29/(t + 237.697) \quad t \text{ in } °C$$

Temperatures calculated at selected pressures:

Pressure (mm Hg)	Boiling Point (°C)
760	31.56
400	14.43
200	−1.78
100	−16.04
60	−25.49
40	−32.44
20	−43.31
10	−53.10
5	−61.94
1	−79.54

9. Index of refraction

10°C	1.43062
15°C	1.42777
20°C	1.42468

10. Absolute viscosity

Temperature (°C)	Viscosity (cps)
−20	0.4478
0	0.3939
20	0.3302

11. Flash point (Cleveland open cup) — 5°F
    (Tagliabue closed cup) — 55°F
12. Explosive limits in Air (28°C) — 7.3–16.0%
13. Auto-ignition temperature — 1058°F

TABLE 24-2. TYPICAL ANALYSIS OF VINYLIDENE CHLORIDE MONOMER

Per cent monomer (excluding phenol)	> 99.6
Chloroacetylene (C_2H_2 equiv.)	10 ppm
Trans 1,2-dichloroethylene	< 0.1%
1,1-Dichloroethane	0.2%
Chloroform	< 0.1%
Phenol inhibitor (% by weight)	0.5–0.8%
Water	< 50 ppm

Preparation

Vinylidene chloride monomer, which contains 73.14 percent of chlorine, may be conveniently prepared in the laboratory by the reaction of 1,1,2-trichloroethane with aqueous alkali:[19]

$$2\,CH_2ClCHCl_2 + Ca(OH)_2 \xrightarrow{90°\,C} 2\,CH_2{=}CCl_2 + CaCl_2 + 2\,H_2O$$

Many early investigators used this general reaction, which later was the subject of more recent patents.[21,30] Other methods for the preparation of 1,1-dichloroethylene are based on bromochloroethane,[27] trichloroethylacetate,[31] tetrachloroethane,[16] and catalytic cracking of trichloroethane.[28] 1,1,2-Trichloroethane may be made from the basic raw materials petroleum and brine by well known reactions involving ethylene and chlorine. Ethylene, made by cracking petroleum, and chlorine, from the electrolysis of brine, combine to form trichloroethane which is converted to vinylidene chloride.

Toxicity

Vinylidene chloride should present no problem from ingestion incidental to its handling and industrial use. If monomer is accidentally swallowed, vomiting should be induced and medical attention obtained immediately. Inhibited vinylidene chloride is moderately irritating to the eyes, causing pain; but permanent damage is not likely. If the eyes become contaminated, they should be flushed immediately with copious amounts of flowing water for about 15 minutes. Medical attention should be obtained immediately. Concentrated phenol inhibitor may cause serious and permanent injury to the eyes.

Liquid vinylidene chloride monomer is irritating to the skin after direct contact of only a few minutes. The inhibitor content of the monomer may be partly responsible for this irritation. Precautions should be taken to prevent skin contact with the monomer; one should wear protective clothing when indicated and wash with plenty of soap and water.

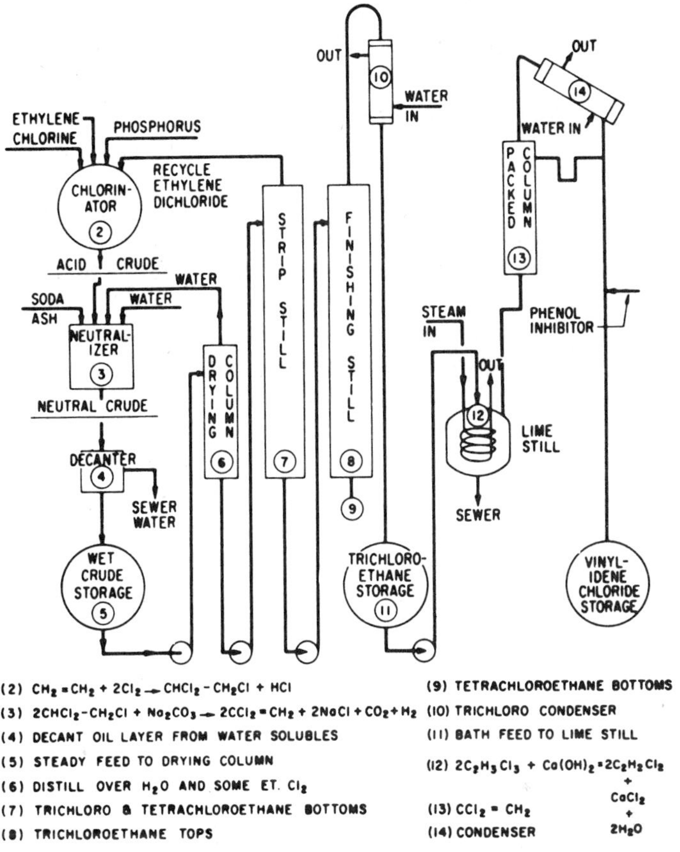

(2) $CH_2 = CH_2 + 2Cl_2 \rightarrow CHCl_2 - CH_2Cl + HCl$

(3) $2CHCl_2 - CH_2Cl + Na_2CO_3 \rightarrow 2CCl_2 = CH_2 + 2NaCl + CO_2 + H_2$

(4) DECANT OIL LAYER FROM WATER SOLUBLES

(5) STEADY FEED TO DRYING COLUMN

(6) DISTILL OVER H_2O AND SOME ET. Cl_2

(7) TRICHLORO & TETRACHLOROETHANE BOTTOMS

(8) TRICHLOROETHANE TOPS

(9) TETRACHLOROETHANE BOTTOMS

(10) TRICHLORO CONDENSER

(11) BATH FEED TO LIME STILL

(12) $2C_2H_3Cl_3 + Ca(OH)_2 = 2C_2H_2Cl_2 + CaCl_2 + 2H_2O$

(13) $CCl_2 = CH_2$

(14) CONDENSER

Figure 24-1. Vinylidene chlorine flow sheet.

The vapor toxicity of vinylidene chloride is slightly greater than that of ethylene dichloride and slightly less than that of carbon tetrachloride. The volatility of vinylidene chloride is such that vapor concentration exceeding 2 percent (20,000 ppm) may occur in instances of large spills on unconfined or unventilated spaces. Its warning properties are inadequate to prevent excessive exposure. The hazard due to inhalation is not unusual, however, and is readily controlled by the observance of precautions commonly taken in the chemical industry. A single exposure for a few minutes to a high concentration of vinylidene chloride monomer vapor (such as 4000 ppm) rapidly produces a "drunkenness" which may progress to unconsciousness if the exposure is continued. Prompt, complete recovery from the anaesthetic effects will result when the exposure is of short duration. A single prolonged exposure or repeated short-term exposures may produce organic injury to the liver and kidneys. The maximum single exposure permitting a reasonably high probability of no injury is about 1000 ppm for up to one hour and 200 ppm for up to eight hours. For repeated exposures, the vapor concentration should be maintained below 25 ppm. Most persons find that 1000 ppm in air has a mild but definite odor; many persons can detect 500 ppm. Handling of the monomer should be conducted in a closed system or with adequate ventilation. Because vinylidene chloride vapors are heavy and may concentrate in low areas, special attention should be paid to low-level ventilation.

If a person should be affected or overcome from breathing vinylidene chloride monomer vapors, he should be removed to fresh air at once and be made to rest and kept warm; medical attention should be obtained immediately. Artificial respiration should be administered immediately if breathing stops.

Fire and Explosion

Vinylidene chloride vapor is flammable at concentrations between 7 and 16 percent by volume in air. Vapors of the liquid monomer, once ignited, burn strongly but not violently. The products of combustion are to be avoided and possible sources of ignition in the area of use should be eliminated or controlled. Vinylidene chloride fires are relatively more difficult to start than are fires with common hydrocarbons such as gasoline or benzene, but vinylidene chloride fires act similarly to these others after a short burning period.

Dry powder, foam, or carbon dioxide extinguishers of the sort recommended for liquid fires may be used to extinguish vinylidene chloride fires. Water spray may be used effectively where cooling is required. Fire fighters should be cautioned about exposure to vinylidene chloride vapors and to other poisonous fumes, smoke or gases which may arise from a vinylidene chloride fire.

Peroxide Compounds

It was believed that vinylidene chloride with the inhibitor removed, in the presence of air or oxygen, forms a complex peroxide compound at temperatures as low as $-40°$ C. The peroxide compound is violently explosive. Reaction products formed with ozone are particularly dangerous. Work currently underway may show this theory to be incorrect and that dichloro-acetylene may have been the culprit.

The decomposition end products of vinylidene chloride peroxides are formaldehyde, phosgene, and hydrochloric acid and can be detected by their sharp odor. Since the peroxide compound may be a polymerization catalyst, the formation and precipitation of vinylidene chloride polymer, when no other catalyst has been added, may also indicate peroxide formation. The presence of peroxides may be confirmed by the liberation of iodine from a slightly acidified dilute potassium iodide solution according to standard analytical procedures. The peroxides are adsorbed on the precipitated polymer and any separation of the polymer by filtration, evaporation or drying will result in an explosive composition. Any dry composition containing more than approximately 15 percent of the peroxide will detonate from a slight mechanical shock or heat. It should be noted that polyvinylidene chloride free of peroxides is not explosive.

Vinylidene chloride which contains peroxides may be purified by washing several times with 10 percent sodium hydroxide at 25° C, or with a fresh 5 percent bisulphite solution. The storage of vinylidene chloride monomer in the absence of air or oxygen is the positive method for preventing the formation of peroxides. The use of a blanket of a nonflammable inert gas (nitrogen) under positive pressure is recommended. Phenol-type polymerization inhibitors which evidently delay or prevent the formation of peroxide compounds are also utilized. Many inhibitors for the polymerization of vinylidene chloride have been described in patents.[5,14,49]

Equipment used for handling or storage of the monomer should be free of any copper or brass because these metals may react with the acetylenic triple bond. Steel containers are satisfactory for inhibited monomer. Nickel is satisfactory for distilled monomer. Steel plugcocks lined with "Teflon*" give excellent service.

POLYMERIZATION

Homopolymerization

Pure vinylidene chloride monomer, free of oxygen, polymerizes very slowly. However, as ordinarily prepared, vinylidene chloride polymerizes readily at temperatures above 0° C. The polymer which forms is insoluble in the monomer and precipitates as a white powder.[38] Oxygen dissolved in the monomer reacts to form peroxides and acid chlorides, thus catalyzing the polymerization reaction. The preferred polymerization catalysts for vinylidene chloride are of the free radical type, e.g., peroxides, potassium peroxysulfate, and 2,2'-azodiisobutyronitrile. In addition to the organic and inorganic peroxygen compounds, other catalysts which have been used successfully for polymerization include organometallic compounds, organic carbonyl compounds, inorganic salts, and inorganic acids.[6,40,50]

*E. I. duPont de Nemours & Co., Inc.

For laboratory polymerization work, benzoyl peroxide has been used frequently in concentrations of 0.05 to 2.0 percent. A typical polymerization curve is shown in Figure 24-2.[41] The straight-line nature of this curve is characteristic. The benzol peroxide-catalyzed polymerizations have little, if any, induction period above 30° C, but they usually show a lower polymerization rate for the first few percent conversion, as illustrated by the enlarged portion of the curve in Figure 24-2. The use of other catalysts or combinations of catalysts causes wide variation in polymerization rates but, in general, does not change the shape of the curves.

Thermal polymerization of vinylidene chloride without added catalyst usually does not occur at a rate sufficiently high to be useful, but photopolymerization has been successful both with and without auxiliary chemical catalysis. Light of wave length less than 4500Å causes polymerization at temperatures as low as –35° C. Many inhibitors for the polymerization of vinylidene chloride have been described in patents.[7,15]

As the polymer is not soluble in the monomer, a solid phase appears during the polymerization of vinylidene chloride. In mass or bulk poly-

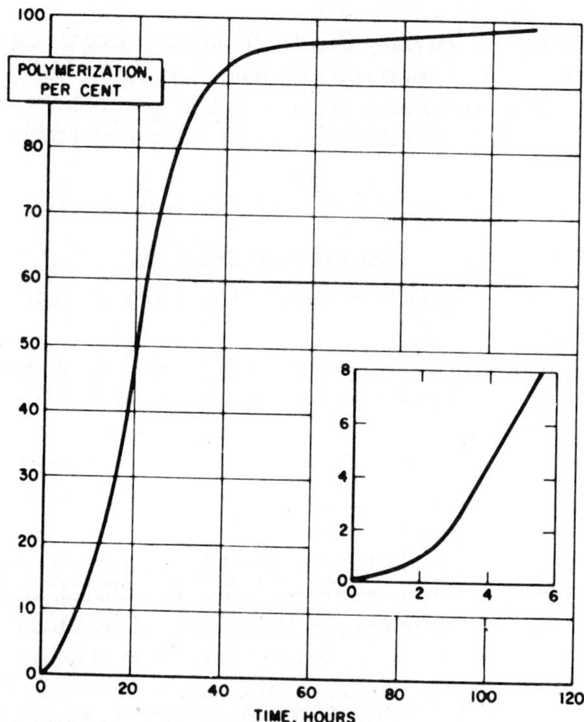

Figure 24-2. Polymerization of vinylidene chloride (catalyst 0.50% benzoyl peroxide; temperature, 45°C; dark).[41]

merizations, where only monomer and catalyst are present, the polymer appears as a flocculant precipitate at low conversion. Between 10 and 20 percent conversion the slurry becomes solid. Above 20 percent conversion, no liquid is visible, and the porous solid becomes harder with increasing polymerization. With proper choice of conditions, the reaction reaches 100 percent conversion to polymer, with the evolution of about 14,400 calories per mole of monomer. Because of the difficulty of controlling the rate of polymerization, and consequently, the physical characteristics of the polymeric product, during mass polymerization it is often desirable to disperse the monomer in a solvent or an immiscible liquid and conduct the polymerization in solution, emulsion, or other dispersed system.[39] The polymers obtained by use of the different polymerization techniques differ only in minor details, whereas the other factors, e.g., cycle time, polymer separation and drying, solvent recovery, and properties control, are important considerations in process selection.

Polyvinylidene chloride $(-CH_2CCl_2-)_n$ has a molecular weight of 10,000 to 100,000 and is a white, porous and nonflammable powder with a softening range of 185 to 200° C and a decomposition temperature of about 225° C. When fused and cooled, the polymer is colorless and nearly transparent. Polyvinylidene chloride possesses two outstanding characteristics: crystallinity, as determined by x-ray diagrams, and insolubility.

Properties of polyvinylidene chloride are listed in Table 24-3. It is insoluble in practically all organic solvents at temperatures below 100° C, and

TABLE 24-3. PROPERTIES OF POLYVINYLIDENE CHLORIDE[41]

Softening temperature, °C	185–200
Decomposition temperature, °C	210–225
Density, at 30°C, g/cu cm.	1.875
Index of refraction, n_D^{20}	1.63

it is unaffected by common reagents at temperatures below 100° C. Fabrication of polyvinylidene chloride by the usual plastic material fabrication techniques is difficult because of its high softening range and its tendency to evolve hydrogen chloride at plastic fabrication temperatures. Plasticizers are often added to polymeric materials to reduce fabrication working temperatures, but the usual plasticizers show a low degree of compatibility with polyvinylidene chloride. These properties, characteristic of the molecule, are not necessarily disadvantages as special techniques can be used for fabrication. However, the unique properties of this polymer may be modified by copolymerization with other monomers to improve the characteristics of narrow processing temperature range, crystallinity, and solubility. The copolymers so derived may be fabricated by the more conventional techniques without any considerable sacrifice in the desirable chemical resistance and permeability characteristics.

Copolymerization

Vinylidene chloride forms copolymers with many substituted ethylenes (particularly the common vinyl compounds), with dienes and their derivatives, and with a number of other unsaturated compounds. Specific examples are described in many patents. Among the more important commercial copolymers of vinylidene chloride are those with vinyl chloride[1,22,51,52] and acrylonitrile.[26] Other compounds which are known to copolymerize with vinylidene chloride are vinyl acetate,[51] styrene,[53] esters of acrylic and methacrylic acid,[13,54] butadiene and its derivatives,[12,44,57] various unsaturated esters,[8,9] unsaturated ethers,[10] and halogen-substituted propenes.[42]

In general, when vinylidene chloride is copolymerized, each mixture of monomers has a different polymerization rate and produces a polymer of different composition. Usually the polymeric product contains a larger proportion of vinylidene chloride than the original monomer composition. Also, the rate of polymerization is slower than for either monomer alone. A typical case is illustrated by Figure 24-3. Although vinylidene chloride and vinyl chloride have nearly identical polymerization rates with 0.5 percent benzoyl peroxide at 45°C, mixtures of the two monomers may have but one-tenth the polymerization rate of either component.

Except in rare cases, the products are true copolymers and contain little or none of the individual homopolymers. Figure 24-4 illustrates the average polymer compositions produced from a number of monomer mixtures. Since the polymer contains more vinylidene chloride than the monomer the vinyl chloride concentration in the monomer increases as polymerization progresses. Consequently, the product of such a copolymerization usually contains a wide distribution of copolymer compositions in addition to the distribution of chain lengths.

The rate of peroxide-catalyzed copolymerization is increased by increasing the concentration of catalyst or by raising the temperature. In practice, a balance between the two is arrived at in order to obtain the desired molecular weight. The thermal and peroxide catalysis of copolymerization is illustrated in Figure 24-5, with 85 percent vinylidene chloride (15 percent vinyl acetate monomer composition). Increased temperature is preferable to higher catalyst concentration as a means of controlling the polymerization rate, which involves consideration of the catalyst half-life. The relationship of viscosity (chain length or molecular weight) to polymerization temperature for these copolymers is illustrated in Figure 24-6. The rapid change of specific viscosity with polymerization temperature is believed due to the chain-transfer effects of vinylidene chloride.

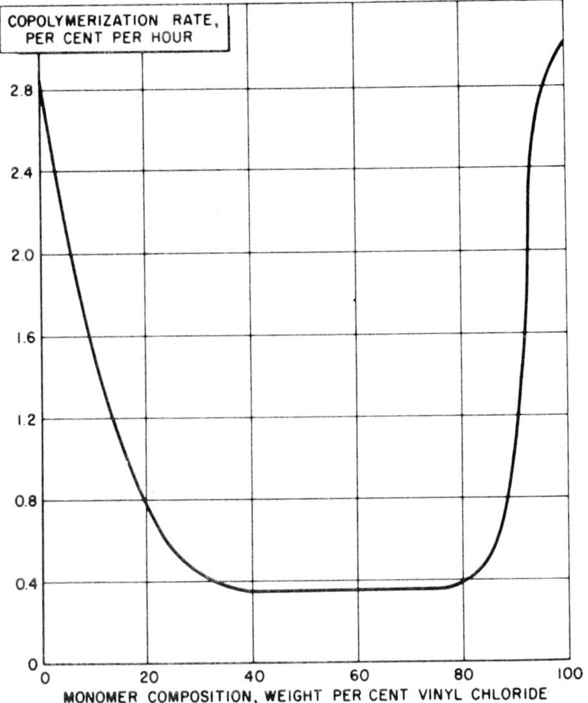

Figure 24-3. Copolymerization rate of vinylidene chloride and vinyl chloride as a function of monomer composition (catalyst 0.50% benzoyl; temperature, 45°C; dark).[41]

The methods used for polymerizing vinylidene chloride, i.e., mass, solution, emulsion, or other disperse systems, are applicable with certain modifications to the polymerization of its copolymers. However, the widely varying polymerization rates and the heterogeneous copolymers introduce important complications in the study and development of both process and products. The balance of the comonomer system, catalyst, process, and temperature affects the solubility, crystallinity, permeability, softening temperature, decomposition temperature, chemical resistance, and processing characteristics.

Vinylidene chloride copolymers may be analyzed by standard analytical techniques. Determination of the chloride percentage is usually sufficient to give the average composition, although the results obtained by this method may be in error by several percent.[3] Chain lengths are comparatively measured by the determination of solution viscosity. Dilute solutions of the polymer in the best solvents (e.g. *o*-dichlorobenzene) at temperatures of 120° C or above are indicated due to the extremely limited solubility of certain copolymers.

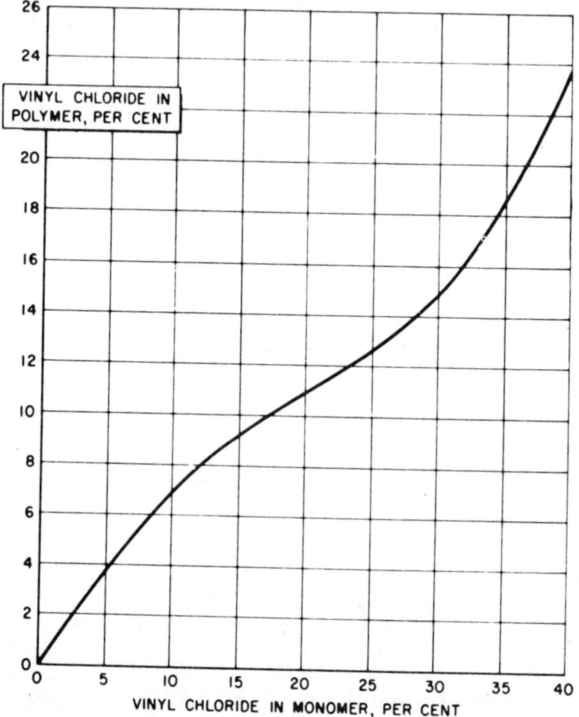

Figure 24-4. Copolymerization of vinylidene chloride and vinyl chloride (catalyst 0.50% benzoyl peroxide; temperature, 45°C; dark; approximately 50% polymerization).[41]

Structure

Polyvinylidene chloride and its copolymers may exist in a *crystalline* state as well as an amorphous state. X-ray diffraction methods indicate a high degree of crystallinity for polyvinylidene chloride, but they indicate a proportionately lesser degree of crystallinity and an appreciable amount of amorphous material for the copolymers of vinylidene chloride. This is the normal crystalline "as-polymerized" state for these polymers. Contributing to the crystallinity of polyvinylidene chloride is its regular molecular structure. As assigned by Staudinger,[47] polyvinylidene chloride shows a head-to-tail configuration, with a serpentine configuration of carbon atoms.

$$\cdots -CH_2-\underset{\underset{Cl}{|}}{\overset{\overset{Cl}{|}}{C}}-CH_2-\underset{\underset{Cl}{|}}{\overset{\overset{Cl}{|}}{C}}-CH_2-\underset{\underset{Cl}{|}}{\overset{\overset{Cl}{|}}{C}}-CH_2-\underset{\underset{Cl}{|}}{\overset{\overset{Cl}{|}}{C}}- \cdots$$

If a polymer or copolymer of vinylidene chloride is melted at a temperature high enough to destroy its crystallites, the material becomes *completely amorphous*. Amorphous polymer, upon being rapidly cooled to a low temperature, remains amorphous. When stored at low temperatures, the polymer remains in the amorphous state for considerable time periods, and storage of the amorphous material at suitable temperatures permits control of the recrystallization rate. On the other hand, when amorphous polymer stands at room temperature or is heated without being remelted, it reverts to the normal state of random high crystallinity. The crystallization induction time for amorphous vinylidene chloride homopolymer is so short that it is difficult to cool it from an elevated temperature fast enough to maintain the amorphous state. The crystallization induction time increases with increasing comonomer concentration.

These two states or modifications of these polymers and copolymers,

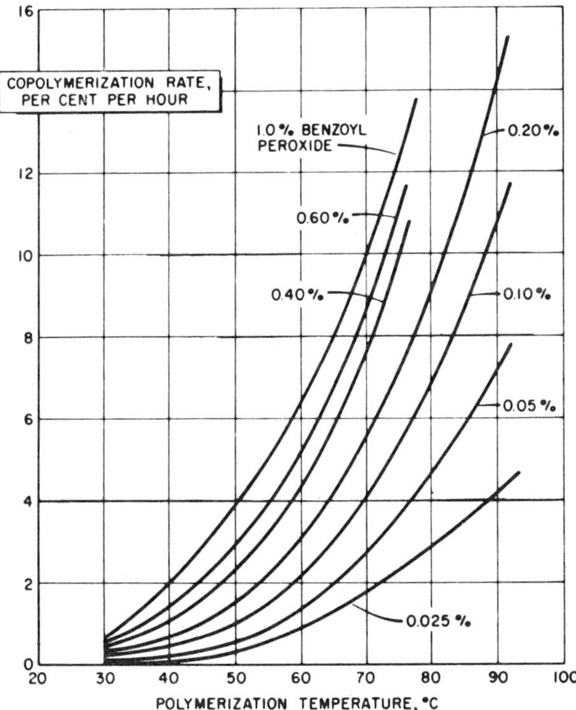

Figure 24-5. Copolymerization rate of vinylidene chloride and vinyl acetate as a function of polymerization temperatures and catalyst concentration (monomer composition, 85% vinylidene chloride, 15% vinyl acetate; temperature, 45°C; dark).

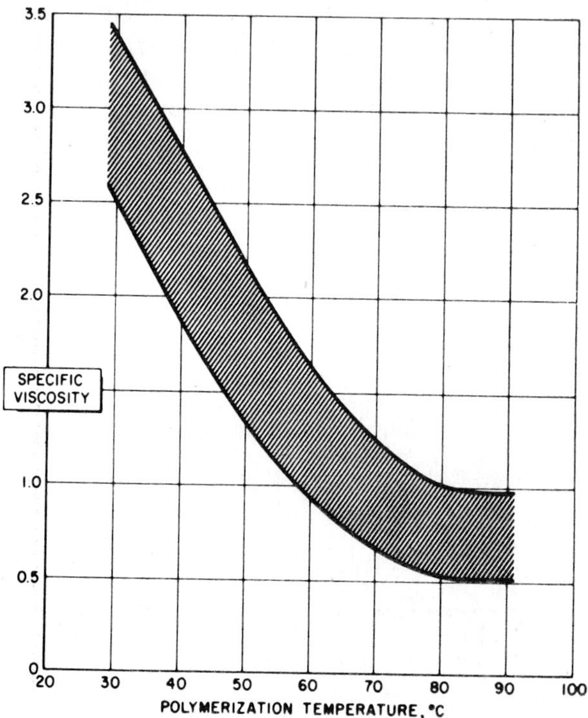

Figure 24-6. Specific viscosity of vinylidene chloride-vinyl acetate copolymers as a function of polymerization temperature (monomer composition 85% vinylidene chloride; 15% vinyl acetate; catalyst, 0.50% benzoyl peroxide; dark; solution, 2% in *o*-dichlorobenzene at 120°C).[41]

together with a third state, the *oriented crystalline* state, have been utilized in developing fabrication techniques. The *oriented crystalline* state is produced by severe mechanical working, as by stretching, of either completely amorphous or partly crystalline polymer. Orientation lines up the crystallites in the direction of stretch and produces a tough, strong, flexible material.

Solution viscosity and osmotic pressure measurements indicate a range of chain lengths of 100 to 1,000 monomer units. As with other high polymers, the chain length is an inverse function of the polymerization temperature and the catalyst concentration. A single chain of polyvinylidene chloride is believed to have the structure of Figure 24-7. With a carbon-valence bond angle of approximately 120° and a carbon-carbon distance of 1.55Å, an identity period of 4.68Å results. This structure is in better agreement with observed data than that suggested by Fuller,[24] a zig-zag arrangement of carbon atoms with the chain shortened by partial rotation.

Determination of the branching coefficient n by the method of Houwink[29] gives $n = 1$. This low value and the marked tendency to crystallize are strong evidence that the chains are not branched. In the amorphous state they are randomly distributed and probably highly curled. Crystallization involves the movement of portions of such molecules into a macromolecular lattice which has no well defined boundaries.

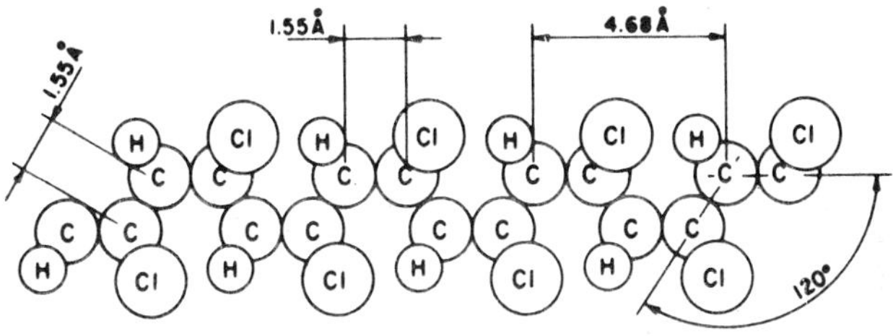

Figure 24-7. Chain configuration of polyvinylidene chloride. (Scale of inter-atomic distances expanded with respect to that of atomic radii.) Each Cl and H atom is one of a pair, of which the other is not shown.[19]

The unit cell of polyvinylidene chloride is monoclinic and contains four —CH_2—C—Cl_2— units. The dimensions of the unit cell are:

$$a = 13.72 \pm 0.015 \text{ Å}$$
$$b = 4.68 \pm 0.01 \text{ Å}$$
$$c = 6.31 \pm 0.01 \text{ Å}$$
$$\sin \beta = 0.8212 \pm 0.015$$
$$\text{volume of cell} = 332.7 \text{ Å}$$

The dimensions of the crystallite are believed to be about 20 to 30 Å parallel to the chain axes, and about 200 to 500 Å at right angles. In the crystalline portions the chains are essentially parallel, although the long-chain molecules undoubtedly traverse both crystalline and amorphous regions. The comparatively low proportion (20 to 40 percent) of crystalline regions in fused and recrystallized polymer, and the high ratio of length to diameter of the molecules, indicate that the amorphous regions are probably the continuous phase. These factors indicate a continuous transition between crystalline and amorphous material, rather than a mesomorphic arrangement or a crystalline modification dispersed in an amorphous region. The crystalline and amorphous modifications each share in determining the ultimate physical properties.

Copolymerization introduces units in the chain which tend to destroy its regularity and, consequently, its ability to crystallize. The introduction of small amounts of some monomers merely results in minor discontinuities in the crystalline regions. The magnitude of the effect thus varies with the structure of the copolymerizing molecule. The vinylidene chloride copolymers become essentially noncrystalline when the monomer mixture contains more than 50 percent vinyl chloride, 20 percent ethyl acrylate, or 15 percent acrylonitrile. Other effects of copolymerization that are related to the decreased crystallinity are reduced softening temperature and increased solubility in organic solvents.

Crystallinity and Orientation

The crystalline homopolymer and copolymers of vinylidene chloride (with the exception of the "as-polymerized" powders) are hard, tough materials resembling ordinary plastics in many respects. In the amorphous state these polymers are soft, rubbery, capable of being mechanically worked, and tend to crystallize on standing at room temperature. The oriented modification is strong, tough, and very flexible. These copolymers possess the most advantageous combination of properties for commercial molding and extrusion. The copolymers which exhibit a low degree of crystallinity due to the particular comonomer combination or concentration tend to have greater solubility in organic solvents and are therefore suitable for solvent-based coating applications.

The crystalline polymers and copolymers of vinylidene chloride have a characteristically narrow softening range. A few degrees above the softening range these polymers have a sharp, quite reproducible crystalline melting point which can be observed by the change in light transmission of a sample heated between crossed polaroids. This melting point probably corresponds to the melting of the most stable crystalline regions. The process of melting is comparable to the solid-liquid transition of crystalline compounds of lower molecular weight. At higher temperatures the molten polymers may be quite fluid. The flow data in Figure 24-8 illustrates the sharp softening point, as measured by plastic flow, of a crystalline vinylidene chloride-vinyl chloride copolymer and compared with polystyrene and ethyl cellulose.

When a normally crystalline copolymer is heated to a temperature sufficient to melt the crystalline portion and then cooled rapidly to a low temperature so that no recrystallization occurs, it remains amorphous and is said to have been super-cooled.[55] When so treated, its tendency to crystallize is a function of many variables, the most important of which are the copolymer composition, the time and temperature of storage, and the presence of addition agents such as plasticizers. Figure 24-9 illustrates the wide variation with temperature of the induction time that elapses before recrystallization begins. This induction period, as measured between crossed polaroids, is a reproducible value for any particular copolymer composition and treatment. Due to the short induction period at the lowest portion of the curves, very rapid cooling of the amorphous material is

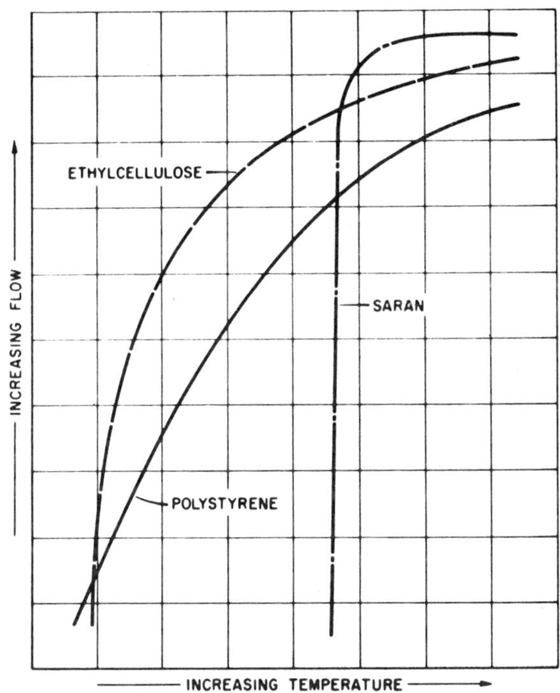

Figure 24-8. Plastic flow versus temperature of several thermoplastics.[53]

necessary to prevent recrystallization from taking place. A storage temperature of under 20° C would be necessary to maintain the material in an amorphous state for an extended period of time. The crystallization induction period above about 110° C is increased by the relatively high thermal energy of the polymer chains. Below about 60° C the increase is due to the high internal viscosity. The effect of certain plasticizers in reducing the crystallization induction period may be attributed to the increased mobility of the chains.

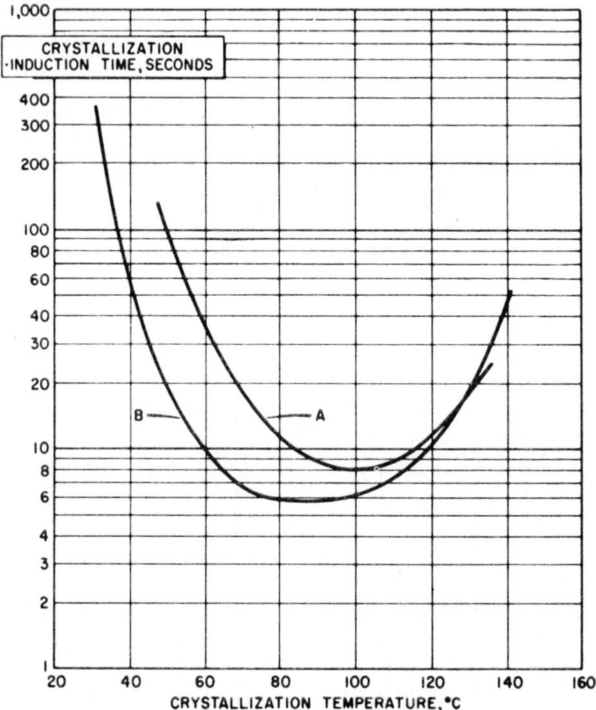

Figure 24-9. Typical crystallization induction period curves for a normally crystalline vinylidene chloride-vinyl chloride copolymer; B, 7% plasticizer; specimen thickness, 0.050 in.[41]

As recrystallization progresses, several marked changes occur in the physical and mechanical properties of the polymer. The most important of these changes are:

(1) A gradual development of a typical x-ray diffraction pattern.
(2) An increase in hardness.
(3) An increase in the force required to produce deformation.
(4) An increase in density.
(5) Evolution of about 3.4 gram-calories per gram of polymer as heat of crystallization.
(6) Variations in electrical properties.
(7) An increase in resistance to the action of solvents.

When either the crystalline or amorphous modifications are mechanically worked, vinylidene chloride polymers and copolymers exhibit orientation. The degree of orientation produced may vary depending on the polymer composition and treatment. If a saran (vinylidene chloride-vinyl chloride copolymer) filament is stretched at a temperature somewhat above the crystalline melting point and quickly cooled, the strand will show slight orientation. Crystalline saran filaments may also be stretched, with the application of comparatively high loads. Filaments so treated will exhibit orientation together with increased strength and flexibility. If an amorphous filament is super-cooled and stretched before crystallization begins, a high degree of orientation results. This technique has been developed commercially to produce fibers of exceptionally high strength and flexibility.[56] The load-elongation curve for a super-cooled filament (Figure 24-10) illustrates that the force required for plastic deformation is nearly constant below 300 percent elongation, but after orientation has occurred, elastic elongation requires a relatively high load. Super-cooled filaments may be

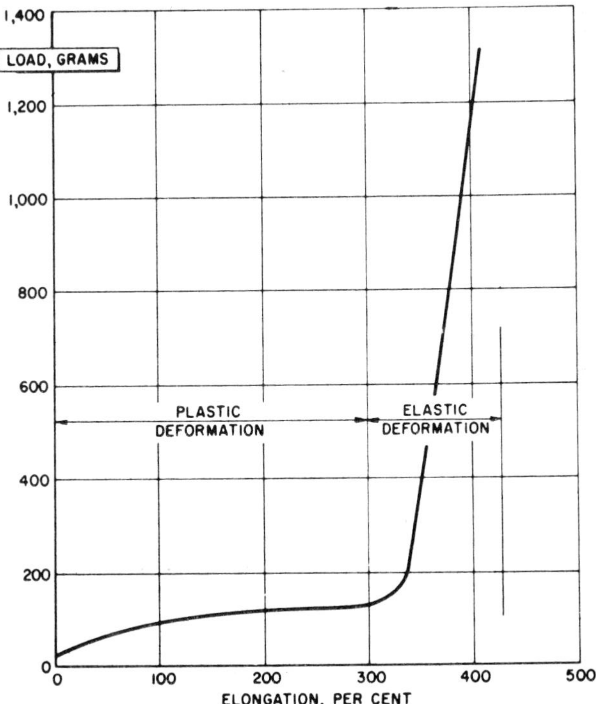

Figure 24-10. Typical load-elongation curve for a super-cooled filament of a normally crystalline vinylidene chloride-vinyl acetate copolymer, plasticized. (Temperature of test 20° C, Scott IP2 testing machine, final diameter of filament 0.011 in.)[41]

stretched to an elongation of 350 to 400 percent, depending on previous treatment. The greatest increase in tensile strength occurs with the last few percent of stretch. Tensile strengths of 100,000 psi have been obtained under optimum stretching conditions, but with the sacrifice of other desirable properties. Under commercial conditions, amorphous filaments of Saran 115, a vinylidene chloride-vinyl chloride copolymer, may be stretched 400 percent; it thereby is changed from the unoriented form, with tensile strength of 8 to 10,000 lb/sq in., low impact and low flexibility, to the oriented form, with tensile strength of 30 to 60,000 lb/sq in., high impact and high flexibility.

Fabrication

Saran (i.e., the homopolymer of vinylidene chloride and the copolymer produced when vinylidene chloride is the major component of the monomer mixture) is adaptable to many of the conventional methods used for the fabrication of plastics. Some modifications of equipment may be necessary to avoid thermal decomposition or to take advantage of the crystalline structure. Among the techniques which have been used with saran are the following:

(1) Extrusion
    unoriented
    oriented
(2) Injection molding
(3) Transfer molding
(4) Compression molding
(5) Calendering
(6) Coating

Extrusion of saran, which is the major fabrication technique, requires machine designs that will permit streamlined plastic flow with minimum opportunity for plastic hold up in the equipment to avoid the thermal decomposition of saran in the hot zones. The thermal decomposition of saran is accompanied by the evolution of hydrogen chloride. Further decomposition may be characterized by a yellow discoloration, and as degradation proceeds, by a reddish color, then brown, and finally black.[33] The final decomposition products are carbon and hydrogen chloride. Therefore, special metals of construction are required for the heated sections of the extruder to resist attack by hydrogen chloride. The metallic salts, e.g., the chlorides of iron, copper, tin and zinc, catalyze this decomposition. The appearance of the first discoloration is evidence of only minor decomposition, and there is very little effect upon the physical properties.

Construction metals recommended for extrusion equipment include "Duranickel" alloy ("Z" nickel) for screws; "Duranickel," "Xaloy" 306, "Hastelloy" D, or Stoody No. 6 alloy for cylinder liners; "A" nickel for nosepieces; and "Duranickel" alloy for dies. "Dowmetal" (magnesium) may be used for experimental dies, but it is too soft for production die use. The nickel alloys and cobalt alloys are not affected by the hydrogen chloride in the degree present in a saran extrusion operation and have no appreciable catalytic effect. They are machinable with care, are hard enough to withstand normal usage, and have a coefficient of expansion similar to that of steel.[20]

Plasticized and stabilized saran formulations are available in powder form for use in extrusion operations. The powder presents no bulk problem with a screw extruder and eliminates a pelletizing operation. Elimination of the pelletizing operation reduces the thermal degradation of the compound as well as the cost of the fabrication step. Mass-colored (pigmented) formulations are available in a wide range of colors. These materials are adaptable to the melt spinning of monofilament. Such monofilament in from five to fifteen thousands diameter has been used widely in the textile field as automobile seat fabric and covering, outdoor furniture, window awning fabric, furniture upholstery, drapery fabric, venetian blind tape, filter fabric, etc.

In the extrusion process, the streamlining of the equipment is necessary to prevent hold up and subsequent polymer degradation. The most satisfactory means for heating saran extrusion equipment is steam. Means for rapid cooling upon shutdown must be provided.

In the extrusion process, saran is heated above its crystalline melting point, where it becomes completely amorphous and very fluid. Because the molten material has no plastic memory, the retention of the desired shape and surface is made possible. When the molten polymer is quenched by immersion in a refrigerated water bath maintained at about 50° F, crystallization is inhibited prior to orientation. When first extruded and cooled in the water bath, the material is soft, weak, and pliable. If it is allowed to remain at room temperature without further treatment, it will gradually harden and partially recrystallize at a slow rate with a random crystal arrangement. By means of heat treatment, recrystallization can be produced at controlled rates. The rate of recrystallization is a function of temperature, as illustrated in Figure 24-11.

The control of extrusion and heat-treatment conditions with saran permits a range of product properties to be obtained (see Table 24-4). A few of the applications for extruded saran include rods for making gaskets, valve seats, ball checks, and medicinal probes; chemically resistant flexible tubing and pipe; and chemical conveyor belts.

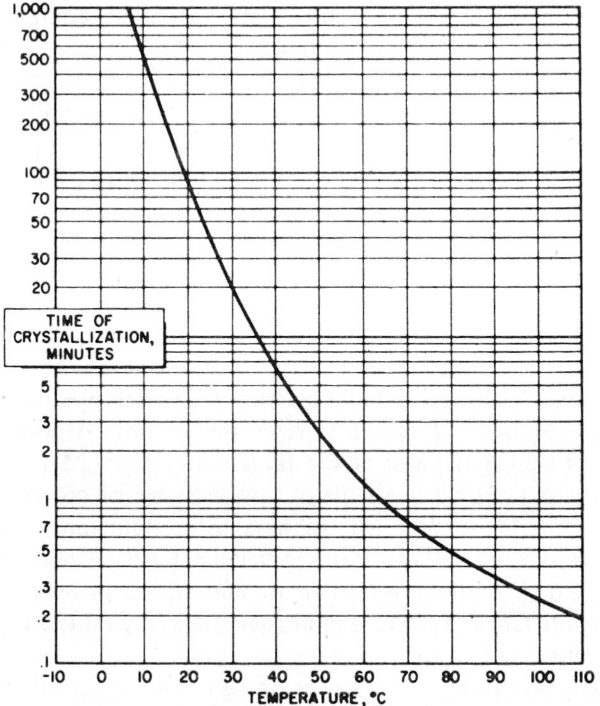

Figure 24-11. Recrystallization time of a typical saran versus temperature.

TABLE 24-4. PROPERTIES OF SARAN

(Range in Properties Covers Several Types of Commercial "Sarans")

Property	Unoriented Forms	Oriented Forms
Specific gravity	1.65–1.72	1.65–1.72
Fabrication data		
Compression molding		
Temperature, °F	250–350	–
Pressure, psi	500–5000	–
Extrusion temperature, °F	300–400	300–400
Injection molding		
Temperature, °F	300–400	–
Pressure, psi	10,000–30,000	–
Mold shrinkage, in./in.	0.005–0.025	–
Mechanical		
Moduli of elasticity (psi)		
In tension, $\times 10^5$	0.55–0.8	0.7–2.0

TABLE 24-4 (*continued*)

Property	Unoriented Forms	Oriented Forms
In compression, $\times 10^5$	0.55–0.95	–
In flexure, $\times 10^5$	0.55–0.95	–
Tensile strength (psi)		
Upper yield	2700–3700	–
Ultimate	3000–5000	20,000–50,000
Compressive properties (psi)		
Yield strength, 0.2% offset	2000–2700	–
Strength at 20% deformation	4500–6100[a]	–
Flexural yield strength	4300–6100[b]	–
Elongation in tension		
At yield, %	15–24	16–23
At ultimate, %	50–240	15–25
Hardness, Rockwell M	50–65	–
Impact strength, Ized notched, ft-lb/in. of notch	0.3–1.0	–
Thermal		
Flammability (Over 0.050 in.), in./min.	self extinguishing	
Flow temperature, °F	240–280	
Heat distortion, at 264 psi, °F	130–150	
Specific heat, Cal/(°C) (g.)	0.32	
Thermal conductivity, Cal/(sec.) (sq cm.) (°C/cm)	3.0×10^{-4}	
Thermal Coefficient of expansion, (in./in.)/°C	19×10^{-5}	
Optical		
n_D^{20}	1.60–1.63	–
Electrical		
Dielectric constant		
At 60 cycles/sec.	4.5–6.0	–
At 10^3 cycles/sec.	3.5–5.0	
At 10^6 cycles/sec.	3.0–4.0	
Dielectric strength, short time, 1/8 in., volts/mil	350	–
Power factor		
At 60 cycles/sec.	0.030–0.045	–
At 10^3 cycles/sec.	0.060–0.075	–
At 10^6 cycles/sec.	0.045–0.065	–
Resistivity, volume, (ohm-cm)	10^{14}–10^{15}	–
Chemical properties		
Water absorption, 24 hr, %	0.1	0.1
Exposure		
Color stability	slight darkening	
Resistance to weathering	slight darkening, otherwise excellent	

[a]No failure, continues to compress
[b]No failure

Saran-lined pipe for the industrial transport of corrosive chemicals is fabricated by swaging to size an oversize steel pipe onto an extruded saran pipe liner. In addition to having good chemical resistance and long life, this pipe is easily installed by conventional tools and requires no special support. Flanged joints and saran gaskets are used to connect lengths of pipe and fittings. Fittings are lined with injection molded saran liners. Saran-lined pipe is not recommended for low temperature use due to the low-temperature brittleness of these formulations.

Crystal orientation is developed by extrusion, followed by supercooling, and subsequently by plastic deformation as by stretching, and heat-treatment. It is thus possible to obtain continuous extrusions of monofilament, tape, or film having exceptional properties. The material leaving the extruder is amorphous, and this condition is maintained by rapid cooling. Stretching the amorphous material produces orientation in a single direction in the extrusion of monofilaments resulting in undirectional properties of high tensile strength, great flexibility, long fatigue life, and good elasticity. These properties are desirable for small monofilaments where the load is along the longitudinal axis. Figure 24-12 illustrates a typical saran

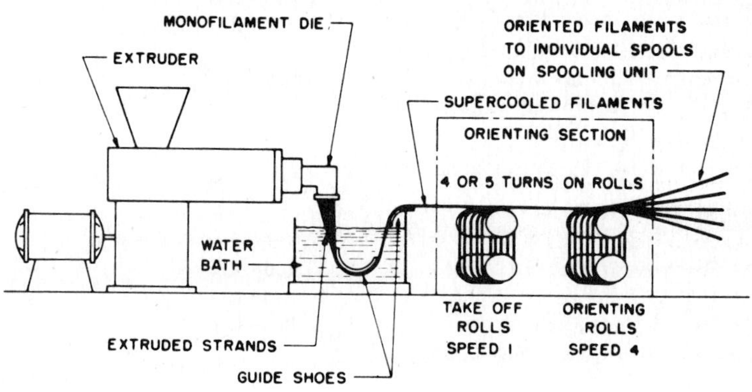

Figure 24-12

monofilament extrusion and orientation train. When a 26-orifice die is used, a typical saran extruder will be capable of extruding saran as 26 monofilaments at 40 pounds per hour, with a linear speed of 450 to 500 feet of monofilament per minute. An average yield of 10-mil monofilament is 5800 yards per pound. These 26 monofilaments may be taken from the supercooling tank as a single "tape," wrapped several times around smooth take-off rolls, and then wrapped several times around orienting rolls which have a linear speed about four times that of the take-off rolls. The difference in roll speeds produces mechanical stretching. In this step, orientation of the crystallites along the longitudinal axis is accomplished, as well as partial recrystallization. The reduction in cross-sectional area is approximately proportional to the elongation. The ultimate size and uniformity of the filaments is governed by the speed and uniformity of the withdrawal rate as well as the difference in roll speeds. Low ratios will not produce uniform filaments. The oriented filaments are then separated and led to individual spools, driven by constant torque motors. Heat-treatment after or during stretching may be used to affect the degree of crystallization desired and thus control the properties of the oriented filaments.

In larger, noncircular sections such as tapes and ovals, the desired degree of transverse orientation can be realized and accompanying transverse properties controlled through the introduction of other factors in the process. For example, a rolling operation incorporated after the quenching step produces some flattening and transverse orientation of the strand with a resultant increase in transverse strength.

Saran monofilaments and multifilaments have many useful properties such as long life, low moisture absorption, chemical resistance, ease of cleaning, fire resistance, insect resistance, flex life, and bright built-in colors. Some of the textile applications have already been mentioned. Other applications include saran fibers in carpeting, fish nets, insect screening, doll hair and manikin wigs.

A variation of the orientation process is used to produce saran film for industrial and household packaging use. Saran film is made by extruding a tube, supercooling this tube, and subsequently orienting it in a continuous process, as shown in Figure 24-13. Air is injected into the supercooled amorphous tube after it leaves the water tank, thus forming an entrapped air bubble in the tube Once the process has been started, the

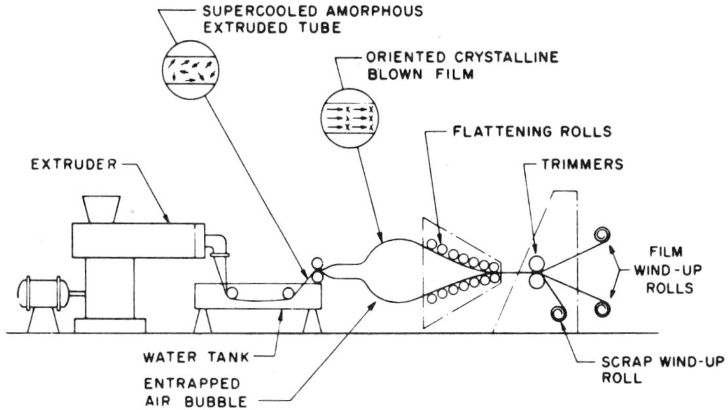

Figure 24-13. Saran film production flow diagram.[19]

entrapped air bubble remains stationary while the extruded tube is oriented as it passes around the bubble. Orientation is produced in the transverse direction as well as in the longitudinal direction, and this creates in the film excellent tensile strength, elongation, and flexibility, while retaining transparency.

The properties of a one-mil saran film are given in Table 24-5. Of special interest are the gas transmission rates and dimensional stability. A one-mil saran film gives the following dry-gas transmission rates at 77° F and 760 mm Hg in cc per 100 sq in. per 24 hours: CO_2, 2.4; N_2, 0.11; O_2, 0.56; H_2, 13.4: He, 33.0; Air, 0.15; Freon F-12, negligible. The im-

TABLE 24-5. PROPERTIES OF A ONE-MIL SARAN FILM

Property	Value
Sp. gr	1.68
Area factor, sq in./lb	16,300
Water vapor transmission, g/100 sq in./24 hr at 100°F. and 90% r.h.	0.20
Water absorption	negligible
Sealing temp., °F	280–300
Burning rate	self-extinguishing
Sp. heat, Btu/(lb)(°F.)	0.32
Thermal conductivity, Btu/(sec.)(sq ft.)(°F/in.)	1.8×10^{-4}
Resistance to heat intermittent	up to 200°F
Resistance to heat, continuous	up to 140°F
Resistance to cold	good flex at 0°F
Tensile strength, psi	7,000–15,000
Elongation, %	20–40
Bursting strength, psi (Mullen)	35
Tear strength, g. Elmendorf	10–20
Folding endurance, M.I.T., 1-kg. load	500,000
n_D^{22}	1.602
Transmission of white light, %	90
Ultraviolet cutoff, A	3000
Transmission of infrared, %	88
Resistance to sunlight	good
Dielectric constant	
At 10^2 cycles	4.9–5.3
At 10^3 cycles	3.9–4.5
At 10^5 cycles	3.4–4.0
Per cent power factor	
At 10^2 cycles	3.5–4.5
At 10^3 cycles	5.2–6.3
At 10^5 cycles	3.9–4.5
Breakdown voltage, volts/mil	3,000–5,000
Surface resistivity, ohms	10^{12}–10^{15}
Volume resistivity, (ohm-cm)	10^{12}–10^{15}

permeability to water vapor and most gases makes saran film an ideal transparent barrier for difficult packaging applications.

Saran films, being highly oriented in the manufacturing process, exhibit some shrinkage when exposed to higher than normal temperatures. This may be desirable in some instances, e.g., in the heat-shrinking of overwraps on packaged items. It may present problems at other times, e.g., in printing and laminating. Pre-shrinking operations can be performed which minimize the residual shrink in the film.

The electronic or dielectric seal is the most satisfactory type for sealing saran film to itself. The bond strength of the electronically fused weld equals the strength of the base film. The heat-sealing temperature of saran film is between 280 and 300° F. Heat sealing with hand or hot-plate sealers will give a "peelable" seal which is satisfactory for sealing airtight overwraps, even though the material may not be actually fused. Impulse sealers, solvent and cement-type seals, and mechanical closures have been used with success. The "clinging" property of Saran wrap household film, due to the combination of flexibility, smooth film surface, and static effects, provides a suitable closure for refrigerator dishes and loose overwraps.

Saran film can be printed by all of the usual methods employed in roll printing, including rotogravure, aniline or Flexographic, letterpress, and so-called roller printing, with the aid of special inks. Modified bag-making machinery is available for the automatic production of saran film bags for the packaging industry.

Injection molding of saran makes possible the production of intricate shapes having properties similar to those obtained by extrusion. As in extrusion, the fabrication equipment involved consists of the conventional-type injection molding machines modified as to contact metals and designs through the heating cylinders. The fundamentals of design include the same contact metals as listed for extrusion, strict streamlining, and the reduction of the thickness of plastic sections in the heating zones. Conventional injection molding die designs and die metals may be used.

The injection molding of saran is unique. With other commercial thermoplastics, the use of cold molds hastens the cooling of the molded parts and shortens the molding cycle. With saran, cold molds produce soft, flexible, amorphous pieces. Rapid hardening of saran is accomplished by heat treatment, through means of heated molds, to induce crystallinity. When injection molding heavy saran sections, heated molds allow the molded piece to retain its heat and recrystallize rapidly. Under these conditions, saran can be ejected from the mold at temperatures as high as 100° C in a strain-free, warp-free, dimensionally stable form. This promotes very rapid cycles with heavy sections, and saran moldings have been produced with quarter-inch sections in a 17-second cycle. Sink marks are usually not a problem in molded saran parts because the outer skin of the molding cools quickly below the melting point, and crystallization takes place rapidly. The crystalline material is stiff and resists shrinkage forces, but if sufficient plastic material is not injected into the mold internal voids are likely to develop. The range of molding temperatures is rather narrow due to the crystalline nature of saran and its thermal sensitivity.

In many instances, saran injection-molded parts are used to satisfy the industrial requirements of such materials as nickel, aluminum, stainless steel, and rubber in applications requiring chemical resistance. Such items as gasoline filters, spinnerets, valves, pipe fittings, balls, containers, and handles have been injection molded for chemical-resistant applications. Articles molded of saran exhibit excellent weld strength. Welds are practically as strong as other portions of the molded part. This characteristic is due to the high fluidity of the plastic material when it is molded, which, in turn, is due to the relatively sharp melting crystalline character. Injection molded saran does not exhibit the high tensile strength of oriented forms. The high tensile strength properties of saran monofilament and film are developed only by orientation.

Saran may be *compression molded* from powder or pellets. This technique for saran differs from that for other thermoplastic materials only in that it requires selection of metals for the molds. Chrome plating is usually most convenient as other noncatalytic metals which can be machined are not well suited to make good mold surfaces because of relatively low hardness and yield point. Molds should be cored for adequate cooling. The conventional type of compression mold is a semi-positive mold.

Saran can also be compression molded from dielectrically heated preforms. The dielectric constant and power factor properties of saran are such that the resin lends itself well to dielectric heating for heat fabrication operations. Dielectric heating minimizes thermal degradation. Large sheets of saran have been fabricated by compression molding for construction of chemical-resistant process equipment.

Saran process equipment may be assembled from compression molded sheet or blocks by welding or cementing. Cementing is not the best method because of the high degree of chemical resistance of saran. However, cementing is satisfactory in some cases. Welding is the preferred technique, and has been used with considerable success. Hot plate welding, at about 400° F, has been used, as well as welding by friction or by radiant heating. However, for the more complex forms, hot gas welding is the most satisfactory. The process is very similar to that employed for gas brazing of metals. The hot gas is usually hot air at a temperature of 400° F to 500° F, blown through a jet in such a manner that it heats the parts to be joined and the saran welding rod simultaneously.

Saran for Plastic Coatings

Saran for plastic coatings may be a copolymer of vinylidene chloride with vinyl chloride, acrylonitrile, or other monomers. Copolymers are prepared for either *latex* or *solvent solution* applications. Saran latexes, such as Saran Latex 122-A15, are mainly used in paper coating and specialty paint applications. These latexes may utilize surface active agents and plasticizers in their preparation so that certain properties such as moisture-vapor transmission or fusion temperature may be affected. Properties of coatings deposited from the aqueous dispersions compare favorably, however, with those obtained from saran lacquer solutions. The saran latexes permit the formulation of high-solids, low-viscosity, fast-drying, low-foaming, solvent-free coating compositions which may be applied by roller, doctor blade, spray, dip, and brush coating. The plasticizers added during production of some of the latexes enables the submicroscopic particles to fuse together during drying to form a film.

Saran latexes deposit smooth, transparent, tough, glossy films on air drying. A high-temperature fusion step at or near the end of the drying cycle will result in improved film properties. Normal aging in light will result in a darkening of the films, which darkening is not accompanied by a great loss in physical properties of the films. Such darkening restricts applications where color retention is the prime requisite.

The vinylidene chloride resins, because of their high chlorine content, produce nonflammable, dense films that have excellent barrier properties. The solvent-deposited coatings are applied normally in dry film thicknesses of four to ten mils by multiple applications. The saran resins such as Saran Resin F-242L, Saran Resin F-120, and Saran Resin F-220 have been used, by solvent coating, satisfactorily for such applications as corrosion-resistant tanker and tank-car linings; coating paper, film and foil; or as a liner for containers. Resistance to various chemical environments varies only slightly with the particular resin used. The copolymers are dissolved or softened by ketones, esters and ethers. The solvents generally used are ketone and ketone-aromatic hydrocarbon mixtures, vapors of which are both toxic and flammable. Applications requiring continuous exposure to temperatures exceeding 150° F are to be avoided as these temperatures may affect the barrier properties of the coatings.

Vinylidene chloride resin solutions, like the vinyls, have poor wetting and penetrating properties and it is necessary to use the best possible method of surface preparation to insure satisfactory adhesion and protection. Blast cleaning of metals should be used wherever possible.

A third application form for coating is chemical resistant or barrier sheet. The copolymer used in this uncured, thermoplastic sheet is termed

"Saraloy." By the use of adhesive systems it has been applied to installation of linings over wood, concrete, or metals, especially where chemical or corrosion resistance is of importance.

Trends

Historically, saran because of its superior properties replaced metals for many critical uses during World War II metal shortages. Such usage continues in chemical resistant applications to the present day. The next major application of the vinylidene chloride copolymers developed almost simultaneously; this was in the textile fiber field, where saran filaments became well known in the automotive seat cover, outdoor furniture, screen and shade cloth, filter fabric, carpeting, and drapery applications. The development of polyolefin fibers has produced a competitive material with some advantages over saran, and several disadvantages. The polyolefin fibers have penetrated some of the filament markets which were formerly held by saran. The saran films, although developed in the World War II period as protective coverings, were soon produced as food packaging films. Saran film, both as a household and commercial packaging film for food applications, continues to be one of the major applications for vinylidene chloride copolymers. In spite of the development and tremendous growth of the polyolefin films, the saran films still provide superior barrier properties for food packaging.

Although the plastic coating vinylidene chloride resins were developed more recently, their production growth has been steady. New uses for the coating resins are being developed, and these materials are now being used for paper and fabric coatings, for tank and container lining, for film and foil coatings, and as a cement additive.

The vinylidene chloride copolymer resins, because of their high chlorine content, have high specific gravities. The specific gravity becomes an economic disadvantage in some applications, but in other applications (such as coatings) the superior barrier properties enable less resin to out-perform other plastic materials. Because of the outstanding and unique properties of the vinylidene chloride copolymer resins, together with the relative ease of fabrication, further growth may be expected.

References

1. U. S. Patent 2,245,742 (1941), Alexander, C. H., and Tucker, H. (to B. F. Goodrich Co.)
2. Bauman, E., *Ann.*, **163**, 308 (1872).
3. Berger, H., *Kunststoffe*, **30**, 35 (1940).
4. Biltz, H., *Ber.*, **35**, 3524 (1902).

5. U. S. Patents 2,121,009–12 (June 21, 1938), Britton, E. C., and LeFevre, W. J. (to Dow Chemical Co.).

6. U. S. Patent 2,206,022 (July 2, 1940), Britton, E. C., and Davis, C. W. (to Dow Chemical Co.).

7. U. S. Patents 2,121,009–12 (1938), Britton, E. C., and LeFevre, W. J.; *Ibid.*, 2,136,347–9 (1938), Wiley, R. M; *Ibid.*, 2,136,333–4 (1938), Coleman, G. W., and Zemba, J. W.

8. U. S. Patents 2,160,940–2 (June 6, 1939), Britton, E. C., Davis, C. W., and Taylor, F. L.

9. U. S. Patent 2,160,946 (June 6, 1939), Britton, E. C., and Taylor, F. L.

10. U. S. Patent 2,160,943 (June 6, 1939), Britton, E. C., and Davis, C. W. (to Dow Chemical Co.).

11. Brooks, B. T., "Non-Benzenoid Hydrocarbons," New York, Chemical Catalog Co., 1922.

12. U. S. Patent 2,029,410 (February 4, 1936), Carothers, W. H., Collins, A. M., and Kirby, J. E. (to Du Pont Co.).

13. British Patent 525,048 (August 20, 1940), Clarke (to Imperial Chemical Ind.).

14. U. S. Patents 2,136,333–34 (November 8, 1938), and U. S. Patent 2,160,944 (June 6, 1939). Coleman, G. H., and Zemba, J. W. (to Dow Chemical Co.).

15. U. S. Patent 2,160,944 (1939), Coleman, G. H., and Zemba, J. W. (to Dow Chemical Co.).

16. Compagnie des Products Chimiques et Electrometallurgiques, French Patent 786,803 (1935).

17. Denzel, J., *Ann.*, **195,** 205 (1879).

18. Ellis, C., "Chemistry of Synthetic Resins," Vol. 2, pp. 1041–2. New York, Reinhold Publishing Corp., 1935.

19. "Encyclopedia of Chemical Technology," Vol. 3, p. 765, Vol. 14, p. 736, New York, The Interscience Encyclopedia, Inc., 1955.

20. "The Extrusion of Saran Monofilament," Dow Chemical Company Plastics Technical Service Bulletin 171–49A, September 1959.

21. German Patent 529,604, I. G. Farbenindustrie (July 15, 1931) (1929).

22. British Patent 477,532 (January 3, 1938), I. G. Faberindustrie A. G.

23. Feisst, W., Doctoral Thesis, Albert Ludwig Univ., Freiberg, Germany 1930.

24. Fuller, C. S., *Chem. Rev.,* **26,** 143 (1940).

25. Goggin, W. C., and Lowry, R. D., "Vinylidene Chloride Polymers," Dow Plastics Technical Service Bulletin, 1945

26. U. S. Patent 2,238,020 (April 8, 1941), Hanson, A. W. and Goggin, W. C. (to Dow Chemical Co.).

27. Henry, M. L., *Bull. Soc. Chim.,* **42** (2), 262 (1884).

28. U. S. Patent 1,921,879 (1933) Hermann, W. O., and Baum, E. (to Consortium fur electro-chem. Industrie Ges.).

29. Houwink, R., "Physikalische Eigenschaften und Feinbau von Naturand Kunstharzen," Leipzig, Akademische Verlagsgesellschaft, 1934.

30. British Patent 534,733, Howell, W. N. (to Imperial Chemical Ind.). (March 17, 1941).

31. Jocitsch, A., *J. Russ. Phys. Chem. Soc.,* **30,** 998 (1898).

32. Kraemer, G., *Ber.,* **3,** 257 (1870).

33. Matheson, L. A., and Boyer, R. F., "Light Stability of Polystyrene and Polyvinylidene Chloride," *Ind. Eng. Chem.,* **44,** 867, April 1952.

34. Natta, G., and Rigamonti, R., *Atti. Accad. Nazl. Lincei., Classe Sci. fis, mat. e nat.,* **24,** 381 (1936).

35. Ostromislenski, I., *J. Russ. Phys. Chem. Soc.,* **48,** 1132 (1916).

36. Regnault, V., *Ann. Chim. Phys.,* **69** (2), 151 (1838).

37. Regnault, V., *J. Pract. Chem.,* **18** (1), 80 (1839).

38. U. S. Patent 2,160,903 (June 6, 1939), Reilly, J. H., and Wiley, R. M. (to Dow Chemical Co.).

39. U. S. Patent 2,160,936 (June 6, 1939), Reilly, J. H., and Russell, C. R. (to Dow Chemical Co.).

40. U. S. Patent 2,160,939 (June 6, 1939), Reinhardt, R. C. (to Dow Chemical Co.).

41. Reinhardt, R. C., *Ind. Eng. Chem.,* **35,** 422 (1943); "Colloid Chemistry, Theoretical and Applied," ed. by Alexander, J., New York, Reinhold Publishing Corp., 1943.

42. U. S. Patent 2,160,947 (June 6, 1939), Reinhardt, R. C. (to Dow Chemical Co.).

43. Sawitsch, V., *Bull. Soc. Chim.,* Seance du 26 Octabre, 239 (1860); *Z. Chem. u. Pharm.,* **3,** 744 (1860); *Jahresber. fur Chem.,* 431 (1860).

44. U. S. Patent 2,215,379 (September 17, 1940), Sebrell, L. B. (to Wingfoot Corp.).

45. Staudinger, H., and Feisst, W., *Helv. Chim. Acta.,* **13,** 832 (1930).

46. Staudinger, H., Brunner, M., and Feisst, W., *Helv. Chim. Acta.,* **13,** 805 (1930).

47. Staudinger, H., "Die hochmolekularen organischen Verbindungen," p. 114, Berlin, Julius Springer, 1932.

48. "Vinylidene Chloride Monomer," Coatings Technical Service Bulletin, The Dow Chemical Company, 1958.

49. U. S. Patents 2,136,347–349 (November 8, 1938), Wiley, R. M. (to Dow Chemical Co.).

50. U. S. Patents 2,160,933–5 (June 6, 1939), Wiley, R. M. (to Dow Chemical Co.).

51. U. S. Patent 2,160,931 (June 6, 1939), Wiley, R. M. (to Dow Chemical Co.).

52. U. S. Patent 2,235,782 (March 18, 1941), Wiley, R. M. (to Dow Chemical Co.).

53. U. S. Patent 2,160,932 (June 6, 1939), Wiley, R. M. (to Dow Chemical Co.).

54. U. S. Patent 2,160,945 (June 6, 1939), Wiley, R. M. (to Dow Chemical Co.).

55. U. S. Patent 2,183,602 (December 19, 1939), Wiley, R. M. (to Dow Chemical Co.).

56. U. S. Patent 2,233,442 (March 4, 1941), Wiley, R. M. (to Dow Chemical Co.).

57. British Patent 533,142 (February 6, 1941), Wingfoot Corp.

25. ZIRCONIUM TETRACHLORIDE AND TITANIUM TETRACHLORIDE

L. W. PIESTER and A. P. MUREN

Pittsburgh Plate Glass Company (Chemical Division)

HISTORICAL

By far, the most important of the titanium and zirconium chlorides are titanium tetrachloride ($TiCl_4$) and zirconium tetrachloride ($ZrCl_4$). $TiCl_4$ has long been used to obtain TiO_2 pigment by hydrolysis. A more recent and important use for $TiCl_4$ has been as a raw material for titanium metal which came into prominence in the mid 1950's when large quantities of titanium were used for military aircraft. The military use has slackened and at present the market for titanium is dependent for the most part on civilian uses. Another new and rising use is the direct oxidation of $TiCl_4$ to form titanium dioxide (TiO_2).

As far as is known, the only important use of $ZrCl_4$ has been for making the metal. The chief use of zirconium metal has been for structural material and components for nuclear reactors and where specialized corrosion problems exist.

The other chlorides of titanium and zirconium will be described very briefly in this chapter. However, because of the importance of the tetrachlorides as compared to the other chlorides, the emphasis will be on the tetrachlorides.

Titanium Chlorides

The titanium poly chlorides were discovered at early dates. George[46] prepared the first $TiCl_4$ in 1825, Ebleman[13] the first $TiCl_3$ in 1847, Friedel and Guerin[19] the first $TiCl_2$ in 1876. The discovery of the monochloride was fairly recent, first having been observed spectroscopically in 1935.[34]

Since the introduction of these compounds, a wealth of information concerning them has been published. Barksdale[1] (67 references) and Skinner, Johnston and Beckett[45] (78 references) both present a summary of the preparation, general chemistry, and physical, chemical and thermodynamic properties of the titanium chlorides.

A very comprehensive review of the physical and thermodynamic properties of many titanium compounds, including the chlorides, was made by Rossini, Cowie, Ellison and Browne.[42]

$TiCl_4$. Pure $TiCl_4$ is a water-white liquid which boils at 136° C. The common method of preparing $TiCl_4$ involves the reaction between chlorine gas and titaniferous materials at elevated temperatures. The pure metal, various alloys, the carbide, the cyanonitride, the dioxide and natural ores have been successfully chlorinated. In the chlorination of the dioxide or the ores, a carbonaceous reducing agent is required to combine with oxygen.

$TiCl_3$. $TiCl_3$ is a crystalline solid that varies in color from dull red to brown, violet or black. At 440° C it decomposes to $TiCl_4$ and $TiCl_2$ while at ordinary temperatures it oxidizes slowly in air. $TiCl_3$ is prepared most easily by the reduction of $TiCl_4$ with metals such as aluminum, antimony, arsenic, magnesium, silver, tin and zinc or hydrogen. Aqueous $TiCl_4$ can also be reduced to $TiCl_3$ with metallic titanium in the presence of hydrochloric acid.

$TiCl_2$. The dichloride is a black solid that decomposes slowly at 400° C in the absence of air. In the presence of air it is unstable at all temperatures. It can be produced by the reduction of $TiCl_4$ with hydrogen, sodium amalgam or titanium and by the thermal decomposition of $TiCl_3$.

$TiCl$. Comparatively little is known of the monochloride. It has been observed from the decomposition of $TiCl_4$ vapor in a discharge tube, but it has not been isolated as such.

Zirconium Chlorides

The preparation, physical and chemical properties of the zirconium chlorides are summarized by Blumenthal with 83 references given.[5] The most important chloride is $ZrCl_4$, although $ZrCl_3$ and $ZrCl_2$ also exist.

$ZrCl_4$. $ZrCl_4$ is a white solid at room temperature which sublimes at 331° F. It can be produced by the chlorination of the metal, the oxide, the fluoride, the pyrophosphate, the carbide, the cyanonitride and zirconium silicate ores. Where oxides are used it is necessary to add a reducing agent to react with oxygen. Various chlorine sources have been used successfully, including carbon tetrachloride, phosgene, sulfur chloride, thionyl chloride, hydrogen chloride or free chlorine.

$ZrCl_3$. The trichloride is a blue-black solid which oxidizes easily in air. It is usually made by the reduction of $ZrCl_4$ at high temperatures. Successful reducing agents have been aluminum, magnesium, iron and zirconium. Hydrogen has not been used with success.

$ZrCl_2$. This material has been produced by the high temperature reduction of $ZrCl_4$ with zirconium metal in the presence of aluminum chloride.

$ZrCl_2$ can also be made by the thermal decomposition of $ZrCl_3$ to $ZrCl_4$ and $ZrCl_2$. The dichloride is very unstable in air or water, undergoing a violent decomposition in either media.

COMMERCIAL PROCESSES—TiCl₄

Ores

$TiCl_4$ is produced according to the reactions:

$$TiO_2 + 2\,Cl_2 + 2\,C \rightarrow TiCl_4 + 2\,CO \qquad (25\text{-}1)$$

$$TiO_2 + 2\,Cl_2 + C \rightarrow TiCl_4 + CO_2 \qquad (25\text{-}2)$$

TiO_2 is supplied in the form of naturally occurring high grade ores or by enriched low grade natural ores. The titanium ore used in commercial $TiCl_4$ plants is usually rutile, which is a high grade ore containing at least 95 percent TiO_2. Many other titanium bearing materials can also be chlorinated. The most important, other than rutile, is naturally occurring ilmenite or beneficiated ilmenite ores such as sorel slag. Of important use for commercial chlorination purposes is leucoxene, a naturally occurring alteration product of ilmenite. Titanium cyanonitride and titanium carbide have also been used.

Source. The rutile used in North America is imported; some comes from India and Africa, but the greatest portion (90 percent) of the world's supply comes from Australia. All of the ore deposits in North America of any commercial importance to the titanium chloride industry, consist of ilmenite and leucoxene. Most of the ilmenite is not high grade ore. Beach sands from Florida, Virginia and Oregon yield ilmenite with 20 to 60 percent TiO_2.[2] The best that can be obtained in sufficient quantity contains 45 to 55 percent TiO_2. Higher grade ilmenite is found outside the North American continent, in West Africa and Travancore, India where ilmenite containing 60 percent TiO_2 is available.

Rutile has been the historical choice of raw material, rather than the low grade domestic ores. Much difficulty is encountered in using a lower grade ore because of the amount of impurities, notably iron oxide, which also chlorinate readily. These impurities not only cause a lowering of $TiCl_4$ productivity from a given volume of reactor and lower chlorine efficiency, but they also cause very difficult problems in recovering pure $TiCl_4$ product. Because of this, domestic ores are usually beneficiated before they are suitable for use in existing processes. However, the cost of processing low-grade ore to yield synthetic rutile has generally been economically unfavorable when compared to the cost of rutile delivered to the United States.

In spite of this fact, a great deal of work has gone into finding suitable and economical methods to beneficiate and use low grade ores. The reasons for this are that the known world supply of rutile is not unlimited, and because an oversea supply of rutile is not an absolutely dependable source in the event of a national emergency. The latter reason concerned the United States Government when the majority of $TiCl_4$ was used to produce titanium for military aircraft. However, concern for a supply of the ore in future years warrants some discussion of beneficiating processes.

Beneficiated Ores. The various methods to up-grade ilmenite generally involve a reduction step to obtain iron and TiO_2, followed by a selective removal. The reduction step can be carried out by roasting the ilmenite with coal, coke, or some other source of carbon which reduces the iron oxides to the metallic state or to a lower oxide. Various selective removal steps can be accomplished such as selective chlorination of the iron at 350 to 1000° C,[12] magnetic or electrofiltration separation methods,[8] or leaching out the iron with hydrochloric acid.[11]

A successful commercial process for beneficiating ore is the electric arc reduction of a mixture of ilmenite and carbon in the form of coal, sawdust or wood chips.[39] In this process, iron is reduced to the elemental molten state. The titanium slag is also molten and two layers form, the slag layer over the iron layer. The process is continuous with several taps per day taken from both layers. With this process, it is possible to obtain ore with up to 76 percent TiO_2 and low-grade pig iron as by-product. This process was actually developed to produce a source of TiO_2 for the pigment industry, and is at present being used for this purpose. It is, however, of importance to the titanium tetrachloride industry where only low grade ores are available.

Much work has been done on the chlorination of domestic ilmenite without beneficiation, and some success along this line is claimed.[14] The difficulties in such a process are numerous; some of these are obtaining a good yield of $TiCl_4$, dealing with the residual molten chlorides of calcium and magnesium in the chlorination bed, and separation of the evolved gaseous chlorides.

A partial list of patents dealing with ore beneficiation is given at the end of this chapter.

Manufacturing

Originally, $TiCl_4$ was made solely for the purpose of upgrading TiO_2 for pigment purposes. For this use, $TiCl_4$ was hydrolyzed to produce a high grade TiO_2 and HCl gas. Very little was required in the way of further processing of $TiCl_4$. However, in order for the $TiCl_4$ to be acceptable for the production of a good grade of ductile metal, further refining of $TiCl_4$ is required. The most damaging impurities are any solids including carbon, dissolved gases such as free chlorine or HCl, dissolved metal chlorides of vanadium, niobium and the like, and dissolved carbon compounds. Most of the impurities from the chlorination are insoluble and can be filtered or settled. The volatile chlorides and oxychlorides present, e.g., $AlCl_3$, $SiCl_4$, $COCl_2$, $TiOCl_2$, $FeCl_3$, HCl, and Cl_2, can be removed from $TiCl_4$ by distillation. Vanadium in crude $TiCl_4$ is probably present as the oxychloride and tetrachloride[3] which boil very closely to $TiCl_4$ and cannot be removed by distillation.

Various methods work satisfactorily to convert vanadium to an insoluble form. Treatment to the insoluble form is always followed by a distillation.[35] Pechukas purified crude $TiCl_4$ by treatment with activated carbon, followed by distillation. The soluble metal salts can be converted to the insoluble sulfides by passing hydrogen sulfide gas into $TiCl_4$[38] or by treating them with hydrogen sulfide in the presence of a heavy metal soap. The vanadium content can be reduced to very low values (less than .001 percent) by mixing the $TiCl_4$ with an alkali metal soap or oil. Another purification process is to reflux crude $TiCl_4$ in the presence of powdered tin or copper, followed by distillation.

The choice of a purification method is a matter of availability of reagent, economics, and disposal of wastes. For an ilmenite chlorination, the waste disposal consideration would be of great importance.

Since the use of $TiCl_4$ as a raw material for the metal has come into prominence, numerous purification patents have appeared. A partial list is given following this chapter.

Figure 25-1 is a simplified flow sheet of a typical commercial $TiCl_4$ plant. The reaction is carried out at a high temperature, either by fluidizing finely divided titanium ore and coke with chlorine or by passing the gas up through a bed of briquettes composed of the ore mixed with coke. The chlorination is highly exothermic. However, the reaction temperature can be controlled by various factors, e.g., the rate of addition of the reactants, and the adjustment of the fineness and/or porosity of the solid reactants, or by mechanical control methods such as cooling coils, etc.

The $TiCl_4$ leaves the reactor as a hot vapor, in the presence of other vaporized metal chlorides and various inert gases such as carbon monoxide, carbon dioxide and phosgene. In the recovery system, the metal chlorides can be totally condensed forming a slurry of $TiCl_4$ and solid metal chlorides. In this case, $TiCl_4$ is separated from the waste solids by settling and decantation. A partial condensation recovery system can also be used for the first portion of the recovery system. In the partial condensation

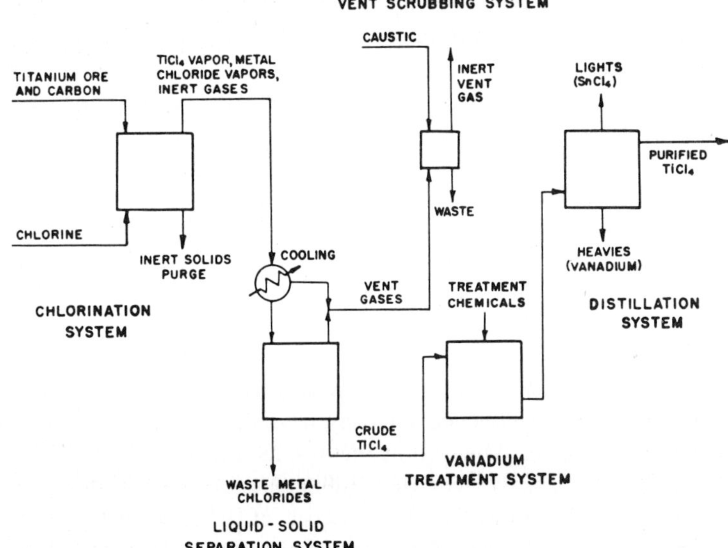

Figure 25-1.   Flow sheet for titanium tetrachloride manufacture.

method, the solid metal chlorides less volatile than $TiCl_4$ are condensed as a powder by means of cooling supplied by vaporizing liquid $TiCl_4$, and with cooling coils. The remaining vapor is greatly enriched in $TiCl_4$ content which can then be totally condensed, giving a liquid relatively free of the metal chloride solids.

The liquid $TiCl_4$ collected in the recovery system contains objectionable quantities of vanadium, silicon, tin and other impurities which must be removed by further treatment consisting of scrubbing with heavy oil or H_2S and distillation in one or more stills. Pure, water-white $TiCl_4$ having the typical impurities shown in Table 25-1 is recovered from the final still.

The ferrous chloride by-product from the chlorination of titanium ores

TABLE 25-1. TYPICAL IMPURITIES IN COMMERCIAL TiCl₄

	Per cent
Iron	.0008 max.
Vanadium	.0008 "
Silicon	.004 "
Tin	.007 "
Carbon	.005 "
Total Residue	.008 "
Boiling range @760 mm	134–138°C

is usually discarded if the ore is rutile. However, in a plant for chlorinating ilmenite or beneficiated ilmenite, the recovery of iron from this residue would be almost essential. Various schemes have been patented for the reduction of the ferrous chloride residue including liquid-phase[36] or vapor-phase reduction with hydrogen,[37] and the oxidation of ferrous chloride to ferric oxide and HCl.

Chlorination

The thermodynamics and kinetics of the chlorination of TiO_2 were studied extensively by Pamfilov and Shtandel.[33] The importance of the addition of carbon is to keep the partial pressure of oxygen at a low value to prevent the reversal of the reaction. Up to 600° C carbon dioxide is formed almost completely, while over 600° C quantities of CO are produced.[4] Maximum efficiency for the reaction is obtained at 850 to 1250° C,[4] which is maintained without the addition of heat. Higher temperatures result in a faster reaction rate and increased yields of $TiCl_4$, but also increase the rate of attack on the reactor lining. Either a fixed or fluid bed can be utilized for the reaction, but in either case, the amount of carbon used is 15 to 35 percent by weight of the ore fed to the system,[4] the exact amount depending on the TiO_2 assay of the ore being chlorinated. In the fluid bed, the particles are sized between 8 and 200 mesh.[25] The finer the fluid bed particles, the faster will be the reaction rate. However, a balance must be struck between the reaction rate and the blow-over rate of very fine particles. For fixed-bed operation, finely divided coke and ore are formed into pillow-block briquettes having the dimensions, $2'' \times 2'' \times 1\frac{1}{4}$ in., as reported in U. S. Patent 2,723,903.

As the reaction progresses, relatively inert materials such as CaO, MgO and their chlorides build up in concentration in the bed.[25] It is necessary to remove these materials to prevent fusion and agglomeration of the solid reactants because of the sticky nature of these inert materials at the temperature at which the reaction is carried out. This is very easily accomplished in a fluid bed by using a continuous purge, while for fixed bed operation, the chlorinator must be periodically shut down and cleaned to remove the accumulated inerts.

Many patents are available describing the production of $TiCl_4$ in both fluid and fixed-bed processes. A representative list is given at the end of this chapter.

USES OF THE TITANIUM CHLORIDES

It is estimated that in 1956, 60,000 tons of $TiCl_4$ was used to manufacture the metal[28] compared to an estimated 25,000 tons in 1959. The decline of $TiCl_4$ used for titanium metal from 1956 to 1959 reflects the reduced consumption of titanium for military aircraft.

TABLE 25-2.   TITANIUM TETRACHLORIDE END USES IN 1959.

Manufacture of Ti	– 25,000 tons (est.)	
Manufacture of TiO_2	– captive use (est. at 75,000 tons)	
In special catalysts	– hundreds of thou- sands of pounds	
Manufacture of $TiCl_3$	– very small tonnage	
Manufacture of alkoxides	– very small tonnage	est. at 1,000 tons
Purification of aluminum	– negligible	
Smokes	– negligible	

Pigment

At present more $TiCl_4$ is used for producing TiO_2 pigment than for producing titanium sponge.   Three general methods of producing TiO_2 from $TiCl_4$ have been used.   In one method, anhydrous $TiCl_4$ is dissolved in water or dilute hydrochloric acid. TiO_2 is then precipitated from the solution by boiling, or by the addition of an alkali.   Another method is to react vaporized $TiCl_4$ with water vapor at 300°C to 400°C to obtain a very finely divided TiO_2.[22]   A third method is the reaction between vaporized anhydrous $TiCl_4$ and dry oxygen or air at 1000°C.   This last method has the advantage of recovering elemental chlorine and avoids the production of extremely corrosive wet HCl.   It is an important method of producing TiO_2 from $TiCl_4$ and will account for greater and greater amounts of $TiCl_4$ usage.   For this reason the oxidation process warrants further discussion.

A highly purified $TiCl_4$ must be used for the direct oxidation compared to the hydrolysis methods, since any impurities present would remain with the product.   There are several patents on the $TiCl_4$ oxidation process, and all are aimed at solving the greatest difficulties, i.e., obtaining control of particle size, obtaining high product quality and obtaining high yields of TiO_2.   A recent patent[47] concerns the details of a burner construction which is quite intricate.   In the process, vaporized $TiCl_4$ and oxygen are contacted in a reaction zone with a flame sustained by a separate inflow of combustible gas such as CO, H_2 or CH_4.   The reaction mixture is added in a number of small jets at a velocity of 5 to 20 meters per second, while the auxiliary combustion gas is added and burned, so its flame completely engulfs each individual jet of reaction mixture in order to sustain the main oxidation.   The purpose of the burner is to permit the reaction mixture to ignite very quickly and uniformly at a temperature of 1000 to 1200°C, which is essential to the formation of properly sized uniform TiO_2

crystals. The velocity of the reaction gas must be held within limits to inhibit formation of agglomerates. A 95 percent yield TiO_2 is obtained.

Another patent describes a process which uses a fluidized bed containing particles of 40 to 60 mesh TiO_2 heated to 1250°C to conduct the oxidation.[41] Again, the prime consideration of the process is particle size control. Another patent[43] describes a process in which $TiCl_3$ is added to the $TiCl_4$—O_2 stream as a nucleating agent to aid in obtaining a rapid reaction rate and thereby better control over particle size.

Titanium Metal

The other major use of $TiCl_4$ is for the production of titanium sponge. The various reduction methods available to obtain titanium metal from $TiCl_4$ are the reduction of $TiCl_4$ with hydrogen, with sodium, with calcium and with magnesium. An important large-scale process for producing titanium sponge has been the magnesium reduction developed by Dr. Wilhelm J. Kroll.[26] The basic chemical reaction for the Kroll process is

$$TiCl_4 + 2\,Mg \rightarrow Ti + 2\,MgCl_2 \tag{25-3}$$

Magnesium is supplied as ingots which are sealed into a vessel having a provision for liquid $TiCl_4$ addition. The entire vessel is placed in a furnace and the magnesium is melted in an inert helium atmosphere. $TiCl_4$ liquid is then dripped on the molten magnesium where it reacts to form titanium and $MgCl_2$. This highly exothermic reaction is controlled at 800 to 900°C by the rate of $TiCl_4$ addition. $MgCl_2$ is also molten at the reaction temperature and, being heavier than molten magnesium, drops to the bottom of the vessel. Solid metallic titanium in the form of a porous sponge is deposited, first on the walls of the vessel and later in a bridge completely covering the magnesium layer. When the batch reduction is completed, the vessel is cooled and the titanium sponge is removed by machining or chipping. The last traces of magnesium and magnesium chloride are removed from the sponge by a high temperature vacuum distillation.

Catalysis

The only other titanium chloride used commercially is $TiCl_3$. It has been used as a catalyst in the production of olefines, alkoxides and plastics. Particular attention has come to the use of $TiCl_3$ for polymerizing propylene. However, because of the expense in obtaining pure $TiCl_3$ it still is not used for this purpose as much as is $TiCl_4$. $TiCl_4$ is used with aluminum triethyl as a catalyst in the Ziegler polypropylene process.[40]

SHIPPING, HANDLING AND SAFETY—TiCl₄

TiCl₄ is classified as a corrosive liquid and is shipped in drums, tank trucks and tank cars in compliance with the Interstate Commerce Commission regulations.

Unloading

Drums are normally unloaded by gravity. While the drums are in a vertical position a suitable unloading valve is fitted to one bung and a Drierite bulb for venting dry air is fitted to the other. The drum is then tipped to a horizontal position for unloading. Tank cars and tank trucks are unloaded through the dome, either by the application of dry air or inert gas pressure or by means of a pump located at ground level. When a pump is used it is still advisable to maintain at least 5 psig pressure on the tanks to keep the pump primed.

On contact with moisture, TiCl₄ reacts to form oxychlorides with the liberation of hydrogen chloride. Consequently, it is imperative that the air or inert gas introduced into the tank have a dew point no greater than –35° F.

At the time of a transfer from the shipping container to a storage tank, vents to the atmosphere require rigid inspection since there is a possibility of plugging by oxychlorides formed by the reaction of the venting vapor and atmospheric moisture. A satisfactory approach is to vent the storage tank back to the shipping container.

Materials of Construction

TiCl₄ may be handled in mild carbon steel equipment under completely anhydrous conditions. Moisture and moist air, however, must be completely avoided. Where maintenance of high quality is desired, it is suggested that the piping system be first flushed with a solvent such as perchloroethylene prior to the dry-air or inert-gas purging. The following construction materials are recommended:

Piping	— all sizes, schedule 40 steel, ASTM A-53.
Flanges	— 150 pound forged steel, weld neck.
Fittings	— all sizes, welded, standard weight.
Union	— flanged.
Valves	— type 304 stainless steel, flanged "Y" pattern with "Teflon" solid chevron ring packing. (Aloyco No. 367Y or equivalent).
Pumps	— packless pump, (Chemco or equivalent).
Gasket material	— Garlock 900, "Teflon," "Kel-F."

Safety

Titanium tetrachloride reacts with moisture to form oxychlorides, oxides and hydrogen chloride with the evolution of considerable heat. The hydrochloric acid formed and the heat generated are the hazardous properties to be considered when handling the material.

Skin contact with $TiCl_4$ should be treated first by the removal of contaminated clothing and the wiping of excess material from the skin. Only after as much of the material as possible has been removed from contact with the skin, should water be applied; then it should be applied in large amounts for several minutes to wash away all residual acid. Eye protection in the form of safety goggles should be used in all areas where $TiCl_4$ is handled. Should the material enter the eye, water should be applied immediately in large quantities. It is only in eye exposure cases that water is applied immediately. Respiratory protection should be the type specified for hydrogen chloride vapors.

In the case of a large-scale spillage, drench the area with copious quantities of water. Small spillages, as might occur in the laboratory, can be handled by spreading sodium carbonate or sodium bicarbonate over the $TiCl_4$. This prevents the escape of hydrogen chloride, and the dry mass can be swept or flushed away with water without fumes.

COMMERCIAL PROCESSES—$ZrCl_4$

Ores

Zircon is the most widely used source of zirconium metal. This ore ($ZrSiO_4$) contains 67.2 percent ZrO_2 and is found in the United States almost entirely in Florida beach sands and in Idaho stream deposits. Practically all of the domestically produced zircon comes from Florida. A large amount of zircon used in the United States is imported from Australia and smaller amounts from India. Baddeleyite, which is zirconium oxide, is imported in small quantities from Brazil.*

Most of the zirconium ores contain $\frac{1}{2}$ to 2 percent hafnium oxide, which is an undesirable impurity in cases where the ore is to be used as a source of reactor grade zirconium via the reduction of $ZrCl_4$. However, where hafnium is to be manufactured as a co-product it is desirable to use ores rich in hafnium. Altered varieties of zircon, such as alvite and cyrtolite contain up to 17 percent hafnium oxide.

*Less than one-quarter of the zircon used in this country goes for the production of zirconium sponge.[28]

Reserves of zirconium minerals, even on the North American Continent, are sufficient for all zirconium metal production in the foreseeable future. Estimates of available ores are fifteen million tons of zircon in the United States; two million tons of baddeleyite in Brazil, plus an undetermined amount of zircon; one million tons of zircon in Ceylon, and five million tons of zircon in India.[15] Zircon reserves in Australia are thought to be high but the actual amount is undetermined.

Ore Treatment

When used for metal production, the most carefully conducted and costly treatment step is the hafnium-zirconium separation. However, several treatment steps are required prior to this separation.

Beach sands are first concentrated by gravity methods and cleaned by magnetic and electrostatic methods. Zircon is a very refractory material both with regard to temperature and chemical digestion, and usually it is treated further before it is chlorinated. In one treatment method used commercially, the zirconium ore is fused at a high temperature with an alkali such as NaOH, and washed with water to remove the excess SiO_2. This mass is treated with sulfuric acid to form soluble $Zr(SO_4)_2$ which is removed from the sand by filtration. Zirconium is precipitated as the hydroxide $[Zr(OH)_4]$ with ammonium hydroxide, washed to remove the sulfates and dried. It is redissolved in nitric acid to form zirconyl nitrate and is then ready for hafnium separation.

Another commercially used treatment process involves heating zircon and coke for 40 hours at 4000° F, using an electric arc furnace. The impurities in this case volatilize leaving zirconium carbide or zirconium oxide. Zirconium carbonitride is also formed by a similar electric arc process. The carbide, oxide or carbonitride is chlorinated to the crude $ZrCl_4$ and $HfCl_4$, and this is followed by the hafnium separation step.

Zircon can also be converted to calcium zirconate which can be readily digested in nitric acid to form the zirconyl nitrate directly.[44] This is done by adding excess lime to zircon and heating to 2900° F where the following reaction takes place:

$$ZrO_2 \cdot SiO_2 + 3\,CaO \rightarrow 2\,CaO \cdot SiO_2 + CaO \cdot ZrO_2 \tag{25-4}$$

This treatment can be aided by the addition of fluorspar to the mixture which lowers the treatment temperature. The reaction is as follows:

$$2(ZrO_2 \cdot SiO_2) + 5\,CaO + CaF_2 \rightarrow 3\,CaO \cdot CaF_2 \cdot 2\,SiO_2 + 2(CaO \cdot ZrO_4) \tag{25-5}$$

The calcium zirconate can be separated by size or by gravity methods. After dissolution in nitric acid or chlorination to $ZrCl_4$, the zirconium material is ready for the hafnium separation step. This step is essential for the production of nuclear-reactor grade zirconium. Even small quantities of hafnium destroy the property of zirconium which makes it useful for reactor construction, which is that of a very low neutron-absorption cross-section. On the other hand, hafnium is sometimes recovered because it has a very high neutron-absorption cross-section which makes it useful as reactor control rods.

A great many separation processes have been described in recent literature. In a method reported by Cox,[10] the solution of zirconium in nitric acid or zirconyl nitrate is extracted by tributyl phosphate diluted with 40 percent (by volume) heptane. The extract of zirconyl nitrate in tributyl phosphate contains less than 100 ppm of hafnium. The zirconium is then precipitated with a base and converted to the oxide. It can be converted to zirconium oxide by redissolving the zirconium precipitate in nitric acid. The resulting zirconyl nitrate is crystallized by flash evaporation of the nitric acid solution. The solid zirconyl nitrate is calcined at 700° C to the oxide.

Zirconyl nitrate can be separated from hafnium by a semi-continuous ion exchange process.[9] The resins, which are the sulfonated polystyrene resins "Dowex 50" and "Zeokarb 225," are first converted into acid form. Zirconyl nitrate is then passed into the first of two columns which contain the two resins in series. There, both zirconium and hafnium are adsorbed. When the resin in tne first column becomes saturated, sulfuric acid is passed through both columns, causing zirconium to elute from the first column. Any hafnium also eluting is caught in the second column. The resins are regenerated (hafnium removal) by 0.5 percent oxalic acid, and reconditioning is done with nitric acid. Silica gel has also been used to separate hafnium and zirconium.[20]

One separation process is based on different rates of chemical reducibility of $ZrCl_4$ and $HfCl_4$[31] with $ZrCl_3$, $ZrCl_2$ or Zr being used as the reducing agents.

Jaywinski and Sisto[24] reacted a mixture of finely divided zirconium and hafnium with CO at 4 to 8 atmospheres and 400° C. The resulting liquid solution of $Zr(CO)_7$ and $Hf(CO)_7$ were separated by distillation. The liquids were reconverted to the metals and CO gas by decomposing the liquids at 50 to 100° C above their respective boiling points.

Another distillation technique is possible if $ZrCl_4$ and $HfCl_4$ are first reacted with $POCl_3$. The reaction products can then be separated in a 50 plate column.[32]

A zirconium-hafnium separation process developed by Carbide and

Carbon Chemicals Corporation at Oak Ridge, Tennessee, was based on the extraction of hafnium thiocyanate in a hexone-thiocyanic acid mixture from an aqueous solution containing zirconium and hafnium.[16,21] This process was used both by Northwest Electrodevelopment Laboratory of Albany, Oregon, and by the Carborundum Metals Corporation of Akron, New York in early zirconium production plants.

In this separation process, zirconium and hafnium in chloride form is fed to an extraction system along with aqueous ammonium thiocyanate. The chloride solution passes countercurrent to a methyl isobutyl ketone and thiocyanic acid mixture, leaving the extraction system with the hafnium removed. The hafnium-free zirconium is then precipitated from solution with sulfuric acid, converted to hydroxide form with ammonium hydroxide and calcined to zirconium oxide. The hafnium-rich solvent is regenerated, first by a sulfuric acid wash to remove metal ions, then by an ammonium hydroxide wash to precipitate thiocyanate ions as ammonium thiocyanate. Both the solvent and the ammonium thiocyanate are recycled to the start of the process.

Chlorination

After the hafnium separation step the zirconium is converted to ZrO_2 for chlorination. For zirconium metal uses (other than for nuclear reactors) the hafnium removal step is not necessary. Especially adapted to this latter application is $ZrCl_4$ obtained from a direct fluid bed chlorination of zircon sand (described in British Patent 759,724). The reaction is as follows:

$$ZrSiO_4 + 4\,Cl_2 + 4C \rightarrow ZrCl_4 + SiCl_4 + 4\,CO \qquad (25\text{-}6)$$

The usual method of $ZrCl_4$ production, however, is the chlorination of hafnium-free ZrO_2 according to the following reaction:

$$ZrO_2 + C + 2Cl_2 \rightarrow ZrCl_4 + CO_2 \qquad (25\text{-}7)$$

The carbon is necessary for the chlorination of ZrO_2 because at no observable temperature does chlorine alone attack zirconium dioxide.[30] In the presence of carbon the chlorination can occur at as low a temperature as 450° F, but higher temperatures give faster reaction rates. Chlorination can be carried out in a fixed bed or fluid bed type reactor. In the fixed bed process,[16] zirconium oxide, carbon, sugar and water are formed into briquettes in a Belgian-roll type briquetting machine. The briquettes, about 1½ in. long by 1 in. wide by ¾ in. thick, are added to the chlorinator continuously by a solids feed mechanism. The bed is brought to temperature by passing an electric current through carbon electrodes imbedded in the briquettes. When the bed is heated sufficiently, it becomes conductive to the passage of current and is heated by its own resistance which supplies heat for the reaction. The chlorinators are operated at a bed temperature of 600 to 800° C.

The chlorination of zirconium carbide or cyanonitride is done conventionally in a packed tower with water cooled walls. A coke fire is started at the bottom of the tower, then when the ore is hot, chlorine is passed through the tower where it enflames. The heat of combustion of the chlorine and ore is sufficient to keep the reaction going without providing heat from an external source. The reaction is

$$ZrC + 2 Cl_2 \rightarrow ZrCl_4 + C \qquad\qquad (25\text{-}8)$$

In all cases, the $ZrCl_4$ vapor is condensed directly to the powder which sublimes at 331° F.

Purification

$ZrCl_4$ can be purified by sublimation[7] which is effective for removal of oxygen compounds and ferrous chloride. However, for ferric chloride separation the sublimation must be done in the presence of a suitable reducing agent such as hydrogen, metallic zinc, cadmium or manganese. This reduces the ferric chloride to ferrous chloride which is much less volatile and easily separated by sublimation.[27]

Another method of purifying the $ZrCl_4$ is to mix it with motor oil which has a carbonizing temperature below the sublimation temperature of $ZrCl_4$. The $ZrCl_4$ heated and sublimed from this mixture is recovered in a highly purified state.[18]

$ZrCl_4$ also can be purified by dissolving it in molten alkali chlorozirconates. The resulting solution is then heated to 500 to 600° F in order to boil out very pure $ZrCl_4$.[23]

Hafnium Tetrachloride

Hafnium tetrachloride can be produced in a manner similar to $ZrCl_4$. The hafnium from the hafnium-zirconium separation step is obtained as a hydroxide filter cake. This is calcined to the oxide at 800° C. The oxide is then briquetted with sugar and carbon in a Belgian-roll type machine, much the same way as the ZrO is briquetted. The hafnium briquettes are 85.5 percent HfO_2, 9.5 percent carbon black and 5 percent sucrose.

The chlorinator is essentially the same as that used for the production of anhydrous $ZrCl_4$. The bed is heated by its own resistance to passage of current (600 amps at 17 volts).[17] HfO_2 briquettes and chlorine are added continuously to the bed while $HfCl_4$ powder is condensed continuously in the overhead gases.

The $HfCl_4$ powder is purified by sublimation in an inert atmosphere which removes most of the objectionable impurities.

USES OF $ZrCl_4$

The $ZrCl_4$ tonnage produced can be estimated at 5600 tons in 1958 and 1959, based on the Bureau of Mines Mineral Market Report No. 2988. This tonnage was used almost exclusively for the manufacture of zirconium sponge. The production of hafnium sponge is not available but this is the exclusive end product of hafnium tetrachloride.

Various methods have been explored to produce zirconium sponge from the tetrachloride. These include reduction with alkali and alkali-earth metals, with magnesium, calcium and sodium—the only ones found to be feasible. High purity calcium is not obtained easily and so reduction with sodium or magnesium remains as the economically possible route with alkali metals. An electrolytic method that produces zirconium powder has been proposed.[27] Reduction with hydrogen at high temperatures has been studied which results in finely divided metal; however, large amounts of pure hydrogen are needed. A method being successfully used commercially to produce zirconium sponge is the Kroll process, which uses magnesium as the reductant.

In the Kroll process the reduction of $ZrCl_4$ to sponge metal proceeds in identical equipment whether the product is to be low hafnium grade zirconium or commercial grade metal containing 2 to 2.5 percent hafnium.

Reduction of the $ZrCl_4$ to sponge metal is carried out in closed retorts, 40 in. ID by 100 in. high, constructed of 310 stainless steel. The retorts are contained in vertical electric furnaces provided with three separately controlled heating zones. The furnace is used for both the purification and reduction steps. In charging the furnace, the required amount of magnesium is placed in a crucible, constructed of type 430 stainless steel. The crucible rests on the bottom of the retort. A batch of $ZrCl_4$ contained in an inconel container is placed on a baffle on top of the magnesium crucible. The retort is evacuated and backfilled with helium. A rather lengthy procedure is then very carefully followed to remove volatile impurities and moisture from the $ZrCl_4$ charge. When the purification cycle is completed after 20 to 30 hours, the temperature of the lower zone is raised to melt the magnesium. When the magnesium is molten, the temperature of the zone around the chloride can is raised to sublime the $ZrCl_4$. The vaporized $ZrCl_4$ reacts with molten magnesium to form the zirconium sponge. The pressure inside the retort is controlled by regulating the temperature of a condensing coil above the $ZrCl_4$ container. When the reaction is completed the reduction crucible contains zirconium sponge, magnesium chloride and unreacted magnesium. The magnesium and magnesium chloride are separated from the sponge by melting and vacuum distillation in separate distillation equipment. The vacuum distillation is done at 960° C and 0.05 microns absolute pressure. After final purification the metal sponge is broken out of the reduction crucible by a hydraulically operated power chisel. The sponge is then crushed and shipped to the point of electric arc melting where the zirconium ingots are produced.

Hafnium metal sponge is made the same way in equipment identical to that used for zirconium sponge.

SAFETY, HANDLING AND SHIPPING

Shipping

$ZrCl_4$ is classified as a corrosive chemical and is shipped in full open-top, rubber-gasketed steel drums, or in rubber hoppers in compliance with ICC Regulations. The drums are unloaded by mechanically tipping them over and emptying the $ZrCl_4$ dust into hoppers. $ZrCl_4$ is a non-free flowing material, and some bumping of the drums is required.

Safety

$ZrCl_4$ reacts with moisture to form oxychlorides, oxides and hydrogen chloride, with evolution of considerable heat. As with $TiCl_4$, the hydrochloric acid and heat generated are the hazardous properties to be considered when handling $ZrCl_4$.

Skin contact with $ZrCl_4$ is treated as is skin contact with $TiCl_4$. First, all contaminated clothing is removed and excess material is wiped from the skin. Only then should water be applied. Eye protection in the form of safety goggles should be used in all areas where $ZrCl_4$ is handled. If material enters the eye, water should be applied immediately and in large quantities. Respiratory protection should be the type specified for hydrogen chloride vapors. Dust masks should be used if necessary to avoid inhalation of the dust. However, any contact where inhalation is a possibility should be avoided.

Handling

Zirconium metal is a powerful getter for carbon, oxygen, nitrogen, hydrogen or carbon dioxide. Because of this property zirconium metal can be a hazardous material. For the same reason application has been found for it in the elimination of residual traces of gases in vacuum tubes. Massive zirconium metal appears not to be active chemically, but this is due to an impervious, invisible, tough, adherent coating of oxide which forms on the metal when it is exposed to air. The metal itself is actually quite active as shown by the fact that dry colloidal zirconium or finely pulverized zirconium will burn in air, nitrogen or CO_2. Dust clouds of zirconium powder in air are explosive; thus zirconium powder is usually shipped in water to avoid fire and explosion hazards. Zirconium acts differently when finely divided than when in the massive form because the surface reaction proceeds at a velocity proportional to the specific surface. At the higher reaction velocity, the evolution of heat is more rapid, and a higher peak temperature is obtained. At the higher peak temperature, there is a qualitatively different behavior than at lower temperatures.[5]

Zirconium sponge can be expected to act in an intermediate fashion between powdered and massive zirconium. However, commercial zirconium sponge is made passive by a controlled amount of air to the retort containing freshly vacuum-distilled sponge; for this reason the hazards of handling zirconium sponge are essentially eliminated. Sponge is normally shipped in a steel pail mounted in a wooden box. The air in the shipping container is displaced with a noble gas such as argon to maintain the original quality of the sponge during long periods of storage.

Zirconium and its insoluble salts appear to have no effects on the human body.[6]

PATENT REVIEW

Ore Beneficiation Patents

U. S. Patents	2,557,455		2,721,793
	2,663,633		2,751,307
	2,664,352		2,752,300
	2,375,268		2,476,453
	2,480,184	German Patent	704,506
	2,680,681	Australian Patent	22,815

TiCl₄ Purification Patents

U. S. Patents	2,344,319		2,560,424
	2,370,525		2,598,898
	2,396,458		2,457,917
	2,412,349		2,512,807
	2,416,191		2,306,184
	2,600,881		2,594,370
	2,543,591		2,530,735
	2,555,361		2,836,547
	2,560,423	British Patent	674,315

Production of TiCl₄ Patents

U. S. Patents	2,681,855		2,723,903
	2,486,912		2,784,058
	2,401,543		2,790,703
	2,401,544		2,701,180
	2,589,466	British Patent	716,681
	2,622,006		

References

1. Barksdale, J., "Titanium, Its Occurrence, Chemistry and Technology," New York, The Ronald Press Company, 1949.
2. *Ibid.*, pp. 15–34.
3. *Ibid.*, p. 321.
4. *Ibid.*, pp. 311–318.
5. Blumenthal, W. B., "The Chemical Behavior of Zirconium," p. 26, Princeton, New Jersey, D. Van Nostrand Company, Inc., 1958.
6. *Ibid.*, pp. 39–45.
7. *Ibid.*, p. 105.
8. Brassert, H. A., U. S. Patent 2,313,044.
9. Bryan, Lister, Duncan, U. S. Patent 2,759,792.
10. Cox, R. P., *Iowa State Coll. J. Sci.*, **31**, 383 (1954).
11. Dawson, Krchma, and McKinney, U. S. Patents 2,127,247 and 2,476,453.
12. Donaldson, K. H., U. S. Patent 2,210,602.
13. Ebelman, J. J., *Am. Chem. Phys.*, **20**(3), 385 (1847).
14. Fetterolf, and McFarland, *Chem. Eng. Progress*, **56**, (5) 768 (1960).
15. Finniston, H. M., and Howe, J. P., Editors, "Progress in Nuclear Energy Series V., Metallurgy and Fuels," New York, McGraw-Hill Book Company, Inc., 1956.
16. *Ibid.*, p. 318.
17. *Ibid.*, p. 337.
18. Frey, Walter, U. S. Patent 2,682,445.
19. Friedel, C., and Guerin, *J. Ann. Chem. Phys.*, **8** (5), 24 (1876).
20. Giffen, R. H., *Iowa State Coll. J. Sci.*, **31**, 422 (1957).
21. Grimes, Barton, Overholser, Blakely, and Redman, United States Atomic Energy Commission, *Y-560*, 64 pages, (1950).
22. Haber, H., and Keebelka, P., U. S. Patents 1,931,380 and 1,931,381.
23. Horrigan, R. V., *J. Metals, 7, Trans. AIME*, 203, 1118 (1955) C. F. British Patent 771,144.
24. Jaywinski, and Sisto, U. S. Patent 2,793,107.
25. Krchma, I. J., U. S. Patent 2,701,180.
26. Kroll, W. J., U. S. Patent 2,205,854.
27. Lindblad, and Pyk, U. S. Patent 2,618,531.
28. "Minerals Yearbook 1956," pp. 1202, Washington, D. C., Bureau of Mines, Department of Interior.
29. *Ibid.*, p. 1364.

30. Morosov, and Korshunov, *Khim, Redkikh, Elementov, Akod, Nouk. S.S.R. Inst. Obshckei i Neorg. Khim.*, (2), p. 102, 1955.

31. Newnham, I. E., U. S. Patent 2,791,485.

32. Niselson, L. A. *Zhur., Neorg. Khim.*, **1,** 2657 (1956).

33. Pamfilov, and Shtandel, *J. Gen. Chem. U.S.S.R.*, **7,** p. 258 (1937).

34. Parker, A. E., and Parker, A. H., *Phys. Rev.*, **47,** 812, (1935).

35. Pechukas, Alphonse, U. S. Patent 2,289,327.

36. Pechukas, Alphonse, U. S. Patent 2,664,352.

37. Pechukas, Alphonse, U. S. Patent 2,663,633.

38. Pechukas, Alphonse, U. S. Patent 2,370,525.

39. "Petrochemical Handbook," *Petroleum Refiner*, **38** (11), 291, (1959).

40. Pierce, Lehighton, Waring, and Fetterolf, U. S. Patent 2,476,453.

41. Richmond, and Bristow, U. S. Patent 2,760,846.

42. Rossini, F. D., Cowie, P. A. Elbson, F. O., Browne, C. C., "Properties of Titanium Compounds and Related Substances," ONR Report ACR-17, U. S. Office of Naval Research, Department of the Navy.

43. Schauman, and Krchma, U. S. Patent 2,691,571.

44. Schoenlaub, Robert A., U. S. Patent 2,721,117.

45. Skinner, G., Johnson, H. L., and Beckett, C., "Titanium and Its Compounds," Columbus, Ohio, Herrick L. Johnson Enterprises, 1954.

46. Vigoroux, E. V., and Arrivant, G. A., *Comp. rend.*, **1907,** 144,485; *Chem. Abs.*, **1907** (1), 1362.

47. Weber, Mutteny, and Frey, U. S. Patent 2,635,946.

26. VINYL CHLORIDE

Anita S. Kastens

Consultant, Union Carbide Chemicals Company

Few basic chemicals can match the magnificent growth performance of that most interesting of chlorine derivatives, vinyl chloride. This monomer is by far the most important starting material of all the vinyl-type resins. Over 900 million* pounds of PVC† were consumed in 1960, compared to less than 400 million pounds just six years earlier in 1954.[5] These materials found their way into hundreds of varied industrial and consumer products, from raincoats and wire insulation, through doll heads and floor tiles, to paints and phonograph records. New developments in resin processing techniques and plastic end uses keep this old work horse intermediate as frisky as a fine chemical with a million pound per year future.

Many thousands of words have been written to describe vinyl chloride, its polymers and copolymers, their many processing techniques, and the endless plastic applications.in which these versatile materials are put to use. One book[90] among others[62,82,100] is outstanding: "Vinyl Resins" by W. Mayo Smith. Excellent review articles and rundowns on the state of the art appear in each January issue of "Modern Plastics" magazine and in the "Modern Plastics Encyclopedia" issued every September by the editors of the publication (since 1942). Other references of general interest include trade publications such as "Bakelite Review," "Monsanto Magazine," and "Dow Diamond."

Vinyl chloride is of commercial interest only as a raw material for its polymer and copolymer derivatives. It is a relatively nontoxic, sweet smelling, and flammable gas at room temperature. It is insoluble in water, but soluble in carbon tetrachloride, ether, ethyl alcohol, and most organic solvents. Its physical properties are listed in Table 26-1.

Chemically, vinyl chloride is a stable compound with little tendency to break down by cracking or hydrolysis. 2,3-Dichloro-1,3-butadiene may be

---

*Actual U. S. Tariff Commission production figure for 1960: 935,508,000 pounds: sales figure: 900,431,000 pounds.

†"PVC" is a term used to designate polymers of vinyl chloride and copolymers that contain over 50 percent of vinyl chloride.

TABLE 26-1. PHYSICAL PROPERTIES OF VINYL CHLORIDE

$$CH_2 =\!\!= CHCl$$

(chloroethylene; chloroethene)

Boiling point, °C	− 13.4
Density at 20°C	0.911
at 25°C	0.901
Flammable limits, % by volume in air	4–22
Flash point (open cup), °C	− 18
Index of refraction at 15°C	1.374
Latent heat of vaporization at the boiling point, Kcal/g mole	5.32
Molecular weight	62.50
Specific gravity, 20°C/20	0.9121
Viscosity at − 20°C, cps	0.3

prepared, however, by heating vinyl chloride at 450 to 650°C for 2 to 10 seconds, followed by rapid cooling.[30] The chlorine atom of vinyl chloride is essentially inactive, but the double bond reactions are normal; for example chlorination of vinyl chloride leads to 1,1,2-trichloroethane. Vinyl chloride is interesting chemically only because of its marked tendency to polymerize with itself or to copolymerize with other monomers. Because these reactions are exothermic and make for slightly hazardous conditions, vinyl chloride must be handled with caution and reacted with adequate safety precautions.

HISTORY

Vinyl chloride was first prepared in 1835 by Regnault.[74] He made it by mixing ethylene dichloride—"the oil of the Dutch chemists"—with an alcoholic solution of potassium hydroxide. Three years later he observed the formation of a white powder when sealed tubes of vinyl chloride stood in the sunlight.[75] In 1872 Baumann[12] polymerized vinyl halides to white solid masses "unaffected by solvents or acids." Ostromislensky[65] was the first to investigate such polymers for commercial exploitation; in 1912 he was issued patents[64] on the production of rubberlike masses from polymerized vinyl halides. The literature before 1930 describes many vinyl polymers, special methods of processing them, and their applications.[1,35,37,42,45,55,66,72,95]

It was not until 1927, however, that resins based on vinyl chloride were experimentally offered, and not until 1933 that they were made in commercial quantities. During that period Carbide and Carbon Chemicals Corporation (now Union Carbide Chemicals Company) offered its first series of "Vinylite" resins.[99] "Vinylite" A series were polymers of vinyl acetate;

"Vinyline" V series were copolymers of vinyl chloride and vinyl acetate; and "Vinylite" Q series were polymers of vinyl chloride. The copolymer resins with vinyl acetate were initially promoted because they were more readily soluble in the common solvents, and accordingly more easily processed for such early uses as can coating (Keg-lined beer cans). With increased processing and plasticizer know-how, however, polyvinyl chloride itself has long since outrun the copolymer derivatives as the major vinyl resin. However, interest in copolymers is again increasing because of certain special performance properties.

In general, vinyl resins have achieved their remarkable growth because of outstanding chemical resistance, ease of processing, mechanical durability, and wide range of color-ability and pattern formation. The widespread potential of vinyl plastics became apparent during World War II when the military consumed almost all resins produced. Because much of the vinyl resin is processed on calenders and molds used in the rubber industry or adapted from such equipment, it was only natural that rubber companies should early turn from processors of vinyl to producers of vinyl (1940). After the War the field was wide open, with new producers as well as old striving vainly to keep production in step with ever increasing demand. Small plants available as package units enticed still more manufacturers to enter the field in the 1950's until now in 1961 there are 24 major suppliers of vinyl chloride and/or polymer[4] (see Table 26-2).

Vinyl chloride resins were well established products in Germany before the War, and much production know-how was gained from studies of the German monomer and polymer operations.[21] This information was utilized freely in the establishment of vinyl resin industries in all the major countries of Europe and in Japan.

PRODUCTION

Monomer Production

Vinyl chloride is usually made commercially from either acetylene or ethylene or a combination of both depending on economics of the specific location.

The U. S. production of ethylene has grown rapidly since 1943 to a 1962 estimated consumption figure of 3.949 billion pounds of which about 2 billion pounds will be produced as merchant ethylene on the Gulf Coast between Sweeney, Texas, and Baton Rouge, Louisiana.

The Gulf Coast merchant price over the past 20 years has shown surprisingly little variation, reaching a peak in 1948 and declining slowly over the past five years.

TABLE 26-2. MAJOR U.S. VINYL CHLORIDE AND/OR PVC PRODUCERS

(1960–61 capacities; est. million pounds/year)[4]

Producer	Monomer Plant	Capacity	Polymer Plant	Capacity
Union Carbide	So. Charleston, W.Va. Texas City, Tex.	260	So. Charleston, W.Va. Texas City, Tex.	260
Goodrich Chemical	Niagara Falls, N.Y. Louisville, Ky. Calvert City, Ky.	240	Niagara Falls, N.Y. Louisville, Ky. Avon Lake, O. Watson, Calif.	260
Dow	Freeport, Tex. Midland, Mich. Plaquemine, La.	200	Freeport, Tex. Midland, Mich.	40
Allied Chemical	Moundsville, W.Va.	150	–	–
Monsanto	Texas City, Tex.	130	Springfield, Mass.	125
Ethyl Corp.	Baton Rouge, La.	150	–	–
	Houston, Tex.	60		
U.S. Rubber	Painesville, O.	70	Painesville, O.	65
Cumberland	Calvert City, Ky.	60	Calvert City, Ky.	55
Diamond Alkali	Deer Park, Tex.	50	Deer Park, Tex.	60
Goodyear	Niagara Falls, N.Y.	45	Niagara Falls, N.Y.	40
General Tire	Ashtabula, O.	30	Ashtabula, O.	45
American Chemical	Watson, Calif.	25	–	–
Firestone Tire	–	–	Pottstown, Pa.	80
Borden Chemical	–	–	Leominster, Mass.	40
			Illiopolis, Ill.	40
Thompson Chemical	–	–	Hebronville, Mass.	75
The Pantasote Co.	–	–	Passaic, N.J.	45
Escambia Chemical	–	–	Pensacola, Fla.	40
Cary Chemical	–	–	Flemington, N.J.	35
Atlantic Tubing and Rubber	–	–	Cranston, R.I.	25
Great American Plastics	–	–	Leominster, Mass.	15
Presto Plastics	–	–	Brooklyn, N.Y.	15
Insular Chemical	–	–	Hicksville, N.Y.	10
Keysor Chemical	–	–	Los Angeles, Calif.	6
Minnesota Mining	–	–	St. Paul, Minn.	5
Totals		1,470		1,381

Year	U. S. Production Billions of Lbs	Commercial 98% grade Gulf Price Ave. ¢ Lb
1944	0.25	4.4
1948	0.35	6.0
1950	1.3	4.25
1952	1.8	4.5
1956	3.2	5.2
1960	3.7	5.0
1962 est.	3.95	4.85

The U. S. acetylene production and price are in sharp contrast to that of ethylene.

Year	U. S. Production Billions of Lbs	Ave. Northeastern Pipe Line Price ¢ Lb
1944	0.378	7.0
1948	0.412	7.3
1952	0.412	7.3
1956	0.620	9.5
1960	0.860	14.0

A new factor is developing in the acetylene picture and it is evident that thermal cracking of hydrocarbons will materially change both the price and location of U. S. acetylene production. Some recently published figures are[81]A

Carbide based acetylene capacity tons	500,000	550,000
Hydrocarbon based acetylene capacity tons	150,000	350,000
Total capacity tons	650,000	900,000

A great deal of interest has been shown in alternate routes to vinyl chloride monomer in the past 30 years and producers are still looking for lower cost routes.

The acetylene process[15,16,41,43,44,57,58,63,69,71,81,92,97,106] may be accomplished either vapor phase or liquid phase over suitable catalysts. The vapor phase process calls for passing a mixture of dry purified acetylene and an excess of anhydrous hydrogen chloride over a bed of catalyst, which is usually mercuric chloride impregnated on activated charcoal pellets. The reaction temperature is maintained between 100 and 250° C. The vinyl chloride is recovered by condensation and distillation and may be stabilized with a small amount of phenol. Unreacted gases are recovered for reuse, and ethylidene chloride is recovered as a minor by-product.

The acetylene process is simply represented by the equation:

$$CH\equiv CH + HCl \rightarrow CH_2=CHCl$$

Each ton of vinyl chloride produced requires 880 pounds of acetylene and 1200 pounds of anhydrous hydrogen chloride. Yields are reported as 80 to 85 percent.[32]

The following description outlines the major features of Naugatuck Chemical's process for making vinyl chloride from acetylene[61] (Figure 26-1):

"Hydrogen chloride gas with a water content of less than 0.2 weight percent is mixed with dry acetylene in a vapor blender to react in equal molar quantities. The hydrogen chloride and acetylene must be dried to avoid excessive corrosion and to hold down the quantity of acetaldehyde formed as a by-product. From the vapor blender the gas mixture is fed to multi-tube reactors. The reactors are connected in banks for parallel flow of gases. Each tube is packed with activated carbon pellets impregnated with a catalyst containing mercuric chloride.

"The reaction must be initiated by heat supplied to the reaction mass, using a circulating heating medium in the shell side of the reactor. Once the reaction is started, it is highly exothermic and the circulating medium used to supply heat is now used as a coolant.

"The temperature in the reactor is controlled to about 200° F. The exact temperature is dependent upon the catalyst age and physical condition.

"Gases from the reactor contain the product vinyl chloride, as well as by-products such as ethylidene dichloride and aldehydes. Small amounts of unreacted hydrogen chloride and acetylene are also present. These gases are cooled before being fed into a combination stripping and fractionating operation. In this operation the unreacted acetylene and hydrogen chloride are removed from the crude vinyl chloride.

"In a fractionating column the crude vinyl chloride is then purified and taken as overhead product. The ethylidene dichloride and the aldehydes are removed from the bottom of this tower. These by-products then go to a pot still, where the residual vinyl chloride is removed from the by-products.

"The pure vinyl chloride product goes overhead and flows to stainless steel storage tanks. A small amount of phenol or tertiary butyl catechol may be added as polymerization inhibitor."

In the manufacture of vinyl chloride from ethylene,[23,18] ethylene is first chlorinated to ethylene dichloride. This in turn is passed through a heated tube which may contain contact catalysts such as pumice or charcoal packed in stainless steel tubes mounted in a cracking furnace. Pressure is 50 psig., temperatures may be as low as 400° C and as high as 950° C, depending on other reaction conditions and the catalyst used. Hydrogen chloride is recovered as a by-product. Yields run about 95 percent. Each ton of vinyl chloride produced requires 3300 pounds of ethylene dichloride, whereas each ton of ethylene dichloride requires on the order of 630 pounds of ethylene and 1600 pounds of chlorine.[32]

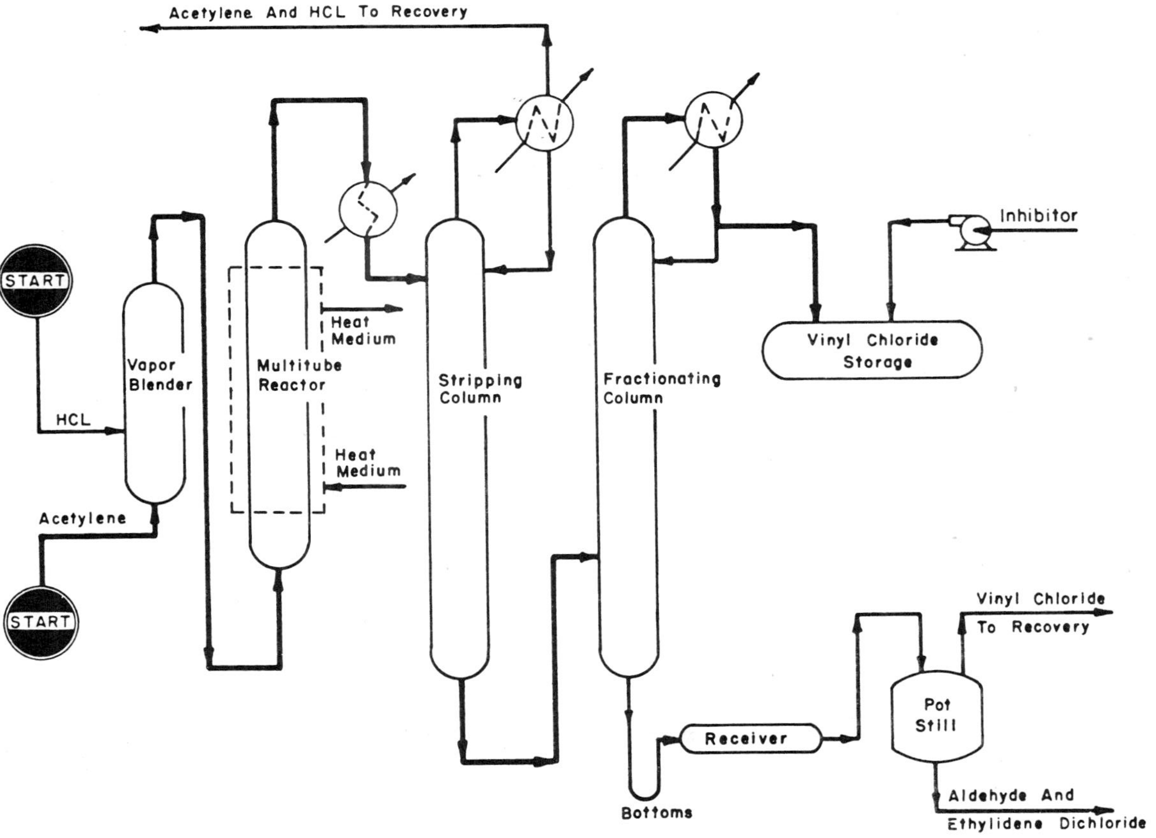

Figure 26-1. Vinyl chloride monomer from acetylene.

Scientific Design's process for making vinyl chloride from ethylene (Figure 26-2) is described as follows:[80]

"Ethylene and chlorine gases are charged to a reactor system containing catalyst and an excess of ethylene dichloride as a solvent for the feed gases. The addition reaction is extremely rapid and goes essentially to completion without any appreciable formation of higher chlorinated compounds.

"The recovered crude ethylene dichloride is carefully purified in a distillation train of two columns. The first column removes traces of unreacted ethylene and any ethane or other low boiling inert materials present in the feed stocks. The second column removes traces of by-product higher-boiling chlorinated compounds which are only created to a small extent in the efficient reaction system, but are present in the recycled dichloroethane.

"The purified ethylene dichloride is then cracked at high temperature in direct fired furnaces to yield a crude mixture containing principally hydrogen chloride, vinyl chloride and unreacted ethylene dichloride. The effluent gases are quenched rapidly for maximum recovery of the desired products. Crude hydrogen chloride gas is discharged to an aqueous scrubber-stripper system for purification. Liquid cracked products are refined in a distillation train which includes two columns—one to remove light ends and the other to remove unreacted ethylene dichloride and other heavy ends formed in the pyrolysis."

Procedures for making both vinyl chloride and vinylidine chloride by the chlorination of ethane are described in the patent literature with patents assigned to both Dow Chemical Company and Ethyl Corporation.[20A,25A]

Ethane is reacted at high temperature in vapor phase with from one to three moles of chlorine in the presence of an inert diluent gas to give a mixture of vinyl and vinylidine chloride. Economics of this reaction depend on an outlet for hydrogen chloride and vinylidine chloride. Some interesting monomer process combinations can be predicted from the use of this process with other more conventional procedures.

Another process for making vinyl chloride through ethylene dichloride[13,14,46]

$$CH_2ClCH_2Cl + NaOH \rightarrow CH_2{=}CHCl + NaCl + H_2O$$

is not commercially economical because of the conversion of one chlorine atom into simple salt.

The particular route any given manufacturer chooses to take to vinyl chloride is increasingly dependent on the economics that govern the ready availability, purity, and price of the basic raw materials and, in the case of ethylene, the easy disposition of the by-product HCl formed. Large-scale producers such as Union Carbide, Monsanto, and Goodrich use both acetylene and ethylene as starting materials. By-product HCl from the ethylene units goes over to the acetylene units for production of more vinyl monomer.[101] Smaller producers rely on neighbors or co-owners to supply raw materials and consume HCl. For example, American Chemical (jointly owned by Stauffer Chemical and Richfield Oil) chlorinates (with Stauffer

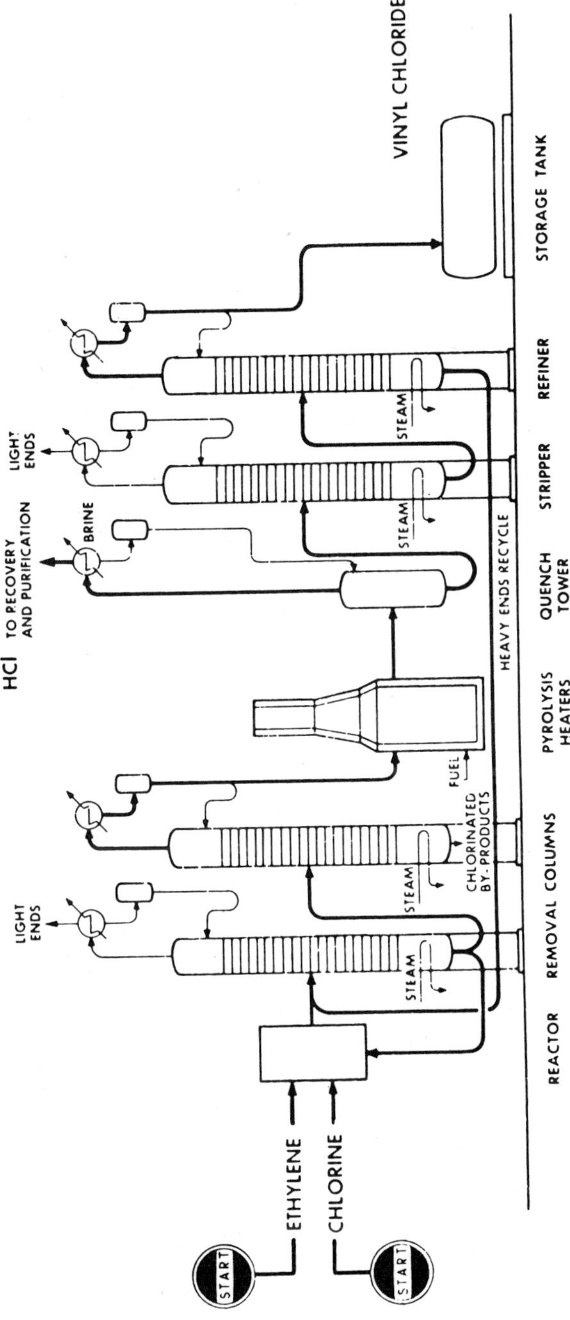

Figure 26-2. Vinyl chloride monomer from ethylene.

chlorine) Richfield's C_2 stream and cracks the resulting ethylene dichloride to monomer and HCl. The by-product HCl is then reacted with ethylene for production of ethyl chloride for conversion into tetraethyl lead. Solvay Process utilizes the excess of HCl produced as a by-product of its chloromethane operations to react with acetylene from an adjacent Union Carbide plant.

Polymer Production

The polymerization of vinyl chloride may be represented by the equation:

$$n{-}CH_2{=}CHCl \longrightarrow \left[\begin{array}{cccccccc} H & H & H & H & H & H & H & H \\ | & | & | & | & | & | & | & | \\ -C & -C & -C & -C & -C & -C & -C & -C- \\ | & | & | & | & | & | & | & | \\ H & Cl & H & Cl & H & Cl & H & Cl \end{array} \right]_n$$

Heat, light, or catalytic agents initiate the reaction. The presence of impurities such as aldehydes, ketones, and oxygen inhibit the reaction.[89] Staudinger's work[93] indicates that the polymers have a simple "head to tail" construction.[34,35] Commercial resins have molecular weights in the general range 50,000 to 150,000.[94]

The commercial polymerization of vinyl chloride to polyvinyl chloride is effected by suspension, emulsion, solution, or bulk techniques.[2,28,36,51,68,86,87,104] Each process yields resins that have particular properties favorable for specific end uses or processing techniques. Bulk polymerization,[26] which yields a particle easily susceptible to plasticization, is used commercially by only one American manufacturer, Union Carbide. Likewise, solution polymerization has had limited industrial utilization and is used on a large scale only by Union Carbide for the production of copolymer resins; the copolymers have excellent properties for solution coating applications. Suspension polymerization is used in by far the greatest volume to produce resins for general use. Emulsion polymerization is relatively new in commercial use, but new plastisol and organosol applications are consuming ever increasing quantities of emulsion polymer.

Suspension Polymerization. Suspension PVC is made in water from which the discrete particles of polymer formed can be easily separated. Reaction temperature, ratio of water to monomer, choice of the suspending agent and catalyst, and possible inclusion of emulsifiers and buffers are variable items.[9,10,40,50,52,59,67,102,103]

Vinyl chloride is fed into water in a reaction vessel, details of which vary considerably both as to size and design. The water-monomer system is vigorously agitated, and the suspending agent is added to stabilize the droplets formed. Catalyst and miscellaneous additives are added last. The reaction temperature is usually kept below 100°C and is set to control the molecular weight and, hence, the physical properties of the resin. The polymerization is complete within 24 hours, although conversion is slow to start and reaches a maximum only after six hours. Residual monomer is recovered and stripped for reuse; end product polymer is washed (usually), dried in rotary driers, and screened.

Gelatin, polyvinyl alcohol, and methyl cellulose are the most commonly used suspending agents. Oil-soluble catalysts include dibenzoyl peroxide, acetyl benzoyl peroxide, and lauroyl peroxide.

Most PVC resin is produced by the suspension method because this process yields a large free-flowing particle that can be readily washed free of major impurities and processes well in conventional types of equipment such as mills, extruders, and calenders. The average particle size varies from 50 to 500 microns with about 150 microns being usual.[7]

Emulsion Polymerization. Again, variations in a basic method are employed for production of emulsion (or dispersion) resins.[17,20,22,24,25,31,38,47,48,49,53,70,73,78,83] Vinyl chloride is added to a vigorously stirred solution of water and emulsifying agent capable of stabilizing both the original emulsion of monomer droplets and the final fine dispersion of solid polymer particles. Protective colloids are added with various inorganic salts that act as buffers to maintain the pH at 8.5. Water-soluble catalysts are used. The temperature is maintained between 50 and 90°C. Unreacted monomer is stripped from the reaction products and the polymer latex spray dried.

Anionic surfactants such as sulfonic and sulfonated hydrocarbons, alkyl naphthalene sulfonic acids, and sulfonated castor oil, are the classical emulsifiers, but they are being displaced by nonionic surface active agents, which minimize the problems in hard water areas. The protective colloids include such materials as polyvinyl alcohol, hydroxyethyl cellulose, and carboxymethyl cellulose. Many catalysts are suitable, including hydrogen peroxide, alkaline persulfates, and calcium and barium peroxides.

Emulsion polymerization yields particles that will disperse in liquid plasticizers to form fluid dispersions called plastisols, or organosols if the dispersions are thinned with other organic liquids such as hydrocarbon diluents or ketone solvents. The average particle size is 0.1 to 2.0 microns.[7] Because the entire polymerizate is recovered the polymers contain the surfactants, catalysts, buffers, etc., used in the polymerization; hence, these resins are used only in end products whose performance is unaffected by these non-polymeric ingredients.

Copolymer Production

Copolymers[19,27,76,77,105] of vinyl chloride and other monomers can be produced by the usual suspension[11] and emulsion[8,84,96] techniques, as well as by bulk or solvent polymerization. Vinyl chloride copolymers tend to have lower softening points and are more soluble than the homopolymers. So-called internally plasticized vinyls, which are merely vinyl chloride copolymers containing minor percentages of other monomers, take advantage of these characteristics. The resins are processed at lower temperatures and dissolve more readily in plasticizers than the pure polymer resins. Vinyl acetate,[29] vinylidene chloride,[79] and acrylonitrile[88] are co-monomers of commercial importance. Copolymers with styrene[98] find widespread application in Germany in floor coverings.

PVC AND VINYL CHLORIDE COPOLYMER RESINS

So many types, grades, and categories of vinyl chloride resins are produced here and abroad that listing them serves no useful purpose. Familiar trade-names include "Bakelite," "Geon," "Opalon," "Exon," and "Marvinol." Physical properties of "Bakelite" vinyl resins (the outgrowth of the original "Vinylite" resins) are listed in Table 26-3 as representative of the various resins available in a well-balanced series.

TABLE 26-3. PHYSICAL PROPERTIES OF "BAKELITE" VINYL RESINS

For Extrusion

Specific gravity	1.20–1.45
Tensile strength, psi	1,100–4,000
Ultimate elongation, %	160–330
Brittle temperature (50% non-failure), °F	−58–+14
Stiffness in torsion, (at 23 °C) psi	
(Apparent mod. of elast.)	560–6,000
Hardness—Durometer Shore Type "A"	58–95
Deformation	15–30
Dissipation factor (at 23 °C, 50 mc/sec)	.01–.07
Dielectric constant	
(at 23 °C, 50 mc/sec)	2.9–3.1
(at 23 °C, 60 mc/sec)	4–9
Dielectric strength (short time), volts/mil	
0.125 in. specimen	400–500
Volume resistivity (at 15.5 °C),	
megohm-cm $\times 10^8$	2.5–5.7
Insulation resistance constant	
(at 15.5 °C), megohms/1,000 ft	730–10,500
Typical uses	spline, gasketing, belting, tubing, cable jacketing, primary insulation.

TABLE 26-3 (continued)

For Molding

Izod impact strength, ft lb/in. of notch	> 10
Tensile Strength, psi	1,000–2,200
Stiffness in torsion (apparent mod. of elasticity), psi 23°C	400–3,000
Elongation in tension, %	150–400
Dielectric strength (short time), volts/mil	340–375
Dielectric constant	
60 cycles	7–10
10^3 cycles	6–9
10^6 cycles	4–5
Dissipation factor	
60 cycles	.07–.11
10^3 cycles	.06–.11
10^6 cycles	.08–.18
Brittle temperature, °F	−40–+14
Specific gravity	1.17–1.42
Hardness—Durometer "A" and "D"	(D) 50–90
Water Absorption, % wt gain	.10–.20

For Calendering & Flooring

Specific density	
(1.0 gm resin/100 ml of methyl isobutyl)	0.550–0.900
(0.2 gm resin/100 ml of nitrobenzene)	0.110–0.250
Plasticizer absorption	excellent
Apparent density (lb/ft^3)	18–35
Specific gravity	1.35–1.40

The commercial resins may be broken down into several categories. General-purpose suspension resins are primarily useful for most molding, extrusion, and calendering operations. Emulsion or dispersion resins are utilized as either plastisols or organosols for slush molding, dip coating, film casting, and foaming. Solution resins are PVC dissolved in solvents, useful for applying thin coatings on paper, fabric, metals. Vinyl latex is PVC dispersed in water. This is again useful for coating operations. Rigid PVC grades are low-molecular weight, comparatively low-melting suspension resins. They are of two types: pure polyvinyl chloride or copolymers containing small amounts of vinyl acetate (Type I) or PVC containing 5 to 20 percent of N-type rubber (Type II) as an additive.

While the vinyl chloride resins are available as over 100 different products, they nonetheless all have in common the desirable properties of the vinyls: chemical inertness, weather resistance, nonflammability, toughness, electrical resistivity, and ease of fabrication. In addition, they are all thermoplastic odorless, tasteless, and nontoxic. These properties, of course, are imparted to the finished products.

Vinyl chloride resins are processed mainly by the usual techniques of molding, extrusion, and calendering.[91] Foaming, casting, blowing, dip coating, and spraying are less important. Before processing, the resins are blended with various additives.

Plasticizers are the major addition to any formulation, being incorporated in general to the extent of 20 to 50 percent by weight of the resin. Plasticizers are essential for softening the hard vinyl to workable materials. Only the rigid grades are not plasticized. The amount of plasticizer used determines the degree of flexibility, and the type determines the physical and chemical properties. Over 100 different plasticizers are available, which may be classified as phthalates, phosphates, low-temperature plasticizers, resinous plasticizers, miscellaneous or special purpose plasticizers, and plasticizer extenders.[33,60] Dioctyl phthalate—both di-2-ethylhexyl phthalate (DOP) and di-iso-octyl phthalate (DIOP)—is by far the most important and most commonly used plasticizing material. The basic characteristic of plasticizers is that the resins are insoluble in them at room temperature, but soluble at the temperatures used for processing.

Various other additives are included in processing formulations. Heat, and sometimes light, stabilizers are added to prevent decomposition. Heat stabilizers are usually metallo-organic compounds. Light stabilizers are generally salicylates and benzoates. Lubricants are commonly added as processing aids. Fillers such as clay or whiting are sometimes incorporated to reduce costs. Both organic and inorganic color pigments may be added to impart color to the basically transparent resins.

APPLICATIONS

Any attempt to describe the many outlets into which PVC is channeled must cover the entire industrial complex. Every industry—building, transportation, textile, clothing, paint, agriculture, appliance, and so on—has adopted vinyl plastics for such widespread application that it is now almost impossible to imagine life without them.

Of the estimated 880 million pounds of PVC consumed in 1960†, approximately 33 percent went into molding and extrusion applications, 24 percent into film and sheeting, 18 percent into flooring, 8 percent into fabric treatment, 5 percent into protective coating, and 12 percent into miscellaneous applications.[6] Film, sheeting, and coatings were the fastest expanding markets in 1960. This is indicated in the consumption pattern for 1960 back through 1956, shown in Table 26-4.[6]

†The estimated 880 million pounds discussed here falls short of the over 900 million pounds actually sold in 1960. Consumption patterns, however, are basically correct.

TABLE 26-4. POLYVINYL CHLORIDE AND COPOLYMER CONSUMPTION, 1955–1960[6]

(In Millions of Pounds)

Use	1955	1956	1957	1958	1959	1960
Film under 10 ml	83	78	78	74	90	96
Sheeting 10 ml and over	51	53	80	74	95	112
Fabric treatment[a]	56	55	55	55	77	67
Paper treatment	8	8	10	9	11	12
Floor coverings	56	66	82	116	155	160
Molding and extrusion	183	205	220	224	306	285
Protective coatings	26	29	32	27	31	40
All other uses[b]	58	74	78	70	109	108
Total	521	568	635	649	874	880

Source: U.S. Tariff Commission, with estimated alterations due to more varied breakdown. Note: Film and sheet resins are impossible to separate accurately but the total of the two as given above is thought to represent a reasonably reliable figure on total amount of resin used for both.

[a]Resin used for laminates with fabric is included in film figure rather than in fabric treatment.
[b]Includes export, some off-grade and various other uses not included in other classifications.

Molding and Extrusion

A breakdown of the outlets in the major volume category—molding and extrusion—is shown in Table 26-5.[6] The largest single market for vinyl resins in this or any other category is as extruded electrical insulation and jacketing. Almost all new wiring in homes and industrial buildings is vinyl coated, as it is in appliances, automobiles, ships, and planes. Power lines and communication wire and cable for telephone, radio, and TV sets are similarly protected. Rubber has been the loser. Rubber has also lost out to vinyls the huge garden hose market, as well as that for flexible tubing for surgical and other purposes. More recently, compounds of rubber and vinyl have been developed which combine some of the better properties of each and have extended the market opportunities for both materials.

Profile extrusion products appear in many forms: refrigerator gaskets, window channeling, automobile and shoe weltings, as well as the hoses and tubing mentioned above. Familiar items processed by slush (from plastisols) and elastomeric molding techniques include: boots, toys, gaskets outlet plugs, battery clips, control knobs.

Vinyls are competing with polystyrene for the growing record market, which has completely abandoned the early shellac records. An 85/15 vinyl chloride/vinyl acetate copolymer is generally used because of its inherent flexibility and lower brittleness, but some low molecular-weight PVC is used also.

Rapidly growing in importance are the rigid vinyls, unique in that no plasticizer is required for processing. Calendered sheets, usually polyvinyl chloride-acetate, appear as automobile, airplane, and motorcycle windshields, as industrial and greenhouse glazing, and as such items as exhaust hoods, tanks, and odd-shaped containers. These sheets can be vacuum formed into most unusual items for use as toys, relief maps, and packaging displays. Extruded vinyl rigid resins are finding a growing market for pipes, ducts, and items of that sort. Fittings such as pipe caps, joints, and angles are fabricated by molding techniques.

Vinyl foams and sponges are a comparatively new outlet for vinyl chloride resins. They are prepared from plastisols that are foamed by mechanical means or by chemical blowing agents. The foams are molded by standard methods into such items as mattresses, seats, cushions, automobile dashboard padding, cold-weather jacket linings, and so on. Vinyl foam backing for carpets and rugs is applied by spraying the foam directly onto the fabric and then curing.

TABLE 26-5. VINYL CHLORIDE RESINS FOR MOLDING AND EXTRUSION[6]

(In Millions of Pounds)

Application	1958	1959	1960
Phonograph records	38	45	45
Slush and elastomeric molding	37	56	58
Profiles	32	50	46
Wire coating	85	105	90
Rigid pipe and shapes	11	20	25
Garden hose[a]	17	18	14
Miscellaneous	5	4	4
Foam and sponge	–	7	3
Totals	225	305	285

[a]A large portion of garden hose is produced from off-grade resin. It is reported here as virgin resin—although it is generally sold to compounders and reported to the Tariff Commission as "all other resins." There is also a good quantity of scrap resin used in garden hose, but that material is not reported in this table.

Film and Sheeting[39]

The primary outlets for vinyl film and sheeting are indicated in Table 26-6.[6] The range of products includes films (under 10 mils thick) and sheets (over 10 mils thick) that are flexible, rigid, or intended for lamination. Most film and sheeting is calendered from suspension resins, but increasing amounts of film under one mil thick are being cast from plastisols.

TABLE 26-6. PATTERN OF CONSUMPTION OF CALENDERED VINYL FILM[6]

(In Millions of Pounds)

Uses	1953	1958	1959	1960
Draperies, bedspreads, kitchen and bath-room curtains	23	13.5	11	11
Yard goods	10	4	4	4
Adhesive-backed film	–	5.5	6	9
Closet accessories	6.5	6.9	8	8.5
Shower curtains[a]	6	9.2	11	11.5
Nursery goods	4	3.8	3.9	4
Baby pants, liners[a]	2.4	6	7	7
Table covers	4	3	3.5	4.5
Appliance covers	3	2.1	{ 8	{ 9
Furniture covers, indoor and outdoor	3	4.7		
Rainwear and sportswear	10	11	7	6
Aprons, including industrial	1.5	1.8	2	2.5
Lamination, quilting	–	11.5	25	27
Wall covering	–	2.5	2.2	2.8
Industrial tape	–	6	8	9
Inflatables	–	7.5	10.5	11
Industrial, agricultural, and miscellaneous	20	21	20	20
Total[b]	93.4	120	137	146.8

This table is intended to include only products made of calendered film up to 10 mils in thickness. Everything 10 mils or over is classified as sheeting. However, it is impossible to eliminate a certain amount of overlapping between film and sheeting and between various classifications of products given above. The figures are given as approximations to show trends. Accurate statistics are not available (see text for discussion).

[a] 1959 figure is down because great quantities of finished goods in these classifications were imported from Hong Kong and Japan.
[b] This total includes film made from imported resin.

Flexible vinyl film is widely used in an almost endless number of products such as bowl covers, crib sheets, garment bags, hat and cap protectors, inflatable toys, cushions and pool markers, kitchen aprons, lamp shades and lamp shade covers, pillow and mattress covers, protective clothing, rainwear, refrigerator food bags, shampoo capes, shower curtains, smocks, sports jackets, table covers, tobacco pouches, umbrellas, vanity seat covers, window curtains and window shades, portable silos, silo liners, covers and caps, pool and well liners, and pressure-sensitive tape. It can be laminated to both sides of delicate lace and plain or figured fabrics for extra toughness and unique creations; or it can be laminated to aluminum foil products to form a tight, moisture resistant wrapper, or it can be applied to food as a snug, water-tight wrapping to guard flavor and freshness.[39]

Flexible vinyl sheeting, used in unsupported form or backed with cloth, appears as hundreds of products: belts, suspenders, key chains, briefcases, book bindings, cosmetic bags, floor mats, furniture upholstery, auto seat covers, rear windows of convertible automobiles, hand bags, hats, shoe uppers, inflatable toys and cushions, industrial aprons, luggage covering, map cases, tobacco cases and wallets. Like vinyl film, the sheeting may be press-polished for a smooth mirror-like surface, or embossed for any variety of special surface textures.

The usefulness of rigid vinyl films and sheets has already been touched on in the molding and extrusion section. A more complete listing of applications would include: advertising displays, advertising novelties, book covers, calling cards, closure liners for cosmetics, drugs, and foods, identification tags and cards, tags for packaging, instruments for calculating, drafting and navigation, name plates, note book covers, plastic bindings, price tags and tag holders, credit cards, formed packages, Christmas decorations and ornaments, and action-type toys. Rigid or semi-rigid laminating film is applied to paper, leather, wood and textiles and appears as a transparent surface coating over printed matter, decorative plaques, lampshades, photographs, and book and menu covers.

Floor Coverings

Vinyl floor coverings have become important rapidly. This market has almost tripled in the last six years, with resin consumption now about 160 million pounds per year. The floor coverings are of three major types: vinyl asbestos tile, flexible all vinyl tile, and vinyl-coated "linoleum"—all available in a variety of clear colors and interesting patterns. They are the ultimate in practicality for schools, homes, libraries, theaters, and public buildings. High-fashion floor styling has appeared with the introduction of metallic-flecked and marbleized vinyl floor materials.

Industry Coatings[56]

Three categories in Table 26-4—fabric treatment, paper treatment, and protective coatings—can all be considered under the general subject of industry coatings. Fabric, paper, and other materials are, of course, laminated with vinyl film and sheeting (see page 000). In addition, however, these materials are treated with surface coatings of a different type: coatings of vinyl solution resins and of plastisols and organosols. These coatings are applied by spray, dip, roller, knife and reverse roll coating methods.

Vinyl solution resins are widely employed in various specialty applications. They are used to protect and strengthen such items as frozen food packages, paper drinking cups, ice cream containers, milk cartons, milk bottle caps, bottle caps and liners, and aluminum foil pouches. Virtually all beer cans today—close to 10 billion a year—are vinyl coated.

Surface coatings based on organosols and plastisols are rapidly growing in importance. They are baked at temperatures ranging from 275 to 400° F in order to fuse to a tough, resilient film. There is virtually a limitless range of formulations for such finishes for use on office furniture, automobile dashboards, business machines and computers, exposed ducts and vent work, school and industrial lockers, outdoor furniture, automobile roofs, softedge glass shelving, stairway bannister runners, tape for industry, ladder treads, "easy grip" handles for scissors, tools, levers and brake wheels, electric motor coils, work gloves, textured bottles, non-skid clothes hangers, table model television sets, and coatings for dishwasher interiors and dish racks. One striking example of a product coated with vinyl is the new line of IBM typewriters, available in a stunning array of colors: azure, royal blue, flame red, lemon yellow, charcoal gray, and blue gray.

Miscellaneous

Vinyl chloride has an interesting outlet as a co-monomer in synthetic fibers. "Vinyon" E is a copolymer of vinyl chloride and vinyl acetate, useful primarily as a nonwoven fiber for tea bags. "Dynel" is a copolymer of vinyl chloride and acrylonitrile. Such fibers are also produced in Japan. These fibers have a number of specialty applications such as "fur" coats and linings, backing for carpeting, carpeting itself, and filter cloths. Filaments of vinyl chloride-vinylidene chloride copolymers, as well as sheeting from such copolymers, are discussed in Chapter 24.

OUTLOOK

Approximately 936 million pounds of PVC (a term used to designate vinyl chloride polymers and copolymers that contain over 50 percent vinyl chloride) were produced in 1960. In 1958, for the first time, sales topped a billion pounds in the general vinyl category which, besides PVC, includes polyvinyl acetate, polybinylbutyral, and vinyl chloride-containing saran, "Dynel," etc. It has been estimated that PVC itself should easily reach one and one half billion pounds production in the 1960's.[6] Capacity for both vinyl chloride and PVC production is now on the order of 1.4 billion pounds.[3] Another 350 million pound capacity is planned or being built.

Overseas production facilities are increasing at an even faster pace. Japan is a major producer, and European nations (Germany, Italy, France, England, etc.) have been turning out PVC almost as fast as the United States. United States 1958 capacity was 490,000 tons per year compared to

European capacity of 414,000 tons per year.[85] These figures mesh into the growth rate patterns of the entire plastics industry: 9.0 percent in 1956/57 and 6.8 percent in 1957/58 for the United States, and 19.2 percent and 16.5 percent for Europe. Solvay is the major European source. Because manufacturing facilities need not be complex, the initial plant investments can be comparatively small, and the fabrication of resins is quite simple, plants are now springing up in much less developed areas of the world. The market for finished goods is almost untapped in such areas.

In the fifties, resins imported from overseas appeared to be cutting heavily into the domestic market, despite a stiff duty on entering PVC. The 1958 U. S. price cut to 23½ cents (see Table 26-7) reflected this foreign invasion, whereas the 1960 price cuts to 18½ cents reflected increasing competition at home.[6] The price of vinyl chloride (about 8 cents/lb) has also been under pressure, even though most monomer is captively consumed at the point of manufacture. Further resin price cuts would effectively make smaller domestic production units marginal operations.

No estimate of ultimate PVC consumption has been valid for more than one or two years. Despite periods of over-production, demand has always caught up with and surpassed supply. Even with competition from polyethylene, polystyrene, and the other plastic materials available, the PVC market is still growing. Thus, the long range outlook remains good.

TABLE 26-7. PRICE HISTORY OF PVC[6]

Polyvinyl Chloride		Vinyl Paste Resin	
Year	Price, cents/lb	Year	Price, cents/lb
1934	0.78	1947	0.40
1935	0.59	1950	0.37
1937	0.56	1950	0.39
1940	0.52	1955	0.34
1942	0.48	1956	0.30
1943	0.44	1958	0.28
1944	0.39	1958	0.265
1945	0.35	1960	0.24
1946	0.33		
1948	0.34		
1950	0.36		
1951	0.38		
1954	0.38		
1955	0.31		
1956	0.275		
1957	0.25		
1958	0.235		
1959	0.22		
1960	0.205		
1960	0.185		

References

1. A. G. fur Anilin-Fabrik., German Patent 362,666 (Oct. 21, 1920).

2. Alexander, C. H., and Tucker, H. (to Goodrich Chemical), U. S. Patent 2,366,306 (Jan. 2, 1945).

3. Anon., *Chem. Eng. News,* p. 26 (May 23, 1960).

4. Anon., *Chem. Week,* **86** (13), 45 (Mar. 26, 1960).

5. Anon., *Modern Plastics,* **32,** 80 (Jan. 1955).

6. Anon., *Modern Plastics,* **37,** 97 (Jan. 1960); **38,** 96 (Jan 1961).

7. Anon., "Modern Plastics Encyclopedia Issue for 1959," p. 167.

8. Arnold, H. W. (to Du Pont), U. S. Patents 2,404,780–1 (July 30, 1946).

9. Baer, M. (to Monsanto), U. S. Patents 2,470,908–11 (May 24, 1949).

10. Baer, M. (to Monsanto), U. S. Patent 2,476,474 (July 19, 1949).

11. Baer, M. (to Monsanto), U. S. Patents 2,492,086–9 (Dec. 20, 1949).

12. Baumann, E., *Ann.,* **163,** 312 (1872).

13. Baxter, G. E. (to Diamond Alkali), U. S. Patent 2,541,022 (Feb. 13, 1951).

14. Baxter, G. E. (to Diamond Alkali), U. S. Patent 2,585,911 (Feb. 19, 1952).

15. Boesler, J., Eberhardt, E., Sandhaas, W., and Stadler, R. (to I. G.), U. S. Patent 2,265,509 (Dec. 9, 1941).

16. Braconier, F. F. A., and Rayet, P. (to Societe Belge de l'Azote et des Produits Chimiques du Marly), U. S. Patent 2,705,732 (April 5, 1955).

17. Britton, E. C., and Le Fevre, W. J. (to Dow), U. S. Patent 2,333,633 (Nov. 9, 1943).

18. Cheney, H. A. (to Shell), U. S. Patent 2,569,923 (Oct. 2, 1951).

19. Christenson, R. M. (to Pittsburgh Plate Glass Co.), U. S. Patent 2,882,251 (April 14, 1959).

20. Collins, H. M. (to Shawinigan), U. S. Patent 2,388,600 (Z v. 6, 1945).

20A. Conrad (to Ethyl Corp.) U. S. Patents 2,765,349-50-1-2 (1956).

21. DeBell, J. M., Goggin, W. C., and Gloor, W. E., "German Plastics Practice," Springfield, DeBell and Richardson, 1946. See also CIOS XXV-20, XXVI-52, XXVII-51, XXVII-85, and XXIX-62; BIOS 104, 811, 927 and 1105; and FIAT 867 and 862.

22. De Coene, R. (to Solvic), U. S. Patents 2,829,133 to 4 (Apr. 1, 1958).

23. de Nie, W. L. J. (to Shell), U. S. Patent 2,474,206 (June 28, 1949).

24. de Nie, W. L. J. (to Shell), U. S. Patent 2,496,384 (Feb. 7, 1950).

25. de Nie, W. L. J. (to Shell), U. S. Patent 2,537,334 (Jan. 9, 1951).

25A. Dirstine, Dance (Dow), U. S. Patent 2,628,259 (1953).

26. Douglas, S. D. (to Union Carbide), U. S. Patent 2,055,468 (Sept. 29, 1936).

27. Douglas, S. D. (to Union Carbide), U. S. Patents 2,075,429 and 2,075,575 (March 30, 1937).

28. Downes, A. W. (to Union Carbide), U. S. Patent 2,345,659 (April 4, 1944).

29. Downes, A. W., and Kernan, J. R. (to Union Carbide), U. S. Patent 2,345,660 (April 4, 1944).

30. Dunn, J. H., Neher, C. M., and Trotter, P. W. (to Ethyl), U. S. Patent 2,635,122 (April 14, 1953).

31. Everard, K. B., and Harris, I. (to I.C.I.), U. S. Patent 2,777,836 (Jan. 15, 1957).

32. Faith, W. L., Keyes, D. B., and Clark, R. L., "Industrial Chemicals," New York, John Wiley & Sons, Inc., 1957.

33. Fedor, W. S., *Chem. Eng. News,* **118** (November 13, 1961).

34. Flory, P. J., *J. Am. Chem. Soc.,* **61**, 1518 (1939).

35. Flumiani, G., *Z. Elektrochem.,* 32, 22 (1926).

36. Folt, V. L. (to Goodrich), U. S. Patent 2,625,539 (Jan. 13, 1953).

37. Frankenburger, W., and Steigerwald, C. (to I. G.), U. S. Patent 1,756,943 (May 6, 1930).

38. Fryling, C. F. (to Goodrich), U. S. Patent 2,356,925 (Aug. 29, 1944).

39. Georgett, E. W., "Vinyl Film and Sheeting for Fabrication and Design," Speech presented before the Commercial Plastics Supply Co., June 26, 1957.

40. Gerhard, J. R., Deegan, C. C., and Fisher, T. W., Jr., (to Firestone), U. S. Patents 2,875,186–7 (Feb. 24, 1959).

41. Halbig, P., U.S. Patent 2,552,425 (May 8, 1951).

42. Herrmann, W. O., and Haehnel, W. (to Consortium fur Electrochemische Industrie), U. S. Patent 1,672,157 (June 5, 1928).

43. Japs, A. B. (to Goodrich), U. S. Patent 2,225,635 (Dec. 24, 1940).

44. Japs, A. B. (to Goodrich), U. S. Patent 2,265,286 (Dec. 9, 1941).

45. Klatte, F., and Rollett, A. (to Chemische Fabrik Griesheim-Electron), U. S. Patent 1,241,738 (Oct. 2, 1917).

46. Koll, R. J. (to Diamond Alkali), U. S. Patent 2,539,307 (Jan. 23, 1951).

47. Kolthoff, I. M., and Medalia, A. I. (to Phillips Petroleum), U. S. Patent 2,647,109 (July 28, 1953).

48. Kolvoort, E. C. H., and Akerman, G. (to Shell), U. S. Patent 2,479,241 (Aug. 16, 1949).

49. Kolvoort, E. C. H., and Akerman, G. (to Shell), U. S. Patent 2,496,222 (Jan. 31, 1950).

50. Kreager, R. M., and Leeson, E. J. (to Goodrich), U. S. Patent 2,812,318 (Nov. 5, 1957).

51. Lawson, W. E., and Werntz, J. H. (to Du Pont), U. S. Patent 1,874,107 (Aug. 30, 1932).

52. Lightfoot, W. J. (to Goodrich), U. S. Patent 2,511,593 (June 13, 1950).

53. Mark, H., Fikentscher, H., Hengstenberg, J., and van Susich, G. (to I. G.), U. S. Patent 2,068,424 (Jan. 19, 1937).

54. Marvel, C. S., *et al., J. Am. Chem. Soc.,* **61**, 3241 (1939); **64**, 2356 (1942).

55. Matheson, H. W., and Skirrow, F. W. (to Canadian Electro Products Company, Ltd.), U. S. Patent 1,725,362 (Aug. 20, 1929).

56. McKnight, W. H., "Vinyls in Modern Industrial Coatings," MS-2007, Union Carbide Plastics Company.

57. Miller, H. S. (to Air Reduction), U. S. Patent 2,448,110 (Aug. 31, 1948).

58. Miller, H. S. (to Cyanamid), U. S. Patent 2,615,054 (Oct. 21, 1952).

59. Morgan, L. B., and Morgan, W. McG. (to I.C.I.), U. S. Patent 2,322,309 (June 22, 1943).

60. Morris, J. J., *Modern Plastics,* **31** (2), (Oct., 1953).

61. Naugatuck Chemical, *Petroleum Refiner,* **38**, 305 (Nov., 1959).

62. Nauth, R., "The Chemistry and Technology of Plastics," New York, Reinhold Publishing Corp., 1947.

63. Niewland, J. A. (to Du Pont), U. S. Patent 1,812,542 (June 30, 1931) and U. S. Patent 1,811,959 (June 30, 1931).

64. Ostromislensky, I., *et al.,* British Patent 6299 (March 13, 1912) and German Patent 264,123 (Jan. 10, 1912).

65. Ostromislensky, I., *Chem. Zentr.,* **1,** 1980, 1983, (1912); *Chem. Ztg.,* **36,** 199 (1912).

66. Ostromislensky, I., U. S. Patent 1,721,034 (July 16, 1929) and (to L. A. Van Dyk), U. S. Patent 1,791,009 (Feb. 3, 1931).

67. Ott, J. B. (to Monsanto), U. S. Patent 2,862,912 (Dec. 2, 1958).

68. Park, H. F. (to Monsanto), U. S. Patent 2,664,416 (Dec. 29, 1953).

69. Perkins, G. A. (to Union Carbide), U. S. Patent 1,934,324 (Apr. 10, 1931).

70. Plambeck, L. Jr. (to Du Pont), U. S. Patent 2,462,422 (Feb. 22, 1949).

71. Plauson, H., U. S. Patent 1,425,130 (Aug. 8, 1922).

72. Plotnikow, I., *Z. wiss. Phot.,* **21,** 117 (1922).

73. Powers, J. R. (to Goodrich), U. S. Patent 2,520,959 (Sept. 5, 1950).

74. Regnault, V., *Ann.,* **15,** 28, 34 (1835) and **15,** 63 (1835); *Ann. Chim.,* **2,** 59, 358 (1835).

75. Regnault, V., *Ann. chim.,* **2,** 69, 151 (1838); *J. Prakt. Chem.,* **I,** 18, 80 (1839).

76. Reid, E. W. (to Union Carbide), U. S. Patent 1,935,577 (Nov. 14, 1933).

77. Reid, E. W. (to Union Carbide), U. S. Patent 2,064,565 (Dec. 15, 1936).

78. Renfrew, A., and Gates, W. E. F. (to I.C.I.), U. S. Patent 2,296,403 (Sept. 22, 1942).

79. Richard, A. P. (to Societe des Glaces et Produits Chimiques), U. S. Patent 2,776,273 (Jan. 1, 1957).

80. Scientific Design Company, Inc., *Petroleum Refiner,* **38,** 306 (Nov. 1959).

81. Schaeffer, E. (vested in Alien Property Custodian), U. S. Patent 2,338,459 (Jan. 4, 1944).

81A. Sherwood, P. Petrochemical Engineer (Oct. 1961).

82. Schildknecht, C. E., "Vinyl and Related Polymers," New York, John Wiley & Sons, Inc., 1952.

83. Schoenfeld, F. K. (to Goodrich), U. S. Patent 2,168,808 (Aug. 8, 1939).

84. Scott, W., and Seymour, R. B. (to Wingfoot Corp.), U. S. Patent 2,348,154 (May 2, 1944).

85. Seidel, H., *Chem. Ind.-International,* p. 6 (March 1960).

86. Seymour, D. C. (to U. S. Rubber), U. S. Patents 2,831,843-4 (Apr. 22, 1958).

87. Shriver, L. C. (to Union Carbide), U. S. Patent 1,938,870 (Dec. 12, 1933).

88. Shriver, L. C., and Fremon, G. H. (to Union Carbide), U. S. Patent 2,420,330 (May 13, 1947).

89. Sittenfeld, M., *Chem. Eng.,* **54,** 129 (Dec. 1947).

90. Smith, W. M., "Vinyl Resins," New York, Reinhold Publishing Corp., 1958.

91. SPI, "Plastics Engineering Handbook," New York, Reinhold Publishing Corp., 1960.

92. Stanley, H. M. (to The Distillers Company, Ltd.), U. S. Patent 2,407,039 (Sept. 3, 1946).

93. Staudinger, H., *Helv. Chim. Acta,* **13,** 805 (1930).
94. Staudinger, H., and Schneiders, J., *Ann.,* **541,** 151 (1939).
95. Stinchfield, R. L. (to Eastman Kodak), U. S. Patent 1,627,935 (May 10, 1927).
96. Strother, C. O. (to Union Carbide), U. S. Patent 2,238,956 (Apr. 22, 1941).
97. Toussaint, W. J. (to Union Carbide), U. S. Patent 1,926,638 (Sept. 12, 1933).
98. Tucker, H. (to Goodrich), U. S. Patent 2,628,957 (Feb. 17, 1953).
99. "Vinylite Resins," Carbide and Carbon Chemicals Corporation, New York, May 1, 1937.
100. Wakeman, R. L., "The Chemistry of Commercial Plastics," New York, Reinhold Publishing Corp., 1947.
101. Weiler, J. F. (to Mathieson), U. S. Patent 2,412,308 (Dec. 10, 1946).
102. Wenning, H. (to Huls), U. S. Patent 2,820,028 (Jan. 14, 1958).
103. Wolf, R. J. (to Goodrich), U. S. Patents 2,564,291-2 (Aug. 14, 1951).
104. Young, C. O., and Douglas, S. D. (to Union Carbide), U. S. Patent 1,775,882 (Sept. 16, 1930).
105. Young, C. O., and Douglas, S. D. (to Union Carbide), U. S. Patent 2,011,132 (Aug. 13, 1935).
106. Young, R. B. (to General Electric), U. S. Patent 2,590,810 (Mar. 25, 1952).

27. METAL CHLORIDES

A. E. Skrzec and Lorraine B. Lowry
Stauffer Chemical Company

INTRODUCTION

While it is technically true that any combination of a metallic element with chlorine will constitute a metal chloride, the industry reserves this general identification to the chlorides of most of the elements of the B sub-group of Groups IV, V, and VI of the periodic table. These are titanium, zirconium, hafnium, vanadium, columbium, tantalum, chromium, molybdenum, tungsten, boron, aluminum, silicon and antimony, all having related uses.

Their primary uses are as catalysts and/or sources of the pure metal, but the number of different processes in which one or more of them are employed as catalysts is almost impossible to catalogue. Until recent years, aluminum chloride was the only one of these chemicals which had significant commercial applications. While it is still of major importance, several of the other metal chlorides are used in large volume, and still others show promise of becoming industrially important. In total, the production of these chlorides required an estimated 60,000 to 70,000 tons of chlorine in 1959.

HISTORY AND USES

Familiarity with this classification of compounds is closely associated with their history and uses. Both are described in the paragraphs that follow.

Aluminum Chloride

The major growth in the use of aluminum chloride, whose catalytic activity Friedel and Crafts discovered in 1877, did not begin until the 1930's, when its use in petroleum refining became important. The moderate decline in sales in the last few years largely has been due to decrease in dyestuff production, to lower refinery runs, and to the use of noble-metal catalysts

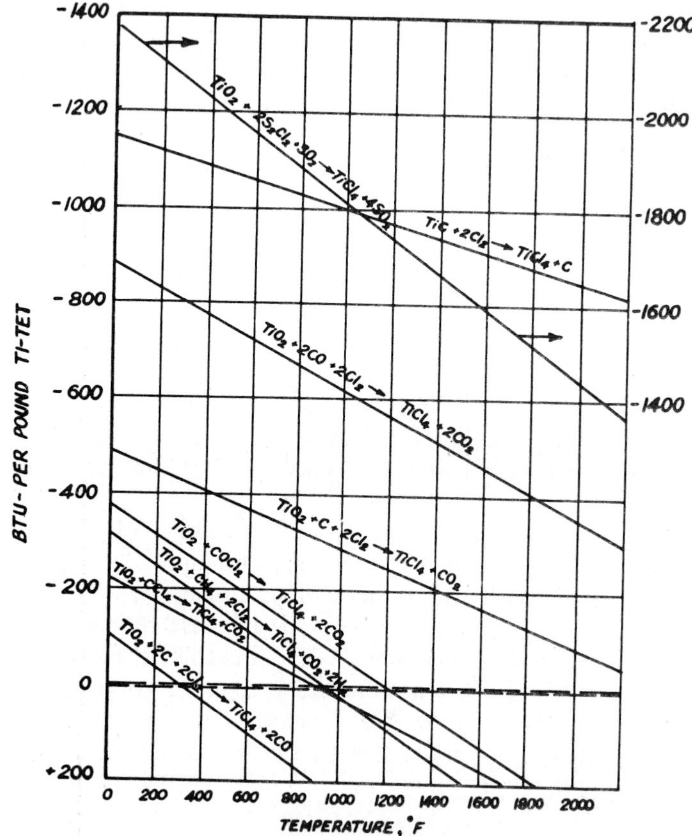

Figure 27-1. Heats of reaction of rutile, $\Delta H_{298°K}$ reactants in a 298°K, products out at reaction temperature.

in place of the chloride in some isomerization operations. As the other uses of aluminum chloride continue to expand, this decline is expected to reverse itself and a steady growth in consumption is expected.

C. A. Thomas, in ACS Monograph 87, "Anhydrous Aluminum Chloride In Chemistry" presents a detailed picture of the uses of this material. However, for the sake of completeness, Table 27-1 lists the estimated end-use pattern of aluminum chloride in 1959.

The "other" category includes end-use areas such as butyl rubber, insecticides, rubber-stabilizers and anti-oxidants, electro-plating, and many others.

Titanium Chlorides

Titanium tetrachloride was first prepared by George in 1825 and somewhat later by Dumas, Wöhler, Merz, Demarcay and Moissan.[1] The more common methods of preparation comprise the direct action of chlorine gas on a great variety of titaniferous materials such as the metal, alloys, carbide, cyanonitride, dioxide and ores, at elevated temperatures.

Anhydrous titanium trichloride is a crystalline violet solid and was first prepared by Eblemen by the reduction of the tetrachloride with hydrogen in a hot tube. This process was improved by Goerges and Stähler, Stähler and Bachran, and Meyer, Bauer and Schmidt, mainly by rapid cooling of the hot gases. It has been developed into a more or less continuous process by Sherfey who refluxed titanium tetrachloride in a stream of hydrogen which was passed over a heated tungsten filament. Billy and Brasseur reported that the trichloride can best be prepared by the action of powdered antimony on titanium tetrachloride at 340° C.

Böck and Moser subjected a mixture of hydrogen and titanium tetrachloride vapor to a silent electric discharge, and a brown trichloride different from the usual violet compound was formed. It readily changed into the violet, but this change was not reversible.

The dichloride is a black solid and is very unstable in air. It was prepared by Friedel and Guerin who reduced the tetrachloride in a stream of hydrogen; Von der Pfordten prepared the compound with sodium amalgam.

The chlorides of titanium have recently become major items of chemical commerce. The output of titanium tetrachloride is considerably greater than that of the trichloride due to its use as a source of titanium metal. It also has become an intermediate in the production of titanium dioxide[5,6,8] pigment. It appears likely that the tetrachloride will remain the primary source of titanium metal. Production of the metal, however, fell from a high of over 17,000 tons in 1957 to 4,600 tons in 1958 and did not increase appreciably in 1959.

Both chlorides usually have been cited together in the literature as catalysts or components of catalyst systems; titanium trichloride has been

TABLE 27-1. ESTIMATED END-USE PATTERN OF ALUMINUM CHLORIDE (1959)

Use	Consumption (%)
Ethyl benzene	16
Ethyl chloride	14
Anthraquinone and other dyestuff chemicals	15
Lubricating-oil refining	20
Dodecyl benzene	21
Alkylation and isomerization catalyst	3
Metallurgical uses (non-ferrous)	3
Pharmaceuticals and perfume	3
Other	5
	100

found preferable in some cases and the tetrachloride in others. The preferred titanium chloride for use in polypropylene catalyst systems is the trichloride; the tetrachloride is the usual choice for linear polyethylene catalyst systems. Catalyst systems for *cis*, 1,4-polyisoprene, *cis* 1,4-polybutadiene, and other polymers probably will incorporate one of these compounds.

The tetrachloride is being employed or has been proposed for coating glass, producing smoke screens and sky-writing, as a component of gaseous flux for brazing and welding, for preparing color lakes, for inhibiting sulfuric acid pickling baths, as a component of a mixture used in grain refining of aluminum alloys, and as a component of flame retardants.

Silicon Tetrachloride

Silicon tetrachloride has a long history of manufacture, primarily because of its use as an intermediate in the production of silicate esters, silanes, silicones and organic silicon compounds. One of its earliest uses was in the preparation of silicate esters for use as waterproofing or sealing agents for stone or masonry. Its newest application is in the preparation of silicon metal for use in the electronic industry.

In 1823, Berzelius discovered silicon tetrachloride when he showed that this compound was formed by the direct union of silicon and chlorine. According to Hempel and Von Haasy, the reaction begins at 280°C and proceeds with incandescence at 480°C. Since its discovery, silicon tetrachloride has been prepared by several methods. Hempel and Von Haasy obtained it by the reaction of chlorine and the sulfosilicates; Troost and Hautefeuille by the reaction of boron trichloride and silica; Stokes by the reaction of aromatic silicon esters and phosphorous oxychloride; Colson by the reaction of mercuric chloride and silicon sulfide; Martin by the reaction of chlorine and silicides at high temperatures. Oersted, Buff and Wöhler, Ebelmen and Schnitzler made the gas by passing a stream of hydrogen chloride or chlorine through a heated mixture of silicon and carbon. This process is the basis for a present-day commercial method of manufacture. Another industrial method used today is based on the work of Warren and De Carli, Hutchins, and Martin. It comprises reacting chlorine and an alloy of silicon, e.g. a 90 percent ferrosilicon mixed with silicon carbide at temperatures from 500 to 1000°C.

Other proposed uses of silicon tetrachloride include production of smoke screens, treatment of methyl methacrylate polymer surfaces to impart mar resistance, heat treatment of molybdenum wires for electrical resistors and various catalytic applications.

Zirconium Tetrachloride

The production of zirconium metal consumes almost all the zirconium tetrachloride output. It is only within the last ten years that this metal has grown to its present production level. The AEC is the largest user since the primary application is in atomic reactors where its corrosion resistance and low neutron capture cross-section is of extreme value.

Zirconium tetrachloride can be formed in a number of ways. It was first prepared by Berzelius by the action of chlorine upon metallic zirconium. The simplest method is to pass chlorine over or through hot zirconium carbide or the cyano nitride. If zircon is used, the reaction temperature is higher and the tetrachloride is removed from the gas stream on a warm condensing surface. Other methods are the method of Smith and Harris, the heating of the oxide with phosphorous pentachloride in a sealed tube at 190° C, and the method of Lely and Hamburger, the action of chlorine and carbon tetrachloride on the oxide at 800° C.

Other proposed uses for zirconium tetrachloride include preparation of zirconium tetrafluoride, production of conductive metal cathode, preparation of magnesium alloys, formulation of anti-perspirants and hair waving preparations, and catalytic applications in processes such as olefin polymerization, petroleum cracking, isomerization, alkylation, and Friedel-Crafts reactions.

Boron Trichloride

The emergence of atomic power development and high energy propellants has focused renewed attention on boron and its compounds. Boron trichloride has become an important raw material for missile fuels. It is also finding increasing use as a catalyst and in the manufacture of boron-containing compounds and boron metal.

As early as 1809, Davy observed that boron burns in chlorine gas but he did not investigate the products of the combustion. In 1824, Berzelius showed that boron trichloride is formed when amorphous boron is warmed in chlorine gas. Several modifications of this method have since been developed. Another method, originally described by Dumas, is the passage of chlorine over an intimate mixture of carbon and boron trioxide at a high temperature. Hoffman reported better yields by substituting iron boride for boric oxide and charcoal. Methods of lesser commercial importance also have been described. Gustavson obtained the trichloride by heating at 150° C in a sealed tube a mixture of boric oxide and phosphorous pentachloride. Moissan found that it was formed when boron sulfide was treated with chlorine or hydrogen chloride at 400° C. In subsequent investigations, Moissan obtained the trichloride by treating calcium boride with chlorine.

Diboron tetrachloride was first isolated by Stock in 1929. It is most conveniently prepared by passing the trichloride vapor through an electric glow discharge struck between mercury electrodes.

Antimony Chlorides

Antimony, the metal, has long been known as a constituent of type metals, solders, fusible alloys and, to a lesser extent, as a medicine. As chlorides, the trichloride and the pentachloride are known and are available commercially. The tetrachloride has been reported.

There is relatively little published information on antimony pentachloride, probably because its commercial applications have been limited. It has been used as a chlorine carrier in organic syntheses and as a catalyst in certain polymerization reactions. The trichloride has found wide use as a catalyst in petroleum processing, in organic chlorinations and polymerizations, in metallurgy as a metal colorant, in the textile industry in the preparation of mordants, and as a reagent useful in the preparation of synthetic Vitamin A.

The trichloride was first prepared in the 17th century by Glauber and Basil Valentine by distillation from a mixture of antimony sulfide with mercuric chloride, salt and clay, or hydrochloric acid. The product, a colorless crystalline solid, with a melting point of 73.4° C, a boiling point of 223° C, and a specific gravity of 3.14, was named *butyrum antimonii* or butter of antimony.[3] It is most conveniently prepared by dissolving antimony trisulfide in hydrochloric acid and distilling off excess acid, after which the chloride is volatilized. Antimony trioxide is converted into the trichloride by the action of chlorine and chlorides of nonmetals and by dissolution in hydrochloric acid. Antimony pentoxide reacts similarly.

A pure solution of the trichloride may be prepared by dissolving antimony oxychloride in hydrochloric acid; crystals may be obtained by evaporation from solution of the trichloride in carbon disulfide or sulfuryl chloride, by sublimation of the trichloride in a current of carbon dioxide, and finally by solidification after fusion.

Many substances of a complex nature have been obtained by different investigators and have been described as oxychlorides of antimony. Studies on the hydrolysis of the trichloride, however, suggested that only two are true compounds, $SbOCl$ and $Sb_4O_5Cl_2$.

Antimony oxychloride is obtained as an amorphous white powder by the hydrolysis of the trichloride at ordinary temperatures. It has also been prepared by heating the trichloride with alcohol in a sealed tube at 160° C.

Tantalum Pentachloride

The commercial availability of tantalum pentachloride is a recent development. Like its sister compound columbium pentachloride, it is used primarily as a source of the metal. Tantalum has excellent properties for use in process equipment and is also used as a raw material for the production of electrical capacitors, switches and film resistors.

It was first prepared by Berzelius in 1815 by the action of pure dry chlorine on heated tantalum metal. Rose and Weber prepared it by heating the oxide and carbon in a current of dry chlorine as did Biltz and Voigt. Others have described the preparation of this compound by the reactions of the oxide with certain chlorides such as phosphorous pentachloride, carbon tetrachloride, carbonyl chloride, and sulfur monochloride.

Ruff and Thomas found that the dichlorides and trichlorides were formed when the pentachloride was reduced by aluminum in the presence of aluminum trichloride by heating at 300°C in a sealed tube. The products were separated by leaching with cold water; the trichloride dissolved leaving the dichloride as a dark olive green powder.

Columbium Chlorides

Moissan and Van Bolton prepared the pentachloride by combination of the elements involved at 205°C. The method used by Rose—one of the commercial methods used today—was that of passing chlorine over an intimate mixture of the pentoxide and carbon at a high temperature. Oxychlorides, which may form if the temperature is too high, were removed by sublimation in a current of chlorine or carbon dioxide. Several other methods of preparation from the pentoxide have been reported. Pennington heated a mixture of pentoxide and phosphorous pentachloride in a sealed tube at 210°C. Demarcay, Camboulives, Ruff and Schiller passed a mixture of chlorine and carbon tetrachloride vapor over red-hot pentoxide; Smith and Hall obtained a very pure product by heating the pentoxide with sulfur monochloride and chlorine at 200°C in a sealed tube, with subsequent distillation. Roscoe obtained the pentachloride by passing the vapor of the oxytrichloride over red-hot carbon.

The other commonly known chloride is the trichloride, a nonvolatile, non-deliquescent black crystalline substance resembling iodine. An early preparation by Roscoe consisted of passing the vapor of the pentachloride through a heated tube. Ott and Stahler tried electrolytic reduction of the pentachloride but found that it did not give rise to trivalent columbium chloride.

Tungsten Chlorides

Roscoe reported the preparation of tungsten dichloride by heating in a bath of molten zinc tungsten tetrachloride in a current of carbon dioxide; or tungsten hexachloride in hydrogen. He described the product as a grey amorphous mass unstable in air.

Bernhardi-Grisson found the trichloride in the electrolytic reduction of a hydrofluoric acid preparation or aqueous alcoholic solution of tungstic acid. The reddish-brown liquid reacted as if it contained a trivalent tungsten salt. Olsson also was able to reduce a solution of tungstic acid to trivalent tungsten by means of tin.

Riche prepared impure tungsten tetrachloride, in voluminous greyish-brown crystals, by partial reduction of the hexachloride. This reaction was also studied by Wöhler and von Borck. Michael and Murphy obtained the tetrachloride by heating tungsten dioxide for many hours at 250° C in a sealed tube with carbon tetrachloride. According to Roscoe, the tetrachloride may be formed by distilling a mixture of the pentachloride and hexachloride in a current of hydrogen or carbon dioxide. The nonvolatile portion is mixed with the volatile portion and the distillation is repeated to convert any dichloride to tetrachloride.

Blomstrand prepared the pentachloride by heating the hexachloride for a long time in hydrogen. Roscoe records that only a small amount is formed by distillation of the hexachloride but reduction always occurs.

The hexachloride has been prepared by several investigators by the action of dry chlorine on heated tungsten. Teclu prepared tungsten hexachloride of high purity by heating a mixture of tungsten trioxide and phosphorus pentachloride in a sealed tube at 170 to 200° C with distillation for the removal of the phosphoryl chloride and subsequent heating in a current of carbon dioxide. Many other laboratory methods of preparation are discussed in the literature.

Vanadium Chlorides

Vanadium dichloride was prepared by Roscoe by passing a mixture of hydrogen and vanadium tetrachloride through a glass tube heated to dull redness. Moissan and Holt observed the formation of the dichloride when vanadium silicide was heated in chlorine; Meyer and Backa obtained it by the action of hydrogen chloride on ferrovanadium at 300° to 400° C. Piccini and Marino obtained an aqueous solution by the electrolytic reduction of the trichloride using a graphite anode.

The method of producing vanadium chloride now in use is that of Roscoe, wherein the reduction of the tetrachloride is completed with hydrogen. This pioneer investigator found that the trichloride is formed when the tetrachloride is merely heated. Other methods of preparation include the heating of the trisulfide in a current of chlorine, the action of dry hydrogen chloride on finely divided vanadium at 300 to 400° C and the passing of carbon dioxide through the tetrachloride at 140 to 150° C.

Roscoe further found that the tetrachloride is formed by heating vanadium nitride to redness in a current of dry chlorine free from air. Other methods include the passing of vanadium oxytrichloride vapors mixed with chlorine through a red-hot layer of charcoal, the passing of mixture of vapor of sulfur monochloride and chlorine over heated vanadium pentoxide and the passing of dry chlorine over ferrovanadium.

COMMERCIAL METHODS OF MANUFACTURE

Theory

A quick estimate of the thermodynamic feasibility of any reaction proposed can be obtained from calculating the net free energy of reaction. Free energy of formation versus temperature data are given generally in a form readily useable by engineers. An example calculation for rutile is given below:

$$TiO_2(s) + 2C + 2Cl_2 \xrightarrow{1000°C} 2CO(g) + TiCl_4(g)$$

$$\Delta F_{(Reaction)} = [2_\Delta F_{F(CO)} + _\Delta F_{F(TiCl_4)} - _\Delta F_{F(TiO_2)}]$$

$$\Delta F_{(Reaction)} = [2(-54) + (-148) - (-170)] = -86 \frac{k.\,cal}{g.\,mole}$$

The large negative value of the net free energy of reaction, shows that conversion to $TiCl_4$ is practically complete if equilibrium conditions are obtained. The kinetics of the reaction, however, are always determined by experimental tests.

Raw Materials

A prospective metal chlorides manufacturer may have a choice of several metal sources. In order of decreasing metal cost, it may be possible to directly chlorinate the metal, its carbide, its oxide or its oxide in combination with other oxides. In general, processing costs increase as the metal source becomes cheaper.

The raw material cost is also affected by factors such as purity, particle size and sieve range, grinding charges, freight and quantity delivered. The cost of most ores will vary directly with the metal content of the ore. As the amount of impurities that can be chlorinated increases in the ore, chlorine wastage increases through formation of unsaleable by-products. Correspondingly, the cost of caustic or lime used in neutralization of these by-products tends to increase. Specifically, such impurities (e.g. calcium, sodium and magnesium) tend to form liquid chlorides at reaction temperatures and result in increased processing costs above those for a high-grade ore.

Although low-grade ores usually show the lowest raw material costs, processing and plant depreciation costs are higher than those for high purity ores. If the plant capacity is relatively small, the optimum unit cost of production well may be obtained by using a high-grade, relatively expensive ore. In fact, use of the up-graded carbides instead of the respective oxides may give cheaper production costs for the chlorides. For large tonnage installations, utilization of the cheapest raw materials available is the preferred, long range objective.

Carbon for the reductive chlorination reactions is usually supplied as calcined petroleum coke. Specifications for the amounts of hydrogen and sulfur contained in the coke are low since each pound of hydrogen consumes thirty-five pounds of chlorine, and each pound of sulfur reacts with about one pound of chlorine. In certain operations involving fluidized-bed chlorinators, coke of a special size range is sometimes specified in order to maintain excellent fluidity. These tight specifications add greatly to the cost of the coke.

Minor raw material charges may include the cost of burning additional coke with oxygen (or air) in order to supply heat to the reactor. Dilute caustic or limewater is required for neutralizing excess chlorine and acid wastes prior to discharge to the sewer or atmosphere.

The over-all raw material efficiencies will be affected chiefly by the reaction efficiency, amount of by-product formation, and processing losses. In a fluidized-bed reactor, for example, elutriation losses of the solids fed to the reactor may be substantial. In order to minimize such losses a careful design of the reactor and its components is necessary, because usually it is neither economical nor practical to recycle elutriated solids to the reactor.

Heat must usually be supplied during the chlorination of an oxide ore—assuming that the reactants enter the reactor at about atmospheric temperatures. If such heat is supplied through other than chemical means, it can still be considered as a raw material. To illustrate, heat requirements for chlorination of rutile and ilmenite are plotted in Figures 27-1 and 27-2 (see pages 806 and 815).

Chlorinating Equipment

Molten Bath. The simplest reactor is the molten bath chlorinator, used for the direct reaction of chlorine gas with a low-melt metal such as

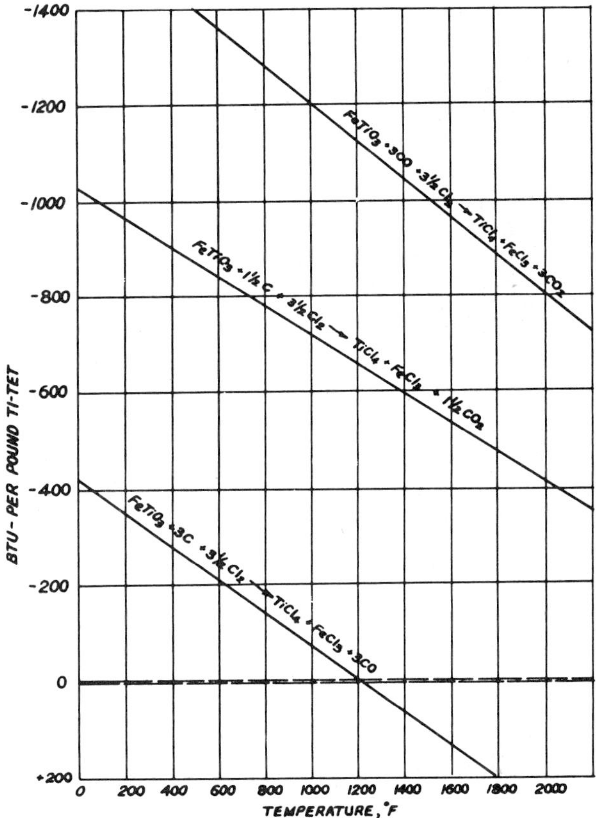

Figure 27-2. Heats of reaction of ilmenite, $\Delta H_{298°K}$ reactants in at 298°K, products out at reaction temperature.

aluminum. The reaction is highly exothermic and the heat liberated from the chlorination is used to melt the aluminum feed. Reactor construction, although requiring refractory, is fairly simple.[11] Chlorine conversion is high and the opportunity for chlorine corrosion quite low.

Moving Bed. The moving or static bed chlorinator for solids is the old standby of the industry. Its good and bad points are usually compared with those of fluo-solid beds which have become quite popular in the past decade. The moving bed reactor offers countercurrent flow of the gas to the solid reactants. This type of flow generally raises reaction efficiency that can be low due to poor gas-solids contact. A big advantage of counter-current flow as provided by the moving-bed reactor is that the sensible heat of product gases is recovered as preheat of solid feed.

Moving beds sometimes utilize feeds in which oxide-carbon mixtures have been sintered or briquetted together. The resultant intimate solids mixture can lead to chlorination temperatures which are 200 to 300° C lower than might otherwise be required. It must be recognized, though, that such feed preparation is costly.

If moving beds can be chlorinated at low temperature an important corrosion advantage is gained, and any heat input requirement is reduced. Going further, lower temperatures can mean a shift in product gas equilibrium from CO to CO_2 and thereby a possible change in the over-all reaction from endothermic-to-exothermic. Moving beds generally are less of a corrosion problem because they do not abrade and sometimes deposit layers of unreacting material on the reactor walls.

Moving beds have the disadvantage of nonuniform flow through the bed, and this leads to channeling of the gas, low chlorine conversion, hot spots, cold spots, bridging, etc. Briquetting often requires nonstoichiometric mixtures in order to get a strong and reactive briquette. In oxide-carbon mixtures, the nonstoichiometry usually takes the form of excess carbon. Continuous discharge of excess carbon, ash, etc. from a chlorinator can be difficult to impossible due to bridging of the bed, corrosion of moving parts, and the problem of gas tight seals. In any event, discharge of solids waste can lead to loss of reactive material.

Fluid bed Reactors. Fluid bed reactors were introduced commercially for the production of metal chlorides in 1954. Due to the intense agitation of fluidized beds, they have the advantage of uniformity. This means that variables such as temperature and reactant concentrations can be controlled, and thereby the rate and composition of product. They permit addition of solids to the reactor and their withdrawal from the bed.[12] On the other hand, fluid beds have poor solid-solid contact, require high temperatures and bring abrasion and chemical reaction right to the vessel wall. The solid feed particles must be sized in a fairly narrow range. These in turn are often reduced in size by reaction and attrition and tend to blow out of the chlorinator with the product gases.

In comparing fluid and moving bed reactors, it should be noted that they utilize different chlorine flows per unit chlorinator cross-sectional area. Fluid chlorinators are operated at one-half to one cubic foot per second per square foot. Moving bed reactors are at speeds well below this.

Recovery and Purification

Regardless of the chlorination step taken, two alternative methods are available for recovery of a condensable liquid-metal chloride product. These involve either a low temperature condensation using refrigeration or absorption-stripping using a minimum of refrigeration. Each system would be followed by the necessary number of continuous packed distillation columns. The chief advantage of the low-temperature condensation system is that the chlorination reaction may be operated independently from the purification system and vice versa. This flexibility is of great importance in a metal chlorides operation in which outages to rod-out plugged lines and equipment are frequent. When using an absorption-stripping system, reactor shutdowns upset column compositions from equilibrium, and considerable time is lost in regaining smooth operation.

If the metal chloride is a solid, it is best removed from the reactor off-gas stream by means of vertical scraped wall condensers. The physical properties of the main product and of the impurities dictate the additional purification methods that might be required. These include pressure sublimations, and the use of eutectic-melt compositions to tie up contaminants.

Instrumentation

Measurement of temperature in a reactor is difficult. In the moving bed, true temperatures cannot be recorded because temperature distribution is not uniform and because of controlling wall effects. In the fluid bed, it is difficult because the refractory thermowells used are corroded readily or are eroded off when placed in the bouncing bed. Since pressure measurements across the fluid reactor are vital, pressure taps are provided. These must be kept open by a steady purge of dry inert gas.

Instruments undergo serious external corrosion in a metal chlorides atmosphere. The sensing elements of the instruments undergo severe service because of accumulation of suspended solids, slimes and sludges. Purge gas for instrument blowbacks and other points in the system should be dry nitrogen or inert gas, to prevent formation of hydrolysis products and subsequent blockages. At certain trouble spots in metal chloride operations, manual control has had to be used as a final alternative in order to achieve satisfactory process control.

Special Metal Chlorides Problems

Solids deposition in reactor off-gas piping and equipment as well as solids dissolved and suspended in the condensed metal chlorides cause frequent blockages. Since these solids vary in properties for each process, there is no universal approach involved in selecting equipment for their removal. The vertical-frame type gas and liquid filters containing large surface areas have worked well in many cases. Vapor and liquid piping are generously oversized and numerous rod-out ports are provided in order to clean out plugged lines.

Where dissolved solids of relatively high vapor pressure are present in the liquid metal chlorides, removal even by evaporation is incomplete. In such cases fouling of heat exchange surfaces and column packing is inevitable. One must make allowances for cleaning the fouled surfaces periodically with water, observing proper safety precautions because of the fumes and acid wastes produced.

Corrosion tests on materials of construction such as refractories are extremely important. These tests can be made in a horizontal tubular furnace in which the sample is subjected to controlled compositions of metal chloride, CO and chlorine. If a fluidized bed reactor is to be used, duplicate samples can be tested simultaneously by placing one in the fluidized bed and the other in the vapor space, in order to determine the severity of service in each location. Typical corrosion test results, obtained under conditions similar to the vapor space above a fluidized bed titanium tetrachloride reactor are shown in Table 27-2. The material which gave the poorest test results, however, has proved to have superior resistance to the erosive action of fluidized bed, and for this reason, it is widely used in commercial reactors.

Process Examples

Good examples of commercial processes used for the production of liquid metal chlorides are those for the manufacture of titanium tetra-

TABLE 27-2. CORROSION TEST RESULTS

Temperature—1100°C.
Gas mixture composition—23% $TiCl_4$, 30% Cl_2, 47% CO

Classification	Code	Alumina (%)	Weight Loss (%) After Exposure of 283.5 hr	533.8 hr	Surface After Exposure
High alumina	A	44.0	0.91	1.53	very slightly chalky
	B	62.5	1.30	2.66	slightly chalky
	C	59.4	1.32	2.83	slightly chalky
	D	59.5	1.52	2.86	slightly chalky
	E	59.5	1.95	3.15	chalky
	F	48.7	2.07	2.50	sandy
	G	44.0	2.82	7.13	very chalky
	H	70.0	3.26	7.10	chalky
	I	42.2	5.44	10.5	slightly chalky
	J	44.0	6.03	9.23	very sandy
	K	45.0	6.19	8.72	chalky
Borosilicate	L		0.46	1.29	slightly chalky
Silica	M		0.50	2.22	very sandy
	N		0.96	3.62	very sandy
	O		3.13	5.34	sandy
	P		5.44	7.01	very sandy

chloride. One process formerly used consists of chlorinating the titanium cyanonitride directly. A variation of this process is the chlorination of a mixture of 50 percent titanium cyanonitride and 50 percent coke-rutile briquettes. The latter has several distinct advantages in that relatively simple equipment can be used, and the heat evolved in the chlorination is sufficient to maintain the reaction temperature. Since rutile is used for both the cyanonitride and the ore briquettes, the initial product is relatively pure titanium tetrachloride. The chief disadvantage of the process is that even with a 50:50 cyanonitride to ore briquette ratio, the cost of raw materials is still high.

The use of ilmenite has two practical commercial features in that it is a low-cost source of titanium dioxide and that its reaction with chlorine liberates more heat than the reaction of chlorine with rutile or titanium dioxide. Part of the saving in cost of the TiO_2 content would not be realized because of the added cost of separating the $TiCl_4$ and $FeCl_3$, the increased condensing problems, and the additional chlorine required. If the handling problems and the by-product $FeCl_3$ can be profitably disposed of or its chlorine content recovered, this would be one of the cheapest processes possible.

A Bureau of Mines process Figure 27-3 consists of reducing briquettes of carbon and rutile at a temperature of 1000° C or higher in a moving bed reactor. The reduction product which consists mainly of titanium suboxides with minor amounts of titanium carbide can be chlorinated at 205 to 540° C. A 79 percent titanium recovery is realized by reduction at 1000° C and chlorination at 500° C. A 96 percent recovery is obtained by raising the reduction temperature to 1200° C.

This process yields a product which can be chlorinated at low temperatures and therefore eliminates the refractory problem. It is a two-step process and consequently entails increased costs of handling materials. The

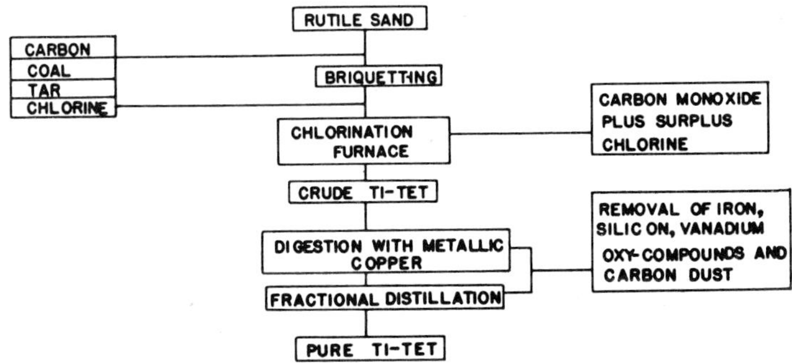

Figure 27-3. Process for production of TI-TET from rutile sand.

initial reduction is difficult to perform economically because of the high temperatures and the low-heat conductivity of the ore and carbon briquettes. The condensation problem is decreased because the product from the chlorination is not diluted with large amounts of non-condensables such as carbon dioxide and monoxide.

A Stauffer patent, U. S. Patent 2,486,912,[2] mentions the use of ferrosilicon to supply the heat necessary to maintain the temperature required for the chlorination of rutile briquettes. A charge containing 0.28 lb silicon/lb titanium dioxide is satisfactory. This method involves the separation of the $SiCl_4$ and $TiCl_4$, but since any chlorination product must be distilled for further purification the removal of the $SiCl_4$ does not present any difficulties. A serious disadvantage of this process is that any oxygen which gets into the system may produce siloxane, and on distillation a siloxane fraction containing 20 to 30 percent titanium tetrachloride is obtained. This results in a loss of titanium tetrachloride and presents a difficult disposal problem.

A simplified flow diagram for a typical fluid-bed process producing a liquid product such as titanium tetrachloride, boron trichloride or silicon tetrachloride is shown in Figure 27-4. The ore, coke and chlorine are fed

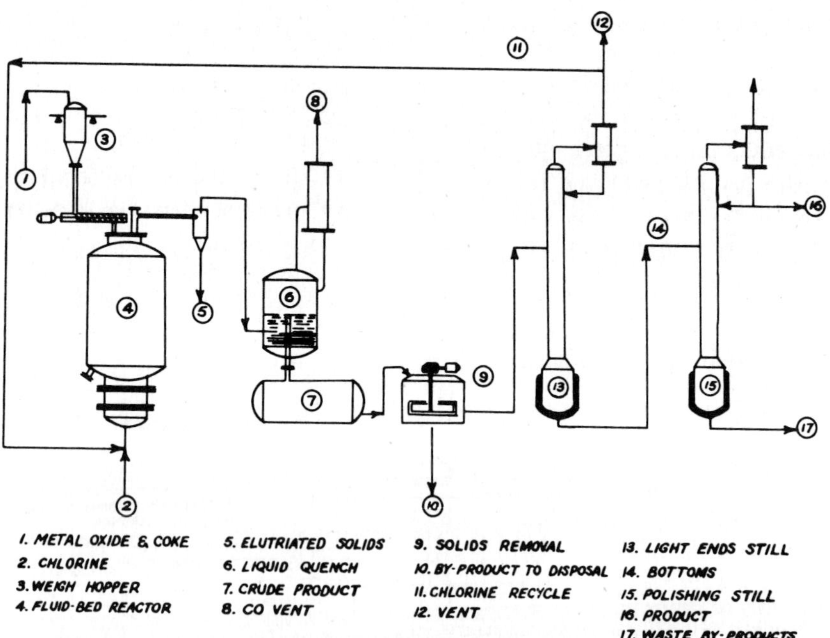

Figure 27-4. Typical metal chlorides process, liquid product.

to a refractory-lined fluidized bed reactor operated usually above 1000° C. The reactor off-gas passes to cyclones or settling chambers where the elutriated solids from the reactor are separated from the off-gas. The gas is then cooled, condensed, and collected as crude product. The solids present in the crude product consist of elutriated material such as coke and ore from the reactor, insoluble metal chlorides such as ferric chloride, and dissolved solids. These can be removed by filtration, centrifugation, or evaporation and drying methods. The solids-free crude is next passed through a distillation train in order to remove soluble by-products. The technical grade product can be sent to a C.P. unit finally for removal of trace impurities, which is usually accomplished by close fractional distillation or chemical treatment.[10]

Figure 27-5. View of a typical metal chlorides plant showing liquid bulk storage facilities.

Figure 27-6 typifies the discontinuous fluid bed used for producing a solid metal chloride such as aluminum chloride from bauxite. The calcined ore is mixed with coking coal and fed directly to a fluid-bed reactor where it is reacted with chlorine at 870° C. At start-up the reactor is heated to temperature with flue gases from an auxiliary heater. These ox-

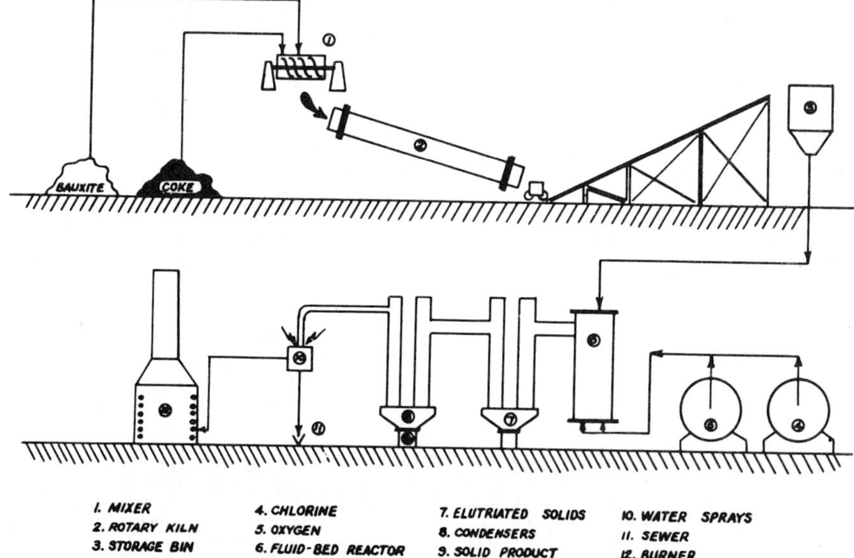

Figure 27-6. Typical metal chlorides process, solid product.[7]

ide chlorinations invariably require heat input which in this case is supplied by burning the coke with oxygen. The oxygen is fed into the reactor as a mixture with the chlorine gas.

Product gases leaving the reactor pass through a vertical U-shaped trap in which any unreacted but elutriated feed is withdrawn. The inside wall of the square cross-section channels are lined with insulating brick and require no additional heating. The solid product is condensed from the gas stream by means of vertical scraped wall condensers. Uncondensable gases such as CO leave the condensers and are burned before their discharge to the atmosphere.

STORAGE, TRANSPORTATION AND SAFE HANDLING

Aluminum Chloride

Since anhydrous aluminum chloride is extremely hygroscopic, caution must be observed in handling it. It forms hydrogen chloride in the presence of moist air. In storage, reaction between moisture and aluminum chloride may take place, causing the surface of the anhydrous aluminum chloride to become glazed with a coating of aluminum chloride hydrate, $AlCl_3 \cdot 6H_2O$. As some of the water of hydration penetrates the still anhydrous salt, reaction occurs, with formation of aluminum oxide and hydrogen chloride.

After a time so much hydrogen chloride may be evolved that the pressure of the gas bursts the container. Hydrogen chloride results in corrosion of iron containers.

Anhydrous aluminum chloride must, therefore, be packed in air-tight sheet iron drums. In technical operation, aluminum chloride is handled most expediently by a totally enclosed screw conveyor from a hopper to the reaction vessel. This prevents exposure to air.

Aluminum chloride is an acidic, nonflammable solid which fumes in moist air. Aluminum chloride should be handled in a well ventilated area and exposure to moisture should be minimized.

Should aluminum chloride come in contact with skin, eyes or clothing, they should be immediately flushed with large amounts of water. Eye exposures should be examined by a physician. Medical attention also should be sought if extensive areas are exposed or if persistent skin irritation follows contact.

Eye protection should be worn; protective clothing and respirators may sometimes be required. Spills should be flushed with copious quantities of

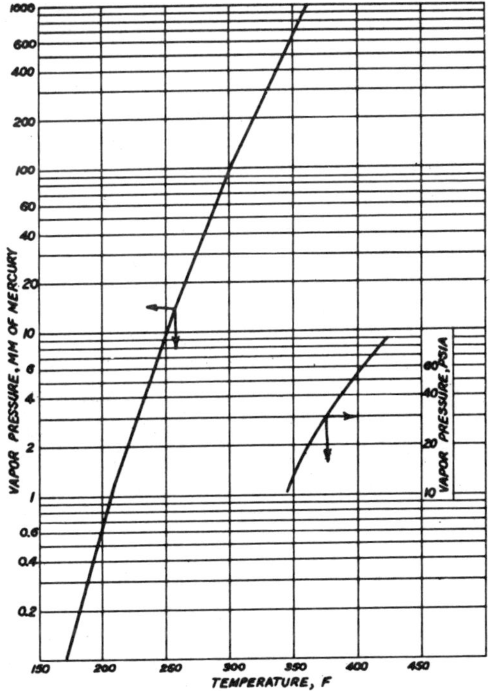

Figure 27-7. Vapor pressure of aluminum chloride.

TABLE 27-3. PROPERTIES OF ALUMINUM CHLORIDE

Formula $AlCl_3$

Molecular wt   133.34 (monomer) 266.68 (dimer)

The dimer exists in the vapor state and often in organic liquids.

	Solid	Gas
Color	white, light yellow, gray	white
Odor	–	pungent, irritating
Density	2.44 g/cc at 25 °C	0.0059 g/cc at 758°C
Melting point	194 °C at 2.5 atm.	
Sublimation point	180.1°C at 760 mm pressure	
Critical temperature		356.4°C
Critical pressure		29.4 atm.
Heat of sublimation	98.9 cal/g at 180.1°C	
Heat of formation	1248 cal/g	1138 cal/g
Free energy of formation	1142 cal/g	
Specific heat		0.1272 cal/g°C.
Heat of solution	26,400 cal for 1 mol. in 900 mols. water at 15.5°C	
Conductivity	vapors are electrical conductors, and the conductivity increases rapidly with rising temperature	

Temperature °C	146	189	209	227	236	245
Specific conductivity × 10^{-6}	0.29	2.6–5.0	0.66	0.86	0.96	1.10
Equivalent conductivity × 10^{-6}	–	–	23	31	35	41

water and then treated like muriatic acid. If possible, the dry powder should be picked up before the addition of water since adding water to aluminum chloride generates heat and acid fumes.

Details on handling are covered in the Manufacturing Chemists Association Safety Data Sheet, SD-62. No shipping regulations are in effect for aluminum chloride; however an M.C.A. warning label is recommended.

Aluminum chloride is shipped in containers that are air-tight sheet iron drums with fully removable heads. A venting screw for pressure release is furnished if requested.

Zirconium Tetrachloride

The material is dusty and in the presence of moisture forms hydrochloric acid fumes. Therefore, a closed system for handling and use is desirable. Ventilation should be adequate. Protective clothing and an acid vapor mask should be worn or readily available. A closed drum which has previously been opened should be re-opened with care. Vapor

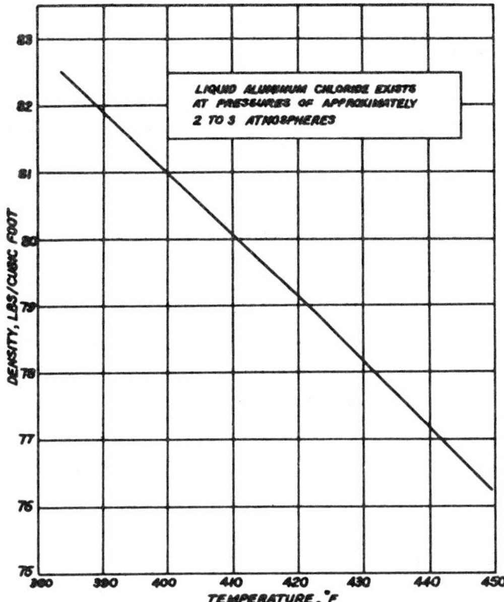

Figure 27-8. Density of liquid $AlCl_3$.

pressure in the drum will depend on how long it has been re-sealed, the temperature during storage and the ambient moisture. In the interest of safety, it is wise to keep drums out of direct sunlight and to avoid long storage in enclosed hot or damp areas.

Zirconium tetrachloride is shipped in 55-gallon drums with full-open heads and lever-lock closures. Shipments are also made in Sealbins of approximately 5 tons capacity, loaded under dry air.

Titanium Tetrachloride

This is a corrosive liquid. When in water solution, it becomes very strongly acid from excess hydrogen chloride formed by hydrolysis. Furthermore, when it comes in contact with the moisture of air, it hydrolyzes to evolve hydrogen chloride fumes, which are very toxic.

This material should be stored in a cool, ventilated place away from areas of acute fire hazard. Containers should be kept closed and carefully stored to avoid mechanical injury.

As long as moisture in any form is excluded, steel drums, tanks and piping are entirely suitable for storing and handling titanium tetrachloride. Water and drainage should be provided to take care of spills promptly.

Silicon Tetrachloride

Silicon tetrachloride has been used in warfare as an irritant gas and to produce smoke screens. In contact with the skin, eyes or mucous membranes of the body, it causes a caustic burning of the ocular and respiratory membranes and its capacity for destroying red blood cells is considerable.

When this material comes in contact with the skin, it causes a whitening of the exposed area followed by blistering. The utmost caution must be exercised in handling. Though highly corrosive to most metals in the presence of water, the anhydrous form does not attack iron or steel. Steel drums and tanks are used for shipping. Adequate ventilation should be provided; also water and drainage to flush spills quickly.

Boron Trichloride

This evolves hydrochloric acid vapors on exposure to moist air. Protective equipment should include rubberized asbestos, acid-proof clothing; safety goggles and/or shield and a "pocket type" cartridge respirator.

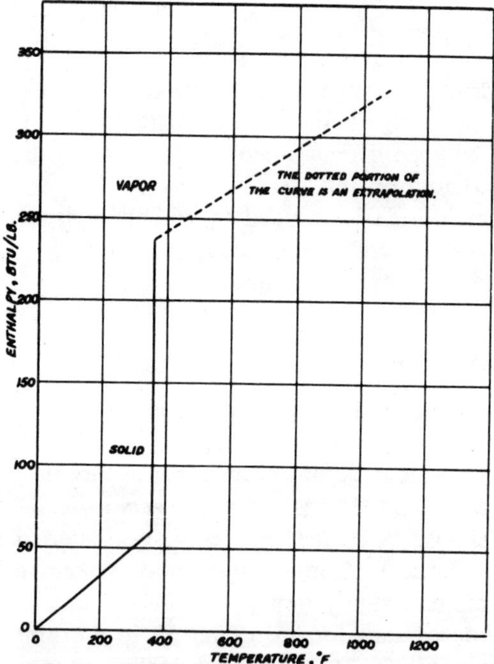

Figure 27-9. Enthalpy of aluminum chloride at atmospheric pressure.

Figure 27-10. Tote-bins for bulk shipment of aluminum chloride and zirconium tetrachloride.

Figure 27-11. Stauffer Chemical Company loading boron trichloride at Niagara Falls, New York.

TABLE 27-4. PROPERTIES OF TITANIUM TETRACHLORIDE

Formula: $TiCl_4$
Molecular wt.: 107.83

	Liquid	Gas
Color	colorless, fuming	colorless, forms white mist in contact with air
Odor	pungent, irritating	pungent, irritating
Specific gravity	1.726 at 20°C	
Boiling point	136.4°C at 760 mm	
Freezing point	−30°C	
Critical temperature		358°C
Coefficient of expansion	0.001086	
Dielectric constant	2.73 at 24°C	
Heat of formation	185 Kg cal	
Specific heat	0.188 cal/g°C	0.129 cal/g°C
Heat of vaporization	79.2 Btu/lb	
Heat of solution	57,870 cals at 17°C for 1 mole $TiCl_4$ in 1600 moles water	

Vapor pressure—temperature − 10 to + 45°C

$$\log P = \frac{A - B}{C} + t$$

$A = 7.09106, B = 1491.4, C = 220.0$

temperature 45 to 225°C

$A = 6.69428, B = 1287.91, C = 201.2$

Boron trichloride is shipped in cylinders weighing 100 and 1800 pounds. It is shipped also in tank cars that are 16 and 30 tons in capacity.

Antimony Chloride

Antimony chloride is a toxic material also known to cause allergic symptoms. Antimony and its compounds are irritating to the skin, often causing dermatitis. They cause severe burns because they are caustic materials.

Vanadium Tetrachloride

Vanadium tetrachloride, in the presence of moisture, forms hydrochloric acid which can, in contact with reactive metals, cause corrosion and release of hydrogen. A build up of sufficient hydrogen in a poorly ventilated area presents a fire and explosion hazard.

TABLE 27-5. PROPERTIES OF SILICON TETRACHLORIDE

Formula $SiCl_4$
Molecular wt.   169.89

	Liquid	Gas
Color	colorless, fuming	colorless, forms white mist in contact with moist air
Odor	pungent, irritating	pungent, irritating
Specific gravity	1.48 at 16°C	
Melting point	−70°C	
Boiling point	57.6°C at 760 mm	
Critical temperature		233.6°C
Refractive index	1.412	
Heat of vaporization	36.07 cal/g	
Heat of formation	−142.65 K-cal/g-mol	
Specific heat	0.132 cal/g°C	
Viscosity	0.472 cent. at 25°C	0.015 cent. at 25°C

Vapor pressure from −70 to +5°C

$$\log P = \frac{-52.23\,B}{T} + C$$

$$B = 30.1,\ C = 7.644$$

TABLE 27-6. PROPERTIES OF ZIRCONIUM TETRACHLORIDE

Formula $ZrCl_4$
Molecular wt.   233.05

	Solid	Gas
Color	white to cream	white
Odor	−	pungent, irritating
Specific gravity	2.80	
Bulk density	4.5 lb/gal	
Sublimation point	331°C	
Heat of sublimation	108.54 cal/g at 331°C	
Heat of formation	−995 cal/g at 25°C	
Heat of fusion	10.51 K-cal/mole at 25°C	
Specific heat	0.128 cal/g °C at 10°C	0.094 cal/g °C

Solubility—soluble in concentrated hydrochloric acid with reaction; in
oxygen-containing solvents such as methyl and ethyl alcohols, ethers, ketones and esters to form products soluble
in the solvent used; dissolves in water with a strong exothermic reaction to form zirconyl chloride ($ZrOCl_2 \cdot 8H_2O$).
Insoluble in carbon tetrachloride, carbon disulfide, benzene and aliphatic hydrocarbons.

TABLE 27-7. PROPERTIES OF BORON TRICHLORIDE

Formula: BCl_3
Molecular wt.:   117.19

	Liquid	Gas
Color	colorless, fuming	colorless, forms white mist in contact with moist air
Odor	pungent, irritating	pungent, irritating
Specific gravity	1.43 g/cc at $0°C$	0.005 g/cc at $13°C$
Boiling point		
Temperature $°C$	12.4 30 50 70 80	
Pressure (atm.)	1.0 1.8 3.2 5.4 6.8	
Critical temperature		$178.8°C$
Freezing point	$-107°C$	
Heat of fusion	4.3 cal/g at $-107°C$	
Heat of vaporization	48.48 cal/g at $12.5°C$	
Heat of formation	879 cal/g	806 cal/g at $25°C$
Free energy formation		0.776 cal/g at $25°C$
Heat of hydrolysis	5.9 cal/g at $25°C$	
Coefficient of expansion	0.0017 at $10°C$	
Surface tension	16.70 dynes/cm at $20°C$	

Viscosity (cs)	1.0	1.1	1.2	1.4	1.5		
Temperature $°C$	12.5	5	−2.2	−13.3	−17.4		
Vapor pressure (mm Hg)	18.0	116	251	760	1368	2432	4104
Temperature $°C$	−60	−30	−15	12.5	30	50	70

TABLE 27-8. PROPERTIES OF ANTIMONY CHLORIDE.

Formula: $SbCl_3$
Molecular wt.: 228.13

	Solid	Liquid
Color	White, pale yellow, gray	−
Odor	−	−
Specific gravity	3.14 at $20°C$	2.67 at $73.4°C$
Melting point	$73.4°C$	−
Boiling point	−	$223°C$
Heat of fusion	13.3 cal/g	−
Heat of vaporization	−	48.5 cal/g
Heat of formation	200 cal/g	−
Specific heat	0.11 cal/g $°C$	−
Dielectric constant	5.4 at $18°C$	33.2 at $75°C$
Conductivity	−	0.00000085 mho at $75°C$

Vapor pressure (mm Hg)	1.1	3.4	29	43	64	92	127
Temperature $°C$	50.3	59.9	120	130	140	150	160
Viscosity (ps)	0.0241	0.0148	0.0108	0.0097	0.0091		
Temperature $°C$	79.1	119.0	159.7	181.2	191.6		

TABLE 27-9. PROPERTIES OF VANADIUM TETRACHLORIDE

Formula: VCl_4
Molecular wt.: 192.78

	Liquid
Color	reddish-brown
Odor	pungent, irritating
Specific gravity	1.816
Melting point	$-26\,^\circ C$
Heat of formation	-141 K-cal/mole
Heat of vaporization	7.9 K-cal/mole
Boiling point	$148.5\,^\circ C$ at 755 mm

Decomposition of this material is accelerated by ultraviolet light, and as decomposition proceeds, enough internal pressure can be built up to cause a flame-sealed glass container to explode; therefore, delays in using this material must be avoided.

In case of inhalation, oxygen should be given and artificial respiration, if necessary. Vanadium tetrachloride should be kept in a flame-sealed glass container, under a second container to exclude light and stored in a closed hood with good ventilation.

Protective clothing must be worn along with safety glasses and face shield when handling. It is advisable to chill the sample below 20°C before opening to reduce the pressure. It should not be stored in con-

TABLE 27-10. SUMMARY OF PHYSICAL PROPERTIES OF METAL CHLORIDES

Formula	Molecular Weight	Color, Cryst. Form. R.I.	Specific Gravity (g/cc)	Melting Point (°C)
$AlCl_3$	133.34	wht, hex, cryst.	2.44	Subl. (188°)
BCl_3	117.19	col. liq.	1.43	-107
$CbCl_5$	270.20	yel-wht, del.	2.75	205
$CoCl_2$	129.85	blue cryst.	3.356	subl.
$CrCl_3$	158.38	violt, pl.	2.76	subl.
$FeCl_2$	126.76	gr-yel. hex.	2.70	677
$FeCl_3$	162.22	bk-brwn, hex.	2.804	307
$HfCl_4$	320.43	wht, cryst.	–	subl.
$NiCl_2$	129.60	yellow	3.544	subl.
$SbCl_3$	228.13	col., rho.	3.14	73
$SiCl_4$	169.89	col., fum. liq.	1.50	-70
$TaCl_5$	358.17	light yel. powd.	3.68	221
$TiCl_3$	154.27	vl. cryst.	–	730
$TiCl_4$	189.73	col. liq.	1.726	-25
VCl_4	192.78	red liq.	1.816	-26
$ZrCl_4$	233.05	wht. lust. cryst.	2.80	subl.

TABLE 27-11. SUMMARY OF THERMAL PROPERTIES[4]

Formula	Heat Capacity $Cp = a + bT + cT^{-2}$			Heat of Formation H_{298} K-cal/mole	Vapor Pressure °K				Heat of Fusion at Melting Point K-cal/mole	Heat of Vap; at Boiling Point K-cal/mole
	a	b	c		$T(10^{-4})$	$T(10^{-3})$	$T(10^{-2})$	Ti		
$AlCl_3$	13.25	28.00	—	−166.8	350	370	394	453	8.48	15.6
BCl_3	17.59	1.80	−2.82	−103.0	160	179	203	286	—	5.7
$CbCl_5$	20	10	—	—	355	377	410	516	6.9	13.1
$CoCl_2$	14.41	14.60	—	− 77.8	780	885	945	1323	14.1	37.6
$CrCl_3$	19.44	7.03	—	−132	800	945	1020	1220	—	56.8
$FeCl_2$	18.94	2.08	—	− 81.5	785	845	935	1299	10.3	30.0
$FeCl_3$	15	6.0	—	− 95.7	445	462	503	592	10.25	6.02
$HfCl_4$	25.5	—	—	−250	400	—	—	590	10.5	24.0
$NiCl_2$	13.10	13.00	—	− 73.0	862	934	1021	1260	18.5	53.8
$SbCl_3$	19.83	—	−1.19	−914	292	319	352	492	3.1	10.4
$SiCl_4$	24.25	1.64	−2.75	−150.1	—	207	235	330	1.85	6.8
$TaCl_5$	21	11	—	−127.4	355	380	412	507	8.2	13.6
$TiCl_3$	20.7	7.4	—	−165.1	900	—	—	—	5.0	33.0
$TiCl_4$	25.45	0.24	−2.36	−181.4	—	256	290	409	2.24	8.35
VCl_4	37	—	—	−141	240	269	300	437	—	7.9
$ZrCl_4$	31.92	—	−2.41	−230	427	460	499	604	10.5	25.29

tainers made of plastic or mild steel or in rubber-lined containers, and exposure to heat during storage is to be avoided.

If appreciable solid precipitate is observed in the sample, it should be discarded by opening it with caution in a hood behind a safety screen and then slowly pouring the liquid into a pail of cold water also in the hood; finally the container should be dropped into the water. If the sample has jelled, it should be placed in a metal container and sent to the chemical dump.

References

*1. Barksdale, J., "Titanium, Its Occurrence, Chemistry and Technology," New York, Ronald Press Co., 1949.

 2. Belchetz, Arnold (to Stauffer Chemical Co., California), U. S. Patent 2,486,912.

*3. Friend, J. N., "Textbook of Inorganic Chemistry," Vol. 6, Part 5, "Antimony and Bismuth," by W. E. Thorneycroft, London, C. Griffin & Co. Ltd., 1936.

 4. Glassner, A., "The Thermochemical Properties of the Oxides, Fluorides and Chlorides to 2500° K," ANL-5750, Superintendent of Documents, Washington, D. C.

 5. Hair, Zera L. (to E. I. du Pont de Nemours & Co.), U. S. Patent 2,784,058.

 6. Krchma, M. J. (to E. I. du Pont de Nemours & Co.), U. S. Patent 2,701,180.

 7. McAfee, A. M., "Cheap Aluminum Chloride," *Chem. Met. Eng.,* 422 (July, 1929).

 8. McKinney, J. J. (to E. I. du Pont de Nemours & Co.), U. S. Patent 2,701,179.

*9. Mellor, J. W., "A Comprehensive Treatise on Inorganic and Theoretical Chemistry," Vols. 5, 6, 7, 9, 11, London, Longmans, Green & Co., 1946, 1947, 1947, 1947, 1948.

10. Nowicke, Edmund S. (to Stauffer Chemical Co.), U. S. Patent 2,543,591.

11. Osborne, Sidney, and Rowland, Jasper (to Hooker Electrochemical), U. S. Patent 2,020,431.

12. Pechukas, Alphonse (to Pittsburgh Plate Glass Co.), U. S. Patent 2,311,466.

13. Quill, Lawrence L., "The Chemistry and Metallurgy of Miscellaneous Materials," New York, McGraw-Hill Book Co., Inc., 1950.

*14. Remy, H., "Treatise on Inorganic Chemistry," Vols. 1 and 2, London, Elsevier Publishing Co., 1956.

15. Saks, N. Irving, "Handbook of Dangerous Materials," New York, Reinhold Publishing Corp., 1957.

*16. Skinner, G. J., Johnston, H. L., and Beckett, C., "Titanium and Its Compounds," Columbus, Ohio, Herrick L. Johnston Enterprises, 1954.

*17. Thomas, C. A., "Anhydrous Aluminum Chloride in Organic Chemistry," New York, Rheinhold Publishing Corp., 1941.

*18. "Thorpe's Dictionary of Applied Chemistry," 4th Ed., Vols. 1, 2, 3, 10, 11, London, Longmans, Green & Co., 1937, 1938, 1939, 1950, 1954.

19. Warburton, D. M. (to Diamond Alkali Co.), U. S. Patent 2,594,370.

*Used generally, without citation, by the author and included for supplementary reading.

28. CHLORINATION OF ALKYLBENZENES

P. E. HOCH

Hooker Chemical Corporation

INTRODUCTION

The important role that chlorinated alkylbenzenes play industrially and academically can hardly be over emphasized. Appreciation of these materials in industrial usage requires more than a casual study of production figures of individual compounds. All of the materials are intermediates in chemical transformations that produce end products in practically every conceivable field. Chlorinated toluenes, for example, are intermediates in producing dyestuffs, perfumes, plasticizers, wetting agents, pesticides, herbicides, germicides, solvents, miticides, resins, lubricants, inhibitors for gasoline, pharmaceuticals, preservatives, cosmetics, mold controlling agents, dyeing assistants, to name a few.

In view of the limited number of producers and the ramification of end uses it is difficult to obtain statistics on market volumes for all of the materials.

In the ensuing paragraphs an attempt has been made to extract from the literature pertinent facts relating to the chlorination of alkylbenzenes. Specific attention has been given to methods of introducing chlorine into the side chain and the nucleus. Thermal and photochemical chlorinations are discussed. A review of catalysts for chlorinations is included. Pertinent chlorinating agents are listed and their applications outlined. The problems relating to orientation, current theories of mechanism of free radical and ionic chlorinations are discussed in detail. The chemistry of various derivatives, industrial usage, and market volumes are also included.

The major emphasis has been placed on those derivatives known to have industrial usage.

CHLORINATION METHODS

Addition Chlorination

Free Radical. Chlorine and benzene react rapidly at moderate to low temperatures in the presence of active light to yield addition products almost exclusively. This observation made in the early nineteenth century was the foundation for a host of industrial products based on benzene hexachloride, the most famous product being the insecticide lindane—the gamma isomer of benzene hexachloride.

In spite of considerable literature on the addition products of benzene there is only sparse reference to studies involving alkylbenzenes. Chlorine addition products of toluene have been prepared at low temperatures in the presence of light.[147] More recently Kharasch and Beckman[106] studied the chlorination of toluene, tertiary butylbenzene and meta-xylene. Toluene and tertiary butylbenzene gave side-chain substitution and addition products at zero degrees in the dark. Light and peroxides catalysed the reaction. These authors reported xylene absorbed chlorine without liberation of hydrogen chloride at low temperatures, but upon warming only nuclear substitution products were isolated. Addition products have been reported with toluene at $-20°$C when a peroxide catalyst is used.[213] A patent describes addition products of ethylbenzene at temperatures of $-30°$C with iodine present.[76] Experimental procedures and characterization of products derived from addition chlorination of toluene, ethylbenzene, dodecylbenzene and tetralin have recently been described[196] in the most informative study of addition chlorination to date. Additions were noted at -40 to $+20°$C with and without light.

The observed tendency for side chain chlorination rather than addition chlorination to take place photochemically at elevated temperatures may be related to the stability of the intermediate free radical. It may be related to the "vicinal" effect noted in the free radical chlorination of aliphatic hydrocarbons.[212,205] At elevated temperatures the tendency for the formation of 1,2-dichloro derivatives is practically nonexistent, and it has been suggested that this is related to the instability of the radical RCH—$CHCl$. The greater stability of the benzyl radical and the high stability of the aromatic ring are undoubtedly involved.

Ionic, Aqueous. One of the interesting means of adding chlorine to the aromatic ring, consists of reacting the aromatic material with chlorine in the presence of water and sodium hydroxide. For example, benzene and chlorobenzene rapidly add three moles of chlorine in the dark at temperatures under $10°$C. Only one percent of a base is necessary. Oxygen does not inhibit the reaction.[58] Water and chlorine in the absence of light give very little reaction. Light, of course, will promote the addition reaction.[84] No work of this nature has been reported for alkylbenzenes. Indications are, however, that alkylbenzenes give mainly nuclear substitution under these conditions.[93]

Radiation. Chlorination in the presence of gamma radiation reportedly gives an excellent yield of addition products with benzene and alkylbenzenes.[116]

Side Chain Chlorinations: Thermal, Non-catalytic

Introduction of chlorine into the side chain of alkylbenzenes as noted above is enhanced in the absence of catalyst or light by increased temperatures. There are many examples of simple thermal chlorinations of alkylbenzenes and chloroalkylbenzenes in the early literature.[19,30,186,105] In most of these procedures chlorine is introduced into the refluxing substrate or into a falling film of the substrate to give mixtures of alpha-chloro products. An extensive study of the thermal chlorination of toluene in the vapor phase (below 250° C) indicated exclusive formation of benzyl and benzal chloride.[128] Another study indicated that at 500° C water and sulfur had little effect on side chain chlorination while iodine, ferric chloride, and ammonia reduced side chain products and increased nuclear substitution. Thermal chlorination of toluene to benzyl chloride has been recently studied in glass-packed columns,[73,185] with surface active catalysts such as carbon, clay,[66] silica and the like. Rather extensive work on thermal chlorinations (150° C) of alkylbenzenes with and without light is reported.[188] Fluorine has been reported as a catalyst at high temperatures.[136] Vapor phase chlorination of benzal and benzyl chloride has been studied and exclusive side chain chlorination is noted.[187]

Air and phosphorus trichloride have been used to effect side chain chlorinations at 120° C.[149]

There have been controversial results recorded in the literature on the thermal chlorination of higher alkylbenzenes.[192] Chlorination of boiling ethylbenzene is reported to yield mainly beta-chloroethylbenzene.[8,75] Mixtures of both the alpha and the beta isomers also have been reported with light favoring the alpha isomer.[70,184] Diethylbenzene has been reported to yield the alpha isomers only.[66] Normal propylbenzene at reflux has been observed to chlorinate in the beta position,[69,77] while isopropylbenzene reportedly chlorinates in the nucleus.

Para-tertiary-butylbenzene chlorinates in the side chain while ortho-tertiary-butyltoluene substitutes in the ring.[182]

H. Goldwhite[80] recently has pointed out that most textbooks represent alphachloro derivatives as the predominant isomer formed, but the above examples indicate this is not the case. Recent work suggests that the site of chlorination does not necessarily depend on the stability of the free radical in the transition state. Evidence indicates that very little bond breaking takes place in the transition state. Resonance stability, therefore, in chlorinations is less important.

Photochemical Chlorination

There are many examples of thermal chlorination in the literature, and in fact some industrially available compounds may be prepared in this manner. The more common means, however, of introducing chlorine into side chains is by the photo mechanism. The use of light is particularly necessary for introducing more than one chlorine on the same carbon.

Photochlorinations have been effected by sunlight, tungsten lights, fluorescent tubes and mercury vapor lamps. The effect of controlled wave length and temperature on formation of specific products and determination of quantum yields has been the subject of extensive study in recent literature.[141,78,68,158,150]

Photochemical chlorinations of side chains are always accompanied by varied amounts of nuclear substitution and higher molecular weight materials.[88,23,208]

Nuclear substitution is promoted by water, trace metals and the like. To minimize the effect of these materials many water-sensitive substances and metal complexing agents have been used in chlorinations. These include phosphorus chlorides, oxychlorides,[72] manganese and arsenic compounds,[156,198,113] benzamides,[125] urea and hexamethylenetetramine[130] in conjunction with trialkyl phosphites.[140]

Sulfuryl chloride, phosphorus oxychloride and arsenic oxide have been suggested as useful materials[65] to minimize thermal decomposition of benzyl chloride.

Catalysts have been reported for photochlorinations which by their nature have a direct effect on the propagation of the free radical process. Thus peroxides,[213,31] diethyl ether or dioxane,[101] dithiazanes and camphor[206] have been suggested for such use.

Special dilution techniques have also been suggested for minimizing side products in photochlorinations.[99]

Chlorinating Agents: Sulfuryl Chloride, Peroxides

The reagent sulfuryl chloride-benzoyl peroxide has been used frequently for introducing chlorine into the side chain of alkyl and chloroalkyl aromatics.[193,56]

Early work with sulfuryl chloride (without peroxide catalyst) revealed nuclear chlorination, some side chain chlorination and considerable sulfonation. Kharasch and Brown[107] were the first to use benzoyl peroxide and sulfuryl chloride and demonstrated the value of this combination in introducing chlorine into the side chain of alkylbenzenes.

The effectiveness of this reagent is demonstrated with toluene. Toluene and sulfuryl chloride at reflux after many hours give low conversions to benzyl chloride. Light promotes the reaction, and good yields are obtained in seven hours. With one hundredth mole percent peroxide catalyst, however, an almost quantitative yield of benzyl chloride is formed in 15 minutes in the dark.

The reagent is of particular value in placing one or two halogens into a side chain. Utility has been found for it in placing side chain halogen in *para*-chlorotoluene, ethylbenzene and isopropylbenzene. The compounds, metaxylene and tertiary-butylbenzene in a normal free-radical chlorination give considerable nuclear substitution.[108] Sulfuryl chloride yields the side chain chlorinated products only.

Other examples in which this reagent has yielded desired side-chain products include the conversions of *para*-xylene to *para*-xylylene dichloride,[157,22] and 2,3,4,5,6-pentachloroethylbenzene to the β-chloroethyl derivative.[165,163]

One halogen can be introduced into ortho and *para*-xylene.[155] Tris-tri-chloromethylbenzene was prepared from chloromethylxylene.[115] Free radical catalysts other than peroxide have been used with sulfuryl chloride. Catalytic amounts of dimethyl-α-α'-azoisobutyrates and α-α'-azoiso-butyronitrile give excellent yields of benzyl chloride.[115a] It is claimed that the nitrogen released aids in sweeping out traces of oxygen and thus is more effective than peroxide.[137]

The reagent does not chlorinate diphenylmethane or nitrotoluene and yields nuclear substitution with methylnaphthalene and fluorene.

Peroxides and azo compounds with chlorine itself allow easy chlorination of side chains.

Chlorinating Agents: Free Radical

N-Chloro Compounds as Chlorinating Agents. There is a considerable amount of literature on N-bromo compounds as brominating agents, especially N-bromosuccinimide, N-bromophthalimide, and N-bromoacetamide. In the presence of peroxide or light, these reagents introduce bromine into the side chain of alkylbenzenes.[44]

The literature is somewhat sparse for studies using N-chloro compounds. Xylenes have been converted to the alpha halogen derivatives with N-chloro succinimide and ultraviolet light.[90] Side chain chlorinations are also reported to take place with N-dichloro derivatives of *p*-toluenesulfonamide where both chlorines are used; peroxide and iodine are cited as catalyst.

The N-chloro compounds may act as a source of positive halogen as well as free radical halogen. For example, N-chloroacetanilide gives nuclear chlorinated products of mesitylene, toluene, ethylbenzene, xylenes, etc. in the presence of acetic acid.[10,11]

Trichloromethanesulfonyl Chloride.[95] Trichloromethanesulfonyl chloride was observed to react with certain alkyl aromatics in a reaction induced by mild photochemical conditions or by catalytic amounts of benzoyl peroxide. Equal amounts of chloroform, sulfur dioxide and alpha-chloroalkyl aromatic are formed. A chain reaction is proposed.

$$Cl_3C\cdot + RH \rightarrow HCCl_3 + R\cdot$$

$$R\cdot + Cl_3C\,SO_2Cl \rightarrow RCl + Cl_3C\cdot + SO_2$$

Toluene, ethylbenzene, *para*-xylene, and *alpha*-chloro-*para*-xylene, are chlorinated in this manner.

Tertiary-butyl Hypochlorite. This reagent is generally used in nuclear chlorinations involving reactive aromatics such as phenols, etc.[2] A more recent study involved halogenation of toluene with *tert*-butyl hypochlorite at 40° C catalyzed with azobisisobutyronitrile or light.[100] The products were benzyl chloride and a small amount of benzal chloride. The reaction is highly exothermic. The effect of various substituents in the *para* position using a Hammett treatment gave *rho* values of –0.83 indicating qualitatively that electron withdrawing groups retard and electron supplying groups accelerate attack of the radical on toluene. Relative reactivities for a large number of alkylbenzenes are recorded.

Liquid Chlorine. Liquid chlorine in the presence of iodine, ferric chloride, or light is a very powerful chlorinating agent.[117,118,119]

Liquid chlorine at atmospheric pressures (–30° C) in the absence of a catalyst will react slowly with alkylbenzenes. In the presence of iodine, however, vigorous nuclear substitution takes place. For example, benzene, toluene, ethylbenzene, cumene and xylene form di- and trichloro derivatives. At 60° C trichlorotoluene can be converted to pentachlorotoluene.

Liquid chlorine quantitatively introduces two chlorines into the side chain of pentachlorotoluene at 30° C in the presence of light.

More surprising and unexpected is the effect of ferric chloride. This catalyst leads to a vigorous reaction with toluene even at low temperatures, *para*-xylene is readily converted to a hexachloro derivative believed to be tetrachloro(bischloromethyl)benzene. Pentachloro or trichlorotoluene at 60° C gives pentachlorobenzyl chloride in quantitative yields. This introduction of chlorine into a side chain with iron as a catalyst is interesting. Butylbenzene is converted into a hexachloro derivative, one chloride entering the side chain.

Another observation is the quantitative cleavage of pentachlorobenzyl chloride to hexachlorobenzene with liquid chlorine (under pressure) using ferric chloride catalyst, at 70 to 80° C.[36]

Electrolytic Chlorinations

Electrochemical chlorination of organic compounds has received sporadic attention in the literature. As with many electro-organic reactions, lack of reproducible results with different investigators presents a confusing picture. Reese[156] summarized the early literature. Both ring and side chain chlorination of toluene have been observed at carbon electrodes; platinum electrodes in the presence of light gave appreciable side chain substitution with toluene. Nuclear substitution of toluene was the dominant reaction in glacial acetic acid. A pressed graphite-sodium chloride-potassium iodide electrode has been used to chlorinate toluene[28] to give only nuclear substitution.

Miscellaneous Chlorinating Systems

Several chlorinating systems resulting in side chain halogenation have been reported in the literature but have received very little study. Aqua regia is reported to yield large amounts of benzyl chloride when the ratio of hydrogen chloride to nitric acid is high.[57]

Considerable side-chain halogenation (20 percent) has been noted in uncatalyzed chlorinations of toluene in ethylene dichloride and carbon tetrachloride with hydrogen chloride present.[7] It is considered most unlikely that hydrogen chloride can serve as chain initiator.[64] An ionic mechanism for addition of chlorine to olefins in general has been proposed.[204] The formation of side chain chlorination in ionic chlorination with alkylbenzene can be explained by a loss of a proton from the alkyl group in the transition state with a subsequent allylic rearrangement of the halogen taking place.

Mechanism

Excellent discussions concerning free-radical chlorination mechanism have appeared in recent texts.[92,82] Extensive experimental evidence is reviewed also by E. W. R. Steacie.[198,194]

The reaction of chlorine with aliphatic and alkyl aromatic is well established as a radical chain process:

$$Cl\cdot + RH \rightarrow HCl + R\cdot$$

$$R\cdot + Cl_2 \rightarrow RCl + Cl\cdot$$

With compounds containing more than one type of carbon-hydrogen bond the products obtained will depend upon the relative rates of reaction of chlorine with the carbon hydrogen bonds available. It has been established that with aliphatic hydrocarbons,[85,152] in the vapor phase the ease of hydrogen displacement decreases in this order: tertiary > secondary > primary. Bond dissociation energies follow the order primary > secondary > tertiary. This bond weakening generally has been attributed to resonance stabilization of the radical in the transition state.[153]

In recent studies of photochemical chlorinations of more complex materials and, more pertinent to the present studies, of alkylbenzenes, it appears that resonance stabilization is perhaps less important than previously thought. Thus, in competitive chlorinations of toluene and *t*-butylbenzene, the C—H bonds of toluene are only slightly more reactive than *t*-butylbenzene, in spite of the large resonance stabilization energy potentially available with toluene.[174,208] The same type of data are reflected in comparing the energy of activation difference between the tertiary hydrogens in cumene and a pure aliphatic hydrocarbon. The energy of activation difference is only 1 kcal., whereas the bond energy difference is estimated to be over 20 kcal.[175] Even more striking have been the observations that the carbon-hydrogen bonds in cyclohexane are more reactive than those in toluene by a factor of 2.7.[174]

The apparent lack of influence of resonance stabilization in free radical chlorinations in the examples above has been attributed to only partial bond breaking of the carbon-hydrogen bond by the chlorine atom in the transition state. If this is the case, factors such as resonance would contribute little to the activation-energy picture. Other data such as lack of steric influence in reactivity of aliphatic hydrocarbons to photochemical chlorination has been cited to support this concept.

The other major factor influencing orientation in free-radical chlorination is the polar effect (inductomeric). The chlorine molecule and the chlorine atom are electrophilic in character, and will tend to substitute hydrogen preferentially at positions of higher electron density.[9] Thus, for example, tertiary hydrogens would be displaced preferentially not only because of stability of the free radical in the transition state but also because alkyl groups would give a maximum electron density by virtue of their inductive effect. Directing effect of groups in free radical chlorinations has been summarized.[38] This existence of the polar factor in free radical chlorination is, perhaps, most adequately illustrated by studying the effect of *meta* and *para* substituents on the side chain halogenation of benzene. The results can be treated in terms of the Hammett equation $\log k/Ko = \sigma \rho$ where k and Ko are rate constants of substituted and unsubstituted toluene respectively and σ a parameter characteristic only of the substituent and represents the ability of the group to attract or repel electrons, and ρ is characteristic of the reaction—in this case free radical halogenation. If then the Hammett σ values (determined for ionization constants for substituted and unsubstituted benzoic acids wherein ρ is taken as unity) are plotted against the log ratio of the rate constants for the chlorination of substituted and unsubstituted toluene, the slope of the line will give the ρ values. Negative values obtained for ρ indicate that the reactions are made more difficult by electron withdrawing substituents.

Several investigators have applied the Hammett treatment to photochemical chlorination using chlorine and sulfuryl chloride and have obtained negative *rho* values.[211,139] The negative values obtained indicate that chlorinations are facilitated by electron donating groups.

There are still other factors that influence orientation in chlorination. Thus, for example, the nature of the chlorinating species and the effect of solvent environment play a role. It has been shown that sulfuryl chloride chlorinations do not give the same distribution of isomers in aliphatic chlorinations as does photochemical chlorination. It has been suggested that the chain propagation for sulfuryl chloride be modified as follows:

(1) $R\cdot + SO_2Cl \rightarrow RCl + SO_2Cl\cdot$ (2) $SO_2Cl\cdot \rightarrow SO_2 + Cl\cdot$

(3) $RH + SO_2Cl\cdot \rightarrow R\cdot + HCl + SO_2$ (4) $RH + Cl\cdot \rightarrow R\cdot + HCl$

In some instances it is suggested that $SO_2Cl\cdot$ is the chlorinating agent, and that being less reactive than the chlorine atom, it exhibits different selectivity. As indicated previously, chlorine atoms are electrophilic in nature and it has been suggested that such atoms can form loose complexes with the aromatic ring.[175,217] Photochemical chlorinations may involve attack by a chlorine-aromatic complex rather than a chlorine atom.

Large solvent effects have been noted in competitive photochlorinations of primary, secondary and tertiary carbon hydrogen bonds.[176,177,178,179] An excellent summary of the solvent effects appeared recently.[181] Aromatic solvents display an effect proportional to their basicity, suggesting the complexed chlorine described above. It has generally been noted that a solvent effect is not as apparent in chlorinations wherein polar factors are controlling, but the solvent effect is very apparent in chlorinations where control by resonance factors is important.

Steric Factors in Free Radical Chlorination

There are many examples of the influence of steric factors in side chain halogenation. It has been observed that no more than two hydrogens can be replaced in a methyl group substituted on a benzene ring when substituted by chlorine in both *ortho* positions.[19,117,88] Perchloromesitylene has been reported prepared from trichloromesitylene but later attempts[88,17] to repeat this work have yielded only the trichlorotris(dichloromethyl)benzene. 2,3,5,6-Tetrachloro-*para*-xylene upon photochlorination yields the bis-dichloromethyl derivative.[132]

Pentachloroethylbenzene chlorinates in the β-position photochemically with chlorine and with sulfuryl chloride.[165,163] Liquid chlorine placed three halogens in the side chain of 2,3,4,5,6-pentachloroethylbenzene, and these are presumably in the β-position.[117] These results are presumably due to steric crowding, though as observed earlier ethylbenzene itself substitutes in the β-position,[80] and it is reported that bromination of 2,3,4,5,6-pentachloroethylbenzene occurs in the α-position.[165]

Photochlorination of monochloroxylenes to the bistrichloromethyl derivatives is achieved with relative ease. Two chlorines in the nucleus virtually prevent formation of the bistrichloromethyl derivative.[33,83]

When methyl groups on the benzene ring are *ortho* to one another only the pentachloro compound has ever been isolated.[189] Thus *o*-xylene gives *o*-trichloromethyl benzalchloride.[155] Pseudocumene gives the octochloro derivative.[189] Durene yields the decachloro compound.[93]

Nuclear Substitution of Chlorine

Alkyl-substituted benzenes can be attacked by chlorine to yield side chain substituted products, addition products of the ring, and nuclear substitution products. As noted above, side chain chlorination is favored by high temperatures and light or a free radical source. Addition is favored by low temperatures and can be initiated by light. The presence of water, however, promotes nuclear substitution at the expense of side chain substitution. Addition chlorination occurs with benzene and chlorobenzene in aqueous systems, but qualitative studies with alkylbenzenes indicate that nuclear substitution is favored over addition. Aqueous chlorinations have received considerable study over the years and a more detailed discussion of history and mechanism is included later. Water serves to enhance the electrophilic character of the chlorine molecule either by favoring charge separation or by aiding in the withdrawal of the chloride ion from a complex in the transition state. From a practical standpoint, the electrophilic character of chlorine is normally enhanced by the so-called chlorine carriers or catalyst.

Chlorinating Catalyst

Although nuclear chlorination of polyalkylbenzenes does take place to a certain extent in the absence of catalysts practical rates are only obtained with their use. The earliest studies of the solution chlorination of alkylbenzenes utilized iodine, antimony, aluminum, ferric and molybdenum halides as catalyst.[19,186] A thorough study of the chlorination of toluene and chlorotoluene was made using an aluminum-mercury couple as catalyst.[51] A lead chloride-ammonium chloride catalyst was claimed to yield only ortho substitution; however, a later attempt to repeat this observation was negative[214] though the later workers did find lead chloride favored ortho substitution at 700° C.

In more recent years, a thorough study of the effect of catalysts on orientation in chlorination studies of toluene has been made.[188] Titanium tetrachloride favors ortho chlorination (73 percent), while antimony penta-chloride favors para substitution (50 percent). *Para* substitution is favored at $-10°$ C, and antimony sulfide is used. Aluminum chloride was considered a very poor catalyst.

In the recent literature, other catalysts have been reported, most of them for use at elevated temperatures. Rubidium and iridium salts in combination with active clays, alumina and platinum and alumina with manganese salts are reported.[35] Boron trifluoride diacetate was used to chlorinate toluene.[221] A catalyst study with hydrogen chloride, chloroacetic acids, stannic chloride, and ferric chloride was reported.[121] Iodine chloride has been observed to catalyse chlorination of polyalkylbenzene.[5]

Aqueous Chlorination

The study of aqueous chlorinations has received a considerable amount of attention in the literature. The major portion of the work has been undertaken to establish the active chlorinating species for a given sub-strate through kinetic studies. The studies are primarily of academic in-terest, though preparative procedures in which the chloronium ion (Cl^+) is believed to take part have been reported.

When chlorine is dissolved in water a complex array of potential chlorin-ating species is possible. These include in order of electrophilic activity Cl^+, ClH_2O^+, Cl_2, Cl_2O, $HClO$, OCl^-, Cl_3^-. If acetic acid is present acyl hypochlorite is also possible. Actually over the past fifty years each of these species have been postulated as the chlorinating agent in a particular study. With such a complexity of chlorinating species possible, it is a small wonder that there have been so many conflicting opinions.

Over the past ten to twelve years the picture has been clarified somewhat through very careful kinetic studies wherein experimental conditions have been chosen to limit the possible active agents. Excellent reviews of the subject have appeared in the literature.[97,60,161]

It is generally conceded now that with solutions of chlorine in water or acetic acid, molecular chlorine is the chlorinating species, rather than hypo-halites, positive chlorine atoms, or other species.[159,160,161] The kinetics are clearly second order for phenol ethers, anilides and, more pertinent to our discussion, alkylaromatics.[12,42,159] Relative reactivities of alkylbenzenes have been established in acetic acid and indeed such studies have contrib-uted considerably to the understanding of the mechanism of chlorination of aromatic systems.[162]

Kinetic studies of chlorinations in aqueous medium are greatly simpli-fied if pure hypochlorous acid is isolated and used in the study. The picture is then made simpler by working with dilute but strongly acid solutions and having silver or mercury ions present to remove chloride ions as they are formed. The high acid concentration favors protonation of hypohalous acid and promotes formation of a positive chlorine as follows:

$$HClO \rightarrow Cl^+ + OH^-$$

The silver ions react with the chloride ion formed in the substitution to prevent formation of molecular chlorine. Such procedures have yielded experimental data establishing beyond reasonable doubt that the positive chlorine (chloronium ion) or its hydrate is the chlorinating agent.

The chloronium ion should be the ultimate in activity and indeed preparative procedures utilizing it have been published.[81] The general procedure consists of dissolving the aromatic material in 85 percent sulfuric acid with silver sulfate present, adding an excess of chlorine in carbon tetrachloride, then shaking thirty minutes. Thus, *para*-nitrotoluene and benzoic acid gave almost quantitative yields of the monochloro derivative.

Kinetic studies with positive chlorine in chlorination of toluene show that its activity and orientation compared to the nitronium ion and the bromonium ion are essentially the same for the *meta-para* positions. (A $\frac{1}{2} m : p$ ratio is essentially constant for all three reagents). Comparative data on *ortho* substitution indicate that the chloronium ion is the least effected by the steric shielding of the methyl group, the order being $NO_2 >$ $Br > Cl^+$.[62] The chloronium ion also shows a higher $o : p$ ratio than chlorination with the chlorine molecule for toluene and *t*-butylbenzene, reflecting the decreased size of the positive ion. A decreased selectivity should be observed for the chloronium ion as has been established for the iodonium ion.

A recent paper has reviewed aqueous chlorination and reported a study of the halogenation of anisole derivatives using hypohalous acid in 40 and 75 percent acetic acid.[197] It is pointed out that actually the chloride ion formed does not significantly effect the kinetics in 75 percent acetic acid. The chlorinating agent is believed to be an acyl hypohalite present in small concentration. These conclusions differ somewhat from results of earlier workers[60] where acid catalysis in chlorinations with glacial acetic acid and hypochlorous acid was observed. The only explanation for these differences is that hypochlorous acid can function in different ways depending on conditions and substrate.

Chlorinating Agents

A variety of reagents other than chlorine itself has been used to bring about nuclear chlorination. As far as is known, none of them have any commercial significance in the chlorination of alkylbenzenes.

The most widely studied reagent is sulfuryl chloride.[56] This material was used by several early workers to chlorinate alkylbenzene, but it was never used effectively; many side products and poor yields resulted. Silberrad[191] discovered that sulfuryl chloride in conjunction with small amounts of aluminum chloride and sulfur monochloride was a very powerful and useful chlorinating agent. Quantitative yields of alkylchlorobenzenes were claimed. The reagent gives somewhat different ratios of isomers than those obtained in direct chlorination. For example, only 2,4-dichlorotoluene is formed. Formation of pentachlorotoluene in quantitative yields without cleavage is claimed.

The reagent has been of particular value in placing four halogens into benzotrichloride with only mild cleavage occurring.[15] More recently trichloro benzotrichloride and 1,4-bistrichlorobenzene have been chlorinated to the perchloro derivatives for the first time, using a large excess of the reagent and a prolonged reaction period. Cleavage was also observed; in fact, hexachlorobenzene was readily formed from perchlorotoluene with the reagent.[17] Tris(trichloromethyl)benzene was uneffected by the reagent while ethylpentachlorobenzene gave incomplete chlorination and cleavage.

The sulfuryl chloride reagent also has been used to place four chlorines in the nucleus of orthoxylene[168,165a] and five chlorines on β-chloroethylbenzene.[165] Cleavage reactions with the reagent using iron in place of aluminum chlorine have been described.[166] Side chain chlorination of ethyl pentachlorobenzene is reported.[163]

Miscellaneous Chlorinating Techniques

Other means of introducing chlorine into aromatic nuclei include the use of liquid chlorine,[117] a very effective reagent with iodine or ferric chloride as catalyst. Electrolytic chlorination has been reported sporadically in the literature. Recently a special carbon-potassium iodide electrode has been used.[28] Phenyl iodonium chloride, nitrogen chloride and ferric chloride itself have been used to chlorinate alkylbenzenes.[44,111]

Chlorinations have also been reported using a water emulsion technique. A 75 percent yield of *ortho*-chlorotoluene is claimed with toluene.[71]

Mechanism

There are excellent text books published in recent years giving fine discussions of the subject of electrophilic chlorinations.[97,92,61]

The non-free radical substitution of hydrogen by chlorine in an aromatic ring is an electrophilic process. The ease of chlorination will increase when a substituent on the ring increases the electron density of the ring. Thus, a substituent such as the amino, hydroxyl, or alkyl group by virtue of its positive inductive and conjugative effects increases the electron density of the ring and facilitates chlorination. A substituent, such as the nitro group which has a negative inductive effect, will withdraw electrons from the ring, making chlorination more difficult. The former substituents will not demand a chlorinating species of high electrophilicity, thus chlorine itself reacts very rapidly with phenol or aniline derivatives; on the other hand nitrobenzene and other substituted benzenes of this type require a highly electrophilic chlorinating agent.

The discussion under aqueous chlorinations mentioned the chloronium ion (Cl^+). This ion should be a most active chlorinating species. There is reasonably conclusive experimental data that this ion takes part in certain chlorinations when proper care is taken to throttle competing side reactions. Such an ion, however, has an experimental basis for existence only in hydroxylic solvents such as water and acetic acid. Practical preparative procedures have been reported and indicate that difficultly chlorinated compounds do chlorinate with relative ease by this method.

Generally, for other than aqueous chlorinations, chlorination is catalysed by the addition of various Lewis acids as noted above. The catalysts which differ in their effectiveness polarize the chlorine-chlorine bond at some stage of the chlorination process.

The exact role of the catalyst, and the stage at which it enters into the chlorination is still not completely clear.

Since the late 1930's and, in fact, in most current text books, the catalyst role was pictured rather simply.[151] The neutral chlorine molecule formed a complex with the catalyst which tended to aid the separation of positive and negative chlorine. The positive end of the complex attacks the aromatic ring in a rate controlling step; then a rapid proton removal occurs with concurrent regeneration of catalyst and loss of hydrogen chloride. The effectiveness of the catalyst in forming such a complex is presumed to have a direct effect on the rate of chlorination.

Over the past twelve years a somewhat different picture of chlorination and the role of the catalyst has been presented. The first step in the chlorination is believed to be formation of an aromatic halogen complex.[161] It is known from spectroscopic and solubility data that chlorine forms loose pi complexes with the pi electrons of the aromatic ring.[145] Such pi complexes have also been demonstrated for hydrogen chloride.[39] There is also ample evidence for sigma complexes. Thus it has been observed that when aluminum chloride is present hydrogen chloride forms an entirely different type complex with aromatic hydrocarbons having physical properties for an ionic structure.[40] It is suggested[61b] that both the loosely held pi complex and the sigma complex may play a role in electrophilic substitutions. The pi complex may occur at the beginning of the reaction path and the deeper rooted sigma complex may actually be the transition state of the reaction.

The catalyst might be pictured than as assisting in the chlorine-chlorine bond cleavage in the complex above by a rate-determining nucleophilic push.

In a chlorinating medium of low polarity such as carbon tetrachloride the catalyst can be the halogen molecule itself or the hydrogen halide.[4] When a polar medium is used the sensitivity to extraneous polar substances is reduced.

The reaction rate of chlorination changes with the solvent. The activation energy rather than entropy appears to be the controlling factor in the rate of chlorination of alkylbenzenes. The differences depend on several factors all of which bear on their effect on the transition complex described above. Thus, the increased rate for chlorinations in trifluoroacetic acid over acetic acid is believed to be due to the increased ability in sharing protons with the negative chloride in the parting complex.[7] The effects of solvents on chlorinations are believed to fit in with the mechanism of preliminary complex of halogen and aromatic substrates.[5,104]

ORIENTATION

The earliest studies of chlorination of alkylbenzenes established the fact that the methyl group gave mainly *ortho* and *para* chlorination products. The relative amounts of *ortho, para* and *meta* substitution and the rate of chlorination are influenced by several factors which will be discussed in the following paragraphs. These factors are:

  (1) the number and type of alkyl groups, size and degree of branching
  (2) the chlorinating species
  (3) the solvent

The methyl group is an electropositive group placing an electrostatic negative charge on the benzene ring by the permanent inductive effect and by the inductomeric effect called into play in the transition state. The inductive effect in a saturated system decreases sharply with distance from the primary pole. In an aromatic ring the inductive effect should be more effective in the *ortho* position less in the *meta* position and least in the *para* position. Studies directed toward measurement of inductive effects are usually made on the *meta* position. The *ortho* position is effected by steric and direct field effects.[67,32,45,27]

In the saturated system the inductive effect is increased with increased branching. Thus, the effect increases in the series $H < CH_3 < CH_2CH_3 < HC(CH_3)_2$. Studies of electrophilic substitution of the benzene ring with such groups attached have indicated conclusively that simple inductive effects do not explain all the facts.[13] Electrophilic substitutions, including chlorination have been studied with toluene wherein the *para* position is substituted with alkyl groups with increased branching.[203] LeFevre[120] summarized these studies and noted that substitution *ortho* to the methyl group was the favored position which is contrary to the expected inductive effect. LeFevre explained this effect on the basis of steric shielding of the positions *ortho* to the branched alkyl groups. Another explanation for these observations was made by Baker and Nathan, based on their concept of hyperconjugation.[13]

Hyperconjugation or "no bond resonance" is the term originally applied to the apparent tendency of the methyl group to donate electrons and enter into conjugative effects. The term has been extended to include any three single bonds when attached to a carbon which is conjugated with a multiple link. Such effects were first used to explain the apparent reversal of the expected inductive trend in displacement studies of alkyl, substituted benzyl and benzhydryl halides. Increased branching on the *alpha* methyl group of the alkyl substituent decreased the rate of displacement contrary to the expected inductive trend.

There has been and still is a degree of controversy in extending the argument of hyperconjugative effects into aromatic electrophilic substitution. Thus, for example, LeFevre's conclusions above may be the correct interpretation especially for the nitration results. It has been demonstrated[52] that in comparing the rate of nitration of the individual nuclear positions of toluene and *tert*-butylbenzene that the most important contribution of the alkyl group is their inductive effect. The *tert*-butyl group, has in fact, a higher rate of nitration.

Hyperconjugative effects, however, are brought into play in bromination and halogenation. Toluene is brominated much faster then *tert*-butylbenzene. *Para* substitution is much higher with tertiary butylbenzene and this is attributed to the steric factor.[162,25,59,218] There has been a difference of opinion in interpreting the rate difference. It has been argued that the tertiary butyl group statistically reduces available positions for attack, this would be reflected in a reduction of over-all rate.[14,61b] Other investigators argue that this should have but little effect, the substitution rate in the *para* position would simply increase.[25,59]

The relatively high rate of halogenation of tertiary butylbenzene compared to that of benzene is not believed to be due solely to the inductive effect, and of course, the lack of *alpha*-hydrogens prevents normal hyperconjugation. The enhanced activity is explained by second order hyperconjugation of the carbon-carbon bonds in the tertiary butyl group.[26]

These concepts have met with heavy criticism.[14]

Bulk or steric effects as well as the hyperconjugative effects thus can greatly influence the position taken by the chlorine atom. The directive effect of the methyl group can also be drastically effected by substituting halogens for the *alpha*-hydrogens. In going from toluene to benzyl, benzal and benzotrichloride, the rate of nitration decreases, and the orientation shifts from 4 percent *meta* to 64 percent *meta*. This effect involves the steric as well as the negative inductive effect of the halogen. Introduction of a chlorine into the ring of an alkyl benzene tends to shift the orientation in favor of the chlorine.[34, 199]

The size and electrophilic potency of the reagent is known to effect the orientation. In chlorinations, the chlorinating agent can be molecular or ionic. Toluene chlorinated with acidified hypochlorous acid wherein the electrophilic reagent is the Cl^+ ion gives a higher *o:p* ratio then chlorinations with molecular chlorine.[62] This effect is believed to be due primarily to the size of the attacking electrophilic species.

This increase in *para* substitution with the less active chlorine molecule may also be partially explained in terms of the electronic demand of the reagent. The hyperconjugative effect would activate the *para* position more then the *ortho* and a less reactive electrophilic reagent could permit this effect. The more highly reactive Cl^+ ion would also be less selective and yield more *meta* substitution.

Catalyst or chlorine carriers discussed previously have been claimed to effect the *ortho-para* orientation in alkyl benzenes. It is quite possible that these observations in many instances are valid. The effect could be related to steric factors or activity of the chlorinating complex.

The role of solvent in chlorination and other electrophilic substitution has received considerable attention in the recent literature.[200a] Kinetic observations for non-catalytic chlorination of toluene in different solvents indicate a significant change in the isomer ratio.[16] Chlorination of toluene in hydroxylic solvents has recently been reported to yield 60 percent *ortho* and 40 percent *para*-chlorotoluene, while in non-hydroxylic solvents such as acetonitrile, nitromethane, ethylene dichloride, the isomer ratio is inverted to 40 percent *ortho*- and 60 percent *para*-chlorotoluene. Chlorination of tertiary-butylbenzene in nitromethane gives a 90 percent yield of the *para*-chloro product, compared with a 76 percent yield of the *para* product obtained in acetic acid.[200]

Rates of Chlorination

Several quantitative studies of the relative rates of chlorination of various alkylbenzenes at various positions in the ring have been made in recent years.[59, 42, 12]

Relative reaction rates for chlorination in acetic acid have been determined as follows.[42,59] Using benzene as unity: toluene 344 (345), *o*-xylene 2.1 × 10^3, (4.6 × 10^3), *m*-xylene 1.85 × 10^5, *p*-xylene 2.08 × 10^3 (2.2 × 10^3). Relative reaction rates where toluene is taken as 100 were reported for: toluene 100, ethylbenzene 84, isopropylbenzene 51 and *tert*-butylbenzene 32. Rate data recently has been reported for mesitylene, durene, isodurene and pentamethylbenzene.[12] These data when referred to benzene as unity give relative rates roughly as follows: mesitylene 4.6 × 10^7, durene 1.36 × 10^6, isodurene 8.5 × 10^7, pentamethylbenzene 2.1 × 10^8. Such data reflect the important electron donating ability of the methyl group.[195,55]

Partial rate factors for chlorinations determined by multiplying the relative reaction rate, and the percentage of the isomer isolated by six, have been accurately calculated for toluene recently.[53,59] These figures can be used to calculate the reaction rates of polymethylbenzenes. Such calculations check observed values within reasonable error. In this manner the relative reaction rates for prehnitene 4.1 × 10^6, hemimellitene 8.37 × 10^5 and pseudocumene 7.34 × 10^5 have been determined.[42]

Solvents can effect reaction rates quite significantly. This fact appears to be related to the ability of the solvent to influence the transition complex. Both the activation energy and the entropies of activation are effected by the polarity of the solvent.[6,46,96]

Cleavage of Alkyl and Chloroalkyl Groups in Chlorination

The cleavage of alkyl or chloroalkyl groups has been noted frequently in the literature both in free radical chlorination studies and in ionic, electrophilic chlorination.

Cleavage in photochemical chlorinations has been noted in polychlorinated alkylbenzenes as well as in simple alkylbenzenes. For example, extensive chlorinolysis was observed in the photochlorination of the tetrachlorinated derivatives of xylene, paracumene, ethylbenzene and diisopropylbenzenes. Cleavage occurred at higher temperatures under forcing conditions.[131,133,134,135] Similar observations have been made in the attempt to force more halogen into tris(dichloromethyl)trichlorobenzene,[88] and bis(trichloromethyl)tetrachlorobenzene.[17]

There are many more examples of the cleavage reaction in electrophilic chlorinations catalyzed by metal salts. Cleavage of the methyl group only occurs after complete ring substitution takes place, and even then it is observed only at elevated temperatures. Paraxylene is converted to hexachlorobenzene at 200° C with an iridium chloride catalyst.[35] Chlorinolysis of the trichloromethyl group is more common. Trichlorobenzotrichloride upon treatment with chlorine in the presence of antimony chloride gives hexachlorobenzene.[19,146]

Cleavage occurs more readily with higher alkyl groups, especially branched chain alkyl groups. Chlorination of paracymene and of cumene in the presence of iodine or ferric chloride results in cleavage of the isopropyl group to yield pentachlorotoluene and hexachlorobenzene respectively.[154] The isopropyl, the ethyl and the tertiary butyl group have been observed to cleave and yield hexachlorobenzene.[166] Cleavage is easiest with the tertiary butyl group.[87,122,122a]

Extensive cleavage of the ethyl group has been observed to occur in treating 2,3,4,5,6-pentachloroethylbenzene and 2,3,5,6-tetrachloro-1,4-diethylbenzene with sulfuryl chloride in the presence of aluminum or iron chloride and sulfur monochloride.[166] All three reagents were necessary to cause cleavage. The product was hexachlorobenzene.

The sulfuryl chloride, aluminum chloride, sulfur chloride reagents have been observed to cleave the trichloromethyl group from perchlorotoluene under forcing conditions.[17]

Liquid chlorine above 80° C in the presence of iron quantitatively cleaves pentachlorobenzyl chloride to hexachlorobenzene.[36]

The cleavage reaction, in ionic chlorination at least, appears to be favored by electron-releasing groups *ortho* or *para* to the leaving group, by steric crowding of the leaving group, and by minimizing possible positions of substitution. Relief of steric strain may in fact play an important part in these displacements. The observed increasing ease of displacement with increased branching of the side chain may also be due to steric factors or it may be related to increased stability of the resulting carbonium ion or free radical as the case may be.

COMMERCIAL PRODUCTS

Benzyl Chloride

The halogen in benzyl chloride undergoes relatively easy displacement reactions with nucleophilic reagents. This reaction at the present time is the most important one from a commercial standpoint. The halogen is displaced readily by such nucleophilic reagents as the hydroxyl ion, cyanide ion, acetate ion, mercaptide ion and nitrogen bases. Benzyl chloride also couples readily in the presence of certain metals. It undergoes alkylation reactions in the presence of Lewis acids such as iron, zinc or aluminum chlorides. In this manner it reacts with itself to form polymeric oils and solids.[86,183,94,112,143,210,164,18]

Markets. Benzyl chloride is supplied by four companies in the United States. The market volume in sales for 1960 is listed at approximately 21 million pounds.[208a] It is used in the manufacture of benzyl alcohols and its esters, benzylamine derivatives, benzyl sulfides, benzyl cyanide, benzyl thiocyanate, and *p*-benzyl phenol, and in the synthesis of various dyes and dye-intermediates, inhibitors for gasoline, etc. Its growth has been good, increasing from approximately 2.5 million pounds in 1940 to 21 million pounds in 1960.

End Products

Benzyl Alcohol. Essentially all of the benzyl alcohol produced in the United States is prepared by the hydrolysis of benzyl chloride. The perfume industry is the largest consumer. The major product is benzyl acetate (1,130,000 lb, 1960)[208a] prepared by direct esterification.[127] Benzyl acetate is used in jasmine and hyacinth scented perfumes, and with benzyl alcohol it has been used as special solvents for casein gelatine, cellulose esters (acetate, nitrate), shellac, etc. Benzyl alcohol also has a minor use as a local anesthetic.

Other Benzyl Esters. Some esters are prepared directly from benzyl chloride and the sodium salt of the acid as well as from the alcohol. These include benzyl cinnamate (3,000 lb, 1960)[208a] used in perfumes and medicinals, and benzyl benzoate[207] which reached a peak production of 2 to 3 million pounds in the period from 1950 to 1952 due to U. S. Army use in miticides and tick eradicants. It is now used in insect repellents, as a solvent in perfumes, in the treatment of scabies and some insecticides at a market volume of 200,000 to 300,000 lb. Benzyl butyl phthalate is a primary plasticizer for various plastics including polyvinyl chloride types. Miscellaneous esters include the salicylate, (sun screening agent and perfume fixer), sebacate, formate, propionate, etc. which are used in soap perfumes and flavoring.[109]

Benzylamine Derivatives

Benzylamine and benzylaniline are both used in the manufacture of certain antihistamines. Substituted amines formed indirectly via benzyl chloride are also found in antihistimines and other drugs.[47] The drugs include Pyrabenzamine, Decapryn, Antazaline, Chlortrimetron, etc.

Benzyl chloride is also used in preparing quarternary ammonium salts. Such materials have found extensive use as germicides especially in hospitals, dairies, barber shops and the like. Quarternary salts such as benzyldodecyl dimethyl ammonium chloride are useful as cationic detergents (1,361,000 lb).[208a]

Benzyl Cyanide

There are at least four manufacturers of benzyl cyanide. It is prepared readily from benzyl chloride either by reaction with an alkali cyanide or copper cyanide. Sales are reported around 554,000 lb.[208a]

Benzyl cyanide is catalytically hydrogenated to 2-phenylethylamine, a valuable pharmaceutical intermediate used in producing a variety of analgesics. It is also hydrolysed to produce phenylacetic acid potassium salt (1,000,000 lb),[208a] an intermediate in pharmaceuticals and perfumes.

Other Uses

Benzyl thiocyanate has established use as a lubricant additive and is used as pesticide to a minor extent.

Benzyl chloride is used to an unknown extent in preparing diphenylmethane, (used in fruit packing) and as an intermediate to benzophenone, a product used in heavy perfumes (geranium)—especially in soaps, in the manufacture of antihistimines, hypnotics, and insecticides.[138] Benzyl chloride is also used in preparing the germicide, preservative and antisceptic parabenzylphenol. It is nitrated, and this intermediate is used in certain dyes including sulfur yellow. Other dyes in which benzyl chloride is an intermediate include Alkali green, Benzyl violet, Fast green, and Chrysaline. Other uses are found in the fields of resins and plastics, gas gum inhibitors, synthetic tannins, rubber accelerators, etc.

Benzal Chloride

There are three suppliers of benzal chloride in the United States. The major portion of benzal chloride is still consumed in the manufacture of benzaldehyde. Actually well over two thirds of the benzaldehyde in the United States is made by simple hydrolysis of benzal chloride. Sales figures are unavailable. Sales for benzaldehyde are close to 2 million pounds.[208a] The benzaldehyde made in this way is contaminated with nuclear halogen and cannot be used in perfumery. The competitive route to benzaldehyde via oxidation of toluene may possibly capture an increasing share of the market.

Benzaldehyde is used in making a variety of dyes of the triphenyl methane types; it is also used in the manufacture of cinnamic and mandelic acid which find use in perfumes and flavorings.

Chlorinated Derivatives. These, converted to the aldehydes, are offered commercially. Uses are suggested in dye stuffs.

Benzotrichloride

The more important chemistry of this reactive material is its tendency to hydrolyse readily with one or two moles of water to form benzoyl chloride and benzoic acid respectively. It also reacts with acids and anhydrides converting them to acid chlorides. A Friedel-Crafts catalyst causes benzotrichloride to condense with aromatic hydrocarbons to yield polyphenyl methanes. It has been nitrated and chlorinated to yield *meta* substitution predominantly. The halogens can be replaced by fluorine.

Industrial Usage. Sales are listed at 256,000 lb (1958) for benzotrichloride. It is an important chemical to certain sections of industry.

Dyestuffs. The largest direct consumer for benzotrichloride is the dyestuff industry. Here use is made of the tendency for condensation with other aromatic systems in the presence of a Friedel-Crafts catalyst. The important dyes include Alizarin yellow A, Quinoline red, and C.I. Mordant green 31. They are primarily of the triphenyl methane type.

Benzoyl Chloride

Partial hydrolysis of benzotrichloride in the presence of catalyst yields benzoyl chloride. The largest industrial consumption of this chemical is in dyestuff manufacture. Relatively moderate quantities are used in preparing benzophenone which is consumed in dyestuff manufacture, in perfumery as a fixative, in pharmaceuticals such as antihistamines (benzhydrol derivatives) and hypnotics. Another derived chemical is benzamide which is a useful intermediate in certain dyes and pharmaceuticals and has use as an inhibitor for linseed oil and as a metal scavenger in chlorination. Benzoyl chloride is also used in manufacturing esters of diethylene and dipropylene glycol which are consumed as primary plasticizers. Other esters such as butylbenzoate are used as dye carriers for synthetic fibers.

A relatively large amount of benzoyl chloride is used in making benzoyl peroxide. This chemical has well established use as a free radical catalyst in vinyl polymerizations, as an oxidizing agent in food processing, as a bleaching agent in oil decolorization.

Benzoic Acid

There are two well established industrial procedures for preparing benzoic acid, namely, pyrolysis of orthophthalic acid in the presence of a chromium catalyst and hydrolysis of benzotrichloride. Other processes include oxidation of toluene. This latter route is the most direct, but apparently side products are difficult to control. There are at least six manufacturers of benzoic acid and sodium benzoate with a listed volume of 10,780,000 lb.[208a] The largest portion of benzoic acid is made by the pyrolytic procedure. Construction of a 12 million pounds-per-year plant for benzoic acid has recently been announced.[103]

Industrially, benzoic acid or the sodium salt finds its largest use as a preservative. It is used for a preservative in foods, textile sizing agents, rubber latex, cosmetics, creams and lotions, antiseptics, dentifrices, tobacco curing agents, soft drinks and syrups, and certain drugs. It finds use as a corrosion inhibitor in glycols and as a dispersing aid for pigments.

Products competitive with benzoic include propionic acid (Mycoben) and hydroxybenzoic acid derivatives (Parasepts).

Nitration of benzoic acid yields *meta*-nitrobenzoic acid, a useful dye intermediate.

Some esters of benzoic have been mentioned earlier and can be prepared from the acid chloride, the acid, or acid salt. Benzyl benzoate is produced in the largest quantities, but other products include 2-naphthyl, amyl, iso-butyl, ethyl and methyl derivatives. All have uses in perfumes and medicines.

Other derivatives such as chlorinated products have a growing use in herbicides. *Meta*-chlorobenzoyl chloride is offered commercially and is presumably made from the acid chloride.

Benzotrifluoride

Benzotrichloride is converted to benzotrifluoride with hydrogen fluoride and a catalyst, or with antimony fluorides.

This material finds use as a dye intermediate; the trifluoro groups improves stability of certain dyestuffs.

The chlorinated and fluorinated derivatives of benzotrifluoride have dielectric properties which render them suitable as transformer fluids and insulating media.

Recently 4-nitro-3-trifluoro methyl phenol has been found effective as a lamprecide.[48,49] Benzotrifluoride is the intermediate for this material.

Many other uses have been suggested in the patent literature and include larvacides, and pharmaceuticals as well as many dyestuffs, fungicides, insecticides, and herbicides.

Chlorotoluenes

The market volume for chlorotoluenes is not listed. A good portion of the chlorotoluenes are made and used captively, especially in the dye and lubricant fields.

Chlorinated toluenes, primarily the mono- and dichloro derivatives, and sometimes mixtures thereof, are used as solvents, dye intermediates, and lubricant additive intermediates.

The solvent use is associated with rubber cement and also other polymers. The market has been erratic and generally declining.

The *ortho*-chlorotoluene and 2,6-dichlorotoluene are used as dye intermediates. They are partially oxidized (either by chlorination and hydrolysis or by direct oxidative techniques) to the chlorobenzaldehydes and then used in preparing triphenylmethane dyes. The benzoic acid derivatives also have use as dye intermediates. A good portion of the *ortho*-chlorotoluenes are prepared from *para*-toluenesulfonic acids. Chlorination of this compound gives the 2-chloro derivative after the sulfonate group is hydrolyzed off. Aurin dyes are prepared from the chlorinated benzaldehydes. Nitration products go into indigo dyes.

Large amounts of chlorinated toluenes are used as intermediates in the high pressure lubricant field. The corresponding chlorobenzyl chlorides are prepared and converted to disulfides. The competitor in this field is dibenzyl disulfide itself.

Other chlorinated derivatives of toluenes are finding increasing use as intermediates in the field of herbicides (chlorinated benzoic acids).

Miscellaneous Chlorinated Products

Chlorinated toluenes other than those mentioned above are offered in small quantities primarily as development chemicals. These include pure *ortho* and *para* isomers of monochlorotoluenes, benzyl chlorides, benzoyl chlorides, benzotrichloride, benzoic acids, benzaldehydes, and also *ortho, meta* and *para* dichloro toluenes, benzyl chlorides, benzotrichlorides, benzaldehydes, benzoyl chlorides, benzoic acids.[79]

Other chlorinated products available commercially in pilot plant quantities derived from xylenes include side-chain chlorinated derivatives such as 1,3-bis(trichloromethyl)benzene, 1,4-bis(trichloromethyl)benzene, 2-(trichloromethyl) benzal chloride, the corresponding acid chlorides, isophthaloyl chloride, and terephthaloyl chloride, as well as phthaladehydric acid. The 1,4-bis(trichloromethyl)benzene has long and continued small use as a dye intermediate.

The acid chlorides are potentially of use in preparing polyesters made via the interfacial polymerization or Schotten-Baumann technique.[54,67] Phthaladehydric acid has many potential uses in dyestufffs, pharmaceuticals, etc.

Bis(chloromethyl)benzene, bis(chloromethyl)xylenes and bis(chloromethyl) durene have been offered as development chemicals in recent years. Such materials have suggested uses in polymers.[126,142]

References

1. Adam, J., Gosselain, P. A., and Goldfinger, P., *Bull. Soc. Chim. Belges,* **65**, 523 (1956).
2. Anbor, M., and Ginsberg, D., *Chem. Rev.,* **54**, 925 (1954).
3. Andrews, L. J., and Keefer, R. M., *ACS Abstracts,* 1 (September 11, 1960).
4. Andrews, L. J., and Keefer, R. M., *J. Am. Chem. Soc.,* **79**, 1412, 5169 (1957).
5. *Ibid.,* **79**, 5174 (1957).
6. *Ibid.,* **81**, 1063 (1959).

7. *Ibid.,* **81,** 1067 (1959).

8. *Anshütz, Ann.* **235,** 329 (1886).

9. Ash, A. B., and Brown, H. C., *Record Chem. Prog.* (Kresge-Hooker Sci. Library) **9,** 81 (1948).

10. Ayad, N., Garwood, R., and Hickenbottom, W. S., *Chemistry & Industry,* 1122 (1955).

11. Ayad, N., Beard, C., Garwood, R., and Hickenbottom, W. J., *J. Chem. Soc.* 2981 (1957).

12. Baciocchi, E., and Illuminati, G., *Chemistry & Industry,* 917 (1958).

13. Baker, J. W., and Nathan, W. S., *J. Chem. Soc.,* 1840 (1935).

14. Baker, J. W., "Hyperconjugation," Oxford, Clarendon Press, 1952.

15. Ballester, M., *Mem. real. acad. cienc. y. ortes, Barcelona,* **29** (7), 3 (1948), *Chem. Abst.,* **45,** 567 (1951); **50,** 238 (1956).

16. *Ibid.,* **49,** 13922 (1955).

17. Ballester, M., Molinet, C., and Castâner, J., *J. Am. Chem. Soc.,* **82,** 4259 (1960).

18. Ballester, M., Molinet, C., and Rosa, J., *Tetrahedron,* **6,** 109 (1959).

19. Beilstein, and Kuhlberg, D., *Ann.,* **150,** 286 (1869).

20. Stanley, G., and Shorter, J., *J. Chem. Soc.,* **246,** 256 (1958).

21. Benoy, G., and Jungers, J. C., *Bull. Soc. Chim. Belges.,* **65,** 769 (1956).

22. Bengelsdorff, I., *J. Org. Chem.,* **23,** 242 (1958).

23. Bennett, G., and Jones, B., *J. Chem. Soc.,* 1815 (1935).

24. "Benzotrifluoride," Hooker Chemical Corp., Bull. 12A.

25. Berliner, E., and Bondhus, R. J., *J. Am. Chem. Soc.,* **68,** 2355 (1946).

26. *Ibid.,* **70,** 854 (1947), *Ibid.,* **71,** 1195 (1949).

27. Berliner, E., and Chen, M. H., *Ibid.,* **80,** 343 (1958).

28. Bionda, G., and Civera, M., *Ann. chim.* (Rome), **41,** 814 (1951).

30. Book, Eggert, *Z. Elekt. Chem.,* **29,** 521 (1933).

31. Boder, H., Edmiston, W., and Rosen, H., *J. Am. Chem. Soc.,* **78,** 2590 (1956).

32. Boyars, C., *Ibid.,* **75,** 1989 (1953).

33. Bradshaw, C. K., Gross, P. M., and Hobbs, H., *Ibid.,* **70,** 1317 (1948).

34. Brimelow, H. C., Jones, R. L., and Metcalfe, T. P., *J. Chem. Soc.,* 1203 (1951).

35. Brintzinger, H., Orth, H., Monatsh, **85,** 1015 (1954), *Chem. Abstr.,* **49,** 15788 (1955).

36. Brodbeck, M. B., (private communication).

37. Brown, H. C., *J. Am. Chem. Soc.,* **75,** 6292 (1953).

38. Brown, H. C., and Ash, A. B., *Ibid.,* **77,** 4019 (1955).

39. Brown, H. C., and Brady, J. D., *Ibid.,* **74,** 3570 (1952).

40. Brown, H. C., and Piersall, W., *Ibid.,* 191 (1952).

41. Brown, H. C., and Russel, G. A., *Ibid.,* **74,** 3995 (1952).

42. Brown, H. C., and Stock, L. M., *Ibid.,* **79,** 5175 (1957).

43. Brown, H. C., and Stock, L. M., *Ibid.,* **81,** 3327 (1959).

44. Buu-Hoi, *Rec. Chem. Prog.,* **13,** 30 (1952).

45. Cachia, M., and Wahl, H., *Bull. Soc. Chim.* (France) 1234 (1956).

46. Callis, M. J., Gintz, F., Goddard, D., and Hebdon, E., *Chemistry and Industry* 1742 (1955).

47. *Chem. Week,* p. 7 (March 3, 1951).

48. *Ibid.,* p. 94, (September 5, 1959).

49. *Chem. Eng. News* p. 40 (July 20, 1959).

50. Claus, A., and Kautz, H., *Ber.,* **18,** 1367 (1885).

51. Cohan, J., and Dakin, H., *J. Chem. Soc.,* **79,** 1111 (1901); **81,** 1324 (1902).

52. Cohn, H., Hughes, E. D., Jones, M. H., and Peeling, M. G., *Nature,* **169,** 291 (1952).

53. Condon, F. E., *J. Am. Chem. Soc.,* **70,** 1963 (1948).

54. Coniz, A., *Ind. Eng. Chem.,* **51,** 147 (1959).

55. Crosseley, A., and Hills, J., *J. Chem. Soc.,* **89,** 882 (1906).

56. Cutter, H. B., and Brown, H. C., *J. Chem. Ed.,* **21,** 443 (1944).

57. Datta, R. J., *J. Am. Chem. Soc.,* **36,** 1011 (1914).

58. Dehn, F. C., U. S. Patent 2,744,145 (1956).

59. De LaMare, P. B. D., and Robertson, P. W., *J. Chem. Soc.,* 279 (1943).

60. De LaMare, P. B. D., Ketley, A. D., and Vernon, C., *Research* (London) **6,** 12 (1953).

61a. De LaMare, P. B. D., *J. Chem. Soc.,* 1290 (1954).

61b. De LaMare, P. B. D., and Ridd, J. H., "Aromatic Substitution, Nitration and Halogenation," New York, Academic Press, 1959.

62. De LaMare, P. B. D., Harvey, J. T., Hasson, M., and Varma, S., *J. Chem. Soc.,* 2756 (1958).

63. Danjam, M., *Chem. Zent.,* **1,** 4928 (1939).

64. Dewar, M. J. S., "The Electronic Theory of Organic Chemistry," Oxford Univ. Press, Amen House, London, E. C. 4, 1949

65. Dory, I., *Chem. Abstr.,* **51,** 14575 (1957).

66. Dresbach, Monroe, U. S. Patent 2,926,201 (1960).

67. Eareckson, W. E., *J. Polymer Sci.,* **40,** 399 (1959).

68. Ellis, U. S. Patent 1,202,040.

69. Errera, G., *Gazz. chim. ital.,* **14,** 504 (1884).

70. Evans, E., Mobatt, E., and Turner, E., *J. Chem. Soc.,* 1159 (1927).

71. Farbenfabriken Bayer, British Patent 691,504.

72. Farthing, A. C., *J. Chem. Soc.,* 3261, 3270 (1953).

73. Firth, J. B., and Smith, T. A., *J. Chem. Soc.,* 339 (1936).

74. Fittig, Hoogenwerff, *Ann.,* **150,** 325 (1869).

75. Fittig, Kiesaw, *Ibid.,* **156,** 246 (1870).

76. Galitzenstein, E., and Woolf, C., U. S. Patent 2,573,394 (1951).

77. Genvresse, *Bull. soc. chim.,* **9,** (3), 219 (1893).

78. Gibbs, Geiger, U. S. Patent 1,246,739 (1917).

79. Godfrey, K. L., U. S. Patent 2,790,819 (1957).

80. Goldwhite, H., *J. Chem. Ed.,* **37** (6), 295 (1960).

81. Gorvin, J. H., *Chemistry and Industry* (London), 910 (1951).

82. Gould, F. S., "Mechanism and Structure in Organic Chemistry," New York, H. Holt and Co., 1960.

83. Gross, P., U. S. Patent 2,394,442 (1946).

84. Gunther, F., *Chemistry and Industry,* **65,** 399 (1946).

85. Haas, H. B., McBee, E. T., and Weber, P., *Ind. Eng. Chem.,* **28,** 333 (1936).

86. Haas, H. C., Livingston, D. I., and Saunders, M., *J. Polym. Sci.,* **15,** 503 (1955).

87. Harvey, P. G., Smith, F., Stacey, M., and Tatlow, J., *J. Applied Chem.,* **4,** 325 (1954).

88. *Ibid.,* **4,** 319 (1954).

89. Head, J. D., and Irvins, O., U. S. Patent 2,748,161 (1956).

90. Hebelynck, M. F., and Martin, R. H., *Bull. soc. chim. Belges,* **59,** 193 (1950); *Chem. Abstr.,* **45,** 2411; **60,** 54 (1951); **46,** 7051 (1952).

91. Hinkel, L. E., Ayling, E., and Bevan, L. C., *J. Chem. Soc.,* 1874 (1928).

92. Hine, J., "Physical Organic Chemistry," New York, McGraw Hill Book Co., 1956.

93. Hooker Chemical Corp. (private communication).

94. Hugel, G., *Chem. Abstr.,* **50,** 8189 (1956).

95. Huyser, E. S., Abstracts of papers, p. 6D, presented at Am. Chem. Soc. meeting, New York, Sept. 11, 1960.

96. Illuminati, G., and Marino, G., *J. Am. Chem. Soc.,* **78,** 4975 (1956).

97. Ingold, C. K., "Structure and Mechanism in Organic Chemistry," London, G. Bell and Sons Ltd., 1953.

98. Ishii, Y., and Yamashito, Y., *Chem. Abstr.,* **49,** 8332 (1955).

99. Irvins, O. D., Head, J. D., and Britton, E. C., U. S. Patent 2,810,688 (1957).

100. Jacknow, B. B., "The Reaction of Tert-butylhypochlorite With Organic Compounds," Columbia Univ., Doctoral Dissertation, 1960.

101. Johnson, A. N., and Brunfeld, P. E., U. S. Patent 2,738,320 (1956).

102. Jonsson, A. E., *Svensk. Kem. Tidskr.,* **67,** 162 (1955).

103. *J. of Commerce* (March 8, 1960).

104. Keefer, R. M., Ottenberg, A., and Andrews, L. J., *J. Am. Chem. Soc.,* **78,** 255 (1956).

105. Kenner, J., and Whitham, E., *J. Chem. Soc.,* 119, 1460 (1921).

106. Kharasch, M. S., and Berkman, M., *J. Org. Chem.,* **6,** 810 (1941).

107. Kharasch, M. S., and Brown, H. C., *J. Am. Chem. Soc.,* **61,** 2142 (1939).

108. King, Merrian, *Chem. Abstr.,* **29,** 6214 (1935).

109. Kirk, R. E., and Othmer, D., "Encyclopedia of Chemical Technology," Vol. 3, New York, Interscience Publishers Inc.

110. *Ibid.,* Vol. 3.

111. Kovacic, P., *J. Amer. Chem. Soc.,* **82,** 1917 (1960).

112. Knust, E., U. S. Patent 2,752,322 (1956).

113. Kress, B. H., British Patent 625,281 (1949).

114. *Ibid.,* 627,509 (1949).

114a. Kretov, A. E., Syrovatko, A. D., *J. Gen. Chem. of The USSR,* **30**(7) 3109, (1960).

115. Kulka, M., *Can. J. Res.,* **23,** Sect. B, 103 (1945), *Ibid.,* 492,592 (1953).

115a. Lane, E. W., U. S. Patent 2,926,202 (1960).

116. Leigh, C. Anderson, Bray, B., and Martin, J., Proc. Intern. Conf. Peaceful Uses of Atomic Energy, (Geneva) **15,** 235 (1956), Harmer, D. E., AECV-3077 (1955), I Rosen, J. P. Stallings, Ind. Eng. Chem. *50,* 1511 (1958).

117. Lawlor, F., U. S. Patent 2,608,532 (1952).

118. Lawlor, F., U. S. Patent 2,608,591 (1952).

119. Lawlor, F., and Brodbeck, M., U. S. Patent 2,608,592 (1954).

120. LeFevre, R. J., *J. Chem. Soc.,* 980 (1933).

121. LePage, J., and Jungers, J. C., *Compt. rend.,* **244,** 2915 (1957).

122. Lerer, M., Fabre, C., and Hugel, G., *Bull. soc. chim.* (France), **173** (1957); (a). *ibid.,* 1235.

123. Ligett, W. B., U. S. Patent 2,654,789 (1953).

124. Loverde, A., Beanblossom, W. S., U. S. Patent 2,566,065 (1951), *Chem. Abstr.,* **46,** 1578 (1952).

125. Loverde, A., U. S. Patent 2,695,873 (1954).

126. Manka, J., Tomaszewski, J., and Wajnryb, M., *Chem. Abstr.,* **50,** 10045 (1956).

127. Manufacturing Chemist 294 (1954).

128. Mason, J., Smale, C., Thompson, R., and Wheeler, T. S., *J. Chem. Soc.,* 3150 (1931).

129. Matthews, F., *J. Chem. Soc.,* **59,** 165 (1891), *Ibid.,* **61,** 104 (1892).

130. Maude, A. H., and Rosenberg, D. S., U. S. Patent 2,847,481 (1958).

131. McBee, E. T., Haas, H. B., Rothrock, G. M., Newcomer, J. S., Clipp, W., Welch, Z. D., and Gochenour, C. I., *Ind. Eng. Chem.,* **39,** No. 3, 384 (1947).

132. McBee, E. T., Haas, H. B., Weimer, P. E., Burt, W. E., Welch, Z. D., Ralib, R., and Speyer, E., *Ibid.,* 387 (1947).

133. McBee, E. T., and Leech, R. E., *Ibid.,* 393 (1947).

134. McBee, E. T., Haas, H. B., Leech, R. E., Hodneth, E., Frederick, M. R., and Bonner, W. A., *Ibid.,* 395 (1947).

135a. McBee, E. T., and Pierce, O. R., *Ibid.,* 397 (1947).

135b. *Ibid.,* 399 (1947).

136. McBee, E. T., and Bettler, J. A., U. S. Patent 2,716,140 (1955).

137. McFord, W. Waters, *J. Chem. Soc.,* 1851 (1951).

138. "Merck Index," 7th Ed., Rahway, New Jersey, Merck and Co., (1960).

139. Miller, B., *Dissertation Absts.,* 15, 2016 (1955).

140. Moyer, R. H., U. S. Patent 2,817,632 (1957) and U. S. Patent 2,817,633 (1957).

141. Myazaki, S., *Chem. Abstr.,* **46,** 11125, 5977, 4915 (1952).

142. Nelson, J. A., U. S. Patent 2,734,045 (1956).

143. Nakamura, Y., *Chem. Abstr.,* **50,** 2093 (1956).

144. Norton, J., U. S. Patent 2,446,430 (1948).

145. Ogimachi, N., Andrews, L. J., and Keefer, R. M., *J. Am. Chem. Soc.,* **77,** 4202 (1955).

146. Page, *Ann.,* **225,** 208 (1884).

147. Pieper, *Ann.,* **142,** 304 (1867).

148. Pines, H., and Mark, V., *J. Am. Chem. Soc.,* **78,** 4316 (1956).

149. Porai-Koshits, A. E., and Porai-Koshits, B. A., *Chem. Abstr.,* **50,** 4880 (1956).

150. Puck, R., and Jungers, J. C., *Bull. soc. chem. Belges,* **60,** 357 (1951), *Chem. Abstr.,* **46,** 4916 (1952).

151. Price, C. C., *Chem. Revs.,* **29,** 37 (1941).

152. Pritchard, J., and Meyers, P., U. S. Patent 2,751,323 (1956).

153. Pyke, J. D., and Dickenson, A. F., *J. Amer. Chem. Soc.,* **77,** 2629 (1955).

154. Quist, W., and Holmberg, N., *Acta. Acad. Oboensis. Math., Phys.,* **6**, (14), 3 (1932); **8**, (4), 30 (1934); *Chem. Abstr.,* **27**, 5726 (1933), *Ibid.,* **29**, 6884 (1935), *Ibid.,* **47**, 110 (1953).

155. Rabjohn, N., *J. Am. Chem. Soc.,* **76**, 5479 (1954).

156. Reese, J., *Chem. Revs.,* **14**, 67 (1934).

157. Reynolds, R. D., and Allen, G. F., U. S. Patent 2,811,486 (1957).

158. Ritchie, M., and Winning, W., *J. Chem. Soc.,* 3579 (1950).

159. Robertson, P. W., and De LaMare, P. B. D., *J. Chem. Soc.,* 279 (1943).

160. Robertson, P. W., Dixon, R., Goodwin, W., McDonald, I., and Scaife, J. F., *J. Chem. Soc.,* 294 (1949).

161. Robertson, P. W., *J. Chem. Soc.,* 1267 (1954).

162. Robertson, P. W., De LaMare, P. B. D., and Swedlund, B. E., *J. Chem. Soc.,* 782 (1953).

163. Robinson, P., and Reid, C., U. S. Patent 2,530,080 (1950).

164. Robinson, K. R., *Chem. Abstr.,* **46**, 952 (1952).

165. Ross, S. D., *J. Am. Chem. Soc.,* **69**, 1914 (1947); **71**, 396 (1949).

165a. Ross, S. D., Coburn, R., Markarian, M., and Schwartz, M., *J. Org. Chem.* **25**, 2102 (1960).

166. Ross, S. D., and Nazzevivski, M. J., *J. Am. Chem. Soc.,* 3146 (1947).

167. Ross, S. D., and Markarian, M., U. S Patent 2,528,445 (1950).

168. Ross, S. D., and Markarian, M., U. S. Patent **2**, 600,691 (1952).

169. Ross, S. D., and Markarian, M., *J. Polym. Sci.,* **9**, 219 (1952).

170. Ross, S. D., and Markarian, M., U. S. Patent 2,631,168 (1953).

171. Ross, S. D., Markarian, M., and Schwarz, M., *J. Am. Chem. Soc.,* **75**, 4967 (1953).

172. Ross, S. D., and Markarian, M., U. S. Patent 2,702,825 (1955).

173. Russell, G. A., and Brown, H. C., *J. Am. Chem. Soc.,* **77**, 403 (1955).

174. Russell, G. A., and Brown, H. C., *Ibid.,* 4578 (1955).

175. Russell, G. A., and Brown, H. C., *Ibid.,* 4031 (1955).

176. Russell, G. A., *Ibid.,* **70**, 2977 (1957).

177. Russell, G. A., *Ibid.,* **80**, 4987 (1958).

178. Russell, G. A., *Ibid.,* **80**, 4997 (1958).

179. Russell, G. A., *Ibid.,* **80**, 5002 (1958).

180. Russell, G. A., *Ibid.,* **70**, 2977 (1957).

181. Russell, G. A., *Tetrahedron,* **8**, 101 (1960).

182. Salibil, J., *Chem. Ztg.,* **35**, 97 (1911); *Bull. intern. acad. Sci. Crocovie* (A) 606 (1910).

183. Schriner, R., and Berger, A., *J. Org. Chem.,* **6**, 305 (1941).

184. Schramm, *Monatsh* **18**, 102 (1887).

185. Scipioni, A., *Ann. Chim.,* (Rome) **41**, 491 (1951).

186. Seelig, *Ann.,* 237, 129 (1887).

187. Sherago, H., and Hobbs, M., *J. Am. Chem. Soc.,* **70**, 3015 (1948).

188. Scherer, P B L 75332, Frames 2247-2256, Fiat Microfilm Reel C-61 PB 17658.

189. Scherer, P B L 75450, Frames 2846, Fiat Microfilm Reel C-61, PB 17658 (1936).

190. Scherer, P B L 75448, Frames 2835-2841, Fiat Microfilm Reel 61, PB 17658 (1936).
191. Silberrad, O., *J. Chem. Soc.,* **127,** 2680 (1925).
192. Skinner, G. S., and Knauss, D. L., *J. Am. Chem. Soc.,* **76,** 1188 (1954).
193. Slough, L. H., *Dissert. Abst.,* **17,** 984 (1957).
194. Smith, A. Noyes, *J. Am. Chem. Soc.,* **55,** 4444 (1935).
195. Smith, L. I., and Mayle, C. L., *Ibid.,* **58,** 8 (1936).
196. Spengler, C., and Hossl, H. O., *Seifen Ol Fette Wachse,* **82** (14), 393 (1956).
197. Stanley, G., and Shorter, J., *J. Chem. Soc.,* 246, (1958).
198. Steacie, E. W., "Atomic and Free Radical Reactions," 2nd Ed., Chapter 10, New York, Reinhold Publ. Corp., 1954. .
199. Stempel, G., Jr., and Greene, C., *J. Am. Chem. Soc.,* **73,** 455 (1951).
200. Stock, L., and Himoe, A., Abstracts of Papers, Am. Chem. Soc. Meeting, p. 1P, New York, 1960.
200a. Stock, L., Himhoe, A., *J. Amer. Chem. Soc.,* **83,** 1936 (1961).
201. Veijola, Vaino, *Suomen Kemistilehti,* **27B,** (11), 74 (1954).
202. Suzuki, M., and Miyazaki, S., *Chem. Abstr.,* **49,** 5133 (1955).
203. Swart, G., and Stempel, G., British Patent 687,267 (1953).
204. Taft, R. N., *J. Am. Chem. Soc.,* **70,** 3364 (1948).
205. Tischenko, *J. Gen. Chem. USSR,* **9,** 138 (1939).
206. Takakashi, T., and Noda, M., *Chem. Abst.,* **46,** 8059 (1952).
207. Thorp, I., and Nattorf, H., *Ind. Eng. Chem.,* **39,** 1300 (1947).
208. Truce, W. E., McBee, E. T., and Alfieri, C., *J. Am. Chem. Soc.,* **71,** 752 (1949).
208a. U. S. Tariff Commission Reports 1960.
209. Vanderlinden, *Ber.,* **45,** 231 (1912).
210. Valentine, L., and Winter, R., *J. Chem. Soc.,* 4768 (1956).
211. VanHeldon, R., and Kooyman, E. C., *Rec. trav. Chim.,* **73,** 269 (1954).
212. Vaughan, W. E., and Rust, F. F., *J. Org. Chem.,* **5,** 449 (1940).
213. Veijola, V., *Chem. Abstr.,* **49,** 15753 (1955), *Ibid.,* **51,** 10398, *Ibid.,* **50,** 4064 (1956).
214. Vermeylen, *Bull. soc. chim.,* **16,** 1937 (1922).
215. Walling, C., and Miller, B., 126th meeting ACS, p. 15R, New York, September 1954.
216. Walling, C., *J. Am. Chem. Soc.,* **81,** 1485 (1959).
217. Walling, C., and Miller, B., *Ibid.,* 79, 4181 (1957).
218. Werber, J., U. S. Patent 2,697,123 (1954).
219. Wertyporoch, *Ann.,* **493,** 163 (1932).
220. Hahn, W., *Chem. Abstr.,* **50,** 13711 (1956).
221. Wulff, C., and Steinbrink, H., German Patent 825,397.

29. SIMPLE ORGANIC COMPOUNDS CONTAINING FLUORINE

Ralph C. Downing

"Freon" Products Division
E. I. du Pont de Nemours & Co., Inc.

SCOPE

Organic compounds containing fluorine have been known for many years. General interest in these products has accelerated in the past two decades, and hundreds of fluorinated compounds have been synthesized and studied. At present, commercial production is limited to relatively few products. However, new developments and applications seem certain to lead to increased production in the future. This chapter is not a general review of fluorine chemistry but will include some comments on the commercial and manufacturing aspects and will be limited primarily to the following compounds:

(1) Compounds containing one or two carbon atoms
(2) Compounds in commercial or semicommercial production
(3) Compounds generally involving chlorine or chlorinated compounds in the preparation.

Several good reviews of this field have been published[2,7,10,11,14,26,27] and are suggested as sources of further information.

DESCRIPTION OF PRODUCTS

Physical Properties

A few physical properties of some of the more common organic compounds containing chlorine and fluorine are listed in Tables 29-1, 29-2, and 29-3. Most of these products have been studied extensively, and in many cases further information is available in the literature or from commercial manufacturers. The compounds have been divided into groups in order to simplify the listing of physical properties. Some products containing one carbon atom are included in Table 29-1. Table 29-2 includes

TABLE 29-1. PHYSICAL PROPERTIES OF ONE-CARBON COMPOUNDS CONTAINING CHLORINE AND FLUORINE.[8]

Code Number	11	12	13	14	21	22	23
Name	trichlorofluoromethane	dichlorodifluoromethane	chlorotrifluoromethane	tetrafluoromethane	dichlorofluoromethane	chlorodifluoromethane	trifluoromethane
Chemical formula	CCl_3F	CCl_2F_2	$CClF_3$	CF_4	$CHCl_2F$	$CHClF_2$	CHF_3
Molecular weight	137.38	120.93	104.47	88.01	102.93	86.48	70.02
Boiling point at 1 atm, °C	23.8	−29.8	−81.4	−128.0	8.9	−40.8	−82.0
Freezing point, °C	−111	−158	−181	−184	−135	−160	−160
Critical temperature, °C	198.0	112.0	28.9	−45.5	178.5	96.0	25.9
Critical pressure, psia	635	596.9	561	542	750	716	701
Critical volume, cc/mol	247	217	181	140	197	164	133
Liquid heat capacity at 30°C cal/(gram) (°C)	0.209	0.240	0.25 (−30°C)	0.27 (−80°C)	0.256	0.335	0.328 (−30°C)
Flammable or explosive	No	No	No	No	No	No	No

TABLE 29-2. PHYSICAL PROPERTIES OF SATURATED TWO-CARBON COMPOUNDS CONTAINING CHLORINE AND FLUORINE.[8]

Code number	112	113	114	152a	142b
Name	1,1,2,2-tetrachloro-1,2-difluoroethane	1,1,2-trichloro-1,2,2 trifluoro-ethane	1,2-dichloro-1,1,2,2-tetrafluoroethane	1,1-difluoro ethane	1-chloro-1,1-di-fluoroethane
Chemical formula	CCl_2FCCl_2F	CCl_2FCClF_2	$CClF_2CClF_2$	CH_3CHF_2	CH_3CClF_2
Molecular weight	203.85	187.39	170.93	66.05	100.50
Boiling point at 1 atm, °C	92.8	47.6	3.6	-24.8^{15}	-9.2^{15}
Freezing point °C	26	-35	-94	-117^{15}	-131
Critical temperature, °C	278	214.1	145.7	113.5^{15}	137.1^{15}
Critical pressure, psia	500	495	474	652^{15}	598^{15}
Critical volume, cc/mol	370	325	293	181^{15}	231^{15}
Liquid heat capacity at 30°C cal/(g) (°C)	0.205 (21.1°C)	0.218	0.238	*	*
Flammable or ex-plosive	no	no	no	yes	yes

TABLE 29-3. PHYSICAL PROPERTIES OF UNSATURATED TWO-CARBON COMPOUNDS CONTAINING CHLORINE AND FLUORINE[8]

Name	Chlorotri-fluoro-ethylene (CTFE)	Tetra-fluoro-ethylene (TFE)	Vinyl fluo-ride (VF)	Vinylidene fluoride (VF$_2$)
Chemical formula	$CClF{=}CF_2$	$CF_2{=}CF_2$	$CH_2{=}CHF$	$CH_2{=}CF_2$
Molecular weight	116.48	100.02	46.04	64.04
Boiling point at 1 atm, °C	-26.8[14]	-76[14]	-72	-85.7[15]
Freezing point °C	-157.5[3]	-142.5[22]	-160	-144[15]
Critical tempera-ture, °C	107[4]	33.3[22]	54.7	30.1[15]
Critical pressure, psia		572[22]	760	643[15]
Critical volume cc/mol	212	174[22]	144	154[15]
Flammable or explosive	yes	yes	yes	yes

some two carbon saturated compounds and table 29-3 two carbon unsaturated products.

Effect of Fluorine on Properties

The presence of fluorine in an organic molecule tends to increase the stability and decrease the toxicity. It has much less effect on the boiling point when substituted for H than the other halogen atoms. As a general rule, compounds containing fluorine are nonflammable, relatively low-boiling materials with high density, low viscosity, and low surface tension. This unusual combination of properties has contributed to their usefulness in a number of different applications.

The effect of fluorine on the boiling point is illustrated in Figure 29-1 for a number of derivatives of methane. A comparison with the effect of chlorine is especially interesting. Carbon tetrachloride boils at about 77° C. As chlorine is replaced by fluorine, the boiling point is progressively lowered, and tetrafluoromethane boils at $-128°$ C which is just 34 degrees higher than the boiling point of methane itself. Substituted methanes containing less fluorine boil higher than tetrafluoromethane, undoubtedly because of a higher degree of association in the liquid.

The change in liquid viscosity when chlorine is replaced by fluorine is illustrated in Table 29-4. When measured at the same temperature the viscosity becomes progressively lower. This difference may be due primarily to the decrease in boiling point. At the same reduced temperature, the

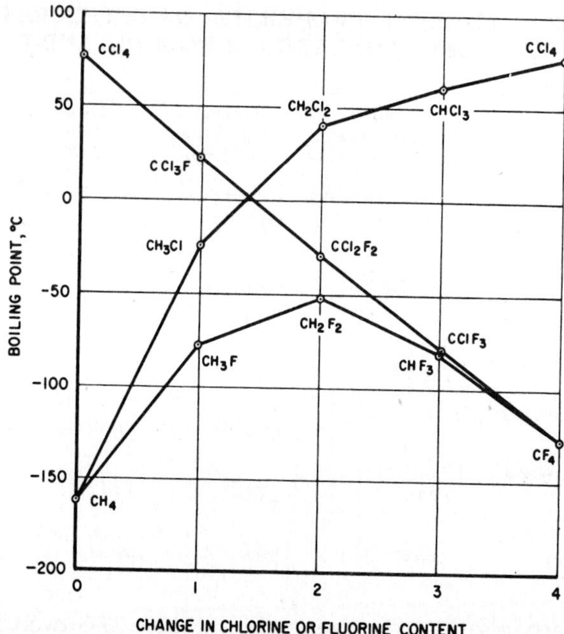

Figure 29-1. Boiling points of several halogenated
methanes.

viscosities of the compounds illustrated in Table 29-4 are quite similar as
shown in the last column.

As a further example of the effect of fluorine, differences in the toxicity
based on acute inhalation tests are given in Table 29-5. It is interesting to
note that in these tests dichlorodifluoromethane had slightly less effect on

TABLE 29-4.   COMPARISON OF LIQUID VISCOSITY DATA OF METHANES.[8]

Compound	Boiling point, °C	Critical temp., °C	Temperature, °C corresponding to $T_R = 0.527$*	Liquid viscosity, cps. 20°C	Liquid viscosity, cps. $T_R = 0.527$
CCl_4	76.8	283.2	20	0.96	0.96
CCl_3F	23.8	198.0	−24.9	0.44	0.72
CCl_2F_2	−29.8	112.0	−70.2	0.26	0.86
$CHCl_3$	61.3	262.9	20	0.56	0.56
$CHCl_2F$	8.9	178.5	−26.2	0.35	0.48
$CHClF_2$	−40.8	96.0	−71.3	0.24	0.57

$$*T_R = \frac{\text{Temperature, °K}}{\text{Crit. temp., °K}}$$

TABLE 29-5. EFFECT OF FLUORINE ON TOXICITY IN ONE- AND TWO-CARBON COMPOUNDS

Compound	Underwriters' Laboratories rating*(30)
CCl_4	3
CCl_3F	5a
CCl_2F_2	6
$CClF_3$	6[†]
CF_4	6[†]
$CHCl_3$	3
$CHCl_2F$	4-5
$CHClF_2$	5a
CHF_3	6[†]
CCl_2FCClF_2	4-5
$CClF_2CClF_2$	6

*A number of products have been tested for toxicity by the Underwriters' Laboratories and divided into groups based on their effect on guinea pigs. Group 1 includes very toxic materials such as sulfur dioxide while group 6 contains compounds with very low toxicity. Carbon dioxide is in group 5a.

[†]Estimated rating based on preliminary testing.

the test animals than CO_2 which is generally considered harmless. The presence of fluorine in the molecule, however, is not always a guarantee of stability and low toxicity. When the molecular structure permits rapid hydrolysis, loss of HF, or the formation of reactive centers, toxicity can be extreme. For example, monofluoro acetic acid has long been used as a rat poison. More recently it has been shown that perfluoroisobutylene is very highly toxic.

The extreme electronegativity of the fluorine atom produces exceptionally strong bonds with carbon and accounts for many of the unusual properties of fluorinated compounds. Interatomic distances generally reflect the strength of the chemical bond. Some examples of substituted methanes are shown in Table 29-6. It can be seen that the carbon–fluorine distance is considerably less than the carbon–chlorine distance, although more than that of carbon–hydrogen. As more fluorine is introduced into the molecule, the carbon–fluorine bond becomes shorter, thus accounting in part for the increased stability of the more highly fluorinated compounds.

NOMENCLATURE

Some of the chlorofluoro derivatives of methane and ethane were commercially developed as refrigerants. In order to simplify their commercial use a numbering system was developed by Midgley and Henney in 1929 and extended by Park.[26] The numbers were incorporated in the registered trademarks of "Kinetic" Chemicals, Inc. and later acquired by Du Pont. The numbering system was donated to the industry by Du Pont for general use. However, the original trade name "Freon" was not abandoned and is still the registered trademark of the Du Pont Company.

The numbering system for halogenated derivatives has been incorporated in Standard B79.1 of the American Standards Association and is now in general use in nearly all industrial applications. This nomenclature was recommended for adoption by a working committee of the International Organization of Standardization in October 1960 and is expected to be officially adopted throughout the world soon.

TABLE 29-6. INTERATOMIC DISTANCES IN SUBSTITUTED METHANES, A

	C-F	C-Cl	C-H
CH_3F	1.39[21]		1.10[1]
CH_2F_2	1.36[21]		
CHF_3	1.33[21]		
CF_4	(1.30)[21]		
$CHCl_2F$	1.41[32]	1.73[32]	1.09[32]
$CHClF_2$	1.36[32]	1.73[32]	1.09[32]
CCl_2F_2	1.35[21]	1.74[21]	
$CClF_3$	1.33[21]	1.75[21]	

In the numbering system the first digit on the right is the number of fluorine (F) atoms in the compound.

The second digit from the right is one more than the number of hydrogen (H) atoms in the compound.

The third digit from the right is one less than the number of carbon atoms in the compound. When this digit is 0 it is omitted from the number. If the compound is unsaturated, the number one (1) is used as the fourth digit from the right.

The number of chlorine (Cl) atoms in the compound is found by subtracting the sum of the fluorine and hydrogen atoms from the total number of atoms which can be connected to the carbon atoms.

In the case of isomers, each has the same number. The most symetrical isomer is indicated by the number without any letter following it. As the isomers become more unsymetrical, the letters a, b, etc. are added.

The use of this systematic nomenclature rather than chemical names or arbitrary trade names has eliminated considerable confusion in discussions and descriptions of these compounds.

Manufacturers have adopted individual trade names which are used in conjunction with the number system. For example, trichlorofluoromethane may be known as "Freon-11," "Genetron" 11, etc. A partial list of trade names and manufacturers is given below.

Trade Name	*Company*	*Country*
"Freon"	E. I. du Pont de Nemours & Co., Inc.	USA
"Genetron"	Allied Chemical & Dye Corp.	USA
"Isotron"	Pennsalt Chem. Co.	USA
"Ucon"	Union Carbide Chem. Co.	USA
"Arcton"	Imperial Chemical Industries, Ltd.	England
"Isceon"	Imperial Smelting Co. (Sales) Ltd.	England
"Forane"	Societe d'Electro-Chimie d'Electro Metal-lurgic et des Acieries Electriques d'Ugine	France
"Flugene"	Pechiney	France
"Frigen"	Farbwerke Hoechst A.G.	W. Germany
"Heydogen"	Chemische Fabrik von Heyden A. G.	W. Germany
"Frigedohn"	V.E.B. Alcid Fluorwek Dohna	E. Germany
"Algofrene"	Montecatini	Italy
"Col-flon"	Nitto Chem. Co., Ltd.	Japan
"Daiflon"	Osaka Kinzoku Kogyo Kabushiki Kaisha	Japan
"Fresane"	Uniechemie N.V.	Netherlands
"Eskimon"	Unknown	Russia
"Flurion"	Productos Quimicas "Industrial" Comas Ing.	Spain
"Algeon"	Fluoder S.R.L.	Argentina
"Fration"	La Fluorhidrica	Argentina

HISTORICAL

A few highlights in the development of fluorine and compounds containing fluorine are mentioned below to illustrate the progress in this field.

Calcium fluoride (or fluorspar) and its use as a flux was described by Agricola about 1530. Many years later, in 1771, its composition was described by Scheele. Impure hydrofluoric acid was obtained from fluorospar probably as early as 1670. However, the completely anhydrous acid, necessary for commercial production of fluorocarbon compounds, was first prepared by Fremy in 1856 by the distillation of potassium difluoride.

The search for elemental fluorine gas covered a period of more than 70 years and many different attempts and methods. One of the great moments of chemistry occured in 1886 when fluorine was first obtained by Moissan in Paris. It was prepared by the electrolytic decomposition of hydrogen fluoride, and this is still the only known method of isolating this reactive chemical. The first successful electrolysis was made with impure hydrogen fluoride which happened to contain enough potassium fluoride to make it conducting. A second experiment before a committee from the Academy of

Sciences using purified hydrogen fluoride failed until Moissan recognized the necessity of adding potassium fluoride.

The isolation of elemental fluorine aroused interest in compounds containing fluorine and lead to considerable activity in this field in the latter part of the 19th century. Earlier in 1881 Young[35] had prepared an organic fluoride by adding HF to amylene. Beginning about 1888 and continuing into the first part of the 20th century there was considerable activity in the preparation of organic fluorine compounds, principally by Swarts in Belgium, Meslans in France, and Ruff in Germany. A number of different metal fluorides were used with and without HF to replace other halogens with fluorine. Silver fluoride was used extensively to prepare fluorinated derivatives of the lower hydrocarbons. The salts of other metals such as antimony, tin, mercury, lead, etc. were used in these earlier studies. In most cases, however, reaction was slow or incomplete. About 1890 a significant improvement was discovered by Swarts. This reaction, which bears his name, includes the addition of a small amount of pentavalent antimony to SbF_3 when it is used as a fluorinating catalyst. This addition causes complete and rapid exchange of fluorine with other halogens and is the basis for many commercial processes today.

Interest in fluorine chemistry continued at a rather slow pace until 1930 when Midgely and Henne[16] developed commercial procedures for the preparation of chlorofluoro derivatives of methane and ethane. Henne continued exhaustive studies of these compounds. He developed the use of HgO and anhydrous hydrogen fluoride as a versatile fluorinating system which could be used to make derivatives not easily prepared by other methods. In the course of his work Henne prepared and characterized most of the chlorofluoroderivatives of the lower aliphatic series.

During World War II a great deal of work was done in the preparation of fluorinated products for use in the separation of uranium isotropes. Since then, studies in the field of fluorine chemistry have accelerated in many directions. Many products of theoretical and commercial significance have been developed.

METHODS OF PREPARATION

In the following paragraphs a number of methods of preparing compounds containing fluorine are described. For the most part this discussion is limited to methods involving chlorine or chlorinated starting materials. The order of discussion corresponds approximately to commercial importance of the method. The reactions outlined in Figure 29-2 illustrate the formation and relationship of some of the chlorofluoro derivatives of methane and ethane.

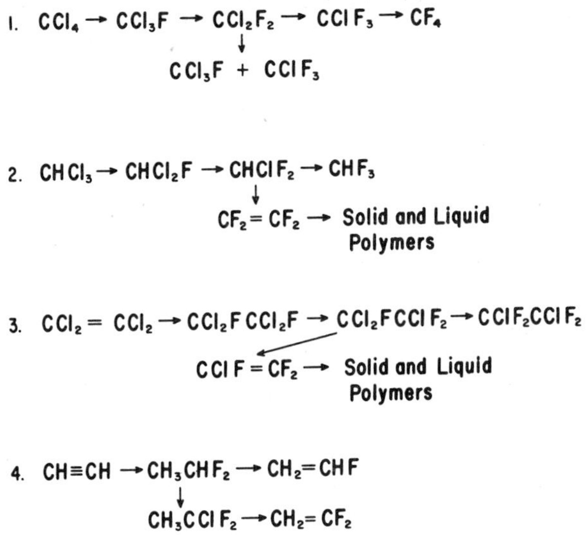

1. $CCl_4 \rightarrow CCl_3F \rightarrow CCl_2F_2 \rightarrow CClF_3 \rightarrow CF_4$
$$\downarrow$$
$$CCl_3F + CClF_3$$

2. $CHCl_3 \rightarrow CHCl_2F \rightarrow CHClF_2 \rightarrow CHF_3$
$$\downarrow$$
$$CF_2=CF_2 \rightarrow \text{Solid and Liquid Polymers}$$

3. $CCl_2=CCl_2 \rightarrow CCl_2FCCl_2F \rightarrow CCl_2FCClF_2 \rightarrow CClF_2CClF_2$
$$CClF=CF_2 \rightarrow \text{Solid and Liquid Polymers}$$

4. $CH{\equiv}CH \rightarrow CH_3CHF_2 \rightarrow CH_2=CHF$
$$\downarrow$$
$$CH_3CClF_2 \rightarrow CH_2=CF_2$$

Figure 29-2. Outline of preparation of some fluorocarbons.

Use of Metal Fluorides

Metal fluorides have been used to prepare organic fluorocarbons by replacing chlorine and other halogens since the earliest studies of fluorine
chemistry. A large number of different fluorides have been studied with
varying degrees of success. The most reactive and versatile are the fluorides
of mercury, silver and antimony but other metals, e.g., potassium, zinc,
copper, cadmium, lead, cobalt, manganese, etc., have been used, at least in
laboratory preparations. The effectiveness of each agent may be different
under different conditions or with different products. An interesting comparison of the effect of various agents was made by Tewksbury and Haendler in the vapor phase fluorination of benzotrichloride at 225° C.[28]

No benzotrifluoride was found with fluorides of lithium, potassium,
calcium, magnesium, aluminum or manganese. The amount of benzotrifluoride obtained with other agents is shown in Table 29-7.

Often, in laboratory work, the metal fluorides are prepared separately,
and the fluorination is performed in a subsequent step. This would not be
practical in commercial manufacture, however, and in practice the metal
salts are used as catalysts with the fluorine being continuously supplied by
hydrofluoric acid.

The following random observations are listed to illustrate the scope and
problems of fluorination with metal fluorides.

TABLE 29-7. FORMATION OF BENZOTRIFLUORIDE WITH
VARIOUS FLUORIDES

Fluoride	%
ZnF_2	70
SbF_3	60–65
PbF_2	45
CuF_2	44
BiF_3	29
CoF_2	18
CdF_2	18
NaF	15

Iodine is most easily, bromine moderately easily, and chlorine least easily replaced in organic compounds. However, side reactions are more likely to occur when iodine is replaced. For this reason and because of the lower cost and greater availability of organic chlorides, they are used as starting materials in all commercial manufacture.

As fluorine is introduced into the molecule, further fluorination becomes more difficult. The preparation of completely fluorinated products requires extreme conditions of temperature or the use of elemental fluorine or salts in higher than normal valence states.

Fluorination is easier with compounds containing more halogen atoms. For example, CCl_4 is easier to fluorinate than CH_3Cl.

Raising a metal salt to its highest valent state increases its fluorinating effectiveness, and the increase in valency can be produced by a halogen other than fluorine. Swarts discovered this effect about 1890 with antimony, and it forms the basis for most commercial fluorinations. Later, Henne developed a similar relationship using mercurous fluoride and chlorine.

The use of hydrogen fluoride with mercury salts requires considerably higher temperatures than with antimony salts.

Whalley[33] has reported that stannic chloride is superior to antimony pentachloride for replacing the chlorine atoms in compounds of the type $R\,CCl_2R'$.

Chlorine atoms adjacent to olefinic bonds are often more easily replaced than in other positions.

Fluorine can be added to double bonds using PbO_2 and HF. An example is the preparation of $CHClFCCl_2F$ from $CHCl{=}CCl_2$.

The use of metal fluorides to replace organic chlorine has been studied extensively, but much more work will be needed to provide a clear understanding of the mechanism and to explain the reactivity of different metal salts. Parameters of temperature, pressure, dilution, etc. are all important in determining the course and extent of the reaction.

Addition of Hydrofluoric Acid to Unsaturated Compounds

The addition of HF to acetylene is assuming commercial importance in the preparation of fluorinated ethanes and ethylenes. By the use of suitable conditions and catalysts the reaction can be controlled to add either one or two mols of HF. Catalysts such as CuCN, AlF_3, Hg salts, etc. in vapor phase reactions at elevated temperatures promote the formation of $CH_2=CHF$. Liquid phase reaction at temperatures of 0 to 25° C produces quantitative yields of CH_3CHF_2 with catalysts such as FSO_3H, BF_3, and nickel shavings. Thermal cracking of CH_3CHF_2 to eliminate HF is an alternate route to $CH_2=CHF$ with generally better yields and no catalyst problems. Addition of HF to higher alkynes occurs more readily than to acetylene, and usually no catalyst is needed.

Chlorination of ethyl fluoride or ethylidene fluoride gives various derivatives depending on conditions. The product, CH_3CClF_2 is of special importance as an intermediate in the preparation of $CH_2=CF_2$.

The addition of HF to double bonds follows Markownikoff's rule, and the fluorine atom goes to the carbon with the least number of hydrogens on it. The ease of addition depends on the structure of the unsaturated compound. In general, addition is more difficult if other halogen atoms are present. For example, HF adds with difficulty to $CH_3CH=CBr_2$ but rapidly to $CH_3CH=CClF$ and $CH_3CH=CF_2$. The use of catalysts such as BF_3 aids the reaction. HF can be added to $CCl_2=CCl_2$ and $CHCl=CCl_2$ using BF_3, while without a catalyst there is no reaction. Higher temperatures also help.

Disproportionation

The heating of simple organic compounds containing chlorine and fluorine can often be made to result in products containing both more and less fluorine than the original. For example, CCl_3F forms CCl_4, CCl_2F_2 and $CClF_3$ in good yields.[20] Murray prepared $CClF_3$ in 96 percent yield from CCl_2F_2 or CCl_3F.[19] Miller heated $CClF_2CClF_2$ and obtained CCl_3CF_3 with some minor amounts of other isomers.[17]

This type of disproportionation reaction offers a means of preparing more highly fluorinated compounds which might be difficult to form by direct replacement of chlorine. Aluminum and chromium salts have been effective catalysts. Attempts to disproporionate more complex molecules generally give a variety of products.

Chlorination

Chlorofluorocarbons can be prepared by chlorinating partially fluorinated compounds by standard methods. Henne and his co-workers have studied this subject extensively from the standpoint of orientation and directive influence of the fluorine atoms. The —CF_3 group generally deactivates atoms on adjacent carbons. Another generality is that chlorination tends to continue on the carbon where it starts. When these two tendencies collide, the stabilizing effect of —CF_3 is apparently overcome as illustrated in the following reactions.[12]

$$CF_3CH_2CH_3 + Cl_2 \rightarrow CF_3CH_2CCl_3$$

$$CF_3CHClCH_3 + Cl_2 \rightarrow CF_3CCl_2CH_3$$

Electrochemical Methods

The production of fluorocarbons and their derivatives by an electrochemical process was originally developed by Simons.[25] Hydrocarbons or organic acids, amines, alcohols or compounds with other functional groups are used as raw materials. The starting compound is dissolved in liquid hydrogen fluoride in an electrolytic cell using nickel anodes and iron cathodes. The operating voltage is too low to produce elemental fluorine, and fluorocarbon products are obtained directly. Although not involving chlorine, this process is mentioned here since it has attained commercial stature and is a versatile method of producing a great variety of fluorinated materials.

The preparation of chlorofluorocompounds by a similar process has been described by Wolfe.[34] Using methylene chloride as a starting material, CCl_2F_2, $CHClF_2$ and $CHCl_2F$ were obtained. $CHClFCCl_2F$ and $CClF_2CCl_2F$ were obtained from $CHCl{=}CCl_2$. The temperature of electrolysis and the amount of agitation, if the starting material was not soluble in liquid hydrogen fluoride, affected the result.

Direct Fluorination

Fluorination with elemental fluorine is difficult to control, but a great deal of progress has been made in studying methods of using it. Moissan[18] treated methane, chloroform and carbon tetrachloride with fluorine and in each case obtained tetrafluoromethane as the chief product. Fluorination of carbon tetrachloride vapor by Ruff and Keim[23] using a cobalt catalyst produced CF_4 and various chlorofluoromethanes. Bigelow and co-workers[29] have studied the use of fluorine under a variety of conditions. Vapor phase fluorination of ethyl chloride gave $CClF_3$ and CF_3CCl_2F.[5] As a general rule the use of elemental fluorine is an expensive and often difficult process, and when other methods are available they are preferred.

Other Methods

Organic compounds containing fluorine are subject to the manipulations and reactions common to organic chemistry. Many different techniques and variations in procedure for making fluorinated compounds are described in the literature. The methods outlined above are intended as illustrations of some of the processes and problems and no attempt has been made to include all of the possibilities.

COMMERCIAL MANUFACTURE

Most commercial methods of producing organic compounds containing chlorine and fluorine are based on the Swarts reaction. Antimony is used as the catalyst, and hydrogen fluoride and the starting material (e.g. CCl_4 or $CHCl_3$) are fed continuously into the reaction pot. A small amount of chlorine is also added to form some pentavalent antimony necessary for an efficient and quantitative reaction. The amount of chlorine used is related to the difficulty of the fluorination. Strict control of the chlorine, and therefore of the pentavalent antimony in the catalyst, is necessary to produce the desired degree of fluorination. Temperatures and pressures in the reaction chamber must also be closely controlled.

A typical flow diagram is shown in Figure 29-3. Hydrogen fluoride can be generated as needed or can be stored separately. The fluorinated products are continuously removed from the top of the reaction vessel through a fractionating column. By-product hydrogen chloride is removed by distillation, and the organic products are scrubbed to remove traces of acids.

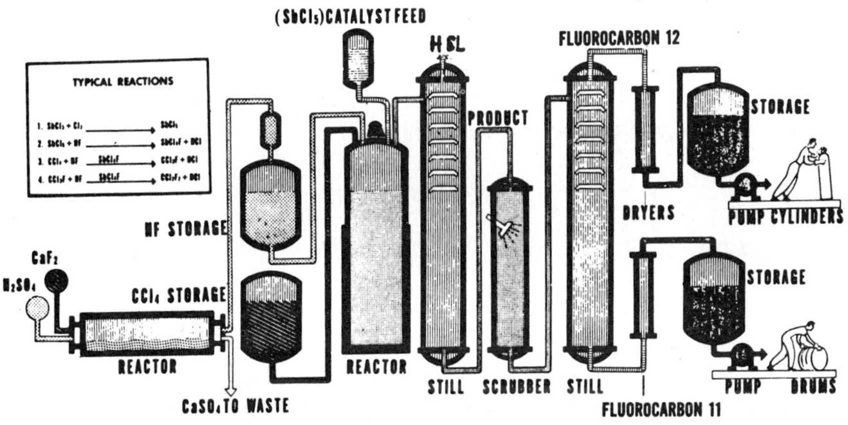

Figure 29-3. Typical flow diagram of a fluorinated process.

After thorough drying with silica or alumina, the product is stored until final packaging and delivery to the consumer.

An exceptionally pure grade of hydrogen chloride is obtained as a by-product in the fluorination and can often be used directly in other chemical processes.

A vapor phase reaction at elevated temperatures has also been used to prepare fluorinated products. The organic chloride and hydrogen fluoride are passed continuously through a heated tube containing the catalyst. Treatment of the product is similar to that described above.

The preparation of one-carbon products by chlorinolysis of fluorinated ethanes has been reported but is of doubtful commercial importance at present. For example, severe treatment of difluoroethane with chlorine will break the carbon-carbon bond to form CCl_2F_2 and CCl_4.[6]

An important and rapidly developing side of fluorocarbon production lies in the field of polymers, both plastic and elastic. The monomers from which some of the polymers are made commercially are obtained by Reactions (2), (3) and (4) in Figure 29-2.

Tetrafluoroethylene is obtained by pyrolysis of $CHClF_2$ at high temperatures in a metal tube with HCl as a by-product. Complete polymerization gives "Teflon" tetrafluoroethylene resin, while partial polymerization or telomerization produces stable liquids and oils.

The reduction of CCl_2FCClF_2 with zinc and alcohol forms $CClF{=}CF_2$ (CTFE) which can be polymerized to "Kel-F" resin. Here also, partial polymerization forms useful oils and greases.

The addition of HF to acetylene is the starting point for two commercially interesting monomers. Difluoroethane can be pyrolyzed at high temperatures with the elimination of HF to form vinyl fluoride. If it is first monochlorinated and then pyrolyzed, HCl is removed and vinylidene fluoride is formed. Both of these monomers are currently used in the preparation of polymers with valuable properties, either by themselves or when copolymerized with other olefines.

APPLICATIONS

General

The estimated production of fluorocarbon products in the United States in 1960 is shown in Table 29-8. The five major products which form the backbone of the refrigerant and aerosol propellent industries still account for a high percentage of the total fluorocarbon market. The 1.7 percent of fluorocarbon production shown in the table under "all others" includes compounds discussed elsewhere in this chapter and, in addition, such other products as oils and greases and fluorinated derivatives with functional

TABLE 29-8. ESTIMATED U. S. PRODUCTION OF ORGANIC
FLUOROCHEMICALS IN 1960[31]

Code number	Formula	Production Millions of pounds	%
11	CCl_3F	72	24.2
12	CCl_2F_2	166	55.7
22	$CHClF_2$	40	13.4
113	$C_2Cl_3F_3$	6	2.0
114	$C_2Cl_2F_4$	9	3.0
All others	–	5	1.7
Total		298	

groups such as alcohols, ethers, esters, etc. These latter products are in limited production but have established specialty applications based on various outstanding properties.

In most cases, the applications of fluorocarbons depend on the unique properties associated with the presence of fluorine in the molecule. These properties include low toxicity, nonflammability, low surface tension, low viscosity, relatively low boiling point compared with molecular weight, high density and unusual stability. The cost of preparing fluorinated products is considerably higher than that of preparing other halogenated compounds so that distinctly superior performance must be demonstrated to overcome the cost disadvantage.

At present, one product, dichlorodifluoromethane, accounts for more than half of the entire fluorocarbon production. However, this picture will undoubtedly change in the future, especially if developments in fluorinated plastics and elastomers continue to fulfill their earlier promise.

The approximate distribution of fluorocarbon applications is listed in Table 29-9. Use as aerosol propellents is now slightly ahead of refrigerant use, and the aerosol industry is still growing rapidly. However, both the refrigerant and propellent fields are fairly well established and defined. Any new burst of growth must come from the other applications mentioned in Table 29-9, or perhaps from some future developments not now known.

TABLE 29-9. ESTIMATED DISTRIBUTION OF
FLUOROCARBON APPLICATIONS[7]

Application	Total fluorocarbon production (%)
Aerosol propellents	50
Refrigerants	44
Other uses (plastics, films, elastomers, coatings, lubricants, textile treating agents, etc.)	6

Refrigeration

The commercial development of fluorocarbon products is based on the work of Midgely and Henne[16] and their application as refrigerants. The selection of chlorofluoro derivatives of methane and ethane as refrigerants was the direct result of a search for materials without the flammable and toxic characteristics of earlier refrigerants. The development of the fluorocarbon products has made possible the tremendous growth of home refrigeration and air conditioning and the air conditioning of public buildings where large numbers of people gather, such as theaters and auditoriums. This growth would have been considerably hampered with the older refrigerants which would have required special installation precautions to avoid the hazards of fire and toxicity in the event of a refrigerant leak.

At present, more than 90 percent of new refrigeration and air-conditioning equipment uses a fluorinated refrigerant. Ammonia is still used to some extent in ice making plants, frozen food plants, ice skating rinks, etc., but even in these commercial applications, the newer refrigerants are gradually finding a place. The one- and two-carbon compounds containing chlorine and fluorine offer a number of different products of interest as refrigerants. They include the five major products shown in Table 29-8 as well as other less commonly used compounds. From these products it is possible to select a suitable refrigerant for nearly every refrigeration and air-conditioning application. The list ranges from Refrigerant 113 (trichlorotrifluoroethane) boiling at 48°C to Refrigerant 14 (tetrafluoromethane) boiling at –128°C. The properties of this family of fluorinated products are similar with respect to stability, nonflammability, low toxicity, high density, etc. However, each product is a different chemical compound, and the properties of each are slightly different. Such factors as oil solubility and effect on gasketing and insulation materials may vary considerably from one refrigerant to another. This difference must be recognized when designing equipment for use with a particular refrigerant.

The requirements of the modern refrigeration industry demand a degree of refrigerant purity not often found in organic compounds produced on a large scale. Organic purity generally exceeds 99.9 percent. Water is limited to a maximum of 10 ppm by weight. The air specification for the commercial products is less than 1.5 percent by volume in the vapor phase. Since air is not very soluble in the liquid, the total air concentration is exceedingly small. The residue on evaporation is limited to 0.01 percent. There must be no acidity or detectable chloride concentration.

The need for physical and thermodynamic property data has led to extensive studies and measurements by the manufacturers. For the commercial products, complete information is available for vapor.pressure, liquid and vapor density, heat capacity, viscosity, thermal conductivity, etc. Equations were developed to represent the experimental measurements. The heat content or enthalpy and the entropy have been calculated by thermodynamic relationships. Extensive tables of these properties have been prepared to meet the requirements of refrigeration design engineers.

Large air-conditioning equipment is based on the use of Refrigerant 11 and to a lesser extent on Refrigerant 113 in centrifugal compressors. These units range in size from about 300 to 600 tons. One ton of refrigeration is equal to 12,000 Btu/hour. This rate of heat exchange is necessary to melt one ton of ice in 24 hours. Refrigerant 12 is also used to some extent in this type of equipment in hermetically sealed systems.

For capacities ranging downward from about 300 tons, Refrigerant 22 or Refrigerant 12 is generally used in reciprocating compressors. With single-stage compression, temperature down to about $-40°$ F can be obtained. For lower temperatures, compound compression can be used with these refrigerants in two separate compressors or in different cylinders in the same compressor. Refrigerant 22 is used in nearly all window air-conditioning units where size is important. The properties of Refrigerant 22 are such that 60 percent more capacity can be obtained from a compressor than when Refrigerant 12 is used. The lower boiling point and higher operating pressures of Refrigerant 22 cause more moles of refrigerant to be transferred with each stroke of the piston in the compressor.

TABLE 29-10. REFRIGERANT APPLICATIONS

Refrigerant	Application
11, 113	in centrifugal compressors for industrial and commercial air conditioning and for industrial process water and brine cooling
12	household and commercial refrigeration and air conditioning, refrigerators, frozen food cabinets, ice cream cabinets, food locker plants, water coolers, room and window air-conditioning equipment; generally used in reciprocating compressors ranging in size from fractional to 800 horsepower and in rotary compressors in the smaller sizes
22	all types of household and commercial refrigeration and air-conditioning applications with reciprocating compressors; the properties of Refrigerant 22 permit the use of smaller equipment than with Refrigerant-12 and it is often preferred where size is a problem
114	in large industrial process cooling and air conditioning with multistage centrifugal compressors and to some extent in small refrigeration systems with rotary compressors.

Instead of compressing Refrigerant 22 or Refrigerant 12 in two stages, temperatures down to about –70° F can be obtained in single-stage compression using bromotrifluoromethane. This material is used principally as a fire extinguishing agent but also to a small extent as a refrigerant.

A cascaded system of refrigeration is often used for producing low temperatures. Two separate systems are used with one serving to cool the condenser of the other. For example, Refrigerant 22 can be used in one system to produce temperatures of about –40° F. In a separate system chlorotrifluoromethane (Refrigerant 13) can be condensed at this temperature and produce temperatures down to about –125° F when it evaporates. By using tetrafluoromethane (Refrigerant 14) in a third system, temperatures down to about –200 to –250° F can be produced.

Aerosol

The modern aerosol industry stems from a war time development of Goodhue and Sullivan.[9] Seeking a way of protecting front-line soldiers from lice, mosquitos and other insects, they pressurized a solution of insecticide with dichlorodifluoromethane. When this was sprayed through a suitable nozzle, a fog of insecticide of true aerosol proportions was formed. The result was a very effective method of insect control. The self-contained, pressurized spray method of packaging and dispensing has been extended to many other useful products, ranging from hair treatments and room deodorants to paints and shaving cream. More than three hundred different products have been commercially developed.

Although most of the sprays produced by these self-pressurized packages are not true aerosols, the term has been appropriated to describe the products and the industry which has grown up around them.

The rapid growth of the aerosol industry is an indication of the utility and convenience of the self-pressurized method of application. The production data in Table 29-11 is from annual surveys by the Aerosol Division of the Chemical Specialties Manufacturers Association.

**TABLE 29-11. ANNUAL PRODUCTION OF ALL
NON-FOOD AEROSOL PACKAGES**

(in million units)

1953	140
1954	185
1955	240
1956	320
1957	390
1958	470
1959	575
1960	670
1961	796

The aerosol industry was founded on insect sprays, and these products still account for a good share of the market. However, hair preparations and shaving lather have each surpassed them in popularity. The distribution of various product types in 1959 and 1960 is shown in Table 29-12. Some of the groupings include a number of different products and more detailed information is available from the CSMA. Some examples of the variety of materials now packaged as aerosols include the following:

(1) *Household Products:* Snow, glass cleaners, leather dressings, waxes and polishes, starches, metal cleaners, water repellents, etc.

(2) *Personal Products:* Colognes and perfumes, medicinals and pharmaceuticals, dental cream, shampoos, sun tan preparations, hand lotions, powders, etc.

(3) *Miscellaneous:* De-icers, veterinary and pet products, automotive products, lubricants, etc.

TABLE 29-12. DISTRIBUTION OF AEROSOL PRODUCT TYPES

	Total Production (%)	
	1959	1960
Hair sprays	16.0	18.4
Household products	15.6	18.1
Shaving lather	14.6	11.2
Insect sprays	13.2	11.1
Coatings and finishes	10.5	11.1
Room deodorants	12.0	11.0
Personal products	13.6	10.5
Miscellaneous	4.5	8.6

The materials in aerosol formulations can be classified in general as active ingredient, solvent and propellent. The active ingredient is the foundation of the entire formulation, and if it is not effective, no amount of research will turn it into a good product. However, in many cases effectiveness can be increased or applications can be developed which are difficult or impossible by other methods.

The term solvent is rather loosely used in aerosol formulating. Its principal functions include (a) producing miscible or compatible formulations; (b) increasing their effectiveness; (c) lowering the pressure. As a general rule all of these effects occur at the same time, although the principal reason for adding the solvent may have been limited to one or two. In addition to its primary functions, the solvent must be stable in storage and must not react with the container, valve components or other materials present. It must be safe to use both from the standpoint of toxicity and from effect on surfaces exposed to the spray.

Many different solvents are used in aerosol formulations, including methylene chloride and trichlorofluoromethane. The former is an excellent solvent and usually can be present only in limited amounts without affecting the valve or other materials. The latter has moderate solvent activity and most often is added as a pressure depressant. It can be considered as part of the propellent and usually is sold as a mixture with dichlorodifluoromethane (Propellent 12).

The propellent force in some aerosol products is supplied by a compressed gas such as air or nitrogen. In a majority of formulations, however, the propellent is a liquefied compressed gas which provides a constant pressure until nearly all of the contents have been discharged. Dichlorodifluoromethane (Propellent 12) is the basic propellent for most products. Since the vapor pressure of Propellent 12 is too high (70 psig at 70° F) for many applications, it is often mixed with trichlorofluoromethane (Propellent 11) or dichlorotetrafluoroethane (Propellent 114). These mixed propellents are available from manufacturers in whatever composition is most suitable for a particular formulation.

Comprehensive reviews of the aerosol industry and products have been published and are recommended for further information.[24,13]

Plastics and Elastomers

Plastics and elastomers containing fluorine in general are (a) resistant to chemical attack, (b) stable at elevated temperatures and (c) good electrical insulators. These properties make them useful under extreme conditions where other materials fail. The list of products in Table 29-13 is representative of the commercial materials available but may not necessarily be complete.

Blowing Agents

The use of moderately low-boiling fluorinated compounds as blowing agents in the production of plastic and elastic foams presents an interesting variation from other applications of these products. Not only does the foam have improved properties in many cases when the fluorocarbons are used, but the process is less expensive. This unusual situation is best illustrated by foamed polyurethanes. Originally, foaming was produced by carbon dioxide gas generated by the action of water on the expensive isocyanate component. The use of fluorocarbons is not only cheaper, but the thermal conductivity of the resulting foam is considerably less with consequent improvement in the insulating value. The fluorocarbon is dissolved in one of the components and the heat of reaction causes it to vaporize.

TABLE 29-13. SOME FLUORINATED PLASTICS AND ELASTOMERS

Monomer	Product	Manufacturer	Comment
$CF_2{=}CF_2$	"Teflon" TFE fluorocarbon resin	E. I. du Pont de Nemours & Co., Inc.	first commercial fluorocarbon plastic; used in gaskets, piston rings, wire insulation, electrical insulation, in providing antistick surfaces and many similar applications
$CF_2{=}CF_2$ + $CF_3CF{=}CF_2$	"Teflon" FEP fluorocarbon resin	E. I. du Pont de Nemours & Co., Inc.	similar to "Teflon" but easier to mold and fabricate
$CClF{=}CF_2$	"Kel-F" plastic	Minnesota Mining and Manufacturing Co.	used for molding electrical and electronic parts, insulation, gasketing tubing, etc.
$CClF{=}CF_2$	"Halon"	Allied Chemical & Dye Co.	new product
$CH_2{=}CF_2$	RC 2525	Pennsalt Chemical Co.	new product
$CH_2{=}CHF$	"Tedlar" polyvinyl fluoride film	E. I. du Pont de Nemours & Co.	exceptional resistance to weathering
$CClF{=}CF_2$ + $CH_2{=}CF_2$	"Kel-F" elastomer	Minnesota Mining and Manufacturing Co.	used in gasketing chemical resistant liners, insulation, etc.
$CH_2{=}CF_2$ + $CF_3CF{=}CF_2$	"Fluorel" elastomer	Minnesota Mining and Manufacturing Co.	used for gasketing, seals and hose
$CH_2{=}CF_2$ + $CF_3CF{=}CF_2$	"Viton" fluoroelastomer	E. I. du Pont de Nemours & Co., Inc.	high temperature elastomer with good resistance to oils and fuels

In this application, the principal agent is trichlorofluoromethane (Blowing Agent 11), but dichlorodifluoromethane and trichlorotrifluoroethane have also been used, and in some cases mixed agents are selected for special purposes.

In other types of foams, fluorocarbons can often be used to replace the more expensive organic blowing agents such as azo or nitroso compounds. Considerable growth in the application of fluorocarbons as blowing agents is expected.

Solvents and Cleaning Agents

The fluorocarbon solvents lie above hydrocarbons and below chlorinated products in solvent power. They have a selective activity, which permits their use for removing oil, grease and dirt without harm to plastics, elastomers, or metals. This property together with their nonflammability, low toxicity and stability has led to their use in a number of specialty applications. Fluorocarbon solvents are being used in metal degreasing, in cleaning complicated electronic equipment without disassembling it, and in cleaning other special equipment such as gyroscopes, electronic printed circuits, magnetic tape, etc.

Fire Extinguishing Agents

The principal fluorinated fire extinguishing agent is bromotrifluoromethane. This agent is used for engine fire protection on many commercial aircraft. Replacement of the older carbon dioxide system with this new agent has been estimated to save enough weight to allow one additional paying passenger. It also is used for marine and home applications and by the army for truck and tank protection. It is considerably more effective than carbon tetrachloride on either a weight or volume basis and at the same time presents no toxic hazard.

Other fluorinated products also exhibit fire extinguishing activity but have found little commercial application.

Miscellaneous Applications

The major uses of fluorocarbon products have been discussed in preceeding paragraphs but they are gradually finding places in many other specialty applications. The list includes: lubricants and greases, dielectric fluids, heat-transfer fluids, leather and textile treating agents, power fluids, etc. The fluorocarbon field is growing and can be expected to have a bright future.

References

1. Andersen, F. A., Bak, B., and Brodersen, S., "Normal Vibration Frequencies of CD_3F; Structure of CH_3F and CD_3F from Infrared and Microwave Spectra," *J. Chem. Phys.*, **24**, 989 (1956).

2. Barbour, A. K., "Industrial Aspects of Fluorine Chemistry," *Chemistry & Industry*, 958 (July 1, 1961).

3. Booth, H. S., Burchfield, P. E., Bixby, E. M., and McKelvey, J. B., "Fluorochloroethylenes," *J. Am. Chem. Soc.*, **55**, 2231 (1933).

4. Booth, H. S., and Swinehart, C. F., "The Critical Constants and Vapor Pressures at High Pressure of Some Gaseous Fluorides of Group IV," *J. Am. Chem. Soc.*, **57**, 1337 (1935).

5. Calfee, J. D., and Fukuhara, N., *et al.*, *J. Am. Chem. Soc.*, **62**, 267 (1940).

6. Calfee, J. D., and Smith, L. B., U. S. Patent 2,417,059 (1948).

7. *Chem. Eng. News*, 92 (July 18, 1960).

8. E. I. du Pont de Nemours & Co., Inc., Technical Bulletins, "Freon" Products Division.

9. Goodhue, L. D., and Sullivan, W. N., U. S. Patent 2,321,023 (1943).

10. Haszeldine, R. N., and Sharpe, A. G., "Fluorine and Its Compounds," New York, John Wiley & Sons, 1951.

11. Henne, A. L., in "Organic Reactions," Vol. II, New York, John Wiley & Sons, Inc., 1944.

12. Henne, A. L., and Whaley, A. M., *J. Am. Chem. Soc.*, **64**, 1157 (1942).

13. Herzka, A., and Pickthall, J., "Pressurized Packaging (Aerosols)," London, Butterworths Publications Limited, 1958.

14. Lovelace, A. M., Postelnek, W., and Rausch, D. A., "Aliphatic Fluorine Compounds," ACS Monograph No. 138, New York, Reinhold Publishing Corp., 1958.

15. Mears, W. H., Stahl, R. F., Orfeo, S. R., Shair, R. C., Kells, L. F., Thompson, W., and McCann, H., "Thermodynamic Properties of Halogenated Ethanes and Ethylenes," *Ind. Eng. Chem.*, **47** (7), 1449 (July 1955).

16. Midgely, T., and Henne, A. L., "Organic Fluorides as Refrigerants," *Ind. Eng. Chem.*, **22**, 542, May 1930.

17. Miller, W. T., *J. Am. Chem. Soc.*, **72**, 705 (1950).

18. Moissan, H., *Ann. Chim. Phys.*, **19** (6), 272 (1891).

19. Murray, W. S., U. S. Patent 2,426,637 (1947).

20. Pailthorp, J. R., U. S. Patent 2,694,739 (1954).

21. Pauling, L., "The Nature of the Chemical Bond," 3rd Ed., Ithaca, New York, Cornell University Press, 1960.

22. Renfrew, M. M., and Lewis, E. E., "Polytetrafluoroethylene," *Ind. Eng. Chem.*, **38**, 870 (1946).

23. Ruff, O., and Keim, R., *Z. Anorg. Allgem. Chem.*, **201**, 245 (1931).

24. Shepherd, H. R., "Aerosols: Science and Technology," New York, Interscience Publishers Inc., 1961.

25. Simons, J. H., *J. Electrochem. Soc.*, **95**, 47 (1949).

26. Simons, J. H., Ed., "Fluorine Chemistry," Vol. I, New York, Academic Press, Inc., 1950. *Ibid.,* Vol. II, 1954.

27. Tatlow, J. C., and Sharpe, A. G., "Advances in Fluorine Chemistry," Vol. I, New York, Academic Press Inc., 1960.

28. Tewksbury, C. I., and Haendler, H. M., *J. Am. Chem. Soc.,* **71,** 2336 (1949).

29. Tyczkowski, E. A., and Bigelow, L. A., *J. Am. Chem. Soc.,* **77,** 3007 (1955).

30. Underwriters' Laboratories, Chicago, Ill. Reports MH 2375, MH 2630 and MH 3134.

31. U. S. Tariff Commission Report, "Synthetic Organic Chemicals: United States Production and Sales of Miscellaneous Chemicals, 1960."

32. Weissman, H. B., Meister, A. G., and Cleveland, F. F., "Substituted Methanes. XXVIII. Infrared Spectral Data and Assignments for $CHCl_2F$ and $CHClF_2$ and Potential Energy Constants and Calculated Thermodynamic Properties for $CHCl_2F$, $CDCl_2F$, $CHClF_2$ and $CDClF_2$," *J. Chem. Phys.,* **29,** 72 (1958).

33. Whalley, W. B., U. S. Patent 2,452,975 (1948).

34. Wolfe, J. K., U. S. Patent 2,806,817 (1957).

35. Young, S., "Note on the Formation of an Alcoholic Fluoride," *J. Chem. Soc.,* **39,** 489 (1881).

INDEX